IIIB	IVB	VB	VIB	VIIB	0.1785†
					1s² Helium

5 10.811 (3) — 4275 / 2365 / 2.34 — **B** — 1s²2s²p¹ — Boron

6 12.011 ±4,2 — 5100* / 3825* / 2.26 — **C** — 1s²2s²p² — Carbon

7 14.0067 ±3,5,4,2 — 77.344 / 63.15 / 1.251† — **N** — 1s²2s²p³ — Nitrogen

8 15.9994 -2 — 90.188 / 54.8 / 1.429† — **O** — 1s²2s²p⁴ — Oxygen

9 18.99840 -1 — 85.0 / 53.55 / 1.696† — **F** — 1s²2s²p⁵ — Fluorine

10 20.1797 — 27.10 / 24.55 / 0.900† — **Ne** — 1s²2s²p⁶ — Neon

13 26.98154 (3) — 2740 / 933.5 / 2.70 — **Al** — [Ne]3s²p¹ — Aluminum

14 28.0855 4,2 — 2630 / 1683 / 2.33 — **Si** — [Ne]3s²p² — Silicon

15 30.97376 ±3,5,4 — 553 / 317.3 / 1.82 — **P** — [Ne]3s²p³ — Phosphorus

16 32.066 ±2,4,6 — 717.82 / 392.2 / 2.07 — **S** — [Ne]3s²p⁴ — Sulfur

17 35.4527 ±1,3,5,7 — 239.18 / 172.17 / 3.214† — **Cl** — [Ne]3s²p⁵ — Chlorine

18 39.948 — 87.45 / 83.95 / 1.784† — **Ar** — [Ne]3s²p⁶ — Argon

IB	IIB

28 58.6934 2,3 — 3005 / 1726 / 8.90 — **Ni** — [Ar]3d⁸4s² — Nickel

29 63.546 2,1 — 2840 / 1356.6 / 8.96 — **Cu** — [Ar]3d¹⁰4s¹ — Copper

30 65.39 — 1180 / 692.73 / 7.13 — **Zn** — [Ar]3d¹⁰4s² — Zinc

31 69.723 (3) — 2478 / 302.92 / 5.91 — **Ga** — [Ar]3d¹⁰4s²p¹ — Gallium

32 72.61 4 — 3107 / 1211.5 / 5.32 — **Ge** — [Ar]3d¹⁰4s²p² — Germanium

33 74.9216 ±3,5 — 876 (subl.) / 1090 (28 atm.) / 5.78 — **As** — [Ar]3d¹⁰4s²p³ — Arsenic

34 78.96 -2,4,6 — 958 / 494 / 4.79 — **Se** — [Ar]3d¹⁰4s²p⁴ — Selenium

35 79.904 ±1,5,7 — 331.85 / 265.95 / 3.12 — **Br** — [Ar]3d¹⁰4s²p⁵ — Bromine

36 83.80 0,2 — 120.85 / 116 / 3.75† — **Kr** — [Ar]3d¹⁰4s²p⁶ — Krypton

46 106.42 2,4 — 3240 / 1825 / 12.0 — **Pd** — [Kr]4d¹⁰ — Palladium

47 107.868 1 — 2436 / 1235.08 / 10.50 — **Ag** — [Kr]4d¹⁰5s¹ — Silver

48 112.41 2 — 1040 / 594.26 / 8.65 — **Cd** — [Kr]4d¹⁰5s² — Cadmium

49 114.82 3 — 2350 / 429.78 / 7.31 — **In** — [Kr]4d¹⁰5s²p¹ — Indium

50 118.710 4,2 — 2876 / 505.12 / 7.31 — **Sn** — [Kr]4d¹⁰5s²p² — Tin

51 121.760 ±3,5 — 1860 / 903.91 / 6.69 — **Sb** — [Kr]4d¹⁰5s²p³ — Antimony

52 127.60 -2,4,6 — 1261 / 722.72 / 6.24 — **Te** — [Kr]4d¹⁰5s²p⁴ — Tellurium

53 126.9045 ±1,5,7 — 457.5 / 386.7 / 4.93 — **I** — [Kr]4d¹⁰5s²p⁵ — Iodine

54 131.29 0,2,4,6 — 165.1 / 161.39 / 5.90† — **Xe** — [Kr]4d¹⁰5s²p⁶ — Xenon

78 195.08 2,4 — 4100 / 2042.1 / 21.45 — **Pt** — [Xe]4f¹⁴5d⁹6s¹ — Platinum

79 196.9665 3,1 — 3130 / 1337.58 / 19.3 — **Au** — [Xe]4f¹⁴5d¹⁰6s¹ — Gold

80 200.59 2,1 — 629.88 / 234.31 / 13.55 — **Hg** — [Xe]4f¹⁴5d¹⁰6s² — Mercury

81 204.383 3,1 — 1746 / 577 / 11.85 — **Tl** — [Xe]4f¹⁴5d¹⁰6s²p¹ — Thallium

82 207.2 4,2 — 2023 / 600.65 / 11.35 — **Pb** — [Xe]4f¹⁴5d¹⁰6s²p² — Lead

83 208.9804 3,5 — 1837 / 544.59 / 9.75 — **Bi** — [Xe]4f¹⁴5d¹⁰6s²p³ — Bismuth

84 (209) 4,2,6 — 1837 / 527 / 9.3 — **Po** — [Xe]4f¹⁴5d¹⁰6s²p⁴ — Polonium

85 (210) ±1,3,5,7 — 610* / 575 / — — **At** — [Xe]4f¹⁴5d¹⁰6s²p⁵ — Astatine

86 (222) 0,2 — 211.4 / 202 / 9.73 — **Rn** — [Xe]4f¹⁴5d¹⁰6s²p⁶ — Radon

110 (269) — — / — / — — **Uun** — [Rn]5f¹⁴6d⁸7s¹* — (Ununnilium)

111 (272) — — / — / — — **Uuu** — [Rn]5f¹⁴6d⁹7s¹* — (Unununium)

The A & B subgroup designations, applicable to elements in rows 4, 5, 6 and 7, are those recommended by the International Union of Pure and Applied Chemistry.

64 157.25 3 — 3545 / 1585 / 7.90 — **Gd** — [Xe]4f⁷5d¹6s² — Gadolinium

65 158.9253 3,4 — 3500 / 1629 / 8.23 — **Tb** — [Xe]4f⁹6s² — Terbium

66 162.50 3 — 2840 / 1685 / 8.55 — **Dy** — [Xe]4f¹⁰6s² — Dysprosium

67 164.9303 3 — 2968 / 1747 / 8.80 — **Ho** — [Xe]4f¹¹6s² — Holmium

68 167.26 3 — 3140 / 1802 / 9.07 — **Er** — [Xe]4f¹²6s² — Erbium

69 168.9342 3,2 — 2223 / 1818 / 9.32 — **Tm** — [Xe]4f¹³6s² — Thulium

70 173.04 3,2 — 1469 / 1092 / 6.97 — **Yb** — [Xe]4f¹⁴6s² — Ytterbium

71 174.967 3 — 3668 / 1936 / 9.84 — **Lu** — [Xe]4f¹⁴5d¹6s² — Lutetium

96 (247) 3 — 1620 / 13.5 — **Cm** — [Rn]5f⁷6d¹7s² — Curium

97 (247) 4,3 — 14* — **Bk** — [Rn]5f⁹7s² — Berkelium

98 (251) 3 — 1170* — **Cf** — [Rn]5f¹⁰7s² — Californium

99 (252) 3 — 1130* — **Es** — [Rn]5f¹¹7s² — Einsteinium

100 (257) 3 — 1800* — **Fm** — [Rn]5f¹²7s² — Fermium

101 (258) 3 — 1100* — **Md** — [Rn]5f¹³7s² — Mendelevium

102 (259) 2,3 — 1100* — **No** — [Rn]5f¹⁴7s² — Nobelium

103 (262) 3 — 1900* — **Lr** — [Rn]5f¹⁴7s²p¹ — Lawrencium

VWR Scientific Products
Science Education
P.O. Box 5229, Buffalo Grove, IL 60089-5229
1-800-727-4368 FAX 1-800-676-2540

Catalog Number WLS-18806

Side 1

ENGINEERING MATERIALS TECHNOLOGY

Structures, Processing, Properties, and Selection

FOURTH EDITION

JAMES A. JACOBS
Professor of Technology
Norfolk State University

THOMAS F. KILDUFF
Professor Emeritus
Thomas Nelson Community College

Upper Saddle River, New Jersey
Columbus, Ohio

Library of Congress Cataloging-in-Publication Data

Jacobs, James A., 1942-
    Engineering materials technology: structures, processing, properties, and selection /
    James A. Jacobs, Thomas F. Kilduff.— 4th ed.
        p. cm.
    Includes bibliographical references and index.
    ISBN 0-13-014280-8
    1. Materials. I. Kilduff, Thomas F., 1924- II. Title.

TA403 .J26 2001
620.1'1—dc21   00-055056

Vice President and Publisher: Dave Garza
Editor in Chief: Stephen Helba
Acquisitions Editor: Debbie Yarnell
Associate Editor: Michelle Churma
Production Editor: Louise N. Sette
Production Supervision: Carlisle Publishers Services
Design Coordinator: Robin G. Chukes
Cover Designer: Thomas Borah
Production Manager: Brian Fox
Marketing Manager: Chris Bracken

This book was set in Times Roman and Univers by Carlisle Communications, Ltd.
It was printed and bound by R.R. Donnelley & Sons Company. The cover was
printed by The Lehigh Press, Inc.

10 9 8 7 6 5 4 3 2 1
ISBN 0-13-014280-8

As authors we are indebted to the many people whose constructive suggestions and cooperation helped bring this book to completion. However, most of all, we dedicate this book to our wives, Martha and Virginia, and daughters, Sherri, Tammi, Jeanene, Liz, and Suzy, to all of whom we owe a major debt for their understanding, unfailing support, optimism, and tangible support as we toiled for many years.

Contents

Preface

Advances in materials science and engineering have had a profound effect on the world's technological development. The present-day globalization of the world's economy is driven by high technology, and the research and development of materials is in the forefront of our changing world society. As a consequence many people can benefit from a study of engineering materials technology. Our first edition was written for students in engineering and industrial technology programs. Since then we have learned that many other people found our text refreshing and useful to them, including professors, secondary school teachers, and industrial trainers with the responsibility to teach materials science and technology subjects.

Because materials technology is so pervasive in our readers' lives, we continually encourage them to become "materials observers" and point out the relevance of this important subject as seen in everyday products and systems. Another effective teaching tool for learning, in evidence since the original text, is the use of integrated common themes: 1) relationships among structure, properties, processing, and applications of materials, 2) the importance of consideration of the total Materials Cycle in material synthesis, selection, processing, and economics, 3) the need for environmentally friendly design and manufacturing of systems and products, 4) new and/or improved technologies that influence many aspects of engineering materials technology such as micro/nanotechnology, recycling, surface engineering, smart materials, intelligent structures, and biomimetrics, 5) encouragement for the reader to explore the many resources and databases outside this book including the use of the Internet, and 6) the role of standard practices in all aspects of design, testing, processing, manufacturing, selection, and applications of materials.

In a rapidly advancing society, people must have a sufficient technical background to understand the basis for those new technologies thrust upon them on nearly a daily basis. But this understanding is only a first step. To make a living using these new technologies will be a necessity. All of us will be called upon to be better users of these technological developments which will require the use of less of the earth's dwindling resources accompanied by a reduction in pollution of the planet earth. Accomplishing these broad objectives calls for not only intelligent citizens but intelligent consumers. Throughout the book we point to efforts to strike a balance between improving materials, processes, products, and systems and those impacts on our environment with approaches such as "green design," "green manufacturing" and "green materials." Building on the drive for "green design and manufacturing" for a sustainable environment we have included extensive coverage of product and system Life Cycle Analysis as related to the total Materials Cycle.

We have added a number of new features to this edition of the text, such as the *EMT WEBLINKS* in each module. These links will help students view dramatic images, animations and movies, probe databases, and observe R&D at corporations, universities, and government laboratories. They also permit the learners to explore numerous examples of the rapid advancements in the fields of material sciences and engineering, and in some cases engage in tutorials and other interactive exercises. Also, a color section provides vivid images of a wide array of example materials, which are among the many new graphics that help illustrate concepts.

Knowledge of new, advanced materials in all members of the family of materials is extensive. In addition, problems incurred in actually applying these materials are also brought to the attention of our readers. The span of time and the development efforts expanded on bringing new advance materials to the production phase are described. The development of technologies that will permit these materials to be processed economically in industry and the ever-present need for compromise in applying these new materials to safe and reliable products and structures is examined. The above statements are sufficient justification for more study of the role of processing and fabrication of today's engineering materials. Near-net shaping, rapid solidification, rapid prototyping, thermal spraying, hydroforming, and tailored blanks are but few of such subjects. We also examine materials as found in building construction and civil engineering technologies. Corrosion gets expanded coverage, including electrochemistry principles, methods of corrosion prevention, and better design procedures, due to the recognized fact that all materials are subjected to corrosion under service conditions.

To present the fundamentals of materials science on a level appropriate to learners who have not completed formal courses in chemistry, physics or mathematics beyond the uses of algebra is uppermost in our mind. As a consequence the terminology is defined while using a maximum of graphics such as equations, graphs, tables, and micrographs. Illustrative problems requiring calculations are numerous whenever a law or principle can be expressed as an equation. Emphasis is consistently placed on the proper use of units in solving equations using both the English and International System (SI) of units. Substitution of units as well as numbers in equations is stressed as being essential for arriving at a correct answer. Problems in materials science, in general, require a degree of creative analysis of concepts coupled with research into databases in order to solve them. Ample opportunity for this type of problem-solving endeavor is provided by problems contained in each module's Self Assessment section. Many of these types of questions can serve as useful class discussions.

ORGANIZATION

The book is organized in self-contained modules that consist of:

- Module Outcomes that present what the reader should be able to do after studying the module;
- Pause & Ponders that present case studies or applications of materials developments, aimed at sparking the reader's interest and providing a reason to learn the content;
- Solved problems and numerous illustrations and examples representing a full range of products, systems, and technologies that support concepts within the module's topic;
- Applications & Alternatives that reinforce the concepts learned by illustrating current use of the information;
- Self-Assessments that provide open-ended essay questions, problems to solve, and objective questions;
- References & Related Readings that provide source information, plus a listing of useful handbooks, journals, periodicals, and similar data; and
- Listings of Experiments and Updates from past National Educators Workshops, which can be obtained by using information in the Instructor's Manual.

The experiments and demonstrations listed in each module have been tested and thoroughly peer-reviewed and offer some excellent activities to supplement the text. The answers to all the problems and questions for the Self-Assessments are available in the Instructor's Manual, which offers a test bank from the Self-Assessments and other useful supplements.

The book's organization—including numerous tables in modules, and extensive appendices, coupled with key terms highlighted in bold face—make it comprehensive enough to serve as a Ready Reference Book on materials for the immediate future and beyond.

Please become familiar with the Contents and Index section. We have taken great care to fully develop these features in order to help you make the many cross-references required of this wide-ranging discipline of engineering materials technology. With the Contents you can locate many useful aids in the Appendix such as Greek symbols, SI/US Customary conventions and conversions, tables of properties, ASTM abbreviations, hardness to strength conversions, trade names, material selection guides, and more. Within modules you will find useful visual aids such as our hardness scale comparison figure. All of these features are cross-referenced in the Index. This book should be a useful guide or ready reference, much like *Machinery's Handbook,* which you should keep at your side as you engage in engineering/industrial problem solving.

NEW: UPDATES AND OTHER RESOURCES

As creator of the National Educators Workshop series (NEW: Update) in 1986, Jim Jacobs has co-directed these annual workshops which aimed at providing emerging concepts on engineering materials, science, and technology with emphasis on laboratory experiments. Major sponsors of these workshops and related research were the National Aeronautics and Space Administration (NASA), National Institute of Standards and Technology (NIST), and Department of Energy (DOE). Faculty from community colleges, four-year colleges, and major universities from across the United States, Canada, and other nations gathered at these sessions to share ideas on more effective materials education. Many of the leading corporate and governmental agencies with materials science and engineering laboratories provided updates at the workshops on the full range of materials developments. The seminars presented at these workshops provided us with an opportunity to learn of the latest improvements. Much of the fine work of all the contributors to NEW:Updates has been compiled into a useful CD-ROM, *EMSET 2 (Experiments in Materials Science and Engineering 2)* to complement the book as well as other courses. This CD-ROM, now in its second edition, has over 350 experiments and demonstrations as well as numerous color images, web links and other instructional aids that should serve as a valuable supplement to a variety of courses.

Since our last edition the Internet has exploded and technical CD-ROMs have become readily available. The Materials Science and Technology Gateway website, *http://MST-online. nsu.edu,* (housed at Norfolk State University with links to excellent sites) is one example of many Internet resources you will find useful in addition to the many URLs listed in the book.

The enclosed demonstration CD-ROM of MaterialTool by Wroblewski and Vanka, provides another useful supplement. This demo disk provides a sampling of the user-friendly, interactive software that catalogues over 1,000 materials and manufacturing processes—all fully supported with clear text descriptions, images, animations, property tables, hyperlinks, and references.

ACKNOWLEDGMENTS

Until you write a textbook, especially one as extensive as *EMT,* you cannot fully appreciate the work involved. We have been most fortunate to receive so much input from so many sources. From those hundreds of educators, researchers, engineers, scientists, and students who participated in the National Educator's Workshops series, we gained valuable insights and resources that helped us to keep our material current and appropriate.

We thank our colleagues in community colleges and four-year institutions who provided their suggestions. The thorough review by Mitchel Lifland of Eastern Kentucky University and Patricia Olesak of Purdue University served as important guides for *EMT4.* Also, we used valuable input from Dale Wilson and his students from Tennessee Technological University, students at Norfolk State University, Tidewater Community College, and Norfolk Ford Assembly Plant. We welcome feedback from readers of our work and invite you to contact us through Prentice Hall.

Illustrating a text of this nature is a very difficult job. We appreciate the assistance of the numerous colleagues, societies, companies, and governmental laboratories, most of whom are listed with each figure's caption, for providing the valuable information and illustrations. We extend a special thank you to Luther Gammon of Boeing Materials Technology who provided a wealth of photomicrographs for this book, and for *EMSET 2*. He also provided an entire short course on "Microscopy of Fiber-reinforced Polymer Composites" for *EMSET 2*. From *Advanced Materials & Processes,* editors Margaret Hunt and Donald Baxter were very helpful in supplying photographs and illustrations that improve readability.

Our detailed index resulted from the thorough work done by Tammi Thompson. The editors and support staff listed on the copyright page gave us needed support and understanding as we worked to meet deadlines.

We especially wish to acknowledge invaluable assistance from the following people for their various input over several editions:

Ed Widener—Purdue University

Clint Bertrand—Texas A&M University

Jim Nagy—Erie Community College

Michael Wehrein—Lethbridge CC, Alberta, Canada

Rhonda Housley—Spartanburg Technical College

Carl Metzloff—Erie Community College

Larry Helsel—Eastern Illinois University

Sarah Joy Nichols—Cape Fear Community College

Richard Cowan—Ohio Northern University

Seth Bates—San Jose State University

James A. Clum—SUNY—Binghamton

Joseph Neville—Wentworth Institute of Technology

Kathleen Kitto—Western Washington University

David A. Smith—Lehigh University

James DeLaura—Central Connecticut State U.

Wayne Elban—Loyola College

Steve Piippo—Richland High School

Roy Bunnell, Mike Schweiger, Irene Hays, Denis Stracher, and Eugene Eschbach—Battelle, Pacific Northwest National Laboratories

James E. Gardner & Charles E. Harris—NASA Langley Research Center

Laurie A. George—American Dental Association, Health Foundation

Robert Berrettini—Editor of JME

Hugh Baker—ASM International

Irwin Poston—General Motors Technical Center

Thomas C. Holka & Bennie S. Bailey—Ford Motor Company

Leonardo E. Boulden—Howmet Corporation

Douglas Craig—Director of Metals and Ceramics at Oak Ridge National Labs

Jonice Adams and Anna Fraker—Materials Science and Engineering Laboratories—NIST

Louis Nenninger, II,—Ceramic Engineer for Owens-Illinois

Sankar Sastri—New York Technical College

Peter Beardmore—Director of Ford Motors' Materials Research Staff

George G. Marra—Deputy Director, U.S. Forest Products Laboratory, Madison, Wisconsin

Charles V. White—GMI Engineering and Management Institute

Cindy Neilson—Ceramic Engineer, IBM

Walter E. Thomas—Wester Carolina University
Frank J. Rubino—Middlesex County College
Howard Hull—New York City Technical College
Charles Flanders—Texas A & M University
Mario J. Restive—Mohawk Valley Community College
Richard I. Phillips—Southwest Missouri State University
Peter Route—Wentworth Institute
Lynda L. Anderson—SCCC
Pravin Raut—Savannah State College
Alden P. Gandreau—Hudson Valley Community College
William Wait—SUNY-Oswego
and many others.

Module

1

Engineering Materials Technology

After studying this module, you should be able to do the following:

1–1. Recall the nature of the broad field of engineering materials technology and the contributions of materials science and engineering to materials developments, processing, and manufacturing.

1–2. Explain the significance and consequences of the five stages in the materials cycle. Recount how Life Cycle Analysis promotes green design and manufacturing for a sustainable environment.

1–3. Describe the characteristics of the ideal engineering material and discuss obstacles to changing from one material to another.

1–4. Cite examples of recent developments in materials science and engineering that have affected technologies such as information, transportation, production, medical, and entertainment.

1–5. Discuss compromise and trade-off in materials selection and other materials selection factors, including tools used for selection.

1–6. Define algorithm and describe examples of using algorithms to solve an engineering problem.

1–7. Synthesize the personal and societal values gained from the study of engineering materials technology in terms of career preparation, consumerism, and citizenship.

1–8. Use EMT WEBLINKs to gain insight into current developments and evolving content in engineering materials technology.

PAUSE & PONDER

A World of Stuff

Our world is full of "stuff" ranging from cellular phones to roller blades. The *stuff* of that stuff is materials such as wood, metal, plastic, or glass. You cannot have your stuff without materials *stuff*. The material stuff gets cooler by the day, because of the rapid advances in materials science and technology (MST). This Pause & Ponder section aims to stimulate your interest in the science and technology that makes all our stuff possible. Some of the terms used may not be familiar to you, but they will be defined later in the book. This section and the rest of this book should make you a smart materials observer. As a result you will become a better consumer of stuff and a wiser user of materials.

As you will read in Module 2, we live in a period of time that may be labeled the "Materials Age." Much of the explosion of new technology is possible largely due to new and improved materials and processes. For example, the evolutions of telephones and watches (Figure 1-1) owe their progress to advances in semiconductor and liquid crystal (LCD) technology. The phone itself is but one component of a complex telecommunications system involving satellites and transmission lines made of metals, glasses, and plastics.

Figure 1-1 Multifunction cell phone: speaker phone, three-way calling, caller ID, pager, two-way radio, plus offers Internet access. Small size and expanded capabilities lend to the fact that MSE produced improved methods of doping silicon with metal atoms at scales down to 0.5 micrometers (0.5 × 10^{-6} m).

Figure 1-2 (a) Natural polymers such as cotton in jeans and synthetic polymers such as polyester in the girls' shirts and acrylic lenses in sunglasses are the stuff of our world. Synthetic polymer, polyurethane plastic wheels on skates give improved traction, shock absorption, and lighter weight over the older steel. (b) Protective helmets worn by Amanda, Elizabeth, and Jennifer are also made of a tough synthetic polymer, ABS plastic which forms the outer shell while a lightweight, urethane dual-density foam acts as an inner cushion that protects their heads with comfort.

Materials science and engineering (MSE—defined later) is a relatively new field of study as compared with such disciplines as physics or civil engineering. However, MSE with its multidisciplinary nature is fostering improvements in all our stuff from artificial skin for burn victims to frames for bicycles. So MST and MSE provide new materials for our stuff, and of equal importance, better means to produce (manufacture) both the raw materials and finished products. The cell phone shown in Figure 1-1 functions with speaker phone, three-way calling, caller ID, a pager, two-way radio, plus offers Internet access. It owes its small size and expanded capabilities to MSE-produced, improved methods of doping silicon with metal atoms at scales down to 0.5 micrometers (0.5×10^{-6} m). Also, advances in materials and processes technology have provided for smaller, lighter, and even flexible batteries that last longer, thus allowing greater functions in cell phones and watches, as well as for computers and other devices. Much of the credit for miniature and micro scale devices resulted from our missions in space led by NASA.

Designers of our stuff work, using new and improved materials, make changes in existing products such as the dual-wheel skates or totally new designs such as in-line wheel skates seen in Figure 1-2a. Polyurethane plastic wheels on skates have the advantage of improved traction, shock absorption, and lighter weight over the older steel and rubber wheels. The in-line designs improve maneuverability over dual-wheel design. Other features of the skates that add to comfort and use for roller hockey are the polymer memory foam in the ankle area; 70 millimeter, super-high rebound hockey profile wheels; and precision bearings. In Figure 1-2b, notice the protective helmets worn by the skaters. These helmets are examples of how plastics have been able to bring safety to skating, biking, and other sports. A combination of a tough ABS plastic outer shell and a lightweight, urethane dual-density, foam inner cushion protects their heads with comfort.

Metallurgy, the study of metals, is one field on which materials science and engineering was based. Through the centuries, metals in their many combinations as alloys have been developed into a key material for engineering. Much of our stuff is made of metal, and we often see iron and steel in a rusted state as shown in the steamship's engine in Figure 1-3a. Years of saltwater exposure have badly deteriorated the engine. Conversely, the propeller in Figure 1-3b has survived a similar ocean environment but still maintains much of its original luster. The use of brass and bronze for materials subjected to saltwater often solves problems of corrosion, but at a cost. Brass cost much more than iron and steel.

Figure 1-3   (a) Rusted cast-iron engine from steamship *George Farwell,* which sank off of Cape Henry in Virginia Beach, Virginia, in 1906. Years of saltwater exposure have badly deteriorated the engine. (b) On the other hand, the brass alloy used to make the propeller withstands the saltwater environment.

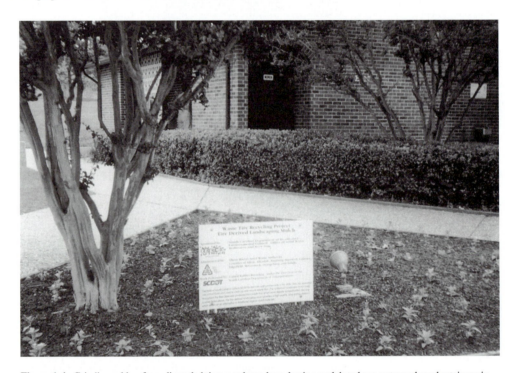

Figure 1-4   Grinding rubber from discarded tires produces long-lasting mulch to keep out weeds and retain moisture for plants while keeping tires from polluting our environment.

Much of our stuff has a limited useful life and must be discarded. Automobile tires are an example of a product that has a service life ranging from 30,000 to 100,000 miles. As a consequence, annually we discard hundreds of thousands of tires, thus creating huge problems in terms of solid waste. Tires are dumped into large piles and create a number of environmental problems ranging from polluting streams, to creating breeding pools for mosquitoes, to being ignited into dangerous fires. Figure 1-4 shows one of the many attempts at recycling used tires. By grinding up the rubber (a polymer) in tires it can serve as a very durable mulch to keep out weeds and retain moisture for plants. Do you know of other ways to recycle tires to keep them out of our landfill dumps?

Figure 1-5 (a) Early food containers of glass and steel use the common ceramic and metal materials of the late nineteenth and early twentieth centuries. (b) Advances in materials technology brought on new containers made of two-piece aluminum and steel cans and "barrier" packages made of laminated composites of plastics, aluminum, and paper. These new containers are lightweight, will not break, and offer long shelf life for their contents.

We often take for granted the stuff we eat without giving much thought to how we keep our food safe and fresh for consumption. Canning with glass (a ceramic material) and "tin cans" has been used for more than a century as seen in some of the early food containers from the beginning of the twentieth century in Figure 1-5a. Improvements in most forms of container materials have been dramatic in recent decades (Figure 1-5b). Many tin cans, actually made of tin-free steel, are now made quite thin and of only two pieces (top and body) versus the three-piece soldered or welded can. Aluminum, although more costly than steel, is used for most beverage cans. Can you imagine why? Many forms of polymers including paper and plastics, often in laminated form with aluminum foil, are used as packaging to keep beverages, such as juice and milk, fresh without refrigeration. You will learn more about these types of composite containers later in the book. *Stop* for a minute and look at various kinds of cans to observe their construction and materials. Then look at the construction of other containers, such as your toothpaste tube; tear it apart and you will see its laminated composite construction that uses polyethylene, paper, and aluminum.

Artists use many kinds of materials to form their creations. Stone (a natural ceramic) of various types (Figure 1-6) has provided a long-lasting medium for art. Through the ages stone art and architecture have provided society with timeless works that live on for centuries after

Figure 1-6 Since early times artists have chosen various types of stone (a natural ceramic) such as marble, granite, and limestone for their media of expression.

their creators are gone. Unfortunately, many of these ancient works of art are threatened by pollutants in the air that result from harmful emissions from gasoline-powered vehicles and a variety of emissions from factories. This dilemma provides a challenge both to protect the works of art and improve our technology for a safer environment for art and living things. Therefore, we must develop improved methods of materials processing if we want to have cool stuff and a healthy environment. Perhaps you will contribute to those improvements.

Throughout this book as you learn basic and advanced concepts about engineering materials technology, you will read about connections to applications in your everyday life, as well as, trends for new technology that may affect you. Your newfound knowledge of materials technology may lead you into a career path that will involve you in selecting and applying materials improvements.

This Pause & Ponder sets the tone for this book. Your study should help you to (1) develop an awareness of the importance of engineering materials in everyday life; (2) recognize society's dependence on materials; and (3) appreciate the value of a knowledge of engineering materials technology for you as a consumer, citizen, and member of the technological workforce.

Many forward-thinking businesses seize on advances in materials technology to propel their corporations beyond their competitors. Their engineers can select from ***designed materials*** rather than settling for **off-the-shelf materials** (materials with limited structure and properties). Throughout this module and the text, we will cite many examples of unique applications of materials and processing technology that set the pace for superior products and systems in fields such as transportation, construction, communication, recreation, health, and entertainment. The examples will help you to develop important competencies in engineering materials technology that will serve you well.

PAUSE for a moment and PONDER this brief discussion of materials and processes. As you read this module, reflect on the discussion to help in comprehending the new concepts presented. We have designed this book to provide you with an understanding of how materials science and engineering fit into the many aspects of technology and society. Industries such as automotive, aircraft, sports, computer, and entertainment rely heavily on materials technology. The illustrations, case studies, problems, and other topics were selected to give you a feel for the significance of both engineering materials and technical problem solving required in modern industry and society. Enjoy!

 EMT WEBLINK*

MIT's World of Materials—*http://www-dmse.mit.edu/wom/*

Materials Science and Technology—*http://MST-Online.nsu.edu*

1.1 ENGINEERING MATERIALS TECHNOLOGY

Materials are the matter of the universe. These substances have properties that make them useful in structures, machines, devices, products, and systems. The term *properties* describes behavior of materials when subjected to some external force or condition. For example, the tensile strength of a metal is a measure of the material's resistance to a pulling force. The *Family of Materials* (which will be explained in more detail in Module 2) consists of four main groups of materials: *Metals* (e.g., steel), *Polymers* (e.g., plastics), *Ceramics* (e.g., porcelain), and *Composites* (e.g., glass-reinforced plastics). The materials in each group have similar properties and/or structures, as will be described later.

Engineering materials is a term often used loosely to define most materials that go into *products* and *systems.* A telephone is a product that would be part of a telephone system composed of many telephones, wires, fiber optics, switches, computers, and so on. Engineering materials can also have a more specific meaning that refers to materials whose structure has been designed to develop specific properties for a given application. For example, *engineering plastics* are

> those plastics and polymeric compositions for which well-defined properties are available such that engineering rather than empirical [trial and error] methods can be used for the design and manufacture of products that require definite and predictable performance in structural applications over a substantial temperature range. (ASTM, 1990, p. 167)

In other words, engineering plastics such as polycarbonates and acetals could replace more *conventional engineering materials* such as steel and wood because their properties are competitive for structural components such as piping, cams, and gears. On the other hand, general-purpose plastics, such as polystyrene and vinyls, do not possess the properties to carry heavy loads but serve as packaging, upholstery, and so on.

The field of *materials engineering* "deals with the synthesis* and use of knowledge [structure, properties, processing, and behavior] in order to develop, prepare, modify, and apply materials to specific needs" (National Research Council, 1989, p. 20). Materials engineers have become very much in demand as we seek to improve the efficiency of products.

Materials science and engineering (MSE) has become a major field of study, one critical to many other fields. As defined by a National Academy of Sciences study, MSE involves the generation and application of knowledge relating the *composition, structure,* and *processing* of materials to their *properties* and *uses.* The "science" focuses on discovering the nature of materials, which in turn leads to theories or descriptions that explain how structure relates to composition, properties, and behavior. The "engineering," on the other hand, deals with use of the science in order to develop, prepare, modify, and apply materials to meet specific needs. The field is often considered an engineering science because of its applied nature. Materials

---

(*EMT WEBLINKS listed throughout the book will link you to Internet sites related to topics being covered. The dynamic nature of the Web brings frequent changes, so some of these sites, while available at the time of writing, may not be up now.)

*Several terms, such as *synthesis* and *process,* introduced in this module will be developed later in the book. Our thorough index allows easy cross-reference of most terms.

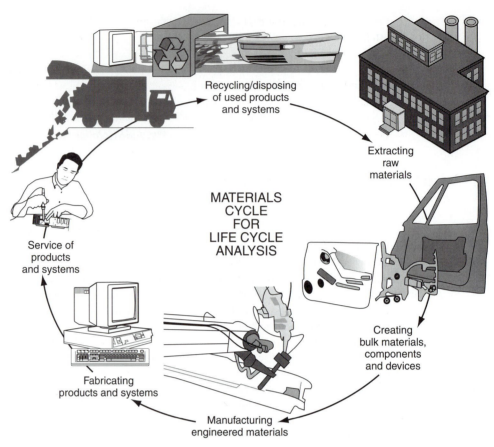

Figure 1-7 Materials Cycle—Life Cycle Analysis

science and engineering is interdisciplinary or multidisciplinary, embracing areas such as metallurgy, ceramics, solid-state physics, and polymer chemistry.

Engineering materials technology (EMT) covers fields of applied science related to
materials, materials processing, and the many engineering specialties dealing with materials
such as research and development, design, manufacturing, construction, and maintenance.

Many new processes have evolved for the manufacture of engineering materials.
Through these processes, coupled with the design of engineering materials, we now enjoy the
benefits of superior engineering materials.

1.2 THE MATERIALS CYCLE—LIFE CYCLE ANALYSIS

To better understand engineering materials technology, it is useful to view the materials cycle.
The *materials cycle* can be broken into phases as shown in Figure 1-7. An explanation of each
of the stages listed in the figure will provide you with insight into the importance of materials,
how they affect our lives, and their total impact on society. As explained later in this module
and discussed throughout the book, the Materials Cycle allows for **Life Cycle Analysis (LCA)**
of products to determine their impact on the global environment. Each phase of the cycle presents a challenge for improved efficiencies in dealing with products regarding materials and
processes in terms of cost and environmental impacts.

1.2.1 Extracting Raw Materials

The earth has provided us with the basic ingredients for producing an unlimited variety of materials. The basic building blocks of these materials are the 106 known chemical elements that
comprise the *periodic table of elements,* shown in inside front cover Table 10-1 of the appen-

dix. Some of these elements, in the form of *solid materials,* are readily available to use, such as gold. Pure carbon in the crystalline form of diamond, the hardest known material, also takes little processing. But other elements are locked in the earth, ocean, or atmosphere.

It is often necessary to mine ore to obtain raw materials. Aluminum (Al), for example, is the most abundant metal; it comprises more than 15% of the earth's crust. However, when it is extracted from the earth, Al is combined with oxygen (O) to form alumina (Al_2O_3). It takes large amounts of electricity to *extract* Al from bauxite (Al_2O_3) ore. It takes 95% less energy to produce aluminum by *recycling* scrap aluminum products.

Synthesis involves transforming gases, liquids, and solid elements by chemical and physical means, where atoms and molecules are combined to form solid materials. An example of synthesis is production of a bulk material, such as nylon. The synthesis of polymers from ***raw materials,*** including coal, petroleum, water, and air, yields substances for producing *bulk materials* such as bulk nylon. Synthesis requires constant research by scientists to improve synthesis techniques.

While also dealing with atoms and molecules, ***materials processing*** "includes control of structure at higher levels of aggregation and may sometimes have an engineering aspect" (National Research Council, 1989, p. 224). Materials processing will yield bulk materials such as nylon tubing, individual components such as ceramic jet nozzles, devices such as semiconductors for computers, structures such as automobile frames, and systems such as fiber-optic communications systems. All the products and systems seen in the photos of this module resulted from materials processing. Lyle H. Schwartz (1996) coined the term ***scitech*** to describe the complex intertwining of science with technological applications such as materials processing and manufacturing.

Materials processing has historically been considered technology rather than science. Generally, trial and error was used to develop the technology rather than scientific inquiry and theoretical analysis. But more recently, applications in fields such as applied chemistry, solid-state physics, and fracture mechanics have caused a merging of science with engineering and the development of the discipline of materials science and engineering.

The syntheses phases of materials technology have been considered more basic science involving scientists such as metallurgists, polymer chemists, and nuclear physicists. However, the development of advanced ceramics and advanced composites as well as that of electronic materials has caused a melding of disciplines. Many universities now offer degrees in materials science and engineering because of the need to ensure that the synthesis phase incorporates considerations for the processing of raw materials. These degrees might be considered *scitech* degrees.

One example of emerging materials technology is ***shape-limited synthesis,*** a new method developed to produce materials that blend synthesis with processing by beginning with one of the chemical agents already in the form of the final shape (see Module 8). This technology contrasts with the conventional procedure of first synthesizing liquids, particles, powders, or pellets, which are then processed by casting or molding into the final shape. These techniques are used in the production of advanced ceramics and composites.

During the raw materials extraction phase, there must be a concern for the by-products of refining and synthesis. Are gases released to the atmosphere that may have adverse effects on people, animals, vegetation, and even on other materials? Is the water used in separation processes returned safely to rivers and streams, or into the ground? Does the waste contain heavy metals or toxic chemicals that would be dangerous to the animal food chain (insects, birds, fish, cattle, then humans)? New techniques of Life Cycle Analysis should promote reduced impacts of materials on the environment.

1.2.2 Creating Bulk Materials, Components, and Devices

Bulk materials are the products of synthesis, materials extraction, refinement, and processing. There are many bulk materials with which you are familiar, such as fir plywood, sheet steel, acrylic tubing, window glass, copper wire, and concrete. Bulk materials are usually made in large quantities through continuous processing and then supplied to manufacturers of components and devices.

Components include gears, electrical wires, screws, nuts, jet engine turbine blades, brackets, levers, and the thousands of constituent parts that go into many products and systems. ***Devices,***

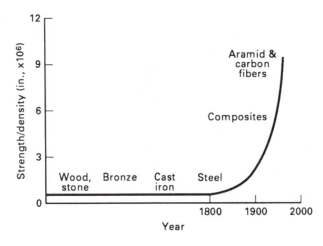

Figure 1-8 Creating bulk materials and components using innovative techniques yields large gains. Modern materials have 50 times higher strength-to-weight ratios than cast iron and other early engineering materials. (Oak Ridge National Laboratory)

which include microprocessors, resistors, switches, and heating elements, are usually more complex than components and are designed to serve a specific purpose. ***Products*** are individual units, such as roller-blade skate sets, television sets, chairs, and telephones. ***Systems*** are an aggregate of products, components, and devices. For example, a telephone system is made up of millions of products (e.g., telephones, microwave transmitters, and computers), components (e.g., optical fiber, copper wire, and lasers), and devices (e.g., switches, relays, and microprocessors).

The source of bulk materials varies by material type. Some materials go through all stages from extraction of raw materials to production of finished parts by the same producers, but this would be an exception. Normally, specialists are involved in the various stages. For example, production of bulk polymers (plastic, rubber, paint, adhesives) is usually done by materials makers known as *polymer manufacturers*. These manufacturers are often located near oil refineries because most plastics are petroleum based (see Figure 6-5). The flow diagram for polymer production shows that the manufacturer begins with raw materials such as crude oil, natural gas, trees, and cotton to make chemical compounds such as ethane, trichloromethane, hydrogen fluoride, and ethylene chloride. From these raw materials, monomers (single molecules) are produced and include ethylene, methylmethacrylate, and vinyl chloride. Through heat, pressure, and the addition of chemicals (catalysts), the monomers are polymerized to form long-chain polymers. The final polymer (resin) is a bulk "virgin material" ready to go to the next stage of production, which is normally handled by ***fabricators***. Polymer manufacturers may sell monomers to ***processors;*** they polymerize the monomers into bulk plastics such as acrylic sheet or PVC [poly (vinyl chloride)] tubing. The processors then sell their bulk plastics to fabricators, who then produce engineered materials such as acrylic lenses for automobile brakelights.

Figure 1-8 depicts how advances in engineering materials technology have helped to improve properties of engineering materials. As the curve shows, traditional materials such as stone and cast iron had very low strength-to-density ratios (*specific strength*). In the latter half of the twentieth century, specific strength has increased in engineered materials to over 10×10^6 in. for advanced composites, compared to less than 0.5×10^6 in. with traditional, "off-the-shelf," engineering materials of wood, stone, and metals (National Research Council, 1989, p. 20). Figure 1-9 shows some applications of advanced ceramic and advanced composites, new breeds of materials, for engines.

Pre–industrial revolution production techniques meant that production was done manually and individual components were cast or cut and *wrought* (shaped with heat and force) from bulk materials. Next, the bulk materials were turned into finished "customized" products; this required considerable skill and labor.

Figure 1-9   More efficient engine components made of *advanced ceramics* and *advanced composites* replace *traditional materials.* Clockwise from upper left: silicon-nitride piston pin, two titanium valves, rapid solidification (RS) aluminum valve, silicon-nitride valve, metal-matrix composite (MMC) piston, RS aluminum connecting rod, and two titanium valve spring retainers. *(Advanced Materials & Processes)*

Mass production, which evolved through the industrial revolution, made it possible to produce large numbers of parts using machines that reduced the amount of human labor required. More jobs were created as demand for affordable products grew in developed countries. The materials science and engineering (MSE) field was in a formative stage as metallurgy began to evolve as a science aimed at better understanding the nature, properties, and processibility of metals and their alloys. Then, during the middle 1900s, MSE began to grow in importance as society began to demand more from technology; conventional materials could no longer meet the needs for such materials properties as higher temperatures, greater strength-to-weight ratios, and directional properties. It became necessary to engineer materials as well as products and systems.

1.2.3 Manufacturing Engineered Materials

This phase may involve bulk materials such as plastic sheets or may create a bulk material, component, or product. The new generation of engineering materials are often *designed materials;* they have been *engineered* to provide designated properties. In other words, instead of designers selecting from a list of available materials, they may specify the desired properties for their needs, and then rely on materials engineers and technologists to create materials to suit the need. *Advanced composites* are examples of engineered materials because engineers and technicians determine how reinforcing fibers should be aligned to withstand the stresses that a product will encounter under service conditions. These composites may be made of plastic resins that can withstand higher temperatures than general-purpose plastics. *Advanced ceramics,* one class of engineered materials, may be tougher than the normally brittle ceramics. Figure 7-22 shows the technique for manufacturing an engineered ceramic: Chemical vapors of silicon carbide form a matrix as they are infiltrated into reinforcing fibers of silicon carbide for net-shape advanced ceramic composite parts.

Figures 1-10 and 1-11 depict the nature of manufacturing technology for engineered materials. For example, advancement of composite science and technology involves an integrated program, as seen in Figure 1-10, where materials science couples with the studies of engineering mechanics (study of forces in a materials system), durability, and engineering

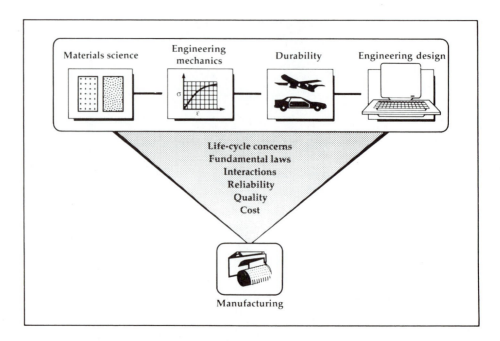

Figure 1-10 Manufacturing engineering materials. An integrated program combines traditional strengths with work in manufacturing processes to yield an expanding science base, new engineering methodologies and practice, and a unified-life-cycle approach to manufacturing. *(Advanced Materials & Processes)*

Figure 1-11 Left: Engineered graphite epoxy framed tennis racquet provides directional properties for a large sweet spot (at arrow) and considerable power. Right: Force (at arrow) improperly exerted on the frame, instead of the strings, caused the racquet to fail catastrophically.

design; these studies then converge with issues of product life-cycle, fundamental laws (e.g., physics and chemistry), how material components interact, the quality and reliability of the materials system, and finally, how cost is affected by the system. Figure 1-11 shows an example of an advanced composite product that has resulted from the systems approach to manufacturing. The tennis racquet on the left has a pointer indicating the sweet spot of the racquet. The graphite fibers within the epoxy matrix, racquet frame were "engineered" to

Figure 1-12 Manufacturing engineering materials with improvements. An engineering thermoplastic ABS resin, used in these interior door panels, elevates this grade of Cycolac® so that it is a competitor with the popular polypropylene plastic. Advantages of the ABS include reduced glare for customer acceptance, and molded-in color, which is preferred by processors because of the potential to eliminate painting, lower component cost, reduce emission of VOCs (volatile organic compounds), and enhance recycling. (G.E. Plastic)

provide the proper directional properties to withstand impact from a tennis ball in the sweet spot region. An equivalent amount of force improperly applied to the racquet in another region, as seen by the pointer in the upper right, resulted in racquet failure. The directional properties of the materials in the frame were engineered only for normal impact from proper use of the racquet.

Many new processes have evolved for the manufacture of engineering materials, as shown in Figure 1-12. At each stage of the materials cycle, we must look at various issues, ranging from materials and production costs, to environmental impacts, to consumer acceptance. Note from Figure 1-12 how producers of engineering materials strive to address issues such as lowering cost, reducing processing steps, recycling materials, and improving air quality (reducing VOCs [volatile organic compounds] improves air quality). ***Intelligent processing of materials (IPM)*** is an evolving technology that uses computer modeling of processes and sensors put in place to monitor and permit control of processing. IPM yields better quality and more reliable products.

Another group of engineered materials is ***smart materials,*** a term referring to a variety of liquids and solids that have the ability at a predetermined condition to sense stresses and respond to alter their properties. For example, smart glass will darken when an electrical current is passed through a laminated grid in the glass. Electrical current can also thicken certain fluids. Optical fibers, metal fibers, and electrorheological (ER) fluids inserted into solids such as plastic composites, aluminum sheet, or concrete can sense stresses and cracking to provide early warning of probable failure. These fibers might be used in engineering applications such as insertion into aircraft wings and highway bridges to provide early warning of failure. Fibers could also serve to change the stiffness of automotive springs, helicopter blades, or golf clubs. Smart products using piezoelectric devices include smart skis and bicycle shock absorbers.

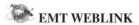

EMT WEBLINK

Smart skis—*http://www.acx.com/cool_smartski.html*
Smart bicycle shocks—*http://www.acx.com/cool_shock.html*

In the search for smart materials, ***biomimicking*** has been employed to study nature and attempt to mimic its wonders. Examples of biomimickry include efforts to reproduce the way that spiders produce very strong fibers and the procedures used by mollusks in building their shells, which may lead to better techniques for fabricating integrated circuits for computers and microprocessors.

1.2.4 Fabricating Products and Systems

Once engineering materials have been manufactured, they are assembled into many useful products and systems. New fabricating techniques are evolving through improvements in manufacturing engineering. ***Manufacturing engineering*** is the study of techniques to turn bulk materials into finished products and systems. Once a field of empirical methods in which processes were handed down from generation to generation, with improvements coming gradually as a result of observation and analysis, manufacturing has recently become more involved, with detailed study and application of theoretical principles. College programs leading to degrees have sprung up for manufacturing technology and manufacturing engineering at levels from associate degrees through doctorates in engineering.

Automation is the common element today in manufacturing, with less manual labor involved. Computers, sensors, robotics, machine vision, adaptive control, and artificial intelligence are applied by manufacturing engineers to perform the manufacturing processes once carried out by humans. Improved quality, smaller lot sizes, more product options, and reduction in price have been the benefits of the improved technology. Fabrication of products and systems is done in a safer work environment, where much of the handling of parts is done by machines. **Just-in-time** (JIT) techniques that rely on computer assistance keep raw materials and parts moving with a minimum of warehousing. People who wish to work in manufacturing are expected to be well educated because of their newer role as *problem solvers* rather than laborers.

A clear understanding of the total materials cycle is now necessary by the entire technological team of crafters, technicians, technologists, engineers, and scientists. Computer-integrated manufacturing (CIM) is a technological team approach to manufacturing that places the materials engineers and technologists together with the design engineers, technicians, and manufacturing engineers, plus the environmental engineers and even marketing personnel for a ***systems approach*** to products. New laws regarding the environment and natural resources are intended to ensure that the technological team use Life Cycle Analysis on all stages of the materials/product cycle to have appropriate safeguards that protect the environment as well as present and future generations. See Figure 1-12 for an example.

The change in the nature of fabricating products has meant numerous changes in the workplace. Manufacturing jobs for crafters and production workers (skilled, semiskilled, and unskilled labor) have decreased drastically. Jobs call for educated people who can design products and manufacturing systems; set up, monitor, troubleshoot, and service manufacturing equipment; and serve as liaison (go-between) among various stages of the total materials/product cycle to ensure that quality is always improving and customer needs are met.

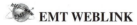

EMT WEBLINK

Manufacturing—*http://www.manufacturingiscool.com/*

1.2.5 Service of Products and Systems

Shifts in manufacturing have resulted in a larger service workforce and a smaller manufacturing workforce. The complexity of products makes it harder for the average person to make repairs on his or her own products. Special diagnostic equipment is used to analyze everything from automobiles to robots to appliances.

The demand for better quality in products and systems has resulted in improved, long-term warranties. Manufacturers are very interested in analyzing materials that fail so that they can improve materials engineering and product design. Proposed laws, and some laws already on the books, involving Life Cycle Analysis place responsibility for disposal of materials/products on the manufacturer once the service life is over. ***Design for assembly*** places emphasis on designed products that lend themselves to easy assembly by robots and other automated equipment. ***Design for disassembly*** is a concept that places recycling at the beginning or design stage of the materials cycle to ensure that waste going into municipal landfills will be minimized.

1.2.6 Recycling/Disposal

The last stage of the materials cycle can become the first stage through the resurrection of material when recycling is employed. Most materials can be recycled. However, it is very difficult for manufacturers to develop a full materials cycle that will ensure recycling. Examples of industries with successful recycling programs are steel and aluminum. These materials have been recycled successfully because of the tremendous savings gained through the production of bulk materials using recycled materials. It has taken legislation to force recycling, even of aluminum, as well as of glass and paper. This recycling has been accomplished by requiring deposits on beverage containers to provide financial incentives for people to return cans and bottles.

Federal, state, and local laws have put mandates on recycling by restricting the amount of municipal solid waste that can be placed in landfills. Clean air and water regulations have restricted the amount and type of waste that can be incinerated or dumped into the ocean. But much remains to be accomplished to develop the proper attitudes and habits among our citizens if we are to make the total materials cycle efficient and thus protect the environment and natural resources for future generations. Data from Life Cycle Analysis will promote wiser decisions for recycling/disposal.

One example of the problems involved relates to materials for packaging. Manufacturers of polystyrene packaging (often improperly called styrofoam) worked with McDonald's and other fast-food restaurants to head off complaints based on the fact that most plastics are not biodegradable. Citizen groups have called for elimination of the polystyrene clamshells used to package hot dogs, hamburgers, and other sandwiches, as well as for drink cups. Manufacturers set up a recycling system to reclaim the polystyrene. Because of the pressure and fear of losing business, most fast-food chains moved back to paper packaging before the recycling system could be fully implemented. Was this the best action?

First, we should recognize that paper will *not* biodegrade in the modern landfill. Modern landfills are designed to stop materials from breaking down, because they can then leach into the groundwater. So going to paper packaging of foods might actually create more solid waste because this type of paper is difficult to recycle. Complex issues are involved in decisions regarding the use of plastics versus paper or glass. Most of our local, state, and federal legislators lack technical backgrounds and therefore are not qualified to decide some of these complex issues. It is therefore vital that citizens prepare themselves with a knowledge of materials technology to provide input to the legal issues. Recycling and disposal of materials is one area that will continue to be a major issue.

1.2.7 Green Manufacturing and the Materials Cycle

Green manufacturing, a major global effort, seeks to make all phases of the materials cycle environmentally friendly; thereby, placing the least amount of stress on our delicate balance of nature. Materials issues are a key to green manufacturing. For example, the Partnership for a New Generation Vehicle (PNGV), discussed later in this book, aims to make a "green vehicle" with 50% mass reduction in body and chassis by using ultra-high-strength steel, reinforced plastics, aluminum, and magnesium.

Achieving the goals of green manufacturing requires a beginning-to-end view of a product in the manner we depicted with the materials cycle in Figure 1-7. ***Lifecycle analysis (LCA)*** looks at all factors involved with a product by making an inventory of the inputs (e.g., raw materials and energy) and outputs (e.g., products and by-products). **(LCI)** a listing of raw materials, bulk materials, gas emissions, product waste, recyclables, etc. LCA results in a **Life Cycle Inventory.** Using the LCI to evaluate possible impacts, such as depletion of old-growth trees and water pollution, should allow a thorough interpretation of the factors to minimize harmful results. LCA thereby provides a perspective on the impact of manufacturing.

While the need to comply with laws that regulate manufacturing's impact on the environment is evident, manufacturers also recognize the need to demonstrate to the public that they are environmentally responsible. The push for green manufacturing resulted from a variety of forces, but today most major manufacturers and their industries seek to be seen as environmentally friendly.

The auto steel industry has built up a good infrastructure of recycling, so that most of the steel from cars goes through salvage yards and is returned for new steel. In fact, the system is so well established that modern steelmaking requires scrap steel. See Figure 5-47, A Flowline of Steelmaking. Aluminum, newsprint paper, and bottle glass are other examples of successful recycling. But we have a long way to go to reduce waste which fills landfills with many of our engineering materials. As we see in Figure 1-7, recycled materials can make great substitutes for raw materials, but it takes a major effort to create cost-effective systems.

Plastics offer a challenge in our need to reduce products that are buried as wasted materials instead of being recycled. Many plastics lack an efficient recycling infrastructure. Discussed in Module 6, certain packaging plastics such as polyethylene terephthalate and high-density polyethylene are having some success with recycling. Nylon makers have begun a recycling system to recycle nylon carpet to keep it out of landfills. Recycling of the carpet yields a basic raw material, caprolactam, used in making nylon 6. The recycling of nylon carpet not only saves the raw material (oil), but also saves energy for processing. This is often true; recycling aluminum, for example, can save 95% of the electricity required to process bauxite ore into aluminum ingot.

Mistakes costly to the environment and society have been made with mining, well drilling, and tree harvesting. Mine tailings left behind when mining ore are washed by rain, causing the leaching of concentrated minerals that might be harmful to plants and animals in the streams, rivers, and lakes. Oil drilling can result in harmful spillage. Clearcutting of trees can result in severe soil erosion and can upset wildlife.

On the other hand, if we did not extract these raw materials from the earth, civilization would not benefit from the valuable resources harvested, people would not have employment in the related fields, and we would not have a thriving commerce that offers a good quality of life. In other words, we have made trade-offs, some conscious, some not. For example, the plastics industry is a comparatively young industry that evolved during a time when oil and gas seemed limitless. There was not much knowledge or concern about pollution, global warming, the ozone layer, or the scarcity of landfill space for municipal solid waste. Plastics offered us sanitary packaging, great strength-to-weight ratios in materials, improved design flexibility, lower-cost products, and other advantages over the traditional materials such as metal, wood, paper, and ceramics.

Now that we recognize such problems as limited resources, energy waste, harmful by-products of materials processing, and materials disposal, the expectation is that everyone will apply knowledge of materials to ensure that our natural resources are best utilized and that engineering materials technology safeguards the environment.

1.3 MATERIALS SELECTION

We are surrounded by materials and we rarely think about how these materials are selected. Why was your desk made of solid wood, plywood, or plastic-laminated particle board? Why have so many plastics replaced steel and zinc in automobiles? What is the controversy about using foamed polystyrene plastic to package fast-food?

While you might take for granted the materials that make up your products, you can be sure that the designers did not. People who design homes, cars, aircraft, clothing, furniture, and other products or systems devote a lot of attention to the selection of the materials they use. Material selection might make or break a company. But how do the designers make that selection to arrive at the best material?

Did you realize that there are hundreds of different types of plastics? They range from very soft to quite hard, inexpensive to very expensive, and transparent to opaque. Wood, too, is available in numerous varieties, ranging from the very soft, lightweight balsa wood used for model airplanes to a group of birch and beech trees that are so heavy and hard that they are sometimes called ironwood. Lignum vitae, one of the hardest and heaviest woods, has been used for bearings and pulley sheaves on ships since ancient times.

Metals in combination with other metals or nonmetal elements, known as *alloys,* include many varieties of steels (iron and carbon), aluminum alloys, brass (copper and zinc), and hundreds of others. Steel, the most common production metal, is found in the bodies of cars, as railroad-train wheels, and for piano wire.

Some glasses, such as most drinking glasses, are very delicate and cannot stand much temperature change, but thermally hard glasses are used as windows and spacecraft on furnaces. Cement is one type of ceramic, as is glass. Clays are ceramic raw material; they are used for making dishes, toilets, and spark plug insulators.

With the nearly limitless range of materials available to the designer or architect, how do they make a materials selection for products and buildings? What selection criteria are most important?

1.3.1 The Ideal Material

The engineer, technologist, technician, or architect searches for the ideal materials to suit the designated need. What is an *ideal material?* Among other characteristics, we can list the following for the ideal material.

1. Endless and readily available source of supply
2. Cheap to refine and produce
3. Energy efficient
4. Strong, stiff, and dimensionally stable at all temperatures
5. Lightweight
6. Corrosion resistant
7. No harmful effects on the environment or people
8. Biodegradable
9. Numerous secondary uses

It is a very complex process for the designer to find the ideal material for a specific product.

Selection of materials for musical instruments has been a challenge for centuries as can be seen in the wide application of materials used with the piano. Figure 1-13 pictures a modern, conventional piano made of such metals as cast iron for the sound board, brass foot pedals and hinges, and various steel alloys for the piano wire, and mounting posts. The piano uses several natural polymers including a walnut cabinet, cotton, felt hammers, and plastic keys (formerly made of the natural material ivory, which is now illegal to use).

With over 100,000 engineering materials from which to choose, selecting the best material for a given application or part involves an understanding of the many relationships that exist between selection and a few important variables: (1) selection and design, (2) selection and performance characteristics of materials (including properties and structures of materials), and (3) materials selection and manufacturing processes. A cost variable is inherent in all these variables.

Figure 1-13 Materials of Entertainment. The piano is an example of materials selection involving such properties as appearance (beautiful wood colors and grain pattern), durability (repeated striking of hammers on strings), and acoustics (musical sound quality).

1.3.2 Selection and Compromise

Compromise is the rule, not the exception, in materials selection. For example, ***space-age materials*** such as graphite/epoxy composites and Kevlar (aramid fiber) epoxy composites possess strength-to-weight ratios three to five times greater than those of steel. So why don't automakers use graphite epoxy and aramid epoxy composites to make car bodies, axles, and drive shafts? Wouldn't use of these composites allow for much more fuel-efficient vehicles? Yes, the space-age materials would be much more fuel efficient—but because these materials can cost 15 to 40 times as much as steel, an $11,000 Ford might cost $300,000. Who could afford to buy it?

Ford Motor Company experimented with their popular Taurus model by building a prototype chassis that reflects designers' keen interest in using reinforced-plastic composite technology for improved vehicles. Instead of using the more exotic fiber-reinforced plastic (FRP) composites such as graphite epoxy, they used glass-reinforced plastics (GRP). In the Taurus prototype, they were able to use only 8 or 10 GRP parts to replace over 400 stamped steel parts. A GRP chassis would save about 30% of the car weight because a lighter chassis means downsizing engines, brakes, and other components. Additional benefits come through more efficient manufacturing techniques. The Partnership for a New Generation Vehicle (PNGV) was formed among the big three U.S. automakers and the federal government to explore alternative materials and designs.

Throughout this book, we provide numerous examples of alternative designs, materials, and processes from many fields of technology. These examples help to illustrate the complexity of materials selection. As pointed out before, materials selection is not simply a matter of making decisions about cost and materials properties.

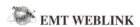 **EMT WEBLINK**

NASA photos and animation—*http://oea.larc.nasa.gov/Photo/photo.html*

Figure 1-14 Failure turned to success in the form of Post-It note pads. An alert 3M Company research realized that a weak adhesive had value for bookmarkers and other temporary notes as seen in the wide variety of Post-It products.

1.3.3 Obstacles to Change

Switching from traditional materials such as steel and concrete to newer materials such as plastic-based composites seems a simple, straightforward approach for the contemporary designer. The newer materials are often superior, but sometimes there are complications. Often, lack of experience with a new material causes hesitation by designers. Departures from tried-and-true materials may be costly, as was the case with plumbing pipes. With a shortage of copper and the health threat of lead–tin alloyed soldered water pipes, there was a quest to find substitute materials. The answer seemed to lie in polybutylene oxide plastic pipe, which relied on metal/plastic compression fittings. The compression fittings seemed to be an excellent substitute for the highly skill-oriented sweat soldering required of copper pipes. Unfortunately, builders and municipal water departments that switched to the polybutylene soon learned that this new water piping system had its problems.

With time, many of the joints began to leak. The designers who selected the new pipe were led to believe that fabricating with compression joints was a rather simple task. The makers of the polybutylene system claimed that failures were due to improper installation of the compression joints. The users of the new piping blamed the problem on oxidizing polymers (plastics) and faulty aluminum compression rings. Manufacturers involved with polybutylene pipe paid out millions of dollars in claims to replace leaking pipes and make repairs from leaks.

This controversy focuses attention on a leading problem of all new materials: It requires time before both designers and fabricators gain sufficient experience to become comfortable with them and the associated processes required to make products or systems. This problem is exacerbated when human life might be in jeopardy, such as when designing for aircraft. As a consequence, new materials and processes are usually slower to enter the marketplace than might be expected. These issues are all a part of engineering problem solving. Materials selection is a problem-solving issue that requires an algorithm for its solution.

While there are often obstacles to change, sometimes changes occur by chance or *serendipity*. There are some noted examples in the history of materials developments in which a successful new material resulted from accident or an apparent failed development. The Post-It Notes system from the 3M Company is an example of a very successful product now available in many forms (Figure 1-14) all because an effort to develop a strong adhesive failed and resulted in a weak adhesive. So the adhesive that did not stick resulted in "temporarily permanent" bookmarkers that had just the right amount of adhesion to stick to paper and other surfaces without damaging them.

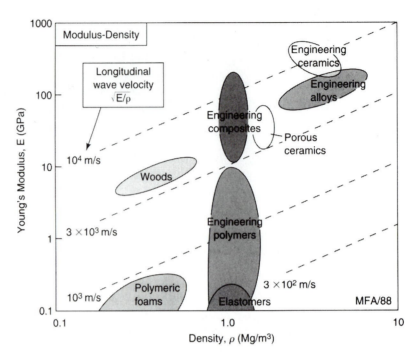

Figure 1-15 Graphical tool for materials selection. The materials selection chart depicts a plot of Young's modulus and material density. Boundaries encircle the clustering plots of these two properties to allow comparisons of groups of materials. (Ashby, Michael F., "Materials in Mechanical Design," *Journal of Materials Education.* Vol. 15, pp. 143–166, 1993.)

 EMT WEBLINK

3M Company innovations—*http://3m.com/*

1.3.4 Algorithm for Materials Selection

Engineering requires clearly stated, unambiguous steps for problem solving. ***Algorithms*** are well-defined methods for solving specific problems. Computer programs are written after an algorithm has been developed to lay out clearly the steps that the program is to solve. For example, you could write a simple algorithm to calculate the strength required of a light pole to withstand the pushing forces (compression) from a light fixture. A much more complex algorithm would be required to select a piston connecting rod for an internal-combustion engine. The first problem requires only the selection of a material of suitable size/strength to hold up the light fixture, and almost any material would suffice as long as it was sufficiently strong and pleasing to the user. On the other hand, a connecting rod will undergo many types of mechanical stress, ranging from compressive to tensile to torsional to gravity forces, in addition to thermal stress from the combustion chamber. How does the designer match component requirements with available materials?

   1.3.4.1 Selection tools.   To aid in the creation of materials selection algorithms, databases must be available to answer questions on material suitability. A materials ***database*** involves tables listing properties of materials, such as tensile strength, hardness, corrosion resistance, and the ability to withstand heat. Thousands of reference books are available with such data. Much of these data are computerized to allow easier access.

   Certain graphical techniques aid the designer in materials selection. In Figure 1-15 a graphical plot of two important properties shows a relationship of stiffness (Young's modulus) to weight (density) of major groups of materials. For example, ellipses encircle the stiffness-to-density ratio, which represents those of most engineering ceramics and engineering alloys. By encircling major materials groups, a designer can see that both groups of materials possess high stiffness but at the expense of high density. On the other hand, engineering composites, as a group, possess stiffness that approaches or may equal that of engineering ceramics and alloys; however, composites have the advantage of much lower density. Similar graphical plots are possible for a variety of properties ratios that allow the narrowing of materials as design requirements narrow the choices.

1.3.4.2 Properties of materials.   Periodicals, websites, and handbooks can provide current data and performance criteria that involve structural (load-bearing) materials:

1. Strength (tensile, compressive, flexural, shear, and torsional)
2. Resistance to elevated temperatures
3. Fatigue resistance (repeated loading and unloading)
4. Toughness (resistance to impact)
5. Wear resistance (hardness)
6. Corrosion resistance

Technical society and manufacturer's references present values for the performance criteria (properties) for metals, polymers, and ceramics, with updates on newer materials such as aramid fibers, zinc aluminum alloys, and superalloys.

Various periodicals, such as *Advanced Materials Processes, Machine Design* and *Modern Plastics,* have annual materials selectors that provide general information on properties for a long list of materials. There are also many handbooks, such as the multiple-volume *Metals Handbook* and the series of *Engineered Materials Handbooks* that cover nonmetals; both are published by ASM International.

Tables as found in Module 10, covering representative materials, provide comparative data on selected ceramics, metals, and nylon. The designer must use such data to determine if a material has appropriate physical, mechanical, and chemical properties to withstand the service conditions to which a part will be subjected. For example, from Table 10-9 we learn that nylon is an engineering thermoplastic and is used for some bearing applications. By turning to Table 7-2, Typical Properties of Selected Engineering and Technical Ceramics, which compares selected plastics, metals, and ceramics, you will note that nylon has a melting point of 215°C and a maximum service temperature of 422 K (149°C). On the other hand, alumina (used for spark plugs and computer modules) has a melting point of 2050°C and a maximum service temperature of 2222 K. Alumina is also much harder than nylon: 9 Mohs versus 2 Mohs. The modulus of elasticity [given in megapascal (MPa)] is a comparison of stiffness; alumina is very stiff (modulus of 34.5×10^4 MPa) and brittle, whereas nylon is flexible (modulus of 0.33×10^4 MPa). From the comparison of hardness and temperature resistance, it is clear why ceramics are selected for furnaces, ovens, and other applications where high abrasion and high heat are present. On the other hand, nylon is selected for rope, tubing, and gears and bearings in electrical appliances.

The two tables discussed above provide general data on properties for a simple comparison. Selection of specific materials requires many more detailed specifications. General databases from handbooks will provide much detail, but the final selection often requires that materials manufacturers supply their own properties database for their product lines. While databases are imperative in the initial selection steps, there are other factors that complicate materials selection.

1.3.4.3 Materials systems.   Materials rarely exist in isolation, without interacting with other materials. Rather, a combination of materials are selected to complement one another. In a successful *materials system,* each component is compatible with the others while contributing its distinctive properties to the overall characteristics of the system of which it is a part. A state-of-the-art telephone is a good example (Figure 1-1). The casing might be a tough

ABS plastic, which houses a microchip (a solid-state ceramic device) that provides memory and sound-transmission capabilities. Copper leads join the circuitry together. There might be a battery and a ceramic light-emitting diode to show when the battery is low. The acid in the battery must be isolated to prevent corrosion, and the copper leads must be insulated so that they do not short out. Each component is made of materials that meet the demands of the physical and chemical environment normally encountered when using the system.

1.3.4.4 Additional selection criteria.

Existing specifications have a lot of influence on the choice of material. These specifications or "standards" are used when redesigning an improved model of the product. When the materials selection algorithm results in selection of a new material, it might not be covered by current specifications from such standardization agencies as the National Institute of Standards and Technology, Underwriters' Laboratory, fire departments, American Society for Testing and Materials, or the U.S. Food and Drug Administration. The Occupational Safety and Health Act (OSHA) of 1970 sets forth conditions of safety that must be met by those involved in the manufacture or use of goods and services in the United States. It might take considerable time for these agencies to alter their specifications to include the new material or they might not approve its use.

Availability is another concern of the designer. Will the material be easily available in the quantities and sizes required by the production demands? Also, will it be available in the shapes required? Aluminum extrusions, for example, are available in many varieties of standard shapes, such as round, oval, and square tubing. In the past, designers were limited to existing materials such as metal alloys, woods, or concrete. Now, it is possible to start from scratch at the synthesis stage to have materials engineers design a materials system to provide properties to meet the expected needs.

Processibility, the ease with which raw materials can be transformed into a finished product, is of paramount concern. Much of the current focus is on low-energy processing. Companies may have difficulty processing the new material on existing equipment. Can they afford to invest in new equipment? Today, the reverse question is usually asked: Can we afford not to use the new material and process? If we do not, the competition might make the change and run us out of business with their superior product. Many new technologies are now available.

Near-net-shape production involves incorporating numerous separate parts into a single, integrated assembly, thus saving overall production costs. This new technology is receiving considerable attention from various industries. Net-shape and near-net-shape processing aims to improve product reliability while reducing costs of materials and processing. One such example, the Super Plug™, a patented door hardware module (see Figure 1-16), replaced the 61 separate, stamped and formed steel, molded plastic, and rubber door parts with one plastic/glass composite system made of XENOY® polycarbonate/polyester blend with 30% glass fiber. The single system not only saves materials and processing costs but also provides manufacturing cost savings at assembly and yields overall weight reduction for fuel savings for the life of the vehicle. Another picture of the module is shown with the door in Figure 6-21.

The near-net-shape concept is a growing trend because it is consistent with the new thinking about design/manufacturing; that is, the concept design, engineering analysis, materials selection, and processing are now orchestrated in a team approach. In traditional practice, design was handled by an engineering group. It sent the drawings to manufacturing; that department was responsible for determining how to produce the part. Often, there was conflict in the demands that each group made on the other. The new CAD/CAM technologies force an integrated approach. Materials selection is central to the design process and has great bearing on manufacturing techniques. ***Expert systems*** as a part of computer-based ***artificial intelligence*** are becoming available to help make selections from the increasing number of variables and levels of optimization inherent in materials.

Plastics, ceramics, and composites have become increasingly competitive because engineering materials and near-net-shape products and parts are suitable for plastics such as polystyrene, nylon, polypropylene, polycarbonate, and glass-filled epoxy. The metal industry

Figure 1-16   (a) G.E. Plastics Super Plug™ is an example of net-shape processing. The glass-reinforced plastic composite provides savings in both production and operating costs over the current door module (b), which is made of stamped steel and uses up to 61 parts. (G.E. Plastics)

has also brought out new processes and refined older ones to simplify and integrate parts for near-net shapes.

Quality and performance are two aspects that achieve consumer satisfaction. The high cost of most durable goods and the competition for customer acceptance has resulted in extended warranties. Materials selection must ensure that parts will not rust, break under repeated stress, or fail to perform in any other way for the predicted service life of the product.

Consumer acceptance includes many factors beyond excellent quality and high performance; there are also societal aspects. Society as a whole as well as governmental agencies are requiring a closer look (LcA-LcI) at manufactured products. Any product has to be considered in terms of its total life cycle. What are the results of the processing methods? Are polluting gases being released into the environment, or are toxic metals and chemicals being flushed into our rivers and streams? During use, does the product safeguard our health? At the end of the product's useful life, how can it be disposed of safely? Municipal solid waste is a hidden product cost that we pay in the form of higher taxes and a poorer quality of life. In a previous example, fast-food restaurant chains moved away from polystyrene packages because the public believed that these plastic containers were more harmful to the environment than paper packaging. Soft drink manufacturers are moving toward *reusable* plastic bottles. Currently, poly(ethylene terephthalate) (PET) is used for 2-liter soft drink bottles; it is a good recycling plastic. But a more durable bottle of polycarbonate can be rewashed and refilled just as many glass bottles are now handled. Why hasn't polycarbonate been adopted?

Design for disassembly has become a theme in much of product design by major corporations. Europe, which has a higher degree of ecological concern, has led the way. With the desire to facilitate recycling, manufacturers of small appliances and durable goods are establishing procedures to ensure that products can be broken into components for easy sorting prior to recycling. Among the procedures are reducing the variety of plastics, adding labels to plastics for easy identification of plastic type, and eliminating screws and adhesives so that parts will disassemble easily.

Newer computer software programs offer design that make products easier to fix. *Design for Service* programs takes its place alongside previous software programs called *Design for Assembly* and *Design for Manufacturability.* The new programs help product designers consider repair issues early in the design stage. Objectives of the program include making repairs less costly and extending the functioning life of products. Environmental issues such as recycling are directly addressed by this new computer software, which may have customers fixing products rather than tossing them out. In addition, this software augments previous software that addresses the need for disassembly of a product for whatever reason.

More often than not, *cost* is the primary selection criterion that will determine the final choice of material. In other words, if several materials have the specified physical, mechanical, and chemical properties, and are suitable for the processing technique selected, the lower-cost materials would be the logical choice. Determining cost is not as simple as it may seem. For example, a variety of plastics, including PET [poly(ethylene terephthalate)] and HDPE (high-density polyethylene), have replaced glass as containers for soft drinks, milk, and juices. Although the initial cost of plastic may be greater, the plastic bottles provide savings due to their toughness (less breakage) and the savings in shipping (PET and HDPE are much lighter than glass bottles).

The traditional "stick framed" home (Figure 1-17a) uses 2×4 studs (sticks) and a wall material such as sheet rock for the inside and brick veneer and perhaps vinyl siding. Recycled plastics provide an alternative to the very old and inefficient stick framed house. General electric has developed factory built composite wall systems (Figure 1-17b) combine a corrugated engineered-wood by-product with a low-density reinforced PPO [poly(phenylene oxide)] foam and recycled polystyrene, which is covered with a fire-retarding glass-reinforced plastic. This type of wall panel can cut down on building costs and promote recycling while improving overall quality.

Product liability is civil (as opposed to criminal) liability of the manufacturer to an ultimate user for injury resulting from a defective product. *Caveat emptor* (let the buyer beware) was once the rule. Today, numerous liability laws are in effect. For those involved in materials, particularly materials selection in the design process, the trend is for courts and juries to identify members of the design team as being responsible for some fault in a product liability action. Therefore, it is imperative to obtain and use the latest information about materials selection, particularly materials' long-term characteristics.

Figure 1-17 (a) Most contemporary homes in the United States use the "stick framed" system for walls. This system is hundreds of years old and does not take advantage of newer approaches to engineering walls. (b) A developmental wall system combines an experimental corrugated strand product with low-density polyphenylene oxide foam and reclaimed hamburger packages. The panel is covered with a glass-reinforced plastic that is an excellent fire retardant. (General Electric Plastics)

1.3.4.5 Materials selection in this book. Many of the illustrations and tables in this book provide examples of materials selection. The Self-Assessment at the end of each chapter will provide practice in materials selection as well as testing other knowledge gained from the module text. **STOP:** You are also encouraged to constantly observe materials applications in your everyday life. Keep a *materials journal* in which you write materials applications as "observations." Next to the observations section, place a section entitled "analysis"; there, record the positive or negative aspects of the observed use of materials. For some observations, state how you could make materials substitutions to improve on the designer's choice. Discuss these with others. As you progress through the book, your observations and analyses will become more sophisticated. This valuable knowledge will serve you well as a consumer, as an intelligent citizen, and in your career.

1.4 TECHNOLOGICAL LITERACY

Our society continues to become more complex. Corporate "downsizing," company restructuring, and similar trends that aim at fewer workers for greater profits place people in jeopardy of losing their jobs. To avoid being a victim of technological advancements, we must be prepared to change jobs several times during our working lifetime. To be agile in the job market, one must become ***technologically literate,*** that is, understand the language and concepts of technology to understand new technological advances. Technological literacy is not only important from a career standpoint, it is also required to function in our ever-changing technological society.

 EMT WEBLINK

MatScience Library—*http://msewww2.engin.umich.edu:591/default.html*

1.4.1 The Materials Consumer

Each year, an increasing number of newly developed materials are substituted for more familiar materials with limited properties. Today, 40% or more of a manufactured item's price represents materials cost. To be an informed and intelligent consumer requires a basic understanding of materials. The selection of a product can be improved by a greater understanding of the nature and properties of the materials used in the product. After a product has been purchased, the problem of failure, sometimes by the *abuse* or *misuse* of the product, can be lessened by such knowledge. Learning about the structure of materials, hence how materials behave, should permit an intelligent analysis of a failure and possibly pinpoint its source and cause. A knowledgeable consumer stands a much better chance of success in demanding remedial action from both manufacturers and retailers of faulty products than does one with less knowledge of the behavior of materials and a poor technical vocabulary with which to explain such behavior.

Wood is our oldest building material and possesses unique structure, warmth, and beauty, but perhaps more important, the ability to renew itself makes it equal in importance to the newer materials (Figure 1-18). With complex structures (Figure 1-18a) and wide varieties, wood as a natural composite, has many desirable properties. Still, materials technology always looks for improvements and wood often finds stiff competition, for example, in a new design for drumsticks (Figure 1-18b). In 1850, about 90% of U.S. energy came from wood, both in the raw state and refined into charcoal for fuel in glass and iron making. By 1900, wood accounted for only 20% of the nation's fuel and continued to decline. Wood use as a fuel is now on the rise in both home and industry. Industries such as lumber and paper make extensive use of wood residues, including waste chemical by-products and bark. Wood does not figure as a major fuel for meeting the void left by depleted fossil fuels, but it holds promise as a substitute for more energy-intensive materials.

In construction materials for similar phases of building, the comparison of fuel required for the preparation of building products shows that wood requires the least energy, plastic nearly six times, and steel about eight times the amount of fuel needed for wood products. Concrete, glass, and other ceramics also show great promise as future materials for building and manufacture because the raw materials are as vast as the sands of a desert. Scarcity of old-growth trees and competition for high-grade wood to make lumber, furniture, and so on, has resulted in the development of new composites that may be superior to wood for many construction and nonbuilding applications. Consumers can make choices among products made

Figure 1-18 Wood's complex structure. (a) Electron microscope reveals microstructure. (Dr. Wilfred A. Côté, SUNY College of Environmental Science and Forestry) (b) Improving on wood—a new composite material for entertainment technology. The drumstick combines the strength and vibration absorption efficiency of an aluminum core with the strong sound-making ability of the cover molded of a thermoplastic polyurethane resin. The resin is hard enough to produce strong sound and not wear out, but soft enough not to damage the drum sets. When the covers wear out, they can be replaced for less than half the cost of wooden drumsticks. The first major endorsement of the new advanced high efficiency alloy drumstick (AHEAD) was by a heavy metal rock star, drummer Matt Sorum of the band Guns 'N' Roses. *(Advanced Materials & Processes)* (c) A wide range of natural and synthetic polymers. There are natural polymers including the teak laminated cabinet, leather boots with rubber soles, cotton jeans, and skin and hair of the Reid and Connor. The chair is made of such synthetic polymers as polyurethane and nylon fabric. The carpet of natural fiber wool Berber covers a plywood (composite wood) floor. Teak, used on the cabinet, is a beautiful wood, but quite expensive. Laminating 1 millimeter of teak over particleboard (made of wood chips) makes beautiful furniture and is available at a more economical price.

of new composites, as well as, natural and synthetic polymers. A knowledge of the properties of materials will add in decision making (Figure 1-18b and c).

1.4.2 The Intelligent Citizen

The technology of materials also provides us with the necessary knowledge to make decisions based on personal values relating to political, social, and ecological issues. *A better informed citizen is a better citizen,* who is much needed in today's changing technological society in which the great issues over energy and materials resources are being debated and voted on. Citizens who comprehend key concepts of materials science and technology, such as the Materials Cycle and Life Cycle Analysis, will guide good legislation and participate in green manufacturing practices such as reduced resources consumption and recycling.

Long-range industrial research and development of new alloys and nonmetals, intensification of programs to conserve and reclaim metals, enlargement of the search for and development of new domestic sources, and the utilization of ocean resources are all affected by issues that find a source in the political, social, economic, or ecological spheres within our society. The fact that the United States is dependent on foreign sources for most metals (with the exception of iron and copper), plus the negative effects of our high rate of consumption of the world's known reserves, places greater emphasis on a good working knowledge of the technology of materials so that citizens can guide their government representatives in making the correct decisions on matters that will have a lasting effect on our lives and living standard.

APPLICATIONS & ALTERNATIVES

As consumers demand better products, the technological team of technicians, scientists, engineers, crafters, and technologists will improve the materials and processes for better design and higher quality. One example of applying new materials and processing technology is seen in the Optimum Compaction Low Void (OCLV) bicycle in Figure 1-19.

To produce a world-class competition bike, the U.S. company Trek Bicycle Corporation set out to design an agile, lighter, and more durable frame. The uniquely constructed one-piece main frame consists of 65% carbon fiber strands that have been compacted in an epoxy resin matrix with a process that yields less than 1% void (air pockets) in the finished composite frame. Trek advertises the OCLV carbon frames as the world's lightest, stiffest, and strongest. Of their frames made of ZX aluminum alloy, Trek states that the alloy achieves desirable properties by combining aluminum, with its low density, with magnesium for tensile strength and zinc to improve corrosion resistance.

Attached to the rear of the OCLV frame is a 6061 T aluminum alloy rear triangle that is tungsten inert gas (TIG) welded for a light and stiff structure. Trek maintains that human crafters provide the precision and passion for bike assembly that robots do not possess. It is easy to accept the passion of humans, but many will question whether humans perform with greater precision than robots.

As you see in Figure 1-19, the OCLV model bike uses various materials, including parts made of rubber, aluminum, steel, and plastic, that complement the carbon composite frame. As with many newer product designs, adhesive bonding plays a role in joining dissimilar materials. Advanced bonding technology (ABT) relies on epoxy resin to glue components into alignment to achieve the designed geometry without the need for post-fabrication adjustments, as required in other fabrication techniques. Teflon-impregnated composite bearings are also bonded into the rear aluminum triangle and main frame to eliminate maintenance and lubrication.

The Trek bicycle and its innovations portray the aggressive nature with which forward-thinking businesses seize on advances in materials technology to propel their corporations beyond their competitors. Their engineers selected from ***designed materials*** rather than settling for **off-the-shelf materials** (materials with limited structure and properties).

What materials systems can you recall? Watch for uses of materials systems as you read newspapers and magazines or observe products because often an idea used in one field can be modified and employed in another. Can you see applications for materials systems in other fields that use the principles discussed here?

SELF-ASSESSMENT

1-1. Give two examples to describe where failure to consider the total materials cycle had harmful results.

1-2. How have the changes in use of materials and manufacturing techniques affected the workforce in modern industry?

Figure 1-19 Advances in engineering materials technology allow for improvements in many products. Advanced composites developed by the aerospace industry are now found in many types of sporting equipment. The OCLV frame for the performance competition racer, seen here, uses carbon fiber composite technology similar to that used for stealth fighters and the space shuttle. (Trek Bicycle Corporation)

1-3. Name a major obstacle to changing from one material to another.

1-4. What must a designer do to make a material selection since there is rarely one ideal material for a given application?

1-5. After all other considerations are made, what is very often the major selection criterion that determines the final choice of material?

1-6. What is a materials database? Name examples of available databases?

1-7. What advantage does net-shape processing offer?

1-8. Because environmental and social factors place emphasis on ensuring that materials be recycled instead of becoming municipal solid waste, what is a new design technique that will aid in recycling?

1-9. Give three reasons why technological literacy is important to you.

1-10. Define the terms (**a**) *engineering materials technology;* (**b**) *materials science;* (**c**) *materials engineering.*

1-11. Give your own example of a materials system.

1-12. In an age of exotic materials, why is wood (one of the oldest materials) still valuable?

1-13. In the section Pause & Ponder, we asked you to begin to take note of materials. Name several examples of materials applications you can see now.

1-14. How does Life Cycle Analysis promote green design, green manufacturing, and green materials usage for a sustainable environment?

OBJECTIVE QUESTIONS

1-1. Recall the Pause & Ponder and Applications & Alternatives in light of what you have read in this module. Which statement is MOST TRUE about materials technology?

    a. Materials science and engineering usually follows the lead of developments in other technologies.

    b. Industry and the consumer are quick to accept new materials and processes innovations.

    c. Developments in materials science and engineering have allowed for innovations in most other technologies.

    d. Existing specifications and laws make it easy to introduce new materials in fields such as building construction and commercial aircraft.

Questions 1-2 through 1-5: Choose the BEST TERM to match the definition.

   a. engineering plastics, **b.** engineering materials, **c.** engineering materials technology, **d.** materials, **e.** materials science, **f.** materials engineering, **g.** conventional materials, **h.** materials science and engineering.

1-2. A term that loosely defines most materials in products and systems.

1-3. A field that deals with developing, preparing, modifying, and applying materials to a specific need.

1-4. A broad term that covers fields of applied science related to all aspects of materials, materials processing, and the many engineering specialties dealing with materials such as those used in manufacturing and construction.

1-5. A field that involves the generation and application of knowledge relating the composition, structure, and processing of materials to their properties and uses.

1-6. Which concept of the materials cycle has been LEAST AFFECTED by recent trends:

   a. Concern for the entire cycle that includes effects of harmful by-products and unrecycled old products

   b. Ability to design materials to meet ever-increasing demands of technology rather than use "off-the-shelf" materials

   c. Need for a more highly educated workforce to support new manufacturing techniques

   d. Desire to achieve the maximum properties from any given material

1-7. Transforming gaseous, liquid, and solid elements by chemical and physical means, where atoms or molecules are combined to form solid materials.

   a. Wroughting   **b.** Synthesis   **c.** Analysis   **d.** Processing

1-8. Control of structure at higher levels of aggregation to yield bulk materials.

   a. Wroughting   **b.** Synthesis   **c.** Analysis   **d.** Processing

1-9. Composites produced with innovative processes that yield high strength and stiffness plus weight reduction.

   a. Advanced composites   **b.** Fiberglass   **c.** Reinforced concrete   **d.** Hydrocarbons

1-10. Manufacturing engineering has evolved to a sophisticated field of study. Which two aspects of fabricating products and systems contribute most to the new approaches of this field?

   a. Systems approach   **b.** Chemistry   **c.** Automation   **d.** Mathematics

1-11. What approach to design reflects a new concern for disposing of products after their useful life?

   a. JIT   **b.** Design for assembly   **c.** Design for disassembly   **d.** CIM

1-12. Which is *not* an obstacle to acceptance of new engineering materials and processes?

   a. Lack of experience by designers and fabricators

   b. Concern for safety with new techniques

   c. Need for compromise between favorable properties and cost

   d. Concern for keeping up with one's competition

1-13. What factor in materials selection usually dominates the final choice?

   a. Cost   **b.** Processibility   **c.** Recyclability   **d.** Weight

1-14. Which term describes well-defined methods for solving specific problems, such as those used to develop an approach to materials selection?

   a. Algorithm   **b.** JIT   **c.** Design   **d.** CIM

1-15. Which of the following best fits the category of a material system?

   a. Gold   **b.** Battery   **c.** Plastic spoon   **d.** Titanium

1-16. Name the method used during design and manufacturing to determine the impact of a product or system on the environment.

   a. Analysis   **b.** TQM   **c.** Life Cycle Analysis   **d.** Just in Time

REFERENCES & RELATED READINGS

AMATO, IVAN. *Stuff: The Materials the World is Made Of.* Avon Books, 1997.

AMERICAN SOCIETY FOR TESTING AND MATERIALS. *Compilation of ASTM Standard Definitions,* 7th ed. Philadelphia: ASTM, 1990.

ASHBY, MICHAEL F. "Materials in Mechanical Design," *Journal of Materials Education.* Vol. 15, pp. 143–166, 1993.

ASM INTERNATIONAL. *ASM Engineered Materials Handbook Desk Edition,* 1995.

ASM INTERNATIONAL. *ASM Materials Engineering Dictionary,* 1992.

ASM INTERNATIONAL. "Green Manufacturing," *Advanced Materials and Processes.* January 1997, pp. 25–28.

BALL, PHILIP. *Made to Measure: New Materials for the 21st Century.* Ewing, NJ: Princeton University Press, 1997.

CALLISTER, WILLIAM D., JR. *Materials Science and Engineering: An Introduction,* 5th ed. New York: John Wiley & Sons, Inc., 2000.

CANBY, THOMAS Y. "Advanced Materials: Reshaping Our Lives," *National Geographic,* December 1989, pp. 746–781.

HUMMEL, ROLF E. *Understanding Materials Science: History-Properties-Applications.* New York: Springer-Verlag, 1998.

NATIONAL RESEARCH COUNCIL. *Materials Science and Engineering for the 1990s: Maintaining Competitiveness in the Materials Age.* Washington, D.C.: NRC, 1989.

ROGERS, CRAIG A. "Intelligent Materials Systems: The Dawn of a New Materials Age," Center for Intelligent Materials and Structures, Virginia Polytechnic Institute and State University, 1992.

SME. "New Nylon 6 Grades Environmentally Friendly," *Molding Systems,* October 1999, pp. 8, 9.

SCHWARTZ, LYLE H. "Industry, Government Cooperation: New in a Changing World," *Advanced Materials & Processing,* February 1996, pp. 110–116.

3M Company website. *http://3m.com/profile/looking/1980.html,* October 21, 1999.

Periodicals

Advanced Materials and Processes	*Manufacturing Engineering*
Automotive Engineering	*MIT's Technology Review*
Machine Design	*Popular Science*

Module
2

Nature and Family of Materials

After studying this module, you should be able to do the following:

2–1. Using the materials and civilization time line, give an example of how materials technology developments sparked new technology in the Scientific Age.

2–2. Use the periodic table of elements and illustrations to determine and/or calculate the atomic structure of elements, ionization potential, and electronegativity.

2–3. Apply the SI rules for prefixes in expressing multiple quantities.

2–4. Use sketches and explanations to describe the microstructure and macrostructure of solid-state materials plus bonding forces within materials.

2–5. Discuss the significance of fullerenes and the polymorphs of carbon.

2–6. Name the five main groups that make up the family of materials; cite examples of materials within each group; and sketch the typical microstructure for ceramics, metals, polymers, and composites.

2–7. Describe intelligent materials systems (smart materials) and discuss some goals of materials science and engineering (MSE) for this class of materials.

2–8. Use EMT Weblinks* to further explore the nature of materials.

*EMT Weblinks listed throughout the book will link you to Internet sites related to topics being covered. The dynamic nature of the Web brings frequent changes, so some of these sites, while available at the time of writing, may not be up now.

PAUSE & PONDER

Recalling our study of history, materials technology played a major influence on societies. Periods of civilization were named for the materials that dominated each era: from the earliest time of primitive people in the Stone Age, through the Bronze Age and the Iron Age. The era we live in has been given many names, including the Scientific Age, the Information Age, the Space Age, and even the Materials Age. Figure 2-1 provides a historical perspective of the concurrent development of civilization and materials technology. This figure, provided by the Franklin Institute's "Future Materials" exhibit, allows a graphical comparison of the relative importance of materials through the ages. Note that the materials are placed into four groups: metals, polymers, composites, and ceramics. In this module, we will provide an in-depth look at these groupings as we look at the family of materials.

The Stone Age, which includes prehistory, was so named because the evolution of civilization began to make great strides with the development of tools. Naturally occurring flint allowed for shaping, and people discovered that flint and other stones could be joined with sticks to produce tools and weapons. Clay dug from the earth was formed and dried into pottery. People of this period (which might be more accurately called the ceramic, hide, and wood age) used the naturally occurring "engineering materials" for tools.

Throughout history humans have been initiating chemical reactions to produce new substances from natural materials as in the production of pottery from clay; but many of these reactions were discovered accidently. **Chemical reactions** are made up of two main ingredients or substances. One group of substances called **reactants** may consist of one or more existing substances. The second category of substances called **products** are the new substances produced by the reaction. Stated another way a chemical reaction is a process in which atoms in the reactant molecules are rearranged to form the molecules of the products. The reaction can be represented in chemical shorthand by a **chemical equation** using the symbols for the elements as found in the periodic table. These equations tell us more than just what substances react together and what substances are produced. They show us how many atoms or molecules of each reactant takes part in the reaction as well as how many atoms or molecules of each product are formed. In other words, a chemical equation is a good bookkeeping device that keeps track of the atoms, molecules, and elements involved in the reaction. Ordinary chemical equations are symbolic statements to show both qualitatively and quantitatively what has taken place. *Qualitative* denotes a description of something whereas *quantitative* refers to a measurement of how much of something is present. These equations involve primarily the outer shell electrons (valence electrons) of the atoms. Throughout this book the various chemical reactions such as corrosion of materials (Module 4) can be explained by writing chemical equations. Most forms of corrosion are electrochemical in nature; that is, they involve the flow of electrons. These interchanges of electrons can be easily represented by using chemical equations which facilitate understanding of the reactions of the environment upon materials.

The Bronze Age represented another major leap forward as "materials technology" brought about the first alloying, when copper and tin were melted together to yield a superior metal useful for shields, knives, ornaments, cups, and urns. Note on Figure 2-1 that ceramics, still widely used, were also advancing during the Bronze Age with the development of glass and composite bricks (clay and straw). The Old Testament, among other ancient sources, contains numerous materials-related quotations such as Deuteronomy 8 in which Moses describes the Promised Land as "a land whose stones are iron, and out of whose hills you can dig for copper." It is evident that the people could be enticed by such a description because of their knowledge of materials which a large proportion of the people must have possessed in those early times. Once iron was developed, around 1100 B.C., it became of great importance to so-

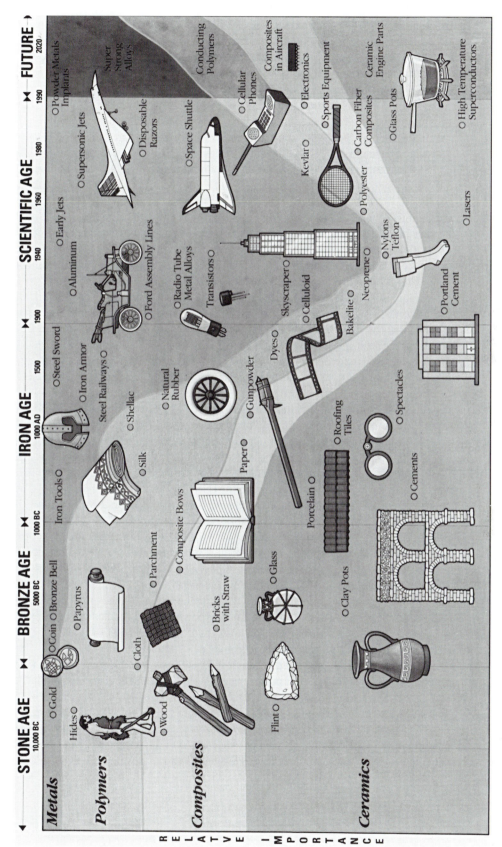

Figure 2-1 Material and civilization time line. (From Future Materials exhibit, Franklin Institute Science Museum)

ciety. Iron's abundance and fine engineering properties proved superior to bronze. With the eventual alloying of iron with carbon and other elements, steel became the superior engineering material; to this day steel has been dominant, although often challenged. During the Iron Age and into the Scientific Age, thousands of iron and steel alloys were developed by metallurgists to meet the demands of the industrial revolution; then mass production received a great boost from the Ford assembly line.

Aluminum, the third most abundant element in the earth's crust, remained a rare and expensive metal until 1886, when an American, Charles Martin Hall, and a Frenchman, Paul Heroult, independently devised a practical process for electrolytic production of aluminum. Hall discovered the aluminum process about a month before Heroult according to his sister's comprehensive laboratory reports. This process is complicated by the fact that aluminum oxide has a melting point of 2060°C, which is too high for it to be conveniently and economically used for electrolysis in the molten state; however, the process can be represented basically by a simple chemical equation, $2Al_2O_3 \rightarrow 4Al + 3O_2$, which shows the aluminum oxide on the left as the reactant and the new substances (products) produced, namely the pure aluminum (Al) plus oxygen (O) on the right of the arrow. The arrow represents, among other things, the addition of energy in the form of heat by passing an electric current over the aluminum mineral reactants (bauxite and cryolite). Approximately 10 times as much energy is required to produce a ton of aluminum as is needed to produce a ton of steel. Still used today, the Hall-Heroult process is the largest consumer of electricity in the United States, making the recycling of aluminum products most worthwhile. Now iron and steel had competition from a lightweight, corrosion-resistant metal. Plastics and rubbers, including Bakelite, Celluloid, Neoprene, nylon, and Teflon (Figure 2-1), were also being developed, which could substitute economically for natural materials such as ivory, tortoiseshell, wood, and silk. Experimentation with combinations of metals (alloys) aimed at meeting the needs of communications, transportation, and construction. Building tall skyscrapers with new types of steel supported the growth of large cities. Soon the radio vacuum tube was improved upon with transistors, and eventually, electronics benefited from the microchip as transistors were reduced in size and combined with other circuit components to produce integrated circuits made with semiconductor materials. It is exciting today to follow the application of new electronics materials to computer technology in areas such as artificial intelligence (AI), which mimics human reasoning, and virtual reality (VR), which provides vast new dimensions to visualization using computer graphics.

Now as we move into the future with results of research and development from the Scientific Age or Age of Materials, technology is beginning to shift away from the dominance of iron and steel. Note in Figure 2-1 that the lines separating the relative importance of metals, polymers, composites, and ceramics are altering again. Advanced materials such as the fiber Kevlar used in aircraft and spacecraft, high-performance ceramics for engine parts, and high-temperature ceramic superconductors are causing the zones of relative importance for ceramics and composites to widen. Continued use of powder metals, super-strong alloys, and conducting polymers shows that the polymer and metal technologies will maintain their importance but not the dominance of recent eras.

In the 21st century, improved materials—the advanced materials and the smart materials—will be called on for use in new products and systems. These materials will be produced through scitech as materials engineers work with manufacturing engineers and materials scientists to invent new materials processing. As you study the nature and family of materials, keep in mind the historical perspective of materials and recognize that technology can cause many shifts in the relative significance of materials. It is important for you to understand the similarities and differences in the main groups of materials and to keep abreast of new developments.

 EMT WEBLINKS

Franklin Institute Science Museum—*Http://sln.fi.edu*
Smithsonian Institution—*Http://www.si.edu*

2.1 NATURE

Through an understanding of the nature and structure of materials, one can predict how materials should behave when exposed to certain forces and environments. The nature of materials results from their composition and structure, which determine their properties. In this module, we will present basic concepts of atomic and molecular structure that lay a foundation for the discussion throughout the book on the nature, properties, and uses of materials. The ability to grasp these fundamental materials technology concepts not only allows you to deal with today's materials and processes, but also prepares you for new developments in materials.

2.1.1 Internal Structure

The structure of materials follows the structure of a building because both develop through the joining of smaller units. Particles of earth form bricks, bricks are stacked together to form walls, walls form rooms, and rooms make a building. *Matter* is anything that has mass and occupies volume. It exists in one of four physical states: solid, liquid, gas, or plasma. *Plasma* exists only under drastic conditions. At very high temperatures, matter is shredded into positively charged particles. These charged particles move about in a cloud of negatively charged electrons. On the surface of the sun, hydrogen and helium gases are changed into plasma by the heat of the sun (6000°C). A *state* of matter is the condition of a substance, plasma, gas, liquid, or solid at a particular temperature. A *solid* is a sample of matter that has a fixed volume and a fixed shape. A *liquid* is a sample of matter that has a fixed volume but assumes the shape of the container it occupies. A *gas* is a sample of matter that has neither fixed volume nor fixed shape. Any substance may assume any of the four states of matter under varying conditions, in particular, as the temperature changes. This last statement describes the convertibility of these states of matter which is a property of most substances. Remember that we are dealing with the same substance in all four states (see Figure 3-56 for a graphic picture of three states of water).

The terms *mass* and *weight* are often used interchangeably. *Mass* is the amount of matter in an object. Mass is a scalar quantity (no direction or sense) which signifies that it possesses only a magnitude. It is the term used to express an object's resistance to change in its state of motion or rest. The greater that resistance the greater the mass. On earth the force that brings about a change in motion in an object is the force of gravitational attraction between the earth and the object. This gravitational force is called *weight,* a vector quantity which signifies that, to properly describe it, its magnitude as well as its direction must be defined. The magnitude and direction of the weight depends on the mass of the object, the mass of the earth, and the distance of the object from the center of the earth. The SI unit for measuring the mass of an object is the kilogram (kg) whereas the SI unit for measuring force is the newton (N). Remember, if an object such as the lunar lander is moved to the moon, its mass stays the same as what it was on earth but its weight would change because it would be subjected to a different force of gravitational attraction. When two or more *atoms* are joined by covalent bonds, the unit of matter formed is called a *molecule.* An *element* is a substance that cannot be broken down any further by chemical reaction. Its atoms all have the same atomic number (A), to be defined. They occur in nature as mixtures of their isotopes. Some elements such as phosphorus have only a single naturally occurring isotope. Examples of elements are carbon, oxygen, or sulfur. Elements combine to form compounds. An example of a compound is NaCl. A *compound* is a substance that can be broken down into elements by a chemical reaction. Molecules combine through chemical reactions to form polymeric substances known as *polymers.* Elements, molecules, and compounds exist in many forms, such as a gas, liquid, or solid, as described.

2.1.1.1 Atomic structure.   An *atom* is the smallest particle of an element that possesses the physical and chemical properties of that element. It is the smallest particle of an element that can enter into a chemical change. Atoms are the basic building blocks of matter. The average diameter of an atom is only about 10^{-10} meter. It takes more than 10^6 atoms edge to edge to make the thickness of this page.

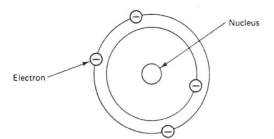

Figure 2-2 Planetary model of an atom.

What the atom really looks like or how the charged particles distribute themselves and move within the atom is not exactly known. Consequently, scientists resort to various models of an atom to help explain its characteristics. One model, the planetary model, appears to be the most popular in explaining the nature of an atom (see Figure 2-2). However, this model is far from being the most accurate. The atom is pictured as a sphere with a very dense center, the nucleus, containing the protons and neutrons around which negatively charged particles (the electrons) are moving in circular or elliptical orbits, much like the motion of the planets around the sun. The diameter of the nucleus is about 1/10,000 the size of an atom, but it contains more than 90% of the mass of the atom. If one represented the nucleus as a golf ball, the electrons would revolve in a sphere with a radius of about 1½ miles from the nucleus.

For most atoms and their isotopes the nucleus is spherical and stable. In studying *radioactivity* (i.e., the spontaneous disintegration of a nucleus), it was determined long ago that the nuclei of all the isotopes of elements with atomic numbers greater than 83 were "unstable"; that is, they decompose into nuclei of other elements by giving off alpha particles, electrons, or gamma rays. However, atoms with atomic numbers between 51 and 67, and more specifically europium (Eu) and holmium (Ho) containing 63 and 67 protons, respectively, are unstable and have "highly deformed" nuclei. These two elements belong to a group of elements called the *lanthanides* or rare earth elements that can be found in Period 6 but separately listed below the main table. Argonne National Laboratory in 1998, studying europium and holmium, determined that they engage in a rare form of radioactivity called *proton emission.* The rate of that ejection depends on the shape of the nucleus, which was determined to be football shaped. For further information on radioactivity see Module 7. Another model of the atom, sometimes sketched in three dimensions, attempts to show the spaces within which the electrons are confined as fuzzy volumes. The atom can be thought of as a sphere of empty space with 90% of its mass located in its center or nucleus.

For our purposes, the nucleus consists of protons and neutrons. A *proton* is a particle of matter that carries a positive electrical charge equivalent to the negative charge on an electron. *Neutrons* are uncharged particles in the nucleus with a mass nearly equal to the proton's mass. The *atomic number* Z of an element is equal to the number of protons in the nucleus. The number of protons in the nucleus equals the number of electrons in the atom in the balanced, neutral, or equilibrium state. Elements in the periodic table (see Table 10-1A) are in order according to their atomic numbers (A). An element may be defined as a substance whose atoms all have the same atomic number (**Z**).

All atoms of an element contain the same number of protons, but these atoms can contain different numbers of neutrons. In other words, these atoms are different, but their positive and negative charges are identical. Such atoms with different numbers of neutrons are isotopes. Another way of describing an *isotope* is by saying that atoms of an element that have different masses are isotopes. For example, hydrogen has 3 isotopes (see Figure 2-3). Tin has 10 naturally occurring isotopes.

The total number of protons and neutrons in the nucleus is the *mass number of the nucleus* (**A**). Protons and neutrons are referred to as *nucleons.* These nucleons are composed of still smaller particles called *quarks,* which might be considered the fundamental building

Figure 2-3 Isotopes of hydrogen. 99.985% of hydrogen is in the form of protium. Deuterium represents only 0.015% of hydrogen. Tritium is an unstable isotope of hydrogen.

Figure 2-4 Subatomic parts of the oxygen atom, A.N. 8. This atom is shown in an equilibrium, balanced, or electrically neutral state.

blocks of all materials. The mass of an atom is concentrated in the nucleus while the electrons account for most of the *volume*. *Mass*, or simply the quantity of matter of a substance, is expressed in the SI system of units using kilograms (kg). The mass of a hydrogen atom is 1.6735×10^{-27} kg or using grams 1.6735×10^{-24}g. Such small numbers are inconvenient to use; therefore, atomic mass is expressed in several other ways, including *atomic mass units*. An atomic mass unit (u) is defined as being equal to 1/12 the mass of a single carbon-12 atom. The mass of one C atom is defined as exactly 12 u. Another way to display or convey the meaning of the average mass of an atom is to express the mass relative to 1/12 the mass of a C atom. This *relative average atomic mass* (or *average atomic mass*) has no units. For example, the average mass of an oxygen atom (with three isotopes) is 15.9994 times 1/12 the mass of a C atom, so the relative average mass of an O atom is 15.9994. This atomic mass is often called *atomic weight*, but this is a misnomer because weight is different from mass as previously explained. (Refer to the key for the elements in Table 10-1A, for use of the term *atomic weight*.)

The standard notation for expressing the composition of the nucleus is to write the atomic number (A) at the lower left (a subscript) of the symbol of the element and the mass number (Z) of its nucleus as a superscript at the upper left. For example, a carbon atom is expressed as C indicating that the C atom has 12 nucleons in the nucleus of which 6 are protons. From that we know that the carbon atom has 12–6 = 6 neutrons. All atoms have zero charge overall. Its nuclear charge (+Z) must be balanced by the total number of the surrounding electrons each of which has a negative charge of –1. This nuclear charge (Z) differentiates the atoms of an element from those of another. An *element* may also be defined as a substance whose atoms all have the same **atomic number** (A). Figure 2-3 shows this notation below the subshells for the isotopes of hydrogen. Hydrogen exists in three isotopic forms. Tritium is the unstable form; that is, it decomposes spontaneously into nuclei of other elements. All but two of the elements with atomic numbers less than 83 have one or more stable isotopes. Elements with atomic numbers 43 and 61 have no stable isotopes.

Electrons.   In summary, an atom (see Figure 2-4) of an element is made up of a dense core, the nucleus, consisting of protons and neutrons surrounded by even smaller particles called

Figure 2-5 Positive ion or cation.

Figure 2-6 Negative ion or anion.

electrons. The electrons carry a negative (–) electrical charge much like the charge at the negative terminal of a battery. Those electrons that occupy the outermost ring or shell from the nucleus are *valence electrons. Valence* is the capacity of an element to combine with other elements. For example, one oxygen atom combines with two hydrogen atoms to produce the water molecule H_2O. Oxygen is said to have a valence of 2. All elements in a group in the periodic table have the same primary or principal valence. For example, sodium, potassium, and the other elements in Group 1 have a valence of 1. For Groups V to VIII the principal valence is equal to 8–*n* where *n* is the group number. Chlorine in Group VII has a valence of 8–7 = 1. Some elements have more than one valence or valence number. If the atom is in a balanced or equilibrium state (see Figure 2-4), it possesses the same number of protons as electrons, and the electrical charge carried by the atom is zero. The negative (−) charge of the electrons is balanced by the positive (+) charge of the protons in the nucleus.

The process of pulling away (removing) or adding valence electrons from a balanced or neutral atom is called *ionization.* The atoms with unbalanced electrical charge are called *ions* (see Figures 2-5 and 2-6). The tendency of atoms to lose their valence electrons and the energy required to remove these electrons will be discussed in the treatment of the periodic table and its underlying concepts.

Electronic Structure.   Many new theories were developed between 1900 and 1930 that laid the basis for a new branch of physics called *quantum mechanics.* One theory was the idea that light is quantized, that it is made up of discrete amounts of energy. In a similar manner, we consider our present-day model of an atom with its electrons as having both wavelike and particlelike characteristics. The position of these electrons in relation to the nucleus of an atom, according to this new theory, must be described in terms of probability distribution (electron cloud) rather than the precisely determined values of classical theory. Four parameters called *quantum numbers* characterize an electron as to its size, shape, and spatial orientation. Shells or principal energy levels are specified by a principal quantum number *n,* where *n* = 1 through 7 or K, L, M, N, O, P, and Q (see Figure 2-7a). The second quantum number, 1, signifies the sublevel, subshell, or orbital (s, p, d, f). The number of energy states for each subshell is determined by the third quantum number m_1. For the s subshell, there is a single energy state; for p, there are three; for d, there are five; and for f, there are seven. The fourth quantum number m_s relates to the electron spin moment, for which there are two values possible (+½ and −½), one for each of the spin orientations. Spin momentum controls the manner in which electrons fill their subshells and also the magnetic properties of materials (see Module 9).

(a)

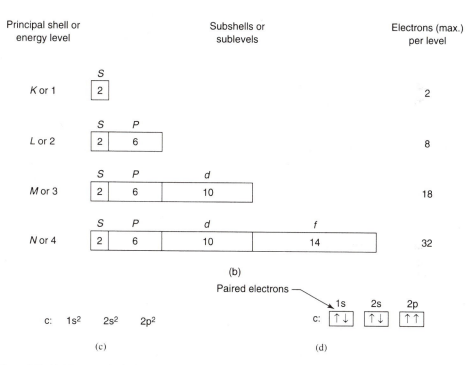

(b)

Paired electrons —

c: 1s²   2s²   2p²

c: 1s [↑↓]   2s [↑↓]   2p [↑ ↑]

(c)   (d)

Figure 2-7   (a) Electron principal energy levels or orbits. (b) Block diagram of the first four energy levels showing their respective sublevels or orbitals along with the maximum number of electrons per energy level. *Orbitals* are used to describe the locations of electrons within the energy levels. They refer to volumes of space around the nucleus of an atom where an electron could probably be found. An orbital can be occupied by one electron or by two paired electrons only. An s orbital is spherical in shape; p orbitals have identical dumbbell shapes oriented around the x-, y-, and z-axes; and d orbitals take the shape of a four-leaf clover. (c) Electron configuration for the element carbon (C), atomic number A.N. 6 in Period 2 of the periodic table, shows three different ways of representing the location of its six electrons.

Three basic rules define the order in which electrons fill their subshells or establish their electron configurations. One says that the electrons occupy their subshells or orbitals in such a way as to minimize the energy of the atom. A second rule states that only two electrons may exist in the same orbital, and these electrons must have opposite spins. The last rule says that when orbitals of equal energy are being filled, the electrons initially occupy these orbitals singly. The result of this rule is that electrons try to get as far apart from each other as possible. To accomplish this, they favor occupying empty orbitals of equal energy as opposed to pairing up with other electrons in half-filled orbitals. As a result, atoms tend to have as many unpaired electrons as possible. Iron (Fe), atomic number 26, has six electrons in the 3d orbital of which four are not paired together. The sixth electron fills in the 3d orbital pairs with one of the previous five. In Fe, the unpaired inner electrons are responsible, in part, for the ferromagnetism which Fe exhibits. **Paired electrons** in the inner shells produce no resultant magnetic dipole moment. For further discussion of ferromagnetism and magnetic domains see Module 9. The orbital diagram is used to demonstrate electrons filling orbitals. Figure 2-7c shows three ways to depict the electron configuration of the element carbon (C). This figure is similar to Figures 2-7a and b because each subshell is broken down into individual orbitals of equal energy. Electrons are shown as arrows. An arrow pointing up corresponds to one type of spin ($\frac{1}{2}$) and an arrow pointing down to the other ($-\frac{1}{2}$). Electrons in the same orbital with opposing (opposite) spins are said to be paired ($\uparrow \downarrow$). The electrons in 1s and 2s orbitals are paired. Electrons in different, singly occupied orbitals of the same subshell have parallel spins (arrows pointing in the same direction). This orbital diagram represents the carbon atom in its most stable or ground state. The order in which all electrons fill subshells, and thus minimize the energy of the atom, as established by experiment, is (with but a few exceptions) as follows:

1s, 2s, 2p, 3s, 3p, 4s, 3d, 4p, 5s, 4d, 5p, 6s, 4f, 5d, 6p, 7s, 5f, 6d, 7p

There are other techniques or methods used to determine the order of filling; they are not discussed in this text.

The energy levels in atoms have been given letter names starting with K, the first and lowest energy level (closest to the nucleus), and proceeding outward to Q, the seventh level, where the electrons have the highest energy. Each level can hold only a certain number of electrons at any one time. To determine the maximum number of electrons at any energy level, the relationship $2n^2$ can be used. The number of the energy level is substituted for n. For example, the third or M energy level can contain a maximum (2×3^2) or 18 electrons. Figure 2-7a contains a simple sketch of all energy levels drawn concentric with the nucleus of an atom. Some energy levels are listed along with the maximum number of electrons permitted at each level in Figure 2-7b.

The manner in which the electrons of an atom distribute themselves in the ground, or lowest energy, state is the **electron configuration** of the atom. Learning the electron configuration of atoms permits us to predict what the atoms will do in the presence of other atoms. In other words, the number of electrons in an atom's outermost energy level (the valence electrons) is the key that controls the chemical properties of an element. As an example, hydrogen with an atomic number of 1 in the ground state has its one electron in the K level. Carbon has six electrons; two fill the K level and the remaining four are in the L level. Atoms that have a completed energy level are chemically inert; that is, their atoms do not react with other atoms. These atoms do not attract other electrons, nor do they readily lose their outer electrons. Hence, these atoms possess atomic stability. Helium has a complete K energy level with its two electrons. Some atoms have the tendency to build up the number of electrons in their levels to eight. (This tendency is known as the **octet rule** or the **rule of eight.**) Once they arrive at eight, the next electron goes into the next higher level. Those atoms that have only one electron in their outermost energy level are extremely reactive; they tend to lose their electron. Examples of such elements are lithium and sodium.

Probing deeper into the electron world, scientists have determined that the energy levels (shells) are divided into sublevels (orbitals) that also contain a maximum number of electrons. Table 2-1 contains an example of sublevels for selected elements, which are given lowercase-letter names.

The K level has only one sublevel, the s sublevel. Using a block diagram, the energy levels could be represented as shown in Figure 2-7b. The electrons fill these sublevels in a particular way. The first 18 elements of the periodic table build up their permitted number in a routine manner. Once the K level reaches its maximum of two electrons, the electrons begin filling up the L level until a maximum of eight electrons is reached, as with neon (Ne). Then

TABLE 2-1 ELECTRON CONFIGURATIONS OF SELECTED ELEMENTS

Element	Atomic number	K 1 s	L 2 s p	M 3 s p d	N 4 s p d f orbits	} orbits orbitals
Oxygen	8	2	2 4			
Sodium	11	2	2 6	1		
Aluminum	13	2	2 6	2 1		
Silicon	14	2	2 6	2		
Titanium	22	2	2 6	2 6 2	2	
Arsenic	33	2	2 6	2 6 10	2 3	

the third energy level from the nucleus (M level) with its three sublevels (s, p, d) begins filling to its maximum of 18. This point is reached with the element argon (Ar, atomic number 18). Note that both neon (Ne) and argon (Ar) have a maximum possible number of electrons in their outermost energy levels. These elements are classified as chemically *inert gases;* that is, they are very stable elements and do not react with other elements. Neon (Ne), with its 10 electrons, has an electron configuration: Ne: $1s^2 2s^2 2p^6$. Note that it has a full L, or 2, level of eight electrons (octet rule). This neon core of electrons can be represented as [Ne]. Argon, atomic number 18, can be represented in a similar way as [Ar]. Again, note that its outermost level (M, or 3) has eight electrons. To represent the electron configuration of titanium (Ti, atomic number 22) showing the order in which the electrons fill the orbitals and using the argon core symbol, we write: Ti: [Ar] $3d^2 4s^2$. The longer version of the Ti electron configuration follows in the next paragraph. Table 2-1 contains other elements' electron configurations.

Using a shorthand notation, the electronic configuration of oxygen is ^8O: $1s^2, 2s^2, 2p^4$. Note that the L energy level has only six electrons, but the p sublevel could contain a maximum of eight electrons. As a result, oxygen tends to accept two additional electrons to form a stable outer energy level. By definition, oxygen is an acceptor of electrons and hence is classified as a *nonmetal.* A *metal* such as aluminum, with its electronic configuration ^{13}Al: $1s^2, 2s^2, 2p^6$, $3s^2, 3p^1$, has a stable L energy level below the farthest energy level, the third or M shell. The three valence electrons in the M level are close to those in the L level and possess similar bonding energies, which results in these three electrons having a tendency to stay with their atom. In addition, to achieve stability, aluminum would need the addition of five more electrons, which is unlikely to occur. Hence, aluminum is classified as a metal. One final example is titanium (Ti), with the electronic notation ^{22}Ti: $1s^2, 2s^2, 2p^6, 3s^2, 3p^6, 3d^2, 4s^2$. This element is chosen to show that the filling of electron levels is somewhat irregular. Note that two electrons have filled the 4s sublevel, whereas the 3d sublevel is far from being filled, with only two electrons (see Figure 2-7b). For an example of the use of this notation, refer to Section 3.4.9.

Illustrative Problem

Write the electronic configuration for the element sulfur.

Solution

Step 1. Determine the chemical symbol and atomic number for sulfur from the periodic table.

S   atomic number 16

Step 2. Determine the total number of electrons in the sulfur atom.

16

Step 3. Write the order in which the orbitals are filled (see paragraph 2.1.1.1).

1s, 2s, 2p, 3s, 3p, 4s, 3d, etc.

Step 4. Just below the orbitals in step 3 write the number of electrons per orbital for each orbit and level (refer to Figure 2-7).

2 2 6 2 6 2 10

Step 5. Write the configuration:

$$S = 1s^2\, 2s^2\, 2p^6\, 3s^2\, 3p^4$$

Step 6. Check by adding the number of superscripts, which represent the number of electrons in each orbit, to arrive at the sum of 16.

On the other hand, let us look at some elements that have only one electron in their outermost energy level. Lithium (Li), atomic number 3, with two electrons in the K level and only one electron in the L level, is extremely reactive. The same can be said of sodium (Na) and potassium (K). For elements with atomic numbers greater than 18, starting with potassium (K), the routine followed by the next 18 elements in filling their electron levels is somewhat different (see order of filling in this paragraph).

Through an understanding of the electronic structure of atoms, the properties of elements can be ascertained and the bonding of atoms of elements together to form molecules, compounds, or alloys can be explained. Additionally, these electrons determine the size of atoms, the degree of electrical and thermal conductivity of materials, and the optical characteristics of materials.

When atoms take part in chemical reactions, there is an energy change or transfer, usually resulting in the release of energy in the form of heat. This energy change brings about a rearrangement of the electrons, and the end result is an electron configuration with lower energy than any of the reactant elements. Only the outermost (valence) electrons are rearranged. The electrons in the inner energy levels (or shells) are too stable, or shielded from this activity. More will be said about this phenomenon when chemical reaction and bonding of atoms are discussed later in this module.

Periodic Table.   Elements are the basic building blocks of all macroscopic matter. Just as musicians experiment with musical notes to produce new music, materials scientists experiment with new combinations and arrangements of these elements to produce new and exciting materials that may become part of a technological device. Each has its own chemical symbol, a shorthand abbreviation of its chemical name, and *each has its own specific properties.* It is important to reiterate that elements with the same number of valence electrons have similar properties. Scientists have long recognized that some elements had similar properties while others, such as sodium and chlorine, had very different properties. All versions of the modern periodic table arrange the elements in order of increasing atomic number, which usually matches the order of increasing atomic mass.

Refer to Table 10-1A for the following discussion. The horizontal rows, called ***periods,*** are numbered from 1 to 7. The vertical columns, called ***groups,*** are numbered in two different ways. If only the main groups (also referred to as representative elements) are numbered, the roman numerals I to VIII are used. If the transition element groups are also numbered, then the arabic numbers 1 to 18 are used. The 10 elements that fall between groups IIA and IIIA in Periods 4, 5, and 6 (a total of 18 elements) are known as ***transition elements.*** Their properties are attributable to the delayed filling of the next-to-last or outermost energy level. Period 6 contains 14 additional elements between lanthanum (La) and hafnium (Hf), which are called lanthanides or rare earth metals. Nd-Fe-B magnets are one example of their use. About 70,000 tons are used worldwide in a year. The total number of elements in this period is 32. The 14 elements in period 7 that form a similar series are called actinides. To give the table a convenient size and shape, these two series of transition elements are set off separately at the bottom of the table. Each ***period*** of the periodic table corresponds to the filling, or at least the partial filling, of the energy levels. The period in which an element appears, therefore, tells us how many electron energy levels are in that atom. The ***group number*** tells us how many electrons (the valence electrons) are in the outermost energy level. Elements within a group have similar properties. Group I or IA elements with only one electron in their outer energy level are the most active metals. Group VIII or VIIIA elements known as the noble gases or inert gases have their outer energy levels completely full of valence electrons. As a consequence they are mostly inactive; that is, they do not get involved in making compounds with other elements. Argon (Ar) is

EMT WEBLINKS

Periodic Tables of the Elements
Kiwi Web—*http://www.chemistry.co.nz/periodic_tables.htm*
Periodic Table of Comic Books—*http://www.uky.edu/~holler/periodic*
Periodic Table—*http://www.periodictable.com*

used in gas welding to provide a shield against the intrusion of oxygen into the weld metal. Xenon (Xe), atomic number 54, in the form of short-lived radioisotopes, is playing a role in an automatic sensory device strategically located in some 80 different locations throughout the world to help police the Comprehensive Nuclear Test Ban Treaty by detecting small underground nuclear explosions. The sensor automatically samples large volumes of the atmosphere for the presence of any noble gases, principally xenon, that might escape from an underground explosion because they are inert and do not react with soil.

The elements can be classified broadly into three basic groupings (metals, metalloids, and nonmetals) because they have certain common properties. *Metals* have certain characteristic properties: They are solids at room temperature (with the exception of mercury), they are malleable and ductile, and they conduct electricity. *Nonmetals* (a total of 17 elements) are poor conductors of heat, they are mostly insulators, and they tend to be brittle and fracture easily. *Metalloids,* or *semimetals,* have some properties like the metals; that is, they conduct electricity and other properties like nonmetals. They consist of seven elements (B, Si, Ge, As, Sb, Te, and At) adjacent to the zigzag border between the metals and the nonmetals. Semimetals are brittle and are poor conductors of heat and electricity. Silicon and germanium, for example, are called semiconductors and they are playing a major role in our technological world (see Module 9).

The atomic radius of an atom (size of atom) decreases as one moves from left to right across a period. As one moves down a particular group of elements, the size of the atoms increases. Ionization, defined previously, involves the loss of a valence electron by an atom. The energy required to accomplish this is known as the *ionization potential.* The first ionization potential is a measure of the energy needed to remove the first electron from the atom in its ground or equilibrium state. In general, ionization potentials follow the periodic relationship and decrease as the atomic numbers of the elements increase in a given periodic group. The lower the ionization potential, the greater the metallic character. For all elements, the ionization potential is usually listed in the periodic table expressed in electron volts (eV), or other energy terms, such as calories per mole. For example, it takes 13.6 eV of energy (313 kcal/mol) (shown as 13.598 in Table 10-1B) to free the electron from the hydrogen atom. A detailed discussion of the ionic bonding between sodium and chlorine to produce a molecule of sodium chloride (NaCl) can be found in paragraph Ionic Bonding (Electron Swapping).

This binding energy increases as the size of the atom decreases. In other words, the farther away an electron is from its nucleus, the smaller the force of attraction between it and its positively charged nucleus. Metallic elements on the far left of the table readily ionize, whereas atoms of nonmetals located on the right of the table more readily share or accept additional electrons and become negative ions. The observation to be made is that the tendency of atoms to ionize depends on their relative positions in the periodic table. *Electronegativity* is the degree to which an atom attracts electrons. It is measured on a scale from 0 to 4.1, as shown in Table 10-1B. The nonmetallics, which accept electrons to produce negative ions (anions), are the most electronegative elements, and Groups VIB and VIIB contain the most electronegative elements, with a value of 3.98 for fluorine (F). Francium (Fr) has the lowest value, 0.7.

Electropositive elements are metallic in nature and give up electrons in chemical reactions to produce positive ions (cations). The most electropositive elements are found in Group 1A. Another term that describes this phenomenon is *electron affinity,* which is a measure of the attraction between the electron and the nucleus. Figure 2-8 is a graph based on the Pauling electronegativity scale showing the percentage of ionic bonding (character) in a compound

Figure 2-8 Percent ionic character of a single bond plotted as a function of the difference in electronegativities of the two bonded atoms.

versus the difference in electronegativity between two elements in the compound. If the difference between the electronegativity of two elements is greater than 1.7, the bond is ionic. The higher the electronegativity value for an element the greater is the ability of that element to attract electrons to itself. The greater the difference in electronegativities between two atoms forming a bond the more polar is the bond (see *polar covalent bonding* or Covalent Electron Sharing).

Illustrative Problem

Determine the type of bonding in the compound MgO.

Solution: Using Table 10-1B, the electronegativity for oxygen is found to be 3.44 and for magnesium, it is 1.31. Entering a difference of 2.13 in the graph shown in Figure 2-8, the percent ionic bonding is about 68%, leaving 32% of the bonding covalent. A more accurate solution is obtainable using data in the small table at the top of Table 10-1B.

2.1.1.2 Molecular structure and bonding.

We mentioned previously that atoms join together to form molecules and that molecules combine to form various compounds. A **compound** is a substance containing two or more elements combined in fixed proportions. Compounds lose the characteristics of the elements; for example, sodium Na combines with chlorine Cl to make the compound sodium chloride (NaCl), as described in the next section.

In some molecules, the atoms are of the same element. Fluorine (F) exists as a diatomic molecule under normal conditions because two fluorine atoms combine to form one fluorine molecule. All the Group VIIB elements form diatomic molecules. In this section, we will gain further understanding about how the 100 or so elements combine or chemically bond to form not only more than 3 million organic compounds, but the various inorganic and bulk metals such as iron or copper.

Molecules and compounds are formed by chemical bonding. **Chemical bonding** can be explained simply as the end product of the interaction of the electrical forces of attraction and repulsion between oppositely charged or similarly charged particles of matter. Chemical bonds are formed by the electrons in the atom's outermost energy levels or regions. In our study of these electron energy levels, we have learned that, generally, atoms that have eight electrons in their outermost orbit are very stable (in equilibrium state). This tendency to start filling the next-higher level once an atom has eight electrons in an energy level is known as the *octet rule,* described previously. Atoms with fewer than eight outermost electrons attempt to seek this stable condition, the electron configuration of the nearest noble gas (Group VIII) in the periodic table, by bonding with each other so that each atom can attain this stable configuration of eight electrons (valence electrons) in its outermost energy level. One way that atoms can achieve this stability is by sharing electrons with other atoms that also need eight electrons in their outermost energy levels. For purposes of study, we can divide the various types of bonding into two groups, strong or *primary bonding,* and weak or *secondary bonding.* In the first group, we include covalent, ionic, and metallic bonding. The second group includes weak atomic bonds, electric dipoles, polar molecules, and the hydrogen bond. The forces that produce secondary bonds are known as *van der Waals forces.* It is emphasized that atoms of one material do not

$$
\begin{array}{c}
H \\
| \\
H - C - H \\
| \\
H
\end{array}
\qquad
N \equiv N
$$

N_2 molecule   **Figure 2-9**   (a) Single bond. (b) Triple covalent bond.

bond solely by one particular bonding mechanism. One type of bond can predominate in a material, but in many materials the various bonding types are represented and mixed-bond types of materials are produced.

   Polar Covalent Bonding. Covalent Electron Sharing, or Electron-Pair Bonding. The sharing of electrons between two or more atoms is known as ***covalent*** or ***shared electron-pair bonding.*** As an example, fluorine (F) atoms have a total of nine electrons: two in the inner energy level and seven in the outermost level. Two fluorine atoms can share one of the electrons with the other to form a single covalent bond, depicted with a single dash representing the bond. Each (–) is a pair of shared electrons (see Figure 2-9a). Bonding of the elements is indicated with the small bar (—). Each bar indicates a covalent bond. Covalent bonding gets its name from the fact that a pair of valence electrons is part of the electron structure of each atom bonding the two atoms together. In general, organic compounds (complex compounds containing carbon and/or hydrogen) are formed by covalent bonding. A ***triple covalent*** bond is one in which two atoms *share three* of their electrons with each other. Figure 2-9b illustrates this with nitrogen (N). Nitrogen needs eight electrons to fill the outer energy level. It has only five. Hence, it shares three with another nitrogen atom, forming a triple bond.

   Organic chemistry is the chemistry of carbon compounds. Of all the millions of known chemical compounds, more than 95% are compounds of carbon. The simplest organic compounds are the ***hydrocarbons (HCs),*** which contain only hydrogen and carbon. ***Alkanes,*** also known as saturated HCs, are a series of related compounds that contain only single bonds. Figure 2-10 shows the first three compounds in this series: methane, ethane, and propane. They all have a continuous, open-ended chain of carbon atoms and possess similar chemical properties. Each bar in Figure 2-11 represents the single bond of a hydrogen atom with a carbon atom to form the compound methane ethylene. Ethylene is the first member of a family of HCs called ***alkenes.*** The structure of ethylene as shown in Figures 2-10 and 2-12 is the basic compound for producing polyethylene. More than half of the production of ethylene goes into the manufacture of polyethylene. In the form of ethylene glycol, it is also the major component of most brands of antifreeze. In these figures, the C represents the carbon atom and the H represents the hydrogen atom. Note in Figure 2-10 that, as the molecular weight increases, the compounds change from a gaseous state to a liquid and finally to a solid as more and more molecules bond to their chains of molecules. Each alkene contains a carbon-to-carbon double bond. Acetylene is the first member of the ***alkyne*** family that contains a triple bond. Collectively, alkenes and alkynes are called unsaturated HCs because they can add more hydrogen atoms to form saturated HCs or *alkanes.* One unusual feature of unsaturated HCs is that they can also add to each other to form very large molecules called polymers (discussed in Module 6).

   Benzene and similar compounds such as toluene are referred to as ***aromatic hydrocarbons*** because some of the first benzenelike compounds had strong aromas. Today, any compound that contains a benzene ring or has similar properties to those of benzene is called *aromatic.* Figure 2-13 shows many ways of representing benzene. All possible structures have double bonds and some have triple bonds, but benzene does not react that way chemically. Because the three pairs of electrons that would form double bonds are not localized (they seem to lie above and below the plane of the carbon atoms), the popular way of representing benzene is with a circle within a hexagon. The circle represents the six unassigned electrons and the hexagon represents the ring of six carbon atoms. This ring of electrons resists being disrupted,

Compound name	Pictorial	Molecular weight	Chemical equation	Structural diagram
Methane		16 Gas	CH_4 (single bond)	
Ethane		30 Gas	C_2H_6 (single bond)	
Propane		44 Gas	C_3H_8 (single bond)	
Ethylene		28 Gas	C_2H_4 (double bond)	
Acetylene		26 Gas	C_2H_2 (triple bond)	
Paraffin		225 Waxy solid	$C_{16}H_{33}$ (single bond)	
Polyethylene		1402 Solid	$C_{100}H_{202}$ (single bond)	

◯ Carbon   ○ Hydrogen

Figure 2-10 Hydrocarbons.

providing the benzene molecule with a stable structure. Aromatic HCs are all liquids that are lighter than water. They are used mainly as solvents.

The chemical formula (Figure 2-13) may be considered as a one-dimensional description of a chemical compound or molecule, the structural description as a two-dimensional view, and the pictorial as a three-dimensional view. Covalent bonding occurs between atoms of nonmetallic elements such as hydrogen, fluorine, chlorine, and carbon, and in molecular compounds such as water, carbon dioxide, ammonia, and ethylene. By sharing electrons, the atoms achieve sta-

Figure 2-11 Covalent bonding of methane.

Figure 2-12 Covalent bonding of ethylene.

Figure 2-13 Benzene or benzene ring.

bility through the creation of interatomic electrical forces, which bond the atoms together to form molecules and compounds. Research has shown that double or triple bonds are stronger than single bonds between the same two atoms. Carbon forms four single covalent bonds.

The many forms (*polymorphism*) of carbon serve to illustrate the array of properties achieved by variations in bonding and location of atoms. Carbon exists as diamond, the hardest natural material known, and graphite, an effective lubricant. Graphite is the stable form of carbon at temperatures up to 3000°C and pressures of 10^4 atmospheres (one atmosphere = 14.7 pounds per square inch) and higher. Diamond is the stable form of carbon at very high pressures (see sect. 7.3.3.4 for discussion of natural/synthetic diamond). *Diamond* has a cubic symmetry (Figure 2-14a), with perfect covalent bonding that joins all atoms into a single molecule. *Graphite* develops into sheets with good covalent bonding, but between sheets van der Waals bonds are weak and shear easily, thus allowing the sheets to slide past each other (Figure 2-14b). *Fullerene* and *buckyball* are the two common names for the *buckminsterfullerene* molecule. Figure 2-14c depicts the recently discovered carbon form ^{60}C, which draws its name from Buckminster Fuller, inventor of geodesic domes, because this complex molecule has a soccer-ball shape, composed of 20 hexagons and 12 pentagons. Fullerene fibers, also known as *nanotubes,* are carbon fibers with tubular structures. They are potentially 30 to 100 times stronger than steel but only one-sixth its weight. The fibers measure about 1 nanometer in diameter.

The discovery of fullerenes (^{60}C molecules; Figure 2-14c) was announced by Richard E. Smalley in 1985. The scanning tunneling microscope, discussed in Module 3, allowed the visual proof of the discovery. This new polymorph of carbon holds much promise as materials scientists and engineers unfold the techniques for processing it. The two other polymorphs of carbon are diamond and graphite (Figures 2-14a and b). The sooty substances, which are

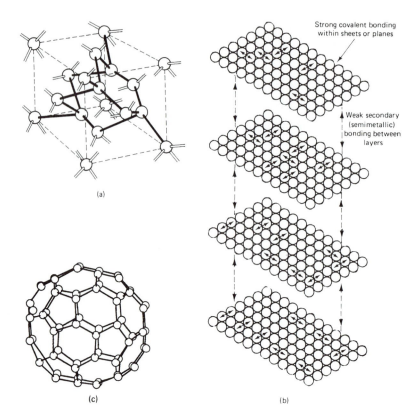

Figure 2-14 Polymorphism of carbon. (a) Diamond's cubic structure: each carbon atom forms strong covalent bonds with four other carbon atoms to develop a tetrahedron in the same manner as silicon and germanium; diamond cutters use a sharp instrument and a sharp blow to split crystal along cleavage planes (111) to produce perfect smaller jewels. (b) Layered structure in graphite: van der Waal bonding between layers allows easy cleavage into sheets, thus providing good lubrication properties. (c) C-60 or fullerene molecules have a spherical shape, like a soccer ball, composed of 60 carbon atoms covalently bonded together.

generated at about 2500°C, can be produced as helixes much like DNA, the basic building block of all living cells. The helix form of fullerene is another candidate for making resistance-free electrical superconductors. On the other hand, adding sodium atoms to ^{60}C spheres will yield electrical insulators.

Among other applications of fullerenes are superstrong fibers for construction materials or thin diamond films for hard-coat cutting tools and electronic circuitry. When joined with ammonia atoms, buckyballs become magnetic. Steel is an alloy of iron, carbon, and other elements. When the ^{60}C tubular molecules were alloyed in steel, greater strength was achieved while using conventional steelmaking techniques.

Fullerenes can be expected to play a key role in many materials applications because this unique substance provides alternatives in numerous phases of materials science and engineering. Rapid development of ^{60}C will mirror those of many other new materials.

Materials science and engineering have the benefit of powerful new tools, such as the scanning tunneling microscope to aid in unlocking the secret of structure. Buckyballs are much like charm bracelets that allow one to attach many charms of different types. ^{60}C permits the combination of numerous elements and molecules that will keep chemists and materials scientists at work for many decades producing new materials.

Ionic Bonding (Electron Swapping).   Instead of sharing electrons, some elements actually swap or ***transfer*** electrons to other elements. This type of bonding is called ***ionic.*** In general, inorganic compounds bond ionically. Elements having outer electron levels that are

Figure 2-15   Ionic bonding (electron swapping) of a sodium chloride molecule. Upon bonding both atoms have the equivalent of eight valence electrons in their outer shells. The orbitals within the shells are not shown in this figure. The subshells are not shown.

almost full or almost empty tend to gain or lose electrons to complete their outer energy levels. A classic example of ionic bonding of two elements is sodium (Na) and chlorine (Cl). From the periodic table, we learn that neutral sodium has 11 electrons, one of which is in the outer energy level. This single valence electron is weakly attracted to the positive nucleus. Chlorine (Cl) has 17 electrons, leaving 7 electrons in the outer or third energy level. As you recall, chlorine needs 8 electrons in its third energy level to attain stability. Since each atom attempts to attain equilibrium (lowest energy state) by completing its required number of electrons for each energy level, the sodium atom gives up its single electron through the action of a strong driving force of the chlorine atom to form a stable outer shell (Figure 2-15). This results in a temporary unstable condition that is soon changed by the sodium atom, which has now become a positive ion (***cation***) due to the loss of one of its electrons. The positively charged sodium atom (cation) interacts with the chlorine atom, which is now a negative ion (***anion***) due to the gain of the electron from the sodium atom. This interaction is a strong electrical (electrostatic) force between oppositely charged ions forming the ionic bond, which produces the molecule of sodium chloride (NaCl), common salt. Each of these elements has its own characteristic properties. Sodium is an active metal and chlorine is a poisonous gas. Yet combined as an ionic molecule, they have the well-known properties of common salt. Sodium and chloride ions are also different from their respective neutral substances (atoms). The greenish-yellow element chlorine, whose atoms have seven valence electrons, is very active. Chloride ions, in the formation of NaCl, have satisfied the octet rule with eight electrons in their outermost shells, and thus are stable. Otherwise, Cl^{-1} ions are extremely active, reacting with metals to produce corrosion. All these bonds are electrostatic in origin, so that the chief distinction among the various ways in which atoms bond to form molecules is in the distribution of the electrons around the atoms and molecules. ***Metals,*** defined later, can be characterized by the tendency of their atoms to lose their valence electrons. An ***active metal*** is one that loses its electrons quite readily. Similarly, an ***active nonmetal*** is one whose atoms readily accept electrons. We also know that nonmetallic elements have a larger number of valence electrons than do metallic elements.

    Metallic Bonding (Electron Swarming).   Metals contain one, two, or three valence electrons. These are shielded from the strong attractive forces of the positive nucleus by the inner electrons and thus they bond to the nucleus relatively weakly. Consequently, when in the company of other metal atoms, these metal atoms will lose their weakly held valence electrons, which, in turn, enter a common free-electron cloud, swarm, sea, or gas. These free or delocalized electrons can then move in three dimensions. Once these electrons leave their atoms, the atoms become positive ions (cations). An electrostatic balance is maintained between the cation and the ***electron cloud,*** which results in the cations arranging themselves in a three-dimensional

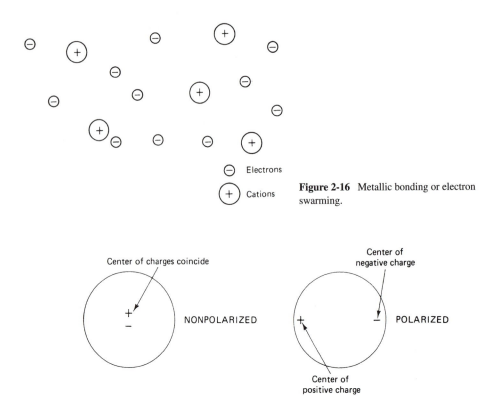

Figure 2-16 Metallic bonding or electron swarming.

Figure 2-17 Schematic of a nonpolarized and a polarized molecule.

pattern as shown in Figure 2-16. This electrostatic balance is the glue that bonds the metallic structure. The ever-moving free-electron cloud acts like a matrix surrounding the positive metallic ions and provides rigidity. If these ions were represented as spheres, their space arrangement in bulk metal could be likened to a box full of table tennis balls.

Secondary Bonding. The van der Waals bonds, which are much weaker forces of attraction, produce bonding between atoms and between molecules. Group VIIIA elements (inert gases) have a full complement of valence electrons in their outer levels. Thus, they have no inclination to lose, gain, or share electrons. At ordinary temperatures, they remain as single atoms (monatomic). Only at very low temperatures will these gases condense. It is the presence of van der Waals forces that permits this condensation. A similar situation occurs with covalently bonded molecules that have achieved equilibrium by sharing electrons. The van der Waals forces also permit these molecules to condense at low temperatures. All molecules consist of distributions of electrical charge. When two molecules are in close proximity, these electrical charge distributions interact and create intermolecular forces. The van der Waals forces of attraction between molecules have their origin in the forces of attraction of the nucleus (positive) of one molecule for the electron charge (negative) of a neighboring molecule, which form fluctuating dipoles in the two molecules. Such forces increase with an increase in the number of electrons per molecule because the electrons in the large molecule are more readily polarized by these forces.

In a *polar molecule* or *electric dipole,* the charges are polarized; that is, both the positive and negative charges are localized within the molecule. Figure 2-17 illustrates these two situations. A hydrogen molecule is a simple example of a *nonpolar molecule.* This diatomic molecule, in which both atoms are of the same element and are covalently bonded together, is sketched in Figure 2-18. Each atom shares the bonding electrons equally, producing an electrical charge distribution that is symmetrical about a line joining the two nuclei.

Any diatomic molecule, such as hydrogen chloride (HCl), in which the atoms are different from each other is polar (Figure 2-19a). The negative pole is located at the atom that is

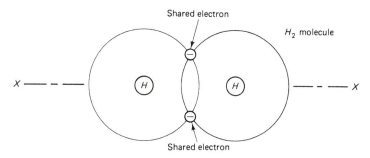

Shared electron

H_2 molecule

H

H

Shared electron

(a) Both atoms share two electrons equally forming a single covalent bond to complete each atom's outer energy level (K level).

180°

X

X

(b)

Figure 2-18   (a) A nonpolar diatomic hydrogen molecule. (b) A linear hydrogen molecule.

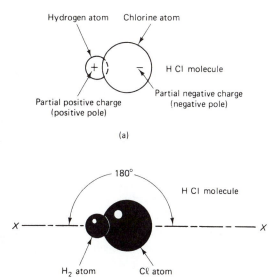

Hydrogen atom   Chlorine atom

+

H Cl molecule

Partial positive charge
(positive pole)

Partial negative charge
(negative pole)

(a)

180°

H Cl molecule

X

X

H_2 atom   Cℓ atom

(b)

Figure 2-19   (a) A polar diatomic molecule of hydrogen chloride (HCl). (b) A linear hydrogen chloride molecule.

more electronegative (attracts electrons more strongly). In a polar molecule, the electrons are not shared equally. The chlorine atom, being more electronegative, attracts more electrons than the hydrogen atom does, which causes the shared electrons to be associated more closely with the chlorine nucleus. Figure 2-19b is a three-dimensional sketch of this molecule showing that, like the hydrogen molecule, it has an electrical charge distribution with a symmetry around the line x–x.

A good example of a polar molecule containing more than two atoms is water (H_2O), sketched in Figure 2-20. Oxygen has a greater attraction for electrons than does hydrogen. As a result, the four electrons being shared in the two covalent bonds remain closer to the oxygen atom. This unequal sharing of electrons has the effect of creating a partial positive charge

(a) Three-dimensional   (b) Two-dimensional

(c) View showing centers of polarity   **Figure 2-20**   Various sketches of a water molecule (H_2O).

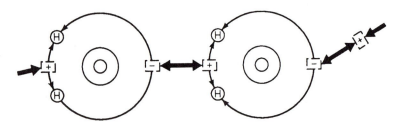

Figure 2-21   Hydrogen bond (bridge) in liquid water.

between the hydrogen atoms and a partial negative charge on the oxygen, which produces a *polar covalent bond.* The three-dimensional sketch (Figure 2-20a) of the water molecule with the oxygen atom in the center of a cube and with two hydrogen atoms occupying two of the eight corners shows a bent molecule (not symmetrical), with an angle between the two hydrogen atoms of about 105° (Figure 2-20b). Most covalent bonds are polar. The asymmetrical charge distribution on the water molecule results in the formation of regions of positive (+) and negative (–) charges located at a maximum distance from each other (Figure 2-20c). If the charge distribution were symmetrical, the positive (+) and negative (–) charges would cancel each other, as both centers would be at the center of the molecule. Thus, each center of charge can exert an attractive force on an adjacent charge of opposite sign. This statement explains, in part, how water molecules bond together to form liquid water. As pictured, the hydrogen atoms bond on one side of the oxygen, leaving the protons of the hydrogen atom farther from the nucleus of the oxygen atom.

The covalent bonds between the hydrogen and oxygen atoms in a water molecule, being polar, result in an asymmetrical charge distribution. The oxygen atom, having a slight negative charge, attracts a positive hydrogen atom belonging to an adjacent water molecule and forms a ***hydrogen bond,*** sometimes called a ***hydrogen bridge.*** This bond is the strongest of the secondary bonds. Figure 2-21 shows one schematic for representing the hydrogen bond between two water

TABLE 2-2 SUMMARY OF ATOMIC BONDING

Type of bond	Number of electrons "shared"	Types of atoms involved	Remarks
van der Waals (molecular)	0	Same	Weak electrostatic attraction due to asymmetrical electrical charges in electrically neutral (as a whole) atoms or molecules
Ionic	1 (or more transferred)	Different	Strong electrostatic attraction
Covalent		Same or different	Electron pair "revolves" in common orbit about both nuclei, one atom supplying one electron
Normal	2		
Coordinate	2	Different	Electron pair "revolves" in common orbit about both nuclei, one atom supplying both electrons, for other atom supplying none (quite rare)
Metallic	∞	Same	General attraction of a very large number of positive (metallic) ions for a dispersed cloud of electrons

Source: C.O. Smith, *The Science of Engineering Materials,* 2nd ed., Englewood Cliffs, N.J.: Prentice-Hall, 1977.

molecules in the liquid phase. The relatively high boiling temperature as well as other extraordinary properties of water are attributable to the bonding action of the hydrogen bridging.

As do all bonds, the hydrogen bond affects the properties and behavior of materials, particularly polymeric materials. In thermoplastics, the hydrogen bond joins long chainlike molecules to each other. These relatively weak bonds can be easily loosened or broken by heating, permitting flow to take place. This explains why thermoplastics can be converted by heating to a soft, flexible state and then back to rigid, solid plastic material upon cooling.

Our discussion of chemical bonding is well summarized in Table 2-2, which shows one secondary bond and the three primary bonds. Further discussion of these chemical bonding mechanisms comes up in dealing with the various groups within the family of materials, seen in Table 2-3.

2.1.2 Solid State

A *solid* is a sample of matter that has a fixed volume or size and a fixed shape that it retains indefinitely without any need to confine it. Picturing the atom as a sphere of microscopic size, atoms that comprise a solid are packed closer together than the same atoms that comprise a liquid or gas. We can say that solids do not flow like liquids nor do they expand like gases. The atoms of a solid do not possess the kinetic energy or the motion they would possess if they were in a liquid or gas. These atoms are quite subdued and do not act as independent units, but move with their neighboring atoms. As we will see, solids can be grouped into several categories, such as metals, plastics, ceramics, and composites. The attractive forces between atoms in a solid are much greater than between atoms in a liquid or gas. As such, these forces may position the atoms in some orderly geometric arrangement to form a *crystalline structure* (as in most metals) in which the atoms vibrate, rotate, and oscillate around fixed locations, maintaining a minimum dynamic equilibrium between adjacent atoms. From this, we can conclude that the geometric arrangement of atoms is related to its energy content. If the arrangement is orderly, this implies a minimum energy content. On the other hand, a disorderly, random distribution of particles (producing an *amorphous structure,* as in many plastics) is an indicator of motion. Hence, unbalanced forces are acting on the atoms, and thus an increase in internal energy by the system of particles is necessary to sustain such motion of particles from a small energy demand for vibration to a larger energy requirement for translational motion. Finally, another indicator of energy is temperature. The higher the temperature of the system of particles, the higher the disorder of the particles will be and the more energy that will be possessed by the system. With solids, we know that most of the motion of the atoms is confined to a vibrational type.

2.1.3 Processing and Structure

Throughout this book, you will learn how internal structure is altered as a result of processing. If materials are allowed to cool slowly as they transform from the liquid to the solid form, they will achieve equilibrium, which results in their natural internal structure and a given set of properties. However, if the cooling process is accelerated or interrupted, an entirely different structure and set of properties often results. For example, in casting molten metals to produce a desired shape, the rate of cooling is normally around 1 kelvin per second. But with use of a rapid-solidification processing technique, the molten metal may be cooled at a rate of 10^8 kelvin per second, which results in disorderly (amorphous) structure and unique magnetic properties not available in the normally cooled metal. Module 3 covers the rapid solidification process.

Structure can also be changed when materials are in the solid state. If the proper ingredients have been mixed together during material formulation, it becomes possible to reheat the solid, then vary the rate of cooling to obtain a variation in properties such as hardness or electrical conductivity. The process of bending or compressing a solid material will alter the internal structure in a manner that will possibly increase the hardness and improve strength.

2.2 FAMILY OF MATERIALS

There is nearly a limitless variety of materials. With so many materials, how can one be expected to understand them? The method used in this book is to consider all materials as members of a big family. Materials that possess common characteristics are then placed into their own group within the family. Even though overlaps exist in the grouping system, it is easier to understand materials when relationships are identified.

As seen in Table 2-3, four groups, *metallics, polymers, ceramics,* and *composites,* comprise the main groups; a fifth group of *other materials* is used for materials that do not fit well into the four groups. Each group then divides into subgroups. Some examples of each subgroup are listed to illustrate common materials in the subgroup. Figure 2-22 illustrates the variety of materials used in one of the communications systems: telecommunications. Within any system such as transportation, communications, or construction, numerous *materials systems* exist. These systems must pull materials from the groups within the material's family to meet their special needs. The wise materials user recognizes the many options available because of the compatibility between groups and then proceeds to develop a materials system that meets a specific need.

2.2.1 Other Systems of Grouping

The grouping system used in this book follows closely grouping systems common to engineering. Slight differences exist in other grouping systems. For example, one system might divide materials into metals, polymers, ceramics, glass, wood, and concrete. Some systems divide all materials into three groups: organics, metals, and ceramics. Then they distinguish between natural and synthetic materials. Others group materials as crystalline or amorphous. Regardless of the system, some inconsistencies develop because of the nearly limitless variety of materials. Still, the advantages of grouping outweigh the limitations.

In this section, we will present the system of grouping for this book and study briefly the general nature of each group and subgroup. Detailed study of each group follows in subsequent modules.

2.2.2 Metallics

Metallics, or metallic materials, include metal alloys. In a strict definition, *metal* refers only to an element such as iron, gold, aluminum, and lead. (See the discussion of metals and nonmetals in Section 2.1.1.1.)

TABLE 2-3 FAMILY OF MATERIALS

Group	Subgroup	Examples
Metallics (metals and alloys)	Ferrous	Iron
		Steel
		Cast iron
	Nonferrous	Aluminum
		Tin
		Zinc
		Magnesium
		Copper
		Gold
	Powdered metal	Sintered steel
		Sintered brass
Polymerics	Human-made	Plastics
		Elastomers
		Adhesives
		Paper
	Natural	Wood
		Rubber
	Animal	Bone
		Skin
Ceramics	Crystalline compounds	Porcelain
		Structural clay
		Abrasives
	Glass	Glassware
		Annealed glass
Composites	Polymer based	Plywood
		Laminated timber
		Impregnated wood
		Fiberglass
		Graphite epoxy
		Plastic laminates
	Metallic based	Boron aluminum
		Primex
	Ceramic based	Reinforced concrete
		CFCC
	Cermets	Tungsten carbide
		Chromium alumina
	Other	Reinforced glass
Others	Electronic materials	Semiconductors
		Superconductors
	Lubricants	Graphite
	Fuels	Coal
		Oil
	Protective coatings	Anodized aluminum
	Biomaterials	Carbon implants
	Smart materials	Shape memory alloys
		Shape memory polymers

2.2.2.1 *Metallic* defined.   The definition used for a metal will differ depending on the field of study. Chemists might use a different definition for metals than that used by physicists.

- *Metals* are elements that can be defined by their properties, such as ductility, toughness, malleability, electrical and heat conductivity, and thermal expansion.
- *Metals* are also large aggregations (collections of millions of crystals composed of different types of atoms held together by metallic bonds; see Figure 2-16). Metals usually have fewer than four valence electrons (electrons in the outer orbits of their atoms), as opposed to nonmetals, which generally have four to seven. The metal atom is generally much larger than the atom of the nonmetal.

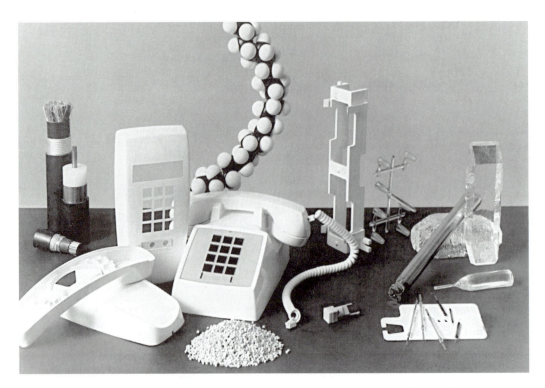

Figure 2-22 Family of materials in communication. One of the major factors influencing performance of telecommunications equipment is performance of materials used in its construction. Plastics (polymers) for outdoor, undersea, and underground cables; magnetic alloys (metals); fiber optics; synthetic single crystals (ceramics); and fiber-reinforced rubber (composites) are five principal areas in which materials research and development discoveries greatly influence telecommunications. (Bell Laboratories)

- *Alloys* consist of metal elements combined with other elements. Steel is an iron alloy made by combining iron, carbon, and some other elements. Aluminum–lithium alloys provide a 10% saving in weight over conventional aluminum alloy.

2.2.2.2 Types of metallics.

While metals comprise about three-fourths of the elements that we use (see metal groups in the periodic table), few find service in their pure form. There are several reasons for not using pure metals; pure metals may be too hard or too soft, or they may be too costly because of their scarcity, but the key factor normally is that the desired property sought in engineering requires a blending of metals and elements. Thus, the combination forms (alloys) find greatest use. Therefore, *metals* and *metallics* become interchangeable terms. Metallics are broken into subgroups of ferrous and nonferrous metals.

Ferrous.   *Ferrous* is a Latin-based word meaning "iron." Ferrous metals include iron and alloys of at least 50% iron, such as cast iron, wrought iron, steel, and stainless steel. Each of these alloys is highly dependent on possessing the element carbon.

Steel is our most widely used alloy. Sheet steel forms car bodies, desk bodies, cabinets for refrigerators, stoves, and washing machines; it is used in doors, "tin" cans, shelving, and thousands of other products. Heavier steel products, such as plate, I beams, angle iron, pipe, and bar, form the structural frames of buildings, bridges, ships, automobiles, roadways, and many other structures.

Nonferrous.   Metal elements other than iron are called *nonferrous* metals. The nonferrous subgroup includes common lightweight metals such as titanium and beryllium and common heavier metals such as copper, lead, tin, and zinc. Among the heavier metals is a group of white metals, including tin, lead, and cadmium; they have lower melting points, about 230° to 330°C. Among the high-temperature (refractory) nonferrous metals are molybdenum,

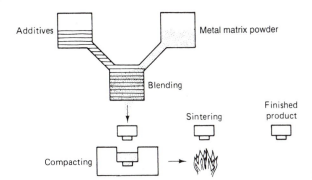

Figure 2-23 Powdered metal.

niobium, tantalum, and tungsten. Tungsten has the highest melting point of all metals: 3400°C. Metal alloys other than iron are called nonferrous alloys. The possible combinations of nonferrous alloys are practically endless.

   Powdered metals.   Alloying of metals involves melting the main ingredients together so that on cooling, the metal alloy is generally a nonporous solid. Powder metal is often used instead solid metal because it is undesirable or impractical to join the elements through alloying or to produce parts by casting or other forming processes. ***Powdered metal*** is sometimes called ***sintered*** metal. As shown in Figure 2-23, this process consists of producing small particles, compacting, and *sintering* (applying heat below the melting point of the main component). The squeezing pressure with added heat bonds the metal powder into a strong (sometimes porous) solid. Powdered metals can be ferrous, nonferrous, or a combination of ferrous and nonferrous elements with nonmetallic elements. Many powdered metal parts are specifically designed to be porous, for example, bearings, cam followers, and medical implants. The ability to control porosity (density) also allows for control of the modulus of elasticity.

2.2.3 Polymerics

Polymeric materials basically are materials that contain many parts. ***Poly*** means "many" and ***mer*** comes from *meros,* which means "parts." A ***polymer*** is a chainlike molecule made up of smaller molecular units (monomers). The ***monomers,*** made up of atoms, bond together covalently to form a polymer that usually has a carbon backbone.

   Figure 2-10 shows a simple sketch of an ethylene monomer. The atoms of carbon (C) and hydrogen (H) form covalent bonds. In Figure 2-10, many of the ethylene monomers have joined into a polyethylene polymer. Thousands of polyethylene polymers join together to form polyethylene plastic. The same process of polymerization is responsible for the formation of human-made rubbers and plastics; natural fibers; wood; rubber; animal bone; skin; and the tissues of humans, animals, and insects.

   2.2.3.1 Plastics.   The term ***plastic*** is used to define human-made, polymer resins containing carbon atoms covalently bonded with other elements, along with organic and inorganic substances. The word *plastic* also means "moldable" or "workable," such as with dough or wet clay. Plastic materials are either liquid or moldable during the processing state, after which they turn to a solid. After processing, some plastics cannot be returned to the plastic or moldable state; they are ***thermosetting plastics*** or ***thermosets.*** *Thermo* means "heat," and *set* means "permanent." Common thermosetting plastics include epoxy, phenolic, and polyurethane. Other plastics can be repeatedly reheated to return to the plastic state; they are ***thermoplastics.*** Examples of thermoplastic plastics are acrylics (e.g., Plexiglas or Lucite), nylon, and polyethylene. Although at present most plastics are produced from oil, they can also be made from other organic (carbon) materials, such as coal or agricultural crops, including wood and soybean. As our limited supply of oil is depleted, the major sources of polymerics will change.

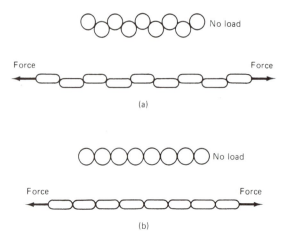

No load

Force   Force

(a)

No load

Force   Force

(b)

Figure 2-24 Elastomer structure. (a) Bond straightening. (b) Bond lengthening.

2.2.3.2 Wood.   Of all the materials used in industry, *wood* is the most familiar and most used. Wood is a natural polymer. In the same manner that polymers of ethylene are joined to form polyethylene, *glucose* monomers polymerize in wood to form **cellulose** polymers ($C_6H_{10}O_5$). Glucose is a sugar made up of carbon (C), hydrogen (H), and oxygen (O). Cellulose polymers join in layers with the gluelike substance **lignin,** which is another polymer.

2.2.3.3 Elastomers.   Prior to World War II, most rubbers (*elastomers*) consisted of natural rubbers. Today, synthetic or human-made rubbers far exceed the use of natural rubbers. An **elastomer** is defined as any polymeric material that can be stretched at room temperature to at least twice its original length and return to its original length after the stretching force has been removed. Some stretch to over 10 times their original length. Elastomers are able to store energy, so they can return to their original length and/or shape repeatedly.

Elastomers have a molecular, amorphous structure similar to that of other polymeric materials. This *amorphous,* or shapeless, structure consists of long coiled-up chains of giant molecules (polymers) that are entangled with each other. Adjacent polymers are not strongly bonded together. As seen in Figure 2-24a, when a tensile force (pulling force) is applied, these coils straighten out (**bond straightening**) and snap back like springs to their original coiled condition on removal of the force. When this same type of force is strong enough, not only do the polymers increase in length as a result of the bonds lengthening between individual atoms, they also lengthen (**bond lengthening,** Figure 2-24b) due to the unwinding coils, resulting in a temporary structure that approaches crystallinity, or a more orderly structure. Remember that with crystallinity comes strength.

To further increase the strength of the elastomers, the process of **vulcanization** is used to form the necessary cross-links (strong bonds) between adjacent polymers (to be discussed later in Section 6.2.3). Vulcanization is a chemical process (invented by Charles Goodyear in 1839) that produces covalent bonding between adjacent polymers with the help of a small amount of sulfur. Cross-links tie the polymers together to produce a tough, strong, and hard rubber for many uses in industry, such as in automobile tires.

2.2.3.4 Other natural polymers.   A most amazing natural polymer is human skin, which has no equal substitute. Animal skin or hide in the form of fur and leather has limited industrial use because synthetic materials have been developed that offer greater advantages to the designer than those of the natural polymers. Medical science continues to study such natural polymers as bones, nails, and tissues of human beings and animals to synthesize these materials for replacement when they are damaged due to injury or illness. Bioengineering and biomechanics are newer fields that integrate engineering and medicine to solve material problems in the treatment of humans.

(SiO₄ ion)

(a)

(b)

Figure 2-25 Ceramic structure.
(a) Silicate tetrahedron. (b) Single chain of
silicate.

2.2.4 Ceramics

Ceramics are crystalline compounds combining metallic and nonmetallic elements. *Glass* is grouped with ceramics because it has similar properties, but most glass is *amorphous.* Included in ceramics are porcelain such as pottery; abrasives such as emery used on sandpaper; refractories (materials with good resistance to heat) such as tantalum carbide, with a melting temperature of about 3870°C; and structural clay such as brick. Ceramics, including glass, are hard, brittle (no internal slip), stiff, and have high melting points. Ceramics primarily have ionic bonds, but covalent bonding is also present. Silica is a basic unit in many ceramics. The internal structure of silica has a pyramid (tetrahedron) unit as diagrammed in Figure 2-25a, which shows both the graphical diagram and the chemical equation. These silicate tetrahedrons join into chains, with the graphical diagram shown in Figure 2-25a. In Figures 2-25a and b, note how the larger oxygen (O) atoms surround the small silicon (Si) atom. The silicon atom occupies the space opening (interstice) between the oxygen atoms and shares four valence electrons with the four oxygen atoms. Chains such as shown in Figure 2-25b are extremely long and join together in three dimensions. The chains are held together by ionic bonds, whereas individual silica tetrahedrals bond together covalently. Silica is combined with metals such as aluminum, magnesium, and other elements to form a wide variety of ceramic materials.

2.2.5 Composites

By strict definition, a *composite* is a material containing two or more integrated materials (constituents or phases), with each material keeping its own identity. Normally, combining of the materials serves to rectify weaknesses possessed by each constituent when it exists alone. By this strict definition, many natural materials exist as composites; for example, wood is a combination of cellulose and lignin, but wood and natural materials are not classified as composites. Some familiar composites include plywood, laminated dimes and quarters, and shoe soles.

While most of the groups of the family of materials could be classed as composites because of the way they are placed in service (such as painted steel or case-hardened steel), the composite classification commonly refers to materials developed to meet the demands in building, electronic, aerospace, and auto industries. With an ever-increasing use of composites, they are truly the material of today and the near future because composites can be designed to be stronger, lighter, stiffer, and more heat resistant than natural materials or to possess properties required by technology that are not available in a single material. Composites allow the designer to select the right combination of materials to perform safely at the lowest cost.

The subgroups of composites shown in Table 2-3 include polymer based, metallic based, ceramic based, cermets, and others. It is also possible to classify composites by their structure. Composite structures include layers, fibers, particles, and any combination of the three. *Layered composites,* shown in Figure 2-26, consist of *laminations* like a sandwich. The laminations are usually bonded together by adhesives, but other forces could be used, such as those provided by welding.

Fibers and particles are integrated into composites by suspending them in a *matrix* or by the use of *cohesive* forces. The matrix is the material component, such as plastic, epoxy cement,

Figure 2-26 Typical composites. (a) Layered composite. (b) Particle composite. (c) Fiber composite.

rubber, or metal, that surrounds the fibers or particles. **Cohesive** forces involve the molecular attraction of one constituent to the other.

Examples of layered composites (Figure 2-26a) include plywood, laminated boards such as particleboard (wood chips) covered with plastic sheets, thermocouples (different metals welded together), safety glass (plastic sandwiched between sheets of glass), cardboard, and Alclad (copper alloy core that gains corrosion resistance from outside covers of aluminum). **Particle composites** include concrete (gravel, rock, and sand in Portland cement matrix), and particleboard (wood chips in resin matrix), as shown in Figure 2-26b. **Fiber composites,** such as reinforced plastics (FRPs), include fiberglass (glass fibers in a polyester plastic matrix), hardboard (vegetable fibers in an adhesive matrix), boron fibers in an aluminum matrix, reinforced glass (wire mesh in a glass matrix), and graphite epoxy (graphite fibers in an epoxy plastic matrix), as shown in Figure 2-26c. This reinforcing phase is usually stiffer and stronger than the other phase (the matrix) and serves as the load-carrying constituent. In general for fiber composites, one phase acts as a reinforcement of a second phase.

2.2.6 Other Materials

Another group is used to include materials that do not fit well into the groups discussed previously. Comparing the materials in the "other" group with the rest of the family of materials, Table 2-3 reveals that most come from the metallics, polymers, ceramics, and composite groups, but their applications justify a separate study. Some are discussed in Module 9.

Materials systems as described in Module 1 open up a whole new opportunity for grouping. Each major group of our family of materials has some materials from a newly evolving subgroup known as intelligent material systems.

Intelligent material systems, smart materials, or ***smart structures*** are materials and material systems designed to mimic biological organisms and offer the ultimate material system that can place control and feedback into a material structure. These materials take their cue from biological elements such as muscles, nerves, and bodily control systems that adapt to environmental changes. Just as alchemists long yearned to turn lead into gold, so too, materials scientists and engineers seek to endow materials with abilities such as (1) making controlled adaptations to changes in stress or heat, (2) making self repairs, and (3) providing feedback information on conditions that may have caused a material failure.

Science fiction such as the movie *Terminator II* has long foretold of "morphing" (changing shape and structure) or ***biomimetics*** (materials mimicking living tissue). As foretold in science fiction, materials designers continue to evolve material structures that do allow for modification to unpredictable environments. Also, feedback provides information to designers and users throughout the life of the material or structure. An example seen later in Module 6, a smart house, uses acrylic windows that change from clear to translucent white with the changing of outside light conditions. The transformation results from liquid crystals embedded in the glazing. Another feature available for smart houses includes new exterior walls that change from light colors that reflect heat in the summer to dark colors that absorb heat in the winter.

Monitoring systems using piezoelectric devices (explained in Module 9) are embedded into or adhesively bonded onto structures such as bridges, aircraft skin, building walls, and similar load-bearing structures. Connected to recording instruments, these intelligent structures provide information to help predict material/structural failures caused by aging, earthquakes, storms, or other stresses.

Nitinol, an alloy of nickel and titanium, illustrates how "smart materials" can be tailored to provide unique properties. The alloy is a ***shape memory alloy (SMA).*** SMAs, through heat treating, can be trained to hold a shape that they will always "remember," even if drastically bent out of shape. As seen in Figure 9-52, an eyeglass frame temple made of an SMA can be wrapped around one's fingers and then snapped back into its original shape. Through training, it is possible to set the shape memory so that it will change from a new shape back to its designed shape with application of a desired temperature, with $\pm$ 1°C accuracy. Nitinol is available in a variety of forms, including wire, sheet, ribbon, and springs. It has high operating temperature, abrasion and corrosion resistance, and high tensile strength. The nitinol alloy offers a variety of potential uses, ranging from moving robotic fingers by application of electrical currents to stress/strain sensors imbedded in bridges and other structures.

Biological structures have evolved over the millennia, and we can learn from their traits and have begun to imitate nature with designs of synthetic materials. Some of the design goals for smart materials and intelligent structures follow:*

1. Cost-efficient, durable structures whose performances match demands on the structure.
2. Change properties, color, shape, and manner to handle external physical loads to repair damage or make repairs from damage.
3. Possess the five senses of smell, taste, hearing, sight, and touch.
4. Allow structures to learn, grow, survive, and age with grace and simplicity.
5. Transmit information back to designers and user.
6. Incorporate adaptive features and intelligence to reduce mass and energy needs.
7. Allow for specification of materials and structural requirements to arrive at designs that are affordable while fulfilling design objectives.

The few examples of intelligent material systems cited here and in other parts of this book reveal "the dawn of a new materials age," as explained by Rogers (1992). You should watch for this evolution of new materials. The concepts learned here will allow you to comprehend the innovations.

*List adapted from Craig Rogers, 1992.

Figure 2-27 The first Boeing 777 lifting off for its maiden test flight. (The Boeing Company)

APPLICATIONS & ALTERNATIVES

This module introduced the family of materials grouping system as a means to help organize your thinking about the infinite variety of engineering materials. The Boeing 777 in Figure 2-27 represents a new breed of airplane, the latest in the application of engineering materials technology. Several illustrated discussions throughout this book will focus on the 777 and the innovative application of design, materials, and manufacturing techniques used on this new generation of airplane. With an estimated development cost of $5 billion and an average per plane cost of $100 million, the 777 relied on newer approaches to product design, development, and manufacture.

An approach to design/manufacturing, ***concurrent engineering,*** is a highly coordinated team approach that commences when a project begins. All parties (designers, manufacturers, suppliers, and customers) provide input throughout the project. Another useful tool is ***simulation,*** which uses computers and testing equipment to replace prototypes, live test beds, and other traditional approaches to development and saves time and money. Another approach involved *extensive component testing* to ensure reliability of each component, such as engines or landing gears, often done prior to assembly on the aircraft.

The 777 was designed to meet projected twenty-first century air travel patterns that use planes smaller than the Boeing 747 but larger than the 767. Designed with more reliable and very large engines, the 777 twin engine is built for transcontinental flight. Figure 2-27 shows the first Boeing 777 lifting off for its maiden test flight.

Pause for a moment and think about each of the main groups of the family of materials. How many applications can you think of for each main group that was used on this aircraft? Later in the text, we will show you some innovative applications of materials on the 777.

*Two excellent case study sources on the 777 are the PBS video series *21st Century Jet* (1 800 828-4PBS) and *21st Century Jet: The Making and Marketing of the Boeing 777* (Karl Sabbagh, Scribner).

SELF-ASSESSMENT

2-1. From the time line (Figure 2-1) discussed in the section Pause & Ponder, what principal group of our family of materials is regaining prominence as we move into the future? Explain why.

2-2. Using the time line in Figure 2-1, give an example of how a material development sparked a new technology in the Scientific Age.

2-3. How many elements have been discovered to date?

2-4. How many protons are in the nucleus of the fluorine atom?

2-5. What is an isotope? How does the mass of the deuterium isotope of hydrogen compare with the protium isotope of hydrogen?

2-6. Explain why the mass of a helium atom (Z = 2) is four times the mass of the simplest hydrogen atom (Z = 1), which consists of a single proton.

2-7. Fill in Table 2-1 for the element gallium. Write the shorthand notation for the electronic configuration of iron. Write the electron configuration for the elements oxygen and boron.

2-8. How many neutrons does oxygen-18 have in its nucleus?

2-9. The formula $2n^2$ is an easy way to determine the maximum possible number of electrons in any energy level of an atom. Using the formula, calculate the maximum number of electrons in the P energy level.

2-10. The key word in describing ionic bonding is *transfer*. What key word can be used to describe covalent bonding?

2-11. What is the name of the recently discovered carbon structure that resembles a soccer ball?

2-12. Give examples of ionic, covalent, and metallic bonding.

2-13. Express the average interatomic distance in units of nanometers and inches.

2-14. What is the relationship between processing and structure?

2-15. In which family group belong the materials with large chainlike molecules that consist of smaller molecules or monomers? Are they generally amorphous or crystalline?

2-16. When a metallic contains at least 50% iron, what is its classification?

2-17. What is the name for a metallic consisting of metal elements and other elements? Give three examples.

2-18. Define *composites*. List three composite structures.

2-19. In the ceramic grouping, most materials are crystalline, but which one has an amorphous structure?

2-20. List three examples of lightweight nonferrous metals or alloys, three examples of heavier nonferrous metals, and one example of a refractory metal.

2-21. What is the subgroup of metallics in which metal particles are joined by pressure and sintering heat?

2-22. What subgroup covers human-made polymers containing carbon atoms covalently bonded together with other elements? List the two divisions of this polymeric material and give an example of each.

2-23. Name three natural polymeric materials.

2-24. How do elastomers differ from other polymeric materials? How can the strength of rubber be improved?

2-25. How does the energy shortage threaten the supply of plastics and rubber?

2-26. What is the prime purpose of selecting a composite material over material from the other family groups?

2-27. List three materials that were grouped in the "other" family group of materials.

OBJECTIVE QUESTIONS

2-1. The largest consumer of electricity in the United States is the refining of the metal _____.
  a. Steel   **b.** Copper   **c.** Aluminum

2-2. A _____ is a substance that cannot be broken down any further by a chemical reaction.
  a. Molecule   **b.** Compound   **c.** Element   **d.** Atom

2-3. A compound is a substance that can be broken down by chemical reactions into _____.
  a. Elements   **b.** Molecules   **c.** Atoms

2-4. Protons and neutrons are known as _____.
  a. Quarks   **b.** Electrons   **c.** Neutrinos   **d.** Nucleons

2-5. Another name for an electron shell is _____.
  a. Quantum   **b.** Principal energy level   **c.** Energy state   **d.** Ion

2-6. Two electrons can exist in the same orbital if they have _____.

    a. The same energy   **b.** The same polarity   (**c.**) Opposite spins

2-7. The maximum number of electrons in the L, or second, energy level is _____.

    a. 2   **b.** 18   (**c.**) 8   **d.** 0

2-8. Inert gases can also be called _____.

    a. Stable   **b.** Group 5 elements   (**c.**) Noble gases   **d.** Active

2-9. The core electrons in neon (Ne) number _____.

    a. 2   (**b.**) 10   **c.** 18   **d.** 6

2-10. Horizontal rows in the periodic table are called _____.

    a. Groups   **b.** Transition elements   (**c.**) Periods   **d.** Metals

2-11. Metalloids consist of how many elements?

    a. 6   **b.** 10   (**c.**) 7

2-12. The electronegativity of iron (Fe) is _____.

    (**a.**) 1.83   **b.** 3.98   **c.** 0.93

2-13. What is the principal valence of Group VIII elements?

    a. 16   **b.** 6   (**c.**) 8   **d.** 12

2-14. $^{16}_{8}O$, $^{17}_{8}O$, and $^{18}_{8}O$, are the isotopes of oxygen. What is the atomic number of oxygen? How does the number vary among the three isotopes?

2-15. A distinction that sets intelligent structures apart from other advanced materials is:

    a. Cost per unit volume is often higher than traditional materials.

    (**b.**) Ability to adapt to changes in environment.

    c. Requires more exotic processing.

    d. Benefits from superior properties offset material cost.

REFERENCES & RELATED READINGS

AMATO, IVAN. *Stuff: The Materials the World Is Made Of.* New York: Harper Collins, 1997; Avon Books (Bard), 1998.

ASHBY, M. F. "Technology of the 1990s: Advanced Materials and Predictive Design," *Philosophical Transactions of the Royal Society of London,* Vol. A322, 1987, pp. 393–407.

ASIMOV, ISAAC. *Building Blocks of the Universe.* New York: Abelard, 1957.

BELLMEYER, FRED W. *Synthetic Polymers.* Garden City, NY: Doubleday, 1972.

BETTS, JOHN E. *Physics for Technology.* Reston, VA: Reston, 1964.

BROWN, THEODORE L., AND H. EUGENE LEMAY, JR. *Chemistry: The Central Science,* 2nd ed. Englewood Cliffs, NJ: Prentice-Hall, 1981.

CHRISTIANSEN, G. S., AND PAUL H. GARRETT. *Structure and Change: An Introduction to the Science of Matter.* San Francisco: W.H. Freeman, 1960.

COTTERILL, RODNEY. *The Cambridge Guide to the Material World.* Cambridge: Cambridge University Press, 1989.

CRAIG, DOUGLAS F. "Structural Ceramics," *National Educators' Workshop, NEW: Update 91,* Oak Ridge National Laboratory, Oak Ridge, TN, November 12, 1991.

EASTERLING, KEN. *Tomorrow's Materials.* London: The Institute of Metals, 1988.

EWEN, DALE, AND LERAY HEATON. *Physics for Technology Education.* Englewood Cliffs, NJ: Prentice-Hall, 1981.

GAMOW, GEORGE A. *Mr. Tomkins Explores the Atom.* New York: Macmillan, 1945.

GIANCOLI, DOUGLAS C. *Physics.* Englewood Cliffs, NJ: Prentice-Hall, 1980.

HUMMEL, ROLF E. *Understanding Materials Science: History, Properties, Applications.* New York: Springer-Verlag, 1998 ($59.95).

NATIONAL RESEARCH COUNCIL. *Materials Science and Engineering for the 1990s: Maintaining Competitiveness in the Age of Materials.* Washington, DC: NRC, 1989.

ROGERS, CRAIG A. "Intelligent Materials Systems: The Dawn of a New Materials Age." Center for Intelligent Materials and Structures, Virginia Polytechnic Institute and State University, 1992.

SASS, STEPHEN L. *The Substance of Civilization: Materials and Human History from the Stone Age to the Age of Silicon.* New York: Ariade Publishing, 1998 ($24.95).

SHACKELFORD, JAMES, AND WILLIAM ALEXANDER. *The CRC Materials Science and Engineering Handbook.* Boca Raton, FL: CRC Press, 1991.

SMITH, CHARLES O. *Science of Engineering Materials.* Englewood Cliffs, NJ: Prentice-Hall, 1977.

Periodicals

Adhesive Age

Advanced Materials and Processes

Ceramics Bulletin

Chemical and Engineering News

Chemical Engineering

Discover

Journal of Materials Education

Machine Design

Manufacturing Engineer

NASA Technology Briefs

Plastic World

Popular Science

Scientific American

Technology Review

Module

3

Processing and Structure of Solid Materials

After studying this module, you should be able to do the following:

3–1. Discuss the relationships between processing and structure of materials.

3–2. Recall the synthesis and processing capabilities of contemporary materials technology at the various scales (macro, micro, nano, and atomic).

3–3. Recall techniques and instruments used to determine material structures, including the capabilities of the optical, electron, and scanning tunneling microscopes.

3–4. Identify characterization components of substrate and coating on a material specimen, including oxide inclusion, substrate roughness, void, pore, crack, cohesive strength between particles, particles, phases, and grains. Explain qualitative and quantitative material characterization.

3–5. Use sketches, unit cell structures, and standard notations to explain and solve problems related to crystallography. Determine the unit cell structure of given elements from a periodic table of elements for metallurgists.

3–6. Explain the alloying and doping process of materials in terms of solutions, alloying elements, packing factors, crystal imperfections, and impurities, and use appropriate sketches. Cite specific practical applications for alloying and doping.

3–7. Using cooling curves as an aid, recall the effects of solidification in pure metals, alloys, and nonmetals.

3–8. Recognize Greek symbols, and determine their meaning and proper pronunciation.

3–9. Apply SI rules of usage, make conversions of U.S. customary units to SI (and vice versa), and use both for computations.

3–10. Use EMT Weblinks* to seek current developments and emerging technology related processing and structure*

*EMT Weblinks listed throughout the book will link you to Internet sites related to topics being covered. The dynamic nature of the Web brings frequent changes, so some of these sites, although available at the time of writing, may not be up now.

PAUSE & PONDER

Understanding the natural structure of materials allows for manipulation of structures to achieve results different from those that would normally occur. Normally, metals must be heated to the liquid state before they can be formed into complex shapes. Less complex shapes can be achieved in metals by heating them to a point in which they become *plastic,* or formable. These processes include extrusion, drawing, bending, and so on. Some softer metals and thinner, harder metals can be cold formed at room temperature.

Thixotropic (read "thicks-oh-tropic") *metals* exist at a semisolid state; they liquefy while under shearing stress and then solidify when left standing. Jello is semisolid but, of course, much softer than a thixotropic metal. It is possible to turn metal alloys into the thixotropic state by stirring metal vigorously as it cools from the liquid state en route to the solid state. When metal cools under *equilibrium* conditions (slow, natural change from liquid to solid) as shown in Figure 3-1, crystals form: first, by *nucleating* (seed crystals begin to form), then with further cooling growing to form *dendrites* (tree-shaped structures), and finally growing into *grains* or *crystals.* If the metal is stirred vigorously while in a state between liquid and solid (much like the mush of melting snow, which is part ice crystals and part water), the dendrites will form spherelike shapes (Figure 3-2) rather than the normal treelike dendrites to produce a *semisolid material (SSM).*

Injection molding, shown in Figure 3-3, utilizes a screwing action that forces thixotropic magnesium through a nozzle into a mold or die made of very hard tool steel or similar die stock. The mold has had a cavity formed into it in the reverse shape (female) of the part to be produced. Injection molding is a process often used for molding plastic and ceramic parts, but not metals. Dow Chemical has developed a process for injection molding of magnesium parts that is possible through the use of thixotropic alloys. *Casting* and *molding* are shaping operations in which the starting material is a heated liquid or semifluid. When referring to metals, the term *casting* is normally used, whereas *molding* is the common term used for plastics. Both processes are used in glassworking. Figure 5-61a shows the sequence of operations in the investment casting of titanium alloys.

In Figure 3-3, you will notice that cold magnesium pellets (in a thixotropic state) are fed through a hopper into an argon atmosphere, then into the heated barrel of the injection-molding machine. Argon is an *inert gas* and, as such, protects the magnesium from oxygen in the factory atmosphere. The shearing action of the screw, acting on the magnesium pellets, transforms them into a thixo-extrudate at 580°C. When enough of the extruded magnesium slurry (solid and liquid) col-

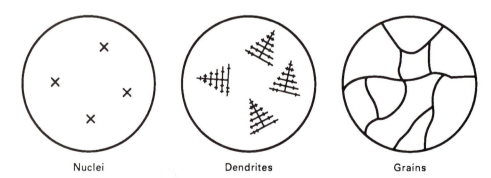

Nuclei   Dendrites   Grains

Figure 3-1 Stages of a metal cooling through equilibrium: first nucleates, then forms dendrites, and finally cools into a crystalline solid of many metal grains.

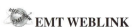 **EMT WEBLINK**

Thixtropic injection molding processes with image gallery—*http://www.thixomat.com*

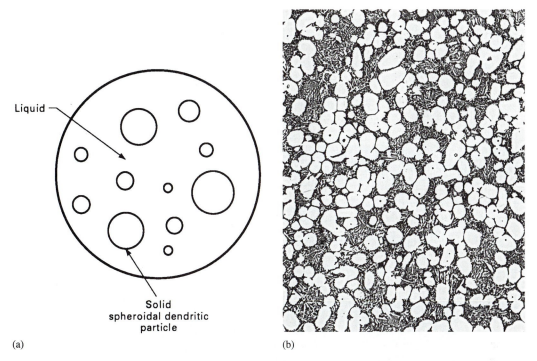

Liquid

Solid
spheroidal dendritic
particle

(a)   (b)

Figure 3-2 Semisolid, thixotropic alloys form nearly spheroidal dendrites. (a) Diagram of SSM structure.
(b) Photomicrograph reveals how continuous stirring during solidification of semisolid material (aluminum casting
alloy) results in a microstructure of large equiaxed solid particles within the formerly liquid eutectic mixture. The
solid particles agglomerate during the SSM processing, affecting flow behavior. (ASM International)

lects in the heating zone chamber, it is injected by the force of the high-speed shot system, which
pushes against the thrust bearing, causing the reciprocating screw to inject slurry through the non-
return valve past the nozzle into the mold cavity. The molded SSM part is the equivalent to a part
produced through melting and casting in the liquid state. The injection process offers advantages
over die casting: improvements in mold cycle time, less downtime, and improved part quality.

These processing techniques illustrate the desire to apply technology that provides an
end product of the highest quality with as few steps as possible. In this module, we provide information on the basic structure of materials and how structure is altered by processes and synthesis. There is a never-ending quest for improved techniques that can form and treat materials so that the ideal material can be produced for the perfect product or system. The quest
requires constant study to keep up with new developments. This module can prepare you to
deal with today's materials and processes and to adapt to coming changes.

3.1 SYNTHESIS AND PROCESSING

Solid materials are products of the conversion of raw materials into bulk materials. With many
polymers, synthesis is a part of raw materials production. Newer materials, such as artificially
structured materials, are also synthesized. The result of raw materials processing or synthesis
with conventional means provides us with familiar structures such as polycrystalline metals,
amorphous plastics, and crystalline ceramics. *Artificially structured materials,* which are synthesized on micro, nano, and atomic scales, have engineered unique structures that may be ultrapure and/or possess electronic circuitry and optical properties based on a special design. In
the modules on metals, polymers, ceramics, and composites, you will learn that a newer process,
such as rapid solidification of metal, provides an amorphous structure rather than the natural
polycrystalline structure. Chemical vapor deposition (CVD) and molecular beam epitaxy
(MBE) are recent synthesis techniques that allow for the production of "designed materials."

Thixotropic magnesium injection molding

(a)

(b)

Figure 3-3 (a) Dow Chemical's thixotropic magnesium injection molding process. (b) SSM magnesium alloy parts formed by thixotropic injection molding. (ASM International and Thixomat Corp.)

TABLE 3-1 COMPARISON OF OUR CONTEMPORARY CAPACITY TO PREPARE NEW MATERIALS OR COMPOSITES ARRANGED AS A HIERARCHY OF SIZE

Scale	Capacity of the technology	Type of materials synthesized or processed
Macro (cm–m)	Excellent	Materials for very large structures have been optimized (e.g., reinforced concrete, airframe alloys)
Micro (mm–μm)	Very good	Lab scale and small objects (GRP to graphite–epoxy composites, transparent Lucalox, composite transducers, etc.)
Nano (0.1–10 nm)	Primitive but new	New materials designed and made (intercalates, glass-ceramics, toughened zirconia, nanocomposite desiccants, etc.)
Atomic (0.1 nm)	Poor	New materials found by serendipity (penicillin, Teflon, magnetic garnets, 1:2:3 superconductors, etc.)

Source: Adapted from Roy (1992).

Many *"beam" processing* and synthesis techniques have evolved with the use of laser, ion, and electron beams to produce micro- and nano-scale structures, as shown in Table 3-1.

Manufacturing processes are traditionally used to convert the bulk synthesized or processed materials into structural shapes and final subassemblies of products. Manufacturing processes will usually change the structure of the bulk material, thereby requiring a variety of "secondary processes" to ensure that the final product has the necessary structure. *Near-net-shape* and *net-shape processing* aim to achieve a final product or specified structure through the initial synthesis or raw material processing, thereby avoiding secondary processes.

Aluminum serves as an ideal example. Although aluminum is the most plentiful metal in our earth's crust, it requires enormous amounts of electrical and chemical energy to extract it from bauxite ore. Great savings are realized by recycling, which makes it practical and desirable for all concerned. Often, however, bulk aluminum must go through many *secondary processes,* such as machine cutting, welding, grinding, anodizing, and heat treating, to achieve a finished product. Using near-net-shape processing to go from the raw aluminum to the final product, while achieving the desired structure and properties, would save even more energy. In Module 8, we will explain how pressureless, infiltration processes produce near-net-shape metal-matrix composite (MMC) parts.

Table 3-1 provides a comparison of contemporary synthesis/process methods based on the scale (macro, micro, nano, or atomic) of the technique employed. Note that as the scale gets smaller, the capacity of the technology diminishes. From Figure 2-1, showing the time line of materials development, the immediate future of materials technology is classified as the nanocomposite era (defined later in this module). The ability to develop improved instrumentation to explore and manipulate matter at the nano and atomic scales will determine the winners of the race to place products in the marketplace. All developed countries and many developing countries are investing heavily to improve or maintain their competitiveness in the races for materials such as those listed in the nano and atomic categories for use in products such as high-definition television (HDTV), superior computer technology (neural networks, multimedia, artificial intelligence, virtual reality), improved transportation (magnetic-levitation trains and surface seacraft), better communications and entertainment technology, and smart homes and smart highways.

With instruments such as the scanning tunneling microscope (described later), materials scientists may view and manipulate the nano and atomic levels for better understanding and control of structure and improved synthesis and processing. At the nano and atomic scales, it is becoming possible to produce new structures such as *nanocomposites*—materials with atoms and molecules arranged into unique phases to provide designed properties. We will see later in Figure 3-13 that *nanoscale mixing* will provide properties such as the following (Roy, 1992):

1. *Chemical/structural:* control of phases formed, lower reaction and sintering temperatures, control of microstructure, control of morphology via solid-state epitaxy
2. *Mechanical:* fivefold increase in strength as size goes from 100 to 20 nm; superhard (> diamond) materials
3. *Optical:* tenfold increase in luminescence
4. *Electrical:* change in fundamental conduction
5. *Magnetic:* newer super-, para-, and ferrimagnets

3.1.1 Nanotechnology

Nanotechnology provides the potential to design and manipulate material structures at the atomic level. Electronic components such as transistors, capacitors, and diodes with dimensions less than 0.25 micrometers are emerging from the nation's laboratories. Units such as micrometers are less useful as units of measure. The next smallest unit of measure is the *nanometer (nm),* defined as one billionth of a meter. To emphasize the magnitude of a nanometer, the diameter of most atoms is 0.1 to 0.4 nm. The range of the diameter of human hair is 40 to 120 μm, which expressed in nanometers would require multiplying by 1000. (See Figure 3-12 for a comparison of synthetic ceramic fibers with human hair.) This new technology is being investigated for use in manufacturing. Starting at the molecular level and assembling any number of molecules of any size, material can be constructed with perfect precision. This attempt has been given the name of *molecular manufacturing, nanofabrication manufacturing,* or *molecular nanotechnology.*

Nanotechnology is defined as the processing of materials into microelectromechanical structures and devices whose sizes vary from hundreds to hundredths of a micrometer (mm). Combining science and technology, nanotechnology builds at the atomic and molecular levels. The materials involved can include diamond films, organic films, semiconductors, metal films, dielectrics, ferroelectric films, and piezoelectric films. Processing techniques involved include ion implantation, optical and e-beam lithography, plasma and wet etchings, and physical and chemical vapor

EMT WEBLINK

> MicroScapes-too small for the human eye, a series of photographs (Silicon Dreams, The Edge of Light, Solid States, Symmetry in Motion, Memories, Glass and Apparitions) taken through high power microscopes—*http://www.lucent.com/microscapes/*

deposition (CVD). These processes are producing flat panel displays, microelectronics, optoelectronics, and solid-state sensors. Many of these topics are covered throughout this book.

3.2 DETERMINING MATERIAL STRUCTURE

The kinetic theory of matter describes atoms in a solid as being at their lowest energy state. In this state, their motions have slowed considerably. These atoms of solids are so close to their neighboring atoms that they can no longer act as independent particles. They vibrate about their equilibrium positions, depending on the degree of remaining kinetic energy (thermal energy) they may possess. They can also move in conjunction with their neighboring atoms. But they can and do move. In the discussion about bonding in Module 2, we learned that two or more atoms can chemically join each other, depending on their electronic structures. We also learned that atoms, ions, and molecules are of different sizes.

The internal structure of solids determines to a great extent the properties of the solid or how a particular solid material will perform or behave in a given application. The study of the internal structure is pursued in various ways using x-rays as well as electron, proton, or neutron beams to disclose crystal size, crystal structures, crystal imperfections, bonding types, spacing of adjacent atoms and adjacent planes, and the different atoms in a solid. A brief knowledge of some of the more important tools in the field of spectroscopy and microscopy that permit a still deeper probe into the internal structure of materials is vital to a greater understanding of the effects of structure on the behavior of natural and human-made materials. Figure 3-4 shows photomicrographs of varying magnification produced with the aid of various instruments used in materials science.

3.2.1 Spectroscopy

White light, when refracted (light rays undergo a change in direction when passing from one medium to another), is broken up into a number of different colors and forms a spectrum. Each spectral color corresponds to a particular wavelength. Every chemical element (as well as its atoms) produces a characteristic spectrum, which may be detected and measured by a spectroscope. Spectroscopy using such an instrument studies substances through an analysis of their spectra. In other words, *spectroscopy* permits the precise measurement of the wavelengths of atoms' radiations, which are discrete and distinct for each atom. wWhen atoms are exposed to high energy from an outside source, such as infrared, atomic absorption, arc-spark, or glow discharge, they become excited and their electrons move to higher energy levels. Becoming unstable, the electrons tend to return to their lower energy levels. The interchanges of energy between radiation and matter take place in discrete units called *quanta* (discussed in Modules 4 and 9). Under excitation an atomic substance will reach a state of dynamic equilibrium in which some atoms are releasing energy at the same rate that it is being absorbed by others. An atom loses energy when its excited electrons transit to a lower energy level, giving off a *photon* of light. The wavelength of this photon is related to the magnitude of the change of energy. Such transitions, when observed by the spectroscope, produce atomic spectra (Figure 3-5) with discrete and distinguishing wavelengths corresponding to the various electron transitions between energy levels. These studies permit scientists to increase their knowledge of atoms, electronic structures, and various bonding mechanisms.

3.2.2 X-Ray Diffraction

One of the most useful tools in the study of crystal structures of solids is *x-ray diffraction.* An x-ray diffractometer is used in this process. X-rays have wavelengths about equal to the diam-

(a)   (b)   (c)

(d)   (e)   (f)

Figure 3-4 Photomicrographs of structural features of some representative materials. (a) Steel (SAE/AISI 1045) 1000×. (NASA) (b) Steel (SAE/AISI 1045) SEM 2550×. (NASA) (c) Wood (maple) 415×. (U.S. Forest Products Lab) (d) Plastic (polyethylene) dendrite crystals. (Dr. Philip H. Geil—U. of Illinois) (e) Glass ceramic: crystals growing in amorphous glass, 17,900×. (Corning Glass Works) (f) Fiber metal composite. (NASA)

Figure 3-5 Atomic spectrum of sodium in the visible region. (Adapted from D. C. Giancoli, *Physics,* Prentice-Hall, Inc., Englewood Cliffs, N.J., 1980.)

eters of atoms (10^{-10} meters) or about the same length as the spacing between atoms (or ions) in solids. When these x-rays are directed at a solid with a crystalline structure, the waves are refracted. The equation, known as *Bragg's law,* $n\lambda = 2d \sin \theta$, relates the wavelength, λ, to the distance between planes of atoms, d, and the glancing angle between the incident beam and

Figure 3-6 (a) Scattering of x-rays by atoms in parallel planes. (b) Constructive and destructive interference of waves. (Theodore L. Brown and H. Eugene LeMay, Jr., *Chemistry—The Central Science,* 2nd ed., Prentice-Hall, Inc., Englewood Cliffs, N.J., 1981.)

EMT WEBLINK

IBM Research—*http://www.research.ibm.com/topics/serious/materials*

the plane of atoms, θ. Figure 3-6a shows a sketch of this relationship. If the incident beam strikes the planes of atoms at some arbitrary angle, the reflected beam may be nonexistent, because the refracted rays from the atomic planes will be out of phase and produce destructive interference (Figure 3-6b) and cancellation of the reflected beam. It has been determined that,

EMT WEBLINK

NIST Gallery—*http://www.nist.gov/public_affairs/gallery/gallery.html*

at a particular angle θ, the reflected beam will be in phase and produce constructive interference (Figure 3-6b). Mathematically, this can be expressed as $\lambda_1 = n\lambda_2$, where $n = 1, 2, 3, \ldots$; the distance traveled from the different parallel planes of atoms represents an integral number (n) of wavelengths. Referring to Figure 3-6a, this statement may be written as

$$ACA' = BEB' - n\lambda \qquad \text{where } n = 1, 2, 3, \ldots$$

Knowing the wavelength of the incident beam of x-rays, the glancing angle can be measured experimentally, and solving Bragg's law for $d = n\lambda/2 \sin \theta$, the interplanar distance can be calculated. Figure 3-7 shows a photographic plate on which the constructive waves produce a series of dots, indicating that the x-rays are scattered from crystals at only certain angles. All the various angles at which diffraction occurs are determined by measurements on the photographic film. By studying the directions of diffracted x-ray beams, as well as the intensities produced by this powerful tool, much can be learned about the crystal structure of solids.

3.2.3 Optical Microscopy

The human eye can distinguish between two points in space only if the two points are approximately 0.1 mm apart at a distance of 25 cm. Microscopic objects perhaps only 0.5 μm apart with the eye focused at the normal reading distance of 25 cm can be seen as an image magnified by a factor of 200 or more. The compound microscope fulfills this requirement by having

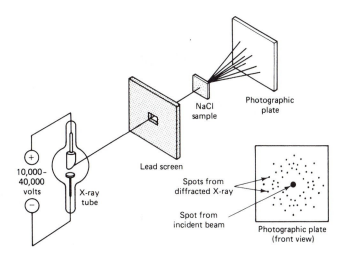

Figure 3-7 The x-ray diffraction pattern for NaCl and the experimental method by which it is obtained. (T. L. Brown and H. E. LeMay, Jr., *Chemistry—The Central Science,* 2nd ed., Prentice-Hall, Inc., Englewood Cliffs, N.J., 1981.)

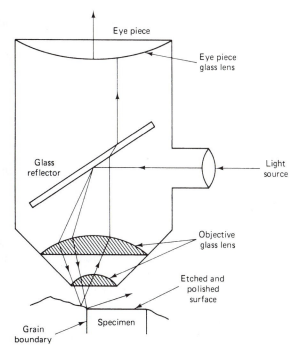

Figure 3-8 Schematic of an optical system of a metallographic microscope detecting a grain boundary in a crystal.

two lenses, the objective and the eyepiece, to give two stages of magnification. The quality of the microscope image is determined by the magnification, degree of diffraction, distortion due to aberrations of the lenses, and contrast. ***Metallographs*** are optical, metallurgy microscopes capable of photographing the images of solid specimens (see Figures 3-8 and 3-9). Opaque specimens, such as metallurgical samples, are typically observed with a reflection microscope in which light is reflected from the sample. Specimens often show very poor contrast between adjacent areas, even though these areas differ in chemical composition or morphology. As a result, most metallographic specimens are acid-etched to make specimen details visible. The ***etching*** reacts with the grain boundary areas at a different rate than the grains themselves. When

Figure 3-9 Photograph of a metallograph. (Buehler Ltd.)

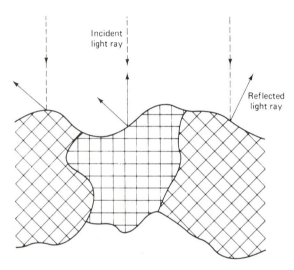

Figure 3-10 Sketch showing identification of individual crystals by reflected light rays.

incident light rays strike these areas with their different crystal orientations, as well as their difference in rate of etching, the reflected light travels back to the eye in different amounts (see Figure 3-10). If most of the incident light is reflected back to the eye, the grain will appear bright. Other crystals reflecting less light will appear darker. A *photomicrograph* of a specimen taken through a metallograph (Figure 3-11) shows the details of the microstructure, including the grain boundaries. Photomicrographs of various materials are shown throughout the book.

3.2.4 Electron Microscopy

The maximum magnification produced by optical microscopes is about 2000×. This limitation is imposed by the wavelengths of visible light, which limit the resolution of minute details in the specimen being observed. *Resolving power* is a term used to describe the ability of a lens to reveal detail in an image. Optical microscopes (discussed previously) do not have the ability to discern details smaller than 10^{-6} m. To observe objects in the 10^{-10} m range, high-energy electrons with wavelengths around 4×10^{-11} are used. These electrons are similar to visible light because they both are forms of electromagnetic radiation. By using magnetic lenses to focus the electrons (see Figure 3-12a) their resolving power can be increased greatly to give magnifications over 200,000×. Remember that metallic specimens are opaque to electrons (i.e., electrons are absorbed by the metallic atoms).

Figure 3-11 A photomicrograph of low carbon steel. (Buehler Ltd.)

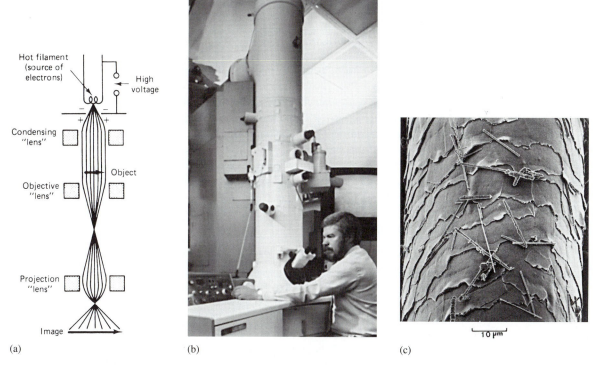

(a)   (b)   (c)

Figure 3-12 Transmission. (a) Electron microscope. The squares represent magnetic field coils for the "magnetic lenses." (Douglas C. Giancoli, *Physics—Principles with Applications.* © 1980, reprinted with permission of Prentice-Hall, Inc., Englewood Cliffs, N.J.) (b) Photograph of a transmission electron microscope (TEM). (c) TEM micrograph comparing whiskers of ceramic (SiC) to a human hair (ORNL).

The *transmission electron microscope (TEM)* (see Figure 3-12b) has been a driving force for the development in electron optics earlier in this century. In brief, an unfocused electron beam is incident on a very thin sample and, by using the transmitted electrons, magnifying projection optics form an image of the sample on photographic film with a resolution of a few billionths of a meter. Because the electron beam passes through the specimen, details of the internal microstructure are made observable. However, solid materials are highly absorptive to electrons and specimens must be prepared in the form of a very thin foil. The transmitted beam is projected onto a fluorescent screen or photographic film so that the image may be viewed. Whiskers are short, crystalline fibers with outstanding strength that have an average diameter of about 0.05 μm. These whiskers are shown in Figure 3-12c along with a human hair (the larger dark object).

The *scanning electron microscope (SEM)* scans the surface of a specimen with an electron beam and the reflected (back scattered) beam of electrons is collected and displayed at the same scanning rate on a cathode ray tube. The image on the screen contains the surface features (with good depth of field) of the specimen, which may be photographed. The major requirement for the specimen is that it be electrically conductive. Accessory equipment allows analysis of the chemical composition of very localized areas of the sample.

An example of the use of electron microscopy can be found in the research and development of synthetic crystalline materials produced by ***molecular beam epitaxy (MBE).*** These materials not found in nature have extremely thin layers of atoms deposited alternately on a semiconducting substrate or base. By controlling the chemical composition and thickness of the layers, various crystalline structures are formed with varying electrical, mechanical, and optical properties. The MBE method (described in Module 9) would not be possible without the prior development of electron microscopes with small wavelengths (0.2 nm) and large magnifications, which permitted scientists to verify that their efforts were indeed producing a crystal with extremely thin atomic layers of alternating composition. Atomic layers as thin as one ten-millionth of a centimeter were determined to exist in these new human-made materials.

EMT WEBLINK

Micro Worlds with Tutorials—*http://www.ibl.gov/MicroWorlds/*

Knowing that (1) changes in the internal structure of solids are accompanied by a change in energy, and (2) the release of this energy in the form of heat (***exothermic***) or the absorption of this heat (***endothermic***) can be detected and measured leads to another technique that increases our knowledge about the internal structure of solids. ***Differential thermal analysis (DTA)*** techniques produce temperature patterns that can be interpreted to obtain information about the various structural changes that solids undergo as a result of the application of different external forces. Finally, even the vibration of atoms in a molecule can be detected using infrared spectroscopy. Such techniques, for example, are used to identify unknown polymeric materials.

3.2.5 Other Instruments

The ***scanning tunneling microscope (STM)*** is a newer form of electron microscope invented in the 1980s by two Swiss scientists, Binnig and Rohrer, at IBM that uses a very fine metal tip to scan across a sample at a height of a few atomic spacings. At such spacings, a tunneling current occurs when only a few volts are applied between the tip and sample. This tunneling current is sensitive to the tip-to-sample spacing and is used to measure and control the height of the tip. In this way, scanning micrographs of the surface with a height resolution of about one atom and lateral resolution of about one to three atomic spacings can be produced. Also, the electron energy distribution of the tunneling current depends on the electronic properties of the sample and is used to select or view different types of surface atoms. This instrument made it possible to see the new polymorph of carbon known as *fullerenes.*

Scanning probe microscopy (SPM) involves many related technologies, all of which operate by scanning a fine probe tip over a sample surface (Figure 3-13 shows a schematic of the probe and surface) to generate a high-magnification ($10^9 \times$), three-dimensional image. Piezoelectric ceramics, discussed in Module 9, control the probe motions in the *x, y,* and *z* directions with nm resolution (see Figure 3-13c). SPM can be used under air, liquid, or vacuum conditions. What is noteworthy is that the sizes of these pieces of equipment range from palm- to desk-size. Their importance is so momentous because SPMs are the connecting bridge between our world and the world of the atom and the molecule. Scientists who are now crossing this bridge are engaged in developing a new technology (discussed in various sections of this book) called by various names: *nanotechnology, molecular manufacturing,* or *molecular nanotechnology.* Regardless of the name, this new technological revolution will someday bring us the capability to manufacture materials as nature does by assembling individual atoms (see Figure 3-13f) and/or molecules to construct whatever material is needed. A few examples of available techniques using SPM are listed following:

Figure 3-13   (a) Atomic force microscope has a resolution of 10×10^{-10} and a scan range of 100^{-6} m. (Park Scientific Instruments) (b) Scanning tunneling microscope (STM) schematic diagram. (c) Enlarged diagram of probe. (d) Enlarged photo shows tungsten probe tip 0.2 nanometers wide to probe conductor and semiconductor materials.

Atomic force microscopy (AFM)—a cantilever probe senses electrostatic interactions with the specimen (see Figure 3-13). It can work with samples in their natural state.

Lateral force microscopy (LFM)—frictional response between the cantilever probe and the surface is plotted.

Modulated force microscopy—maps the variation in sample compliance (hardness) due to changes in cantilever modulation.

Electric force microscopy (EFM)—measures the electrostatic force of attraction or repulsion between the tip and the sample.

Magnetic force microscopy (MFM)—measures the magnetic force of attraction or repulsion between the tip and sample.

Scanning thermal microscopy (SThM)—uses an electrically resistive probe to map thermal conductivity or temperature variations between the tip and the sample.

(e)

(g)

(f)

Figure 3-13 (e) Schematic showing how tungsten probe moves along and scans silicon surface. (f) Image from STM of gallium arsenide (GaAs) atoms. (g) Nanoscale composite mixing—top: STM image of silicon surface; middle: a voltage pulse of +3 volts applied between probe tip and surface led to formation of a mound of silicon atoms (large white spot) surrounded by a black matlike area; bottom: a second voltage pulse of +3 volts leads to removal of the mound, then with the probe in a different position, application of pulse −3 volts led to a cluster (white spot) being deposited on the surface, leaving a large empty area (large black spot). (IBM Research)

New techniques are continually being developed. An example is the SPM with an SEM. This technique combines the best features of both microscopes into a powerful instrument in the study and characterization of materials. The scanning ion-conductance microscope (SICM) and the laser force microscope (LFM) are helping to advance synthesis of high-temperature superconductors and to improve the understanding of all materials.

In 1951, the *field-ion microscope* was introduced. It can provide direct images of individual atoms in a solid surface. It achieves its high magnification from a radial projection of ions from the tip of a sharp needle to a fluorescent screen. The needle, called a "field emitter," or more commonly, a tip, has a radius of curvature at the end of around 10 nm. Ions are generated above the tip surface in a process known as *field ionization.* The tip is placed in a vacuum chamber and a low-pressure "imaging gas" is introduced. An electric field is also applied. The imaging gas atoms in the vicinity of the tip are ionized by an electron tunneling from the tip surface to the specimen surface. The positive field ions formed follow the electric field lines away from the tip surface to the fluorescent screen, where they produce image spots that are representations of individual atoms. From the symmetry of the pattern of spots, the Miller indices (Section 3.4.3) of the observed planes of atoms can be determined. The field-ion microscope is similar in design to the *field-emission microscope* invented in 1936.

Only a few of the many sophisticated instruments and techniques allow scientists today to measure the features of materials, define their properties, manipulate their structure at the atomic and molecular levels to increase scientific knowledge, assist in the manufacture of

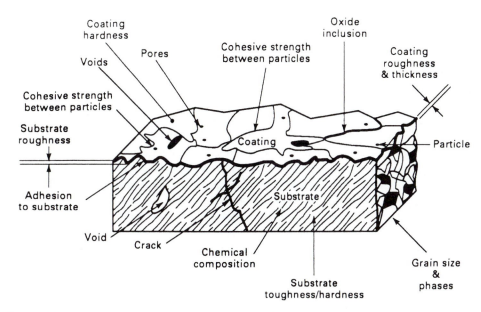

Figure 3-14   Characterization components of a substrate material and coating.

products, and contribute to the process of analyzing applications. As stated earlier regarding synthesis and nanoscale mixing (Table 3-1, and illustrated in Figure 3-13), nanoscale technology is achieved with processing/syntheses using probe microscopes, beam processing (MBE, laser, etc.) at scales of 50 nm or less. These techniques are carried out on advanced metals, ceramics, and composites.

 EMT WEBLINK

> Oak Ridge National Laboratory-STM and AFM-Animated Image Gallery—*http://www.ornl.gov/img/afm.html*

3.3 MATERIALS CHARACTERIZATION

Just as humans have fingerprints unique to each person, materials have unique characteristics. ***Characterization*** allows the identification and analysis of unique properties of a material's microstructure, revealed as defects, crystal structure, and impurities, as well as macrostructural properties such as surface roughness. Characterization may use some of the same tools and techniques as those used for destructive or nondestructive testing and evaluation (see Module 4), but characterization focuses more on structure and composition. The analysis can be qualitative or quantitative; it can be used to provide chemical or physical characteristics of the material. ***Qualitative characterization*** has been practiced since early times as crafters examined wood, metal, or stone to determine the suitability of that material for the tool or product they would craft. Figure 3-14 shows some of the characteristics for analysis of a material's substrate and coating.

   Quantitative characterization results from the ability to link computers with analysis instruments and to capture large amounts of data to make a comparison against a standard. Quantitative characterization is a recent technology that has evolved as a result of the need for improvement over older evaluation techniques that could not provide the precision required by advanced materials. Very hard materials or materials such as ceramic fibers in softer polymer composite matrices are often difficult to prepare for analysis by conventional metallographic techniques. Different means of preparing the same specimen for analysis can yield different results. For characterization techniques to be useful, they must provide true micro/macrostructural information. On the specimen shown, polishing with silicon carbide paper or fixed diamond platens seems to indicate voids in the coating. However, examination of the same specimen after diamond lapping revealed that it was the polishing medium that caused pullout of the coating.

To overcome problems of inaccurate characterization, both improved specimen preparation and analysis may be required. For example, digital analysis is now possible for images obtained with microscopes and x-rays, or generated from acoustic (sound) devices. Digital analysis provides data on features such as porosity, grain size, and surface flaws. ***Artificial intelligence (AI),*** the use of computers to solve problems using methods similar to those used by humans, has been employed to make quantitative characterization and statistical analysis of structures that would be too tedious for humans. Advances in digital analysis have come on the heels of methods to enhance optical and other imaging techniques with technology such as machine vision, which feeds special data into image analyzers. Improved methods of spectrometry, such as flame atomic absorption, allow chemical characterization of trace elements in high-performance alloys down to a few parts per million (ppm). These analyses permit the use of chemically characterized alloyed parts to be qualified for use in very critical applications, such as aircraft gas turbines.

3.4 STRUCTURE OF SOLID MATERIALS

In simplified form, structure is the "cause" that produces the properties ("effect") possessed by a material. In fact, there are many levels of structure, from the atomic to the macroscopic, each with its own chemical composition and distribution of its structural components (i.e., atoms, molecules, crystals, and phases, or what can be called the basic building blocks of materials). All these material structures give rise to an infinite array of materials, each possessing its own unique set of properties. All who work with materials or use them must constantly be aware that subjecting a material to external forces and/or conditions (impact, changes in temperature, corrosive conditions, radiation) will, in the majority of instances, have an effect on its structure with attendant change in its properties. Some of these changes can be handled quite satisfactorily by the material if the design process was correct in every respect. A final observation of the awareness of the many structures in a material helps us understand the role of material scientists, who are constantly working to develop materials with different structures or with different combinations of structures with the right mix of properties to help solve our technological problems.

Definitions of solids most always contain the phrase "definite volume and shape," together with the fact that solids maintain their shape to a varying extent when subjected to external mechanical forces. Solids occur in two *basic* forms: crystalline and amorphous. A true ***crystalline solid*** possesses an ordered, three-dimensional, geometric arrangement that repeats itself. A metal would be representative of a crystalline solid (though not perfectly

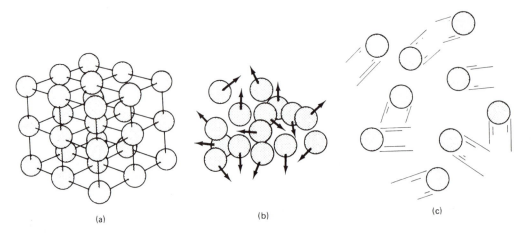

(a)   (b)   (c)

Figure 3-15 How atoms are arranged in (a) a crystalline solid, (b) a liquid, (c) a gas. (D.C. Giancoli, *Physics,* Prentice-Hall, Inc., Englewood Cliffs, N.J., 1980.)

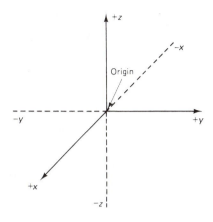

Figure 3-16 A standard axis system.

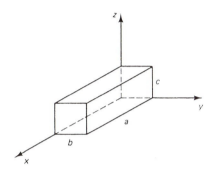

Figure 3-17 A box aligned on an axis system.

crystalline). An ***amorphous solid,*** on the contrary, contains no repetitious pattern of atom locations to any extent. The classic example of an amorphous solid, glass, is sometimes referred to as a supercooled liquid because of the random nature of its atomic arrangement (see Figure 7-41). Figure 3-15 illustrates the different atomic arrangements of three states of matter, emphasizing the different degrees of disorder and closeness of their positions in relation to each other.

In our treatment of ***crystallography*** (the study of crystalline structures), we will limit ourselves to the orderly arrangement of the atoms in their microscopic world. In so doing, we represent atoms, ions, or molecules essentially as spheres of varying sizes occupying points at various distances from each other in space—hence, the need for an ***axis system.*** Such a system is shown in Figure 3-16.

The arrowhead on each axis points in the positive direction. The negative directions, therefore, are opposite to the arrowheads. The x-, y-, and z-axes are, in this case, perpendicular to each other, with their origins coinciding at any convenient point. A box with dimensions *a,* *b,* and *c* is sketched in Figure 3-17, using this axis system. If this solid figure were a cube, these dimensions would be of the same magnitude. However, in one important crystal system they are not equal in magnitude. Another interpretation of these letters (or *lattice parameters*) is that they represent distances from the origin to points on the axes. The letter *b* would, for example, represent the perpendicular distance from the origin to a point on the y-axis.

The three axes in Figure 3-18, which are infinite in length, form three mutually perpendicular planes that are also infinite in size. A box is drawn using these axes as sides to help illustrate some basics concerning crystal systems. The x–y horizontal plane forms the base of the box. The top of the box is contained in a plane parallel to the x–y plane. A portion of the x–y plane contains the rear (as looked at by the viewer) end of the box, and the front end would be contained in a plane parallel to the y–z plane. There are countless planes parallel to these three primary (x–y, y–z, and x–z) planes, which can be referred to using a standard technique called ***Miller indices*** (see Section 3.4.3).

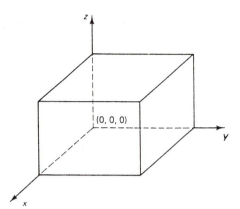

Figure 3-18 A three-dimensional box aligned on an axis system with an origin at the lower-left rear corner.

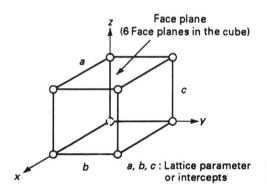

Figure 3-19 A simple crystal lattice unit cell with equal intercepts *a*, *b*, and *c*.

3.4.1 Unit Cells and Space Lattices

In our study of crystals, the axes system just described (called a ***simple cubic crystal system***) is modified slightly for the sake of uniformity. The standardized axes system shown in Figure 3-19 retains the x-, y-, and z-axes. The box is now called a *unit cell* (defined later). The angles between the principal planes are named α (Greek letter alpha), β (Greek letter beta), and γ (Greek letter gamma). For example, in Figure 3-20, α is the angle measured in degrees between the x–y and x–z planes; angle β, between x–y and y–z planes; and angle γ, between the x–z and y–z planes. The sides of the box (unit cell), labeled *a, b,* and *c,* are the lattice parameters in the x, y, and z directions, respectively. These distances are also known as ***intercepts.*** To describe a particular axes or crystal system adequately, all six of the preceding dimensions are needed.

    The term ***unit cell*** is used to describe the basic building block or basic geometric arrangement of atoms in a crystal. You can compare a unit cell to a single brick in a brick wall. Knowing that atoms are located at each corner of the single brick, it is easy to picture the atomic structure of a crystal, with such a unit cell repeating itself in three-dimensional space.

    If you repeat the unit cell in all three dimensions, you create a crystalline structure with a definite pattern. This larger pattern of atoms in a single crystal is known as a ***space lattice*** or ***crystal lattice.*** A space lattice is three sets of straight lines at angles to each other, constructed to divide space into small volumes of equal size, with atoms (ions or molecules) located at the intersections of these lines or between the various lines. We must remember that the lines and points in a space lattice are only imaginary. The lattice concept is used to show the positions of atoms, molecules, or ions in relation to each other. Atoms may be represented by circles, spheres, table tennis balls, or tennis balls located either at the intersection of these lines or between these lines. A part of a space lattice is sketched in Figure 3-21, with small spheres rep-

 EMT WEBLINK

MIT's World of Materials-unit cells—*http://www-dmse.mit.edu/wom/fcc.html*

Parameters of unit cell

a, b, c : Edges of unit cell
α, β, γ : Angles between **a, b, c** or
the planes containing
a, b, c

Figure 3-20 Angles between planes and intercepts in a
unit cell.

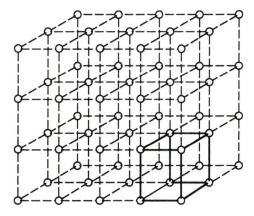

Figure 3-21 A space or crystal lattice with a simple
cubic unit cell outlined.

resenting atoms. We must also remember that the actual atoms in solids are located as close to
each other as possible, thus attaining the lowest possible energy level. Two atoms closest to
each other would be represented by two spheres touching each other. The closer the atoms are,
the denser the solid.

3.4.2 Crystal Systems

Mostly pure substances form crystals in the solid state. In this section, we emphasize crystal
systems composed of atoms and ions, but the structural particles of crystalline solids can be
atoms, ions, or molecules. Solid methane, CH_4, a molecular solid, has a face-centered cubic
(fcc) structure, which means that there is a CH_4 molecule at each corner and at the center of
each face in its unit cell. The forces acting among these structural particles may be metallic
bonds, interionic attractions, van der Waals forces, or covalent bonds. Most metals, a good
many ceramic materials, and certain polymers crystallize when they solidify. Most metals have
a cubic close-packed, hexagonal close-packed, or body-centered cubic lattice structure. There
are some exceptions. Tin (Sn) and indium (Im) form body-centered tetragonal structures. (Figure 3-25 shows a bcc tetragonal unit cell.) Table 3-2 indicates this tetragonal structure as XX
showing, among other things, that this structure is not a common one for pure elements in the
periodic table. Zirconia (ZrO_2), most important in improving toughness (see Module 7), forms
a tetragonal unit cell structure. Martensite, a form of tough steel, gains its hardness and
strength from its hard, brittle, tetragonal phase that is supersaturated with carbon.

The cubic system's basic structure has been demonstrated in Figures 3-16 to 3-20, assuming that the sides of the boxes described are all equal in length ($a = b = c$). It is possible
for atoms to form three different patterns or arrangements with this cube. These patterns, of
which there are a total of 14, are known as ***Brevais lattices.*** Each pattern can be described in

TABLE 3-2   PERIODIC SYSTEM OF THE ELEMENTS FOR FERROUS METALLURGISTS

PERIODIC SYSTEM OF THE ELEMENTS

Adapted Primarily for Ferrous Metallurgists

Atomic size factors (in parentheses) are % smaller (−) or larger (+) than gamma (FCC) iron at 75 F. Lattice environment (Coordination No.) is taken into account; CN is 12 except 6 for interstitials H, B, C, N & O. Groups VI, VIb, VII & VIIb form ionic compounds with the metals. Atomic size is based largely on work of W. Hume-Rothery and associates and L. Pauling (Some values, such as those for H & O, are approximate). Alloying valences are those of Pauling.

0	Ia / I	IIa / II	IIIa / III	IVa	Va	VIa	VIIa	VIII	VIII	VIII	Ib	IIb	IIIb	IVb	Vb	VIb	VIIb
He-2 FCC (Others)	Li-3 (+23) ● BCC* HCP†	Be-4 (−11) ● HCP* BCC	B-5 (−29) △ XX											C-6 (−34) △ XX ⊗	N-7 (−36) △ XX ⊗	O-8 (−33) △ XX	F-9
Ne-10 FCC	Na-11 (+50) ⊗ BCC* HCP	Mg-12 (+27) ⊗ HCP	Al-13 (+14) ● FCC											Si-14 (+7) ⊗ XX	P-15 (+2) ⊗ XX	S-16 (+1) ⊗ XX	Cl-17 XX
Ar-18 FCC	K-19 (+86) ⊗ BCC	Ca-20 (+56) ⊗ FCC* BCC	Sc-21 (+29) ⊗ HCP* BCC	Ti-22 (+16) ● HCP* BCC	V-23 (+6) ● BCC	Cr-24 (+1) ● BCC	Mn-25 (+1) ● XX* FCC†	Fe-26 (0) ● BCC* FCC	Co-27 (−1) ● HCP* FCC	Ni-28 (−1) ● FCC	Cu-29 (+1) ● FCC	Zn-30 (+6) ● HCP	Ga-31 (+12) ⊗ XX	Ge-32 (+9) ⊗ XX	As-33 (+11) ⊗ XX	Se-34 (+11) ● XX	Br-35 XX
Kr-36 FCC	Rb-37 (+97) ⊗ BCC	Sr-38 (+71) ⊗ FCC* HCP†	Y-39 (+42) ⊗ HCP* BCC	Zr-40 (+27) ⊗ HCP* BCC	Cb-41 (+15) ◐ BCC	Mo-42 (+10) ● BCC	Tc-43 (+8) ● HCP	Ru-44 (+6) ● HCP	Rh-45 (+6) ● FCC	Pd-46 (+9) ● FCC	Ag-47 (+14) ⊗ FCC	Cd-48 (+20) ⊗ HCP	In-49 (+25) ⊗ XX	Sn-50 (+23) ⊗ XX	Sb-51 (+27) ⊗ XX	Te-52 (+27) ⊗ XX	I-53 XX
Xe-54 FCC	Cs-55 (+112) ⊗ BCC	Ba-56 (+76) ⊗ BCC	La-57 (+48) ⊗ HCP* FCC†	Hf-72 (+26) ⊗ HCP* BCC	Ta-73 (+16) ● BCC	W-74 (+11) ● BCC	Re-75 (+9) ● HCP	Os-76 (+7) ● HCP	Ir-77 (+8) ● FCC	Pt-78 (+10) ● FCC	Au-79 (+14) ● FCC	Hg-80 (+25) ⊗ XX	Tl-81 (+36) ⊗ HCP* BCC	Pb-82 (+39) ⊗ FCC	Bi-83 (+35) ⊗ XX	Po-84 (+40) ⊗ XX	At-85
Rn-86	Fr-87	Ra-88	Ac-89 (+49) ⊗ FCC														
Alloying Valence	1	2	3	4	5	6	6	6	6	6	5.56	4.56	3.56	2.56 Note 2	1.56 Note 2	(2) Note 3	(1) Note 3

H-1 (−58) ⊗ XX ▲

Note 1: The rare-earth (lanthanide, 58-71) and actinide (90-103) series are omitted.
Note 2: Valence is 4 for C; 3 for N and P.
Note 3: (1) and (2) are not alloying valences.

STRUCTURE

BCC – BODY CENTERED CUBIC
FCC – FACE CENTERED CUBIC
HCP – HEXAGONAL CLOSE PACKED
XX – NOT BCC, FCC OR HCP — USUALLY MORE COMPLEX
*– STRUCTURE AT 75 F
†– ALSO FCC   ‡– ALSO BCC

SUBSTITUTIONAL SOLID SOLUTIONS

● FAVORABLE SIZE FACTOR: 0 TO ± 13%
◐ BORDERLINE SIZE FACTOR: ± 14 TO ± 16%
⊗ UNFAVORABLE SIZE FACTOR: > ± 16%

INTERSTITIAL SOLID SOLUTIONS

▲ FAVORABLE SIZE FACTOR: > (−40%)
△ BORDERLINE SIZE FACTOR: (−30) TO (−40) %
▵ UNFAVORABLE SIZE FACTOR: < (−30%)

TYPE OF GAMMA IRON (FCC) FIELD IF ALLOYED WITH IRON

GAMMA LOOP, LIKE Cr
LIMITED GAMMA LOOP, LIKE B
OPEN GAMMA REGION, LIKE Ni
LIMITED GAMMA REGION, LIKE C

Prepared by O. O. Miller

Copyright © 1965, 1966 by The International Nickel Company, Inc.

[Handwritten margin notes:]
FCC – FACE CENTERED CUBIC
BCC – BODY CENTERED CUBIC
HCP – HEXAGONAL CLOSE PACKED   } 3 MOST IMPORTANT

● Visible atom
◌ Hidden atom

Figure 3-22 A simple cubic unit cell showing atoms only at the corners of the cube.

8 atoms (one at each corner)

1 atom (at center of cube)

Figure 3-23 A body-centered cubic unit cell.

detail by using a unit cell as a reference. The *cubic axis system,* with its mutually perpendicular axes and equal lattice parameters, has already been described. The cubic coordinate axis system allows us to form three basic unit cells by placing atoms, ions, or molecules at various lattice positions.

The simple ***cubic unit cell*** (Figure 3-22) consists of eight atoms located at each corner of the cube. This structure does not exist in nature because the space in the center would need to be filled by other atoms. The point to be made is that this representation of atoms in a solid shows only the location of atoms. It must be remembered that if you represented these eight atoms with hard rubber balls and arranged them in accordance with this simple cubic unit cell, all eight atoms would be touching each other.

Another unit cell (Figure 3-23) is known as the ***body-centered cubic (bcc).*** It is similar to the simple cubic unit cell, but contains an additional atom located in the center of the cube. The third type of unit cell formed from the cubic axis system is the ***face-centered cubic (fcc).*** One atom at each corner and one in the center of each of the cube faces make up the complement of atoms. There is no atom at the center of the cube (Figure 3-24).

For both the bcc and fcc structures, the same type of atoms must fill both the center/face-centered locations, as do the cell corners. If, in fact, a different type of atom fills the center position than the cell corners in a body-centered cubic structure, the structure is correctly called a simple cubic (sc) structure. An alloy made up of about 50% copper and 50% zinc forms such a structure.

The ***tetragonal crystal system*** has similar unit cells to the cubic, but the sides are not equal. As an example, the ***body-centered tetragonal (bct)*** crystal lattice unit cell is shown in Figure 3-25. Tin forms a tetragonal unit cell. The tetragonal is similar because the axes are all normal to each other. The difference lies in the length of the intercepts. The x- and y-intercepts

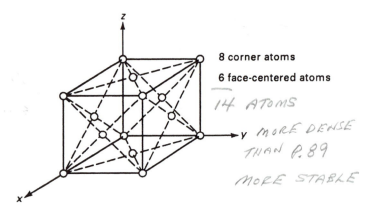

8 corner atoms

6 face-centered atoms

14 ATOMS

→y MORE DENSE THAN P.89

MORE STABLE

Figure 3-24   A face-centered cubic unit cell.

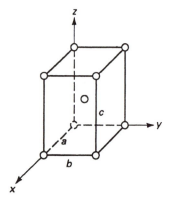

Figure 3-25   A body-centered tetragonal crystal lattice unit cell.

have the same magnitude. The z-intercept is larger than the x- or y-intercept. Martensite, a combination of iron and carbon that is contained in a hard steel, has its atoms of iron and carbon in a tetragonal lattice structure (Module 5).

The *hexagonal crystal system* (Figure 3-26a) can best be described using three axes (a_1, a_2, and a_3) in the x–y plane 120° apart and a fourth axis (z) at 90° to the x–y plane. The intercepts along the three axes in the horizontal plane are equal in length ($a = a = a$), but the fourth intercept, labeled c, is of a different length. This unit cell is made up basically of two parallel planes (top and bottom basal) separated by a distance equal to the dimension c. The atoms shown in the figure trace out a right hexagonal prism. Each of these two places can be divided into six equilateral triangles, with each side equal to the intercept a (Figure 3-26b).

Atoms of solid materials do not form the purely hexagonal unit cell as in Figure 3-26 because they cannot satisfy equilibrium conditions by being so far apart. In other words, they would be unstable. Consequently, they form the hexagonal unit cell, called *close-packed hexagonal (cph)*, as shown in Figure 3-27, with its three mid-plane atoms at a distance of c/2. Zinc, titanium, and magnesium form cph unit cells.

In summary, there are a total of seven crystal systems, of which three were mentioned. Two examples of some of the unit cells from the orthorhombic and monoclinic crystal systems are shown in Figure 3-28. Not discussed are the triclinic and rhombohedral systems. Using these seven axes systems, atoms can form 14 patterns in space (Brevais lattices), of which five have been mentioned and described. As an example, using unit cells for the cubic crystal system, atoms can form three different patterns: simple cubic (Figure 3-22), bcc cubic (Figure 3-23), and fcc cubic (Figure 3-24). Table 3-3 lists the intercepts and the angles between axes for five of seven crystal systems with which the earth's elements form their particular atomic structures.

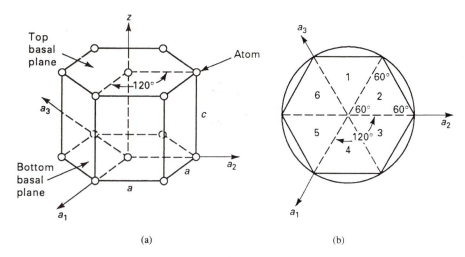

Figure 3-26   (a) A hexagonal crystal lattice unit cell. (b) Top plane of unit cell showing six equilateral triangles.

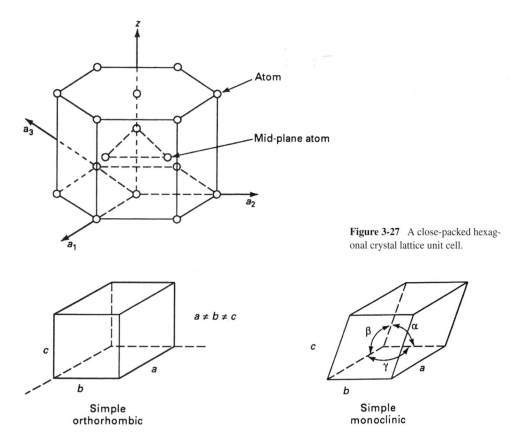

Figure 3-27   A close-packed hexagonal crystal lattice unit cell.

Figure 3-28   Orthorhombic and monoclinic unit cells.

3.4.3 Crystallographic Planes, Miller Indices, and Crystal Directions

The amazing thing about crystallographic structures is that any variation in the arrangement of atoms or the stacking of the planes formed by the atoms has a pronounced effect on the properties of the solid material. Atoms within the solids tend to move in coordination with each other. Solids tend to deform in the directions along the planes that are the most closely packed because it is easier for an atom (or ion) to move in these close-packed planes. Planes of atoms

TABLE 3-3   CRYSTAL SYSTEMS SUMMARY

System	Intercepts	Angles between axes
Cubic	$a = b = c$	$\alpha = \beta = \gamma = 90°$
Tetragonal	$a = b \neq c$	$\alpha = \beta = \gamma = 90°$
Hexagonal	$a = a \neq c$	$\alpha = \beta = 90°; \gamma = 120°$
Monoclinic	$a \neq b \neq c$	$\alpha = \gamma = 90° \neq \beta$
Orthorhombic	$a \neq b \neq c$	$\alpha = \beta = \gamma = 90°$

move, or slip, in parallel directions much like two adjacent cards in a deck of playing cards. This movement of atoms in a solid results from the application of some type of external force, such as a compressive or tensile load. The degree of movement of planes of atoms gives an indication of the ductility and strength of the particular material. Therefore, it is important to know something about how the principal crystal structures differ in the location of the atoms in their space lattices/unit cells and something about the planes in these structures. The bcc crystal structure is not a close-packed structure since the atoms could be packed closer together. Of the volume of the bcc unit cell, 68% is occupied by atoms and the remainder is empty space. The atomic packing factor (APF) (see paragraph 3.4.6) for a fcc unit cell is 0.74 indicating that the atoms are packed as close together as possible. Contrast this close-packed structure with 74% of the volume of the fcc unit cell occupied by atoms to the bcc structure which is not close-packed.

Slip, the displacement of atoms, is easiest to produce on close-packed planes because less force is needed to move the atoms from one position to the next adjacent one.

First, let us consider how atoms (or ions) move in a plane. A simple experiment can show this phenomenon. Take a number of pennies and line them up in two rows, all touching, with an equal number in each row. If you push one row with your finger in one direction parallel to the row and do the same with the other row, but in the other direction, each row will move easily. Next, if you line up the objects in less than a perfect arrangement with maybe only one of the objects projecting into the other row and then push the rows as before, the result is very little movement in the direction of the pushing force and considerable movement in many directions. Later you will learn that metals and metal alloys that arrange themselves with fcc unit cells exhibit the most ductility mainly because they possess the greatest number of close-packed planes in the greatest number of directions. Also, if not parallel, these planes will intersect and movement of atoms may be hindered or stopped completely at some stage.

The number of atoms in any atom plane depends on the stacking sequence of the planes. The atom planes in these crystal lattices that have the largest number of atoms are called ***close-packed planes.*** Representing atoms (or ions) as spheres, the closest packing of atoms occurs when each atom has 12 other atoms of equal size around it.

Of all the crystalline materials found in nature, the most important fall into only three stacking sequences. The stacking of four atom planes in a particular order creates the fcc arrangement of atoms. Of course, in a solid material there would be an infinite number of planes stacked in this particular way. Fcc metals such as Al, Cu, and Si therefore have four sets of nonparallel, close-packed planes in their unit cells, as discussed later. Bcc unit cells for elements such as Cr, Li, or Mo contain no close-packed planes. Cph unit cells are formed using a different sequence of stacking. As a result, they have two hexagonal bases (basal planes), as shown in Figure 3-26.

Numerous exercises can be developed to help gain better visualization of these structures. In so doing, remember that no adjacent planes can be identical. Any deviation from the stacking sequences mentioned are considered *stacking faults.*

In the bcc crystal system, we note in observing the bcc unit cell in Figures 3-23 and 3-29 that, by use of a cutting plane through this cell, the plane containing the most atoms would be the one that passed through four of the eight corner atoms and the one in the center of the cube. All figures are expanded views. Actually, the atoms are much closer. This diagonal plane

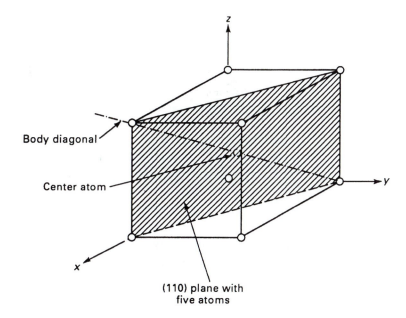

Figure 3-29   A bcc unit cell showing (110) plane.

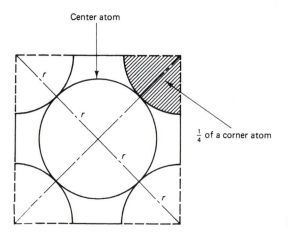

Figure 3-30   A bcc unit cell showing (110) plane.

could be a (110) plane, as sketched in Figure 3-29. It should be noted that, although these planes contain the most atoms, the atoms are not as closely packed as they could be. A view of this diagonal plane from a position at right angles to it, as in Figure 3-30, shows that the spaces are not occupied by atoms, the diagonal has a length of $4r$ (r is the radius of each atom), and the atoms touch each other along the bcc unit cell body diagonal.

In the *fcc unit cell,* the planes with the most atoms can be represented by plane (111) as sketched in Figure 3-31 or seven other planes with the same number of atoms. Observe that the plane contains three face-centered atoms and three corner atoms. Figure 3-32 shows these atoms all in contact with each other with a minimum of unoccupied space between them. With atoms of the same size, six atoms are touching each atom in such a plane.

The *cph unit cell* readily tells us the most closely packed planes of atoms are the basal planes and all planes parallel to them. This means that a metal with a cph crystal lattice can slip in only one direction. In other words, this metal would be brittle. The bcc or fcc metals, having a greater number of slip planes, can slip in many directions and are thereby classified as ductile metals. Recall that, at this stage, we are talking about the atomic structure of millions of atoms in one crystal. The vast majority of solid materials are polycrystalline

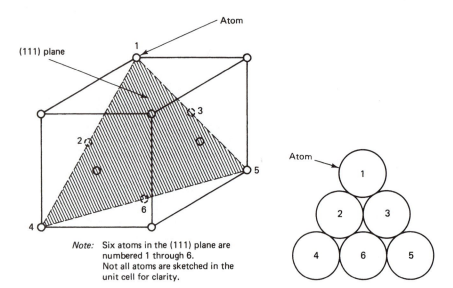

Note: Six atoms in the (111) plane are
numbered 1 through 6.
Not all atoms are sketched in the
unit cell for clarity.

Figure 3-31 (111) plane in fcc unit cell.

Figure 3-32 Normal view of diagonal plane (111)
in fcc unit cell showing six numbered atoms.

(i.e., composed of millions and millions of individual crystals whose axes are oriented in random directions). This random orientation could result in a blocking effect; that is, a crystal would stop the slippage of atoms in any adjacent crystal. More details of the movement of atoms will be discussed later.

3.4.3.1 Miller indices for atomic planes and directions.

Because atoms move, it is appropriate to be able to describe the location of atoms in a unit cell as well as the direction of their movement. The sites or locations of atoms and/or points in a unit cell are described by their axial coefficients expressed in unit-cell dimensions. These coefficients (also called coordinates or intercepts) are the respective distances along the x-, y-, and z-axes from the origin of the axis system. By convention, the origin is located at the lower-left rear corner of the unit cell (Figure 3-18). However, it may be moved temporarily to another position if it is so indicated. In Figure 3-18, for example, the location of the atom shown at the origin of the axis system would have the coordinates 0,0,0, as indicated. The atom above the origin on the z-axis in Figure 3-19 at a distance of 1 unit length would have the coefficients 0,0,c or 0,0,1. In Figure 3-24, the face-centered atom in the x–y plane is located by the coefficients ½, ½, 0. Note that the distances making up the coordinates are separated by commas. The planes in a unit cell are designated by their *Miller indices*.

The procedure for determining the Miller indices of a plane is illustrated in the following problems.

Illustrative Problem

In Figure 3-33, two planes *A* and *B* are shown with their intercepts. The four-step procedure for determining the Miller indices of plane *A* is as follows:

Step 1. Record the intercepts in order of the x-, y-, and z-axes.

x-axis	intercept 1
y-axis	intercept ½
z-axis	intercept ⅓

Step 2. Clear any fractions by taking the reciprocals of the intercepts.

x-axis	$1/1 = 1$
y-axis	$1/½ = 2$
z-axis	$1/⅓ = 3$

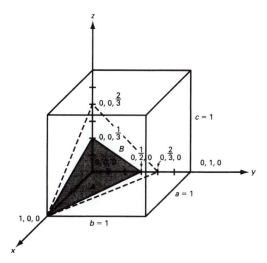

Figure 3-33 Miller indices for atomic planes for a cubic unit cell. Plane *A* is shaded; plane *B* is drawn with dashed lines.

Step 3. Clear any fractions arrived at in step 2 by multiplying by a common multiplier.

no fractions

Step 4. Record the numbers (indices) in parentheses with no commas separating the numbers.

(123)

Illustrative Problem

Determine the Miller indices for plane *B*.

Step 1. Intercepts are:

x-axis	1
y-axis	$\frac{2}{3}$
z-axis	$\frac{2}{3}$

Step 2. Taking reciprocals:

x-axis	$1/1 = 1$
y-axis	$1/\frac{2}{3} = \frac{3}{2}$
z-axis	$1/\frac{2}{3} = \frac{3}{2}$

Step 3. Clearing fractions:

common multiplier = 2

$$1 \times 2 = 2 \qquad \tfrac{3}{2} \times 2 = 3 \qquad \tfrac{3}{2} \times 2 = 3$$

Step 4. Record numbers (indices):

(233)

Illustrative Problem

Determine the Miller indices for the face plane in Figure 3-19 that parallels the x–z plane at a distance, b = 1.

Step 1. Intercepts:

x-axis	∞ (plane intercepts x-axis at infinity)
y-axis	1
z-axis	∞ (plane intercepts z-axis at infinity)

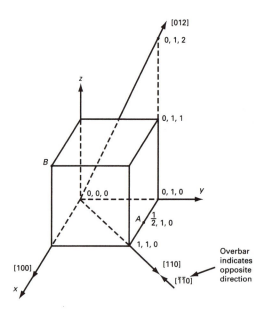

Figure 3-34 A cubic unit cell with some intercepts and crystal directions indicated. Note use of square brackets and the omission of commas to express Miller indices of direction.

Step 2. Reciprocals:

x-axis	$1/\infty = 0$
y-axis	$1/1 = 1$
z-axis	$1/\infty = 0$

Step 3. No fractions.

Step 4. (010) indices.

Figure 3-29 shows the (110) plane in a bcc unit cell, and Figure 3-31 shows a sketch of the (111) in a fcc unit cell.

3.4.3.2 Crystal directions.

Many materials are anisotropic; consequently, their properties are not the same in all directions. To specify directions in a crystal structure, Miller indices are used. As with Miller indices for specifying points and planes in a crystal-lattice system, there is a procedure for determining the Miller indices for directions. The first step is to determine the coordinates of two points that lie in the particular line of direction. The first point, sometimes called the "head point," is farthest from the origin. Using the origin as the second point simplifies the procedure. The second step is to subtract the second point from the first point. The third step calls for clearing of any fractions to obtain indices with the lowest integer values. The fourth step is the writing of the indices in square brackets without commas. Negative integer values are indicated by the use of a bar placed over the integer. The procedure is illustrated in the following problem (refer to Figure 3-34).

Illustrative Problem

In Figure 3-34, find the direction for a line passing through point *A* with coordinates $\frac{1}{2}$, 1, 0 and the origin of the unit cell.

Step 1. Axis system, defined in Figure 3-34

Step 2. Head point coordinates $\frac{1}{2}$, 1, 0

Tail point coordinates $- 0,0,0$
_____
$\frac{1}{2}$, 1, 0

Step 3. Clear fractions (multiply by 2) and reduce to lowest integers: 1 2 0

Step 4. Enclose in brackets without commas: [120]

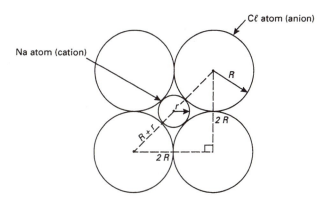

Note: Two Cℓ anions located along the axes
perpendicular to the page are not
shown for clarity.

Figure 3-35 A face plane in NaCl crystal.

In more complex crystal systems, the determination of directions requires temporary re-location of the origin to another point in the unit cell to simplify the procedure. If the properties of a crystal measured along two different directions are identical, the two directions are termed *equivalent.* Examples of equivalent directions are the face diagonals or the body diagonals in a cubic crystal. Equivalent directions are referred to as a *family of directions.* For example, the directions for the face diagonals (a family of directions) is specified by using spread or angle brackets: $\langle 1\ 1\ 0 \rangle$. This family consists of 12 directions, which can be arrived at by taking all the various ways (permutations) of expressing the three digits, 1, $\bar{1}$, and 0.

3.4.4 Coordination Number

To describe how many atoms are touching each other in a group of coordinated atoms, the term **coordination number** (CN) is used. The CN is the number of neighboring atoms that each atom has immediately surrounding it. Note in Figure 3-27 that each upper and lower basal plane of a cph unit cell contains an atom at its center. Each atom touches six atoms in its own plane, plus three atoms above and below in adjacent planes. Consequently, the CN for these atoms would be 12. The number of nearest atoms is dependent on two factors: (1) the type of bonding, and (2) the relative size of the atoms or ions involved. In our discussion of bonding, for example, we learned that valence electrons determine the type of bonding as well as the number of bonds an atom or ion can have. Carbon (C) in Group IV has four covalent bonds and therefore a CN of 4. The Group VII elements, such as chlorine (Cl), form only one bond (CN is 1). The relative size of the atoms determines how many neighboring atoms will touch another atom. Ionic bonding involves ions of different charges, hence, different sizes. The limiting factor in this case is the ratio of the size (radii) of the combined atoms of ions. The minimum ratios of atomic (ionic) radii produce various CNs. During ionization, atoms decrease in size when they change to cations and increase in size as they form anions.

Figure 3-35 represents the five ions occupying one of the six faces of the fcc unit cell for NaCl. The Na ion is just the right size to fit between the Cl ions at the corners of the unit cell. Thus, the ions are closely packed, with each cation separated from other cations by a layer of anions. Each cation and each anion are shared equally by six oppositely charged ions. Therefore, a CN of 6 describes this geometric arrangement. The center-to-center distance between the Na ion (radius = r) and its six neighboring Cl ions (radius = R) is determined to be the hypotenuse of an isosceles right triangle (two legs or sides of equal length). This triangle is outlined in Figure 3-35 with its sides labeled 2R. Using Pythagorean theorem $x^2 + y^2 = z^2$ produces

$$(2R)^2 + (2R)^2 = (2r + 2R)^2$$

Taking the square root of both sides gives the expression

TABLE 3-4   MINIMUM RADII RATIOS FOR CNS

CN	$\dfrac{r}{R}$
3	≥ 0.155
4	≥ 0.225
6	≥ 0.414
8 (bcc)	≥ 0.732
12 (cph or fcc)	1.0

r, radius of smaller atom; R, radius of larger atom; $\geq$, greater than or equal to.

$$\sqrt{2}(2R) = (2r + 2R)$$

Solving for the radius (r) of the smaller ion in terms of the larger ions:

$$r = (\sqrt{2} - 1)R = 0.414R$$

These data show that the smallest ion that will just fill the spaces between six larger ions or that just touches six ions must have a radius that is at the minimum 0.414 times the size of the larger ions. This same result can be obtained by using a proportion between the isosceles triangle in Figure 3-35 and the standard 45–45–90 degree right triangle.

$$\frac{1}{2R} = \frac{\sqrt{2}}{(2r + 2R)}$$

Cross multiplying:

$$r + R = \sqrt{2}R$$

Solving for r:

$$r = \sqrt{2}R - R = R(0.414)$$

Table 3-4 reveals that, to have a CN of 6, the ratio of r/R must be at least 0.414.

In the case of sodium chloride (NaCl), r/R is 0.95/1.81 = 0.520. This being greater than 0.414, NaCl qualifies for a CN of 6. Note that, as the difference between r and R decreases, higher CNs are possible. A CN of 12 is the maximum, which occurs when the atoms (ions) have the same radius and the ratio becomes 1. In other words, as the r gets smaller than R (radius of surrounding atoms), the fewer neighboring atoms can make contact or touch the smaller atoms. Table 3-4 lists the minimum radii ratios for some common CNs.

With a CN of 12, each atom has contact with 12 other atoms. Each atom in a fcc or cph unit cell meets this description, provided that their radii are of similar size.

3.4.5 Allotropy/Polymorphism

It is of interest to list some common materials and show their crystal structure and melting point to illustrate the concept of the movement of atomic planes within a solid. Table 3-5 contains several elements that exist in more than one crystal structure depending primarily on the temperature. This phenomenon is known as **allotropy** or **polymorphism** (*poly,* "many"; *morph,* "shape"). Actually, an allotropic material, after changing to one structure, can reverse the phenomenon and return to its previous structure; a polymorphic material does not possess this reverse phenomenon. Over one-fourth of the elements are allotropic. Steel owes its extensive use to this property. The cost of steel is relatively cheap when compared with the many advanced materials that are competing with steel in today's marketplace. With the recent advances made

TABLE 3-5   CHARACTERISTICS OF SELECTED METALS

Element (symbol)	Crystal structure (20°C)	Atomic radius (10^{-10} m)	Melting point (K)
Beryllium (Be)	cph	1.14	1562
Aluminum (Al)	fcc	1.43	933
Titanium (Ti)	cph	1.46	1945
	bcc 877°C		
Iron (Fe)	bcc	1.24	1810
	fcc 910–1399°C	1.27	
	bcc 1399–1538°C		
Cesium (Cs)	bcc	2.62	300
Tungsten (W)	bcc	1.37	3680

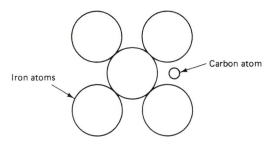

Iron atoms

Carbon atom

Atomic radius C, 0.077 nm
Atomic radius Fe, 0.127 nm

Figure 3-36   A face plane of an fcc unit cell for iron showing a carbon atom located in an interstice.

in the research and development of lightweight, high-strength steels, steel is a great competitor especially in the transportation industry. Because of allotropy, steel can be produced from iron. When iron is heated to above 910°C, its structure changes from bcc to fcc, allowing for a much greater absorption of carbon atoms (2% maximum). Figure 3-36 shows one carbon atom in an fcc unit cell of iron.

In general, we now know that any allotropic change in an element means different properties. All the properties of steel change almost instantaneously at 910°C. The best example of a change in properties with crystal structure is that of carbon (C). One polymorph of pure carbon is graphite—a black, greasy, low-strength material. A second polymorph is diamond, the hardest naturally occurring substance. A third naturally occurring polymorph is the recently discovered fullerenes (see Figure 2-14). A final example is zirconia (ZrO_2), a ceramic used to toughen alumina (Al_2O_3) and described in detail in Module 7 (see Section 7.4). ZrO_2 is polymorphic with a fcc structure at high temperature. As it cools, it changes first to a simple tetragonal structure, followed by a further change to a monoclinic polymorph. Figure 3-37 is a sketch of the change in volume of ZrO_2 as it goes through these transformations on being cooled. These changes require the consumption of energy that would otherwise be used to help a crack propagate through the ceramic alumina. The end result is alumina toughened by this addition of zirconia.

3.4.6 Volume Changes and Packing Factor

In discussing crystal structure changes, we mentioned that every change in atomic structure brings changes in properties of the solid. One of these changes is volume. When steel transforms from a bcc to fcc structure, it decreases in volume. The explanation for this phenomenon involves the density of atoms in the various unit cells. *Density* is the ratio of the mass to the volume of a substance, which stays constant provided it is nonallotropic. With mass measured in kilograms (kg) and volume in cubic meters (m^3), density has the units of kilograms per meter cubed (kg/m^3). Other typical units are g/cm^3.

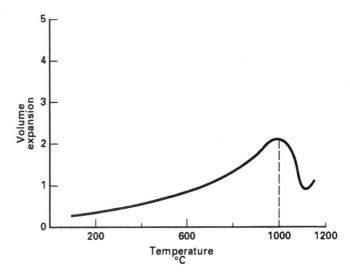

Figure 3-37 Volume expansion curve for ZrO_2 emphasizing the effects of the polymorphic transformation on the rates of expansion and the abrupt changes in volume.

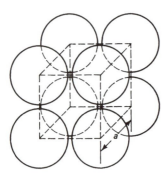

Figure 3-38 A simple cubic unit cell.

The **atomic packing factor (APF),** or *packing factor (PF),* is the ratio of the volume of atoms present in a crystal (unit cell) to the volume of the unit cell. In calculating the volume of an atom, we assume the atom is spherical. The difference between the PF and unity (1) is known as the *void fraction,* that is, the fraction of void (unoccupied or empty) space in the unit cell. Using the *simple cubic crystal unit cell* sketched in Figure 3-38, with atoms of equal radius (R) located at each corner, the volume of the cell occupied by the eight atoms is equivalent to one atom. Each corner atom contributes one-eighth of its volume. Therefore, the volume of one atom is $\frac{4}{3}\pi R^3$. The (100) face plane of this simple cubic unit cell shows four corner atoms touching each other (see Figure 3-39). The relationship of the radius R of an atom to the lattice parameter or edge length a of the unit cell is $2R$. The volume of the unit cell is a^3. The volume of atoms is $(\pi/6)a^3$. Thus,

$$PF = \frac{\text{volume of atoms}}{\text{volume of unit cell}} = \frac{(\pi/6)a^3}{a^3} = \frac{\pi}{6} \cong 0.52$$

Solving for the PF in terms of the radius R produces the same results. The void factor is therefore $1 - 0.52 = 0.48$. What the calculation tells us is that only about half (52%) of the space in the simple cubic unit cell is occupied by the atoms. This is too inefficient, so atoms of metals do not crystallize in this structure. Remember, the closer the atoms come to each other, the less energy they have and the more stable is their structure.

The *bcc unit cell,* as you recall, is quite similar to the simple cubic unit cell with the addition of one atom in the very center of the unit cell. Therefore, the bcc unit cell contains the equivalent of two atoms.

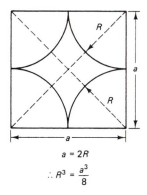

$a = 2R$

$\therefore R^3 = \dfrac{a^3}{8}$

Figure 3-39 (100) face plane of simple cubic unit cell.

For the *fcc unit cell,* there are four net atoms. Each of the eight corner atoms contributes one-eighth of an atom. Each of the six face atoms contributes one-half. The total is $1 + 3$, or 4, atoms. Notice that the atomic radius (R), or a as used in the formula, cancels out in all these calculations, which tells us that the PF is not dependent on the radius of the spheres being packed if all the atoms are of the same size.

The fcc structure has the maximum PF for a pure metal. The cph structure also has a PF of 0.74. Finally, we should note that the coordination number varies directly with the PF. As an example, the $CN_{bcc} = 8$ and $PF_{bcc} = 0.68$; the $CN_{fcc} = 12$ and $PF_{fcc} = 0.74$.

In our study of metals, we will see that pure iron will change its structure on heating from bcc to fcc at 910°C. Knowing the PF for both these structures will lead us to the conclusion that iron will contract in volume as it is heated above 910°C. This change in structure forms the basis for the production of steel as well as the heat treatment of steel.

3.4.7 Crystal Imperfections

So far in our study of crystal structures we have dealt with atom arrangements that are perfect in every way. These structures have **long-range order** because the orderly arrangement of atoms extends throughout the entire material, forming a regular gridlike lattice or pattern. To grow a perfect crystal requires laboratory conditions, and even with such a controlled environment we have achieved limited success. Of course, this lack of perfection in the microstructure of materials used by people is far from being all bad. If it were not for imperfections of many kinds in solid materials, these solids would not possess the properties that we desire them to have. An example would be the heat-treating process used with high-carbon steel to change the properties of the steel to suit certain conditions demanded by our present technological age. Without imperfections in the crystalline arrangement of atoms, these processes would be severely limited in, if not incapable of, changing the structure and hence the properties of steel. The whole semiconductor industry owes much of its existence to the imperfections in bonding arrangements of the atoms' outer shell electrons. Therefore, it is now essential to delve into the several imperfections of a solid's atomic structure so that we can learn how to take advantage of such disorder in the atomic structure.

It is important to point out here that we are still talking about crystalline materials and not amorphous (noncrystalline) materials. We know that **amorphous** materials have no regular atomic structure. Known as **short-range-order** materials, their order is limited to an atom's nearest neighboring atoms. For example, water has short-range order due to its covalent bonding between the hydrogen and oxygen atoms. Even with its secondary bonding (hydrogen bridging), the water molecules are randomly joined. Glass and polymers also have short-range order. This is not to say that amorphous materials are of little use. Many materials are being produced by processes such as powder metallurgy and rapid solidification that are amorphous in structure and that possess remarkable properties needed for some specific application.

Crystal imperfections fall into two categories: those involving impurity atoms and those in which there is some disorder in the atomic structure brought about by something other than impurity atoms.

3.4.8 Crystal Impurities

The word *impurity* comes from the use of the word when referring to the small percentage of copper in sterling silver that distinguishes sterling from pure silver. One talks about the impurities in copper that reduce its conductivity. In many instances, these impurities are purposely added to improve a material's properties and/or reduce its cost. Such a solid, called an *alloy,* is a material composed of two or more elements, at least one of which is an element that possesses metallic properties. **Wrought alloys** are those that can be mechanically deformed. Cast alloys are those alloys whose brittleness is such that they cannot be cold worked.

Brass is an alloy consisting of copper to which has been added some zinc. The addition of zinc has a great effect on the hardness, strength, ductility, and conductivity of the pure copper. Our objective in the following discussion is to explain in simple terms why adding these impurities (the zinc atoms) to the copper atom produces such differences in the properties between the pure metal and the alloy.

Before proceeding, definitions of the words *solution, solvent, solute, mixture, alloy,* and *diffusion* are required. A *solution* is a homogeneous mixture of chemically distinct substances that forms a phase. A *phase* is defined briefly as a physically distinct material that has its own structure, composition, or both. Uses of phases in solids are discussed in this module. The atoms or molecules of one substance are uniformly distributed throughout the other on a random basis. The substance present in the greatest proportion is the *solvent.* The other substance or substances present are the *solutes.*

A *mixture,* on the other hand, is a material that has no fixed composition and contains more than one phase. The components (substances) can be identified and separated by physical means. Thus, a mixture of sugar and salt crystals is not homogeneous nor is it random. One can readily see the two distinct crystalline phases and, with some patience, segregate one crystal phase from the other. However, if you dissolve sugar and salt in water in a dilute concentration, they form a liquid solution. In this situation, the salt and sugar lose their individual identities by dissolving in water.

Air is an example of a solution of many gases dissolved in another gas. A similar situation occurs in solids, producing *solid solutions.* A *solid solution* is simply a solution in the solid state that consists of two kinds of atoms combined in one type of space lattice. In other words, when two elements are soluble in each other in the solid state, they produce a solid solution. *Alloys,* then, are a combination of a metal and one or more other elements forming either a mixture or a solid solution. Steel is a mixture of iron with a bcc structure and cementite. Brass is a solid solution because its single-phase structure is all fcc; it is a solid solution of copper (solvent) to which some zinc (solute) has been purposely added. The zinc atoms are diffused into the atomic structure of the copper on a random, uniform basis. *Plastic alloys* are blends of polymers or copolymers with other polymers or elastomers; they are also called *polymer blends.* Styrene-acrylonitrile is a plastic alloy. To describe the amount of material in a mixture or an alloy, an expression is used that contains "wt" (for weight) and "%" ("for each hundred"). As an example, in describing monel metal later in this module, the expression 70 wt% Ni states that 70 grams of nickel (Ni) are present in 100 grams of this alloy of nickel and copper.

Diffusion comes from the Latin verb meaning "to pour out" or "to spread out." In materials science, it means the intermingling in solid materials of atoms (in metals), ions (in ceramics), or molecules (in polymers). This active movement of particles is fairly well understood through our experiences with gases and liquids. A bottle of perfume uncapped in one part of a room will soon disclose its presence to people in a distant corner of the room by the diffusing of perfume atoms through the air in the room. Salt or sugar dissolved in water will diffuse throughout the water and form a solution in which the salt or sugar atoms will be evenly distributed throughout the water.

Our main preoccupation in our present study of materials is with solids. Diffusion takes place in solids, too. We know that the individual atoms and molecules in a gaseous or liquid state are relatively far apart, offering little opposition to other atoms migrating through them. In solids the atoms are held tightly and close together. If a metallic crystalline solid were formed with a perfect crystalline structure, atoms would find it impossible to move about. But in our study of imperfect crystal structures, we will learn that point defects are the rule and not the exception. These defects are one main reason why atoms of a solid can actively move about within the atomic structure of the solid. Combine the presence of vacancies (the absence of an atom or atoms in a lattice site) with the fact that each atom possesses sufficient energy to cause it to vibrate about its position in the lattice structure and it is fairly easy to visualize why certain atoms possessing higher average energy in the crystal structure can break their bonds and "jump" from one lattice site to one that is not occupied (a vacancy) in the lattice structure. Once the atom moves to a new site, it leaves behind another vacancy. In other words, the atom exchanges positions with a vacancy.

The phenomenon of diffusion in solid materials is especially important in understanding the manufacturing and functioning of semiconductor materials; the carburizing of steel in surface or case hardening; the production of metal alloys including steel, the primary alloy of iron; and the heat treating of aluminum alloys, called precipitation hardening. ***Precipitation*** is a change in the solid state (change in phase) brought about by diffusion. How this precipitation is controlled to produce strong aluminum alloys is discussed in Module 5. These metallurgical processes are discussed elsewhere in this book.

In polymeric materials, diffusion is aided by defects in the molecular structure similar to those in metals. The diffusion of a penetrant into a solid polymer is of great importance. If the penetrant is a gas, the gas may permeate through the solid (even glass). Plastic materials can absorb liquids, gases, or vapors. Such permeation frequently causes swelling, a change in the chemical structure, or environmental stress cracking. Polypropylene (PPO) stress cracks when exposed to mineral oil. This diffusion of a fluid through a porous body is described as ***permeation.*** To permeate is to pass into, pass through, penetrate, or spread through something, such as ink passing into and spreading throughout a blotter. The term is used to describe the action of atoms diffusing through a crystal structure. Polymer films are designed to prevent the diffusion of gases and water vapor into foods that have been wrapped in the film. Liquid solvents may permeate a polymer and produce a softening of the polymer, color changes, and odors.

The ***doping*** (alloying with a very small amount of alloy addition) of silicon with phosphorus in the alloying of silicon to increase its electrical conductivity is another diffusion process that depends on point defects in the silicone structure. Many sintering processes, such as powder metallurgy and the large number of welding and brazing processes for the joining of two metals by local ***coalescence*** (to grow, unite into a single body), depend on the transport of atoms by diffusion. Most surface modification treatments depend on atomic diffusion (see Section 5.9).

Because the internal energy possessed by atoms is related to temperature, an increase in temperature will increase the rate of diffusion. Of course, if there are no vacancies in the structure, very little, if any, diffusion will occur.

Another factor that affects the diffusion of atoms, molecules, or ions is the type of bonding of the matrix atoms (i.e., strong bonding takes more energy to break the bonds). An example is high-melting-point solids. Smaller permeating atoms stand a better chance of diffusion through a structure of larger atoms, as in the case of carbon atoms diffusing through a structure of iron atoms. Our study of the microstructure of solids tells us that a lattice structure that contains loosely packed atoms (less dense) will offer less resistance to diffusing atoms than one whose structure contains tightly packed atoms. Finally, diffusion depends on time. This translates usually into allowing sufficient time for diffusion to take place at some higher temperature, as in an oven or furnace.

In summary, diffusion is the process depended on by material scientists to change the microstructure of solids and thus vary the properties of the many solid-state materials in use

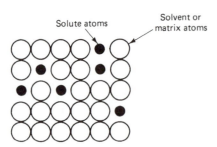

Solute atoms

Solvent or matrix atoms

Figure 3-40 Substitutional solid solution.

by our society. Having completed a brief background discussion of solvents, solutes, solid solutions, alloys, and the very important phenomenon of diffusion, we will return to the subject of crystal imperfections. The first of two categories of such imperfections in a crystal structure involves, as mentioned, impurity atoms, which produce two types of solid solutions, substitutional and interstitial solid solutions.

3.4.9 Substitutional Solid Solutions

In a substitutional solid solution (Figure 3-40), the solute atoms replace some of the solvent atoms in a crystal structure of the solvent. Using brass as our solid solution, up to about 40% of the copper atoms can be replaced by zinc atoms. This is possible because the atoms of copper and zinc are much alike. Their atoms are about the same size (the atomic radius of the copper atom is 1.278×10^{-10} m and of zinc, it is 1.39×10^{-10} m). Their electron configurations are comparable:

$$^{29}Cu: 1s^2, 2s^2, 2p^6, 3s^2, 3p^6, 4s^1, 3d^{10}$$
$$^{30}Zn: 1s^2, 2s^2, 2p^6, 3s^2, 3p^6, 4s^2, 3d^{10}$$

Their crystalline structures are both fcc with a CN of 12. Another good example of a solid solution is Monel. Monel is a solution of copper in nickel (about 70% Ni and 30% Cu). The range of solubility goes from practically no nickel to almost 100% nickel. Again, these two elements have a common crystalline structure (fcc), and the atomic radius of nickel is 0.1246 nm. Therefore, Monel is a substitutional solid solution. Another interesting fact about this type of solid solution is that atoms may fill only one type of site in the lattice structure of the solvent atoms. For example, in the alloy of copper and gold, the majority of copper atoms occupy the face-centered sites and the gold atoms, the corner sites of the face-centered cubic unit cell. An *ordered substitutional solid solution* is formed. As a rule, two distinct elements may form a substitutional solid solution if the sizes of their atoms do not differ by more than 15%. Of course, there are further restrictions on the degree of solubility brought about by any differences in their crystal or electron structure. The solute or impurity atoms, although of similar size, may be larger or smaller than the solvent atom, which will produce only a slight distortion in the lattice structure. A Cu-Zn phase diagram shows the beta (β) phase with Cu and Zn atoms randomly dispersed at lattice points. Upon cooling, the phase changes to an ordered structure (β'), which is still bcc but the Cu atoms are located at the corners and the Zn atoms are at the center of the unit cells. In most cases the addition of solid solution elements tends to decrease ductility as strength increases. However, the addition of Zn (up to about 30 wt%) increases ductility along with strength (46,000 psi).

3.4.10 Interstitial Solid Solutions

The second type of solid solution formed by impurity atoms is the *interstitial* (Figure 3-41). If the impurity atoms take up sites in the lattice structure that are normally unfilled or unoccupied by the pure (solvent) atoms, they form an interstitial solid solution. These normally unfilled voids or vacant spaces are called *interstices.* In the fcc unit cell, we know there is a

Figure 3-41 Interstitial solid solution.

Figure 3-42 Point defects (vacancies).

relatively large interstice in the center and smaller interstices near each corner atom. It is worthwhile to point out that steelmaking is made possible because of the formation of an interstitial solid solution. First, we know that iron is allotropic. At temperatures below 910°C, iron is in the bcc form. Above that temperature, the bcc structure changes to fcc to accommodate a higher energy level of the atoms. In the fcc structures, carbon atoms can form in the interstices of the iron unit cell. At temperatures below 910°C, the bcc structure contains no room for the carbon atoms to fit between the iron atoms. This fact forms the basis for many of the heat-treating procedures used to produce a multitude of steels with the many different properties required by our technological society. We will discuss heat treating in detail later in the book.

3.4.11 Crystal Defects

The second category of crystal imperfections or lattice defects, a disorder of the crystal structure, is brought about by some mechanism such as thermal agitation of the crystal during its formation, the effects of gravity, or the result of high-energy radiation. Such deviations from the perfect crystal are classified for purposes of explanation as *point defects, line defects,* and *area defects.* Actually, defects occur in all combinations.

Point defects (1 or 2 atom positions), the simplest and best understood, affect only the small volume of the crystal surrounding a single lattice site. One such point defect is a ***vacancy,*** that is, the absence of an atom at a lattice site in the otherwise regular crystal (see Figure 3-42). As a result, the electronic bonding of the adjacent atoms is disrupted, which changes the effective radii of these atoms. This weakens the crystal. If sufficient vacancies were produced by heating of a crystalline solid, the crystal structure would lose its long-range symmetry and order, resulting ultimately in porosity or a change to a fluid. It is important to note that these local imperfections in the crystal structure produce a disequilibrium that has a great effect on the important properties of crystalline solids such as density, mechanical strength, diffusion, and electrical conductivity. By themselves, point defects do not affect strength as much as they affect diffusion—the migration of atoms.

Another point defect is called an ***interstitial defect*** or ***interstitialcy.*** This is produced by the presence of an extra atom in a void, the space between normal lattice positions. This interstitial atom may be added specifically as an alloying element or it might be an impurity atom indigenous to the solid. Other impurity atoms may deposit themselves in a lattice position reserved for atoms of the solid. Regardless of how they are formed, these point defects produce local aberrations in the atomic arrangement in the crystal, which produce varying degrees of

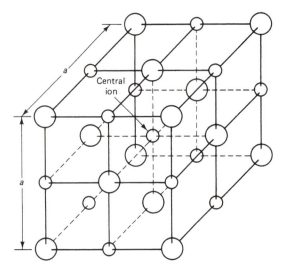

o - Na ion
O - Cl ion

Figure 3-43   A unit cell of NaCl.

local disorder in the bond structure and energy distributions in the solid. It is worthwhile to mention that disorder can occur below the atomic level (i.e., at the subatomic level). The imperfections in the electronic structure of atoms exist also. This fact has been capitalized on by the semiconductor industry, which produces materials with varying electrical properties. Further treatment of point defects in the electronic structure of atoms will be included under the topic of electronic bond theory of nonmetallic crystalline solids. It should be evident now that imperfections in the atomic, ionic, and electronic structure of solids do not all have a negative effect on a solid's properties. Some imperfections can improve certain properties; others may degrade some properties. All imperfections do not affect the strength of solids.

So far we have discussed point defects in terms of metallic crystals, with the exception of the brief mention of nonmetallic crystals with the "imperfections" in their electronic structures. As we know, all metals crystallize when they solidify, provided that they are given enough time to form their inherent structures. So do most ceramics and some polymeric (plastic) materials. In the case of ceramic materials, the crystal structures are more complex than those of metals, and the bonding is ionic, covalent, or a combination of metallic and covalent. With ionic bonding the atoms, of course, become ions. In the ionic solid sodium chloride (NaCl), each Na^+ cation finds itself surrounded by six equally spaced Cl^- anions, and each Cl^- anion is surrounded by six Na^+ cations (CN = 6). Figure 3-43 is a sketch of a model of the unit cell for NaCl, upon which the crystalline structure is built. Each unit cell has four sodium ions and four chlorine ions associated with it. In this model the very center of the cube is occupied by a sodium ion. Each of the other 12 sodium ions is shared by three adjacent unit cells so that each contributes one-fourth of an ion to the unit cell sketched. The total number of sodium ions, therefore, is $1 + \frac{1}{4}(12) = 4$. The 14 chlorine ions distribute themselves in a similar fashion and contribute a total of 4 chlorine ions to the unit cell under discussion. Each set of ions forms an fcc structure with equal numbers of sodium and chlorine ions. The radii ratio of sodium to chlorine is 0.54, with the Na ion being almost half the size of the Cl ion. This difference in size of the ions places restrictions on other ions that could replace them. Furthermore, the replacement ions must have the same number of exterior (valence) electrons. With these restrictions in mind, point defects in ionic or ceramic crystals are the rule rather than the exception. A vacancy may consist of pairs of ions of opposite charge. An interstitialcy would consist of a displaced ion located at an interstitial site in the lattice structure.

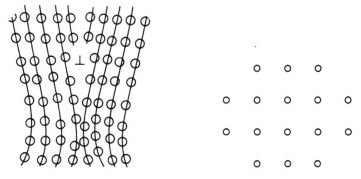

Figure 3-44 Line defects (dislocation).   **Figure 3-45** Undeformed crystal.

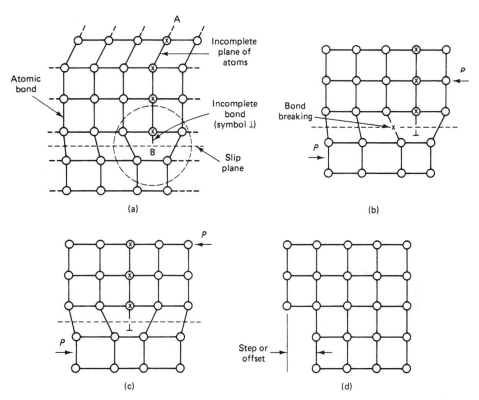

Figure 3-46 Edge dislocation movement (slip). (a) Dislocation—incomplete row of atoms above slip plane.
(b) Shear force (*P*), causing dislocation to move. (c) Dislocation moved one row to the left. (d) Dislocation
reaches surface of crystal, producing plastic strain (deformation).

The second type of imperfection, the (one-dimensional) *line defect,* is also known as a
dislocation (see Figure 3-44). A ***dislocation*** is a linear array of atoms along which there is some
imperfection in the bonding of the atoms. An undeformed crystal lattice is represented by Figure 3-45. Figures 3-46a and b show what appears as an extra or incomplete plane of atoms (*A–B*)
that causes distortion of the crystal structure. The two-dimensional representation also has
atoms behind each atom, shown as open dots. The atoms in the area circled by a dashed line represent a center or core of poorly bonded atoms that extend back into the material along a line
normal to the paper. This line defect is known as an ***edge dislocation,*** whose symbol is (⊥). A
force *P*, acting as shown, would cause the rows of atoms to move in the direction of the force,
each row moving one at a time much like dominoes striking each other (Figure 3-46c). As each

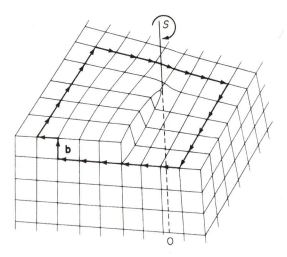

Figure 3-47   A screw dislocation. (Arthur L. Ruoff, *Materials Science* © 1973. Reprinted by permission of Prentice-Hall, Inc., Englewood Cliffs, N.J.)

row of atoms moves, the next follows in turn until all planes of atoms in the area have been displaced sufficiently to make all planes continuous, as shown in Figure 3-46d.

This line of local disturbance represented by the five atoms circled in Figure 3-46a, which may extend to the boundaries of the crystal, is a region of higher energy. As stated previously, this region contains the line defect or dislocation. The dislocation plane contains this line. Above this line the atoms are under a compressive stress, while those atoms below the line are experiencing a tensile stress. These bonding forces are not as strong as in a perfect crystal lattice, which permits a relatively small shear force to break the bonds, allowing the dislocation to move. The bonds re-form after the dislocation passes.

The successive passage or slipping of planes of atoms has been likened to the sliding of a large, heavy rug. To move the rug requires a large force; however, if you make a wrinkle in the rug and push the wrinkle a little at a time, the rug can be moved. The small movement of the rug to make the wrinkle can be thought of as the slipped portion of the rug; the other portion of the rug, the unslipped region. The wrinkle is the dislocation that separates these two regions.

The displacement of atoms (*slip*) is in a direction perpendicular to the dislocation line and/or plane. Where the direction of slip is parallel to the dislocation line, the line defect is called a ***screw dislocation,*** denoted by the symbol ↻. This is depicted in Figure 3-47. As with point defects, many line defects are actually combinations of edge and screw dislocations, producing curved dislocation lines or loops that start and end within the crystal. Dislocations originate during crystallization or plastic (inelastic) deformation. More will be said about dislocations when the topic of plastic deformation is treated later in this book.

However, before leaving dislocations, the following observations are made. The ideal crystal structure would contain no deformations. Experience with human-made, near-perfect crystals or **whiskers** has indicated that such whiskers contain great strength. The reason why a relatively little force is able to deform a crystal structure with strongly bonded atoms or ions is that only a few atomic bonds need to be broken and re-formed when dislocations are present in the crystal structure. In view of our present inability to produce near-perfect crystalline solids, the problem resolves into one of determining how to control this movement of dislocations by hindering the movement (strengthening the solid) or facilitating it (temporarily weakening the solid for some purpose).

In summary, line defects have a great deal to do with the strength of a solid. An abundance of them will cause a mutual interference in their movement through a crystal, preventing the planes of atoms from slipping, thereby strengthening the material. The presence of a few dislocations increases the ductility of a crystalline solid.

Area defects (interfacial or two-dimensional) are the third type of imperfection and exist in the form of grain boundaries (Figure 3-48). As each crystal grows it establishes its own axis system, on which the atoms/ions orient themselves. Adjacent crystals with their differently ori-

(a)

Nucleation and
dendrite formation

Grains forming
and orienting

Complete
solidification

(b)

Figure 3-48 (a) Grain boundaries.
(b) Grain growth.

ented lattice structures close in on each other. The last atoms to take up position in a crystal find
it more difficult to occupy normal lattice sites. Consequently, a transition zone is formed that is
not aligned with any of the adjoining crystals. The atoms making up the grain boundary possess
greater disorder, and hence greater energy, than their counterparts within the crystals themselves. Furthermore, the atoms are less efficiently packed together. These factors signify that the
atoms in the grain boundaries are ready to act as sources of new crystal formation (nucleation
sites) once the right conditions are met. Second, they assist in the diffusion of atoms through
the solid. Third, they offer resistance to the movement of dislocations and therefore modify the
strength and the ability of materials to plastically deform. Fourth, they act as sinks for vacancies. A solid with a large number of individual crystals (also called **grains**) has more grain
boundaries than a solid with a lesser number of grains. Fine-grained material will, at normal
temperatures, be stronger than coarse-grained material. Figure 3-48 is a two-dimensional sketch
showing grain boundaries with different orientations of grains and grain growth.

Remember that a solid contains crystals, each having the same lattice structure. What is
different between the individual grains is the orientation of that structure within each grain. A
technique for revealing the details of the grain structure of solids uses reflected light through
a metallographic microscope (Figure 3-49).

3.5 PHASE DIAGRAMS

Phase or equilibrium diagrams serve as maps for finding one's way through the many solid-
state reactions that occur in materials. Equilibrium conditions may be defined as slow heating
and/or cooling of the many materials to permit any phase change to occur. **Equilibrium,** from

Figure 3-49 Identification of individual grains by reflected light.

a material's standpoint, has been defined as a condition of minimum energy. It should be noted in pure terms equilibrium is a condition that is neither reachable nor attainable; but for practical purposes we can treat processes as if equilibrium has been reached. This condition might be compared to a metal spring that is neither stretched nor compressed. Once the spring is stretched or compressed, a force is set up inside the spring that tends to return the spring to its free-length position. The spring stores up the work done to stretch or compress it in the form of potential energy. Anywhere there is disorder in the crystalline structure of a solid (grain boundaries, imperfections, etc.) is a place of high energy similar to the example of the spring under action of some external force. The magnitude of the energy difference is proportional to the magnitude of the disorder.

To describe completely the conditions of equilibrium for a particular system, three externally controlled variables of temperature, pressure, and composition must be specified. Normally, phase diagrams record the data when the pressure is held constant under normal atmospheric conditions. Consequently, phase diagrams especially are graphical representations of a material system under varying conditions of temperature and composition. To determine the data necessary to plot these many phase diagrams, specialized equipment and techniques are used. Cooling curves for a pure metal, a pure iron, a metal alloy, and a nonmetal are sketched in the following pages to illustrate the phase changes that occur, if any. Note that for pure iron, three of the four phase changes occur while the iron is in the solid state. The α-iron stage existing at the higher temperatures is nonmagnetic.

A *cooling curve* is a graphical plot of the structure of a material (usually a pure metal or metal alloy) over the entire temperature range through which it cools under equilibrium conditions. Figures 3-50b and c show the structures that the particular metal or alloy is in at any particular temperature and time. Three structures are indicated for a pure metal or alloy: (1) liquid phase (melt), (2) a combination structure consisting of both a liquid and solid phase, and (3) a final solid phase near or at room temperature.

For the temperature to remain constant, as in Fig. 3-50a, between points b and c, there must be a balance between the heat withdrawn from the metal and the heat supplied. In this instance, the metal is giving up heat (latent heat of fusion), which balances the heat removed. The result is a net change of zero, as measured by a thermometer. As a metal cools from a higher to a lower temperature, its electrons require less thermal energy as a result of the relatively more ordered, more dense, and more bonded positions occupied by the atoms. As the electrons surrender their excess thermal energy (in the form of heat), they move closer to their nuclei and slow their motions to match the energy level of their new positions (*equilibrium* position). Cooling curves for nonmetals such as glass or plastics (Figure 3-51) show no clearly defined melting points.

Samples of an alloy are taken at various temperatures and quickly cooled to capture the structure at that particular temperature. Metallurgical microscopes using visible light study the grain structure, including the size, shape, and distribution of grains, with a maximum magnification of about 2000×. X-rays, gamma rays, and electron beams probe the crystal structure with great resolution and magnification. Electron microscopy permits the detailed chemical analysis of selected areas within the solid material, as well as the character of an individual

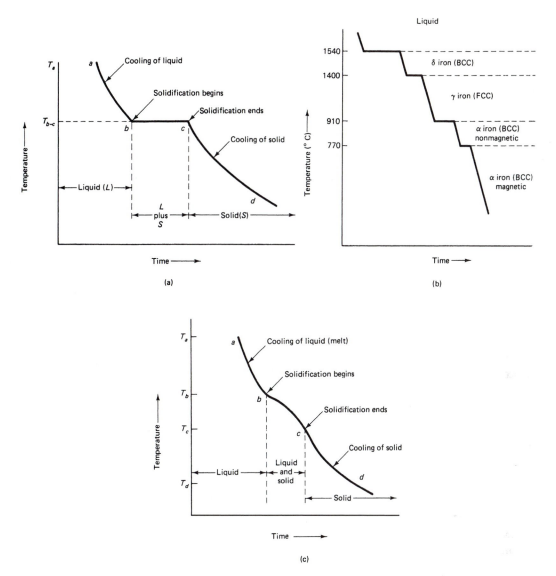

Figure 3-50   (a) Cooling curve for pure metal: (1) Heat pure metal to point T_a (liquid phase or melt); (2) cooling of liquid metal $a - b$; (3) at point b, pure metal starts to *precipitate* out of solution; (4) point c, *precipitation* (pure metal completely solid); curve from b to c straight horizontal line showing constant temperature $T_b - T_c$ because thermal energy absorbed in change from liquid to solid; (5) more cooling of solid pure metal from c to d and temperature begins to fall again; change in volume follows same pattern. (b) Cooling curve for pure iron. (c) Cooling curve for a metal alloy: (1) Two metals heated to point a (liquid phase or melt, with both metals soluble in each other); (2) cooling of alloy ($A + B$) in liquid phase; (3) point b, solidification begins; (4) point c, solidification complete (both metals in solid form); sloped $b - c$ due to changing from liquid to solid over range of temperature T_b to T_c because metals A and B have different melting/cooling temperatures; (5) further cooling from c to d of solid-state metal alloy.

dislocation. The data collected are used to construct various representations, particularly of the structure of the alloy under varying conditions of temperature, pressure, and composition. Most, if not all, of these graphical representations indicate the structure of a system under *equilibrium conditions*. Therefore, sufficient time must be allowed for atoms to diffuse and thus establish equilibrium conditions at any particular pressure, temperature, and composition. In practice, equilibrium conditions are normally departed from either because of the length of time involved (and attendant cost) or purposely to produce a solid phase that is in a ***nonequilibrium*** condition.

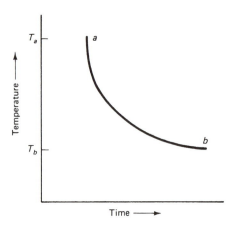

Figure 3-51 Cooling curve for a nonmetal. No phase transformation occurs. Nonmetallic heated to temperature T_a and allowed to cool to room temperature T_b. With no phase changes, continuous plot of temperature and time produces a smooth curve.

In our previous discussion we learned that atoms possess higher energies at elevated temperatures than at some lower temperature. This higher energy permits the atoms to diffuse more quickly. At lower temperatures, this kinetic energy is lessened. One can see that this reduced atomic mobility, coupled with the disequilibrium energy of a crystal structure at positions of high-energy content, presents a broad spectrum of possibilities for the diffusion (precipitation) of atoms producing a phase transformation in the solid state of a system.

The phase diagram for water makes a good departure point for an understanding of phase diagrams for systems comprising more than one component. The single component case of water is the compound H_2O. The system under study would then contain a definite amount of H_2O. The one-component diagram for a definite amount of water shown in Figure 3-52 is a graphical plot of pressure along the vertical axis versus temperature along the horizontal axis. These variables are expressed in SI units. This curve represents the physical changes that water (H_2O) undergoes. Note that, regardless of the ***physical changes*** (liquid to solid, i.e., water to ice), there is no alteration of the chemical composition of the water. Only its structure is changing. On the other hand, ***chemical change*** always alters a material's composition. Observe in Figure 3-53 how a change in temperature chemically changes the composition of the metal alloy plotted along the horizontal axis.

The pressure–temperature fixed composition diagram shown in Figure 3-52 actually contains three curves. The fusion curve is the line that separates the solid from the liquid phase. The curved line separating the liquid from the vapor phases, called the ***vaporization curve,*** is a plot of the various pressures and temperature combinations at which the liquid and vapor phases coexist in equilibrium with each other. The same can be said of the sublimation curve separating the solid from the vapor phases. ***Sublimation*** is the process by which a solid changes directly to its vapor state without passing through its liquid phase. Each curve forms a boundary between two different phases.

> Curve T–F, the fusion curve, represents where the liquid and solid phases are in equilibrium. In other words it is a graph of the freezing point versus pressure. At one ***atmosphere (atm),*** the freezing point of water is 0°C. At this pressure, water is a liquid if its temperature is between 0°C and 100°C, a solid if below 0°C, and a vapor if above 100°C.
>
> Curve S–T, the sublimation curve, represents where the solid and vapor phases are in equilibrium. It is a graphic plot of the sublimation point versus the pressure. Carbon dioxide, for example, in the solid phase (called dry ice) sublimates at atmospheric pressure.
>
> Curve T–V, the vaporization curve, is a plot of those points where the liquid and vapor phases are in equilibrium. It is therefore a graphic of the boiling point versus pressure.
>
> Curve T–F is unusual in that it has a negative slope (slopes upward to the left). For substances that normally contract upon freezing this curve would have a positive slope

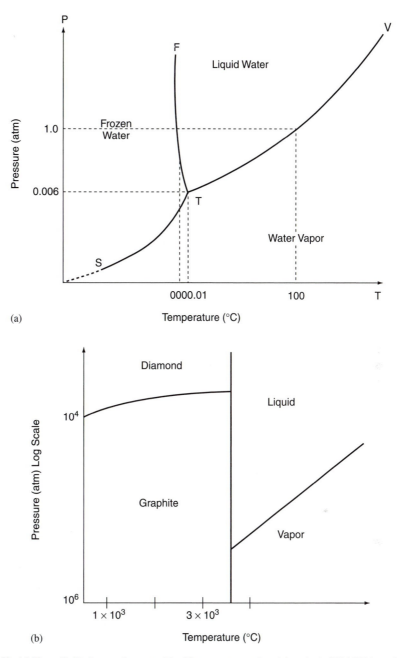

Figure 3-52 (a) Phase (P–T) diagram for water. The SI temperature at the triple point is 273.16 K (note that scales are not linear). (b) Phase diagram for carbon. Graphite is the stable form of carbon at temperatures up to 3000°C and pressures of 10^4 atmospheres and higher. Diamond is the stable form of carbon at very high pressures. See Figure 2-14 showing the polymorphs of carbon.

(sloping upward to the right). In the case of water, it expands upon freezing which requires a lower temperature at a higher pressure to cause the liquid to freeze. When water freezes, the molecules form a hexagonal lattice structure. This open structure accounts for water's unique property of expanding on freezing and being less dense as a solid than as a liquid. At 4°C this structure begins to form and at 0°C it is complete. The volume of a given amount of water decreases as it is cooled from room temperature until its temperature reaches 4°C. Below 4°C the volume increases and therefore the density decreases.

Figure 3-53   A set of hypothetical cooling curves for an alloy system phase diagram. If cooling curves for several combinations of two metals (metal alloy) are plotted as shown here, a phase or equilibrium diagram forms. Points *b*, in cooling curves where solidification begins, are connected to form a smooth line (*liquidus* line). Points *d*, where solidification is completed for each alloy, are also connected by a smooth line (*solidus* line). More discussion of phase diagrams for metals and ceramics follow in the text.

Special note should be made of the point where all three curves intersect. This point, known as the ***triple point,*** is where all three phases coexist. For water the triple point occurs at a temperature of 273.16 K (~0°C) and a pressure of 610.47 N/m^2.

In our study of metal alloys, not only do we have more than one component, but we want to study the solid-state reactions while varying the amounts of these components. To accomplish this, several approaches are used. Fortunately, in practice, most metallurgical processes are conducted at normal atmospheric conditions, which simplifies matters considerably.

If the cooling curves for many combinations of two metals (metal alloy) or ceramics were plotted, as in Figure 3-53, a phase, or equilibrium diagram, would be formed. Points *b* in the cooling curves, at which solidification began, can be connected to form a smooth line, or *liquidus.* Points *d,* at which solidification is complete for each alloy, are also connected by a smooth line, a *solidus.* There are various types of phase diagrams of metal alloys, depending on whether the two components are soluble, insoluble, or partially soluble in each other in the liquid or solid phase.

The ***liquidus*** is the name of the curve that represents the temperatures at which solidification begins. It marks the lower boundary of the liquid phase, and thus it is a solubility curve that shows the limit of solubility of the components *A* and *B* in the liquid. The ***solidus*** is the curve that passes through all points at which solidification is completed. It marks the upper boundary of the solid solution phase and therefore it is the solubility limit of both components *A* and *B* in the solid solution. Between these curves, where a two-phase slushy region of liquid plus solid exists, the process of solidification takes place. It is evident that the points where the liquidus and solidus curves meet represent compositions of 100% *A* or 100% *B* (i.e., pure metals). Between these two points are an indefinite number of varying compositions of the two pure metals *A* and *B*. Metal *A* could represent Cu and metal *B,* Ni.

3.5.1 Phases

A ***phase*** is a homogeneous part or aggregation of material that differs from another part due to difference in structure, composition, or both. The difference in structures forms an interface

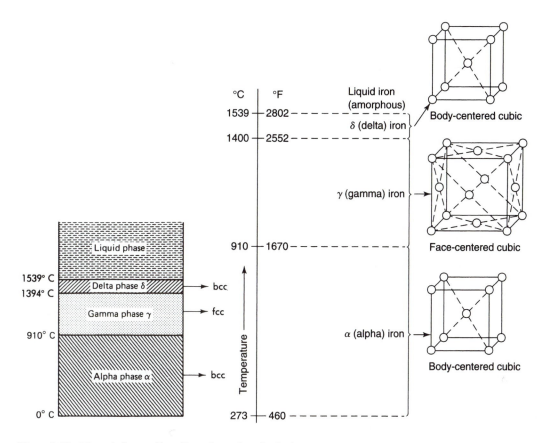

Figure 3-54   Allotropic forms of iron (three phases: bcc, fcc, bcc).

between adjacent or surrounding phases. Some solid materials have the capability of changing their crystal structure with varying conditions of pressure and temperature. These varying conditions cause these materials to change phase. Water can exist as a gas, liquid, or solid. Note that each of these general phases contains the same *components* (i.e., the same basic chemical elements or chemical compounds). The single component present in water is the compound H_2O, made up of hydrogen and oxygen.

Water in the solid phase can have different phases because it forms different crystal structures under different conditions of temperature and pressure. At the triple point of water (see Figure 3-52a) all three distinct phases can coexist with identical compositions, each having unique atomic structures and properties.

Being allotropic, like many metallic elements, iron can have different crystal structures and hence different phases (see Figure 3-54). At room temperature iron has a bcc structure. Heated to above 910°C (1670°F), iron's structure transforms to fcc. These allotropic or polymorphic forms are also referred to by Greek letters: α (alpha), γ (gamma), and δ (delta). Then, on heating above 1394°C (2541°F), iron's structure changes back to bcc.

It is worthwhile to pause briefly to recall that *phases,* being physically distinct with their own characteristic crystal structure, must of necessity possess different properties. No elements found in nature are 100% pure. During the processing of metals, additional impurities are introduced unintentionally, due partly to the high costs involved in trying to eliminate the impurities.

What are the effects of adding other substances to a pure element? The results produced can be many and varied. One result of this addition of impurities might be the formation of three different solid phases. Second, solid solutions may result. (A ***solid solution*** is simply a solution in the solid state that consists of two kinds of atoms combined in one type of space lattice.) A third possibility is that one or more compounds could form. Finally, two or more of

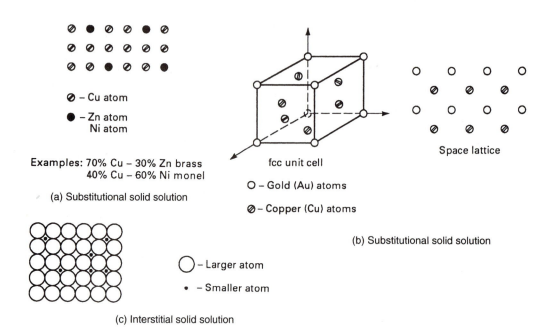

Figure 3-55 (a) Disordered (random) substitutional solid solution. Cu and Zn atoms occupy lattice sites in random. (b) Ordered substitutional solid solution in which Au and Cu atoms occupy restricted sites in the space lattice.

the preceding results could coexist, depending on the prevailing conditions of pressure, temperature, and degree of concentration of the components of the system. Some examples that illustrate the diversity of the results follow. An fcc structure of solvent Cu atoms will substitute Ni atoms in a solid solution of copper and nickel, known as a ***substitutional solid solution.*** The substitution of Zn atoms for Cu atoms in an fcc structure forms brass, a disordered or random substitutional solid solution (see Figure 3-55a). If the ordering were 100%, a compound would be formed. The alloy Cu–Au (Figure 3-55b) has Cu atoms occupying the face-centered sites and Au atoms occupying the corner sites of the fcc unit cell.

Ordered solid solutions form generally at lower temperatures, or they come into existence when a disordered solid solution becomes unstable at a lower temperature. The fundamental lattice structure may or may not change during this particular transformation. It is important to point out again that this modification in the arrangement of the atoms brings on an alteration in physical properties. In some instances the ordered arrangement is harder and has greater electrical conductivity than the disordered arrangement.

Another type of crystal structure is formed when some atoms take up positions between the regular solvent atoms, that is, in the vacant spaces between lattice points. These solute atoms and the solid solutions formed are ***interstitial.*** The carbon atoms in ferrite (alpha iron, or alpha ferrite) occupy the interstices (spaces) in this bcc allotropic form of iron. For solute atoms to form interstitial solid solutions (Figure 3-55c), they must be small enough to fit into the interstices between the normal lattice points.

It is time to ask, is it possible to determine ahead of time whether an element will form a separate phase or a solid solution when added to another element? In other words, can we predict the degree of solubility occurring in certain systems? Certainly, this lies at the heart of metals technology in developing alloys (solid solutions) that possess ductility for subsequent forming operations or two-phase materials with greater strength and hardness than those of the pure substances alone. A set of general rules known as the Hume-Rothery solubility rules provides this guidance, and like all such statements is not to be considered foolproof.

3.5.1.1 Solid Solution Strengthening.

By following the ***Hume-Rothery solubility rules*** set forth below, we can produce solid solution alloys that result in solid solution strengthening. As with all rules, however, they are not foolproof. For example, Cu and Ag meet

(a)   (b)

Figure 3-56   Solid solutions. (a) Monel (Ni, 67.5%; Cu, 30.18%). (b) Brass (Zn, 70%; Cu, 30%). (Buehler Ltd.)

all the requirements set forth by the rules but they have limited solid solubility. Solid solution alloys have greater strength than the pure metals. The difference in atomic size between the solvent and the solute atoms will produce greater strength, because there is a greater disruption of the initial crystal or lattice structure making slip more difficult. The larger the amount of the alloying or solute atoms, the greater the strengthening. See paragraph 8.6.4.2 for a discussion of dispersion strengthening of composite materials when the limit of solubility may be exceeded. The large size difference between the atoms also produces a disruption of the initial lattice.

Rule 1.   The difference in atomic diameters between the solvent atom and the solute atom should be less than 15%. ATOMS MUST BE OF SIMILAR SIZE. Note that smaller atoms generally produce greater strengthening than larger atoms. For example, Be in Cu produces greater strength properties than Al when added to Cu.

Rule 2.   The metals must have the same crystal or lattice structure.

Rule 3.   The atoms of the metals must have the same valence. Such elements do not form compounds. Near neighbors in the periodic table such as Fe and Co tend to form solutions in one another. If the valence difference is significant, then the metals may form compounds not solutions.

Rule 4.   The atoms of the metals must have approximately the same electronegativity. With different electronegativities the tendency is to form compounds (note the case of Na and Cl to form NaCl).

A pause to discuss these rules with regard to other previous definitions of solid solutions is worthwhile at this time. Assume two metals are soluble in each other to some extent. If the size difference in their atoms is less than 8% and the other conditions are satisfactory, there is almost complete solid solubility. In other words, the two metals are soluble in each other in all possible proportions. Examples of this are Monel (Ni–Cu) and brass (Cu–Zn) (Figures 3-55a, 3-56a and b). Referring to Table 10-1B, copper's atomic radius is 1.57 Å (1Å $= 10^{-10}$m) and nickel's is about 1.62Å. Expressed in nanometers, these radii are 0.157 nm and 0.162 nm, respectively. Both have an fcc structure. Note their position in the periodic table. They are elements 29 and 28, respectively. The fcc Cu–Ni alloys can range from near 0% Ni to almost 100% Ni.

There are some 25 alloys used in industry that are based on this mutual solubility of copper and nickel in all proportions. These nickel alloys, known as **_Monel alloys,_** contain about 30% copper and have combinations of high strength and good corrosion resistance. Some high-strength Monels, in addition to being nonmagnetic, are equivalent to heat-treated steels having tensile strengths approaching 200,000 psi. In contrast to pure Cu, which has relatively low strength, the alloys of Cu and Ni permit much higher strength because of the interaction between the different atoms in the solid solution producing increased resistance to slip.

Another common alloy, **brass,** is a solid solution of Cu and Zn (Figure 3-56b). Zinc's atomic radius is about 0.153 nm. It is element 30 in the periodic table. Cu and Zn atoms differ in size by less than 15%. In this system, Zn atoms can replace Cu atoms up to a maximum of 50% and produce disordered substitutional solid solutions. We can add tin (Sn) to Cu; Sn, the fiftieth element of the periodic table, has a radius of about 0.146 nm. This indicates that we are approaching the 15% limit. In fact, only about a maximum of 10% of the Cu atoms can be substituted by Sn atoms to produce the single-phase alloys known as **bronzes.**

The term *bronze* was originally used to describe copper–tin alloys. Now it is used for any copper alloy that contains up to approximately 12% of a principal alloying element. It is incorrectly applied to some alloys that are special brasses. Architectural bronze, a trade designation for an alloy that contains 57% Cu, 40% Zn, and 3% Pb, is actually a brass.

In summary, the effects of solid solution strengthening are realized in the properties of the material, which is consistent with the basic premise that a change in the structure of a material results in a change in its properties. Some of these effects that are produced in solid solution alloys as opposed to the pure metals are as follows:

1. Tensile strength and yield strength are greater than in the pure metals.
2. Hardness is also increased.
3. Ductility is usually decreased, as well as electrical conductivity.
4. High-strength capabilities are increased, as well as resistance to creep at high temperatures.

For an interstitial solid solution to form, the added atoms must be sufficiently small to fit into the interstices between the solvent atoms. From our study of atoms and the periodic table, we know that the atom's radius is measured from the center of the atom to the outermost electron. We also know that this diameter decreases as we move from left to right in a period of the table (Table 3-2). Taking these facts into consideration, the five elements with radii less than 0.1 nm are hydrogen (H), carbon (C), boron (B), nitrogen (N), and helium (He). A classic example is the interstitial solid solution formed by C in Fe in the production of steel. Pure Fe has a bcc structure at room temperature. This phase is called α (alpha) iron. In this form the Fe is relatively weak and incapable of being shaped for commercial use. Above 912°C, iron, being allotropic, changes to an fcc structure known as γ (gamma) iron. From our study of unit cells of the various lattice structures, we know that the interstices between atoms in a bcc structure are quite small. In the fcc unit cell, a relatively large interstice exists in the center. Carbon added to α-Fe allows only 0.025% of the C atoms to take up positions in the iron bcc structure. If, instead, the α-Fe is heated to above 912°C and then the C atoms are added, the γ-Fe will accommodate up to 2% of the C atoms. This interstitial solid solution is given the name **austenite.** The formation of steel and cast iron will be treated extensively later when phase equilibrium diagrams and heat treating are discussed.

In studying the formation of compounds in liquid or gaseous states, it may be fair to conclude that most chemical compounds are nonmetallic. Second, they involve the exchange or sharing of electrons. Third, one of the elements involved must have a positive valence and the other a negative valence (oxidation states). Fourth, the algebraic sum of these oxidation states must be zero in the compound formed. Experience indicates that two metals show no inclination to join together in a chemical way to produce compounds. The latter statement is true for metals in the liquid and gaseous states. We are now discussing solid-state reactions. In solids, two metals can combine to form **intermediate alloy phases.** Their compositions are intermediate between the two pure metals, with crystal lattices differing generally from those of the pure metals. The intermediate phase can occur alone or accompanied by the pure metal or their solid-solution phases. These later phases are then called terminal phases.

Many intermediate phases have great technological importance in the strengthening of solid materials. They will be eventually called on, as an example, in a heat-treating process to

 EMT WEBLINK

Centre for Materials Science—*http://www.cms.fy.chalmers.se/*

precipitate out at some lower temperature to impede the slippage of planes of atoms over one another, thus producing an increase in strength and hardness of steel. (Intermediate phases are discussed in more detail in Section 3.5.2.)

The liquid (or molten) state of a metal may consist of a single pure metal. It may comprise a solution of two or more metals that are completely soluble in each other. A third possibility is that the liquid state represents two *immiscible* metals. Finally, it could contain two insoluble liquid solutions. Note that the word ***insoluble*** is synonymous with ***immiscible.*** Substances that are not soluble (that are insoluble) are immiscible. Water and oil have no solubility; that is, they are insoluble, or immiscible.

Very few metals are insoluble in each other in the liquid state. Aluminum and lead come close to being immiscible in both the liquid and solid states. In the solid state they solidify into two separate layers with a clear line of contact showing no appreciable degree of diffusion. Most metals are soluble in each other to some degree in their liquid phases. When these metals are cooled under equilibrium conditions from their liquid phases, they solidify and produce various solid phases. Assuming that they are completely soluble in the liquid phase, on cooling they may be:

1. Completely miscible in the solid state
2. Insoluble in the solid state
3. Partly soluble in the solid state

Other products may be formed, such as intermediate phases. Also, if the liquid phases are not completely soluble, other results may be produced in the solid state. To understand why and how these various results occur requires a good knowledge of the construction and use of *phase diagrams.*

The process of solidification begins with the formation of seed crystals or embryos from the liquid. Seed crystals, embryos, or nuclei (do not confuse these with the nucleus of an atom) subsequently grow into full crystals. For these nuclei to form, ***supercooling*** or undercooling of the liquid phase is necessary; that is, a temperature decrease below the equilibrium temperature is needed to make the phase transformation proceed at a measurable rate. This supercooling is an indication of the removal of the latent heat of solidification. Again, we can call on our example of the spring that, once deformed, stores up energy within itself. Supercooling is comparable to the amount the spring is deformed. For a nucleus to form—a process called ***nucleation***—energy is needed to create the nucleus by overcoming surface tension effects. Nucleation can be aided by the presence of impurity atoms, which are almost always present. As each of these nuclei grows larger, it forms dendritic, or treelike, crystals with a random distribution of crystallographic planes relative to one another (see Figure 3-57). These crystals, called ***grains,*** are surrounded by regions of high energy known as ***grain boundaries.*** If many nucleation sites are formed, a fine grain size in the solid state is the result. To obtain this fine grain, a minute amount of some element is purposely added. The particles of the element added act as impurities, on which the nuclei can form. In other words, the average grain size is proportional to the degree of supercooling.

3.5.2 Phase Diagrams (Equilibrium Diagrams) for Metallic Systems

Building on our knowledge of phase diagrams presented in Section 3.5.1, we will describe "road maps" that guide and direct those involved in the development, fabrication, heat treatment,

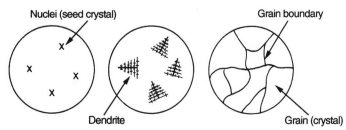

Figure 3-57 (a) Grain formation.

(a)

(b)　　　　　　　　　　　　　　　　　　　(c)

Figure 3-57　(b) Dendrites of "tough pitch" copper, dendrites outlined by copper oxides (200×). (Buehler Ltd.)
(c) Grains of "high-carbon" steel. (American Iron and Steel Institute)

and design of alloys using some simple diagrams to illustrate their main features. Before beginning this task, note that various rules must be observed to ensure correctness of the diagrams. These rules are derived from Gibbs' formulations developed over 100 years ago, the most important of which is the ***Gibbs phase rule.***

The construction of phase diagrams is based on the Gibbs phase rule, which is expressed in equation form as

$$P + F = C + 2$$

where P is the number of phases in equilibrium; F is the variance or number of degrees of freedom, or the number of variables such as pressure, temperature, or composition that can be varied without affecting the number of phases in equilibrium; and C is the number of components (elements, compounds, or solutions) in a particular system. The number "2" in the equation stands for temperature and pressure, the two variables that can be allowed to change.

The phase diagram for water is plotted in Figure 3-58, with the pressure as ordinate and the temperature as abscissa. This is a ***unary phase diagram,*** meaning that it consists of one component—in this instance, water (H_2O). The series of curves in the diagram show the division between the three phases in which water can exist depending on temperature and pressure conditions.

Point A in the liquid region is defined by a certain combination of pressure and temperature. The number of components (C) at this point is 1, a unary diagram. The number of phases (P) is 1 because only liquid water exists in this region. The phase rule tells us that the number of degrees of freedom is

$$F = C - P + 2$$
$$= 1 - 1 + 2$$
$$= 2$$

The "2" means that, within the limits of the liquid phase, the pressure, temperature, or both can be changed and the phase would still be liquid. No phase changes occur and equilibrium is maintained.

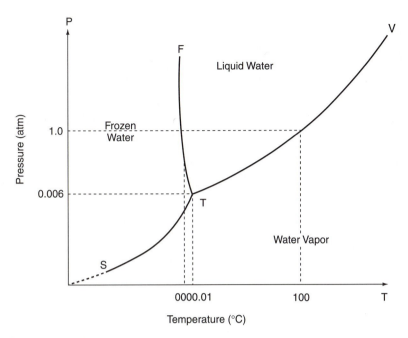

Figure 3-58   Unary phase diagram for water.

Point B is on the fusion curve (T–F), the boundary between the liquid and solid phases of water. The number of components has not changed and therefore $C = 1$. The number of phases is $P = 2$ because anywhere along this curve, liquid and solid phases of water coexist in equilibrium. Using the phase rule to find the number of variables that can be changed without changing any phases in equilibrium, we have

$$F = C - P + 2$$
$$= 1 - 2 + 2$$
$$= 1$$

This single degree of freedom means that if the pressure is changed, the temperature must also be changed to stay anywhere on that boundary line. We can no longer change either the pressure or the temperature independently and stay on the boundary line where liquid and solid phases of water coexist in equilibrium.

Point T, at the intersection of the three curves, is known as the triple point of water. Here the three phases of water coexist. The phase rule tells us that the number of degrees of freedom for this position on the phase diagram is

$$F = C - P + 2$$
$$= 1 - 3 + 2$$
$$= 0$$

Zero degrees of freedom means that we cannot change pressure or temperature. If we did, the three phases would no longer coexist in equilibrium. It also means that the pressure and temperature are fixed for this condition to exist. The word *invariant* is also used to describe this particular point.

Point V, which is known as a ***critical point***, means that no matter how high the temperature and pressure rise beyond this point, the vapor phase will never change to a liquid.

Materials science deals primarily with the solid and liquid phases of materials. Second, the pressure variable has only a small effect on materials. Third, metallic phase diagrams are

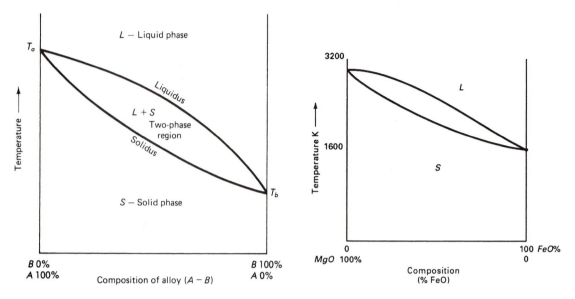

Figure 3-59   Phase diagrams of metals and ceramics. (a) Metals *A* and *B* completely soluble in each other in both liquid and solid phases. (b) Phase diagram representing two ceramic materials completely soluble in each other in the liquid and solid states.

Figure 3-60   Phase diagram for metals *A* and *B* completely insoluble in each other in both liquid and solid phases (hypothetical).

usually limited to showing the interactions between only two components and thus are known as **binary diagrams.** For example, the iron–iron carbide diagram or the phase diagram for brass has only two components, copper and zinc. Consequently, phase diagrams deal with temperature changes versus changes in the composition of materials. This means that the phase rule can be simplified as follows:

$$P + F = C + 1$$

We may now wish to know what the *actual composition of the two-phase region* (Figure 3-59) is for a particular alloy at a specific temperature. Figure 3-59a shows a typical phase diagram for two metallic elements, and Figure 3-59b shows ceramic materials that are completely soluble in each other in both liquid and solid phases. The vertical axis at the left represents a pure metal *A* with a melting point of T_a. The vertical axis at the right represents metal *B* and has a melting temperature less than metal *A*. The upper curved line of the cigar-shaped *L* + *S* region is the liquidus and the lower line is the solidus. The pure metals and any composition of the two are in the liquid (melt) phase upon heating above the liquidus. Upon cooling to the solidus, they solidify. When an alloy is at a temperature above the solidus but below the liquidus, it exists as part liquid and part solid in a two-phase region labeled *L* + *S*. Figure 3-60, for com-

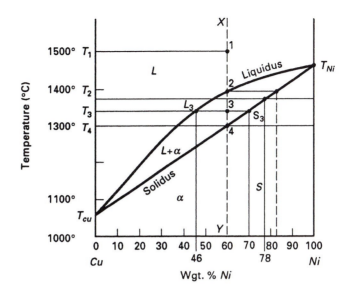

Figure 3-61   Cu–Ni phase diagram. Alloy *X–Y* is 60% Ni–40% Cu.

parison purposes, is a hypothetical phase diagram for two metals that are completely insoluble in each other, both in the liquid and solid phases.

Using an alloy with a composition of 60% Ni–40% Cu, we will plot this composition on Figure 3-61 to determine first its chemical composition at any given temperature. Vertical line *X–Y* represents our alloy. At point 1 the alloy is at temperature T_1 in the liquid solution phase. Upon slow cooling, point 2 is reached at a temperature of T_2. Our alloy is now entering the two-phase region. Part of our alloy is liquid and part has formed a solid solution. To determine the composition of these two phases at any point, such as at 3, a horizontal temperature line called a ***tie-line*** is drawn through point 3. This tie-line will cross both the liquidus and solidus curves at points labeled L_3 and S_3, respectively. From both these points, drop vertical lines to the abscissa (horizontal axis) to determine the composition of these two phases at the temperature T_3. The liquid phase formed so far has a composition of 54% Cu and 46% Ni, determined by proceeding from point 3 along the tie-line to the liquidus L_3, then vertically down to the horizontal axis at the point labeled 46% Ni. Similarly, the composition of the solid phase is found to be 70% Ni–30% Cu. This time we proceeded along the tie-line to the solidus at S_3 and vertically downward to the horizontal axis to read 70%. Note that the tie-line drawn across the two-phase region ties in or connects the adjacent phases. In this instance, the solid phase is on the right and the liquid phase is on the left of the two-phase region. This is an excellent way of identifying any two-phase region. As the alloy continues to cool to temperature T_4, observe, using the same procedure, that both phases continue to increase in the percentage of copper. Upon reaching the solidus at temperature T_4, the last liquid, very rich in Cu, solidifies at the grain boundaries. Through diffusion of the atoms, all the solid solution will be at an overall composition of 60% Ni–40% Cu. The solid solution is designated α solid solution.

The tie-line procedure plays an important role in identifying two-phase regions in a phase diagram. In Figure 3-62, a phase diagram that roughly resembles the Al–Si phase diagram shows the two-phase regions labeled. Greek letter symbols are commonly used to identify those solid phases first to form. By drawing a tie-line across these regions and knowing what phases are present at the extremities of the tie-line, the two-phase region can readily be labeled. Note the complete solidus line, showing that any phase below it is a solid phase. Lines *BE* and *DF* are solvus lines, or ***lines of maximum solubility*** of the terminal or first-formed solid solutions. Note the reduction in the limits of solubility of these solutions to absorb other phases as the temperature decreases.

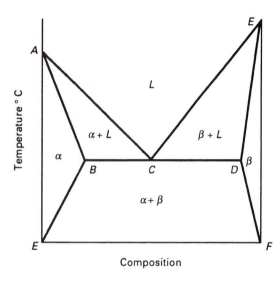

Figure 3-62 Two-phase regions of a phase diagram showing partial solid solubility.
ACE—liquidus
ABCDE—solidus
BE and *DF*—solvus lines
C—eutectic point

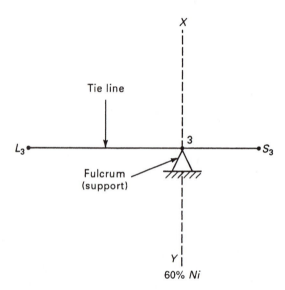

Figure 3-63 Tie-line from Figure 3-61.

   The next problem to be tackled is to determine the *relative amounts of liquid and solid existing at a specific temperature in a two-phase region.* So far we know that at temperature T_3 (see Figure 3-61) two phases exist: a solid phase with a composition of 70% Ni–30% Cu and a liquid phase with a composition of 54% Cu–46% Ni. How much of the liquid has solidified upon being slowly cooled to the temperature T_3? To determine this percentage, we use what is known as the ***lever rule*** (also known as the ***inverse-ratio law,*** as we shall soon demonstrate). The law gets its name from treating the tie-line as though it were a lever supported (fulcrum) at the point where the vertical line representing our alloy *X–Y* crosses the tie-line. Figure 3-63 is an enlarged sketch of the tie-line in Figure 3-61.

   The complete tie-line from L_3 to S_3 represents the total weight (100%) of the two phases present at temperature T_3. The vertical line (*X–Y*) representing our alloy cuts the tie-line into two parts or lever arms; one part forms L_3 to 3, and the second forms 3 to S_3. The length of the first lever arm, L_3 to 3, represents the amount of solid phase present; the right lever arm, 3 to S_3, represents the amount of liquid phase present. Note that the left arm or part contains L_3 on the liquidus line; yet it represents the amount of solid phase. The right lever arm contains a point S_3 on the solidus, but this arm represents the amount of liquid phase present. Thus, the

Figure 3-64 Tie-line from Figure 3-63.

lever rule states that these two lengths are inversely proportional to the amount of the phase present in the two-phase region. Observe that point 3 is closer to the solid-phase region, and therefore the quantity of the liquid phase is greater than the quantity of the solid phase at that temperature. The lengths of these two parts or lever arms of the tie-line are measured in units of composition.

Now let's apply this lever rule to our alloy 60% Ni–40% Cu. Figure 3-64 is another sketch of the L_3-3-S_3 tie-line in Figure 3-61 (a repeat of Figure 3-63). The tie-line's length in terms of composition is 24, or $70 - 46 = 24$, leaving out the percent sign. The length of L_3 to 3 (left lever arm) is 14, leaving 10 for the length of the right arm. Next ask yourself: What percentage of the entire line is the right lever arm? To find the answer, you would take the length of 3 to S_3 and divide it by the total length of the tie-line, L_3 to S_3, and multiply the result by 100%.

$$\frac{10}{24} \times 100\% = 42\% \quad \text{(liquid formed)}$$

Using the same procedure, we find that the left part (left arm) represents the percentage of solid formed:

$$\frac{14}{24} \times 100\% = 58\% \quad \text{(solid formed)}$$

Let's take another look at Figure 3-61 to note another metallurgical process that is taking place: ***diffusion.*** For our purposes, note that the solid formed at T_2 consists of a composition of about 82% Ni–18% Cu. Assuming slow equilibrium cooling, our alloy will ultimately solidify completely, with a composition of about 60% Ni–40% Cu. What this tells us is that sufficient time must elapse for the atoms to migrate through the solid phase to satisfy the principles of equilibrium; that is, given sufficient time, atoms will diffuse away from places of high concentration such that a uniform distribution of atoms in the crystal structure is once again achieved. The reheating of a solid metal to give atoms greater energy to diffuse or move toward equilibrium (hence more uniform composition) is termed ***homogenization.*** Homogenizing eliminates the segregation of phases, such as coring, that form during solidification. Such segregation leads to nonuniform properties. Annealing, specifically normalizing (see Module 5), is an example of a homogenization process involved in the heat treating of metals.

Research into the diffusion process discloses that the rate of diffusion slows as the difference in concentration of different atoms is reduced. Second, diffusion in a solid slows with reduced temperature. Therefore, in practice it is extremely difficult to achieve equilibrium cooling. Also, slow cooling produces large grain size, which is usually undesirable. Faster cooling rates are the rule, not the exception, and they prevent complete diffusion. The end result is that the initial crystals formed are of one composition. As the crystal grows in a dendrite fashion by atoms attaching themselves to the original crystal, the composition of the total crystal changes. The original higher-melting-point central portion is surrounded by lower-melting-point solid solutions. This condition is known as ***coring*** or ***dendritic segregation.*** Segregation is always present to some extent when metals are melted and subsequently solidified in the making of steel (steel ingots) and in the casting of metal parts. A metallurgical process that attacks the problem of cored structures will be discussed later.

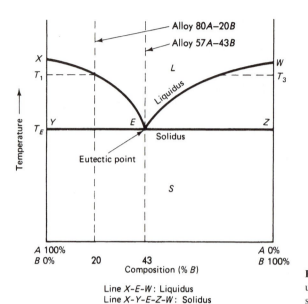

Line *X-E-W*: Liquidus
Line *X-Y-E-Z-W*: Solidus

Figure 3-65 Metals *A* and *B* completely soluble in the liquid phase and completely insoluble in the solid phase (hypothetical).

Illustrative Problem

Determine the relative amounts of liquid and solid existing at temperature T_3. (See Figure 3-61.)

Solution   Using the lever rule and the previously determined compositions of the two phases existing at that temperature, draw a horizontal tie-line at $T_3 = 1336°F$. The complete line (Figure 3-63) from L_3 to S_3 represents the total weight (100%) of the two phases present at this temperature. The vertical line (X–Y) representing the alloy cuts the tie-line into two parts or lever arms. Lever arm L_3 to 3 represents the amount of solid phase present. Lever arm 3 to S_3 represents the amount of liquid phase present. Note that these two lengths are inversely proportional to the amounts of phase present. The length of arm 3 to S_3 is a bit smaller than arm L_3 to 3, hence, one would expect the amount of solid to be a bit larger than the amount of liquid. Referring to Figure 3-64, the estimated lengths of the arms are determined to be L_3 to 3 = $60 - 46 = 14$, and 3 to $S_3 = 70 - 60 = 10$.

Solid phase: $\dfrac{14}{24} \times 100\% = 58\%$

Liquid phase: $\dfrac{10}{24} \times 100\% = 42\%$

Illustrative Problem

Determine the composition of alloy X–Y (40% Cu–60% Ni) at point 2 in Figure 3-61.

Solution   The tie-line through this isotherm (at T_2) intersects the solidus at about 60% Ni–40% Cu giving the composition of the liquid phase. The solid phase at T_2 is 82% Ni–18% Cu.

Illustrative Problem

Determine the composition of the last remaining liquid phase of alloy X–Y in Figure 3-61 just prior to the alloy completely forming a solid solution.

Solution   The tie-line constructed through point 4 ($T_4 - 4$) intersects the liquidus. Reading vertically downward, the percentage composition of the alloy X–Y at this temperature immediately prior to forming a solid solution is about 67% Cu–38% Ni.

To continue our study of phase diagrams, the next step is an understanding of the *eutectic reaction*. The word **eutectic** is taken from the Greek and means "to melt well." Figure 3-65 labels the intersection of the liquidus and solidus as the eutectic point (*E*). The temperature

Figure 3-66   Two-phase lamellae microstructure. (Buehler Ltd.)

that corresponds to it is labeled T_E, which is the lowest temperature at which a liquid solution will remain completely liquid. The alloy 57% *A*–43% *B* has the lowest melting point of any alloy in the system *AB*. Not only does it have the lowest melting temperature, but this eutectic, a mechanical mixture, solidifies completely at this temperature rather than over a range of temperatures. Alloy 80% *A*–20% *B* begins to solidify at temperature T_1 and completes its solidification at temperature T_E. The two-phase region to the left of the eutectic point is composed of metal *A* and liquid. The two-phase region to the right of the eutectic point is made up of metal *B* and liquid. Once they reach the solidus, all these alloys will solidify as separate phases. Thus, the region below the solidus consists of a mixture (not a solution) of two solid phases. This eutectic reaction for the eutectic alloy, in our example 57% *A*–43% *B,* can be written in equation form as follows:

$$L \rightleftharpoons S_1 + S_2$$

The double arrow indicates reversibility. Cooling the liquid phase produces two solid phases (S_1, S_2). Heating the two solid phases produces a liquid (melt) (*L*). For the lowest melting composition of the system *AB,* the eutectic composition 57% *A*–43% *B,* its two solid phases will be completely distinguishable from each other when viewed under a metallurgical microscope. The parallel wavy-line pattern formed, as viewed by the microscope, is known as a lamellar type (Figure 3-66). It can be compared to the stripes of a zebra.

Referring to Figure 3-65, alloys to the left of the eutectic mixture are known as ***hypoeutectic*** mixtures. Those mixtures to the right of the eutectic are known as ***hypereutectic.*** Those alloys closer to the eutectic composition will contain more eutectic mixture in the solidified alloy. All phase diagrams drawn depicting the complete insolubility of two metals in the solid state have been labeled as hypothetical (see Figure 3-60). Most metals are soluble in each other to some degree. An alloy that comes close to being completely insoluble is aluminum silicon (AlSi); see Figure 3-67. Silicon is only slightly soluble in aluminum. Figure 3-67 is a sketch of the Al–Si phase diagram. This family of alloys forms the basis for a number of commercial alloys. Figure 3-68 is a photomicrograph of an Al–Si alloy.

Our interest now turns to the more realistic situation in which the metals are to some degree soluble in each other in the solid state. This situation is depicted in Figure 3-69. Notice that this diagram is basically the same as the hypothetical phase diagram, Figure 3-65, with the addition of points *Y* and *Z* and the lines extending from these points. Also note the similarities of Figures 3-67 and 3-69. Lines *Y*–*S* and *Z*–*T* are *solubility curves,* also known as ***solvus lines.*** Before learning more about these lines, note that there are three areas of single-phase solutions, three regions of two-phase solid solutions, and one eutectic mixture.

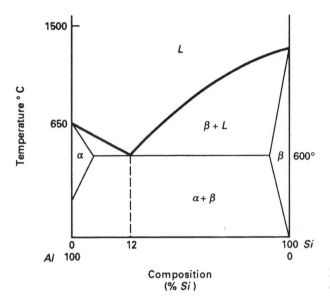

Figure 3-67 Aluminum–silicon phase diagram.

Figure 3-68 Microstructure of Al–Si alloy as cast ×250, unetched. (Buehler Ltd.)

The solid alpha (α) phase is composed mostly of metal *A* with some metal *B*. The solid beta (β) phase is rich in metal *B* along with some metal *A*. The α phase, which is called a terminal phase, differs from pure metal *A* by the presence of some metal *B*. Using a tie-line, the two-phase region between α phase and the liquid (L) phase is labeled α + *L*. The main solid phase, α + β, is determined in the same manner. Figure 3-69 shows the typical shape of a phase diagram where *A* and *B* are partially soluble in each other. As the temperature increases, the more solid (composition *B*) can be dissolved in the liquid. Note that the solvus lines end at points *Y* and *Z* in Figure 3-69. Point *Y* is the limit of solubility of metal *B* and solid solution. At the lowest temperature the degree of solubility is a minimum (98% *A*–2% *B*); that is, only 2% *B* is soluble in *A*, but as the temperature increases this degree of solubility of *B* in *A* increases to a maximum limit (75% *A*–25% *B*).

Now it is time to follow a particular alloy (65% *A*–35% *B*) and explain briefly the phase transformations that take place as it is cooled from somewhere in the liquid phase. At point *l* (Figure 3-69) on vertical line *X–W* and temperature T_l, the liquidus line is reached. Solid solution α is beginning to form. As the alloy cools, this solid solution becomes richer in metal *B* until the solidus is reached at point *P* and temperature T_E. The last liquid to solidify has a composition of 50% *A*–50% *B*. Assuming that diffusion has kept pace with nucleation and growth of crystals, the solid phase has a composition of 75% *A*–25% *B*. Applying the lever rule, we could determine the relative amounts of these two phases present. Note that the last liquid to

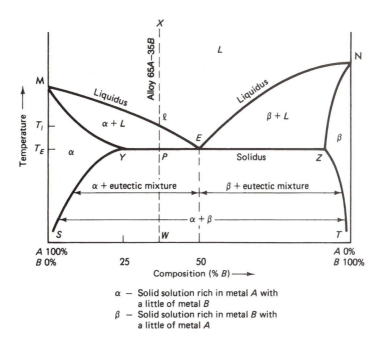

Figure 3-69   Typical eutectic-type phase diagram: metals *A* and *B* completely miscible in the liquid phase and partially miscible in the solid phase.

solidify has the eutectic composition and therefore begins to solidify and form a eutectic mixture consisting of alternate layers of crystal of α phase and crystals of β phase. At this point we are just below temperature T_E. Here is where the solvus lines come into the action. If you were to draw in the tie-line for this temperature, it would be pretty close to the line *Y–Z*. As the temperature continues to lower, the tie-lines that you might draw get longer and longer. Add to this observation the fact that the solvus line *S–Y* shows the maximum solubility of metal *B* in metal *A* at various temperatures, and that this solubility decreases as the temperature is reduced to room temperature. Metal *B* is in solution in the α solid phase. Therefore, the excess β phase must precipitate out of solution. At room temperature our alloy will consist of α phase with some excess β phase precipitated within it, plus the eutectic mixture. This eutectic mixture is also made up of both α and β phases.

3.5.2.1 Intermediate phases.

The phase diagrams discussed up to this point are binary alloy phase diagrams, such as copper-nickel, or binary eutectic phase diagrams similar to Figure 3-69, which could easily represent a lead–tin alloy system. In both these types of phase diagrams, there are only two solid phases α and β, called terminal solid solutions, that exist over a range of compositions near the extremities of the diagram. Other alloy systems contain ***intermediate solid solutions*** or ***intermediate phases.*** These phases are formed at other than the two extremes in composition. The copper–zinc system (not shown) has, in addition to its two terminal phases, four intermediate phases or intermediate solid solutions. None of these four phases extends to the extremities of the phase diagram.

In some solid systems discrete intermediate compounds are formed. They are composed of a metal and a nonmetal that have distinct chemical formulas. Intermediate compounds formed in metallic phase diagrams are known as ***intermetallic compounds*** or ***intermetallics.*** They have a narrow range of composition (see mullite in Figure 7-12), while some have a fixed composition that is represented on a phase diagram as a single vertical line (see alloy *AB* in Figures 3-70 and 3-71). Their atomic bonding can be either mixed ionic/metallic or mixed covalent/metallic. Their properties include brittleness coupled with good high-temperature resistance and a high strength-to-weight ratio. Consequently, intermetallics Ni_3Al and Ti_3Al are finding applications in the aerospace industry. Aluminum casting alloys often contain

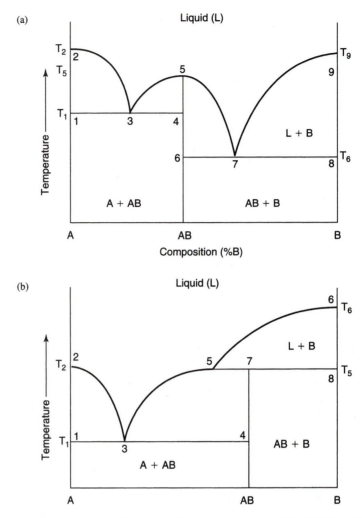

Figure 3-70 (a) Binary phase diagram with an intermediate, congruent intermetallic (*AB*). *Note:* Points 3 and 7 are eutectic points. (b) Binary phase diagram with an intermediate, incongruent intermetallic (*AB*). *Note: AB* never reaches the liquidus. Point 5 is a peritectic point.

intermetallic phases between grains which are prone to form because of the low solubility of the many alloying elements in aluminum. Refer to the phase diagram for mullite (Figure 7-12) and paragraph 8.2.3 for use of intermetallics as matrix materials for metal matrix composites.

These compounds follow the rules set forth in *stoichiometry* (pronounced "stoy-key-om-eh-tree"). Intermetallics are stoichiometric compounds with characteristic metal atom ratios such as CuZn, AlLi, Ni_3Al, Al_3V, AlSb, Mg_2S, and Ti_3Al. Stoichiometry is the calculation of the quantities of chemical elements or compounds involved in chemical reactions. Chemical equations balance the starting materials (reactants) needed to produce a certain amount of product. In addition, the cost of the starting materials determines the cost of the final product. Stoichiometry allows one to calculate the amount of starting materials necessary to obtain the correct amount of product. In ceramic materials, stoichiometry is used to define the ionic state for ionic compounds, wherein there is an exact ratio of cations to anions as predicated by the chemical formula. A ceramic is nonstoichiometric if there is any deviation from this exact ratio. Using calcium chloride as an example, each calcium ion (cation) has a +2 charge (Ca^{2+}) and associated with each chlorine ion (anion) is a single negative charge Cl^{-1}. Thus, there must be twice as many Cl^{-1} anions as there are Ca^{2+} cations in the compound with the formula for

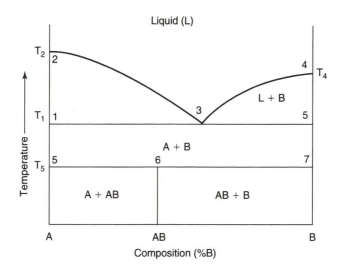

Figure 3-71   Binary phase diagram with an intermediate, dissociating intermetallic stable only over a limited temperature range. *Note: T_5 represents the upper limit of stability upon heating.*

calcium chloride, namely, $CaCl_2$. The positive–negative attraction is what holds the ions together. In this compound a Ca atom gives up its two outermost electrons, transferring one electron to each chlorine atom. The calcium atoms become the positively charge Ca^{2+} ions (cations). Each chlorine atom has an extra electron, producing a negatively charged chlorine, Cl^{-1}, ion (anion). In other words, a stoichiometric ceramic is one in which all crystallographic lattice positions are filled according to the normal chemical formula. The chemical formula of a compound indicates the ratio of cations to anions, or the composition that results in the balancing of the charges.

   In general, intermetallics possess strong ionic/covalent bonding, which gives them nonmetallic properties such as poor ductility and electrical conductivity. An intermetallic might contain but a single component. Therefore it is represented, on a phase diagram, as a vertical line and labeled by a chemical formula such as NiAl or MgSn.

   Congruent phase changes occur when one phase changes into another phase at constant temperature (isothermal) and without any change in chemical composition. Pure metals and some intermediate compounds change phase congruently. In Figure 3-70a, the congruently melting compound is a solid from room temperature up to its melting point (T_5). In other words, a congruent intermetallic melts immediately once it reaches the liquidus line. An ***incongruent*** melting compound (Figure 3-70b) would not melt directly but would decompose into a liquid plus one of the pure metals. In ceramics some compounds are not stable through the complete temperature range. In other words, the compound may form only over a limited temperature range (Figure 3-71). The word ***dissociating*** is used to describe these compounds. Both congruent and incongruent compounds are formed in ceramics.

   One last solid-state reaction will be discussed. The ***eutectoid*** reaction takes place within the solid state. The suffix ***oid*** means "resembling" or "like." Therefore, the eutectoid reaction is like the eutectic reaction. As you recall, the eutectic reaction involves the transformation of a liquid phase into two solid phases ($L \rightleftharpoons S_1 + S_2$). The eutectoid reaction then is the transformation of a solid phase (S_1) into two new solid phases (S_2 and S_3). An appropriate equation describing this reaction is

$$S_1 \rightleftharpoons S_2 + S_3$$

Figure 3-72 shows the many similarities of this type of reaction to the eutectic reaction (see Figure 3-69). In industry the eutectoid reaction is most important because it forms the basis for

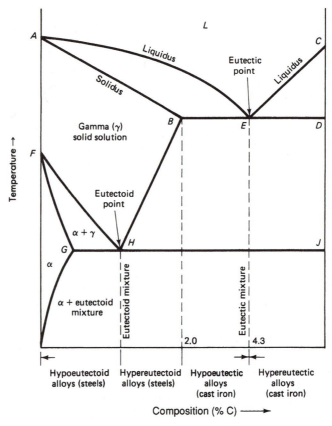

Figure 3-72   Phase diagram showing a eutectoid reaction.

heat treatment of many metallic materials. Point *E* is the eutectic point. Point *H* is the eutectoid point located well below in the solid region of the diagram. As shown, the gamma solid solution at point *H* transforms into the eutectoid mixture, made up of two solid phases. The prefixes ***hypo-*** and ***hyper-*** are attached to particular compositions of metals on either side of the eutectoid mixture in a fashion similar to those compositions related to the eutectic composition. This very crude diagram will be refined and studied in more detail in Module 5. At that time the diagram will be called the iron–iron carbide phase diagram—the heat treater's main road map to the processing of steel and cast iron.

   3.5.2.2 Multiphase metal alloys.   So far our discussion has involved solute and solvent atoms that have the same type of crystal structure and thus form relatively pure ***homogeneous*** mixtures or solutions. Now we advance to alloys that are classified as mixtures of more than one phase that differ from each other in their composition or structure. Under the microscope, the boundaries between phases can be seen in some metal alloys (Figures 3-11 and 3-73). Solder is an example of this type of multiphase metal alloy, with the metals tin and lead present in two separate phases. Steel is a more complex multiphase mixture of different phases, some of which are solid solutions. There are many more ***multiphase metal alloys*** than single-phase solid solution alloys, due primarily to the greater flexibility of their properties, with more variables to control to produce differing properties. In other words, the properties of multiphase metal alloys such as steel depend not only on the phases present, but on the structure, amount, shape, distribution, and orientation of these phases (Figure 3-73). The ability to control these phases allows us to produce desired properties in metals. For example, low-carbon ductile steel is used on auto bodies, and harder and more brittle steel is used in metal files.

Figure 3-73 Microstructure of a multiphase metal alloy. Two different compositions (A and $A + B$) in the solid state. 65 Al–Cu alloy as cast $\times 250$, $Fe(NO_3)_3 \cdot 9H_2O$. (Buehler Ltd.)

3.5.2.3 Phase or structural transformations.   Just how can we bring about these transformations to produce a metal alloy with desirable properties? First, let's consider the atomic world once more. We know that a few metal atoms can change their structural arrangement as a result of a change in temperature. Remember the name of this phenomenon? *Allotropy.*

We know that there is a definite limit of solubility of one solid material in another and that this limit also depends on temperature. If this limit is reached, we see atoms coming out of solution to form a multiphase metal alloy. Remember the name of this phenomenon? *Precipitation.*

We know that atoms can move about in a solid just as cigarette smoke moves in a room, but certainly not as fast. Remember this phenomenon? *Diffusion.* And the rate at which the atoms move varies with the temperature. The atoms move at faster rates because they have more energy. What energy? *Thermal energy.* Where does this energy come from? From an external source of heat. The more heat, the more the energy and the greater the movement. By now you should have a pretty good idea about the answer to the next question. In looking over the various ways that we can change atomic structure in a metal alloy, what is the one common quality or characteristic in all of them? Yes, it is thermal energy.

3.6 SINGLE-CRYSTAL CASTINGS

Ordinary metal castings produce polycrystalline structures. These many crystals or grains of random orientation possess weak atomic bonds across grain boundaries since impurities gather at boundaries. Techniques to improve across-boundary strength have included improved mold material, alloying and producing refined **equiaxed** castings. Then casting technology worked to control the direction of grain boundaries and finally eliminate them for certain applications, as discussed in the Pause & Ponder of this module. **Directional solidification** (Figure 3-74) produces long grains growing continuously and parallel from one end of a casting to the other, with the longitudinal growth designed to be in the direction to withstand greatest stress. The next step involved selection of one grain and control of its solidification into a **single-crystal** casting (**monocrystal,** Figure 3-74). The monocrystal eliminates the need for boundary strengtheners, which depress melting points and limit high-temperature use; allows high treatment temperatures; and improves high-temperature corrosion resistance. The monocrystal turbine blade seen in Figure 3-74 represents the growing acceptance for high-temperature use of single-crystal castings. Space-shuttle turbine blades for the main engines were originally cast as directional solidified and then as single crystal.

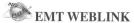 **EMT WEBLINK**

Materials Science and Technology Gateway Website with links to many other sites— *http://MST-Online.nsu.edu*

Figure 3-74 Photograph of metal castings of turbine engine airfoils showing the progress in grain control from equiaxed (right), to directionally solidified (left), to mono crystal (center). (Howmet Turbine Components Corporation)

3.7 STRUCTURES RESULTING FROM NONEQUILIBRIUM CONDITIONS

A new breed of material, Metglas, developed in the late 1970s by Allied Corporation, is an amorphous metal alloy. It is produced by bringing a molten metal alloy at about 1000°F into contact with rapidly moving and relatively cool substrate (or chill block) in a continuous casting process. The drastic quenching operation is similar to that used in conventional solid-state thermal processing of metals such as steel to transform austenite to martensite. It differs because the cooling is so rapid that for most metallic glasses the minimum rate is 10^5 kelvin/second (K/s). To ensure a uniform cooling rate at all points within the metal, one dimension of the liquid layer must be kept to a minimum. Thus, foil, wire, or powder are the forms currently produced. The magnitude of the cooling rate (about 1 million degrees per second) allows but 1 millisecond (ms) for the metal to solidify. The end result is that atomic diffusion is prevented, which in turn precludes any nucleation and growth of crystals. A glassy state consisting of one chemically homogeneous phase, with atoms packed in a random arrangement similar to that of glass or liquid metal, is produced. Alloy compositions consist of transition elements such as Fe and Ni with small percentages of metalloids (B, C, Si, P). Also, mixtures of transition elements such as B and Ni are common. Initial applications of this material are in vacuum furnace brazing of engine parts for turbines in which the older, less desirable brazing transfer tape is replaced by Metglas brazing foils.

The ductility of Metglas foil is greater than that of transfer tape. Metglas foil is capable of being folded back on itself without fracture. This permits it to be converted to various preforms, as well as punched to exact shape to conform with various brazing-joint designs. Many new techniques have been developed since the introduction of Metglas. Known by the name **rapid-solidification processing,** these techniques employ large departures from local equilibrium in the formation of homogeneous structures. Powders for rapid solidification are produced mainly by *inert gas atomization* and *centrifugal atomization.* Both processes produce spherical particles in the size range 20 to 100 μm in diameter by employing cooling rates up to 10^6 K/s. Other processes are *melt spinning* and *self-quenching,* which produce ribbons/filaments that are pulverized into gritlike powders (Figure 3-75). Powder consolidation can be accomplished by *hot isostatic processing (HIP), hot extrusion,* or *dynamic compaction.*

The materials produced by rapid solidification are finding applications in many areas, including as reinforcing filaments in ceramic-matrix composites and as surface-modification materials (surface alloying) for hard facing of tools, dies, valve seats, and turbine blade tips.

(a)

(b)

Figure 3-75   (a) A jet of molten metal is solidified rapidly when it strikes a liquid-cooled rotating drum. (b) This striking photo shows a melt-spinning device at the National Institute of Standards and Technology. Glowing metal, melted by coils at top right, hits a whirling wheel and flies off as rapidly solidified ribbon. Time photography produced multiple images. (Photo courtesy of NIST.)

By adjusting the cooling rate, a variety of new materials are produced with highly desirable mixes of properties. Amorphous magnetic materials with atomic structures that impose no obstacles to the movement of magnetic domain boundaries, low hysteresis losses, and high permeability (see Section 9.3) have been developed for various electronic devices such as sensors and motors. New fine-grained homogeneous crystalline materials (Al and Mg alloys) for use as tough tool steels and nickel-based superalloys are now available. This new technology makes it possible to produce alloys that could never be produced with conventional ingot-casting techniques. The highly alloyed powders are sintered and pressed into bulk form. In some processes the powders are pressed into "green" compacts and sintered without ever coming into contact with the air (see Sections 5.10 and 7.3.1).

Ferrous amorphous metals are "soft" magnetic materials, meaning that they possess no preferred orientation of crystals and hence no easy direction of magnetization. In practical terms, these new materials have a vast potential for use as metal cores in electrical transformers in which billions of kilowatt-hours are lost in overcoming resistance to changes in the direction of magnetization of the presently used silicon metal alloy cores. Other uses for such a family of materials are being developed by a multitude of industrial concerns.

3.8 LABORATORY ACTIVITIES AND CLASSROOM DEMONSTRATIONS WITH PROCESSING AND STRUCTURE

The *Experiment in Materials Science, Engineering and Technology (EMSET) CD-ROM* provides numerous experiments and demonstrations related to processing and structure as result of a cooperative effort by scores of educators and materials specialist from industry. See Prentice-Hall's website for ordering information on the latest edition of *Experiments in Materials Science (CD-ROM), at www. prenhall.com.*

APPLICATIONS & ALTERNATIVES

In the early 1980s, using techniques such as rapid solidification, new structural arrangements of atoms in solids were discovered for the first time. One was thought to be impossible based

Figure 3-76 Quasicrystal (five-fold symmetry) of Al and Mg at 400×. (NIST)

 EMT WEBLINK

National Institute of Standards & Technology—*www.nist.gov*

on the existing laws of crystallography. Known as *quasicrystals,* whose atomic structure remains a mystery to this day, they have spurred the development of three theories, which will ultimately lead to a theory of quasicrystalline structure. The end result will be a theory that will give direction to materials scientists in the development of this new member of the subgroup of materials with a new ordered structure and far-reaching electrical properties. One method for producing these new materials is to melt together aluminum and manganese and subject them to rapid solidification. Such a shock treatment produces a variety of novel structures or phases (Figure 3-76). Quasicrystals do not form unit cells as do regular crystalline materials. Nor are they amorphous, like glass. Instead, they form an ordered structure somewhere between the two recognized structures.

Where will quasicrystals lead? What new material structure is waiting around the bend? Your education must never stop. You must continue to prepare for new applications and alternatives of materials technology.

SELF-ASSESSMENT

3-1. From the section Pause & Ponder, what goals are sought in using metal casting such as thixotropic magnesium injection molding?

3-2. What are the synthesis/process capabilities of contemporary materials technology at the various scales (macro, micro, nano, and atomic)?

3-3. Explain the difference between a metallurgical microscope and a metallograph.

3-4. In an x-ray diffraction analysis, the x-ray source used produces an x-ray with a wavelength of 1.32 $\times 10^{-10}$ m. The angle of maximum diffraction (θ) is found to be 10°40′ with n of 1. What is the interplanar distance between the atomic planes that produced this diffraction?

3-5. Why are electron microscopes operated within a vacuum?

3-6. Glass lenses are used to focus the beam of light in an ordinary optical microscope. What types of lenses are used to focus electron beams? Explain why such lenses affect an electron beam.

3-7. Define the word *spectroscopy.* Is it possible to have a spectrum of something other than sunlight? If so, name some spectra other than light spectra.

3-8. What are some of the new instruments that allow visualization beyond the capabilities of optical and electron microscopes?

3-9. Name an example of qualitative characterization and one of quantitative characterization.

3-10. What distinguishes a cooling curve for a metal from a cooling curve for a nonmetal?

3-11. What is the volume of a piece of iron having a mass of 864 g?

3-12. Under what conditions can atoms or ions form an fcc crystal structure in which each atom is in contact with 12 other atoms or ions?

3-13. What is the coordination number (CN) for the fcc structure formed by ions of sodium and chlorine producing the chemical compound NaCl?

3-14. Table salt and water both form crystals. What, if any, is the difference in the manner in which these crystals are held together as a solid material?

3-15. Using a ball model (rubber or polystyrene balls), construct an fcc structure to show the closest-packed atom planes (potential slip planes). How many such planes are there?

3-16. A single crystal of zinc, like many other single-metal crystals, is fairly ductile. There is nothing to prevent slippage of planes of atoms over one another. Explain why some of these metals, like zinc, become brittle when they are in the polycrystalline state, although they are ductile in single-crystal form.

3-17. What pressure is normally used in constructing a phase diagram?

3-18. What line on a binary diagram marks the upper limit of the solid solution phase?

3-19. Can a ceramic material be represented by a phase diagram?

3-20. What are the Miller indices for the basal plane (base of the box) in Figure 3-26?

3-21. What is the crystal direction for a line from the origin passing through point *A* in Figure 3-34?

3-22. Consult a table of properties of materials and record the melting-point temperatures of several common materials. Explain why there are differences in such temperatures when these solids change from a solid to a liquid.

3-23. What holds the atoms (ions) together in a compound such as NaCl?

3-24. Compare the ionic radii of negative ions with the radii of positive ions for the same elements. Tin, sulfur, or lead would be likely candidates. Explain the differences, if any.

3-25. The atomic radius of copper is 1.278×10^{-10} m. Express this radius in terms of nanometers (nm) and in inches (in.).

3-26. Whenever the density of a material is mentioned or recorded in a table, the temperature is also noted. What is the explanation for this?

3-27. What is the CN for a cph unit cell?

3-28. What is the maximum number of spheres that can surround and be in contact with a given sphere, provided that all of them are the same size?

3-29. Is there some relationship between the linear coefficient of thermal expansion for a solid and its melting point? List both the coefficient and the melting-point temperature for a metal, a ceramic, and a polymer and note the differences.

3-30. The highest atomic-packing factor is obtained when atoms are of the same size. What is its magnitude?

3-31. The ideal shear strength of a perfect crystalline structure of iron has been calculated to be around 10^{10} N/m^2. This theoretical strength has never been attained. "Whiskers" of iron grown in a near-vacuum in a laboratory approach this value. With the new capabilities provided by the NASA space program, discuss the future ramifications of being able to grow large, flawless crystals that are undistorted by their own weight as they form in space laboratory conditions.

3-32. Observe a piece of galvanized steel. Describe and sketch the zinc coating as you observe it with the naked eye.

3-33. In the cubic axes system, the measures of the angles α, β, and γ are how many degrees?

3-34. Sketch an fcc unit cell.

3-35. Using Table 3-2, determine the crystal structure of lead.

3-36. The rapid quenching of austenite in the heat treatment of steel produces a supersaturated and distorted crystal structure known as body-centered tetragonal (bct). Sketch the unit cell for such a structure.

3-37. What are the most likely directions for slip to occur in cph crystals? Sketch the unit cell.

3-38. Describe the solid state in terms of kinetics and energy.

3-39. Why is lead more ductile than tin?

3-40. What steps can be taken to reduce slip in metal crystals?

3-41. Obtain a rubber band. Stretch it as far as it will go without breaking while holding it in contact with your moistened lips. Observe the temperature change. Hold the band in this stretched condition for about 30 seconds. Note any change in temperature. Release the band suddenly to its original unstretched length and touch it to your lips. Note the temperature change. Rubber is an elastic polymer when stressed. The stressed molecules align themselves and local crystallization occurs. When the stress is released the molecules return to their original arrangement. Determine if energy is absorbed or released when an amorphous material crystallizes. Record all your observations and comments.

3-42. Mark the following statements as true (T) or false (F).

(a) Single-phase materials have a single-crystal structure.

(b) Single-crystal structured materials are single-phase materials.

3-43. What is the packing factor of pure iron when it is in **(a)** the bcc structure and **(b)** the fcc structure?

3-44. What happens to the atomic radius of pure iron as it is heated from room temperature (RT) to 910°C?

3-45. If the APF of a material increases when heated, what affect does this have on its volume?

3-46. Using Figure 3-35, sketch an isosceles right triangle described in the accompanying text, label its sides, and derive the equation $(r = R)^2 = 2R^2$.

3-47. Explain the difference between a short-range ordered material and a long-range ordered material. Give an example of each.

3-48. Diffusion of atoms through a solid takes place by two main mechanisms. One is diffusion through vacancies in the atomic structure. Describe another method of diffusion.

3-49. Name two possible substances formed by alloying two or more metals.

3-50. Would chromium atoms be a likely substitute for aluminum atoms when forming a substitutional solid solution? Ruby rods are doped with chromium atoms to convert the well-known ceramic (CrO_3 in Al_2O_3) into a laser material.

3-51. Disorder in the arrangement of atoms in solids is limited to the atomic level. Cite an example that refutes this statement.

3-52. Grain boundaries in crystalline solids play an important role in the movement of dislocations through a solid. Note at least three actions that take place in the grain boundary transition zone in a crystalline material.

3-53. List some examples of allotropic solids.

3-54. Metals are classified as crystalline materials. Name one metal that is an amorphous solid and name at least one recent application in which the use of it is saving energy or providing greater strength and/or corrosion resistance.

OBJECTIVE QUESTIONS

3-1. What type dendrites form in thixotropic metals?
    a. treelike   **b.** spheroidal   **c.** BCC   **d.** FCC

3-2. Ability of a material to exist in different space lattices.
    a. Allotropic   **b.** Crystalline   **c.** Solvent   **d.** Amorphous

3-3. Contemporary synthesizing/processing capabilities are best at which level of the materials technology scale?
    a. Atomic (0.1 nm)   **b.** Macro (cm—m)

3-4. Amorphous metals develop their microstructure as a result of _____.
    a. Dendrites   **b.** Directional solidification
    c. Slip   **d.** Extremely rapid cooling

3-5. In an alloy, the material that dissolves the alloying element is _____.
    a. Solute   **b.** Solvent   **c.** Matrix   **d.** Allotrope

3-6. Methods used to determine unique properties by identifying and analyzing (both quantitatively and qualitatively) features such as voids, roughness, grain size, coating, hardness?
    a. Characterization   **b.** Slip   **c.** Diffusion   **d.** Synthesis

3-7. The Pause & Ponder focused on changing structure through processing. Which processes produced the improved equiaxed grains?
    a. Investment casting   **b.** Shell molding
    c. Thixotropic molding   **d.** Injection molding

3-8. What advantage comes from equiaxed grains?
    a. Lower processing costs   **b.** Uniform strength
    c. Directional strength   **d.** Reduced density

Select the *best* match from the following techniques and instruments for questions 3-9 to 3-14.
    a. Spectroscopy
    b. X-ray diffractometer
    c. Optical microscope
    d. Electron microscope
    e. Scanning tunneling microscope
    f. Nanotechnology

3-9. Uses magnetic lenses to focus electron beams that provide over 200,000× magnification.

3-10. Provides magnification of about 2000× and is a relatively low-cost instrument.

3-11. A general term for techniques that utilize the differences in wavelength spectra produced by chemical elements.

3-12. A method to manipulate structures from hundreds to hundredths of a micrometer.

3-13. Used to measure the angles of electrons glancing off material specimens.

3-14. Uses very fine probes that move over the surface of a material and provide images of individual atoms. e.

3-15. Characterization technique involving visual clues to crafter or inspector.

    a. Nanoscale   **b.** Quantitative   **c.** Qualitative   **d.** Artificial

3-16. Characterization technique involving instruments and methods such as AI.

    a. Nanoscale   **b.** Quantitative   **c.** Qualitative   **d.** Artificial

3-17. What distinguishes a cooling curve for a metal from a cooling curve for a nonmetal?

    a. Phase transformations occur with nonmetals.

    b. No phase transformations occur with nonmetals.

3-18. What is the volume of a piece of iron having a mass of 864 g?

    a. 10.8 cm^3 **b.** 109.78 cm^3 **c.** 20.8 cm^3 **d.** 200.6 cm^3

3-19. What is the coordination number (CN) for the fcc structure formed by ions of sodium and chlorine producing the chemical compound NaCl?

    a. 6   **b.** 8   **c.** 14   **d.** 16

3-20. What pressure is normally used in constructing a phase diagram?

    a. 100 psi   **b.** Depends on material

    c. Ambient   **d.** Normal atmospheric pressure

3-21. What line on a binary diagram makes the upper limit of the solid solution phase?

    a. Liquidus   **b.** Eutectic   **c.** Eutectoid   **d.** Solidus

3-22. A ceramic material cannot be represented by a phase diagram.

    T (true)   **F** (false)

3-23. What are the Miller indices for the basal plane (base of the box) in Figure 3-26?

    a. (0001)   **b.** (001)   **c.** (111)   **d.** (101)

3-24. What is the crystal direction for a line from the origin passing through point *A* in Figure 3-34?

    a. [110]   **b.** [222]   **c.** [120]   **d.** [001]

3-25. What holds the atoms (ions) together in a compound such as NaCl are electrostatic forces between _____ .

    a. Atom and ion   **b.** Covalent bonds

    c. Electrons and nuclei   **d.** Neutrons

3-26. What is the CN for a cph unit cell?

    a. 5   **b.** 8   **c.** 12   **d.** 24

3-27. What is the magnitude of the highest atomic-packing factor when atoms are of the same size?

    a. 0.74   **b.** 0.81   **c.** 21   **d.** 81

3-28. How many directions of slip are there in cph systems?

    a. 3   **b.** 4   **c.** 5   **d.** 6

3-29. Single-phase materials have a single-crystal structure.

    T (true)   **F** (false)

3-30. Single-crystal structured materials are single-phase materials.

    T (true)   **F** (false)

3-31. Diffusion of atoms through a solid takes place by two main mechanisms. One is diffusion through vacancies in the atomic structure. Another method of diffusion is _____ .

    a. Cold   **b.** APF   **c.** Substitutional   **d.** Interstitial

3-32. Two possible substances formed by alloying two or more metals are solid solution and _____ .

    a. Substance   **b.** Interstitial   **c.** Mixture   **d.** Compound

3-33. Traditional ceramics have poor tensile strength due to what?

    a. Ionic bonding within grains

    b. Metallic bonding within grains

    c. Poor bonding across grain boundaries

    d. Good bonding across grain boundaries

3-34. Grain boundaries _____ movement of dislocations through a solid.

    a. Improve   **b.** Inhibit   **c.** Do not affect

3-35. Iron can be alloyed with carbon because it is _____ .

    a. Crystalline   **b.** Amorphous   **c.** A mixture   **d.** Allotropic

3-36. Metals can be cooled only to crystalline solids.

    T (true)   **F** (false)

3-37. Solid crystals that do not form unit cells as do regular crystals.

    a. Allotropes   **b.** Twins   **c.** Quasicrystals   **d.** Alloys

REFERENCES & RELATED READINGS

ASM International. "Advances in Characterization and Testing of Engineering Materials," special issue, *Advanced Materials and Processes,* November 1989.

ASM International. "Characterizing Thermal Spray Coatings," special issue, *Advanced Materials and Processes,* April 1992.

Davis, J. R., ed. *ASM Materials Engineering Dictionary.* Materials Park, Ohio: ASM International, 1993.

Drexler, K. Eric. *Nanosystems: Molecular Machinery, Manufacturing, and Computation.* New York: John Wiley & Sons, Inc., 1992.

Frederick, P. S., N. L. Bradley, and S. C. Erickson. "Injection Molding Magnesium Alloys," *Advanced Materials and Processes,* November 1988, pp. 53–56.

Fulrath, R. M., and J. A. Pask. *Ceramic Microstructures.* Melbourne, FL: R. E. Krieger, 1976.

Hall, Cecil E. *Introduction to Electron Microscopy.* Melbourne, FL: R. E. Krieger, 1983.

Ingersol Rand, *Compressed Air,* October/November 1994, p. 16.

Loveland, R. P. *Photomicrography: A Comprehensive Treatise,* Vols. 1 and 2. Melbourne, FL: R. E. Krieger, 1981.

McDonald, Alan F. "Scanning X Microscopy," *National Educators' Workshop, NEW: Update 89, Standard Experiments in Engineering Materials, Science, and Technology,* NASA Langley Research Center, Hampton, VA, October 19, 1989.

McLean, D. *Mechanical Properties of Metals.* Melbourne, FL: R. E. Krieger, 1977.

Mallardi, Joseph L. *From Teeth to Jet Engines.* Howmet Corporation, 1992.

Roy, Rustum. "New Materials: Fountainhead for New Technologies and New Science," *NEW: Update 92,* Oak Ridge National Laboratory, Oak Ridge, TN, November 12, 1992.

SEMTC. *Navy Manufacturing Letter,* February 1995, p. 6.

Stephens, Peter W., and Alan I. Goldman. "The Structure of Quasicrystals," *Scientific American,* April 1991, pp. 44–53.

Wickramasinghe, H. Kumar. "Scanned-Probe Microscopes," *Scientific American,* October 1989, pp. 98–105.

Periodicals

Advanced Materials & Processes

Journal of Materials Education

Module

4

Properties, Degradation, Failure, and Selection of Materials

After studying this module, you should be able to do the following:

4–1. Use EMT Weblinks to Internet sites for the most current developments related to properties and materials selection.

4–2. Use diagrams, explanations, and calculations to determine physical, mechanical, chemical, and thermal properties of materials.

4–3. Determine the names and types of standard mechanical tests for the properties they measure.

4–4. Explain the types and uses of industrial standards in design and manufacturing, and the role of technical societies and organizations involved in developing standards for engineering materials.

4–5. Analyze service conditions (stresses and environments) to determine properties critical to the ability of parts or products to survive the conditions.

4–6. Explain reasons for the degradation and failure of materials and the role of failure analysis, factors of safety, post-process thermal treatments, and corrosion protection.

4–7. Recall purposes and methods for nondestructive testing, inspection, quality control, and quality assurance.

4–8. Relate properties of materials and the effect of cost in materials selection criteria.

*EMT Weblinks listed throughout the book will link you to Internet sites related to topics being covered. The dynamic nature of the Web brings frequent changes, so some of these sites, while available at the time of writing, may not be up now.

PAUSE & PONDER

How does one tell the difference between the materials that are used to produce the many products we use in our daily lives? The kitchen table, the fishing rod, the automobile tire, or the roof of the house in which we live—each material has a list of identifying features by which it can be identified. These features are known as properties or characteristics. Producing a myriad of materials with the many different properties required by society demands constant application of new science and technology. Many, if not most, of the new materials that are "engineered," as described in Module 1, take on their new properties as a result of the application of heat or cold which, in turn, changes in the materials' atomic structure.

Advances in technology provide new tools to materials engineering and manufacturing to improve the confidence with which to predict the limits of failure or product reliability of their products. Failure to do so in this world economy means a loss of market for one's products and consequent failure in the marketplace.

An example of adoptions of new technology or the use of technologies not previously applied is the case with cryogenics. *Cryogenics* is the science that deals with the production of very low temperatures and their effects on the properties of materials. Various objects are being treated cryogenically, for example, golf balls, brass instruments, cutting tools, firearms, and electronic circuit boards. These materials are subjected to a temperature of –320°F which is the temperature of liquid nitrogen. This treatment changes the crystal (atomic) structure such that it enhances the performance of the material. Cryogenics has seen limited use with engineering materials for nearly 20 years and its use with all types of materials is ongoing. Cryogenic surgery, or cryosurgery, uses the same cold temperature of liquid nitrogen but its purpose is to selectively destroy human tissue by freezing it. Other uses of these cold temperatures are whole-body freezing and storage of human cells as a future source of new tissues or organs.

Figure 4-1a shows the application of cryogenic processing to engine parts for race cars. As seen in the drawing of the car, cryogenic treatments have resulted in reduced wear in such components as bearings, brakes, camshafts, transmissions, and differentials. Cylinder blocks do not warp, piston rings seal better, piston rods exhibit less flexing, and brake rotors better resist cracking. The photos 4-1b show some of the wear-resistant and toughened parts.

This module provides an introduction to properties of materials and the environments these properties must counteract in order to meet their service needs.

 EMT WEBLINK

American Society for Testing and Materials—*http://www.astm.org/*

4.1 PROPERTIES AND ENVIRONMENTAL VARIABLES

The properties of a material are those characteristics that help modify and distinguish one material from another. Taken as a whole, these qualities define a material. All properties are observable and most can be measured quantitatively. Properties are classified into two main groups, physical and chemical properties. *Physical properties* involve no change in the composition of the material. Density, strength, and hardness are examples of such properties. *Chemical properties* are associated with the transformation of one material into another. Iron rusts when it combines to produce an iron oxide through a chemical reaction. Physical properties are, in turn, arbitrarily subdivided into many categories. These subdivisions bear names such as *mechanical, metallurgical, fabrication, general, magnetic, electrical, thermal, optical, ther-*

Camshaft
Cryogenics: Reduce wear on lobes, bearings, reduce breakage.
Coatings: reduce wear on lobes, bearings.

Valve Train
Cryogenics: Valve Springs last longer, lose less spring force. Reduced rocker arm breakage, longer valve life, less seat erosion, less stem and insert wear. Reduce pushrod flex.
Coatings: Reduce temperature of valves and springs.

Cylinder Heads
Cryogenics: Reduce warping, increase resistance to detonation damage.
Coatings: Higher thermal efficiency, less heat loss.

Transmission-Differential
Cryogenics: Reduce wear and breakage on bearings, gears, shift forks, synchronizers.
Coatings: Reduce wear on gears.

Brakes
Cryogenics: Reduce wear on rotors and pads. Reduce rotor warping and cracking.
Coatings: Dissipate heat faster. Reduce heat absorption by calipers to keep brake fluid from boiling.

Engine Block/Crankshaft
Cryogenics: Reduce cylinder wear. Reduce block shift under racing loads. Reduce crank wear and breakage.
Coatings: Reduce wear on crank journals.

Wheel Bearings
Cryogenics: Reduced bearing wear.

Connecting Rods
Cryogenics: Reduce rod breakage.

Pistons
Cryogenics; Less skirt wear, better ring sealing more resistance to detonation damage. Reduced wrist pin wear. Reduce distortion.
Coatings: Heat reflecting coating reduces crown temperature, reduces detonation. Lubrication coatings reduce piston skirt scuffing.

(a)

(b)

Figure 4-1 (a) Cryogenic processing coupled with coatings of engine parts in race cars has increased the life of high-performance race cars three to five times. (b) Improved wear and tougher engine components allow them to endure increased stresses. (Courtesy Controlled Thermal Processing, Inc.)

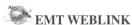 **EMT WEBLINK**

Materials Science and Technology Gateway Website—*http://MST-Online.nsu.edu*

monuclear, and ***electro-optical.*** Regardless of the name of the subdivision, physical properties result from the response of the materials to some ***environmental variable,*** such as a mechanical force, a temperature change, or an electromagnetic field (which includes all the radiation in the electromagnetic radiation spectrum). For purposes of this book, physical properties will be divided primarily into mechanical, thermal, electrical, magnetic, and optical.

4.2 MECHANICAL PROPERTIES

In selecting a material for a product such as a piston in an internal combustion engine, a designer is interested in properties such as strength, ductility, hardness, or fatigue strength. These are some mechanical properties of a material. ***Mechanical properties*** are defined as a measure of a material's ability to carry or resist mechanical forces or stresses. Additionally, mechanical properties of a material affect how the material can be worked. For example, very low hardness can be detrimental to a metal's machining performance. High ductility causes a metal to tear as the chip is formed, resulting in a poor finish. Mechanical properties rank highest in importance when selecting materials for many applications. Terms such as durability, formability, rigidity, strength, and toughness are often used to describe these properties. However, three tests supply the most useful information for most applications—the tensile, impact, and hardness tests. Durability can be assessed in part by the hardness test. Strength can be described by the results of tensile testing, and impact testing provides adequate data for determining the rigidity of a material. When any matter is at rest, the atomic or molecular structure is in equilibrium. The bonding forces in this structure resist any attempt to disrupt this equilibrium. One such attempt may be an external force or load.

Stress (to be defined) results from ***forces*** such as tension, compression, or shear that pull, push, twist, cut, or in some way deform or change the shape of a piece of material. Often, this deformation is so minute that only delicate instruments can detect it. Figure 4-2 shows a universal testing machine used to apply loads to material specimens and, with appropriate instrumentation, to detect minute deformation of materials under load.

Fracture or ***rupture*** occurs when a material or structure separates into two or more pieces. Most often this would be termed a failure of the material. ***Failure*** of material is defined as the inability of a material, structure, or component to perform its designed function. The failure criterion used in design is different for different materials. For concrete, the criterion of failure is crushing strength or compression. For elastomers, the failure criterion is the tear strength. For composite materials it is tensile strength. For ductile metals, in particular, the design criterion is yield strength. Metal strength is affected by many factors such as the amount of cold working and strain hardening. When work is done on metals its strength is directly affected, producing a range of values for the metal's yield strength. Metal in an annealed condition would exhibit a low value while a strain hardened metal would be approaching its tensile strength. Depending upon the design criteria, a failure may occur prior to fracture. Beams that are designed not to deflect beyond a standard amount under normal loads such as floor beams may be termed a failure if they exceed this stated amount. In this case no fracture has occurred. An aluminum desk chair that collapses and buckles is a failure without fracturing. The standard tensile test, the source of nearly a dozen important properties of engineering materials, produces experimental data needed to design structures and machines. The data are obtained by fracturing a specimen of material intended for a specific design project.

4.2.1 Stress

Stress is defined as the resistance offered by a material to external forces or loads. It is measured in terms of the force exerted per area [pounds per square inch (psi)]. Normal stress is that applied perpendicular to the surface to which it is applied, i.e., tension or compression. The corresponding SI units are newtons per square meter (N/m^2) or pascal (Pa). One pascal (1 Pa) equals 1 N/m^2. Another way of defining it is to say that stress is the amount of force (F) di-

(a)

(b)   (c)

Figure 4-2   (a) Universal testing machine. The unloaded grippers would hold
the specimen and strain indicators seen in (b) and (c). (b) Before stressed and
(c) after breaking. (Tinius Olsen Testing Machine Co.) (d) Photoelastic model
of wrench for analysis of stress. (Measurement Group Inc., Raleigh, NC)

(d)

vided by the area (A) over which it acts. Using σ (the Greek letter sigma) as the universal symbol for normal stress, we say mathematically that

$$\sigma = \frac{F}{A}$$

An assumption is made that the stress is the same on each particle of area making up the total
area (A). If this is so, the stress is uniformly distributed. When a load or force is applied to an
object, we are unable to measure the stress produced by this force in the material. What we do

$\Delta\ell = \delta_\ell = \ell - \ell_0$ Total axial deformation
$\Delta d = \delta_d = d - d_0$ Total lateral deformation

Note: Deformations exaggerated for illustrative purposes

(a)

E — gripped ends, may be threaded, plain, or with
       hole for gripping by machine
W — reduced width to insure specimen breaks in
       middle — round on round specimens and flat
       on flat specimens
G — marked gage length to precisely measure the
       change in length before, during and after test
F — fillet to reduce stress concentrations

(b)

Figure 4-3   (a) Rod under a tensile load. (b) Standard tensile test specimen or gage.

is measure the force, identify the area over which the force acts, and measure it as well. These two quantities can then be used to calculate the stress produced in the material by the previous relationship. With the use of polarized light and models made of photoelastic plastic, it is possible to detect concentrations of stress, as seen in Figure 4-2d.

Illustrative Problem

Calculate the stress on a 10-ft circular rod with a diameter of 1 in. that is in tension due to a pulling force of 50 lb. Sketch in two dimensions the rod showing the force and the area over which the stress is acting (to resist the external force). Express your answer in pounds per square inch (psi).

Solution

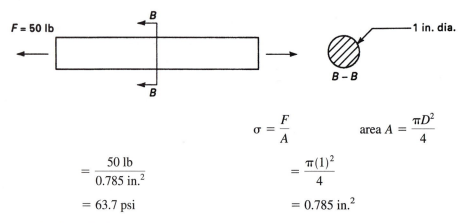

$$\sigma = \frac{F}{A} \qquad\qquad \text{area } A = \frac{\pi D^2}{4}$$

$$= \frac{50 \text{ lb}}{0.785 \text{ in.}^2} \qquad\qquad = \frac{\pi(1)^2}{4}$$

$$= 63.7 \text{ psi} \qquad\qquad = 0.785 \text{ in.}^2$$

4.2.2 Strain (Unit Deformation)

Strain, or *unit deformation,* is defined as the unit change in the size or shape of material as a result of force on the material. Many times we assume that a solid body is rigid; that is, when the body is loaded with some force, the body keeps its same size and shape. This is far from correct. Regardless of how small the force, a body will alter its shape when subjected to a force. In other words, the body will change its dimensions. The change in a physical dimension is called *deformation* (δ, Greek letter delta). Figure 4-3a illustrates a rod of original length (l_0) and original diameter (d_0) placed under a tensile load (F) and elongated (stretched). Its change in length (Δl) equals the difference in the two dimensions ($l - l_0$) in the direction of the force. In this example, Δl is exaggerated for illustrative purposes. This is known as total deformation, δ (Δ, Greek capital letter delta).

The relationship between the total deformation and the unit deformation can be expressed mathematically. Using the symbols δ for total deformation (or Δl), l_0 for Original length, and ϵ (Greek letter epsilon) for unit deformation, we can write the equation $\delta = \epsilon \times l_0$.

Note that the original diameter has been reduced in size and produces a corresponding change in the lateral direction (at right angles to the direction of the load). Note too that in one instance the dimension increased and the other decreased. The change in the length is called a **total axial** or **longitudinal deformation.** The change in the lateral dimension is known as a **total lateral deformation.** The ratio of the total axial deformation to the original length is known as the **unit axial** or **longitudinal strain,** ϵ. The linear units are not cancelled and are kept as part of the term. Therefore, in mathematical terms,

$$\epsilon_{\text{long.}} = \frac{\Delta l \text{ mm}}{l_0 \text{ mm}} = \frac{\delta \text{ mm}}{l_0 \text{ mm}} \quad \text{(longitudinal unit deformation)}$$

$$\epsilon_{\text{lat.}} = \frac{\Delta d \text{ mm}}{d_0 \text{ mm}} = \frac{\delta \text{ mm}}{d_0 \text{ mm}} \quad \text{(lateral unit deformation)}$$

In summary, when a piece of material (a body) is subjected to a load, it will not only deform in the direction of the load (axial deformation), but it will also deform in a lateral direction (at right angles to the direction of the tensile or compression load). The ratio of the lateral unit deformation or strain ($\epsilon_{\text{lat.}}$) to the unit longitudinal deformation or strain ($\epsilon_{\text{long.}}$), given the symbol μ (Greek letter mu), is known as **Poisson's ratio:**

$$\mu = \frac{\epsilon_{\text{lat.}}}{\epsilon_{\text{long.}}}$$

The unit longitudinal deformation is larger than the unit lateral deformation and therefore Poisson's ratio is less than 1. For steel, it is about 0.3.

In our discussion of deformation we demonstrated it by using a tensile load (Figure 4-3a): If a body was loaded in compression, the length of the body would have decreased and its width increased. One major reason for conducting tension and compression tests using standardized equipment and specimens or gages (see Figure 4-3b) is to determine the data needed to plot stress–strain diagrams for the material under investigation so that other meaningful relationships can be determined.

4.2.3 Stress–Strain Diagrams and Hooke's Law

The stress–strain diagram is used to determine how a certain material will react under load. Figure 4-4 is a stress–strain diagram for a low-carbon (mild) steel. Strain values (mm/mm) are plotted along the horizontal axis (abscissa), and along the vertical axis (ordinate) are plotted the stress values (MPa, megapascal).

The straight-line portion of the diagram up to almost the yield point is known as the **elastic region** of the diagram. Within this range of stresses, the material will return to its original dimensions once the load, hence, the nominal stress, has been removed. (Nominal stress is the load divided by the original cross-sectional area. Actual stress can vary due to changes in area caused by lateral deformation.) In the elastic region, each increase in stress will produce a proportionate increase in the strain. This statement is known as **Hooke's law** ($\sigma = E\epsilon$), with E the constant of proportionality (**elastic modulus**—the word *modulus* means a ratio or constant of proportionality). The elastic modulus, also known as Young's modulus, can be obtained graphically by measuring the tangent of the slope angle in the elastic region of the stress–strain curve (see Figure 4-4).

Young's modulus or elastic modulus ($E.$) is defined as the ratio of engineering stress (f) to engineering strain (ϵ) in the linear or elastic region of the stress–strain diagram. It is also a measure of the interatomic bonding forces in a material. The higher the magnitude of these bonding forces the stiffer the material or the higher the resistance of the materials to being

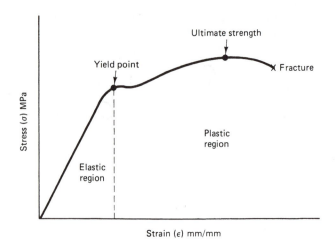

Figure 4-4 Engineering symbols for the sides of the triangle shown in the elastic region of the curve, 2 (theta) is the slope angle, E is the Modulus of elasticity or Young's Modulus and is computed:

$E = \dfrac{\Delta\sigma}{\Delta\epsilon} = \tan\theta$ and Δ is the symbol that means "a measurable change in." Stress–strain diagram.

deformed. When a material is loaded with external forces the material with the highest modulus value experiences the least amount of deformation or strain. This deformation may be elastic or plastic deformation. If the material reverts back to its normal size and shape upon removal of the load it is ***elastic deformation.*** If the applied force or load is removed and the material is permanently deformed (changed in shape), the material is said to have undergone ***plastic deformation.***

Illustrative Problem

Using the same rod as in the Illustrative Problem in Section 4.2.1, determine how much the rod has elongated by the tensile force of 50 lb. Assume that the modulus of elasticity for steel with which the rod is fabricated is 30×10^6 psi.

Solution   Using Hooke's law, $\sigma = E\epsilon$, and solving for ϵ, we find that

$$\epsilon = \frac{\sigma}{E} = \frac{63.5 \text{ psi}}{30 \times 10^6 \text{ psi}} = 2.12 \times 10^{-6} \text{ in./in.} \text{ and}$$

$$\delta = \epsilon \times l = 2.12 \times 10^{-6} \text{ in./in.} \times 10 \text{ ft}$$

$$= 2.12 \times 10^{-5} \text{ in./in.} \times 1 \text{ ft} = 212 \times 10^{-5} \text{ ft}$$

Beyond the yield point, the material will continue to deform, but with less stress than before, because the material has begun to yield. In this region, known as the ***plastic region,*** plastic deformation takes place and when the load is removed the material will not return to its original dimensions. It now has a ***permanent set.*** Note also that it takes less nominal stress to break the metal specimen than it does to reach the ultimate strength. Because the material has yielded, the original cross-sectional area of the specimen has been reduced in size so that less material is available to resist the load.

The *yield point* represents the dividing line or transition from the elastic to the plastic region of the curve. When the stress reaches the yield point, a large increase in strain occurs with little increase in stress. The ***modulus of elasticity, elastic modulus, tensile modulus, Young's modulus, modulus of elasticity in tension,*** or ***coefficient of elasticity,*** given the symbol E, is the ratio of the stress to the strain in the elastic region of the stress–strain diagram.* The ***tensile modulus*** is approximately equal to the ***compressive modulus of elasticity*** within the proportional limit (elastic limit of the diagram). Note that this ratio expresses the slope of the straight-line portion of the curve. Regardless of the name, this modulus is an indication of the

*The SI unit for E is the pascal (Pa).

Figure 4-5   The stiffness of metals versus metal composites. (NASA)

stiffness of the material when subjected to a tensile load. The stiffness of a material is defined as the ratio of the load to the deformation produced. The higher the value of Young's modulus, the stiffer the material, as demonstrated in Figure 4-5, reinforced metal composites possess greater stiffness than nonreinforced metals.

Not all materials produce stress–strain diagrams (Figure 4-6) on which there is a clear indication of the start of yielding as the load is increased. Cast iron is an example. This situation should not be interpreted to mean that cast iron does not exhibit elastic properties under moderate loads. In other words, such materials are elastic with strains returning to zero when moderate loads, hence stresses, are removed. The modulus of elasticity for these materials is sometimes taken as the slope of a tangent to the stress–strain curve at the origin.

4.2.4 Ultimate Strength or Tensile Strength

Ultimate strength or *tensile strength* is the maximum stress developed in a material during a tensile test. It is a good indicator of the presence of defects in the crystal structure of a metal material, but it is not used too much in design because considerable plastic deformation has occurred in reaching this stress. For brittle materials, tensile strength is still a valid criterion. Most grey cast irons are specified by their tensile strengths. Plastic deformation is not all bad. However, in many applications the amount of plastic deformation must be limited to much smaller values than that accompanying the maximum stress. The ultimate shear strength (Section 4.2.7) is about 75% of the ultimate tensile strength.

4.2.5 Yield Strength

Several face-centered cubic materials such as copper and aluminum do not have a well-defined yield point. *Yield strength* is the strain corresponding to the elastic limit. It is the lowest stress at which plastic deformation occurs. For most design purposes, the yield strength is assumed to be the same in tension as in compression. Strain is measured at various points in the engineering stress–strain diagram. The strain corresponding to the elastic limit is called the *yield*

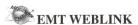 **EMT WEBLINK**

Photos & Animation-LAVA (Langley Animation & Video Archives)-*http://oea.larc. nasa.gov/Photo/photo.html*

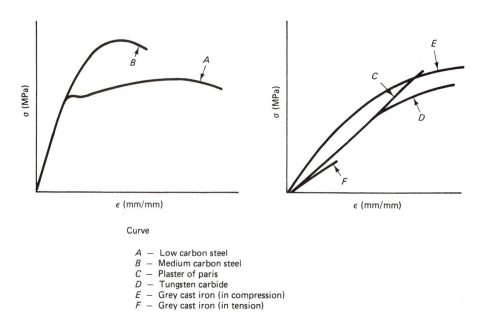

Curve

A — Low carbon steel
B — Medium carbon steel
C — Plaster of paris
D — Tungsten carbide
E — Grey cast iron (in compression)
F — Grey cast iron (in tension)

Figure 4-6 Typical stress–strain diagrams.

point strain (ϵ_{yp}). The elastic limit replaces the yield point in those metals that do not show a yield point on the stress–strain diagram. The stress at this point is the yield strength (F_{ys}). The strain corresponding to the tensile strength is called *uniform strain* (ϵ_u) because the strain to this point is uniformly distributed throughout the specimen or gage cross section. The engineering strain at the fracture point (ϵ_f) may also be used to express ductility of the metal specimen or sample. Low-carbon steel is one of just a few that exhibit a point where the strain increases without an accompanying increase in stress. This poses a problem in deciding when plastic deformation begins for such materials. By agreement, a practical approximation of the elastic limit, called the *offset yield strength,* is used. It is the stress at which a material exhibits a specified plastic strain. For most applications, a plastic strain of 0.002 in./in. can be tolerated, and the stress that produces this strain is the yield strength. This is sometimes expressed as 0.2% strain. The yield strength is determined by drawing a straight line, called the offset line, from the 0.2% strain value on the horizontal axis parallel to the straight-line portion of the stress–strain curve. The stress at which this offset line intersects the stress–strain curve is designated as the yield strength of the material at 0.2% offset. In some cases the offset can be specified as 0.1% or even 0.5%.

Figure 4-7 shows a typical stress–strain curve for an aluminum alloy with no pronounced yield point. The offset, offset line, and point of intersection of the offset line with the stress–strain curve are shown. When reporting yield strength, care must be taken to include the amount of offset as well as the value of the stress at the intersection with the data. In general, the yield strength of metals is much higher than that of other materials. For brittle materials, yield strength differs very little, if at all, from tensile strength. As an example, for class 40 gray iron, both strengths are 40,000 psi.

Ductile materials show a wide difference in these strengths. Figure 4-6 compares typical stress–strain curves for various materials. The yield strength in shear for a ductile material is determined to be about one-half (0.577) the yield strength in tension.

4.2.6 Resilience

The *modulus of resilience* (R), represented by the area under the straight-line portion (elastic region) of the stress–strain curve, is a measure of the energy per unit volume that the material

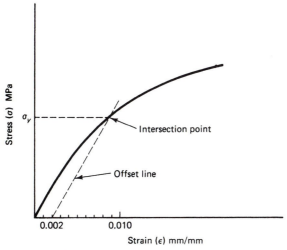

Note: For aluminum alloy 2014-T6 0.2% offset σ_y = 480 MPa.

Figure 4-7 Determining offset yield strength using a stress–strain diagram.

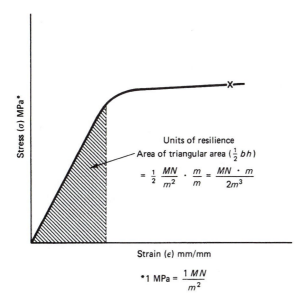

Strain (ϵ) mm/mm

*1 MPa = $\dfrac{1\ MN}{m^2}$

Figure 4-8 Resilience of a material as determined by a stress–strain diagram.

can absorb without plastic deformation. If the unit of volume is one meter cubed, the SI units of resilience are MN · m/m^3. Figure 4-8 shows that this area is a right triangle.

4.2.7 Shear Stress

A second family of stresses is known as *shear stress* or *shearing stress.* The symbol τ (Greek letter tau) represents a shear stress. A shearing force produces a shear stress in a material, which, in turn, results in a shearing deformation. Figure 4-9 shows a shearing load acting. In Figure 4-9a, the shear force F produces an angular deformation of a block of material and not a lengthening or shortening. This deformation is exaggerated for illustrative purposes. In Figure 4-9b, a block of material is subjected to a shear force that, if larger than the shear strength of the material, will shear a section out of the block. A hole in a metal plate can be produced by the action of a punch and hammer that delivers an impact blow, causing the metal in the plate to fracture by shear.

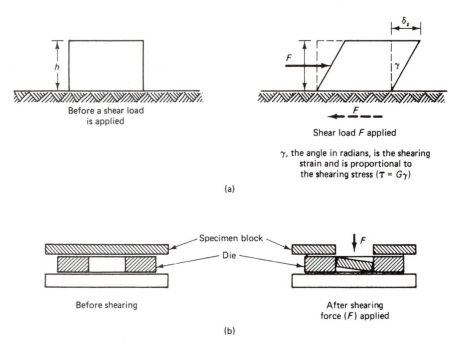

Figure 4-9 Shearing stress and strain. See Figure 4-10 for sheet metal shearing. (a) Before shear load. (b) After shear load.

Shear strain, γ (Greek letter gamma), then is the deformation (δ_s—read as "delta sub s") produced by the shear force F (see Figure 4-9a) divided by the dimension h, or

$$\gamma = \frac{\sigma_\sigma \text{ in.}}{h \text{ in.}}$$

Note that this ratio is also an expression for the tangent (ratio of the opposite side of an angle in a right triangle to the adjacent side of that angle) of the angle labeled γ in radians.

Illustrative Problem

A small pad subjected to a shearing force is deformed at the top of the pad 0.006 in. (see δ_s in Figure 4-9a). The height of the pad (h) is 2 in. What is the shearing strain on the pad?

Solution   Using $\gamma = \delta/h = 0.006$ in./2 in. = 0.003 radian.

Note: γ = is an angular measurement in radians or degrees (π radians = 180°).

A shear force acts parallel to the area over which it acts, producing a shearing stress (τ) and a shear deformation (δ). A torque wrench used to tighten a nut on a bolt produces a shear force on the bolt that acts along a circular path. The atoms along this path resist a tendency to slide past one another. The shear stress, called *torsion,* is discussed more thoroughly later in this module. (Figure 4-2d shows this stress in a model of a plastic wrench, and Figure 4-26 shows a failure produced by torsion.)

A stress–strain diagram can be plotted using shear stress and shear strain. Such a diagram will show a definite straight-line portion (elastic region) in which the shearing stress is directly proportional to the shearing strain. Like the normal stress–strain ratio, the ratio of the two shear quantities, G, is known as the *modulus of rigidity* or *shear modulus* of elasticity. In mathematical terms,

$$G = \frac{\tau}{\gamma}$$

with units of psi or pascals. Finally, we can state, as we did with normal stress and normal strain, that the two quantities can be set equal to each other (Hooke's law), and the preceding equation can be written $\tau = G\gamma$, where G, the constant of proportionality, is the shear modulus. In addition to these simple loads, more complex forces are often imposed on materials, producing corresponding stresses such as a bending force (see Figure 4-18) and a twisting force (see Figure 4-26).

Illustrative Problem

A piece of structural steel with a modulus of rigidity of 12×10^6 psi is subjected to a shearing stress of 15,000 psi. Find the strain on the steel.

Solution   Using Hooke's law and solving for shearing strain yields

$$\gamma = \frac{\tau}{G} = \frac{15{,}000 \text{ psi}}{12 \times 10^6 \text{ psi}} = 0.00125 \text{ rad}$$

4.2.8 Ductility

A material that can undergo large plastic deformation without fracture is called a **ductile material.** A **brittle material,** on the other hand, shows an absence of ductility. Consequently, a brittle material shows little evidence of forthcoming fracture by yielding as a ductile material would do. A brittle fracture is a sudden fracture. A ductile material, by yielding slightly, can relieve excess stress that would ultimately cause fracture. This yielding can be accomplished without any degradation of other strength properties. Figures 4-10a and b show stress–strain curves for both a ductile and a brittle material. Note the difference in the amount of plastic deformation shown by each curve prior to fracture.

Figures 4-10c and d show an application of the plastic deformation of a material undergoing a metal-forming operation called **deep drawing.** Flat sheets of metal (in this case, steel) are drawn and stretched by a punch and die set into the shape of a can. The metal is subjected to a variety of tensile and compressive forces that must be controlled to not exceed the strength of the metal. Other effects of this cold working are discussed in Module 5. If a failure of the metal occurs through tearing or excessive thinning of the can walls, the operation fails. Redrawing, resorted to when containers (shells) are too difficult to draw in one operation, produces longer sides in the container with reduced diameters. Greater can lengths can be produced by multiple drawing that involves drawing, annealing, and then redrawing to a greater depth. The term **annealing,** to be explained further in Module 5, is a heat-treating operation that, among other things, restores ductility to metals after they have been cold worked. Experience gained in deep drawing reveals that **cold-rolled (cold-worked)** sheets of metals with the desired grain orientation contain an **anisotropy** (material properties that differ in direction) conducive to successful drawing operations. In determining this anisotropy, tensile specimens are cut from cold-rolled sheets, subjected to an elongation of 15–20%, and strains are measured to determine the degree of anisotropic properties.

Ductility is measured in either of two ways. In the first method, the **percent elongation** is the ratio of the change in length of a specimen from zero stress to failure, compared to the original length; the quotient is then multiplied by 100%. In terms of a mathematical relationship, the factors above can be written as

$$\% \text{ elongation} = \frac{l_F - l_0 \times 100\%}{l_0} \quad \text{where } l_F = \text{length at fracture, } l_0 = \text{original length.}$$

Illustrative Problem

A test specimen with an extensometer clamped to it is undergoing a tensile test in a universal testing machine (Figure 4-2a–c). Its original length is 2.0 in. and its final length is 2.8 in. What is its percent elongation?

Figure 4-10 Stress–strain curves for (a) ductile and (b) brittle material. (c) Ductile and malleable metals allow complex forming. Photograph of the various stages in deep drawing a tin can. Beginning with a blank of 8.3 to 8.5 in. in diameter, a tin can is drawn with final dimensions of 3¾₆ in. diameter and 4⅜ in. tall. (American Can Co.)
(d) Diagram showing the stages required to produce a two-piece can. (American Can Co.)

Figure 4-11   Photographs of impact specimens of boron/aluminum fiber composites. (NASA)

Solution

$$\% \text{ elongation} = \frac{l_F - l_0}{l_0} \times 100\% = \frac{2.8 - 2.0}{2.0} \times 100\% = 40\%$$

The second method, ***percent reduction in area,*** measures the change in the cross-sectional area of a specimen, compares it to the original cross-sectional area, and multiplies the quotient obtained by 100%:

$$\% \text{ RA in area reduction} = \frac{A_0 - A_F \times 100\%}{A_0}$$

where A_0 = original cross-sectional area, A_F = cross-sectional area at fracture.

Both these measures of ductility do not take into account the nonuniform effect on elongation and area reduction due to necking.

It is customary to consider a specimen that has 5% or less elongation as a brittle material. Brittle materials should not be considered as having inferior strength. Such materials lack the ability to plastically deform under load. Figure 4-11 is a pair of photographs showing the effects of increased ductility on the impact resistance of a specimen of boron–aluminum composite material. Pure metals can undergo elongations of 40–60% without rupture, whereas typical superalloys, because of their different size atoms, have ductilities of less than 20%.

Illustrative Problem

A cylindrical metal specimen undergoes a tensile test. Its original diameter is 10 mm with a gage length of 50 mm. At the end of the test the diameter in the necked region is reduced by 4 mm. Calculate the percent reduction in area and the true strain at fracture.

Solution

Using $A = Bd^2/4$ Original cross-sectional area $(A_0) = Bd^2/4 = 78.54 \text{ mm}^2$

Final cross-sectional area $(A_F) = B(6)^2/4 = 22.21 \text{ mm}^2$

Using %RA $= A_0 - A_F/A_0$ %RA $= 56.33/78.54 \times 100\% = 71.7\%$

Using $\epsilon_f = \ln [100/(100–RA)] = \ln [100/23] = \ln 4.3478 = 1.469$

EMT WEBLINK

Materials Science and Engineering Laboratory-NIST—*http://www.msel.nist.gov*

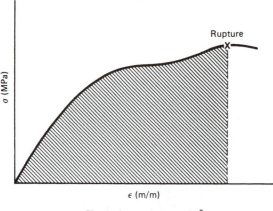

SI units for toughness — J/m^3

Figure 4-12 Modulus of toughness.

4.2.9 Toughness

The ability or capacity of a material to absorb energy during plastic deformation is known as *toughness.* The *modulus of toughness* (T), equal to the total area under the stress–strain curve up to the point of rupture (see Figure 4-12), represents the energy per unit volume of a material required to produce fracture under static conditions.

Toughness can also be expressed in terms of the ease or difficulty in propagating a crack. It can be measured by the amount of energy absorbed by a material in creating a unit area of crack. A tough material would have no defects in its microstructure (not normally attained). Some crack-stopping mechanisms, such as inserting fibers in a metal matrix composite, are used to increase its toughness (Figure 4-11). Having a good amount of ductility makes many materials tough because the plastic flow permitted by the ductility prevents a concentration of stress at the tip of a crack.

Impact is defined as a sudden application of a load confined to a localized area of a material. Exemplified by the striking of a material with a hammer, this relatively quick application of force, as opposed to a slow or static loading of a material, can cause considerable damage to a material that cannot adequately redistribute the stresses caused by the impact.

Ductile materials usually survive impact due to their microstructure, which allows slip to take place. Most metals have good toughness and thus have good impact resistance. Due to their inherent nature as compounds of metals and nonmetals, ceramics do not possess the ability to redistribute stresses and plastically deform. Consequently, they have poor toughness, poor impact resistance, and poor fracture toughness.

Impact tests are also used to give an indication of the relative toughness of a material. Notched-bar impact tests using either of the two standard notched specimens, the *Charpy* or the *Izod,* reveal the material's behavior in sustaining a shock load. A heavy pendulum swinging through its arc strikes and breaks the specimen. By recording the initial and final elevations of the pendulum the difference in potential energy can be calculated. This difference in potential energy is the impact energy absorbed by the specimen during failure. These notched specimens of the material under test have either a keyhole or a V-notch cut to specifications. Figure 4-13a shows a universal impact tester for conducting Charpy, Izod, and tension impact tests on materials. The Charpy impact test, developed by a French scientist in 1905, played a key role in understanding fracture problems of merchant ships during World War II when it was used to determine which hull plates were subject to fracture. Today it is used in many industries including the construction of bridges and pressure vessels. Figures 4-11 and 4-13b show impact failures. Impact testing is considered *dynamic testing,* in contrast to the slower, *static testing,* such as tensile or compression testing. Impact tests serve the building construction industry, which must be concerned with how storms will affect building materials (see Figure 4-14a).

In contrast to the relatively small effect of temperature on the strength and ductility of metals, temperature has a marked effect on the impact resistance of notched bars. The Izod test

(a)   (b)

Figure 4-13   (a) Impact testing machine. (Tinius Olsen Testing Machine Co.) (b) Impact fractures. (John Deere Co.)

is customarily used for nonmetallic materials. V-notched specimens are used to determine the resistance of the material to crack propagation. The impact energy absorbed by a specimen during failure is expressed in foot-pounds (ft lb) or joules (J), both of which are used to describe the toughness of a material. Figure 4-14b shows, in general form, the effects of temperature over a wide range on the impact resistance of metals. Below some critical temperature, a particular metal will experience a brittle failure with little absorption of energy. Cracks normally propagate (grow) most readily along specific crystallographic planes by *cleavage.* Another main path for cracking is an *intergranular* (along the grain boundaries) path particularly when the grain boundaries are weakened by inclusions or segregation. A cleavage fracture is readily recognized by its crystalline or "rock candy" appearance. Another common fracture form produced is the *chevron pattern,* which consists of several crack fronts propagating at different levels in the material from a common origin. A pattern of ridges fans away from the origin. A chevron pattern and its point of origin are visible with the naked eye. These examples illustrate *brittle fracture,* which occurs in high-strength metals or metals that have poor toughness and ductility. In many cases the cause of this type of fracture is impact as opposed to being overloaded. Also some materials experience an inability to absorb low-energy fractures at low temperatures (see this section for more details on toughness of materials and the transition temperature range of materials). Figure 4-15 is a photomicrograph of a cloverleaf deck socket that failed on a new class of U.S. Navy transport ships. These sockets are flame-cut from deck plating and then welded to the decks of such ships. They allow vehicles being transported by the ships to be securely tied to the deck to permit these ships to sail under varying sea conditions to destinations required by national security. Various cracked specimens were analyzed using SEM stereo section fractography (SSF), which uses a scanning electron microscope. This technique is used to carefully evaluate the causes of cleavage initiation (cracking to begin) in the material through simultaneous observation of both the fracture surface and the underlying microstructure. In this case the SSF technique successfully identified the dominant local features which triggered cleavage initiation. The ultimate cause of the failure of the sockets was a combination of overload failure and the transition from ductile to brittle fracture produced by high brittleness in the carbon-rich transformed martensite layer introduced by the cutting flame used in the manufacture of the socket. In addition, high residual stresses were induced in the socket as a consequence of the manufacturing processes. Milling and cutting were substituted for a coining operation and several steps were recommended to reduce the hard and brittle surface layers from the sockets. Further information related to brittle failures can be found in Section 4.3 and in Module 5, Applications & Alternatives. Above some critical temperature, the failures are ductile with a much greater absorption of energy. Between these temperatures there is a zone of temperatures in which these failures are mixed (see Figure 4-14b). In some

(a)

(b)

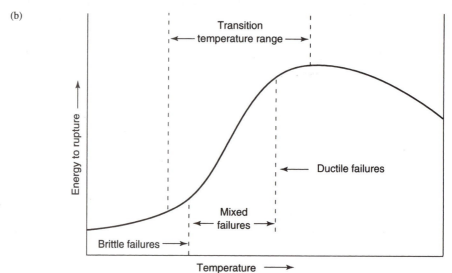

Figure 4-14   (a) Designers of the Hard Rock Cafe located at Universal Studios in Orlando, Florida, chose polycarbonate plastic sheet for the dramatic, covered arched walkway. To ensure that the plastic glazing would endure Florida's hurricanes, the complete window system underwent dynamic impact tests in which projectiles were fired from an air cannon. The plastic glazing passed local building codes for hurricane resistance and met Occupational Health and Safety Administration (OSHA) guidelines for fall-through protection for skylights and other overhead glazing applications. (G.E. Plastics) (b) Variation of temperature versus energy to rupture in impact testing of metals.

Figure 4-15 Top-Cross section of welded cloverleaf deck socket from transport ship. Bottom-Enlargement reveals crack in failed HAZ or heat affected zone (*Advance Materials & Processes*).

steels the ***transition temperature range*** may be abrupt, while with other steels it may extend over a considerable range. Coarse grain size, strain hardening, or the addition of certain elements tend to raise the transition range of temperature, whereas fine grain size, heat treatments, and the addition of certain elements (nickel) tend to enhance notch toughness even at rather low temperatures. Not all materials have a distinct transition temperature. Metals with an fcc crystal structure do not, whereas bcc metals do have such a characteristic.

This knowledge was not known to materials engineers during World War II, when Liberty ships were built with steel that had a high ductility transition temperature; that is, this type of steel could not absorb low energy fractures at low temperatures, with the result that cracks continued to propagate until the ship broke apart. Many of these ships sailing on the Murmansk supply route bringing supplies from the United States to the Soviet Union have never been recovered after sinking somewhere in the North Atlantic. Brittle fractures have plagued the airline industry as well. One example is the Boeing 737 operated by Aloha Airlines in 1988 which lost a canopy without warning in level flight over the Pacific Ocean. Subsequent investigation revealed that the canopy was weakened as a result of extreme corrosion and fatigue of the aluminum alloy used as a skin material. The fatigue analysis concluded that the low stresses developed due to excessive cycles of pressurizing and depressurizing of the cabin. The abnormal amount of cycles was produced by the number of short flights at high altitudes in the mountainous Hawaiian area.

4.2.10 Processing Properties

Malleability, workability, and *formability* are some terms related to ductility that describe, in a general way, the ability of materials to withstand plastic deformation without the occurrence of negative consequences (rupture, cracking, etc.) as a result of undergoing various mechanical processing techniques (see Figure 4-10). Terms such as *weldability* and *machinability* are mentioned here as additional examples used to generally describe the reaction of materials to various manufacturing and/or fabricating processes in industry. In the case of copper alloys (brass, for example) an increase in lead content and hardness improves machinability but reduces ductility (cold workability). An improvement in machinability is always achieved at the expense of cold workability. Increases in copper content increases workability but reduces an alloy's machinability.

Weldability defines the capacity of a metal or combination of metals to be welded, forming a designed structure with weld joints that will perform satisfactorily. Other characteristics included in the term are the ease with which the welding process is accomplished and the absence of weld defects. Sheet metal formability defines the ability of the metal to undergo the desired shape changes without failures such as necking and tearing. In essence, it describes the relative ease with which a metal can be shaped through plastic deformation. Several techniques are used to test the formability of sheet metals.

4.2.11 Flexural or Bending Strength

A *beam* is a structural member that bends or flexes when subjected to forces perpendicular to its longitudinal axis. This axis in beam strength determination is called the beam's *neutral axis (NA)* because no normal stresses are assumed to exist along its length. The neutral axis is the edge view of a neutral plane. Figure 4-16a is a sketch of a *simple supported beam* (a beam supported at its ends with a pin and a roller). The transverse load or force P bends the beam (causes it to deform, i.e., deflect), thus producing both shear and normal stresses in the beam. The load in this example would cause the beam to deflect, as in Figure 4-16b, resulting in normal stresses in compression near the top surface and normal stresses (tensile) at the bottom of the beam. These bending stresses are normal to any cross-sectional area through the beam, which makes their direction parallel to the neutral axis of the beam. The maximum bending stress in tension at failure is called the *bend strength* for ceramic materials and is often referred to as the *modulus of rupture (MOR).* The question of the presence of flaws, their distribution in a ceramic material, and their attendant adverse effects on the strength of the material are not taken into consideration when using this indicator of strength.

Assuming that the beam material is *homogeneous* (same material nature throughout) and *isotropic* (same properties in all directions throughout the beam material), the normal stresses will be at a maximum near the top and bottom surfaces of the beam. These normal stresses (both compressive and tensile) are known as *flexural, fiber,* or *bending stresses.* The *flexure formula,*

$$\sigma = \frac{Mc}{I}$$

relates these stresses (σ) to the *bending moment* (M), the maximum distance (c) from the beam's neutral axis (where bending stresses are zero) to the outer surfaces of the beam, and the rectangular *moment of inertia* of the cross-sectional area of the beam (I). A *flexure test* is performed using a simple-supported beam loaded as shown in Figure 4-16c. The maximum bending stress and deformation (deflection) are recorded for increments of load P. These data are plotted to obtain a stress–strain diagram. The maximum bending stress developed at failure is known as the *flexural strength.* For those materials that do not crack, the maximum bending or flexural stress is called the *flexural yield strength.* A bend test used to determine the ductility of certain materials should not be confused with this flexure test.

(a)

Figure 4-16 (a) Simple-supported beam with a transverse load. (b) Loaded beam showing deformation (deflection). (c) Wood specimen undergoing a static bending (flexural) test and showing evidence of failure due to tension. (Forest Products Laboratory, Forest Service, USDA)

(c)

Illustrative Problem

Referring to the beam in Figure 4-16b, the bending moment (M) produced by the force (P) at the center of the beam is 3530 ft · lb. The beam is a T-beam with a distance of 5 in. (c) from its centroid to the bottom surface of the beam. Calculate the stress at the bottom surface of the beam midway between the supports. The moment of inertia (I) for the beam's cross-section is 136 in⁴. Also indicate in your answer whether the stress is tensile or compressive.

Solution

$$\sigma = \frac{Mc}{I}$$

$$= \frac{3530 \text{ ft} \cdot \text{lb} \times 5 \text{ in.}}{136 \text{ in.}^2} \left(\frac{12 \text{ in.}}{1 \text{ft}} \right)$$

$$= 1560 \text{ psi (tensile)}$$

4.2.12 Fatigue (Endurance) Strength

Service conditions commonly involve many repetitions of applied stress or reversals of stress. We know that when a ***simple beam*** is subjected to a downward-acting, transverse load

AB — Beam
P — Transverse load

Figure 4-17 A simple beam with a concentrated, transverse load.

Figure 4-18 A simple beam deflected by a cyclic, transverse load.

(see Figure 4-17), the material in the top half of the beam will be compressed (the stress s_c is compression) and the lower half of the beam will be subjected to a tensile stress (s_t).* This is demonstrated in Figures 4-18 and 4-19. Roller and pin supports are termed **simple supports** to differentiate them from fixed supports, which can also resist rotational forces. A shaft with a pulley or gear can be compared to the horizontal beam referred to previously. The load or force, in this instance, is a pulley or gear force that causes the shaft to deflect (bend), particularly if sufficient bearing supports are not provided. Now, if a transverse load is applied and removed in some cyclic fashion, the material in the beam or shaft would go from a condition of zero stress to a stressed condition and back to a condition of zero stress. This type of cyclic loading is called *repetitious*. If the stress in the material changes due to the loading from compression to tensile, or vice versa, this is known as **stress reversal.** The latter condition is easier to visualize with a point on the surface of a rotating shaft turning at 1000 revolutions per minute whose bending stress changes from compression to tension 1000 times a minute. In practice, many failures have occurred under such conditions when the stresses developed were well below the ultimate stress and frequently below the yield

Note: σ is the universal symbol for stress. However, lowercase *s* is sometimes substituted for σ, as seen in Figures 4-18 and 4-19.

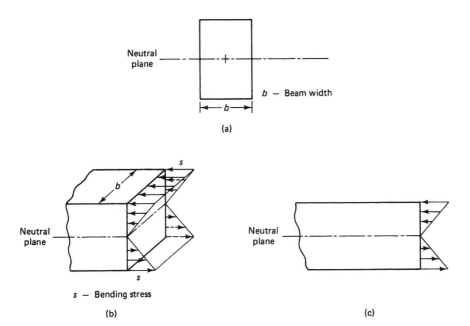

Figure 4-19   (a) End view of beam showing cross-section. (b) Front view (three-dimensional) of a portion of beam showing distribution of tensile and compressive bending stresses acting on a transverse plane. (c) Front view (two-dimensional) showing same stress distribution as in (b).

Figure 4-20   Combined fatigue fracture due to combined bending and torsional loads on an axle. The outer surface was hardened. A crack began at the top surface. The rough surface in the center indicates the last area to fracture instantly when the axle broke. (John Deere Co.)

strength. These failures are called *fatigue* (or *endurance*) *failures* (see Figure 4-20). Fatigue causes over 75% of failures of mechanical parts and components in service.

A fatigue failure starts as a tiny crack whose origin is often traced to an inspection stamp, tool mark, or other defect on the surface. The crack produces a stress concentration that assists in the growth of the crack until eventually the area of material remaining to withstand the stress is insufficient, which results in a sudden fracture (see Figure 4-20). Much empirical research is done using fatigue testing machines to determine the strength of materials under fatigue loading. Machines developed by the National Aeronautics and Space Administration use computers that simulate aircraft or spacecraft flights to test the material until it fails under varying conditions of load and ambient temperatures. The results of these many tests are recorded on semilog or log-log paper to produce *s* (stress)–*N* (cycles) diagrams. Typical **s–N** *diagrams* are shown in Figure 4-21 for a fiberglass composite, in Figure 4-22 for a typical low-carbon steel,

Figure 4-21   s–N diagram for a typical fiberglass composite material.

Figure 4-22   s–N diagram for a typical low-carbon steel.

and in Figure 4-23 for some nonferrous metals and some plastics. Figure 4-24a is a photograph of a typical fatigue fracture showing two very distinct surfaces.

The s–N curve for a low-carbon steel (see Figure 4-22) shows an abrupt break, or *knee*, at which point the curve tends to approach a horizontal line. In other words, with any stress below the fatigue or endurance limit, a low-carbon steel can be cycled continuously without fracturing. Aluminum and other nonferrous materials fracture at relatively low stresses after many cycles. For many ferrous alloys the endurance limit is about one-half the tensile strength of the metal. Furthermore, they exhibit no fatigue limit, which means that there is no stress below which they will not fracture. Therefore, when speaking of the fatigue of nonferrous materials, it is necessary to express both the stress and the number of cycles in describing the life of the material. See Section 4.2.9 for an example of fatigue failure of a Hawaiian aircraft.

It is conservatively estimated that over 50% of the failures occurring under service conditions are fatigue failures. In view of the expense involved in conducting tests, particularly when attempting to duplicate service conditions in a laboratory, a *fatigue* or *endurance ratio*

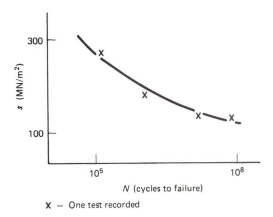

Figure 4-23 *s–N* diagram for nonferrous metal (e.g., aluminum and copper) and some plastics (e.g., nylon).

X — One test recorded

(a)

(b)

Figure 4-24 (a) Fatigue fracture due to progressive fatigue cracking, with cracking progressing across most of the section before the final overload fractures the remaining metal. (John Deere Co.) (b) Sketch illustrating torsional fatigue on a crankshaft (adapted from *Shot Peening Applications,* 7th ed, Metal Improvement Co. Inc., p. 43).

is sometimes used instead. This ratio, varying between 0.25 and 0.45 depending on the material, compares the fatigue limit to tensile strength. The fatigue strength of a bolt after 10^7 cycles is only about 10–20% of the ultimate tensile strength. The shape of the metric threads plus the martensitic microstructure of a high-strength bolt means that fatigue is the limiting mechanism for fasteners.

The *fatigue* or *endurance limit* is the maximum stress below which a material can presumably endure an infinite number of stress cycles (Figures 4-21 and 4-22). *Fatigue strength* is the maximum stress that can be sustained for a specified number of cycles without fracture. In other words, fatigue strength can be any value on the ordinate of the *s–N* diagram. To make it meaningful, the corresponding *N* value must be reported also. In Figure 4-22 the fatigue strength of about 40 kpsi corresponds with an *N* of 10^5 cycles. The fatigue limit, as determined empirically, is generally below the yield strength. Most design stresses are lower than the fatigue or yield strength of a material primarily because of the adverse effects of surface conditions on the strength of materials.

Another term used in describing failures is ***endurance ratio,*** which is the quotient of endurance limit to tensile strength. For many ferrous alloys the endurance limit is about one-half the tensile strength of the metal. In view of the expense involved in running fatigue tests, particularly when attempting to duplicate service conditions in a laboratory, a fatigue or endurance ratio is sometimes used instead. This ratio varies between 0.25 and 0.45, depending on the material.

4.2.13 Creep

Creep is a slow process of plastic deformation that takes place when a material is subjected to a constant condition of loading (stress) below its normal yield strength. Creep occurs at any temperature. However, at low temperatures, slip (movement of dislocations) is impeded by impurity atoms and grain boundaries. At high temperatures, the diffusion of atoms and vacancies permits the dislocations to move around impurity atoms and beyond grain boundaries, which results in much higher creep rates. The word *creep* implies, then, that a material plastically deforms or flows very slowly under load as a function of time. After a certain amount of time has elapsed under constant load, the ***creep strain*** (plastic deformation) will increase and some materials will rupture. This rupture, or fracture, is known as ***creep rupture.*** Aluminum alloys begin to creep at around 100°C. Aluminum aircraft wing panels are creep formed. During ***creep forming,*** a workpiece is loaded against a die or tool for a specified time. The load is normally applied using pressure or vacuum at temperatures around 160°C. The process results in predetermined shapes that require little or no machining. The amount of creep resulting from such operations must be predetermined through numerous creep tests. In addition, springback must be taken into consideration. When the forming or bending force is removed at the end of the deforming operation, elastic energy remains in the deformed part, causing it to recover partially to its original shape. This elastic recovery is called ***springback,*** which is defined as the increase in the included angle of the bent part relative to the included angle of the forming tool after the tool is removed.

Although some materials will creep at low or room temperatures, this type of plastic deformation is usually associated with high temperatures. Polymeric materials creep at room temperature, but this low-temperature creep is called ***cold flow.*** Both steam and gas turbines used for propulsion of ships or the generation of electricity operate, by necessity, at high temperatures over many years. The extremely close fits between turbine blades and their casings prevent the escape of steam or gas past the blades. If these turbine blades were allowed to change their dimensions at any time during their expected service life, during which time they are under large centrifugal loads (rotating at extremely high speeds) at extremely high temperatures for long periods, possibly weeks, the failure would be catastrophic. Consequently, the material from which these turbine blades are fabricated must possess, among other necessary properties, high creep resistance. Continuing research over the years has produced materials for medium- and high-temperature service such as the many titanium alloys and the "superalloys" composed of iron, nickel, and/or cobalt. Module 5 discusses investment casting of such alloys.

Tensile creep tests develop data at a constant level (stress) and at a constant temperature. The amount of creep strain when plotted against time, as in Figure 4-25, indicates that the creep curve formed can be divided into several stages, with the last stage ending with a rupture of the material. The ***creep rate*** is determined, at any point, by the slope of the curve (drawing a tangent to the curve at the point). The creep rate, the slope of the middle portion of the curve, is nearly constant and represents the minimum rate. Once the creep enters the last stage, rupture soon follows. Creep rate is usually expressed in percent creep strain per hour. A typical rate might be $10^{-4}\%$ per hour.

Engineering materials must perform satisfactorily throughout their ***service lives.*** In many applications, service life can extend well beyond 20 years. A determination must be made as to the maximum allowable deformation that can be tolerated during the expected service life of a material before the material is chosen for the particular application.

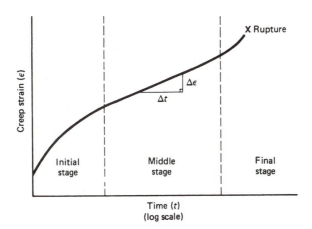

Figure 4-25   Typical creep curve.

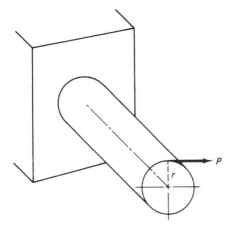

Figure 4-26   Circular rod under torsion.

Research in polycrystalline materials with varying amounts of grain-boundary areas has shown that, at high temperatures, *coarse-grained materials* possess more creep resistance than do fine-grained crystalline materials. At low temperatures, *fine-grained materials* offer more creep resistance. These statements, including the phenomenon of creep itself, can be explained in terms of the movement and stoppage of mobile dislocations (linear arrangements of atoms making up imperfections in the atomic structure of solid materials) throughout the crystal structure. Creep, then, like other properties of materials, is dependent on the structure of materials.

4.2.14 Torsional Strength

Torsion describes the process of twisting. A body such as a circular rod (e.g., a shaft for transmitting power), as shown in Figure 4-26, is under torsion as a result of a force acting to turn one end around the longitudinal axis of the rod while the other end is fixed or twisted in the opposite direction. The material resists this twisting action by generating a similar twist internally. The product of the force P and the radius r perpendicular to the line of action of the force is called a *torque.* The external torque produces both a stress and a deformation of the rod. The stress is classed as a shear stress, which causes the atoms to twist past each other. The *shear deformation* is measured in terms of the angle γ (Greek letter gamma) and the angle θ (Greek letter theta), as indicated in Figure 4-27a. As a result of the torque T, the point C on the surface of the rod moves to point C'. This deformation can be measured by angle γ on the surface of the rod or by angle θ shown on the cross-sectional area of the rod. The angle θ is known as the angle of twist. Both angles are exaggerated for illustrative purposes. Forces can combine to cause failures, as seen in Figure 4-20.

(a)   (b)

Figure 4-27 (a) Circular rod undergoing a torque, with resulting deformation. (b) Torsional fracture. Torsional (twisting) loads produce spiral types of failure. Note the curved line from (A) to (C′) is part of a helix.

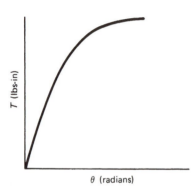

θ (radians)   **Figure 4-28** Typical torque–twist diagram.

A torsion-test machine measures the torque applied to a specimen of material along with the corresponding **angle of twist** (θ). The results can be plotted as a *torque–twist diagram* (see Figure 4-28), which resembles an ordinary stress–strain diagram as obtained by the usual tensile-test procedure. The torsional stress is the shear stress (τ) produced in the material by the applied torque and is calculated using the torsion formula for circular shafts:

$$\tau_{max} = \frac{Tr}{J}$$

where r is the radius of the cross-sectional area and J is the centroidal **polar moment of inertia.** The maximum torsional stress occurs when r is a maximum, as indicated in the preceding equation and illustrated in Figure 4-29. The maximum torsional stress occurring at the outer surfaces of the circular rod is called the **torsional yield strength.** It relates to that point on the torque–twist diagram where the curve begins to depart from a straight line. Torsional yield strength roughly corresponds to the yield strength in shear. The **ultimate torsional strength** or **modulus of rupture** expresses a measure of the ability of material to withstand a twisting load. It is roughly equivalent to the ultimate shear strength. The **torsional modulus of elasticity** as determined by the torque–twist diagram known as the **modulus of rigidity** is approximately equal to the shear modulus or the modulus of elasticity in shear (G).

Illustrative Problem

How much torque (T) is needed to produce a shear stress of 79.6 MPa on the surface of a 40-mm-diameter shaft with a polar moment of inertia (J) equal to 25.12×10^{-8} m⁴?

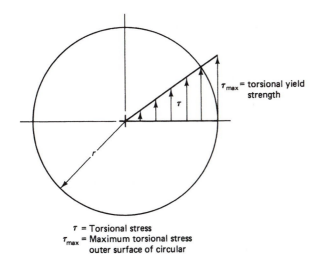

τ = Torsional stress
τ_{max} = Maximum torsional stress
         outer surface of circular
         rod

Figure 4-29 Distribution of torsional stress in a circular rod.

Solution

$$T = Pr \qquad\qquad \tau = \frac{Tr}{J}$$

$$r = \frac{D}{2} = 0.02 \text{ m} \qquad T = \frac{\tau J}{r}$$

$$= \frac{79.6 \text{ MPa} \times 25.12 \times 10^{-8} \text{ m}^4}{0.02 \text{ m}}$$

$$= \frac{79.6 \times 10^6 \text{N/m}^2 \times 25.12 \times 10^{-8}\text{m}^4}{0.02\text{m}}$$

$$= 999.8 \text{ N} \cdot \text{m}_t \cong 1 \text{ kN} \cdot \text{m}$$

4.2.15 Hardness

Hardness is a measure of a material's resistance to penetration (local plastic deformation) or scratching. One of the oldest and most common hardness tests, based on measuring the degree of penetration of a material as an indication of hardness, is the Brinell. ***Brinell hardness numbers*** (HB) are a measure of the size of the penetration made by a 10-mm steel or tungsten carbide sphere with different loads, depending on the material under test. The indentation size is measured using a macroscope containing an ocular scale. ***Vickers hardness numbers*** (HV) employ a diamond pyramid indentor. Otherwise, the two tests are basically similar. ***Rockwell hardness testers,*** using a variety of indentors and loads with corresponding scales, are direct-reading instruments (i.e., the hardness is read directly from a dial). The hardness number, for example, 65 HRC, indicates the reading came from the C scale using a diamond cone indentor and a 150-kg load. It is therefore important in reporting Rockwell hardness readings to include the scale letter so that the person wishing to use the information knows the type of indentor as well as the size of load used in the test. (See Table 4-1 and Figure 4-30.) The National Institute of Standards and Technology (NIST) and ASTM committees have worked on revision to ASTM E18, *Standard Test Methods for Rockwell Hardness and Rockwell Superficial Hardness of Metallic Materials.* The goal of the revision is to arrive at consistent Rockwell hardness (HR) standard values among the makers of hardness testers, test blocks, and users of the machines and blocks worldwide.

Few, if any, tools are subjected to such an elaborate heat treatment as nail-hammers, an item that weighs no more than 20 oz. The striking face and entire claw section including the

TABLE 4-1   HARDNESS SCALES COMPARISONS

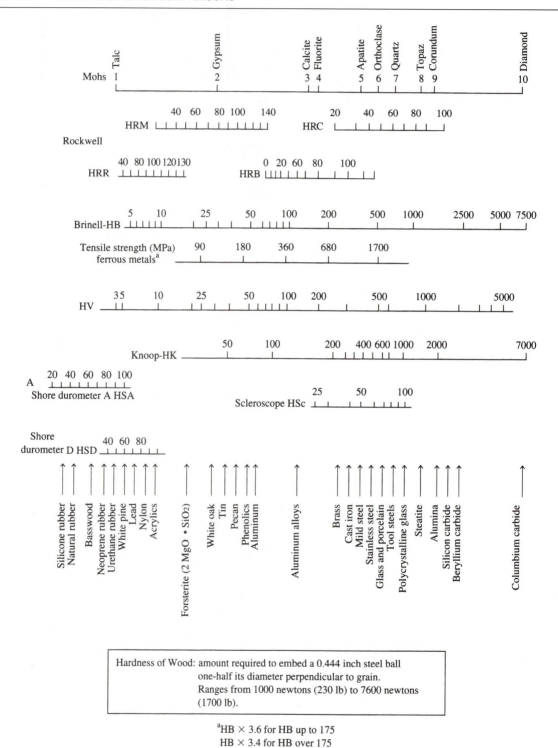

Hardness of Wood: amount required to embed a 0.444 inch steel ball one-half its diameter perpendicular to grain. Ranges from 1000 newtons (230 lb) to 7600 newtons (1700 lb).

[a]HB × 3.6 for HB up to 175
HB × 3.4 for HB over 175

---

[a]HB × 3.6 for HB up to 175
HB × 3.4 for HB over 175

(a)

(b)   (c)

Figure 4-30   (a) Photograph of Tukon microhardness tester. (Measurement Systems Div., Page Wilson Corp.)
(b) Photograph of Rockwell hardness tester. (Measurement Systems Div., Page Wilson Corp.) (c) Photograph of
Air-O-Brinell metal hardness tester with digital readout of Brinell values. (Tinius Olsen Testing Machine Co.)

eye (or crotch) are first hardened to about 62 HRC by induction (see "Surface Hardening" in
Module 5). Because of variations in section thickness of the claw, two different induction frequencies are used. The striking face and the rim around the striking face are tempered by induction, using a coil arrangement that results in 50 to 58 HRC on the striking face and somewhat lower hardness values around the face (called ***ring tempering***). Claws are also induction

Figure 4-30   (d) Microhardness penetrator (Knoop and Vickers) indentations. (Wilson Instrument Division of ACCO) (e) Various standard loads for the Rockwell hardness tester. (Wilson Instrument Division of ACCO) (f) Brale sphero-conical diamond penetrator. (Wilson Instrument Division of ACCO)

tempered. Specified hardnesses of the crotch are 44 to 52 HRC; the claw tips, 47 to 55 HRC. The heads are then bulk tempered in a furnace at 600°F.

   Other tests used to report hardness include a file test (resistance to scratching), a **Scleroscope** HSc test (measures the rebound of a small weight bounced off the surface of the material), and a comparison test that also uses scratch resistance. The latter test compares a material's hardness to some 10 known minerals arranged in order of hardness. Mostly used by mineralogists, the **Mohs scale** classes hardness of all materials between 1 (the hardness of talc) and 10 (hardness of diamond). The scale is based on the ability of a hard material to scratch a softer material (Table 4-1).

Materials such as very thin materials (e.g., coatings, foils, plated surfaces), very brittle materials (e.g., glass or silicon), and very small parts (e.g., gears in a wristwatch) require special care in hardness testing due primarily to their thinness and/or size. Furthermore, laboratory research in materials necessitates hardness testing on a microscale in determining the differences in hardness over the minute area of a single grain of metal or between the middle of a crystal and the grain boundary area. For such purposes a microhardness tester (Figure 4-30a) finds application with loads and indentations that are so small that indentations require microscopic viewing with appropriate scales for accurate measurement.

Table 4-1 provides a comparison of the approximate hardness of a variety of materials using 11 different hardness scales. These readings may be compared to the tensile strengths of steel given in Table 10-11. Some correlation exists between hardness and tensile strength, but it is only approximate. For example, the tensile strength of steel (but not other materials) is about 500 times the HB as listed in the table. As a general rule, the tensile strength of a given ductile metal can be estimated only from the hardness reading within an error of less than 10%.

Illustrative Problem

Given the formula for determining the Brinell hardness number (HB), where P is the applied load (kg), D is the diameter of the steel ball (mm), and d is the diameter of the indentation (mm),

$$HB = \frac{P}{(\pi D/2)(D - \sqrt{D^2 - d^2})}$$

what is the Brinell hardness number if a 10-mm ball with a load of 3000 kg produces an indentation with a diameter of 2.88 mm?

Solution

$$HB = \frac{3000}{\pi(10/2)[10 - \sqrt{100 - (2.88)^2}]}$$

$$= \frac{600}{\pi(10 - 9.576)} = \frac{600}{1.333} = 450$$

4.3 FAILURE ANALYSIS

Up to this point in the module, we have discussed many of the physical properties involved in mechanical failures of materials. In addition, other properties of materials produce failures in materials, such as thermal or chemical. Failures of materials not only cost vast sums of money annually but contribute to loss of human life. Many failures in equipment, structures, and systems are not all attributable to material failures but are the results of other factors not discussed in this book. When a failure occurs, failure analysis is undertaken to help determine whether the failure was due to a design or material deficiency. An examination of the fracture tries to reconstruct the sequence of events and the cause of the fracture. Much information can be determined by analysis of the path a crack makes as it propagates through the material, the origin of the crack, and the cause of the crack initiation (i.e., material flow, oxidation/corrosion, tensile stress overload, thermal shock, etc.). Often, most of these data can be determined from visual observation by a trained investigator. If not, low-power microscopy, scanning electron microscopy, electron probe, and other sophisticated instruments might be needed to analyze the chemical makeup of the fracture surfaces to aid in determining where responsibility for fracture lies. This highly technical area within the broad scope of *failure analysis* is known as *fractography.*

A finding of responsibility is critical not only as an element in liability suits, but is equally important in future design, specification writing, modifications, selection of materials, and the like. A classic example of failure analysis is the investigation of the Hyatt Regency Hotel disaster, in which 113 people were killed and many injured when concrete walkways collapsed in 1981 (Figure 4-31).

(b)

(a)

Figure 4-31   (a) The National Institute for Standards and Technology (NIST) investigated the Hyatt Regency Hotel disaster by using a mock-up and actual samples of box beams that supported the failed walkways. (b) Close-up shows joint pulling loose.

4.3.1 Factors of Safety

In the design of machine parts and structural members, it is the responsibility of the designer to provide a margin of safety, i.e., to ensure that a machine part is safe for operation under normal operating conditions. *Factor of safety* (or design factor, N) is defined as the ratio of ultimate strength to allowable stress. *Allowable stress,* also called *design stress, working stress,* or *safe stress,* is the maximum level of stress that a part will be permitted to endure under operating conditions. In determining the allowable stress, considerations of the type of load on the part (static, impact, fluctuating, reversed, etc.) and the material from which a part is made are paramount. A material's ductility, as well as its strengths and stiffness, must be evaluated because the mode of failure of a ductile material varies dramatically from a brittle material such as gray cast iron, some plastics, or some heat-treated steels. For example, experience warrants that, for fatigue loading (reversal of stress from tension to compression on a repetitive basis), a ductile material be selected with a percent elongation of 10% or higher to ensure against a brittle-type failure. The need for a factor of safety is most evident when knowledge is sketchy about (1) possible unexpected high loads on a structure, (2) the degree of residual stresses within a material due to manufacturing processes, (3) homogeneity of the internal structure of a material throughout its entirety, (4) loading history of the material since its fabrication, and (5) numerous but necessary assumptions used in the analysis and design procedures that can lead to appreciable errors. For ductile metals and plastics under static or steady tensile loading, the yield strength (σ_{ys}) is used as the basis for determining the allowed working stress. In most cases the factor of safety used is 2. Thus, the allowed working stress or design stress (σ_{des}) is calculated:

$$\sigma_{des} = \sigma_{ys}/2$$

For brittle metals such as cast iron, the ultimate tensile strength (σ_{uts}) with a factor of safety of 4 is used:

$$\sigma_{des} = \sigma_{uts}/4$$

The design shear stress or maximum allowed stress is calculated using the materials yield stress and a factor of 2:

$$\tau_{max} = \sigma_{ys}/\,2$$

In dealing with ceramics, a brittle material due to the presence of defects and cracks of varying sizes which will make the tensile strength vary, the design stress will, of necessity, employ probability design methods such as Weibull statistics. Even though ceramics and other brittle materials are quite capable of handling compressive loads, the concern with tensile strength is due to the fact that even ceramics behave as other materials do. When subjected to temperature changes that incur changes in microstructure and induce internal thermal stresses, the result is tension failures with sometimes disastrous results.

Computing the required factor of safety can be quite involved. However, in some cases the factor of safety is specified in codes promulgated by organizations such as the American Institute of Steel Construction (AISC) and the American Society of Mechanical Engineers (ASME), as well as in building codes and company policies as explained in Section 4.6.

4.4 ELECTROCHEMISTRY—ELECTROLYSIS, CORROSION, AND ITS PREVENTION

Electrochemistry is the study of chemical reactions that produce electricity and of how electricity can be used to cause chemical reactions. Electricity drives chemical reactions as in electrolysis and the refinement of aluminum. Oxidation and reduction (redox) reactions (to be explained) involve the transfer of electrons and, by definition, when electrons move there exists an electrical current. An electrochemical cell is a device that uses oxidation and reduction reactions to produce an electrical current. In this technological world the many electronic devices that permeate our existence such as handheld computers, portable radios, watches, and automobiles depend on electricity produced by chemical reactions. Flashlights, pocket calculators, portable computers, and watches are just some of many devices that depend on batteries that use oxidation-reduction (redox) *b* reactions to produce an electric current. Fuel cells depend on oxidation and reduction reactions that convert chemical energy directly into electrical energy. Many industrial processes use electrical energy to generate chemical reactions which, in turn, produce a particular product. An understanding of the working of an electrochemical cell (see Figure 4-32) can be put to good use not only designing materials and structures that can resist corrosion but also in selecting materials that will be subjected to some adverse environmental condition. Electrochemical cells are discussed later in this module. By studying the laws of thermodynamics and the concepts of enthalpy, entropy, and free energy, which are beyond the scope of this book, a greater understanding of chemical reactions can be obtained. *Thermodynamics* is a branch of physics that describes the reversible transformations of heat into other forms of energy and the laws that govern such conversions. It provides a description of these processes in terms of *macroscopic* variables (detectable by our senses).

4.4.1 Chemical Properties and Reactions

Chemical properties are a measure of how a material interacts with gaseous, liquid, or solid environments. The most familiar chemical property is corrosion resistance. *Corrosion resistance* is the ability to resist oxidation. Common examples include the ability of aluminum to resist rusting when exposed to air and moisture, and the ability of rubber to withstand sunlight (ultraviolet rays) without drying and cracking. Many conditions or environments threaten materials. Polluted air is filled with elements emitted from gasoline engines, home furnaces, and industrial plants: These elements combine with materials to cause damage and shorten their service life. The American Lung Association Healthy Home program certifies new homes for good indoor air quality. Stringent requirements include a sealing out of moisture and pollutants

Figure 4-32 Electrochemical cell. This device uses a spontaneous chemical reaction to produce an electric current as in flashlights, pocket calculators, portable radios, and many other devices that rely on batteries. The porous partition allows the movement of ions through the cell but prevents the electrolyte solution for each half cell from mixing. The electrons lost by oxidation of the zinc atoms leave the Zn electrode and flow through the external wire to the copper electrode where they reduce the Cu ions and thus, the Cu atoms coat the surface of the copper electrode.

with a tight, well-insulated building shell and foundation. All air supplied to the house must be filtered. Some of our ancient architectural treasures, such as the Parthenon (an ancient Greek temple) and copper and stone statues, are rapidly deteriorating from corrosive, polluted air. Salty water from ocean spray or road salt used on icy roads promotes corrosion in automobiles. The reaction of some materials to high temperatures and fire is hazardous. Many plastics emit poisonous gases when burned. Sometimes it is desirable for materials to deteriorate or be biodegradable so that the natural environment can break down this material, thus reducing solid waste.

4.4.2 Electrolysis

Electrolysis is the decomposition of a material through the passage of electricity. The electrolysis of water is an excellent example of this chemical reaction brought about by the flow of electrons in which water is broken down into the two gases of oxygen (O_2) and hydrogen (H_2). Figure 4-33 is a schematic of the apparatus needed to decompose water by electrolysis. A few drops of hydrosulfuric acid (H_2SO_4) is added to water to make it more conductive. This acid is an excellent electrolyte which allows for the migration of ions through the water from one electrode to the other. Hydrogen ions are attracted to the cathode. At the cathode they pick up or gain electrons to become H atoms. The external circuit provides the surface of the cathode with electrons, which transfers them to the electrolyte containing the H ions and reduction takes place, forming atoms of H. As atoms, they combine with other H atoms to produce molecules of hydrogen gas (H_2). This gaining of electrons by the atoms is called reduction. At the anode the water molecules (H_2O) that have lost electrons decompose releasing O_2. This is called oxidation. An oxidizing reaction that occurs at the anode results in metal atoms being ionized which then enter the electrolyte. Because the metal ions leave the anode, it corrodes.

Electrons are lost from the metal in the ionizing process and leave the anode through the electrical connection. In this process the electrons are transferred from the electrolyte to the surface of the anode where oxidation takes place. The electric charge is carried through the elec-

Figure 4-33 Electrolysis of water. H^+ ions gain electrons at the cathode and form H atoms. H atoms then form H_2 molecules of gas. At the anode the H_2O decomposes to replace the H^+ ions lost at the cathode and release O_2.

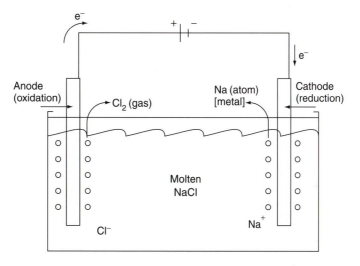

Figure 4-34 Electrolysis of NaCl. At the anode, Cl^- ions give up electrons to the electrode to form Cl atoms which combine to form Cl_2 molecules (gas). At the cathode, Na^+ ions accept e^- from the cathode to produce Na atoms which form liquid Na metal.

trolyte not by the free electrons but by the movement of positive ions toward the negative electrode (cathode) and by negative ions migrating toward the positive electrode (anode). This migration of ions, referred to by some as the internal circuit, is known as ***ionic conduction.*** The external circuit connecting the electrodes with a source of electrical energy contains charge carriers that are the free electrons. The flow of electrons in this circuit is known as ***metallic*** or ***electronic conduction.*** The chemical changes that took place (oxidation and reduction) can be added together and the net effect is represented by the following chemical equation which then represents both the oxidation and reduction reactions that took place.

$$2H_2O \rightarrow 2H_2 + O_2$$

Some examples of electrolysis used in industry are the refining of aluminum from bauxite, the production of chlorine from salt, the production of sodium metal (Na) and chlorine gas (Cl) (shown schematically in Figure 4-34), and electroplating of metals such as chromium and silver. In this last process the object to be plated is made the cathode and the plating material the anode. In ***electroplating*** (see Figure 4-35) this combining of metal ions and electrons (purposely

Figure 4-35 Schematic of electroplating (electrochemical plating). The anode acts as a sacrificial anode supplying the plated metal. Tin plating of steel is done mostly by this method. Metal ions from the anode migrate to the cathode and are reduced to atoms by gaining electrons from the cathode. Chromium provides the hardest plating metal for applications requiring wear resistance.

augmented by an electrical potential) produces a metal that plates out and covers the cathode surface. Electricity can produce chemical change by the process of electrolysis. The synthesis reaction discussed in Section 4.4.1 can be reversed as in the case of molten salt. Molten salt can be changed into sodium, a metal and chlorine (a gas), by passing an electric current through it.

4.4.3 Electrochemical Cell

In our discussion of redox reactions it was not possible to directly observe the transfer of electrons, the formation of ions, or the formation of compounds. By separating these two dependent reactions, the electron transfer can occur through an external circuit and be observed by instruments. Such a device is called an ***electrochemical cell,*** which uses a spontaneous chemical reaction to produce an electric current. This type of electrochemical cell is known as a ***voltaic*** or ***galvanic cell.*** Figure 4-35 is a sketch of an electrochemical cell consisting of two half-cells with the anode (negative electrode) and the cathode (positive electrode) both connected by a wire to provide the external circuit. Both electrodes are dipped into their respective electrolytic solutions that permit the passage of ions from one electrode to the other (the internal circuit), thus completing the entire electrical circuit. Electrons are produced by an oxidation reaction at the anode. They travel through the external circuit to the cathode where they cause the reduction reaction to occur. Anions given off at the cathode move toward the anode and cations move in the opposite direction and thereby complete the circuit inside the cell. The migration of ions results in anions moving toward the anode and cations toward the cathode.

When a metal ion in solution gains electrons at or near the electrodes, the porous barrier keeps the two solutions from mixing without impeding the movement of the ions between the electrodes. The flow of electrons pushed out by the anode flowing through the external circuit produces a potential difference measured by a voltmeter. By using this technology with one hydrogen electrode, standard potentials for the various metals and alloys can be measured and used to construct tables such as the activity series of metals and Table 4-2. An electrochemical cell is capable of defining either electrochemical corrosion or electroplating. These processes differ in their purpose, in the source of the electric potential by which a current is caused to flow, and by what type of chemical reaction occurs at the cathode.

The electrochemical cell illustrated in Figure 4-36 is also helpful in explaining electrochemical corrosion. The cell consists of four basic elements: physically connected electrodes,

ELECTRON CONDUCTOR

CAUSE
(cathode
protected
area)

EFFECT
(anode
corroded
area)

CONTINUOUS LIQUID PATH (electrolyte)

1. **Electrolyte (continuous liquid path, usually water in the form of condensate, salt spray, etc.)**

2. **Cause (cathode—the cause of corrosion— the area through which electricity flows.)**

3. **Electron conductor (in a structure, usually a metal-to-metal contact, e.g. rivets, bolts, spot welds, etc.)**

4. **Effect (anode—the surface or object which corrodes.)**

Figure 4-36 The corrosion circle, a graphic picture of the four essential elements needed for corrosion to occur.

Ions in water (H₂O)

Corrosion of iron (Fe)

(a)   (b)

Figure 4-37 (a) Electrochemical corrosion depicting ionized water; and (b) the corrosion of iron (Fe).

an anode, a cathode, and electrolytes (electrolytic solution) such as water that contains ions. The anode is the electrode that loses or releases electrons as a result of oxidation occurring at that electrode. The cathode is the electrode that gains or receives electrons as a result of reduction occurring at that electrode. Figures 4-37 and 39 graphically represent these conditions. When corrosion occurs two chemical reactions occur simultaneously—oxidation and reduction. These two simultaneous processes are known as a ***redox reaction*** or process. Oxidation occurs at the anode while reduction takes place at the cathode. To prevent or control electrochemical corrosion it is only necessary to eliminate one of the four basic elements. The flow of electrons is from the anode (negative terminal or electrode) through the external wire connection to the cathode (positive electrode or terminal). At the same time the electrical current composed of ions leaves the anode, enters the electrolyte, and proceeds to the cathode where it combines with the electrons that have flown through the external circuit to form some solid metallic material such as rust. Note that the anode is being corroded because of its loss of electrons and the formation of metallic ions. In actual corrosion circumstances, corrosion occurs because the anodic and cathodic areas on the surface of a material are connected by moisture or some other ionic solution that acts as an electrolyte thus providing all four elements needed to complete an electrochemical cell.

Electrolytes are substances that give conducting solutions when mixed in water. Most electrolytes are dissolved in water by dissociation; that is, the molecules break down into ions (see Figure 4-37). Water is a molecular substance that has only a slight electrical conductivity; but when an ionic solid is dissolved in water the ions from the solid separate from each other in the solution and can move about carrying their charge with them. Molten salt is an electrolyte. Figure 4-37 shows polyionic ions in an aqueous electrolyte. In *dry cell batteries* (see Figure 5-60 which describes a nickel–metal hydride (primary cell battery) the electrolyte is a moist paste that fills the space between the electrodes. The alkaline dry cell battery (primary cell) uses potassium hydroxide (KOH) in the form of a gel as an electrolyte. Dry cell batteries are classified as *primary cells* because once the reaction ends (reaches equilibrium) the battery is dead and cannot be recharged. The *lead-acid battery* or *secondary cell* can be recharged by reversing these reactions by passing electric current through them that has a voltage of opposite sign and is larger than the cell voltage. A dilute sulfuric acid solution is the electrolyte. Figure 5-60 is a graphical picture that includes a description of how a typical power cell (secondary cell battery) works. A 12-V battery has six single lead cells connected in series.

Chemical reactions occur when the composition of a material is changed with different physical and chemical properties (rusting of metal, boiling an egg, or tarnishing of silver) from the original material. A chemical reaction is a process in which some chemical bonds are broken, others are formed, and the chemical energy changes as a result of the reaction. These reactions can produce electricity and electricity can cause chemical reactions. From Module 1 physical changes do not change the composition of the materials as in the case of freezing water or the dissolving of sugar in a glass of water. Chemical reactions can be divided into two broad classes: synthesis and decomposition or dissociation. A *synthesis reaction* has two elements combining to produce a compound. Metals react with nonmetals to produce a compound that is generally ionic.

Metals may be considered *donor substances* as they donate or lose one or more electrons in an electrochemical process while nonmetals are sometimes called *acceptor substances* as they accept or gain one or more electrons. For example

$$(1) \qquad 2Na + Cl_2 \rightarrow NaCl$$

The product in this reaction is table salt while the two separate metals, the reactants in the form of elements, are certainly dangerous to humans. The ionic bonding of this compound was discussed in Module 2 along with the terms *ionization, ions, anions,* and *cations*. *Polyatomic ions* are charged particles containing two or more covalently bonded atoms. Examples are OH– (hydroxide ion) and SO_4^{+2} (sulfate ion). Note the hydroxide ion has the atoms of hydrogen (H_2) and oxygen (O_2) covalently bonded together plus an "extra" electron which gives this molecule a negative charge (see Figure 4-37 showing the hydroxide ion in an aqueous electrolyte).

The electric charge on an ion is referred to as its *oxidation state.* This concept will be explained further when discussing galvanic corrosion. If an electric circuit is passed through water the water can be decomposed into hydrogen and oxygen. This *dissociation reaction* can be shown as

$$(2) \qquad 2H_2O \rightarrow 2H_2 + O_2$$

The products formed are two gases.

4.4.3.1 Oxidation-reduction (redox) reactions and electron transfer reactions.
The previous two chemical equations are examples of oxidation-reduction (redox)

reactions. In the past, *oxidation* was defined as the addition of oxygen to an element or compound and *reduction* as the removal of oxygen from an oxide. This was so because many elements combine with oxygen and oxygen is very abundant on the earth's surface in the form of metal oxides. To free these metals from their oxides requires a chemical process called reduction. Iron is separated from its oxide using this process of reduction:

$$(3) \qquad Fe_2O_3 + 3CO \rightarrow 2Fe + 3CO_2$$

Carbon monoxide (CO) removes the oxygen from the iron oxide and is thereby converted to carbon dioxide (CO_2). The iron oxide is said to be reduced to iron. At the same time the CO is oxidized to CO_2. Figures 4-38a and b depict the refining of iron and the rusting of iron. In addition the oxidation of polymeric materials such as old erasers, rubber bands, and automobile tiers is graphically sketched showing the addition of cross-links in the resulting network structure (see Figure 4-38c). Because many other substances can be used to reduce oxides to their elements the definition of oxidation and reduction is broadened from the definition previously given. Today these redox reactions are also known as ***electron-transfer reactions.*** An element is oxidized when it loses electrons because oxidation is the loss of electrons. An element is reduced when it gains electrons because reduction is defined as the gain of electrons. Electrons are transferred from an element that is oxidized to an element that is reduced. Substances that can oxidize other substances are said to be oxidizing agents. Any substance from which electrons can be readily removed such as a metal may reduce other substances and therefore can be defined as a reducing agent. In sum the reaction between a metal and a nonmetal is an electron-transfer reaction in which the metal behaves as a reducing agent and is oxidized and the nonmetal behaves as an oxidizing agent and is reduced.

These two processes, oxidation and reduction, are separated into two parts called ***half-reactions*** for greater understanding. A separate equation represents each process. The two half-reaction equations then can be added together to construct an equation that represents the entire reaction. Using the synthesis equation for forming the compound of table salt from sodium and chlorine can demonstrate this procedure:

$$2Na + Cl_2 \rightarrow 2\,NaCl \text{ (The overall equation)}$$

The half-reaction representing oxidation:

$$2Na \rightarrow 2Na^+ + 2e^-$$

The half-reaction representing reduction:

$$Cl_2 + 2e^- \rightarrow 2Cl^-$$

Note in both half-reactions that ions are formed when combined in the final equation from the compound NaCl in which Na and Cl each are in a balanced electron state. These two elements were used in discussing electrolysis where the two elements are separated from their compound form to produce the metal sodium and the gas chlorine (see Figure 4-34).

4.4.4 Corrosion of Metals

According to the Battelle Institute in Columbus, Ohio, the corrosion of metals amounts to a loss of $300 billion annually. More than 200,000 bridges in the United States are in need of major repairs. Designed to last for 45 years they are failing in 12 to 18 years. Without any protection, most metals will oxidize (corrode). It is estimated that over $150 million is lost to corrosion in products and equipment annually. Corrosion loss has a dual aspect in that as products and equipment corrode they must be replaced with the expenditure of limited material resources and the corroded materials end up in a landfill or are recycled. The production of new equipment to replace

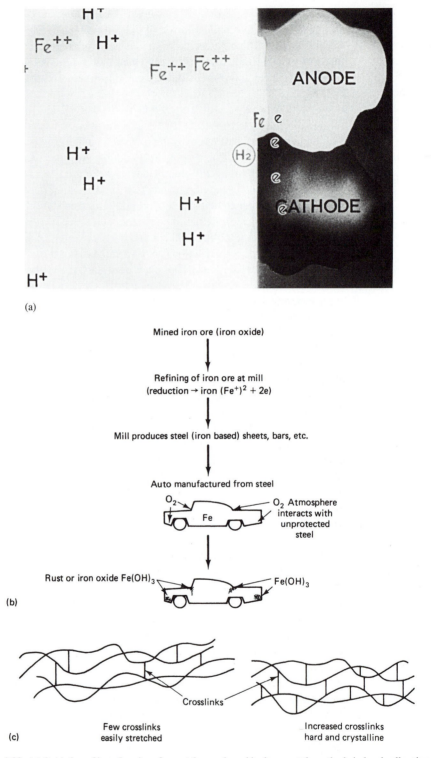

Figure 4-38 (a) Oxidation of iron. Iron ions form at the anode and hydrogen at the cathode in local cell action. (INCO) (b) Cycle of iron and steel seeking natural equilibrium. (c) Typical thermosetting polymer. Increasing number of cross-links produces a network structure.

the corroded also means the expenditure of energy resources that may also produce pollutants harmful to the world's atmosphere as in the case of burning fossil fuels. Corrosion is an electrochemical process. Most metals in nature are found in the form of compounds such as a metallic oxide. Gold is one metal that is found in its metallic form. Once metals are freed from their combined form, such as a metal oxide, they are driven to return to their natural states. This tendency or driving force to return to original form is another way of describing a material's seeking to reach its equilibrium state or state of lowest possible energy. This driving force is the root cause of electrode potential possessed by metals which is described briefly in this module.

Ceramics do not possess this driving force, because they occur in nature as compounds and remain as such while in use; therefore, they do not possess the force that would result in their corroding. That also accounts for ceramics' poor conductivity since their electrons are all locked up in bonding their atoms. Glasses that contain Na or K ions are subject to *leaching* in which these ions diffuse from the surface of the glass into the surrounding medium which is many times a liquid such as water. These ions belong to the alkali metals (Group 1A) in the periodic table which in previous modules were described as being highly active metals having only one valence electron per atom. This fact results in rather weak bonding strength. Polymer materials will corrode but they do so only because of some chemical reaction between their environment and their molecules that comprise these materials.

Corrosion may be defined as the deterioration of a material by interaction between the material and its environment. It is not considered an inherent property of a material because corrosion is the result of the material's reaction with its environment. Remove the environment and no corrosion will occur. The root cause of corrosion is the instability of materials, particularly metals, to revert to their natural states (seeking equilibrium at a lower state of energy). Materials differ in the amount of this driving force or electric potential they possess. Iron rusts because it has a large driving force. This force is described more fully in the study of thermodynamics which involves the study of chemical reactions. This potential (redox potential) is measured for metals and tabulated along with other metals' potentials to determine if a particular metal will corrode in the presence of another metal (dissimilar metals). Metals can undergo an electrochemical reaction that transforms them from the metallic state to the ionic state. In metals, corrosion is accompanied by a loss of material or the formation of a nonmetallic scale. Many environments contain atmospheric air with moisture that provides for ionic conductivity. Metals in aqueous (containing water) environments always disturb the electrochemical equilibrium with the result that electrochemical reactions occur where products are the products of corrosion. That is, corrosion occurs when two or more electrochemical reactions take place on a metal surface. Metal elements or alloys change from a metallic state into a nonmetallic state which results in the lowering of the energy of the system as the metal converts to a lower energy form. Corrosion of steel produces rust, a nonmetallic corrosion product. This combining of the ions and electrons produces a by-product. In the case of electrochemical corrosion this by-product is a solid material such as rust, $Fe(OH)_2$. Corrosion may also be described as the conversion of a metallic element from a zero or lower valence state to a positive or higher valence state. The conversion process is called oxidation. Most metals occur naturally in their oxidized state. The stable state of a metal is the oxidized state.

Some metals lose their chemical reactivity and become inert or passive. This condition, known as *passivity,* is possessed by some metals and their alloys such as Cr, Fe, or Ti. Passivity results in these metals forming a highly adherent but very thin oxide film on their surface. This barrier coating prevents further corrosion by breaking the electrical circuit or by stopping the diffusion of ions as in the case of polarization (to be defined). Stainless steel and aluminum exemplify passivity. Corrosion is affected by increases in temperature of the ambient environment (air, liquid, or solid) as well as increases in the velocity of the surrounding medium. Such increases in the fluid velocity produce still further corrosion of the material called *erosion* (to be discussed further).

Another phenomenon that affects the corrosion of materials is called *polarization.* It describes the condition of many metals when the rate of corrosion decreases with time. This condition is caused when the rate of corrosion is limited by the diffusion of ions within the

Figure 4-39 (a) Stress cells. (b) Intergranular corrosion. (c) Stress cells (anodic regions or anodes) in a bent nail produced by cold working will corrode faster than other areas. A cold-worked steel bar that is bent produces a similar condition.

(a)   (b)   (c)

electrolyte. A decrease in ion concentration at the cathode or an increase in metal ion concentration at the anode will decrease the corrosion rate. The rate of corrosion depends on the cell potential and it, in turn, is changed by the ion concentration. Metals will experience higher corrosion rates if the electrolyte is a moving fluid which increases the ion removal rate, thus reducing the polarization.

In addition to these processes, metals will corrode more easily if they or portions of them are under strain imposed either by some industrial process or by some unexpected application of excessive force. Strained metals are more active (more anodic) than unstrained metals. A bent nail (see Figure 4-39) or dented steel automobile fender will readily show this phenomenon in a graphic way.

4.4.5 Types of Corrosion

A listing of the different forms of corrosion tends to impress the reader of the scope of the problem and to bring home the importance of attending to it early on in the design process. A brief description of these varied forms of corrosion follow and further reference to them can be found throughout the text. *Uniform corrosion* includes general rusting of ferrous metals which leaves a scale or deposit over an entire exposed surface. *Pitting corrosion* results from localized surface defects. One or two examples might illustrate this type of corrosion. A marine organism attached to a metal surface of a boat hull will cause a pit to form which can result in deep, isolated holes under the surface of the metal. Pitting of a chromed automobile bumper occurs when the chromium barrier coating is broken exposing the steel underneath. This type of corrosion is more devastating than uniform corrosion because of the localization of the attack. If a steel tank suffers this type of corrosion it could lead to a hole causing a leak of its contents. Stainless steels and aluminum are not recommended for use with saltwater because they are subject to pitting. Monel and copper alloys are more suitable. *Cavitation corrosion* is the result of high-velocity motion between a metal and a liquid. If the pressure accompanying a high-velocity motion drops to the vapor pressure, the liquid will vaporize and form a vapor cavity at the metal surface. At a slight increase in pressure, this bubble will collapse, causing a concentrated liquid impact resulting in the erosion (metal particles are removed from the metal's surface) and pitting, depressions, and pock marks to appear on the metal surface. *Crevice corrosion* involves the junction of two metals exposed to a corrosive environment. The basic causes have been identified as (1) a difference in oxygen concentration, and (2) a deficiency in metal-ion concentrations. A concentrated solution is one that has a relatively large amount of dissolved solute: For example, a drop of water on a metal surface may have a greater concentration of oxygen near the edge of the drop than in the middle which would aid the corrosion of the metal by the forming of local anodic and cathodic regions. *Fretting corrosion* is produced by vibrations which result in the striking/rubbing at the interface of close-fitting, highly stressed metal surfaces. Splines, keyways, and bearings are examples of where this takes place. *Intergranular corrosion* occurs along grain boundaries (see Figure 4-39b). Grain boundaries such as heavily stressed or cold-worked areas are anodic regions as they have more crystal defects that generate higher energy levels than defect-free crystal regions as in the interior of a crystal or grain. Stainless steels when heated to temperatures between 500° and 925°C for sustained periods become sensitized to this form of attack. The high temperatures cause the precipitation of chromium carbides along the grain boundaries. A chemical reaction occurs between the chromium and carbon atoms, which diffuse to the grain boundaries in the form of particles of chromium carbide. In so doing a chromium-depleted zone is created adjacent to the grain boundaries which is highly susceptible to corrosion because it has lost its former alloying composition. Sensitizing can be brought about by welding these steels. Through proper heat treating, or by lowering the carbon content, this condition can be rectified.

Glasses are degraded by a process called *weathering* that involves the *leaching* (to extract a soluble substance from some materials, to dissolve and be washed away) of ions under high humidity conditions. Degradation of plastics is a chemical interaction with the ultraviolet rays of sunlight. Ceramic materials such as bricks degrade by *spalling* (flake or chip) due to condition of high thermal stress. *Selective leaching,* or *preferential corrosion,* is found in solid solution alloys and results in the preferential removal of one element or component as a consequence of corrosion. The most common example is *dezincification* of brass containing at least 15% Zn when both Zn and Cu are attacked with the Cu ions being replated onto the brass and the zinc is selectively leached leaving only a porous mass of copper in the dezincified region of the metal. Al, Fe, Co, Cr, and other elements are vulnerable to this preferential corrosion. *Erosion* is a form of stress-assisted corrosion. The friction and wear caused by fluid impingement upon the interior surface of metal pipes at locations such as elbows accelerates the corrosion rate.

When metals (especially dissimilar metals) are placed in an electrolytic solution, one metal becomes the anode and the other becomes the cathode just as in a battery. This is *galvanic corrosion.* Cations of one metal (the anode) enter the ionized water to join anions. Left behind are electrons which travel through the metal conductor to the cathode. Figure 4-37b illustrates the electrochemical corrosion of an iron bolt that is connected to a piece of copper, possibly a washer or nut. The positive-charged ions (Fe) enter the water to form bonds with the negative hydroxide ions (OH^-). Metallic bonds are broken to form ionic bonds. Electrons (e^-) left behind in the iron travel through the iron bolt to the copper where they meet positive hydrogen ions (H^+) and become neutral hydrogen atoms (H). The anode (the iron bolt) is pitted as it loses atoms. Corrosion involves the same redox reactions that occur in voltaic cells (batteries) except that they are unwanted. Redox reactions occur between some metals and not others. For example, Cu and Ag produces oxidation-reduction reactions but the combination of Cu and Zn do not. This difference in behavior involves a study of electrode potential, electromotive force, electric charge density, and other concepts beyond the scope of this book.

It is possible to determine how active a metal will be by referring to the electromotive force series using seawater as an electrolytic solution. The metals on the top are stronger oxidizers and thus are active anodes. The bottom metals are less active and become reducing agents (cathodes). Greater galvanic action results from joining metals that have the greater separation on the table. Table 4-2 is a galvanic series in which the different elements and alloys are ranked according to their tendencies to lose their electrons in a particular environment. These series are developed for particular environments such as seawater, freshwater, or industrial atmospheres. They are based upon the electromotive force series which compares the tendencies of a metal to give up its electrons. If a metal in the form of an electrode connected to a standard hydrogen or platinum electrode along with a standard electrolyte has a greater tendency to give up its electrons than does the hydrogen or platinum, then its potential is termed "negative" and the metal is anodic with respect to the hydrogen or platinum electrode. Metals

TABLE 4-2   ORDER IN WHICH DISSIMILAR METALS PRODUCE
ELECTROMOTIVE-FORCE (GALVANIC) SERIES

Greater oxidizing agent ↑	*Electromotive series* (corrosion potential)	Active or Anodic (greatest corrosion) more active ↑
	Magnesium 2+[a]	
	Beryllium	
	Aluminum 3+	
	Zinc 2+	
	Chromium 3+	
	Iron (steel) 2+, 3+	
	Cadmium 2+	
	Tin 2+	
	Nickel 2+	
	Cobalt	
	Lead 2+	
Greater reducing agent ↓	Brass	
	Copper 1+	
	Silver 1+	Noble or Cathodic (least corrosion) more passive ↓
	Titanium 1+	
	Platinum 2+	
	Gold 3+	

[a]Oxidation numbers.

Anodic coatings are more active in the galvanic series than the substrate upon which
the coatings are applied. Noble coatings refer to metal surface coatings that are less
active in the *galvanic* series than the substrate upon which the coatings are applied.

are listed in tables by their tendency to oxidize, to be active, or to be anodic, as in Table 4-2.
Those metals are arranged such that one can determine which of two dissimilar metals will oxidize in the presence of another assuming there is a medium (electrolyte) in contact with both
metals that is capable of transmitting ions (ion exchange). Usually an aqueous solution (containing water) is the most common medium for corrosion to occur. In Table 4-2 Zn is located
high on the list of metals, indicating it is more active or anodic, and more capable of being corroded, than those metals listed below it. Fe is below Zn in the listing, which indicates why Fe
many times is coated with Zn (galvanized metals) to protect the Fe from corrosion. Note that
the metals listed in Table 4-2 have a superscript identified by the asterisk as an oxidation number. *Oxidation number* or *oxidation state (OS)* designates the number of electrons that an
atom loses, gains, or otherwise uses in joining other atoms in compounds. In the oxidation
process the OS of elements increases, and in reduction the OS of an element decreases. The
OS of an atom, by itself, is zero. In the equation for the oxidation of Zn,

$$Zn \rightarrow Zn^{+2} + 2\,e^{-2}$$

The OS of the Zn atom on the left in its free state is 0 and it increases to $+2$ on the right because it has lost 2 e^-. Redox reactions must always occur together. The total number of electrons associated with oxidation must equal the total number associated with reduction. Again
the assumption is that all four conditions must be met for corrosion to occur.

4.4.6 Corrosion Prevention

Corrosion prevention requires large expenditures of capital to prevent the incipient loss of the
world's limited material wealth. Protection from corrosion takes many forms. To prevent corrosion requires a good knowledge of the design of materials, their environments in which materials spend their service lives, and the ability to translate these factors into the proper selection of the best materials for the intended application. *Design* also involves a knowledge of the
many types of corrosion and their causes, including the techniques required to reduce or eliminate corrosion. If the distance between the anode and the cathode is designed to be as large as

possible, then corrosion is lessened. Eliminating the difference in compositions in a material will prevent the setting up of localized cells. Alloying of elements such as aluminum or chromium with steel produces oxide layers that make the metal passive. Conversely, alloying of certain dissimilar metals produces galvanic cells within the metal which enhances corrosion. Two-phase *alloys* have more potential to corrode than a single-phase metal. In steel, ferrite is anodic to cementite which can cause galvanic corrosion. Grain boundaries are regions of higher stress in relation to the interior of the grains. Fine grains are anodic to coarse grains in the same metal. High applied *stressed areas* accelerate corrosion and reduce the time required to produce failure. Such areas can be produced by cold working (see Figure 4-39c).

Stress relieving through heat treatment, structural changes, and design considerations provide more homogeneous metals with nearly equal energy levels. In addition to heat treating metals, alternative techniques are used to produce stress relief. For example, the *Meta-Lax* process with sub-resonance stress relieving is used by auto and marine engine builders. It is effective in reducing the resonant frequency of a material and thus relieving the internal stress buildup in a material as a consequence of a manufacturing process. Magnetic Processing Systems, Inc. has developed a pulsed magnetic treatment that relieves stress and improves fatigue properties of metal parts having diameters up to 300 mm.

Surface roughness decreases corrosion resistance because the rougher the surface the more the chance of collecting entrapped corrosive media. Fatigue failures can be initiated by corrosion. Corrosion fatigue reduces fatigue resistance by initiating cracks and crevices which, in turn, increase crack propagation. A choice of filler wires in *welding,* which are the same or slightly more noble in composition than the base metal, reduces corrosion. The use of insulating washers, bushings, and gaskets in *bolted connections* is a good design practice. Neutralizing or "buffering" the electrolyte by adding inhibitors as is done with antifreeze coolants in automobile radiators has proven to be a successful practice. These inhibitors slow the anode–cathode reactions thereby protecting the cast-iron engines and the brass-copper or aluminum radiators. The avoiding of surfaces that will collect liquids and debris by incorporating drainage or sealing is an example of the many good rules of design that are intended to reduce the opportunity for corrosion to occur. The atmosphere provides excellent electrochemical mechanisms to promote corrosion. Oxygen and ozone (a highly oxidizing allotrope of oxygen) contribute to smog at the earth's surface but provide a protective layer in the upper atmosphere. They also cause the degrading of rubber products. Sulfur dioxide (SO_2) from burning fuel oil and coal, ammonia, hydrogen and hydrogen sulfide, dusts, and salts exist in varying amounts depending on location (marine, industrial, or rural). These active chemicals produce different rates of corrosion for the materials exposed to them. The U.S. Air Force stores its excess aircraft in open storage ranges in Arizona, where the effects of atmospheric corrosion are nil. Table 4-3 shows that materials are tested and then selected because of their corrosion resistance. It is vital to provide steel with some form of corrosion resistance such as adding elements (e.g., chromium) or painting or coating the surfaces. The many different *environmental conditions* under which structures and devices need to operate during their designed operating lives including those new influences and atmospheres that appear almost daily is another category of specialized knowledge that needs to be acquired. The large variety of environments to which a material can be subjected makes corrosion protection quite complex. On the other hand, *polymers* have been developed that are so resistant to our natural environment that they become nearly indestructible and are solid waste pollutants.

4.4.6.1 Protective coatings.
Although aluminum oxidizes quickly, in doing so it forms an oxide film on the surface that prevents oxygen from reaching the metal below. Iron and steel do not oxidize as rapidly as aluminum, but the scale on iron is soft and porous which allows oxygen to penetrate farther into the metal until it corrodes away to nothing. A low alloy steel known as Cor-Ten uses a small percentage (less than 0.5%) of copper to produce a compact oxide; therefore, the oxide coating on Cor-Ten also serves as a barrier and reduces corrosion to less than one-fourth times that of regular steel. Two common *coatings* for high-strength ferrous fasteners are zinc dichromate and a three-layer phosphate-zinc-organic (PZO)

TABLE 4-3 POSSIBLE CORROSION RATES FOR METALS IN THREE GENERAL ATMOSPHERES

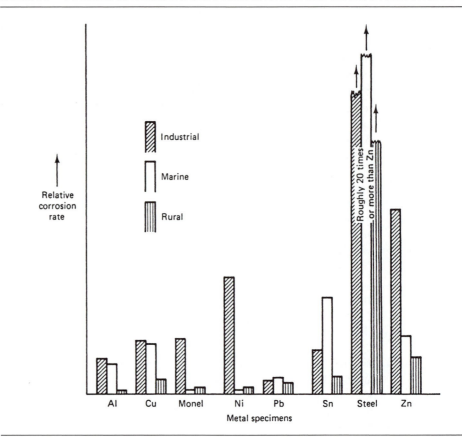

coating. Zinc dichromate consists of an electroplated zinc layer with a chromium conversion top layer. The topcoat is an electrical insulator that resists the flow of electrical current. The zinc layer is anodic, or *sacrificial*, and corrodes preferentially when exposed to a corrosive environment. Zn is used in *cathodic protection*, in which the Zn, the active metal, is called the *sacrificial anode* (see another example of this type of protection also in this paragraph). If a Zn coating on steel is broken, the Fe is still protected and the corrosion product (oxides of Zn) protects the Zn from further corrosion. Such fasteners when tested in a salt spray test, such as ASTM B117, should survive up to 400 hours before red rust appears. The three-layer coating consists of an initial layer of zinc phosphate, which provides a small amount of corrosion resistance, but its main function is to promote the adhesive of other coatings. The second layer is a zinc-rich inorganic resin made of mixed metal oxides. The zinc provides anodic protection to the underlying iron. The final layer is an aluminum-impregnated epoxy resin which protects iron; the aluminum makes the coating compatible with aluminum materials.

If you coat iron with Cu or Sn the iron is protected, provided the coating is not disrupted by cracks or damaged by impact. Once the underlying iron is exposed it will corrode (oxidize) because iron is more active (more anodic) than either Cu or Sn (refer to Table 4-2 to confirm this statement). In using Table 4-2, if one is in the process of designing a structure that calls for connecting two dissimilar metals, find their relative position in the table and identify that metal on the anodic end of the table that will corrode. Alloy additions strongly affect the corrosion potential of metals. Stainless steel is more cathodic than low carbon steel. Copper and its alloys rarely corrode galvanically because copper is cathodic to such metals as iron, aluminum, magnesium, tin, or lead. The further apart two metals are in Table 4-2, the greater the

Figure 4-40   Polluted air not only harms animal life but also accelerates corrosion. Smog-eating radiators use PremAir™ coatings to convert ozone (O_3)—the main component of smog—into breathable oxygen (O_2). (Ford Motor Company)

potential for corrosion when they are connected in the presence of an electrolyte. Suppose instead of Cu or Sn you choose Zn to coat the iron and the coating of Zn is damaged or cracked. The iron substrate is still protected from corroding because the Zn is more active (more anodic) than iron and the Zn will be attacked (oxidized) producing Zn atoms that protect the Zn from further corrosion. Fastening Zn anode plates on boat hulls (steel) or Zn collars on propeller shafts prevents galvanic attack between more noble metals such as steel hulls or bronze propeller shafts. Protective coatings and *surface engineering* receive considerable attention throughout this book. Often the preparation of materials for coatings and application of coatings, plus disposal of gases, liquids, and solids used in coatings, cause harmful effects on the environment. Heavy metals, chlorofluorocarbons (CFCs), and volatile organic compounds (VOCs) have been targeted among the polluters.

A new coating known as PremAir™, developed by Engelhard Corporation and Ford Motor Company, offers the potential for a new system of "smog-eating radiators" (see Figure 4-40). The catalyst system may turn car radiators into smog-eating, air-filtration systems to complement the catalytic converters now used in exhaust systems. This system may have the effect of cleaning all sources of smog from the atmosphere as air flows over vehicle radiators. Not only will this help animal life but it will also reduce corrosion. The developer's testing indicates the system will destroy 97% of VOCs, 90% of nitrogen oxides (NOs), and 96% of carbon monoxide (CO) from the air it contacts. Protective coatings are the most familiar methods for preventing corrosion. Paint, varnish, oil, and a variety of polymeric and ceramic coatings prevent oxygen and moisture from reaching the metal to be protected. Automobile bodies are dipped into protective coating solutions. Zinc-coated steel (galvanized steel) is an example of both protective coating and cathodic protection. Engelhard Corporation uses electrophoresis to coat their automobile radiators.

Most present-day *lubricants* for threaded fasteners are based upon metal powders of lead, nickel, and other heavy metals which are a source of health or environmental risks. These lubricants are also the main ingredients in the production of galvanic corrosion. Dow-Corning Corporation has developed ultrapure, high-temperature, metal-free lubricants made of high chromium, nickel, or molybdenum alloys to overcome the problems of these older lubricants. Galvanic corrosion comes to the forefront again when automakers began to substitute lighter-weight metals in their campaign to produce more efficient automobiles. Bolt-on applications for hoods and trunk lids where aluminum can be substituted for steel presents little problem. The different metals can be kept from making contact with each other by using polymeric materials. When greater use of such metals is attempted, as in the case of using aluminum for welded-in panels such as roofs and floor pans, then a greater problem must be solved: how to join an aluminum roof to a steel body frame. Water, air, and salts will go between the metals and form an electrolyte, causing the two dissimilar metals to act as a battery resulting in destructive corrosion. Texas Instruments has developed a transition metal made of two metals (aluminum and steel) which are joined under such high pressure that they bond at the molecular level. This type of bonding eliminates the galvanic reaction. Such sandwich material is cut into strips and placed between the dissimilar metals to prevent them from making direct contact. The parts would then be welded in the conventional way. This new technique is undergoing testing to integrate it into the assembly line without reducing assembly-line speed.

Corrosion produces cracks on the surface of metal parts which grow with time and produce fractures. Some fractures are catastrophic. Shot peening and nitriding are other processes used to strengthen the surface layers of a metallic part. Another parameter for highly loaded fasteners is the possibility of environmental-assisted cracking (EAC) specifically from *hydrogen embrittlement.* The corrosion protection methods for steel fasteners involve acidic environments, which permit hydrogen to be absorbed into the bulk material and may lead to delayed fractures. For this reason it is important in the manufacture of fasteners to include an additional post-heat treatment to remove residual hydrogen. Titanium wheel lugs have now replaced steel lugs by Ford in the Partnership for a New Generation Vehicle (PNGV). Titanium with its high strength and low density does not need the expensive protective coatings required by ferrous steels. A 9-gram titanium nut replacing a steel nut weighing 16 grams represents a 44% weight reduction. In addition to cutting weight these titanium lugs prevent galvanic corrosion with magnesium wheels. If two metals are to be connected in some aqueous atmosphere, corrosion could be slowed by choosing metals that are close to each other (see Table 4-2). To control or prevent galvanic corrosion it is necessary to eliminate only one of the four essential elements previously mentioned: anode, cathode, electrolyte, or physical contact between the anode and the cathode.

By keeping all surfaces of the materials dry and the humidity below 50%, the effectiveness of the electrolytic substance can be reduced. If this is not possible consideration should be given to electrically insulating one from the other. The relative size of the anode and cathode also plays a role in preventing or slowing corrosion due to galvanic action, a rule known as the *anode–cathode area ratio.* If you make the anode surface large compared with the cathode, corrosion will be slowed or stopped as in the case of a steel bolt joining two aluminum plates. The anode in this case is the two plates and the cathode is the bolt. If you reverse this situation with the steel plate as the anode it would corrode the steel bolt in a rapid fashion. Chromium (Cr) plated steel provides a hard coating especially where wear resistance is desired. Chrome automobile bumpers were once popular because of their appearance; however, if the chrome plating is scratched, the steel substrate will corrode. The small exposed area of steel becomes anodic. The chromium is cathodic to the steel, causing the steel to corrode. This situation violates the rule rendering a small ratio. A summation for this situation is the higher the ratio of the area of the anode to the area of the cathode the less severe the corrosion. Remember the rule: The anodic material is where the current leaves and enters the electrolyte and subsequent to that leaves the electrolyte and enters the cathode. Stopping or slowing oxidation-reduction reactions will stop or slow corrosion.

4.4.6.2 Materials selection.

The subject of *materials selection* which this text stresses comes to the forefront in the battle to save our limited natural resources from the ravages of corrosion. Materials selection must involve the placing of materials into the proper environment to prevent corrosion. The heating of stainless steels up to high temperatures when undergoing heat treating or welding causes a depletion of chromium and a partial loss of corrosion protection. The depletion appears as chromium carbide particulates precipitate along

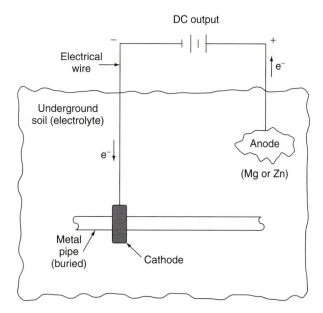

Figure 4-41 Cathodic protection of a buried pipe using a sacrificial anode and a source of electrical current.

the grain boundaries which results in the grain boundary region being susceptible to ***inter-granular corrosion.*** Another technique is alloying the steel with titanium (Ti) or niobium (Nb), called ***scavenger elements,*** which have a greater tendency to form carbides with carbon than does chromium, thus keeping chromium in the solid solution. Measures to prevent ***selective leaching*** involve redissolving the chromium and reducing the amount of carbon below 0.03 wt% C to reduce the carbide formation.

Illustrative Problem

A structure is being designed that has solid metal plates joined with nickel bolts. Electrochemical corrosion is anticipated due to humid environmental conditions (at least 50% humidity). The choice of materials for the solid metal plates, which require considerable tensile strength, is limited to copper and plain carbon steel. Select the best materials for the plates, and explain your reasoning.

Solution   Consulting Table 4-2, the best material is plain carbon steel, as steel is more anodic (active) than nickel. Steel is closer to nickel than is copper, thus reducing the potential for galvanic action between the two metals.

4.4.6.3 Cathodic protection.   Protecting metals from local soil conditions that are unusually corrosive (i.e., with high concentrations of sulfides, chlorides, or hydrogen ions) may necessitate neutralizing the soil by adding limestone, protective coatings or wrappings, or some form of cathodic or anodic protection. Large structures made of steel or iron that are in contact with soils or water such as pipelines or ship hulls can be protected by the use of a sacrificial anode as mentioned in Section 4.4.6.1 (see Figure 4-41). This anode is made from an active metal (anodic to the metal being protected) and connected directly or through an electrical wire to the metal object being protected. Oxidation occurs at the active metal which slowly dissolves in use. The iron or steel object acts as the cathode whose surface acquires electrons from the anodic reaction. Magnesium is one of these metals that finds universal use in cathodic protection. Magnesium drain plugs in auto engines provide cathodic protection: The plug is the anode and the engine block is the cathode. Magnesium anodes in domestic hot water heaters are sacrificial anodes protecting the steel tank through the cathodic protection mechanism. Cathodic protection of a buried steel pipe or structure is accomplished by connecting the pipe or structure (the cathode) to a sacrificial anode of magnesium by a wire conductor with a direct current passing through the soil (the electrolyte in this case) to the pipe or structure and completing the circuit through the wire conductor (see Figure 4-41). By impressing a direct current on a buried steel pipe (the cathode) through the use of a buried graphite anode (the sacrificial anode), the steel pipe is protected by the current passing through the electrolyte (the soil) from the anode to the pipe (cathode).

Shot peening is a cold-work process in which the surface of a metal is bombarded with small media called shot. Each piece of shot acts as a tiny peening hammer that produces a small indentation or dimple on the surface. To produce a dimple, the surface must be subjected to a tension stress which causes the necessary deformation. At the same time, the material below the dimple, in seeking an equilibrium condition, is subjected to a compressive force that attempts to restore the surface to its original shape. The overall effect of the number and overlapping of dimples is the formation of a narrow layer of residual compressive stress at and below the surface. The maximum compressive stress formed is at least as great as half the yield stress of the metal being peened. Surface hardness also increases as a result of the cold working. Compressive stresses are beneficial in increasing resistance to surface fatigue failures (pitting), corrosion fatigue, stress corrosion cracking, galling, and fretting. The accompanying cold working results in work hardening, reduction in porosity, and greater intergranular corrosion resistance. The shot should be as hard or harder than the parts being peened. Media typically used for shot peening are small spheres with no sharp edges or broken particles made from cast steel. Ceramic or glass beads are used where iron contamination on the peened surface is a concern.

Peen forming is used to produce a compound, convex curvature on the peened side in aircraft wing skins to increase resistance to flexural bending fatigue. Hard anodizing of high-strength aluminum improves the fatigue resistance of coated materials when shot peening is used prior to anodizing. A follow-on process that requires shot peening, first developed by Metal Improvement Company, Inc. called ***Chemically Assisted Surface Engineering*** or ***C.A.S.E.,*** extends the surface fatigue life (pitting of the contact faces) beyond that of shot peening alone. First applied to racing gears it has been extended to passenger car transmission gears and boat propellers. Using chemicals this process produces a polished surface (superfinishing) that removes the surface asperities left over by shot peening and provides smooth contact surfaces with no peaks where microwelding can occur and pitting can initiate. Only low micro indentations remain that act as reservoirs for lubricant retention and provide locations for surface micromovement occurring under sliding and rolling pressures.

4.4.7 Corrosion Protection

The large variety of environments to which a material can be subjected makes corrosion protection quite complex. On the other hand, polymers have been developed that are so resistant to our natural environment that they become nearly indestructible and are solid waste pollutants. Some common methods of corrosion protection for metals include cathodic protection, protective coating, stress relieving, insulation, alloying, materials selection, and design. Figure 4-42 shows that materials are tested and then selected because of their corrosion resistance. Many rules of good design are intended to reduce the opportunity for corrosion to occur.

4.4.8 Water Absorption and Biological Resistance

Natural and synthetic polymers are subject to biological attacks and the absorption of water. When left untreated, dried wood will absorb moisture and then serve as a good environment for the growth of fungi (small plants) and insects; they feed on cellulose and lignin and cause deterioration. Unprotected dry wood is also a host for termites and other insects. Some synthetic polymers swell through water absorption, which causes deterioration and provides a good environment for damaging microorganisms. The corrosion of metals through oxidation can result from organisms such as barnacles living on the material. *Biofilms* consist of communities of bacteria that form sticky layers and act as adhe-

(a)

(b)

(c)

Figure 4-42   (a) Severe spalling and delamination of concrete roadway caused by corrosion of embedded rebar (steel reinforcing) initiated by deicing salts. (W.R. Grace & Co.) (b) Plastic-glass composite highway guardrail made of pultruded glass-fiber-reinforced polyester is noncorrosive, maintenance-free, and stronger than similar structures of steel and steel-reinforced concrete; they are also comparable in cost. (*Advanced Materials & Processes*) (c) Corrosion of steels in a marine atmosphere. Left: Low-copper steel. Center: Ordinary steel. Right: Nickel-copper-chromium steel. (The International Nickel Co.)

sives to bond on many substrates such as from human teeth to boat hulls. As the biofilms grow through cell multiplication they form pillars and mushroom-shape structures which cause problems by clogging sewer pipes, fouling machinery, and contaminating drinking water systems. In drinking water pipes the films cause corrosion and disease. To counter biofilms, industry has developed enzymes to dissolve the films and blocking agents to stop their formation.

4.4.9 Wear, Friction, Tribology, and Wear Resistance

The worldwide degradation of engineering materials caused by the insipid attack by chemicals and/or wear amounts to a staggering financial loss and a large drain on the world's limited resources. Previous paragraphs in this module describe corrosion that is the chemical removal of atoms. *Wear,* including some of its various types such as abrasive, adhesive, and erosive, also takes its toll on materials by removing part of the materials by mechanical action using solids, liquids, or gases. Environments such as constantly changing temperatures can cause wear from repeated expansion and contraction. Rates of wear are difficult to measure beforehand, even

Figure 4-43   A friction force (F) generated by a pushing force (P) acting on a solid block resting on a horizontal surface.

though there are many different types of tests for materials under various testing conditions. *Tribology,* the science and technology dealing with interfacing surfaces, focuses on design, friction, lubrication, and wear behavior to avoid problems with mating parts.

Every object moves in accordance with the forces acting on the object. A *force* may be defined as any influence that can cause a change in the state of motion of an object. The magnitude (size) of forces is expressed in units such as pounds (U.S. system) or Newtons (SI system) accompanied by their direction expressed such as up or down, left or right. Forces are vector quantities since they have both a magnitude (size) and a direction (N or S, left or right). If these forces (F) balance each other there is no net force left over and the object does not accelerate (change its motion). If the net force is not zero then this force will cause the object to accelerate (a), producing motion. The acceleration is given by the expression $a = F/m$ in which the mass (formerly called *weight* in the U.S. system of units) of the object is represented by m. Sky divers jumping from great heights out of an airplane fall at a constant, or terminal, speed. Constant speed indicates that divers reach a point in their fall in which they no longer accelerate; that is, their speed no longer changes. The terminal speed, in this case, is approximately 125 mi/hr. What produces this result is the balancing of the forces acting on the diver. The force of gravity acts downward on the diver and the opposing force is friction which is the result of the diver moving through the medium of air. The frictional force produces air resistance.

Friction, or frictional force, exists not only in falling objects in air but also between any object that moves over or through matter. An automobile tire moving over a highway, a fish swimming through water, a tennis racket contacting a ball, a metal piston moving within a cylindrical surface, and a tool insert in contact with a moving metal surface are examples of situations when friction is present. Motion is further defined as relative motion, as it refers to two bodies or objects either in contact with each other such as a brake pad and a disk or one body moving through a medium such as air or water. The only exception in which one can rule out the presence of friction is when the object under study is moving through a perfect vacuum. This environment is difficult to accomplish even in outer space.

Practically all mechanical manufacturing processes involve friction and usually in a negative way. Friction can be considered as either a positive or a negative factor. For instance, it certainly is a positive factor when trying to move an automobile with an inoperable engine by pushing it. People who drive on ice-covered roads also recognize the significance of this statement. Even the simple act of walking would not be possible. In spite of this, friction is looked upon as a negative or undesirable feature that must be overcome by expending energy and financial resources in some form. As a result much effort is spent on designing processes that reduce friction to the greatest extent possible. Friction is a force that impedes sliding (see Figure 4-43). These forces originate at the surfaces of bodies in contact. Irregularities of the surface in contact interlock with each other. Bodies then adhere at the touching points of these irregular surfaces and harder materials tend to "plow out" softer materials with which they are in contact. Without friction we would be unable to walk or drive an automobile along a road. Some 20% of the power developed by an automobile engine is lost to internal friction. Lubrication of rubbing surfaces reduces friction in machinery. Using Figure 4-43, friction (F), a vector quantity, is shown acting on a solid block as sketched in two dimensions sitting on a flat horizontal surface.

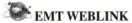 **EMT WEBLINK**

Yahoos Materials Science Site—*http://dir.yahoo.com/science/engineering/material_science*

TABLE 4-4   SOME REPRESENTATIVE COEFFICIENTS OF
FRICTION

Materials	μ_s	μ_k
Glass on glass	0.94	0.35
Steel on aluminum	0.61	0.47
Steel on dry steel	1.20	0.85
Steel on lubricated steel	0.12	0.07
Teflon on Teflon	0.04	0.04

The block represents any object being subjected to friction. The block has mass so therefore it has a gravitational force called weight (W) that acts in a vertical direction. The weight is opposed by the reaction force of the surface on the block called a normal force (N). Since the W force and the N force are equal in magnitude there is no unbalanced force and thus no acceleration or movement in the vertical direction. If we now try to move the block by pushing it to the left with a pushing force (P), the block generates or reacts to this force with a frictional force (F). Note that the direction of the friction force is always opposite to the impending relative motion between the surfaces in contact. When the maximum force of friction is exerted just before the box starts to slide, the relationship between the friction force, the normal force, and the coefficient of friction (u) is $F = uN$. Up to this point the discussion involves the case of static friction and the term *static* could be used throughout. It has been determined experimentally that the coefficients of friction are nearly independent of the size of the contact area between metal surfaces. For polymer surfaces including wood, this statement does not apply.

Thus far we have discussed **static friction,** the frictional force sufficient to prevent relative motion between surfaces. Another form of friction is **kinetic friction,** or **sliding friction,** where there is relative (sliding) motion at the interface of the surfaces in contact. Generally the coefficient of kinetic friction is less than the coefficient of sliding friction for two surfaces (see Table 4-4 for some representative materials and their coefficients). To distinguish between the different coefficients, the subscripts *s* and *k* are used. For example, the coefficient of kinetic (sliding) friction (u_k) for steel sliding on dry steel is 0.48. Note this coefficient is a dimensionless quantity (no units). For glass sliding on glass the coefficient of static friction (u_s) is 0.94.

One major effect produced by friction is wear, or the gradual removal of material from the surface of an object. Manufacturing operations contain numerous examples of wear such as tool wear (saw blades, drilling bits, inserts), erosion in piping, and molds caused by fluid flow or by particles contained in the flowing medium that impinge on the conducting surfaces; and general obsolescence of machinery as a result of the effects of friction over a long period of operation is another example.

Additionally, mechanical properties of a material affect how the material can be worked. For example, very low hardness can be detrimental to a metal's machining performance. High ductility causes a metal to tear as the chip is formed, resulting in a poor finish. Mechanical properties rank highest in importance when selecting materials for many applications. Terms such as *durability, formability, rigidity, strength,* and *toughness* are often used to describe these properties. However, three tests supply the most useful information for most applications, namely, the tensile, impact, and hardness tests. Durability can be assessed in part by the hardness test. Strength can be described by the results of tensile testing, and impact testing provides adequate data for determining the rigidity of a material.

Figure 4-44 shows four common tests to try to determine the rate of wear for some common materials under attack by oxidation, acid corrosion, and erosion by mechanical testing. This figure points out that one factor in the struggle to control wear in materials is to protect their surfaces from these types of attacks. In this case, ceramic (TiC) is used to coat the material (graphite) with a hard surface impervious to chemical attack, wear, or their combination. Further discussion of some of these techniques to provide wear resistance is found in each of the succeeding modules (see Section 7.1.3).

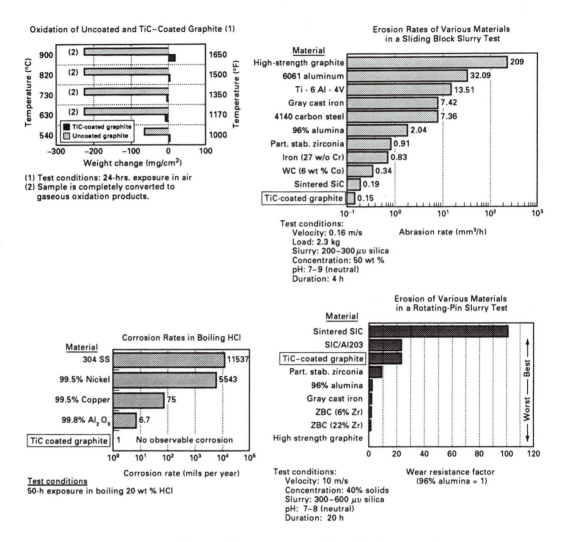

Figure 4-44 Corrosion wear rates for various materials under different source test conditions. Erosion is a corrosion mechanism. (Alanx)

According to developers at Argonne National Laboratory a new *coating* called *near-frictionless carbon (NFC)* or commercially, Slick 001, is much harder and slicker than Teflon. This carbon-based material has a coefficient of friction of only 0.001, which is one-twentieth that of the previous record holder, *molybdenum disulfide.* For comparison, Teflon's coefficient of friction is approximately 0.04. Produced quickly and in large quantities this nontoxic material adheres to a variety of surfaces including plastic. NFC is produced by using a proprietary gas-containing carbon which is converted into a plasma "soup" of atomic elements from which the carbon elements are allowed to drift onto a substrate forming a coating. *Wear resistance* of the new material was tested by sliding a sapphire ball over the coating with a pressure equal to 145,000 psi (normal atmospheric pressure is 14.5 psi). After 5 million cycles the new material lost just 1.0 μm (0.0004 in.) of its thickness. Promising applications are to be found in automobile turbocharger rotors and fuel-injector components, oilless bearings, and spacecraft mechanisms (see Figure 3-17 and Section 7.1.3 for further discussion of these subjects).

4.5 THERMAL PROPERTIES

A knowledge of thermal properties is vital to those who work with materials to make parts that will serve at temperatures other than as fabricated, or that perform some heat-transfer function.

TABLE 4-5   SPECIFIC HEAT (c) FOR SOME SELECTED MATERIALS AT CONSTANT PRESSURE (UNITS OF J/kg · K)

Metals		Ceramics		Polymers	
Ag	230	Al_2O_3	775	PE	2100
Al	900	BeO	1050	PP	1800
Cu	386	Silica	750	PS	1350
Fe	450			PTFE	1050
Ge	322			Nylon	1675
Mg	1020				
Si	700				
Ti	523	Water	4186		
Steel	500	Wood	1700		
Kovar	460	Human body	3470		

For example, when heated deliberately or by hot working, a metal rod exhibits three thermal effects: (1) the rod absorbs heat, (2) it expands, and (3) it transmits heat. Each of these situations has a property that measures and describes the degree to which each effect is present. *Absorption* of heat is characterized by the property's heat capacity (c_p). *Expansion* is usually described by the *coefficient of thermal expansion* (α). *Heat transmission* is identified by *thermal conductivity* (k).

4.5.1 Heat Capacity—Specific Heat

The efficiency of a material to absorb heat is known as heat capacity. **Heat capacity (c)** is the amount of heat required to raise the temperature of a material a given amount. The larger the heat capacity the greater the amount of heat needed to produce a temperature change. Its value depends on both the size of the material and its composition. Using kilocalories (kcal), or 1000 cal, it takes 1 kcal of heat to raise 1 kilogram (kg) of water by 1°C. The mechanical equivalent of heat or the amount of work equivalent to 1 cal of heat is 4.186 joules (J). Thus, 4.186 J = 1 cal or 4.186×10^3 = 1 kcal. The **specific heat** of a material (c_p) is the amount of heat necessary to raise the temperature of exactly one gram of a material by exactly one degree (1°C or 1 K). Therefore, the amount of heat necessary to raise the temperature of a given material is the specific heat times the mass of the material times the change in the temperature. The specific heat of liquid water is much higher than most other materials. Its specific heat is 4.186 J/kg ·°C. Water is a substance that requires a large expenditure of energy to produce a change in temperature. Substances with high specific heat values do not change their temperature appreciably. The amount of heat required to change the temperature of a substance is expressed by the relationship:

$$Q = c_p m \Delta t$$

where Q is the amount of heat expressed in cal or kcal that flows into or out of a substance assuming the pressure on the material stays constant, m is the mass of the substance, and Δt is the temperature change. Solving for c_p gives the specific heat of the substance. Using SI units the units of c_p are kilojoules per kilogram per °C (kJ/kg· °C). Table 4-5 shows the specific heat for some representative materials under constant pressure conditions.

Illustrative Problem

The specific heat of iron is 0.11 kcal/kg·°C. How much heat is required to raise 50 kg of iron from 20° to 100°C?

Solution   Using the expression above

$$Q = mc\Delta t = 50 \text{ kg} \times 0.11 \text{ kcal/kg} \cdot °C \times 80°C = 440 \text{ kcal}$$

TABLE 4-6   COEFFICIENT OF (LINEAR) THERMAL EXPANSION (α) FOR SOME REPRESENTATIVE
MATERIALS

Metals		Ceramics		Polymers	
		Units of $-10^{-6} \times {}^\circ C^{-1}$			
Ag	20	Al_2O_3	7	PE	100–200
Al	25	BeO	9	PP	58–100
Cr	17	SiC	5	PS	60–90
Fe	12	Si	3	PTFE	100
Mg	25	Si_3N_4	2.7	Nylon	6/6 80
Ti	9	Silica 96%	0.8	Epoxy	45–100
Steels	11–17				
Brass	18				
Kovar	5				
Invar	0.9				

Illustrative Problem

Express the specific heat of aluminum 900 J/kg·°C in units of kcal/kg·°C.

Solution

$$900 \text{ J/kg} \cdot {}^\circ C \ (1 \text{ cal}/4.186 \text{ J}) \ (1 \text{ kcal}/1000 \text{ cal}) = 0.215 \text{ kcal/kg} \cdot {}^\circ C$$

4.5.2 Thermal Expansion

Nearly all solid materials expand when heated and contract when cooled. Isotropic materials expand or contract equally in all directions. This temperature deformation results from the changes in the distances between adjacent atoms due to changes in thermal energy. The *coefficient of linear thermal expansion* (α, Greek letter alpha) describes this effect in the following equations for unit thermal strain (ϵ_t) or unit deformation due to temperature change:

$$\epsilon - \frac{\Delta l}{l_0} = \alpha \Delta t$$

where Δl is the change in length of some dimension, l_0 is the original length of that dimension, Δt is the change in temperature, and α, the constant of proportionality, is the coefficient of linear thermal expansion. In expressing the coefficient of thermal expansion the term α $(C^\circ)^{-1}$ is read "the coefficient is expressed per degree Celsius." Another notation, ${}^\circ F^{-1}$, is read in a similar manner. The expression $\Delta l/l_0$ is another way of indicating unit deformation (ϵ, strain) brought about by temperature changes. To determine the total linear deformation (δ) due to a temperature change:

$$\delta_t = \epsilon_t \times l_0$$

The coefficient for thermal expansion (CTE) for lamillar composite materials must match closely the CTE for constituent materials, otherwise the composite material could delaminate at the interfaces. The CTEs for some selected materials are shown in Table 4-6.

In metals there is a close correlation between the coefficient (α) and the melting point (T_m). The lower the coefficient, the higher the T_m value. Knowing the T_m value gives an indication of the strength of the bonding forces. A bimetallic strip (see Figure 4-45) exemplifies the use of the coefficient of thermal expansion in temperature-control relays. With a temperature change, two metals bonded together and having different coefficients will bend, making electrical contact with a switch. If the two metals A and B are thin, the induced stresses cre-

Figure 4-45 A bimetallic strip. Note that temperature t_1 is greater than t_0. After the passage of a discernible amount of time at t_1 bending is produced.

ated by the difference in expansion will cause the metallic strip to bend. The degree of bending can be magnified by the use of a bimetallic coil made of the two materials. The expansion of the coil can be used to break or make an electrical circuit when the desired level of the variable being controlled such as temperature is reached.

E materials, designed and developed by Brush Wellman, Inc., are a family of metal matrix composites made up principally of beryllium (Br) and beryllium oxide platelets. Three grades—E20, E40, and E60—are now produced by blending the beryllium and beryllium oxide powders into a homogenous mixture to create isotropy of physical and thermal properties. Used principally in electronic packaging heat sinks, each alloy can be tailored to contain specific properties by varying the volume percentage ratio of beryllium and beryllium oxide. One property that can be controlled by the proper mixing of constituents is the coefficient of thermal expansion (CTE or COT). The CTE determines the longevity of solder joints on circuit boards. The better the COT match is between joined materials the less strain is imposed on the joining materials such as solder or adhesive bonds. Less strain translates into better fatigue life for the joints and longer life for electronic circuit boards. The CTE for E60 alloy is 6.1 per degree Celsius. The mix is then hot isostatically pressed (HIP) into fully dense blocks for further processing into finished blanks and subsequent machining into components. Applications are found in *avionics* (the branch of electronics dealing with development, manufacture, and use of electronic devices and equipment in aviation and astronautics) to improve the failure rate of electronic components by reducing the vibrations experienced in flight and launch by lessening the dynamic stresses exerted on the solder or adhesive bonded joints of electronic packaged devices. Materials with high elastic modulus (high specific stiffness) have a positive effect on the transmissibility of vibrations. The elastic modulus for this family of materials is about 320 GPa with an average density of about 2.30 g/cc. This new family of lightweight materials was specifically developed to solve such problems as reducing the number of failures of electronic circuit boards as they withstand the rigors of flight.

In the design of materials systems, the deformation of materials due to temperature must be taken into account. Changes in temperature are indications of heat transfer. An example of such a case is the use in a composite structure of two different materials that might have excessive differences in their coefficients of expansion. What is a possible outcome? What force or condition is recognized by ship and road builders who install expansion joints in their structures? What is the cause of the repetitious clinking sound produced by metal wheels running on steel railway tracks? In this discussion so far, all examples indicate that there is no resistance to the expansion of materials due to temperature changes. What happens to the materials if the expansion or contraction is resisted by an external force?

Illustrative Problem

A steel rod 2 m long is fastened at one end to a fixed support and is free to expand or contract. If the ambient temperature rises by 25°C, how much will the rod expand or contract?

Solution

$$\alpha_{steel} = 11.7 \times 10^{-6} \, °C^{-1} \text{ (See Table 4-6)}$$

$$l_0 = 2 \text{ m } \Delta t = 25°C$$

$$\delta_t = \Delta l = l_0 \alpha \, \Delta t$$

$$\delta_t = 2 \text{ m}(11.7 \times 10^{-6} \, °C^{-1})(25°C)$$

$$\delta_t = 0.585 \text{ mm or } 5.85 \times 10^{-4} \text{ m (Total Deformation Expansion)}$$

Polymers have small bonding forces (van der Waals bonds) and consequently larger expansion coefficients than metals. In summary, the covalent and ionic bonded materials have the lowest coefficients (strongest bonding), polymers the highest coefficients (weakest bonding), with metals somewhere between these limits.

4.5.3 Thermal Conductivity

The ability of a nonmetallic solid to transmit heat, its ***thermal conductivity,*** depends primarily on ***phonons*** (quanta of energy). Phonons or "particles" behave somewhat like gas atoms and transfer the solid's lattice vibrations from a region of higher energy (high temperature) to a region of lower energy. They encounter more collisions as they move with the speed of sound through a material as the temperature increases. This results in a lowering of the thermal conductivity with the increase in temperature. Thermal energy and electrical conductivity are both related to a material's atomic structure. Thermal energy can be conducted through a material by two mechanisms: the movement of free electrons and the movement of phonons (lattice vibrations). Using electron band theory this conduction through conductor metals, semiconductors, and insulators is discussed in Module 9. Thermal conductivity of materials has a varied relationship to temperature. In some materials thermal conductivity initially decreases with increasing temperature up to a certain temperature and then increases. Iron and graphite are examples of this type of behavior. With other materials the conductivity increases continuously with increased temperature. Glasses, nylon, and platinum are some examples. In general, metals have comparatively high thermal conductivities while polymers have rather low values. Oriented polymers can have much greater conductivities than unoriented polymers as in the case of filaments of PE. Amorphous ceramics tend to be lower than more crystalline ceramics.

In metals, the free, or conduction, electrons also serve as carriers of thermal energy received from the phonons. Elements added to metals reduce the thermal conductivity, much as porosity and other crystal imperfections do in all materials. Polymers lacking free electrons and having a less crystalline structure (orderly atom arrangement) have poor thermal conductivities. Such a characteristic makes polymers good thermal insulators. A ***coal-based carbon foam*** recently developed by West Virginia University researchers is now being marketed for commercial use. Made from bituminous (high-sulfur) coal, it absorbs shock, is lightweight, strong, and noncombustible. It could be used to insulate Navy ships against fire and protect military vehicles from explosions. Other potential applications are in high-strength parts for the aerospace industry.

Thermal conductivity (k) is a measure of heat flow through a material. It relates heat flow (the flow of heat energy per unit area, per unit time) to the temperature gradient, causing the heat to flow. The term ***temperature gradient*** (in symbols, $\Delta t/\Delta d$) describes a temperature difference per unit distance. Using symbols, the heat flow can be expressed as q/A, where q represents the heat energy in units such as watts, Btu/hr, or calories per second (cal/s); A is the area through which the heat passes, in square units such as square meters or square feet. The change in temperature (Δt) can be expressed in °C, °F, or K. Using these symbols, thermal conductivity can be written

$$\frac{q}{A} = k\frac{\Delta t}{\Delta d}$$

TABLE 4-7   THERMAL CONDUCTIVITIES (k) FOR SOME REPRESENTATIVE MATERIALS

	[unit W/m · K]			
Metals		Ceramics		Polymers
Ag	450	Al_2O_3	35	PE	0.4
Al	300	BeO	220	PP	0.12
Cu	483	SiC	95	PS	0.12
Fe	134	Silica glass	2	PTFE	0.24
Ge	60	SiO_2	1.5	Nylon	0.24
Mg	170			Wood	0.16
Ti	30				
Steel (carbon)	50				
Stainless steel	15				

This relationship shows that thermal conductivity (k) is a constant of proportionality. Solving for k gives

$$k = \frac{q}{A} \cdot \frac{\Delta d}{\Delta t}$$

Substituting SI units into this equation, the units of k are

$$k = W/(m \cdot K) \text{ or}$$
$$k = W \cdot m^{-1} \cdot K^{-1}$$

Note that the superscript, -1, is not an exponent but rather indicates that the base quantity is to be divided as shown in the preceding equation.

Illustrative Problem

An insulating material placed inside the walls of a home is 6 in. thick. The heat flow is calculated to be 4.2 Btu/hr when the inside temperature is 70°F and the outside temperature is 0°F. Determine the thermal conductivity of the insulating material as the heat flow passes through 1 ft² of the wall. In addition to expressing k in units of Btu/(hr · ft · °F), convert your units to SI units using watts, meters, and Kelvin temperature. Using Table 4-7 identify what insulating material was used.

Solution

$$k = \frac{q}{A} \cdot \frac{\Delta d}{\Delta t}$$

$$= \frac{4.2 \text{ Btu}}{1 \text{ ft}^2 \cdot \text{hr}} \cdot \frac{6 \text{ in.}}{70°F} \left(\frac{1 \text{ ft}}{12 \text{ in.}} \right)$$

$$= 0.03 \text{ Btu/(hr} \cdot \text{ft} \cdot °F)$$

Using conversion Table 10-7, we have

$$1 \text{ Btu/ft} \cdot \text{hr} \cdot °F = 1.729 \text{ W/m} \cdot K$$

$$k = \frac{0.03 \text{ Btu}}{\text{hr} \cdot \text{ft} \cdot °F} \left(\frac{\text{ft} \cdot \text{hr} \cdot °F}{1 \text{ Btu}} \cdot \frac{1.729 \text{ W}}{\text{m} \cdot K} \right)$$

$$= 0.052 \text{ W/m} \cdot K \text{ (Polymer Material)}$$

4.5.4 Thermal Resistance

Insulating materials have low thermal conductivities and can retard the transfer of heat. The thermal conductivity of most materials is temperature dependent. In other words, the value changes with a change in temperature. Refractory materials, on the other hand, have thermal conductivities that have a minimum dependence on temperature. Porous materials such as textiles, rock, wood, cork, foamed plastics, and human-made insulating tiles are good insulators partly because of their ability to trap air, which itself has a low thermal conductivity. Furthermore, the entrapped air is free from circulating currents, which aid the transfer of heat.

Numerous applications for insulating materials exist in today's society, particularly in the area of heat (energy) conservation. Wood is a good insulating material. Its largest drawback is its flammability. As it burns it remains strong, with charring action a positive factor in its resistance to burning. The char acts as an insulator material. Once it loses its strength it fails catastrophically. Used in building construction it is often impregnated with fire-retardant chemicals. Further information can be found in Module 6, Section 6.3.2.1 dealing with *flame retardants,* charring, and ablation. Organic materials such as polymers are flammable, but certain materials (additives) such as chlorine, phosphorus, or metallic salts will reduce the chances that combustion will occur or spread. Steel does not burn but it loses strength at high temperatures causing collapse by drooping or otherwise giving way (see Figure 6-52). Insulating materials are used in fire protection of buildings. Mineral fiber is made from spun molten iron slag. Gypsum-based products in the form of blankets or boxes are used to encase building elements. These materials remain inert to fire, do not burn, and do not conduct heat readily. Gypsum is a hydrated sulfate of calcium with the chemical formula $CaSO_4$ $2H_2O$ which occurs in sedimentary rocks. When heated to above 90°C the chemical bonds in gypsum break down and release water which cools a substrate. Cementious-sprayed fire-resistant materials are cement- or ceramic-fiber-based materials commonly referred to as "flung dung," which are sprayed on in liquid form and then left to harden to protect steel beams and columns. Intumescent materials are sprayed, painted, applied as putty, or wrapped in sheets onto a surface to be protected. (*Intumescent* means the material swells under heat.) When this type of material is exposed to high heat it foams up or "blooms." Once foamed the material hardens and remains inert exploiting the foam's air bubbles as an insulating cushion. Also, as they expand, they can stop the spread of fire by sealing gaps and cracks. Endothermic foaming materials developed to protect electrical cables in nuclear power plants are formed into blankets to wrap steel structures and electrical wires. The term *endothermic* is defined as designating or produced by a chemical change in which there is an absorption of heat from the environment. When exposed to intense heat they chemically absorb heat energy keeping the substrate materials cool. Releasing chemically bound water to further cool the surface of the substrate, they combine intumescent and heat-absorbing properties. A new filler material by Dow Corning trademarked "Instill," enclosed within a thin barrier material of metal foil, is evacuated to a high vacuum producing a vacuum-insulating panel with R-value six times as high as traditional glass fiber insulation. One inch of this panel is equivalent to 6 in. of glass fiber. This new fiber material is made of open-cell polystyrene foam which contains open cells that allow all the gas or air within to escape, thus enhancing the heat-blocking effect of the vacuum. Styrofoam® is an example of closed-cell construction. First applications for this new material will be in refrigerated trucks, vending machines, and shipping packages, all of which will produce considerable energy savings. Figure 4-46 shows two vital applications that exemplify several thermal properties mentioned in this brief discussion. See Module 7 for a discussion of thermal shock and thermal shock resistance.

4.6 TESTING, STANDARDS, AND INSPECTION

4.6.1 Testing

Materials are tested to determine their basic properties. In particular, the testing of materials has as its objective the performance of tests to determine numerical values for properties. As stated previously, most properties of engineering materials can be placed into the following

(a)

(b)

Figure 4-46   (a) Thermal resistance. High alumina (Al_2O_3) content spark-plug insulators can withstand thermal shocks from below zero ($-73°C$) to white hot heat (over $1000°C$). Note the frost still on the top of the insulator. (American Ceramics Society) (b) Various thermal insulating materials systems protect space vehicles from the high heat of reentry into the earth's atmosphere. (Corning Glassworks)

major classes: physical, mechanical, chemical, thermal, electromagnetic, and optical. An example of innovative testing techniques are those developed to investigate the high-temperature properties of advanced materials approaching 2200°C (4000°F). Current metallic alloys are capable of withstanding temperatures to only about 700°C (1300°F). These high-temperature techniques fall into two categories: macromechanical testing such as static/cyclic biaxial loading of a hollow specimen, and micromechanical such as a stereo image processing system. *Macromechanical systems* provide data on the strength, fatigue, and tribological properties of materials mainly for engineering design of components, whereas *micromechanical systems* permit detailed experiments for determining and describing specific mechanisms of failure. The terms *testing* and *inspection* are often used interchangeably, but they should not be.

Table 4-8 provides a summary of the general types of tests used to determine the properties discussed in this module. Materials testing is widespread throughout industry and government. Our modern society relies on materials testing of (1) raw materials, (2) product subassemblies or components, (3) products, and (4) systems. Who does the testing? Everyone! Read the caption in Figure 4-47. Think about the liability of Arrow Dynamics, the company who designed and built the Drachen Fire, and Busch Gardens, the company who operates this thrill ride that moves through spiral rolls and travels 60 miles per hour! Every aspect of this amusement must meet rigid quality standards to ensure the safety of riders while protecting the designers, builders, and operators of the ride from product liability. This includes the steel structural pipes and other structural members that make up the track. Also involved are the wheels, bearings, and suspension system that connects the cars to the track. Quality standards extend to the restraint system that holds the "brave souls" in the cars. It includes the composite body that shrouds the cars and makes the seats. But who gets involved in the many tests to ensure this quality?

TABLE 4-8   GENERAL TYPES OF MATERIALS TESTING

Mechanical Tests

Used to evaluate mechanical properties of materials where there is a measure of the material's ability to carry or resist mechanical forces such as tension, compression, shear, torsion, and impact. Tests can be *static,* in which the stress is applied slowly, or *dynamic,* in which the stress is applied suddenly. Testing usually results in permanent damage to the specimen.

Nondestructive Tests

Used to examine materials and components in ways that do not harm the future usefulness and service. They detect, locate, measure, and evaluate discontinuities, defects, and other imperfections to assess integrity, properties, and composition, and to measure geometric characters.

Physical Tests

Used to determine physical properties of materials when there is no change in composition of the material. They include tests to determine density and electrical, optical, thermal, and mechanical properties.

Chemical Tests

Used to determine chemical properties of materials where there is transformation of one material into another. They include tests of the abilities of materials to resist oxidation, corrosion, and rust.

Figure 4-47   Taming the beast. Brave souls challenge the mighty Drachen Fire, the fearsome roller coaster at Busch Gardens in Williamsburg, Virginia. As one of the world's largest roller coasters, it stands 150 feet tall, reaches speeds of more than 60 mph, and features many first-of-a-kind thrill ride elements. Hurtling along 3550 feet of electric-blue colored track, riders are turned upside down six times. (Courtesy Busch Gardens and Arrow Dynamics, Inc.)

4.6.2 Standards and Testing Organizations

Table 4-9 provides categories of agencies, societies, and organizations who (1) develop standards for materials testing and/or (2) conduct materials testing. **Materials testing standards** are rules and procedures established for the testing of materials to ensure objectivity and common practices for making judgments about materials. Just as the standard measurements for

TABLE 4-9   WHO SETS STANDARDS FOR MATERIALS TEST-
ING AND CONDUCTS MATERIALS TESTING?

1. Technical societies
2. Governmental agencies and research centers
3. Private laboratories
4. Manufacturing companies
5. Wholesalers and retailers
6. Consumer groups
7. Individual consumers

the layout of lines on a tennis court ensures that everyone is playing within standard boundaries, it is vital that everyone have clear, standard rules on how to test materials so they have clear meaning. This sounds simple and straightforward, but development of materials testing standards is a complex topic that we will only touch on here. Many people make full careers of developing these standards that the entire civilized world relies on for commerce and safety. CSSinfo, Ann Arbor, Michigan, is a distributor of engineering standards information, and specializes in creating on-line access and distributions systems for industrial standards. CSSinfo has launched a new World Wide Web site devoted to the dissemination of standards information from over 200 worldwide standards developing organizations including ISO, IEC, ASTM, ASME, ANSI, EIA, SAE, API, BSI, CSA, ACI, IEEE, NSF, UL, AAMI, and EN. The site is found at http://www.cssinfo.com and is freely accessible to the public. On-line searching for documents and standards information is free of charge.

Over a two-week period in August 1999, four people died on three separate amusement rides. Additionally, several people suffered heat exhaustion after being trapped upside down for more than an hour on a thrill ride. The incidents in industry took place in spite of claims of good safety records, focused attention on detail, careful fabrication, proper maintenance, thorough inspection, and riders' observations of safety rules on these amusement rides. Attention to these same issues carries over to buildings, transportation vehicles, and most of our products and systems. A need to hold rigid standards to achieve safety of life and property cannot be taken lightly.

4.6.2.1 Technical societies.   These groups of people, made up of technical specialists from private industry, government, and universities plus individuals with common interests, come together and agree on what standards are best for their particular industry. Some examples of technical societies are the Society of Plastic Engineers (SPE), Society of Automotive Engineers (SAE), Society of Manufacturing Engineers (SME), ASM International (ASMI), American Society for Testing and Materials (ASTM), American National Standards Institute (ANSI), and International Standards Organization (ISO). These societies often cooperate in standard development. For example, ASTM publishes the *Annual Book of ASTM Standards,* which involves coordinating 140 main and 2034 subtechnical committees and can include representatives from ASMI, SAE, and ANSI. The *Annual Book* consists of approximately 60 volumes and thousands of standards that serve as guides to materials testing and other related concerns. These ASTM standards are used around the world to set legal criteria and objective means for analysis of materials. For example, ASTM devotes about 40 standards to impact testing. One of these is *ASTM E23, Izod Impact Test of Metallic Materials* (see Section 4.7.9).

4.6.2.2 Governmental agencies and research centers.   Federal, state, and local government agencies become involved in many forms of materials testing. Other testing and development of materials are conducted at regional research centers that are under contract with federal agencies, such as the Department of Energy's Oak Ridge National Laboratory (ORNL) and Pacific Northwest Laboratory (PNL). The National Institute of Standards and Technology (NIST), formerly the National Bureau of Standards (NBS), is a federal agency that falls under

Figure 4-48   Meeting building standards. CoreGuard's one-piece panel incorporates 0.081-in. opaque recycled polycarbonate sheet and 1⅝-in. or 1¼-in. fire-rated gypsum wallboard. The panel provides at least 600 ft/lb of impact strength. (G.E. Plastics)

the jurisdiction of the U.S. Department of Commerce. NIST is deeply involved in developing materials testing standards and conducts a broad range of tests of materials. For example, they set up a test of the structural connection of a concrete walkway that collapsed in a hotel, killing 113 people (see Figure 4-31). NIST was involved in testing the walkway connection because their experts in materials testing could help explain the fault in the walkway system and thus avoid similar disasters and perhaps develop new standards. The National Aeronautics and Space Administration (NASA), another federal research agency, has research centers across the United States that perform materials testing on all sorts of materials relating to both space and aeronautics and also to many associated technologies, such as aircraft runways or satellites.

4.6.2.3 Private laboratories.   Many private laboratories test materials for both private businesses and government agencies. For example, Underwriter Laboratories (UL)® tests and approves appliances, tools, motors, and similar products. In addition to large private labs such as UL, you can find small, private materials companies operating in many communities that test a wide range of materials and products. To assist the designer in selecting a plastic material that is to undergo long-term exposure to elevated temperatures while under load, the UL developed a temperature index (UL 746B) that lists a temperature for each polymeric material. The index temperature is determined by exposing samples to circulating air at various temperatures for 10,000 hours. The temperature that causes a sample to lose 50% of its strength or toughness is the index rating temperature.

Figure 4-48 illustrates the use of standards for building construction materials. The CoreGuard security wall system was designed for high-security areas such as hospital corridors, schools, computer centers, libraries, or anywhere high-impact materials are required. Note the UL® label; the panel system was designed and tested according to UL 263 and ASTM E-119 standards for one-hour and two-hour fire-rated wall assemblies.

4.6.2.4 Manufacturing companies.   Large manufacturing companies rely on their materials testing labs to test all sorts of materials, ranging from raw materials and material stock from outside vendors (to ensure that they meet specifications), to the components and products they make) to be sure they will meet quality standards). For example, they may

Figure 4-49 Fully automated tennis ball compression tester (PCM VersaTest) is computer controlled, featuring a pneumatic pick-and-place machine. (Mecmesin, Inc.)

EMT WEBLINKS for a Sample of Technical Societies and Laboratories

American National Standards Institute—*http://web.ansi.org*

ASM International—*http://www.asm-intl.org*

Materials Research Society—*http://www.mrs.org*

Minerals Metals and Materials Society—*http:www.tms.org*

Underwriters Laboratory—*http://www.ul.com*

Department of Energy-Reports Bibliographic Database—*http://www.osti.gov/html/dra/dra. html*

SAE (Society of Automotive Engineers)—*http://www.sae.org*

test nuts and bolts to ensure that they have sufficient strength, and they may crash test a car to see how the materials and components respond to improved design. See paragraph 4.6.2.3 for information on UL 746B test for plastic materials. Smaller manufacturers may not be able to afford their own testing lab, so they will rely on private labs. Even large manufacturers use private labs such as UL to certify their products. Mecmesin, Inc. of Santa Rosa, California, has designed and manufactured a fully automated pneumatic pick-and-place machine (PCM VersaTest) for testing tennis balls in compression (see Figure 4-49). Pneumatic grips pick a ball from a carousel of 12 balls and place the ball between specially designed platens. A compression test cycle is carried out, the ball is twisted through 90°, and the test is repeated. The ball is automatically returned to the carousel. Two automatic procedures have been developed: one for the ball, and one for preconditioning each ball prior to testing. A report giving test data for each ball and statistics for each batch of balls is automatically printed. This testing system can be adapted for any compression or tension type testing of materials or components within a range of 0 to 550 lb.

Figure 4-50 Ultrasonic-testing setup. (NASA)

4.6.2.5 Wholesalers and retailers.   To ensure product quality, products wholesale distributors and retail stores may have their own lab or use private labs to test samples of products that they will sell. This quality testing is important to ensure reliable products that will keep the confidence of customers. Can you think of examples of retailers and the products they might test?

4.6.2.6 Consumers groups.   Many nonprofit consumer organizations also conduct tests of products. A few months before the Christmas season, you may see consumer groups on television talking about toys considered dangerous as a result of their testing. Consumers Union tests a wide variety of consumer products and publishes the results in various publications, such as *Consumer Reports*. This magazine is a useful resource to learn more about materials and product testing.

4.6.2.7 Individual consumers.   The ultimate testers of materials and products are consumers. Manufacturers try to stay in touch with their customers to ensure satisfaction. As newer materials, such as composites and advanced ceramics, appear on the market in new products, consumer acceptance is of great concern to producers. Even though extensive testing goes into these materials and products, the final and most important testing is by you, the consumer, whose feedback to the manufacturers helps them to improve products as well as to validate their testing procedures.

4.6.3 Nondestructive Testing (NDT), Nondestructive Inspection (NDI), and Nondestructive Evaluation/Examination (NDE)

Inspection differs from testing in its objectives. The ***objectives of inspection*** are to examine parts of materials for the presence of *discontinuities* and *defects*. In other words, inspection estimates the degree to which a product conforms with design specifications. Materials characterization, as discussed in Section 3.3, may be part of the inspection procedure. The American Society for Nondestructive Testing (ASNT) defines a ***discontinuity*** as an interruption in the normal physical structure or configuration of a part, such as a crack or porosity. A discontinuity may or may not be detrimental to the usefulness of a part. A ***defect*** is a discontinuity whose size, shape, location, or properties adversely affect the usefulness of the part or exceed the design criteria for the part. ASNT has five major inspection systems for detecting such discontinuities. These systems are (1) radiographic testing (RT), (2) ultrasonic testing (UT) (see Figure 4-50), (3) eddy-current testing (ET), (4) magnetic-particle testing (MT), and (5) liquid-penetrant testing (PT). In addition to these five major systems, inspection techniques using holography, microwaves, liquid crystals, infrared radiation, and leak testing are available where circumstances dictate. Use of *computer simulation* of materials processing is developing into a reliable means for improving quality processes. Sophisticated simulation allows process engineers to look at most variables prior to building molds for casting and other process components. As is evident from the preceding information, inspection techniques involve tests, most of which are classed as ***nondestructive testing (NDT)*** and ***nondestructive evaluation (NDE).***

NDE was once relegated to the role of inspecting a finished product or a product already in service for defects. Today NDE has been incorporated into the manufacturing cycle (Intelligent Manufacturing) where it provides vital information for feedback into the cycle at various points. This amounts to real-time improvements in the process and greater quality assurance, plus better product quality and reliability. A listing of NDE technologies involved with present-day materials involves ultrasonics, acoustics, holography, microwave, thermography, radiography, tomography, and spectroscopy. Several of these methods or techniques of NDE will be described throughout the text.

Illustrative Problem

A client has recently purchased a new brass lamp. A neighbor asked if the lamp was made of solid brass or plated brass. How can the owner determine the answer to this question?

Solution   Using a small refrigerator type magnet, the owner could place the magnet on the lamp. If the magnet is attracted to the lamp, then the lamp is made of a steel alloy which has been plated with brass. If the magnet fails to adhere to the lamp, then the lamp is made of brass.

PAUSE & LOOK

Look around your room, lab, or office and locate the familiar UL label on an electrical cord or piece of equipment. Discuss the type of material and product testing that may have been done on it. Do some research into UL testing to determine how it helps you, the consumer.

Two terms used in conjunction with a discussion of testing and inspection are ***quality control (QC)*** and ***quality assurance (QA).*** The objective of QC is to determine *statistically* how much testing and inspection are required to ensure that products will meet design specifications and service life expectations. For example, in the production of an automobile engine, each part cannot, because of high cost, be given a complete inspection to see if it meets exact dimensional tolerances, if the metal used is free of all external and internal defects, and if each part has the necessary strength. Rather, a ***statistical sample*** is taken of a batch of parts as they are produced. Through statistical analysis, the sample is determined to be representative (probably the same) as all parts made in the process. Whole engines are also chosen statistically to be samples of the engine-making process. They are tested to failure (destructive test) as a means of providing quality assurance (QA) for the engine itself. QA, the goal of any QC program, is used to refer to the total set of operations and procedures used in manufacturing a product whose goals are conformance of a product to design specifications.

4.7 MATERIALS SELECTION

In the section Pause & Ponder and throughout this book, new materials and processes are described. Often, 10 to 20 years elapse before new materials come into common use. It requires that much time for development of a reliable ***database*** to ensure that the potential (properties, processing

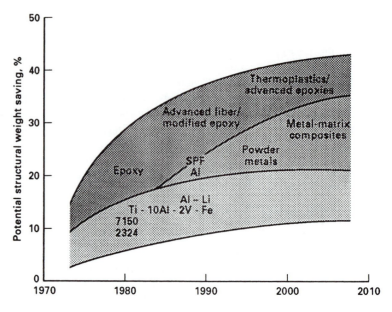

Figure 4-51 Potential weight savings trend for future structural materials. (NASA)

techniques, cost, etc.), as well as any risk assessment, is well known. The normally high cost of committing to production with new materials dictates an *evolutionary* rather than a *revolutionary* procedure for introducing new materials. Computer simulation coupled with rapid prototyping (see Module 6) is helping to shorten the time from concept to reliable materials technology.

Advanced composites serve to illustrate the point. In Module 8, we develop the many benefits available from advanced composites, including the favorable properties studied in this module. However, these materials have been slowly accepted, even in the aerospace market, where their benefits (especially strength-to-weight ratios) are so desirable, as seen in the potential applications shown in Figure 4-51. Among the reasons why aerospace designers have moved cautiously is that material failure in a commercial passenger aircraft carries enormous consequences.

EMT WEBLINKS

GE Plastics Material Selection—*http://www.ge.com/plastics/resins/materialselection/index. html*

Du Pont Industrial Polymers—*http://www.dupont.com/industrial-polymers/index.html*

Another reason is the refinement of new design technology and automation of the fabrication of composites. Costs for advanced composites have declined. Graphite fibers were around $35 a pound in the early 1980s and dropped to around $20 per pound in the mid-1990s. Carbon fiber used as reinforcements ranges in cost from a low of $6 per pound to a high of $200 per pound for high modulus varieties. However, fabrication costs will remain high until automated fabrication of composites, with the help of robotics, is developed further. The section Applications & Alternatives that follows demonstrates why the properties available in advanced composites make them the material of choice when cost is not an overriding consideration. Still, in applications where cost is a major concern, expect to see the trend toward composites grow (Figure 4-51) because their properties make them so attractive that databases, design parameters, and processing/manufacturing technology will quickly advance. In rebuilding the world's transportation infrastructure, low-cost polymer matrices that are environmentally durable and safe will gain ac-

(a)

(b)

Figure 4-52 High-tech testing. (a) Testing aluminum bats to determine the sound produced, "ping," and to increase energy reflectivity or performance. The reflectivity of the bat depends on the properties of the selected aluminum alloy and the wall thickness of the hollow bat. The designer's goal is to increase the "sweet spot" area where the maximum force is transferred to the ball. (b) The curves generated by the impact tester provide a "sweet spot" response. This location (5 in. from the end of the bat) is where maximum energy is transferred from the bat to the ball. (Courtesy GRC Instruments)

ceptance. As you will see in later modules, automobiles, trucks, trains, and oceancraft are also moving toward use of these composites because they offer the right properties to meet new demands related to energy efficiency and environmental impact.

The ASMI Center for Materials Solutions is developing a system to provide the user with an on-line source of materials knowledge. One goal of this program is to manage these case histories and match them with current problems in materials manufacturing and in-service problems. The *materials solutions case histories* will be categorized by materials, including alloy numbers or another suitable designation (aluminum, copper, glasses, etc.), processing (heat treating, surface treatment, casting, etc.), and in-service effects (wearing, failure, corrosion, etc.). The case histories will be published as a guide to solutions of materials problems.

Alloy selection mandates that a matrix be developed, setting forth material requirements based on design criteria and the performance demands made on the material. Using the matrix, various *electronic data banks* may be consulted to find the alloy and the thermomechanical conditions that best satisfy those requirements. A knowledge of how and why material parts fail in service is essential for effective materials selection. Some popular data banks are Alloy Finder, Alloy Digest, and MAPP–98.

As discussed in Module 3, low-energy materials and processes must also be considered, due to cost and environmental issues. Also, in developing countries, which may lack natural resources and sophisticated technology, wood and other polymers and cement and other ceramics will be favored.

Later you will learn how properties of polymers, metals, ceramics, and composites qualify these materials to be selected for critical parts in a variety of products and systems. With a knowledge of the properties of materials and practice in using this knowledge, wise material selection should result.

Figure 4-53 Golf clubs reflect the manner in which sporting equipment makers quickly adopt new materials to benefit from unique properties. Athletes and hobbyist are looking for an edge to improve their performance. Seen on the right, older steel shaft with persimmon headed driver. The putter, iron and 3 wood with aluminum head and titanium face all use graphite epoxy shafts. Sometimes the trend shifts back; persimmon shows up on expensive new woods.

4.8 LABORATORY ACTIVITIES AND CLASSROOM DEMONSTRATIONS WITH PROPERTIES

The *Experiment in Materials Science, Engineering and Technology (EMSET) CD-ROM* provides numerous experiments and demonstrations related to properties as result of a cooperative effort by scores of educators and materials specialist from industry. See Prentice-Hall's website for ordering information on the latest edition of *Experiments in Materials Science (CD-ROM),* at *www.prenhall.com.*

APPLICATIONS & ALTERNATIVES

Sports and recreation are big businesses throughout the world. Both professional and amateur athletes continue the search for the winning edge. College courses offered in materials engineering for sports equipment focus on improving performance through engineered materials and new designs. As engineering materials evolved for the aerospace industry, designers began to apply the new materials to a wide variety of equipment—bicycles, snow skis, race cars, baseball bats, hang gliders, and tennis rackets, just a few of the products that have profited from such innovations.

Advances in technology have provided new tools to materials engineering and manufacturing to improve the confidence with which to predict the *product reliability* (the ability of the product to perform as specified by the designer). Companies may find it very difficult to compete in a global economy if they do not use the new technology and instead rely on only *subjective* (personal experience and feelings) rather than *objective* data (data that can be measured with instruments). Figure 4-52 depicts a high-tech, instrumented materials-testing setup that you might find in a typical manufacturing plant. Because of demand by consumers, quality must be built into products. The way to ensure built-in quality is to test every aspect of the product design and manufacture before releasing it to the public. Quality involves both performance and durability.

A baseball bat should have the proper weight and balance, be able to transfer energy from the batter to the ball efficiently, and be able to withstand the energy transfer from the strongest batter without breaking. To determine the batting quality of aluminum bats, manufacturers evaluated them with a drop-weight impact test, as seen in Figure 4-52a. The falling tip inputs controlled energy to the bat, and a computerized data-acquisition system (wired into the testing machine) generates data curves showing the ratio of energy absorbed to energy reflected. The closer the impact to the "sweet spot," the less energy is absorbed by the bat. ***Sweet spots,*** as found on bats, tennis rackets, badminton rackets, and golf clubs, are the best zones for hitting a ball or birdy to achieve maximum control and distance. Less energy absorption by a bat means the ball should travel farther and with less strain on the batter. The curves shown on the monitor in Figure 4-52b indicate the falling tip has hit 12.5 cm from the fat end of the bat, which is the heart of the sweet spot. Therefore, the load from the impact tip is reflected away with minimum absorption.

Information from computerized materials-testing equipment allows designers to make modifications in design and allows manufacturers to be sure they have obtained their quality goals. Prior to the development of new, computerized materials-testing technology systems, much of the design of sports equipment and many other products was subjective, based on trial and error with no quantitative analysis. In other words, the designer and user had to rely on the look and feel of parts, or maybe the endorsement of some star athlete, rather than actual objective test results. Knowledgeable consumers are beginning to ask for objective evidence like that found in statistics from materials and product testing.

The game of golf is going high tech (see Figure 4-53). Over 26.5 million Americans play golf and spend more than $5 billion a year on clothing and equipment. It is estimated that the exploding population of golfers is around 50 million worldwide. The latest innovations in the game to make it less frustrating involve golf clubs, balls, and teaching techniques. In the early 1990s titanium was used to produce clubs improving the size of the sweet spot (see Section 9.13). Today's high-tech clubs have sweet spots the size of a half dollar. Using lightweight but strong titanium, the club heads are not as heavy which allows the golf club shafts to be longer so to improve the driving distance. Supposedly these new innovations in design may exceed the authorized standards imposed on their construction, because of their illegal springlike effect when the club face flexes upon contact with the ball thus giving the ball greater velocity and driving range. Many players fear that imposing tougher standards would limit further innovations in club design and the use of new materials.

Visit a sporting goods shop and notice the wide variety of golf clubs now available. Read the literature and compare the prices. While in the shop, observe the other equipment to see how materials technology is meeting the desires of sports and recreational enthusiasts.

 EMT WEBLINK

Material Science and Engineering Education forum—*http://www.tms.org:900/*

SELF-ASSESSMENT

4–1. A large percentage of the world's infrastructure—its buildings, roads, bridges, and tunnels—is badly deteriorated, posing a threat to the safety of people using these facilities. What is the main cause of this degradation? Cite one example of how this deterioration of materials is overcome.

4–2. What qualities define a material?

4–3. Define a chemical property.

4–4. Define *stress* and express it in appropriate SI units.

4–5. Which quantity—stress or strain—can be measured directly in a stress–strain tensile test?

4–6. How can you distinguish a ductile material from a brittle material with the aid of a stress–strain diagram?

4–7. Why should the test conditions under which some material property was determined be included with the property?

4–8. Do all materials have a yield point?

4–9. Factors of safety are defined either in terms of the ultimate strength of a material or its yield strength. In other words, by the use of a suitable factor, the ultimate or yield strength is reduced in size to what is known as the design stress or safe working stress. Which factor of safety would be more appropriate for a material that will be subjected to repetitious, suddenly applied loads?

4–10. Stiffness of a material is readily apparent in studying the material's stress–strain curve. Explain.

4–11. Product liability court cases have risen sharply in recent years because of poor procedures for selecting materials for particular applications. Assuming that a knowledge of a material's properties is a valid step in the selection process, cite two examples where such lack of knowledge could or did lead to failure or unsatisfactory performance.

4–12. A material may be tough at room temperature. In technical terms, describe what this means. What happens to this toughness in most materials as the ambient temperature is decreased?

4–13. Make a sketch and fully dimension an Izod impact test specimen.

4–14. At what level of yield stress do most fatigue failures occur?

4–15. From your experience, what typical automobile part requires complete fatigue testing prior to development?

4–16. Identify the origin of most fatigue failures and explain the significance of your answer in terms of handling a metal workpiece in the machining process.

4–17. Torquing a nut on a bolt to specified values using a torque wrench is standard practice in many industries. Explain why.

4–18. A structural steel beam may deflect under its transverse loading. Is this considered bad practice? Is it allowed? Cite a case where deflection of beams (wood or steel) must be limited to a minimum.

4–19. When selecting a material that will be subjected to abrasion, what property would you expect the material to possess to a great degree?

4–20. How does the chemical industry protect its tanks and piping from corrosion due to the action of highly corrosive liquids?

4–21. What would your explanation be for the development of surface cracks on the surface of four-year-old automobile tires?

4–22. Name one method that large automobile manufacturers use to inhibit corrosion of the steel chassis parts of trucks and passenger vehicles.

4–23. A 1-in.-diameter steel rod 3 ft long is used between two immovable (rigid) walls. The rod is welded to the two walls at each of its two ends. The ambient temperature increases by 100°F. The modulus of elasticity for the steel is 30×10^6 psi and the CTE is 6.5×10^{-6}/°F.

    (a) Make a sketch of the above problem using appropriate welding symbols. Note on the sketch that the walls do not move or deform. This sketch will show the rod welded at both ends to the walls.

    (b) Describe what is happening to the rod. What is acting on the rod from the outside? What is happening to the atoms in the steel rod?

    (c) If the rod were free to move, describe its movement and calculate how much it would move and in what direction.

    (d) Make another sketch with the walls replaced by force vectors representing the walls acting on the rod.

    (e) Calculate the stress acting inside the rod.

    (f) Calculate the force on the rod that would be necessary to produce the same stress as calculated in part (e).

4–24. A 3000-kg load using a 10-mm indentor produces a 3.2-mm impression on a ferrous metal specimen that yields a Brinell hardness of 364 (HB = 364). Using the relationship of HB to the tensile strength (TS), determine the approximate TS value for the material. Also, knowing the relationship between TS and the endurance limit (EL), determine the EL value for the material.

4–25. Refer to Table 4-1 and the accompanying text. What prefix would be added to a Rockwell hardness reading taken from a major load of 150 kg with a diamond penetrator?

4–26. Compare the modulus of elasticity with the modulus of rigidity for the same steel and write a statement about the steel's capability to withstand a direct tensile load versus a shearing type of load.

4–27. Thermal resistivity is as important as thermal conductivity. Illustrate both properties by citing an example of each from your own experience or knowledge.

4–28. Using magnetic chucks in a machining operation, how does a knowledge of magnetic properties of metals pay big dividends?

4–29. How does quality control differ from quality assurance?

4–30. Why does a deep drawing operation, as illustrated in Figure 4-10, require so many phases?

4–31. Solid objects are three-dimensional. A bar with a rectangular cross-section under load will deform in how many directions? If it deforms in more than one direction, are the amounts of deformation the same? Which direction would have the greatest deformation?

4–32. Knowing that a common steel has a HB of 207, determine its approximate tensile strength.

4–33. Explain why water is chosen to cool automobile engines. What is the effect of using more than the recommended percentage of antifreeze in the automobile's cooling system?

4–34. Even though heat and work are different forms of the same quality (energy), custom has continued to express them using different units. Name two customary units used to measure heat energy.

4–35. The coefficient of thermal linear expansion may be expressed in units such as m/m/°C. What must you do to convert these units to °F? Would you do the same thing if you were given units of m/m/K?

4–36. Is the coefficient of linear expansion the same for iron and iron alloys (steels)? If not, how do you explain the difference, and what is the implication in designing steel structures?

4–37. Both the coefficient of thermal expansion and the thermal conductivity of most materials are temperature sensitive. Explain this sentence in terms of any changes in their respective values. Can you find a material with a coefficient of thermal expansion that is nearly zero?

4–38. For the major heading Properties, develop a list of at least two properties. Next to each, write a service condition from your own experience in which the property is important.

4–39. Express one cubic inch in terms of cubic centimeters.

4–40. Using Tables 10-7, 10-7A, and 10-11, calculate the thermal mass for type 405 stainless steel in cal/cm^3 · °C (refer to Section 4.5.1).

4–41. When determining if a heat treat oven has sufficient heating capacity, one can calculate the heat required per some stated period of time to raise the temperature of a material a certain number of degrees. Does this calculation represent the complete answer or are there other factors involved?

4–42. The specific heat for iron as stated in Section 4.5.1 is 0.1 cal/g · K. Express the specific heat in units of Btu/lb·°F.

4–43. Calculate the heat required to heat 1000 lb of steel to 400°F in one hour. Assume the specific heat of steel is 0.125 Btu/lb · °F and the ambient temperature is 70°F.

4–44. A steel rod is 4.000 m long (l_0). It is heated from a temperature of 10° to 38°C. The CTE is 12×10^{-6}°C^{-1}. Determine (a) the unit deformation (ϵ_t) and (b) total deformation (δ_t) of the rod due to the change in temperature.

OBJECTIVE QUESTIONS

4–1. Cite one property of a material that cannot be observed.

   a. Conductivity   **c.** None can be observed

   b. Hardness   **d.** All can be observed

4–2. Give two examples of a mechanical property.

   a. Thermal resistance   **b.** Wear resistance   **c.** Hardness   **d.** Strength

4–3. Scissors used in the home cut material by concentrating forces that ultimately produce a certain type of stress within the material. Identify this stress.

   a. Bearing stress   **b.** Shearing stress   **c.** Compressive stress

4–4. Given μ for a material as 0.25 and the strain in the lateral direction as 1.50×10^{-6} mm/mm, what is the axial deformation expressed in inches?

   a. 6×10^{-6} in./in.   **b.** 0.6×10^{-5} in./in.   **c.** Both (a) and (b)

4–5. An aluminum rod 1 in. in diameter ($E = 10.4 \times 10^6$ psi) experiences an elastic tensile strain of 0.0048 in./in. Calculate the stress in the rod.

   a. 49,920 ksi   **b.** 49,920 psi   **c.** 49,920 msi

4–6.   Express the stress in question 4-5 in SI units with an approved prefix.

    a. 0.3442 GPa   **b.** 3.442×10^8 Pa   **c.** 34.42 MPa

4–7.   A 6-ft steel bar is deformed 0.01 in. in an axial direction. What is the unit deformation?

    a. 1.389×10^{-4} **b.** 1.389×10^{-4} in./in.   **c.** 1389×10^{-7} in./in.

4–8.   A shaft supported by thrust bearings placed 4 ft apart will fail if the total deformation exceeds 0.00025 in. What is the maximum allowed strain?

    a. 3.7×10^{-6} in./in.   **c.** 15.6×10^{-5} in./in.

    b. 5.2×10^{-6} in./in.   **d.** None of the above

4–9.   The minimum yield stress for a material is 48,000 psi. What is the factor of safety if the allowed working stress is 24,600 psi?

    a. 0.51   **b.** 1.95   **c.** 3.90

4–10.   Express the yield stress units of question 4–9 in megapascals, the recommended units in the SI system. Use appropriate abbreviations.

    a. 331 MPa   **b.** 331 GPa   **c.** 331 MPA   **d.** All of the above

4–11.   A 1-in.-diameter steel circular rod is subject to a tensile load that reduces its cross-sectional area to 0.64 in.^{2}. Express the rod's ductility using a standard unit of measure.

    a. 18.5%   **b.** 1.85%   **c.** 18.5   **d.** (a) and (c)

4–12.   If a force (push or pull) of 40 lb is used to turn a wrench such that the 40-lb force is applied at right angles to the wrench at a distance of 16 in. from the center of its jaw, what torque is produced?

    a. 640 inch pounds   **b.** 640 #-in   **c.** 640 pound-inches   **d.** (b) and (c)

4–13.   An angle of twist of 0.48° can be expressed as how many radians?

    a. 8.4×10^{-3} **b.** 84 radians   **c.** 0.0084 radians   **d.** (a) and (c)

4–14.   A standard-weight pipe with a nominal diameter of 1 in. has an outside diameter (OD) of 1.315 in. and a thickness of 0.133 in. Calculate its polar moment of inertia. *NONE*

    a. 0.247 in.^{3} **b.** 0.247 in.^{4} **c.** 2.47×10^{-3} in.^{4} **d.** All of the above

4–15.   Determine the bending stress on the top surface of a rectangular beam whose height is 11½ in. if its moment of inertia is 951 in.^{4} and its bending moment is 82,856 pound-inches.

    a. 499.34 psi   **c.** 499.34 #/in.^{2}

    b. 499.34 pounds/sq. in.   **d.** All of the above

4–16.   The specific heat (specific heat capacity) of water is 1.0 kcal $\cdot$ kg$^{-1} \times$ K^{-1}. Write this value using a different set of SI units.

    a. 4184 J/kgK   **c.** 4.184×10^3 J/kg K

    b. 4184 J kg^{-1} K^{-1} **d.** All of the above

4–17.   The specific heat of Alclad 2024 is listed in Table 10-11. Express the specific heat in SI units of kcal/kg°C.

REFERENCES & RELATED READINGS

AMERICAN SOCIETY FOR TESTING AND MATERIALS. *1996 Annual Book of ASTM Standards.* Philadelphia: ASTM, 1996.

AMERICAN SOCIETY FOR TESTING AND MATERIALS. *UNSearch, Version 2.0,* software. Philadelphia: ASTM, 1992.

ASM INTERNATIONAL. *ASM Handbook,* Vol. 1, *Properties and Selection: Irons, Steels and High Performance Alloys;* Vol. 2, *Properties and Selection: Nonferrous Alloys and Special Purpose Materials;* Vol. 8, *Mechanical Testing;* Vol. 11, *Failure Analysis and Prevention;* Vol. 13, *Corrosion;* Vol. 17, *Nondestructive Evaluation and Quality Control.* Materials Park, OH: ASM International, 1990, 1991, 1985, 1985, 1987, 1989.

ASM INTERNATIONAL. "Materials Selection," *Advanced Materials and Processes,* special issue, June 1999.

BERKE, NEAL S., AND LAWRENCE R. ROBERTS. "Reinforced Concrete Durability and ASTM," *ASTM Standardization News,* January 1992, pp. 46–51.

BLAU, PETER, ed. *ASM Handbook,* Vol. 18, *Friction, Lubrication, and Wear Technology.* Materials Park, OH: ASM International, 1992.

BOSICH, JOSEPH F. *Corrosion Prevention for Practicing Engineers.* New York: Barnes & Noble, Inc., 1970.

CUBBERLY, W. H., and RAMON BAKERJIAN, eds. *Tool and Manufacturing Engineers' Handbook.* Dearborn, MI: Society of Manufacturing Engineers, 1989.

DEXTER, S. C., ed. *Biologically Induced Corrosion.* Houston: NACE, 1986.

DIXON, J. I. *Failure Analysis: Techniques and Applications.* Materials Park, OH: ASM International, 1992.

FERRARA, MICHAEL. "Stress Relief Meta-Lax Extends Life of Engine Components," *Turbo & Hi-Tech Performance,* May 1995, pp. 62–65.

JOHNSTON, NORMAN J. "Introduction to High Performance Composites," *National Educators' Workshop, NEW: Update 92,* Oak Ridge National Laboratory, Oak Ridge, Tenn., November 13, 1992.

MCGHEE, DOUGLAS B. "Impacts of New Hardness Standard," *Advanced Materials and Processes,* October 1999, pp. 165–167.

OBERG, E., et al. *Machinery's Handbook,* 25th ed. New York: Industrial Press, 1996.

PROVAN, JAMES W. "An Introduction to Fatigue," *Journal of Materials Education,* Vol. 11, No. 1/2, 1989, pp. 1–105.

RUMBLE, JOHN. "Computerizing ASTM Test Methods," *ASTM Standardization News,* July 1992, pp. 34–38.

Periodicals

Advanced Materials and Processes

ASTM Standardization News

Corrosion

Geotechnical Testing Journal

Journal of Composites Technology and Research

Journal of Materials Education

Journal of Materials Engineering and Performance

Journal of Testing and Evaluation

Materials Performance

Surface Engineering

 EMT WEBLINK

Materials World Modules—*http://mwm.ms.nwu.edu/*

Module

5

Metallic Materials

After studying this module, you should be able to do the following:

5–1. Describe basic alloying, forming, casting, and thermal processing, plus surface engineering methods used by industry to change metals' microstructures and thus vary properties.

5–2. Analyze phase diagrams and photomicrographs to determine thermal processes, basic structures, and properties of ferrous and nonferrous metals.

5–3. Explain the effect on grain size, metal properties (tensile strength, hardness, and toughness) as a result of strain hardening and various thermal processing. Describe how degrees of both hardness and toughness are achieved through thermal treatments.

5–4. Differentiate among ferrous metal alloys, structures, processing, and properties. Cite examples of properties and applications of major classes of ferrous metals and smart metals.

5–5. Differentiate among nonferrous metal alloys, structures, processing, and properties. Cite examples of properties and applications of major classes of nonferrous metals.

5–6. Explain the corrosion resistance mechanism in metals and methods for preventing corrosion in metals and alloys.

5–7. Recall the fabrication techniques for powdered metal products. Explain some of the properties available and applications for powdered metal (P/M).

5–8. Explain changes involving metal structures, properties, and processing used by the metals industry to compete with other materials groups.

5–9. Use tables in the text, other handbooks, and periodicals to compare alloy content, cost, properties, processability, and applications for selecting metallic materials.

5–10. Describe some of the thrusts and barriers to the use of advanced metals and metals processing such as metal matrix composites (MMCs), superplastic forming (SPF), and light aluminum alloys.

5–11. Use EMT Weblinks to seek current developments and emerging technology related to metallic materials.

PAUSE & PONDER

Pause for a moment and think about recent developments relating to cars and to light trucks, which include pickups, minivans, and sport utility vehicles (SUVs). The auto and light-truck industry acts as a major economic player in developed countries. Because these vehicles affect society in so many ways, ranging from ease and comfort of travel to environmental pollution, they continually undergo research and development for improvements.

Have you read anything about "Super Car"? The concept vehicle is a goal of the Partnership for a New Generation of Vehicles (PNGV). This industry/government consortium, composed of the U.S. auto industry and the federal government, aims to improve fuel efficiency to 80 mpg (miles per gallon). Researchers for PNGV seek to simultaneously reduce vehicle weight by as much as 40%, increase engine efficiency by 40–55%, implement regenerative braking, and increase energy storage by 90%.

These goals came from the United States Council for Automotive Research (USCAR). The group was founded in 1992 to facilitate, monitor, and promote a growing number of precompetitive, cooperative research and development programs among DaimlerChrysler, Ford, and General Motors. Many of the research programs organized under the USCAR umbrella are aimed at environmental improvements, fuel economy, and safety for occupants. Super Car focuses on fuel efficiency and will incorporate other aims of USCAR.

Working as a subgroup to USCAR, the United States Automotive Materials Partnership (USAMP) focuses on body materials, which encompass the car's frame and outer skin. The cars in which you ride represent more than a century of evolving automotive technology. Nearly all high-production cars use frames with *unibody,* stamped-steel construction with much of the skin made of sheet steel. The steel industry aims to keep its dominate role in autobody materials as well as other parts for cars; however, steel faces several challenges from plastics, plastic composites, and aluminum. Figure 5-1a depicts an UltraLight *body-in-white* (unibody structure onto which body panels will be affixed). This steel unibody is a result of

Modal analysis
(simulation)

Figure 5-1   (a)UltraLight unibody structure with ghosted styling image. The PNGV-focused project aimed to fully optimize the qualities of steel including a wide range of mass, cost, performance, and safety targets. (American Iron and Steel Institute)

Figure 5-1 (b) Synergy 2010 concept car is a PNGV approach to a family sedan which would be capable of 80 mpg with comfort and convenience and comparable cost to today's sedans. (Ford Motor Co.)

the steel industry's involvement in the PNGV program. Figure 5-1b shows Ford's Synergy 2010 PNGV concept vehicle that uses an aluminum unibody aimed at one-third weight reduction and 40% less aerodynamic drag. The UltraLight and Synergy aluminum unibodies are but two of the numerous approaches being studied by USAMP members for reducing the weight in cars. An aluminum "spaceframe" seen in Figure 5-40 shows another approach to the primary auto structure.

Aluminum offers many advantages that you will learn in this module and from the following EMT Weblinks.

PNGV-USCAR—*http://www.uscar.org*

AISI Autosteel—*http://www.autosteel.org/cars/default.htm*

Aluminum Association Automotive—*http://www.autoaluminum.org*

Aluminum can be processed in many ways like steel and it is about 45% as dense as steel. But sheet aluminum costs about $1.50 per pound compared with $0.30 for sheet steel. Also of great significance, aluminum is about one-third as stiff. These factors hint at the challenges faced by materials competing with steel.

Figure 5-2a shows how current vehicles lose energy from systems problems such as engine, drivetrain, idling, accessories, aerodynamics, braking, and rolling losses. Addressing these losses to achieve the target of 80 mpg will require considerable redesign of current cars. At highway speeds aerodynamic drag accounts for nearly 15% of fuel consumption. An experimental car by GM has a drag coefficient of 0.163 which is less than one-half that of a typical midsize sedan. The lowest drag coefficient in vehicle production, the EV 1 with a drag coefficient of 0.19 has a smooth-shell underbody. The Chrysler ESX11 experimental car—using molded, recyclable plastic panels of reduced weight and cost, a radically new method of assembling the body, and a combined diesel and electric engine to improve fuel economy and lessen emissions—is the most advanced engineering example to date toward the goal of a full-size and yet affordable sedan that will achieve 80 mpg (see rolling losses in Figure 5-2a). Figure 5-2b depicts estimates of percentages of reductions in auto components (comparing current to Super Car weights) to accomplish the 40% weight reduction. Both a change in component design and materials substitutions will be necessary. Figure 5-2b shows how use of materials (in pounds) in cars has changed from 1975 to 1990, and shows projections for materials options beyond the year 2000 to meet Super Car goals. Note an anticipated increase in lighter metals: aluminum, magnesium, and perhaps titanium, along with polymer-based composites and high-strength steel.

Materials substitutions are complex issues that embrace concerns far beyond simple mass reduction. Aerospace technology presently used for many of the candidate materials has proven to be feasible at low production rates for airplanes and spacecraft. This technology must be modified for the much higher production rates required of cars and light trucks. This also means modifying aerospace manufacturing techniques to permit complex molding or

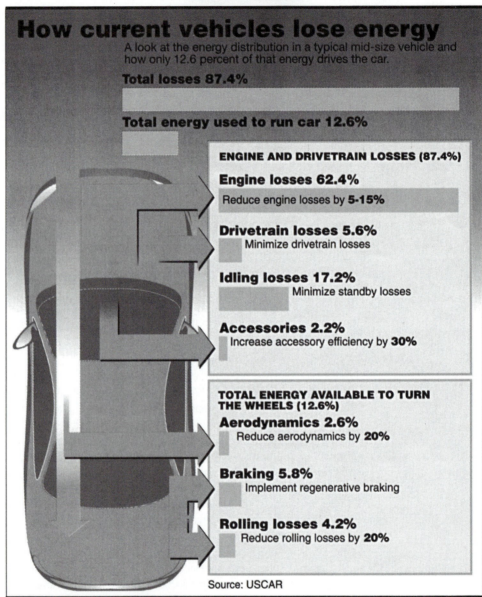

How current vehicles lose energy

A look at the energy distribution in a typical mid-size vehicle and how only 12.6 percent of that energy drives the car.

Total losses 87.4%

Total energy used to run car 12.6%

ENGINE AND DRIVETRAIN LOSSES (87.4%)

Engine losses 62.4%
Reduce engine losses by **5-15%**

Drivetrain losses 5.6%
Minimize drivetrain losses

Idling losses 17.2%
Minimize standby losses

Accessories 2.2%
Increase accessory efficiency by **30%**

TOTAL ENERGY AVAILABLE TO TURN THE WHEELS (12.6%)

Aerodynamics 2.6%
Reduce aerodynamics by **20%**

Braking 5.8%
Implement regenerative braking

Rolling losses 4.2%
Reduce rolling losses by **20%**

Source: USCAR

Figure 5-2   (a) How current vehicles lose energy.

machining, joining of dissimilar materials, and improvements in assembly methods—all at competitive costs in line with consumer demands.

This module will provide you with valuable insights into metals, their structures, properties, methods of processing, and applications so that you can contribute to solving the type of problems faced by USCAR. Enjoy the ride!

 EMT WEBLINKS

United States Council for Automotive Research (USCAR)—*http://www.uscar.org*
ASM International—*http://www.asm-intl.org*

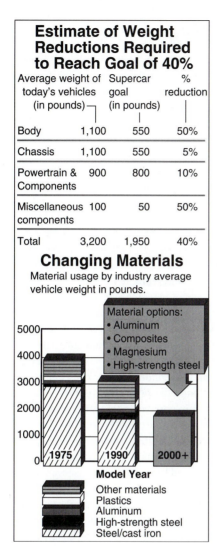

Estimate of Weight Reductions Required to Reach Goal of 40%

	Average weight of today's vehicles (in pounds)	Supercar goal (in pounds)	% reduction
Body	1,100	550	50%
Chassis	1,100	550	5%
Powertrain & Components	900	800	10%
Miscellaneous components	100	50	50%
Total	3,200	1,950	40%

Changing Materials

Material usage by industry average vehicle weight in pounds.

Material options:
• Aluminum
• Composites
• Magnesium
• High-strength steel

Model Year: 1975, 1990, 2000+

Other materials
Plastics
Aluminum
High-strength steel
Steel/cast iron

Figure 5-2   (b) Estimate of spot weight reduction required to reach goal of 40% total weight reduction. (Bob Graham, United States Council for Automotive Research)

Aluminum Association—*http://www.aluminum.org*

American Iron & Steel Institute—*http://www.aisi.org*

Technology Review—*http://web.mit.edu/techreview/*

USCAR—*http://www.theautochannel.com*

Auto/Steel Partnership (A/S P)—*http://www.autosteel.org*

ULSAB—*www.ulsab.org*

Powder Metallurgy—*http://www.mpif.org*

5.1 NATURE OF METALS

Metals have been useful in humanity through the ages because they are "strong" when subjected to the external forces encountered under service conditions, yet they become "soft" enough to yield to a machine cutting tool or to a compressive shaping force. Above a certain temperature, they melt and become liquids capable of being shaped by casting. Only in recent times have we realized that the properties of all types of solid materials, including metals, arise from their atomic architecture, that is, from the manner in which their atoms arrange themselves into a crystalline order, from the number and types of imperfections found in this structure, and from the bonding forces that keep the collection or structure of atoms bound or joined together. Most metals occur in nature as compounds. These compounds must be reduced to free the metal (see Module 4 on oxidation and reduction for a further explanation).

The "softness" quality of metals can be explained by an understanding of the atomic structure and metallic bond of the metal atoms to form a crystalline structure. We recall that the electrons in the metallic bond are free to move about their positive ions in an electron cloud or gas, which acts to glue or bond the ions together. This free movement, within limits, also allows for the movement of the atoms under the influence of an external load. This slight movement, visible only under the most powerful microscopes, is called ***elastic deformation*** or ***elastic strain.*** Once the external force (load), such as a bending force, is removed, the internal electrical forces that cause the atoms to move will decrease, allowing the atoms to return to their normal position; they leave no sign of ever being moved. If you bend a piece of spring steel such as a machinist's rule or vegetable knife, it will return to its original shape, thus experiencing elastic deformation.

If you were not careful and applied too much external force by excessively bending the rule or knife, the atoms might move too far from their original positions to be able to move back again when you released the external force. Consequently, the rule or knife would be permanently bent and no longer fit for use. This permanent deformation is known as ***plastic flow, plastic slip, plastic deformation,*** or ***permanent set.*** When automakers stamp out a metal car body from low-carbon steel in a huge die press, they use this softness quality of metals. The term *cold working* (defined later) is applied to this stamping operation and many other metalworking processes that produce plastic deformation in a metal. Cold-working operations include rolling, heading, spinning, peening, bending, pressing, extruding, drawing, and others. The steel framing of the skyscraper seen in Figure 5-3 consists of structural steel produced by hot working; steel and other metals, shaped by cold working, serve in the construction and furnishing of the typical office building.

The microstructure of metals can be modified in a number of ways. By now we know that this last statement can be interpreted to mean that, through advances in metals technology, we can affect the atomic structure of metals in a precise, controlled manner in the design of metal alloys with the desired properties. Throughout this module much of the text and illustrations refer to the primary metal alloy, steel. However, much, if not all of the text applies to most metal alloys used in industry. In this module we classify, for learning purposes, the basic methods of changing a metal's properties into the following categories: (1) work or strain hardening, (2) thermal processing under equilibrium conditions (solid solution hardening), (3) thermal processing under near-equilibrium conditions (annealing and grain refinement), and (4) thermal processing under nonequilibrium conditions. Of the four categories, only the first does not involve primarily thermal processing. Prior to discussing thermal processing techniques, we need to understand the concepts involved in the internal structure of solids, as presented in Module 3, and have a working knowledge of phase and phase diagrams, covered in the present module.

Work or ***strain hardening*** is a way to change a metal alloy's structure to alter its properties by performing work *(cold working)* on the metal itself. Work is a form of energy. If we can find a way to deliver energy to these metal atoms, we can give them the energy necessary for the atoms to increase their movement and thus their ability to diffuse through the metal structure. Figure 5-4 shows four industrial cold-working methods, and the photomicrographs demonstrate grain structure before and after work hardening.

In our opening remarks about the usefulness of metals, we mentioned that metals have the ability to yield to a mechanical force that can form them into desired shapes. An external, mechanical shaping force is a force that causes a metal to exceed its elastic deformation limit and ***deform plastically without fracturing.*** This force and the deformation produced are the

Figure 5-3 The Steel Triangle, the United States Steel Corporate Center in Pittsburgh. High-strength, low-alloy steel (HSLA), Cor-Ten, forms a tight adherent, dark russet, ferrous-oxide surface coating and requires no other protective coating. The partially completed building exposes steels such as USS "T-1," which is nearly three times the strength of carbon steel and bears great loads as interior columns, and USS Ex-Ten, an HSLA steel that serves as floor beams and some core columns. Much carbon steel finds use in areas not requiring maintenance and high strength. (United States Steel Corporation)

Figure 5-4 Some work-hardening (cold-working) processes.

means for transferring sufficient energy to the atoms to allow them to flow or move plastically. This movement of atoms, one row at a time along planes of close-packed atoms, shifts the positions of the atoms in relation to each other. From our study of crystal structures, we know that the spacing between atoms is not only critical, but varies with the particular crystal structure. Fcc structures have the closest-packed atoms and the greatest number of closest-packed planes of atoms, which take less energy to allow for the slip of the atoms. As more and more slip takes place, more dislocations are produced. The greater the slip, the greater the dislocations and the more distortion of the lattice structure. The end result is that the deformed metal is *stronger* than the original undeformed metal and offers greater resistance to further deformation.

Figure 5-4 also shows that the processing of materials, particularly the operations of rolling and drawing, can affect the orientation of metal grains (crystals). This is another example of the fact that the structure determines (in the main) the properties of a material. In this example, due to the ***preferred orientation*** of the grains, the metals will have different properties in the direction of the orientation than in other directions. This is known as ***anisotropy*** (see Section 5.3.4 for further discussion). Metal grains that are random in their orientation produce ***isotropy,*** the same properties in all directions.

In summary, through cold working we have (1) reduced the metal's ductility, (2) reduced the effectiveness of the metal atoms to slip, (3) reduced movement of dislocations in the structure, (4) created distortion of the lattice structure, and (5) ended up with a stronger metal that requires a greater force or greater amount of energy to deform it further. At the same time, changes in other properties, such as electrical conductivity, have also occurred. This condition of noticeable increase in energy for further deformation or increase in the metal's yield strength is known as ***work*** or ***strain hardening.***

5.2 THERMAL PROCESSING OF STEEL: EQUILIBRIUM CONDITIONS

By controlling the thermal energy, that is, by controlling the supply or removal of heat, we can make metal atoms move. This art and science of controlling thermal energy for the purpose of altering the properties of metals and metal alloys is known as ***thermal processing*** or ***heat treating.***

As with most subjects in materials science, thermal processing concerns a metal's atomic and crystal structure. So far in our study of materials technology, we have accomplished many learning objectives dealing with the "inner workings and hidden mechanisms" in the world of a metal's atoms. Crystal structures, bonding, diffusion, crystal imperfections, dislocations, allotropy, solid solutions, phases, multiphase solid mixtures, solute, solvent, and saturated solid solutions are but some of the many terms that now have meaning for us.

When we studied phases and solid solutions, we used cooling curves for pure metals and metal alloys. These curves are graphic pictures showing the relationships between the variables of time and temperature. Many of these single cooling curves are plotted to give a graphical picture of the possible phases a metal alloy could be in at any particular temperature and at any particular composition. Such phase diagrams are also known as ***equilibrium diagrams.*** *Equilibrium* means that the time variable is not controlled but is allowed to run its course. In other words, *equilibrium* implies that phase changes shown on equilibrium diagrams are produced under conditions of slow cooling with no restraints on time. This allows the atoms that are diffusing through a solid material the time to seek and find the equilibrium position of lowest energy level. Phase diagrams show us the following:

1. The phases that are present for a particular alloy composition and temperature
2. The extent of solubility between two metals
3. The maximum solubilities of each metal in the other
4. The alloy composition with the lowest melting point (eutectic)
5. The melting points of the pure metals making up the alloy

But phase diagrams do not show what happens to a solid system of materials in equilibrium when a change in temperature, pressure, composition, or any combination of these variables occurs. We know that the system will seek another state of equilibrium with its lowered level of energy (free energy). We also know that this process takes time. In many cases this new state of equilibrium is extremely slow, particularly at low or room temperatures. During this time the system is said to be in a nonequilibrium condition or *metastable* state. This metastable state may persist indefinitely. Some steel and nonferrous alloys are in metastable states brought on by heat treatments whose goal is to increase their strength or other desired property. The heat treatments change their microstructures not only in a metastable state during the process, but also in the final or complete solid material.

To learn how thermal processing can produce phase changes, we use iron and its alloy steel as examples of ferrous metals. We will not discuss other industrial materials that are heat treated, such as glass, but leave this important area of study for independent research.

Being an alloy of iron and carbon, steel is also allotropic, existing in several forms. Referring to the steel equilibrium diagram (Figure 5-5c), which is a graphical record of the various phase transformations of steel in the solid state, and using the eutectoid composition, we see that above 727°C steel is in a solid state with a structure called *austenite,* or gamma (γ) iron, a single-phase fcc solid solution. Figure 5-6a and Figure 5-5b is a simpler version of the complete diagram shown in Figure 5-5c. Austenite is characterized by its ability not only to be deformed but also to absorb carbon. The fcc unit cell of iron atoms contains interstices that are large enough for the small carbon atoms to occupy, producing an interstitial solid solution. Many ***hot-working*** operations in industry take place with steel heated to a temperature that produces this austenite phase. Furthermore, the austenite region is the starting point for many of the thermal processes about to be discussed.

At temperatures below 910°C, pure iron changes to a stable phase called alpha ferrite, alpha (α) iron, or ***ferrite.*** This ferrite phase can accommodate up to a maximum of 0.02% carbon (by weight), which produces a solid solution with a bcc structure. The bcc structure contains interstices in its unit cell that are too small to accommodate the carbon atom. Another component in this diagram is ***cementite,*** an interstitial compound with the chemical formula Fe_3C. Cementite is orthorhombic, with twelve iron atoms and four carbon atoms in the interstices among the larger iron atoms in the unit cell. The ratio of iron atoms to carbon atoms is 3 to 1 and it does not change. (Refer to Sections 3.4.10 and 5.8.1.2, and Table 3-3.) *Note:* Cementite is also discussed briefly on page 241. Also called iron carbide, this brittle substance contains 6.67% carbon and has an HB of 700. When steel with the eutectoid composition forms at 727°C, it produces a lamellar two-phase mixture of ferrite and cementite called ***pearlite*** (see Figure 5-6). Using tie-lines across the two-phase regions of Figure 5-6, a hypoeutectoid steel can be described as a mixture of ferrite and pearlite (Figure 5-7), eutectoid steel as having a pearlitic structure, and hypereutectoid steel as a mixture of pearlite and cementite. Such structures can be discerned under a microscope. At this point it should be evident that there are countless alloys of carbon and iron (carbon steels) with varying amounts of carbon, producing a corresponding variety of steels with different properties to serve the demands of industry. Later in this book other elements will be discussed that also play a part in producing numerous specialty steels (alloy steels) that find ever-increasing use.

It is most important to understand that steel is possible only because iron is allotropic and carbon atoms are small enough to fit between the iron atoms in an fcc austenite structure (interstitial solid solution). The many different properties of carbon steel are produced as a result of changes in the amount of carbon and the difference in the abilities of ferrite and austenite to dissolve carbon. Austenite can dissolve almost 100 times more carbon than can ferrite. But austenite does not exist at room temperature under equilibrium conditions, whereas ferrite does. It is worthwhile to note on the iron–iron carbide phase diagrams (Figure 5-5a, b, and c) the locations of wrought iron (almost pure iron), steels, and cast irons (greater than 2% carbon). Most important of all is the realization, first, that everything we have discussed about the allotropic forms of iron, the diffusion of carbon atoms, phase transformations, austenite, ferrite, cementite, and pearlite takes place in the *solid state.* Not once did we get even close to the melting point of iron, 1538°C. Second, all phase transformations took place under *equilibrium conditions.*

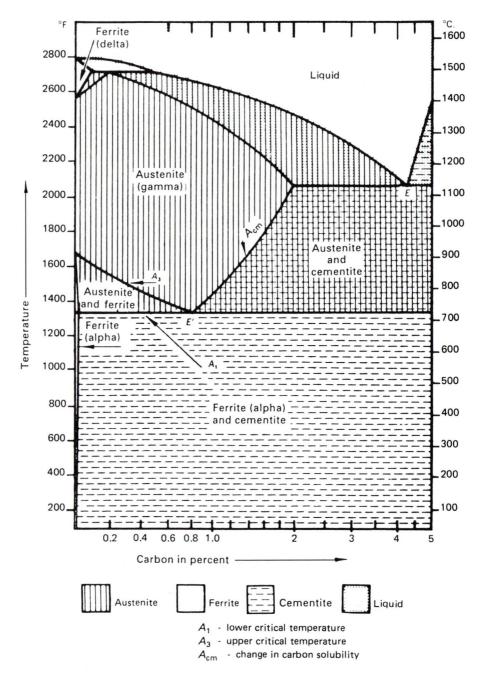

Figure 5-5   (a) Simplified iron–carbon equilibrium diagram. (United States Steel Corporation)

5.3 THERMAL PROCESSING OF STEEL: NEAR-EQUILIBRIUM
CONDITIONS

Several thermal processing techniques used with ferrous metals approach equilibrium conditions, as represented by the iron–iron carbide equilibrium, or phase diagram. By definition, such
processes bring about a change in a material's properties. Therefore, they must affect the material's atomic structure. The first thermal process to be discussed is generally known as ***annealing.***
Annealing processes bring about changes mostly by producing phase transformations that result

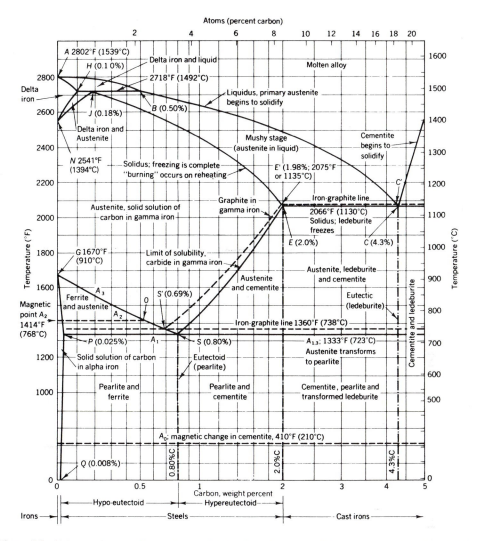

Figure 5-5 (b) Iron–carbon equilibrium diagram. Dashed lines show true equilibrium of iron and graphite. Solid lines show a metastable phase diagram of iron and iron carbide (Fe₃C), also called cementite. The metastable diagram is used in the same manner as a true equilibrium diagram. The distinction between the two is negligible above 2100°C. (ASM International)

in a rearranged, stable atomic structure with less distorted grains. Therefore, the initial step in most annealing techniques is to heat the steel above its critical temperature to form austenite.

Annealing is broken down into a number of related operations whose overall purpose is to reduce hardness, refine grain structure, restore ductility, remove internal stresses left over from some industrial forming process, or to improve machinability or some other property. These operations—full anneal, normalizing, stress relief, process anneal, and spheroidizing—are shown on a simplified equilibrium diagram for steel (see Figure 5-8). A ***full anneal*** consists of heating steel to about 30°C above the line marked A_1 (the eutectoid temperature line), depending on the carbon composition; holding it at that temperature to obtain a homogeneous structure of austenite; and slowly cooling it in a furnace. The result is, for a hypoeutectoid steel, a transformation of austenite to a coarse lamellar pearlite that is soft, stress free, and a fine ferrite. The word ***annealing*** used alone when referring to ferrous alloys implies full annealing. When referring to nonferrous alloys, the word is used to describe a heat treatment designed to soften a cold-worked structure by recrystallization and/or subsequent grain growth. Annealing of age-hardened alloys implies softening of the alloy by a

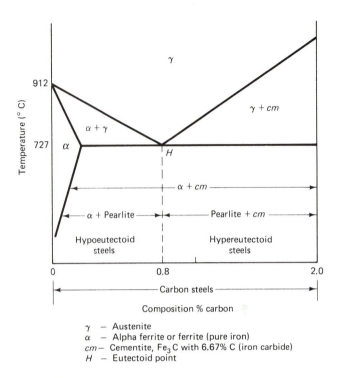

γ — Austenite
α — Alpha ferrite or ferrite (pure iron)
cm — Cementite, Fe_3C with 6.67% C (iron carbide)
H — Eutectoid point

Figure 5-5 (c) Simplified steel equilibrium diagram.

nearly complete precipitation of the second phase. If the annealing is intended for the sole purpose of reducing stresses, it should be designated as stress relieving. Forged parts are annealed to refine the microstructure produced by forging, which is very hard, so that machining operations such as trimming can be performed. Annealing of gray iron, especially small castings that require a large amount of machining, is noted with an example of ASTM class 35 gray iron, which ends up as a class 20 iron due to the reduction in strength as a consequence of the treatment. An example of the annealing treatment taken from the *ASM Handbook* of a heat-resistant casting made of Alloy Castings Institute (ACI) alloy HA, or ASTM A217 iron–chromium heat-treatable alloy used in the oil industry, is cited. These alloys are heated to an annealing temperature of 1625°F, followed by slow cooling in the furnace at about 50°F per hour to below 1300°F. For improved strength these castings are normalized by heating to 1825°F and air cooled to below 1250°F. *Galvannealing* refers to galvanized parts that are given a subsequent heat treatment to convert the zinc coating to an oxide to promote paint adhesion. Steel parts are dipped into molten zinc followed by a high-temperature treatment in an annealing furnace. The result is a zinc-iron alloy coating that contains 9–12% iron. This coating provides better paintability and weldability than a hot-dipped zinc coating of equal thickness.

 Because it is time consuming and expensive (due primarily to the use of furnace cooling), full annealing in many cases is replaced by *normalizing.* Normalizing of steels through using higher temperatures does not require furnace cooling. Instead, all cooling is in still air, and a fine pearlite structure is obtained. Steel is normalized to obtain greater hardness than that of a full anneal. Normalizing does not approach equilibrium cooling conditions as closely as a full anneal. The tensile strength of an annealed alloy is 95,000 psi versus 107,000 psi for the same alloy if normalized. Nodular cast iron castings are usually heat treated to improve their properties. A typical heat treatment to obtain a pearlitic structure is to heat to 1650°F, hold at temperature, cool to 1450°F in the furnace, and then air quench. The quench may be followed by tempering to obtain greater hardness and strength. This additional treatment consists of heating to between 1600° and 1650°F, oil quenching, and tempering. The yield strength for

Figure 5-6 Pearlite. Lamellar two-phase mixture of ferrite and cementite. The white areas in this hypoeutectoid steel are grains of proeutectoid alpha ferrite. The dark areas are pearlite colonies consisting of alternate lamellae of alpha ferrite and Fe_3C (cementite). The resolution of the two phases in the pearlite region depends on magnification and orientation and varies from area to area. (Buehler Ltd.)

Figure 5-7 Slowly cooled, hypereutectoid steel 1000×. White areas are cementite and dark areas are pearlite. (United States Steel Corporation)

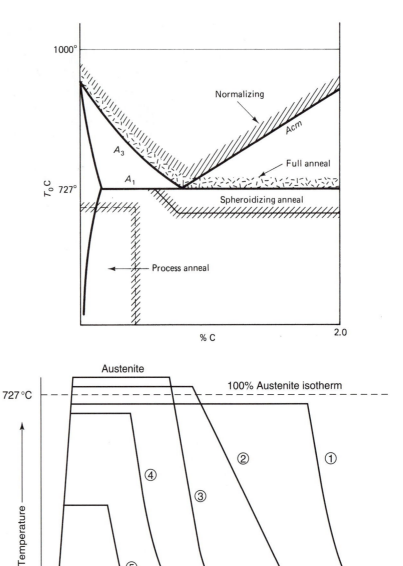

Figure 5-8 (a) Simplified steel equilibrium diagram showing various types of anneals. (b) Schematic of heat-treat annealing processes for steel: 1) spheroidize, 2) full anneal, 3) normalize, 4) process anneal, and 5) stress relief.

normalized and tempered nodular cast iron castings is 76,000 psi, compared to 124,000 psi for normalized only, and the HB increases from 255 to 321. Normalizing of steels through using higher temperatures (about 60°C above the A_3 line; see Figure 5-8) does not require furnace cooling. Instead, all cooling is in still air, and a finer and more abundant pearlite structure is obtained than from a full anneal. However, its main purpose is to homogenize the microstructure, particularly if further hardening is required. Heating to high temperatures is critical. If too high a temperature is used or if the steel is held for too long at a high temperature, the most likely result is undesirable grain growth. In general, slow heating is more desirable, particularly when working with highly stressed materials (i.e., cold worked), to avoid distortion. Parts with varying cross-sections (thick and thin sections) call for slow heating.

     A *stress relief anneal* requires temperatures around 600°C, which are below the critical temperature at which austenite begins to form upon heating. The primary objective of this

Figure 5-8   (c) Spheroidized 1045 steel, 400×. (Buehler Ltd.)

technique is to relieve residual stresses in all metals and steels as a result of a welding, cold-working, or casting operation. These stresses are eliminated or reduced, even though there is no change in the metal's microstructure.

Process anneal is allied with cold working, in which the metal, after heating, is cooled slowly in a furnace down to room temperature. The cooling phase distinguishes this process from normalizing. The end product has higher ductility and lower strength than if normalized. Process annealing is used to "soften" metals, particularly steel sheet and wire products, for further cold working. An electrical transmission line made of copper would become brittle after several drawing operations and, without a process anneal, would fracture. This technique, as well as stress relief, does not involve a phase transformation (no austenizing). The structure of the metal involved, however, is affected. A process called recrystallization is involved, which will be explained later in this module.

Spheroidizing anneal is used to improve the machinability of high-carbon steel. Hypereutectoid steels provide good wear resistance. A spheroidizing anneal helps to toughen them by providing more ductility. The hard and brittle cementite network present in hypereutectoid steel also makes machining to close tolerances difficult. Heating the metal for a longer duration near the critical temperature, followed by slow cooling, is one technique used to produce a spheroidal or globular form of cementite in the ferrite matrix. This entire structure is called spheroidite or spheroidized pearlite; it has a lower hardness, higher ductility, and higher toughness than does the original metal (Figure 5-8).

Most of the above processes may be used to improve machinability but the carbon content of the steel usually is the determining factor. Hypoeutectoid steels with less than 0.3% carbon are normalized, those with up to 0.6% carbon are annealed, and spheroidizing is reserved for steels with 0.6–1.0% carbon.

5.3.1 Grain Refinement

Grains that have become plastically deformed as a result of cold working can be given enough energy through thermal processing to permit an orderly rearrangement of the atoms to take place with less deformed grains. The thermal process involved is annealing, and a certain stage of annealing in which new crystals form, known as recrystallization, is worthy of separate comment.

Recrystallization is the formation of a new, strain-free grain structure from that existing in a cold-worked metal upon heating the metal up through a critical temperature for a suitable period of time. When referring to cold-worked metals (mostly all nonferrous metals), annealing can be described as consisting of three stages: recovery, recrystallization, and grain growth.

The recovery stage (see Figure 5-9), the lowest-temperature region, provides the deformed atoms with sufficient energy to rearrange themselves, thus reducing distortion and its attendant stress. The properties recover toward their original values with no visible change in their microstructure. The recovery is due to the cancellation or annihilation of the point defects and dislocations in the deforming metal. A stress-relief anneal, as described previously, requires temperatures in the lowest-temperature region of the recovery stage. As temperatures increase, a point is reached where the nucleation of new crystals appears in the microstructure. These new crystals have the same structure and composition as those of the original undeformed crystals before the metal was cold worked. The grain boundaries and slip planes with the maximum distortions caused by the cold working contain the majority of these new crystals, and eventually they grow and spread throughout the metal structure. A short time after the appearance of the new crystals, the properties of the metal change rather drastically. Hardness and strength drop off, while ductility increases. As the temperatures increase, the region of grain growth is reached, about 0.3 to 0.6 of the melting point of the metal in Kelvin, which points out that control of grain size can be exercised to some extent by a wise choice of annealing temperatures. Recrystallization temperatures vary with metals and with the degree of cold working. The number and size of crystals that develop in the recrystallization region re-

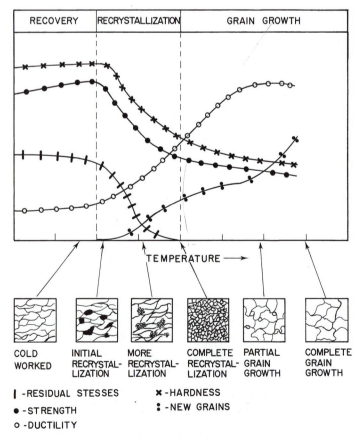

Figure 5-9 Cold-work annealing treatment.

sult from the amount of cold working done to the metal. If a metal is not cold worked prior to heating, recrystallization will not occur and the grains will grow in size. To transfer sufficient energy to the deformed crystals, a minimum plastic deformation of about 7% is required before any change in grain size takes place.

From our knowledge of cold working, we know that a highly cold-worked metal contains very little capacity for further plastic deformation. Consequently, the metal is heated to a temperature that permits the atoms to realign and diffuse to a more stable position. The temperatures over which a marked softening occurs (a drop in tensile strength) are in the recrystallization zone. This range or zone of temperatures falls between one-third and one-half of a metal's melting temperature expressed in absolute units (Kelvin). The average recrystallization temperature is not the same as the critical or transformation temperature. It depends on the amount of cold working, as well as the duration of the heating. The more a metal is cold worked, the more energy the atoms possess. Therefore, they need less energy transfer from an external source than does a less-cold-worked metal. The recrystallization temperature is used to distinguish between cold working and hot working of metals. The cold working of copper at 95°C is at a higher relative temperature than the hot working of zinc at 20°C, because the recrystallization temperatures of pure copper and zinc are about 120° and 10°C, respectively. A point to remember is that these changes do not come about as a result of any phase change. When a metal is not allotropic, several processes are needed to change its properties. First, it must be cold worked to bring about a minimum deformation of the crystal structure. Second, it must be heated sufficiently to produce recrystallization. Third, after slow cooling the emergence of new, stress-free, small, uniformly dimensioned grains will result in the properties desired. Finally, it must be realized that if, while in service, a metal is subjected to a temperature that is higher than its recrystallization temperature, it will recrystallize, lose its strength and hardness, and possibly cause a failure of some magnitude.

5.3.2 Grain Size

Heat treating can alter the size of grains, which produces corresponding changes in properties. A grain, or individual crystal, varies in size from one metal to another. Zinc crystals in galvanized steel sheet and crystals in some brasses can be seen with the naked eye, but these are exceptions to the general rule. Most crystals are visible only with the use of a microscope.

Even within one metal, the grain size can vary from one region to another. Grain size has a direct influence on the properties of the metal. For example, as grain size *decreases,* the yield strength of the metal increases. Another fact is that fine-grained metals are stronger and tougher than coarse-grained metals under low or room temperatures. At high temperatures, the reverse is true; coarse-grained metals are stronger and tougher than fine-grained metals. Grain-size ratings, as contained in the American Society for Testing and Materials (ASTM) specifications, are made by comparison with a standard chart of sizes numbered from 1 to 10. When a match is arrived at, the grain size is then designated by the number corresponding to the index number *(n)* of the matching chart. These matching standard-grain-size charts represent a microstructure projected at a magnification of 100×. The comparison method described above (ASTM E112) is an estimation at best.

Steel is ordered on the basis of either fine grain or coarse grain. Grain size must be specified for metals that are to be cold worked. If the grain size is too coarse, the worked surface will be uneven, and if too fine, the metal will be too hard to work satisfactorily. Several different methods, in addition to the ASTM comparison method, are used in industry to estimate grain size.

Temperature alone causes grain growth and not grain size reduction or refinement. If you wish to reduce the size of grains to refine them, you must destroy the original grains and produce new ones with the desired size. One way of doing this is by heating an allotropic metal sufficiently to produce a phase change with its initially small grains. In steel, the fcc grains of austenite are large, but on transforming to bcc crystals, the grains are small. Another way to refine grains is to deform the metal plastically by cold working. This refinement in

Figure 5-10   Different micrographs showing grain boundaries. (Buehler Ltd.)

grain size is not the primary purpose of cold working. However, due to the smaller size of the grains, the cold-worked metal is found to possess properties different from those before it was cold worked.

5.3.3 Grain Boundaries

We know that the atoms that make up the grain boundaries are the last to solidify. As a result, they are not arranged in the same orderly way as are the same atoms in the interior of the crystal. We also know that any time we observe a disordered arrangement of atoms, we look for distortion, increased stress, and resistance to further plastic deformation or slip—all of which result in stronger material in the grain boundaries (Figure 5-10). Arrows point to some of the many grain boundaries shown in Figures 5-10 and 5-11.

The more small grains present in a metal, the more grain-boundary material per unit of volume of metal. The conclusion drawn is that fine-grained metals are stronger than coarse-grained metals at room temperatures. At high temperatures, the heat supplied to the fine-grained metal atoms that make up the grain boundaries increases their energy. This reduces their strained condition, which allows them to spread out and become less densely packed, which in turn results in a decrease in strength. Based on this reasoning and coupled with knowledge gained from experience, coarse-grained metals are required for high-temperature applications where resistance to creep is critical. Can you recall the definition of *creep?* A good example where knowledge is needed is in the selection of the correct metal for the manufacture of turbine blades that revolve at very high speeds and temperatures, with extremely small clearances between the moving tips of the blades and the stationary parts of the turbine.

Cold working results in the formation of more grain boundaries due to the coming together of the large number of dislocations with their disorderly arrangement of atoms. With more grain boundaries come smaller grains; reduced ductility; and increased hardness, tensile strength, and electrical resistance. These conditions describe the terms ***work*** or ***strain hard-***

Figure 5-11   Micrographs showing: (a) hot-worked, then cold-worked, copper specimen (100×); (b) hot-worked copper specimen; (c) 70/30 brass 300×. Reduction: 1) 0%, 2) 5%, 3) 25%, 4) 75%. (Buehler Ltd.)

ening produced by cold working. A good example of cold working is the drawing of copper wire. ***Drawing*** has two meanings in materials engineering. In heat treatment, it means tempering (to be defined). In cold working, the process of drawing refers to the forming of a recess in sheet metal with dies (see Figure 4-10) and the pulling of metal through dies, as in the manufacture of wire. A copper wire transmission line is cold worked to increase its strength so that it can support itself without excessive sagging over a moderate span.

5.3.4 Grain Shape

Cold and hot working produce a distorted grain structure; that is, the shape of the grain is deformed or becomes elongated in the direction of the metal flow. Figure 5-11a and b compares hot- and cold-worked grains of copper. Such deformed metals no longer possess uniform properties in all directions (isotropic). Instead, the metal generally shows higher-strength properties in certain directions, with a corresponding decrease in those same properties in other directions (anisotropic).

To summarize the preceding discussion, the set of photomicrographs in Figure 5-11c illustrates the concepts of cold working, recrystallization, and grain growth. The specimens are from annealed, commercially available 70% Cu–30% Zn nonleaded brass, initially 0.128 in. thick. Reductions were accomplished by small reduction rolling passes while maintaining the strip at room temperature at all times. All photomicrographs of the strip samples are shown at 300× for the best possible comparison of microstructures.

5.4 THERMAL PROCESSING OF STEEL: NONEQUILIBRIUM CONDITIONS (MARTENSITE STRENGTHENING)

By varying the amount of carbon dissolved in the single-phase solid solution of iron and carbon (austenite) and cooling the austenite under equilibrium conditions, the austenite transformed into a multiphase mixture known as carbon steel. Depending on the amount of carbon, steel alloys of ferrite and pearlite or cementite and pearlite are formed, which have different hardness and strength properties. But we now reach a limit. Low-carbon steel (0.25% carbon) has a strength of about 44,000 psi; eutectoid steel (0.8% carbon) has about 112,000 psi. To obtain steel with greater strengths, we must depart from equilibrium conditions primarily in the cooling of the austenite. This departure, in the form of a rapid cooling, is known as *quenching.*

The degree of cooling depends mostly on the quenching medium used. For drastic quenching, water or brine is used. Less severe cooling rates are achieved with such media as oil, molten salt baths, still air, or molten sand. Most liquid media require agitation of the liquid to help reduce the gaseous layer formed adjacent to the metal as a result of the vaporization of some of the liquid in contact with the hot metal, whose temperature is above the boiling point of the liquid quenching medium.

Rapid cooling of steel produces a metastable phase called *martensite* (Figure 5-12), which is defined as an interstitial supersaturated solid solution of carbon in iron having a bct lattice structure. The transformation of martensite results from a three-dimensional mechani-

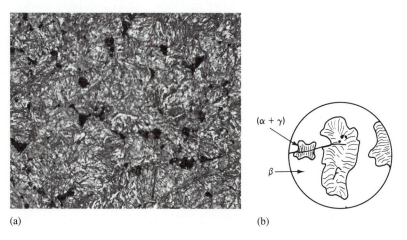

(a)   (b)

Figure 5-12 Micrograph of 1095 steel martensite. (a) 1095 steel. (b) Microstructure of martensite (100×). (Buehler Ltd.)

cal shearing action that distorts the structure, producing hardness and brittleness. Martensite transformations occur in other metals and alloys, such as titanium, lithium, iron, nickel, and copper-aluminum.

To summarize what we have said about equilibrium and nonequilibrium conditions of cooling, our initial step was to heat the basic components of steel (into the *austenite*-forming region) to produce austenite, a single-phase solid solution of carbon in gamma (γ) iron. Once we formed austenite, we could cool it under equilibrium conditions to produce a two-phase mixture of ferrite and cementite called **pearlite.** The carbon in the cementite distorts this structure, producing qualities of strength and hardness. The ferrite contributes ductility. If we depart from equilibrium conditions and cool the austenite more quickly, we can form pearlite, with greater hardness. If we cool it even more rapidly, we produce a transformation product called **martensite,** which is extremely brittle, strong, and hard. It far surpasses the hardness and strength of pearlite. Steel with 100% martensite has limited use due to its sensitivity to fracture by impact. Therefore, we call on another heat-treating process known as **tempering** to heat the martensite to selected levels and to transform some or all of the martensite to other microstructures with differing properties. Tempering is the process of heating the as-quenched steel (in the form of as-quenched martensite) to give it some ductility. This as-quenched martensite is a supersaturated solution of carbon in either a bcc or bct ferrite. The bcc structure is producing the strength of the steel and depends almost entirely on the carbon content of the steel. After tempering, the steel is in the quenched and tempered (Q&T) condition having the as-quenched martensite transformed into stable phases of ferrite and iron carbide (Fe_3C). Tempering is sometimes known by another name, **drawing,** due to the fact that, through controlled heating, hardness can be drawn from the material. Actually, tempering, like any process that supplies energy to atoms through heating, depends on the diffusion of these atoms. The rate of diffusion depends not only on the temperature, but also on the time allowed for diffusion to take place. In allowing the carbon atoms to proceed once more on their paths through the atomic structure, varying microstructures and grain sizes are formed with just the correct hardness and strength desired for some industrial application. One application might require a good amount of toughness with very little brittleness; another might demand greater hardness. As in many situations, we cannot have the best of two worlds; to gain greater strength and toughness, we have to trade a little hardness.

In discussing nonequilibrium conditions in the thermal processing of steel, we make use of an **isothermal-transformation (IT)** or **transformation-temperature-time (TTT)** diagram for a particular steel composition. This diagram is derived by plotting hundreds of isothermal cooling curves for samples of a particular steel composition. The isothermal cooling curve sketched in Figure 5-13 shows the percentage of austenite remaining and the percentage of the austenite transformed phase plotted against time. The isothermal temperatures are all below the critical A_e (or eutectoid) temperature at which the austenite becomes unstable and begins to transform (see Figure 5-14). The data thus collected from these isothermal curves are then used to construct the isothermal-transformation diagram for that particular steel. The diagram shows the beginning and end of the austenite transformation as well as the transformed product at various isothermal temperatures.

This procedure is illustrated in Figure 5-14 using just one set of data. A characteristic S-curve is formed that represents the transformation zone or region at various temperatures. Figure 5-15 is a copy of such a diagram for eutectoid steel, indicating the Brinell and Rockwell hardness readings alongside photomicrographs of the various products formed from austenite at the various temperatures. Figure 5-16, another sketch of a TTT diagram, shows the S-shaped austenite-to-pearlite transformation zone, the austenite-to-martensite transformation zone, and a subdivision of the pearlite into coarse and fine pearlite and bainite, a transformed product of austenite that has characteristics of both pearlite and martensite. In addition, two cooling curves are plotted. The first curve, labeled 1, represents the transformation of austenite to coarse pearlite. Such a curve has a relatively small slope, which indicates a slow cooling rate. The second curve has a much steeper slope and consequently a much faster cooling rate. The latter curve is shown passing to the left of the leftmost portion of the pearlite

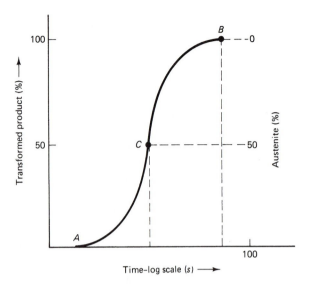

Figure 5-13   Isothermal-transformation curve at temperature $T°C$.

Figure 5-14   Isothermal-transformation diagram (IT or TTT).

transformation zone, known as the knee or nose. As a consequence, all the austenite transforms into martensite.

It should be evident at this point that the goal of thermal processing under nonequilibrium conditions is to produce steels with greater strengths than those produced by solution hardening under equilibrium conditions (very slow cooling). The mechanism used to accomplish this goal is a phase transformation. Such a mechanism is possible because steel is allotropic. In the next few pages we will describe briefly some major thermal processing techniques used by industry to strengthen steel under nonequilibrium conditions.

Figure 5-15   Isothermal-transformation diagram of a eutectoid steel. (United States Steel Corporation)

5.4.1 Conventional (Customary) Heat, Quench, and Temper Process

The previous statements on the formation of austenite, quenching to form martensite, and the final heating to temper the martensite describe the conventional process of making steel parts with desired properties. Study Figure 5-17 for the customary quench/temper cooling rates. The log scale along the bottom indicates time. One serious drawback to this process is the possibility of distorting and cracking the metal as a result of the severe quenching required to form the martensite without transforming any of the austenite to pearlite.

As a metal object is quenched, the outer area is cooled more quickly than the center. Thinner parts are cooled faster than are parts with greater cross-sectional areas. What this means is that transformations of the austenite are proceeding at different rates in a single metal

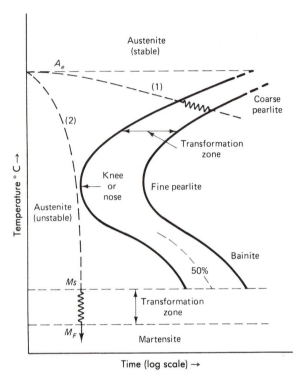

Figure 5-16 IT diagram showing transformation zones, two cooling curves, and products formed.

object. As we cool a metal object, it also contracts and its microstructure occupies less volume. Put these statements together and the conclusion is that extreme care is necessary to prevent undue distortion and/or fracture. Extreme variations in the size of metal objects complicate the work of the heat treater and should be avoided in the design of metal parts. This means that there is a limit to the overall size of parts that can be subjected to such thermal processing.

5.4.2 Martempering

To overcome these restrictions, two other thermal processes are used. The first, *martempering* or *marquenching,* permits the transformation of austenite to martensite to take place at the same time throughout the structure of the metal part. This is shown graphically in Figure 5-17. By using an interrupted quench, cooling is stopped at a point above the martensite transformation region to allow sufficient time for the center to cool to the same temperature as the surface. Then cooling is continued through the martensite region, followed by the usual tempering.

5.4.3 Austempering

A second method of interrupted quenching is called *austempering.* Figure 5-17 shows this process graphically. The quench is interrupted at a higher temperature (200° to 375°C) than for marquenching, to allow the metal at the center of the part to reach the same temperature as the surface. By maintaining that temperature, both the center and the surface are allowed to transform to bainite and are then cooled to room temperature. The advantages of austempering are (1) even less distortion and cracking than marquenching, due primarily to the higher transformation temperatures, and (2) no need for final tempering. However, austempering has the disadvantage of requiring more time, even though it requires no tempering treatment. Also, parts with large sections of thickness cannot be processed. Sections with a maximum thickness of 1/2 in. are cooled sufficiently fast to permit the transformation of austenite to bainite without the formation of pearlite.

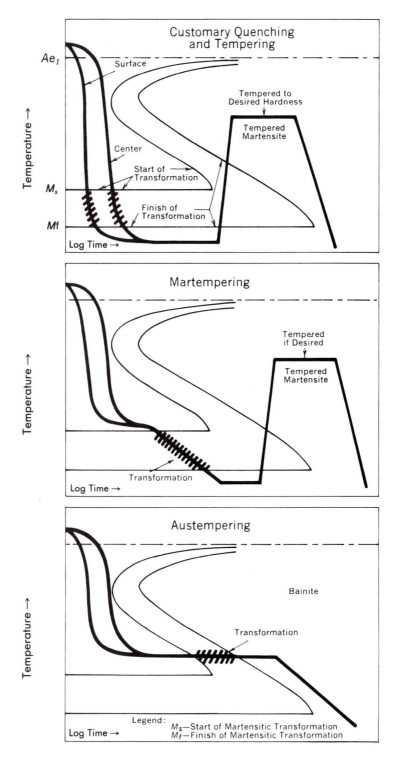

Figure 5-17 Comparison of heat-treating processes. Heating and cooling operations superimposed on a typical isothermal-transformation diagram. The temperature range at which the transformation to the hard products bainite or martensite occurs has been indicated. (United States Steel Corporation)

5.5 PRECIPITATION STRENGTHENING/HARDENING (AGE HARDENING)

Nonferrous metals do not undergo the significant phase transformations possible with steel. Consequently, most thermal processing of nonferrous metals is used to relieve stresses in a single-phase microstructure and/or to produce recrystallization. These techniques do not result in significant strengthening of a metal's structure. Instead, the most effective thermal processing technique for increasing the strength of such metals is *precipitation hardening,* or *age hardening.* Precipitation hardening involves two steps (see Figure 5-18). The first is *solution treatment,* the heating (annealing) of an alloy that exists at room temperatures as a two-phase solid

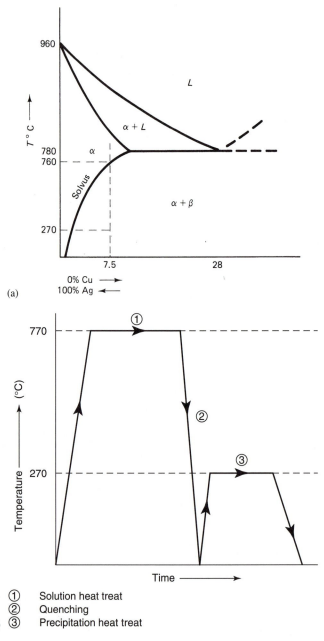

(a)

(b)

① Solution heat treat
② Quenching
③ Precipitation heat treat

Figure 5-18 (a) Schematic of the Ag-rich end of Ag–Cu equilibrium diagram. β phase is pure copper (Cu) plus a little silver (Ag). (b) Temperature/time plot showing heat-treat cycles for precipitation strengthening.

solution to produce a single-phase solid solution at the elevated temperature. The second step is the subsequent rapid cooling *(quenching)* of this single-phase solid solution to form an unstable supersaturated solid solution. Some alloys require a further heating process to expedite the strengthening of the metal.

To explain how this strengthening comes about, it is first necessary to review earlier material in this chapter, specifically that concerning diffusion (page 226) and the effects of solid solubility curves (solvus lines) using Figures 3-67 and 3-69. Such a review will assist the reader in understanding our explanation of precipitation hardening. As stated previously, the partial solubility of one solid metal in another is determined by the slope of the solid solubility curve (solvus line). The solvus must slope so that it indicates there is greater solubility of one metal in another at a higher temperature than at a lower temperature.

Figure 5-18, similar to Figure 3-69, is a schematic of the equilibrium diagram for silver (Ag)–copper (Cu) alloys. At each end (the silver-rich and the copper-rich), the solvus indicates the type of partial solubility desired for precipitation hardening. Figure 5-19 shows the silver-rich end of the diagram (the left end of Figure 5-18), an alloy consisting of about 92.5% Ag and about 7.5% Cu (sterling silver) is shown as a vertical dashed line. At room temperature this alloy exists as a metal consisting of two distinct phases (α and β). Phase α represents a silver-rich solid solution with some dissolved Cu, and β is a solid solution of almost pure Cu with very little dissolved Ag. As the alloy is heated, it crosses the solvus at about 760°C. The β phase dissolves at this temperature and diffuses uniformly to form a solid solution of α. If, then, this same alloy were slowly cooled, the β phase would precipitate out of the α solid solution because the solubility of β (copper) in the α (silver) decreases from about 8.5% at 780°C to less than 1% at room temperature, as shown by the solvus line.

Now if after the solution is heated to 1 around 770°C (see Figure 5-18b), this alloy were cooled quite rapidly (quenched 2), there would be insufficient time to permit the Cu atoms to diffuse, which would produce equilibrium-type precipitation of β atoms. The result is a retained α phase that is unstable, with its excess Cu atoms trying to precipitate out to form the β-phase crystal structure. The unstable, supersaturated solid solution formed is quite ductile.

With the metal in such a crystal structure, it can be worked, straightened if distorted, or machined much more easily than when the metal was in the stable, original form at room temperature. Once the metal is worked, the strengthening process can continue. Other alloys require heating to a temperature below the solvus line to cause this precipitation. Sterling silver can be heated to around 300°C for about 30 minutes and the hardness doubles in value. The increase in strength and hardness is explained by precipitation of the trapped β metal out of the α-metal structure at the grain boundaries and at the sites of impurities in the lattice structure of α metal, which produces resistance to the movement of dislocations. If the precipitation occurs at room temperatures within a few days' duration, it is known as *natural aging.* If additional heating is required, as it is in sterling silver, to cause precipitation and full strengthening, the process is called *artificial aging.* This additional heating/cooling cycle is also referred to as the precipitation heat-treating cycle (Figure 5-18b). At this stage the supersaturated α solid solution is heated to some intermediate temperature (see the 270°C isotherm in Figure 5-19) to increase the rate of diffusion of β participate phase in the form of finely dispersed particles of metal *A* or, in this case, copper atoms.

It is possible to stop the precipitation entirely by lowering the temperature sufficiently through refrigeration. Once the precipitation has produced a saturated normal dual phase at or near room temperature, the metal reaches its greatest hardness and strength. As fine precipitates form, they have a tendency to grow with time, and with natural-aged alloys this growth of the precipitate will result in a decrease in the strength. Such a phenomenon is known as *overaging.* Artificially aged alloys are less subject to overaging because the growing process can be stopped effectively by a simple quench. It must be pointed out that the strength property was used more in this discussion than was any other property. We know that other properties are also affected by distortions of a lattice structure brought about by the precipitation of atoms of another element. The main difference between precipitation strengthening and the thermal processing under nonequilibrium conditions that produces martensite is the formation of a new crystal structure in the transformation of martensite.

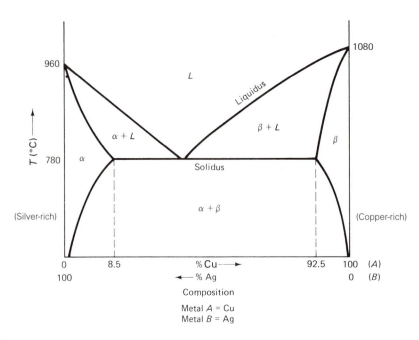

Figure 5-19   Ag–Cu alloys equilibrium phase diagram.

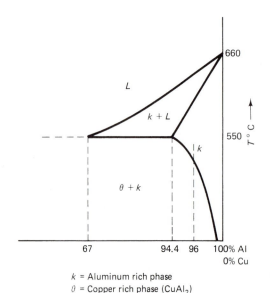

k = Aluminum rich phase
θ = Copper rich phase (CuAl₂)

Figure 5-20   Schematic of the Al-rich end of Al–Cu equilibrium diagram.

To round out our treatment of precipitation hardening, the popular aluminum–copper alloy known as Duralumin (2017) and a part of its equilibrium diagram are used to illustrate the steps taken to bring out the strength of the alloy in industry. Figure 5-20 is a schematic of the aluminum-rich end of the Al–Cu equilibrium diagram. Alloy 2017 contains about 96% Al and 4% Cu. Naturally or artificially aged, this alloy finds many applications, one of which is for aircraft rivets. The vertical dashed line (which intersects the solvus line) in Figure 5-20 represents this particular alloy. Upon solution treatment to a temperature of about 550°C, the single-phase solid solution designated *k* is formed in which all the Cu atoms are diffused into the crystalline structure of the Al atoms. Next it is quenched to form the unstable, supersaturated *k* phase, and the alloy becomes very ductile and can be maintained if it is refrigerated to retard the eventual precipitation of Cu atoms from the *k* phase. Once sufficient energy is

HEAT TREATMENT

Decarburization

Like other carbon bearing high strength alloys, AerMet 100 alloy is subject to decarburization during hardening. Heat treatment should take place in a neutral atmosphere furnace, salt bath or vacuum. Decarburization should be determined by comparing the surface and internal hardness of a small test cube for proper response. Metallographic determination of decarburization is not recommended for this alloy.

Normalizing

AerMet 100 alloy can be normalized by heating to 1650°F (899°C) holding for one hour and air cooling to room temperature. Optimum softening for machining is obtained by following the 1650°F (899°C) normalize with a 16 hour 1250°F (677°C) overage anneal.

Annealing/Overaging

AerMet 100 alloy is softened by using a 1250°F (677°C) overage anneal for 16 hours. The optimum annealed hardness of 40 HRC maximum is obtained following this anneal.

Solution Treatment

The solution treatment temperature range is 1625°F +/−25°F (885°C +/−14°C) for 1 hour. The solution treatment temperature must be monitored by a thermocouple attached to the load.

Quenching

Water quenching is not recommended. Proper quenching practice is essential for AerMet 100 alloy. The alloy should be cooled from the solution treatment temperature to 150°F (66°C) in 1 to 2 hours to develop optimum properties. Individual sections larger than 2" diameter or 1" thick (plate) must be quenched with oil in order to obtain 150°F (66°C) in 1 to 2 hours. Individual sections up to 2" diameter or 1" thick (plate) will air cool to 150°F (66°C) in 1 to 2 hours. The cooling rate of the furnace load must be monitored by a thermocouple attached to the hottest spot in the load to insure that the 2 hour cool to 150° (66°C) is obtained.

Refrigeration

Following cooling to room temperature, to obtain the full toughness capability AerMet 100 alloy should be cooled to −100°F (−73°C) and held for 1 hour. The parts can then be air warmed.

Straightening Treatment

AerMet 100 alloy exhibits minimal size change during heat treatment; however, for some parts, mechanical straightening to compensate for distortion during heat treatment is appropriate.

Prior to straightening, a low temperature stress relief at 350/400°F (177/204°C) for 5 hours following the refrigeration operation will provide an optimal combination of ductility and yield strength for the mechanical straightening operation.

Aging

The standard aging treatment for AerMet 100 alloy is 900°F +/−10°F (482°C +/−6°C) for 5 hours. Parts made from AerMet 100 alloy should never be aged at a temperature below 875°F (468°C).

Effect of Aging Temperature on Hardness—AerMet 100 Alloy

Specimens solution treated using 1625°F (885°C) 1 hour, air cooled, refrigerated −100°F (−73°C) 1 hour.

Aging Temperature	HRC
As hardened	51.0/53.0
875°F (468°C) 5 hrs.	54.5/55.5
900°F (482°C) 5 hrs.	53.0/54.0
925°F (496°C) 5 hrs.	51.0/52.5

TYPICAL MECHANICAL PROPERTIES

Typical Mechanical Properties—AerMet 100 Alloy

Heat treatment—1625°F (885°C) 1 hour, air cooled, −100°F (−73°C) 1 hour, aged 900°F (482°C) 5 hours.

Yield Strength		Ultimate Tensile Strength		% Elongation	% Reduction of Area	Charpy V-Notch Impact Energy		Fracture Toughness K_{IC}*	
ksi	MPa	ksi	MPa			ft-lbs	J	ksi in	MPa M
Longitudinal Orientation									
250	1724	285	1965	14	65	30	41	115	126
Transverse Orientation									
250	1724	285	1965	13	55	25	34	100	100

*Per ASTM E399

Shear strength
ksi 175
MPa 1207

Figure 5-21 Heat treatment and typical mechanical properties of AerMet® 100 alloy steel excerpted from Alloy data sheet for AerMet® 100 alloy. (Courtesy of Carpenter Technology Corporation—Carpenter Steel Division) received through heating to room temperature or beyond, the aging process, and hence an increase in strength, take place. Rivets made from this alloy taken from refrigerated storage can then be driven easily. Once driven, the aging process begins as room temperature is reached, which brings on the attendant growth in strength and hardness. Figure 5-21 describes the heat treatment and typical mechanical properties of AerMet® 100 alloy steel produced by Carpenter Technology Corporation. This alloy is more fully described in Section 5.8 of this module along with Figure 5-34a, a photograph of a McDonnell Douglas F/A-18 carrier-based naval aircraft fighter which uses this alloy in its landing gear. All procedures and vocabulary in this discussion of the heat treating of this alloy steel should pose no problem with students who have studied the thermal processing of steel in this module. As a result, careful reading of this material can serve students well as a test of their learning progress in the area of heat treating.

5.6 HARDENABILITY

Not to be confused with hardness, *hardenability* is the measure of the depth to which hardness can be attained in a metal part using one of the standard thermal processing techniques designed primarily to increase hardness. We know that carbon is vital to steel making. Steels are even classified by their carbon content. Hardness, as an indicator of strength, depends on carbon content. The higher the carbon content, the higher the maximum hardness that can be attained at greater depths within a part, assuming that other factors remain the same. Alloys also play a significant role in producing steels with higher degrees of hardenability. We know that one way to change a metal alloy's microstructure is to purposely add impurities. The impurities are elements that will alloy or join with the main elements or ingredients to form solid solutions—substitutional and/or interstitial. These alloying elements, such as nickel, boron, copper, or silicon, produce effects similar to the effects of impurities found in the original metal when mined and/or those added, out of necessity, during the processing of the metal. The presence of atoms of the alloying element (1) distorts the lattice structure; (2) increases the metal's resistance to plastic slip; (3) hinders the diffusion of atoms past these areas of distortion; (4) effectively increases the brittleness, hardness, and strength of the metal alloy; and (5) increases significantly the metal's hardenability. Adding alloys or increasing carbon content shifts the location of the S-curve (transformation zone) of austenite to pearlite or bainite to the right on the TTT diagram. The result is that a steel need not be quenched as rapidly to achieve the desired hardness. One disadvantage is that alloys are costly, but benefits usually outweigh such costs. One benefit in using a less drastic quench is the reduction in the danger of cracking and/or warping of steel parts, particularly the parts that are, of necessity, complex in shape or size or that require great dimensional stability.

It is essential to point out here the necessity of following proper design standards, which help to reduce such things as residual stresses, warping, and cracking. Such techniques as maintaining, if possible, uniform cross-sections or thicknesses and using fillets and rounds make the job of the heat treater less difficult.

The Jominy end-quench test performed on steels using standardized specimens and procedures determines the necessary nonequilibrium cooling rates to attain a certain hardenability in a particular steel part (see Figure 5-22a). The 25-mm-diameter specimen with a length of 100 mm represents a metal alloy austenized at a prescribed temperature and for a prescribed time. After removal from a furnace it is clamped into a fixture and quenched at a specified flow rate and temperature. After cooling to room temperature, shallow flats are machined along its length and Rockwell hardness readings are taken at prescribed intervals over the first 50 mm. Each hardness reading is plotted to produce a hardenability curve similar to Figure 5-22b. Each steel alloy has its own unique hardenability curve. A steel alloy that has a high hardenability is one that hardens (forms martensite), not only at the surface, but generally throughout the interior. Hence, *hardenability* is defined as the ability of an alloy to be hardened by the formation of martensite as a result of a given heat treatment.

Figure 5-22   (a) Jominy end-quench hardenability test. (b) Hardenability plot of hardness (Rockwell C-scale) versus distance from quenched end of specimen.

Figure 5-23   The ibgNDT Systems Corporation's Preventive Multi-Frequency Tester (eddyliner® P) is shown with some typical automotive parts that can be automatically tested using eddy current testers. (Courtesy of ibgNDT Systems Corporation)

5.7 SURFACE HARDENING AND SURFACE MODIFICATION

In all our discussions about hardening and strengthening metal alloys, our objective was to obtain the same hardness and strength throughout the part that we were heat treating. It would be impractical, if not impossible in many instances, to harden the entire part, that is, the interior and the exterior, to that hardness demanded by wear-resistant surfaces. Even if we could attain this degree of hardness throughout, the part would have little *toughness* (the ability to withstand impact loads). Little application is found in industry for such a part.

    Through hardening is the hardening of a part throughout its cross-section as opposed to surface hardening commonly referred to as case hardening, which only hardens the surface of the part without a change of surface composition from the original steel composition. Figure 5-23 shows an eddy current tester and an array of automotive parts that can be automatically tested to determine their structure, composition, or properties. Included in these test results are a part's hardness, case depth, temper, residual stresses, and structure variations. ***Eddy currents*** are the result of the interaction between a ferrous material and an electromagnetic field. This field is produced by an alternating current flowing in a conductive coil. If a metallic sample is placed near or within this field, then the field of the coil will induce eddy currents which, in turn, interact with the original field of the coil. By observing or measuring the effect of the sample on the coil, much information may be deduced concerning the structure and composition of the sample. Once optimum conditions are established, such as frequency and magnitude of the current in the coil, changes in the electrical conductivity or magnetic permeability of the sample can be detected. Eddy currents do not penetrate deep into a sample. Therefore, flaws at or near the surface are readily detected. Through analysis of the hysteresis loop many characteristics of ferrous materials can be determined (see Module 9 for further explanation of hysteresis loops and magnetic permeability of materials).

Most steel parts require machining, and this operation is best performed when the part is not hard. Once machining is completed and the part is within the required tolerances, it may be hardened on the surface to provide a strong, hard, wear-resistant surface. Figure 5-24a is a case-hardened bar. Figure 5-24b shows a gear with case-hardened teeth.

5.7.1 Case Hardening

Hardening of the surface only is known as **_case hardening,_** and the surface layers that are actually hardened make up the _case_. The inner surfaces that are not hardened are known as the _core_ (Figure 5-24). The first three mentioned case-hardening methods are referred to as chemically modifying methods or diffusion-coating methods. They provide the needed elements that are lacking in the surface layers to permit the hardening reactions to take place. We have learned that carbon is an essential element in steel if we are to harden it (about 0.3% carbon in steel is the minimum for hardening steel). Many hot-working operations result in a removal of the carbon from the surface. This is known as **_decarburization_** (Figure 5-25a).

5.7.2 Carburizing

To harden the steel, the carbon content must be restored. As the carbon content in steel increases, the heat treater's ability to create greater hardness also increases. The purposes of carburizing are to restore lost carbon and/or to increase the amount of carbon in the outer layers of the metal part.

(a)   (b)

Figure 5-24   (a) Polished section of 8620 steel bar carburized to a depth of 0.060 in., measured to a 0.40% carbon content. (Republic Steel Corporation) (b) To combat wear, gear teeth get case hardened so the gear body retains its toughness to handle torsion stresses.

(a)

(b)

Figure 5-25   (a) Decarburizing. Left: Photomicrograph showing "decarb" to a depth of 0.020 in. in the surface of a hot-rolled bar of 1050—magnified 100 times. Right: Same bar after carbon correction annealing. Note the restoration of carbon on the outer surface and also the refinement of grain structure. (b) Carbon correction. Left: Photomicrograph of AISI 5046 before carbon correction—magnified 100 times. Right: Same bar after carbon correction annealing with surface carbon restored. (Republic Steel Corporation)

Basically, the metal part is surrounded by a high-carbon-content solid material or gas; by supplying the necessary energy through high furnace temperatures, the carbon atoms diffuse from this high-carbon-content medium into the crystal structure of the metal part to be case hardened.

After carburization, the steel part is ready for nonequilibrium thermal processing, which transforms the high-carbon-content austenite into martensite. This processing is followed by various quenches and tempers, depending on the degree and depth of hardness as well as the degree of grain refinement desired in both the case and the core (Figure 5-25b).

5.7.3 Nitriding

In our study of the various phase changes produced by heat treating steel, we learned that some iron reacts with carbon to produce a very hard, brittle compound, iron carbide (FeC), that precipitates out during a phase change, giving the steel hardness and strength. Aluminum alloyed with copper forms another compound ($CuAl_2$), with similar results. In addition to Ti, Al, and Fe, there are several other elements, such as chromium (Cr) and molybdenum (Mo), that react with nitrogen (N) in steel to produce compounds called *nitrides* that are also hard and brittle like carbides. The purpose of nitriding is to provide the necessary nitrogen in the steel so that these hard nitrides may form, producing the necessary hardness. The rich nitrogen medium may be ammonia gas. As with carburizing, heat is called on to provide the energy for the nitrogen atoms to diffuse into the steel case. The heat also helps produce the necessary chemical changes, and therefore additional heat treating is not required.

5.7.4 Cyaniding and Carbonitriding

The methods discussed above also provide cases that are rich in carbon and nitrogen. *Cyaniding,* an old process, might be considered as providing both carbon and nitrogen. Cyanide (CN) decomposes in a liquid bath of molten salts containing sodium cyanide, and the nitrogen reacts with iron to form hard iron nitride. The carbon diffuses to assist in further hardening operations, as discussed previously. *Carbonitriding* is very similar to gas carburizing because it uses a medium consisting of a mixture of gases rich in carbon and also nitrogen. The presence of nitrogen in the austenite also permits slower transformation to martensite, and therefore lower temperatures are required, followed by less drastic quenching. These conditions result in less distortion and less danger of cracking.

5.7.5 Flame Hardening and Induction Hardening

Flame hardening and induction hardening do not change the chemical makeup of the steel case. Through the application of heat, the outer layers of the metal part are first transformed into austenite, followed by quenching to transform the austenite into martensite. The steel to be case hardened must contain all the hardening ingredients. Any decarburized surface would have to be removed by machining or grinding prior to induction or flame hardening. *Flame*

(a)   (b)

Figure 5-26   (a) Flame hardening. (b) Water jet cutting allows thick metals to be cut without the thermal effects produced by flame cutting. This Waternife is cutting 3-in. steel plate. (Flow Systems) *hardening* uses oxyacetylene torches as sources of heat, as shown in Figure 5-26. ***Induction hardening*** uses heat generated by the resistance of the metal to the passage of high-frequency-induced electrical currents.

Surface-hardened steel products are truly ***composite materials*** containing two or more integrated components. Reinforced concrete is a composite material whose integrated components are steel rods strong in tension and a concrete matrix strong in compression. They work together in an integrated system to provide strength properties not obtainable by either one acting alone. With surface-hardened steel parts, the integrated components are the hard case and the tough core, together providing superior performance not obtainable by their use alone.

Even steels that are through-hardened, such as fine-pearlite or tempered-martensite steels, are strong and relatively hard as a result of the distribution of fine particles of a second harder phase throughout a soft ferrite matrix (see dispersion hardening). The relatively soft matrix material, the solvent component of an aluminum–copper alloy, though deformable, cannot deform independently of its harder phase. The point is that there is a striking similarity among all human-made materials—steel, aluminum alloys, prestressed or reinforced concrete, fiber-reinforced plastics, or surface-hardened metal products. They all contain integrated components working together to provide the right mix of material properties needed for the almost endless uses in today's highly technological society. We provide further details on composite materials in Module 8.

5.7.6 Other Methods of Surface Modification (Surface Engineering)

Many manufactured metal parts require additional processing of their surfaces to impart certain properties. Zinc-plated steel, jewelry that is plated with precious metals, and gold-plated electrical contacts are but a few such examples. Surface treatments, surface modification, and surface engineering are called upon to improve the following:

1. Wear and erosion resistance
2. Friction and fatigue resistance
3. Stiffness
4. Surface roughness
5. Lubrication (lubricity)
6. Texture, color, or decorative features (appearance)
7. Corrosion protection
8. Solderability
9. Electrical conductivity
10. Frayed/damaged surfaces (rebuilding surfaces)
11. Paintability
12. Weldability

Throughout this book we describe means of modifying or engineering surfaces both for tribological reasons, where interfacing surfaces may cause friction, wear, or erosion, and corrosion prevention. In Section 3.3 you learned about material characterization and may now wish to look back at Figure 3-14 to reexamine surface characteristic features. In the ceramics module you will read about additional surface engineering techniques such as diamond film and titanium coatings for cutting tools and other high friction/abrasion interfaces situations. Similarly, discussions in the polymer and composites modules deal with surface modifications to improve tribological characteristics and aspects of components.

Shot peening subjects a metal surface to a high-velocity stream of small, cast steel pellets (shot) that effectively cold works the surface, inducing compressive stresses into the surface layers which improves fatigue strength. ***Electroplating (electrochemical plating)*** suspends the workpiece in a bath containing a water-based electrolyte solution. The metal to be plated on the workpiece is also suspended in the bath. The workpiece becomes the cathode and the plating metal the anode in an electrical circuit. When the circuit is energized the metal ions from the anode combine with the ions in the solution and deposit themselves on the surface of the cathodic workpiece. A typical thickness of the deposited layer of metal ranges from one or two atomic layers up to a maximum of 0.05 mm. ***Chromium plating,*** also known as ***chromizing, chromidizing,*** or ***chrome plating,*** uses a combination of metals starting with copper followed by a coating of nickel. Chromium coatings are very thin and porous and normally offer little protection against corrosion. Copper is used as an undercoat particularly on zinc-based die castings. See Section 5.9.6 for using cadmium coatings on steel. As the by-products (specifically chromate) of this treatment are, in many cases, harmful to the environment, active research is being undertaken to find secondary uses for these materials. Chromium coatings are used on automobile trim, plumbing fixtures, and appliances. A somewhat similar surface treatment called ***electroless plating*** uses chemical reactions instead of electricity. Nickel is the predominate plating metal. Nickel chloride is chemically reacted with sodium hypophosphite as the reducing agent to produce nickel for plating a workpiece that provides high resistance to wear and corrosion. The hardness of nickel plating is about 500 HV, providing excellent wear and corrosion resistance to the workpiece.

The diffusion process which impregnates the surface of the substrate with foreign elements (atoms) is used in several processes and treatments: ***Carburizing*** and ***nitriding*** (described previously) are such examples. Diffusing of impurity atoms (foreign elements) into the surfaces of silicon chips is used to change the electrical properties of transistors and diodes (see Module 9). ***Ion implantation,*** an alternative to diffusion, embeds foreign elements into a substrate surface using a high-energy beam of ionized particles. ***Hard facing*** is a technique that uses welding to deposit alloys on a metallic substrate. ***Thermal spraying*** uses molten and semimolten coating materials. ***Metallizing*** (or metal spraying) refers to the spraying of metal coatings. ***Porcelain enameling*** is a process that applies porcelain (a ceramic material) to a metal substrate. Anodizing, hard anodizing, physical vapor deposition (PVD), chemical vapor deposition (CVP), and many of these treatments are described in greater detail elsewhere.

Throughout this book we describe means of modifying or engineering surfaces both for tribological reasons, where interfering surfaces may cause friction, wear, or erosion and corrosion protection. In Section 3.3 you learned about material characterization and may now wish to go back and look at Figure 3-14 to reexamine surface characteristic features. In the ceramics module you will read about additional surface engineering techniques such as diamond film and titanium coatings for cutting tools and other high friction/abrasion interference solutions. Similarly, discussions in the polymer and composite modules deal with surface modifications to improve characteristics that are required for some particular application.

Figure 5-27 SEM photomicrograph of crystal-oriented Pb-8Sn-2Cu. "Side view" of deposit shows the pyramidal surface morphology that enhances wettability (defined in Section 6.6.1) by motor oil, which results in improved bearing performance. (*Advanced Materials & Processes*)

1 μm

The automotive industries' never-ending quest to improve engine efficiencies continues to modify materials and materials surfaces. One such example involves the improved tribological aspects of sliding, plain bearings for automotive crankshafts. The modification involves a lead-alloy electroplated overlay. The goal of the improvement was to produce an overlay surface that would best retain lubrication. Thus, an optimum surface is produced by controlling the crystallography of the electroplated alloy deposit. The solution was a designed surface morphology produced by controlling the orientation of the deposits with highly oriented crystal texture in primarily the [200] and [400] planes (refer back to Section 3.4.3). As seen in Figure 5-27, the scanning electron microscope (SEM) photomicrograph reveals a surface of the overlay made up of relatively uniform, pyramid-shaped crystals approximately 1.5 μm across, having bases approximately 2 μm across.

SELF-ASSESSMENT

5-1. To achieve the goal of producing a vehicle that will attain a fuel efficiency of 80 mpg vehicle design must address how current vehicles lose energy. Name six sources of energy loss in present day vehicles.

5-2. In the section Pause & Ponder, we discussed applications of metals on trucks. List three metals that are used in trucks and name their uses.

5-3. Why are metals used so often as engineering materials?

5-4. Name four basic methods for changing a metal's properties.

5-5. What other name does the term *work hardening* go by?

5-6. Using the variable time, describe what the term *equilibrium conditions* means for phase changes in an allotropic metal.

5-7. Describe the differences between the words *eutectic* and *eutectoid.*

5-8. Name four metals, with an application for each, that are out of normal view.

5-9. Name the four methods used to change metal microstructure to achieve different properties.

5-10. Name at least four cold-working operations.

5-11. Name two other terms used to describe permanent deformation of a material.

5-12. Which crystal structure for metals has the closest packed plane of atoms?

5-13. How many types of space lattice does a solid solution contain?

5-14. What is the other name for the inverse-ratio law used in phase diagrams?

5-15. Identify the process that takes place when a solid metal is reheated to allow the atoms to diffuse or move toward equilibrium.

5-16. Copper–zinc phase diagrams contain how many intermediate solid solutions?

5-17. What is the name of the chemical calculation that allows one to determine the quantities of chemical elements and/or compounds involved in a chemical reaction?

5-18. Cite one example of a multiphase metal alloy in which the elements exist in two separate phases.

5-19. When referring to nonferrous alloys, what does the term *annealing* signify?

5-20. What is the usual consequence of heating steel to too high a temperature?

5-21. What heat treating process uses cooling in still air?

5-22. What is the main difference between precipitation strengthening and thermal processing under non-equilibrium conditions that produces martensite?

5-23. Describe what *high hardenability* means.

5-24. A carbon–steel alloy containing 0.20% carbon when cooled from the austenite region crosses the A_3 line in Figure 5-5. What phase begins to form at the austenite grain boundaries at this point?

5-25. Referring to Figure 5-8, determine approximately the temperature to which you would heat a hypereutectoid steel to anneal it fully.

5-26. Name two annealing processes that do not involve phase changes.

5-27. Describe how to reduce the grain size in a piece of metal.

5-28. When grinding metal for sharpening purposes, operators are cautioned not to grind the metal too long; otherwise, the temper in the metal will be drawn. What is meant by this warning, and how do you prevent losing the metal's temper in this situation?

5-29. Figure 5-15 is a diagram of the transformation of austenite in eutectoid steel. How would you re-draw this diagram if the steel were a hypereutectoid steel?

5-30. Hardness and hardenability are related properties. Discuss their differences.

5-31. If a metal part needs to be hard and at the same time possess a high degree of toughness, how are these two properties acquired through thermal treatment?

OBJECTIVE QUESTIONS

5-1. Adding elements to replace base metal atoms in a solid solution.

    a. Interstitial   **b.** Elutriation   **c.** Addition   **d.** Substitutional

5-2. The ability of a metal to exist in more than one space lattice.

    a. Isotopic   **b.** Amorphous   **c.** Allotropic   **d.** Equilibrium

5-3. Heat treatment that draws out some (but not all) hardness to improve toughness.

    a. Tempering   **b.** Spheroidizing   **c.** Case hardening **d.** Annealing

5-4. Heat treatment that relieves stress and produces softer and more ductile metal.

    a. Tempering   **b.** Spheroidizing   **c.** Case hardening **d.** Annealing

5-5. Heat treatment that produces a very hard shell but a tough inner core.

    a. Tempering   **b.** Spheroidizing   **c.** Case hardening **d.** Annealing

5-6. The measure of depth to which hardness can be obtained in a piece of steel.

    a. Hardenability   **b.** Hardness   **c.** Toughness   **d.** Ductility

5-7. Steel that is work hardened.

    a. Hot rolled   **b.** Forged   **c.** Cold rolled   **d.** Plated

5-8. The type of heat treatment process in which the metal is rapidly cooled.

    a. Annealing   **b.** Equilibrium   **c.** Nonequilibrium**d.** Allotropism

5-9. Grain structure that provides good creep strength and toughness at high temperatures.

    a. Fine   **b.** Large   **c.** Polymorphic   **d.** Serrated

5-10. Regions within a metal that are high energy and form as a result of metal cooling into a polycrystalline solid.

    a. Boundaries   **b.** Lattice   **c.** Void **d.** Interstices

5-11. Homogeneous phase with all solute atoms dissolved into the solvent.

    a. Multiphase   **b.** Uniform   **c.** Single phase   **d.** Substitutional

5-12. What is Fe_3C?

    a. Cementite   **b.** Pearlite   **c.** Ferrite   **d.** Austenite

5-13. Permanent change in a metal as a result of force.

    a. Elastic flow   **b.** Elutriation   **c.** Tempering   **d.** Plastic flow

5-14. Metals or alloying elements that are insoluble in each other.

    a. Amorphous   **b.** Immiscible   **c.** Morphologic   **d.** Diffused

5-15. A diagram used to determine quenching points for nonequilibrium thermal processing.

    a. TTT **b.** Alpha   **c.** Gamma   **d.** Jominy

5-16. Metal designation system used to indicate the alloy and percentage of alloy in steel.

    a. AA   **b.** ASTM D2000 **c.** AMS**d.** AISI-SAE

5-17. An element with a radius of less than 0.1 nm.

    a. Mn   **b.** Mg   **c.** C   **d.** Fe

5-18. A complex multiphase mixture of different phases, some of which are solid solutions.

    a. MMC **b.** Steel **c.** Wrought iron

5-19. A two-phase mixture of ferrite and cementite.

    a. Austenite   **b.** Ledeburite   **c.** Pearlite

5-20. Annealing that is intended for the sole purpose of relieving stresses.

    a. Normalizing   **b.** Full anneal   **c.** Stress relief   **d.** Tempering

5-21. Spheroidizing of steels with a carbon content.

    a. Less than 0.1% **b.** 0.3% to 0.6% **c.** 0.6% to 1.0%

5-22. A eutectoid steel transformed from austenite to fine pearlite at about 750°F.

    a. HRC 31   **b.** HRC 44   **c.** HRC 58

Top: New MacGold steel bars are precision hot-rolled by a reducing mill installed within the continuous casting and mill production line. Bottom: The mill line includes a three-strand continuous caster (MacSteel—Fort Smith, Arkansas)

Figure 5-28   A flowline of steelmaking. (American Iron and Steel Institute)

5.8 FERROUS METALS

Iron and its many alloys, including cast irons and a nearly limitless variety of steels, comprise the ferrous metals group. ***Wrought iron*** is a commercial iron consisting of slag (iron silicate) fibers entrained in a ferrite matrix. It contains approximately 2% slag but very little carbon, and is easily shaped by hot-forming operations such as forging. ***Ingot iron*** contains about 0.1% impurities including 0.01% carbon. It is used in applications where high ductility or corrosion resistance is needed. ***Electrolytic iron,*** about 99.99% pure, is used mostly for research.

Figure 5-29   Tapping a heat of steel from a state-of-the-art
electric furnace. (MacSteel)

Pig iron is iron that is tapped from the base of the blast furnace and contains over 4% carbon
plus other elements (impurities). It is converted into gray cast iron by heating, along with scrap,
in a vertical furnace called a cupola. For steelmaking, it is refined into steel in the basic oxygen furnace (BOF) or electric furnace. Even with the wide acceptance of aluminum and polymeric materials, the iron-based alloys dominate all other materials in the weight consumed annually for manufactured products. Ten times more iron (mainly in the form of steel) is used
than all other metals combined. Figures 5-28 and 5-29 show the processes that iron ore, coal,
and limestone undergo in the production of iron. Coke is made from coal; many other products also come from this source.

5.8.1 Cast Iron

As shown in the iron–iron carbon equilibrium diagram (Figure 5-5), cast iron has between 2%
and 4% carbon, compared with less than 2% for steel. Other elements in cast iron are silicon
and manganese, plus special alloying elements for special cast irons. Many cast-iron products
are used as they are cast, but others require changes in properties, which are achieved through
heat treatment of the cast parts.

5.8.1.1 Gray cast iron.   Gray cast iron is a supersaturated solution of carbon existing in a *pearlite* (two-phase structure) matrix. This carbon is mostly in the form of *graphite*
flakes (soft form of carbon known as *elemented carbon*). It is the familiar metal formerly used
as the engine block of automobiles and for other internal combustion engines. Figure 5-30
shows photomicrographs of two "as-cast" gray cast-iron specimens, one of low strength and
one of medium strength. The amount of carbon, 3.2%, exceeds the solubility limits of iron, and
the carbon precipitates out of solution with ferrite (carbon precipitates out in graphite form).
Inoculation with small amounts of iron–silicon alloys helps produce finer graphite flakes, thus

(a)

Figure 5-30 Gray cast iron. (a) The low-magnification photomicrograph at left illustrates the graphite distribution, type, and size. At 1000× (at right), the pearlite colonies and small ferrite grains adjacent to the graphite flakes are clearly distinguishable. The large, round gray particles are manganese–sulfide inclusions. The surface of the specimen contains products of transformation of a faster cooling rate. (b) The distribution of fine graphite flakes in this sample results in an increase in strength of the iron. (Buehler Ltd.)

(b)

improving strength. The graphite promotes machinability and lubricity of this metal. The damping ability of this alloy provides excellent absorption of vibrations and noise, which leads to its selection as piano sound-board frames, engine blocks, and machine tool bases. These combined properties have not only made gray cast iron a popular gearing material, it is also the least expensive of all metallic materials.

The ASTM system of designation for gray iron places it into classes 20 to 60 based on the minimum tensile strength for each class. For example, class 30 would have a minimum tensile strength of 30,000 psi (207 MPa), while a class 60 gray iron would be 60,000 psi (414 MPa). This classification is often preceded by "48" (ASTM A48 class 40), which designates the specification used to determine the mechanical properties of representative samples. Brinell hardness numbers (HB) range from 160 to 200 for ASTM 48 class 20, to 212 to 248 for ASTM 48 class 60.

Machinability is a term used to attempt to describe the ease of machining steel to the size, shape, and surface finish required commercially. From this term comes *machinability rating,* with cold-drawn B1112 grade steel rated at 100%. This standard test for B1112 machining requires turning with a suitable cutting fluid at 180 ft per min under normal conditions. All other steels are rated above or below this 100% level. Such ratings have proved to be undependable primarily due to unintended variations in the chemical content of steel. Further, it must be pointed out that there is a poor correlation between the hardness of steels and their machinability ratings. Gray cast iron is also available alloyed with nickel, chromium, and molybdenum to improve resistance to wear, corrosion, and heat while improving strength. Flame and induction hardening allow for increased surface hardness with a slightly tougher core.

5.8.1.2 White cast iron.

Through slow cooling in sand molds, chilling of specific portions of a casting, and alloying, graphitic carbon is stopped from precipitating out of solution with the ferrite to produce a white cast iron. Most of the carbon exists as cementite instead

Figure 5-31 White cast iron. This specimen shows a hypereutectoid structure of pearlite and massive cementite. The dark areas are pearlite colonies surrounded by a network of cementite. At higher magnification, the alternate lamellae of alpha ferrite and Fe_3C are clearly resolved. (Buehler Ltd.)

of graphite, as is the case with gray cast iron. This is due to the low silicon content (less than 1%). The name *white iron* comes from the white color produced in the fracture surface of the alloys. Figure 5-31 shows a photomicrograph of white cast iron to contrast with the photomicrograph of gray cast iron in Figure 5-30, which reveals graphite flakes. The carbon composition of 3.5% for unalloyed white iron has 0.5% silicon. The structure is an interstitial compound of carbon and iron known as *cementite,* plus a layered two-phase solution of ferrite and cementite known as *pearlite.* The castings are very brittle, with Brinell hardness values from over 444 to 712. White cast iron has very good compressive strength, above 200,000 psi (1380 MPa), with tensile strength around 20,000 psi, and good wear resistance; it finds applications as rolls for steel making, stone and ore crushing mills, and brickmaking equipment. Heat treatment can reduce brittleness, and as with gray iron, other properties are possible with alloying, but both of these add cost to the castings.

5.8.1.3 Nodular or ductile cast iron.

With 3.5% C and 2.5% Si, the addition of small amounts of magnesium (Mg), sodium (Na), cerium (Ce), calcium (Ca), lithium (Li), or other elements to molten iron will cause tiny balls, or *spherulites,* of graphite to precipitate out. As the name implies, the spherulitic structure improves the elongation or ductility while yielding superior tensile strength (150,000 psi, or 1034 MPa) and machinability (similar to gray iron).

ASTM specifications for nodular iron indicate minimum tensile strength, minimum yield strength, and minimum percentage of elongation in 2 in. For example, ASTM 120-90-02 would have a minimum tensile strength of 120,000 psi (828 MPa), minimum yield strength of 90,000 psi (621 MPa), and 2% minimum elongation in 2 in. (50.8 mm). In internal combustion engines, nodular iron finds application as crankshafts, rocker arms, and pistons. It is also used for cast gears, pumps, and ship propellers, and with caustic-handling equipment. It is more expensive than gray iron in terms of weight, but its specific strength makes ductile iron more economical. However, it lacks the damping ability and thermal conductivity of gray iron.

5.8.1.4 Malleable iron.

The annealing of white iron castings causes nodules (large flakes) of soft graphitic carbon to form through the breakdown of hard and brittle cementite (Fe_3C). Two basic types of malleable iron are possible by varying the heat-treatment cycle. *Pearlitic malleable* iron is strong and hard, whereas *ferritic malleable* iron is softer, more ductile, and easier to machine. Malleable iron has 2.2% carbon and 1% silicon. In pearlitic malleable iron, 0.3–0.9% of the carbon is combined as cementite and allows for selective hardening of portions of a casting.

According to ASTM specifications A47-52 and A197-47, three grades are available: 35018, 32510, and cupola malleable iron. The 35018 and 32510 grades are ferritic, with the latter lower in silicon and consequently more ductile. Cupola malleable iron has a higher

TABLE 5-1   PROPERTIES OF THREE CAST-IRON GRADES

Grade	Minimum tensile strength [psi (MPa)]	Minimum yield strength [psi (MPa)]	Minimum elongation [% in 2 in. (50.8 mm)]
35018	53,000 (365)	35,000 (241)	18
32510	50,000 (345)	32,500 (224)	10
Cupola	40,000 (276)	30,000 (206)	5

Figure 5-32 Wrought iron. The matrix of this specimen consists of ferrite grains similar to that of ingot iron. The elongated stringers are inclusions of slag composed largely of FeO and SiO$_2$. At higher magnification, small, dark particles within the ferrite grains are visible. These are finely dispersed impurities, apparent only after etching. (Buehler Ltd.)

carbon and lower silicon content than the other grades, which yields lower strength and ductility. Basic properties for the three grades are shown in Table 5-1.

Applications of the ferritic grades include machined parts (120% machinability rating), automotive power trains, and hand tools such as pipe wrenches that take hard beatings. Applications for the stronger and harder pearlitic malleable iron include parts that require high surface hardness (up to HRC 60 or HB 163-269), such as bearing surfaces on automobiles, trucks, and heavy machinery.

Table 10-11 provides a comparison of properties of cast irons and wrought iron. As shown in the photomicrograph in Figure 5-32, wrought iron is an iron of high purity (less than 0.001 part carbon) with the slag (iron silicate) rolled or wrought into it. The ferrite matrix encloses iron silicate fibers shaped in the direction of rolling, which makes it an easy material to form. It is not a common metal today, but before the development of cast iron and steelmaking, a cruder form of wrought iron served as weapons, tools, and architectural shapes. The Eiffel Tower in Paris was constructed of wrought iron in 1872.

5.8.2 Steel

As the most widely used engineering material, steel is available in an almost limitless variety. Steel contributes to "greening of manufacturing"; it is the most recycled of all materials. Refer to Figure 5-8, and note scrap steel is a basic ingredient in the steelmaking process.

Several groups can be used, such as cast steel and wrought steel. Wrought (hot or cold worked) steel covers the largest group and is the steel most common to consumers. Steel is cast into ingots when it comes from steelmaking processes such as the open-hearth furnace or basic oxygen furnace. These ingots are processed further while in the hot "plastic" state to produce a variety of *wrought* or *hot-rolled steel* (HRS) products, such as bars, angles, sheet, or plate. Further working of HRS sheet or bar stock at below the recrystallization temperature of the steel is known as *cold working* or *cold finishing. Cold-rolled steel* (CRS) is a harder steel because its grains have work hardened. Further classifications of steel are the carbon steels and alloyed steels.

The classification of steels takes a variety of forms. A very common system developed by both the Society of Automotive Engineers (SAE) and the American Iron and Steel Institute (AISI) uses four or five digits and certain prefix and suffix letters to cover many steels and steel alloys. As shown in Figure 5-33a, the numbers reveal the major alloying element, its approximate percentage, and the approximate amount of carbon in *hundreds of 1 percent,* commonly called *points of carbon.* Table 5-2 shows the major groupings of carbon and alloy steels under the SAE-AISI classification. The alloy and carbon contents given in this four- or five-digit system are approximations. Complete specifications are available from SAE, AISI, handbooks such as *Machinery's Handbook,* and Table 10-11. For example, the chromium–steel alloy 5120 has the following composition: C = 0.17 to 0.22%, Mn = 0.70 to 0.90%, Cr = 0.80 to 1.10%, P = 0.040%, and Si = 0.20 to 0.35%. Prefixes can indicate the process in making the steel, such as E for electric-arc furnace; and suffixes further clarify, for instance, H for hardenability guaranteed. Other societies, such as ASTM and ASME (American Society of Mechanical Engineers), have specifications for specialty steels, such as tool steels for dies, cutting tools, and punches or structural bolts and plates. Steel is now classified by ASTM as an *engineered* material.

5.8.2.1 Carbon steel.   This group of steels, also known as *plain carbon* and *mild steel,* dominates all other steels produced and is essentially iron and carbon with other elements that occur naturally in iron ore or result from processing. These elements are held to certain maximum levels: manganese (Mn), 1.65%; silicon (Si), 0.60%; and copper (Cu), 0.60%. Carbon steel may be cast or wrought. Typically, cast steels have more uniform properties since wrought steel develops *directional properties* as a result of rolling it into shape. See Table 10-11.

5.8.2.1.1 Ultrahigh-Carbon Steel (UHCS).   Ultrahigh-carbon steels are hypereutectoid steels ranging in composition from 1.0–2.1% carbon. They have remarkable mechanical properties in spite of their high carbon content. The basis for this characteristic is the ability to achieve desirable microstructures by control of thermo-mechanical processing and chemistry. They are superplastic at intermediate and high temperatures. At room temperatures, they have ultrahigh strengths with good ductility. They also can be made exceptionally hard accompanied by high compression toughness when heat treated to achieve martensitic structures. These steels may well be a rediscovery of Damascus steels—ancient blacksmiths produced this steel for swords at the time of Alexander the Great (325 B.C.). Damascus steels are forged from ultrahigh-carbon steel with beautiful surface markings that to this day are not fully explainable. This type of steel surface marking can be almost duplicated by modern solid-state forging and rolling procedures. With *cladded steels,* already in use, is the cladding of low-carbon steels with stainless steel. *Laminated steel* is an example of the bonding of low- and high-carbon direct cast strips similar to the approach used for ancient swords but with much greater control. Table 10-12 provides a listing of some of these steels with their principal properties.

High- and Ultrahigh-Strength Steel

The UltraLight steel autobody (ULSAB) design used over 90% of these steels in the body structure. ULSAB defines high-strength steel as having a yield strength 210 through 550 MPa and ultrahigh-strength steel as having yield strength above 550 MPa. Material thicknesses range from 0.65 to 2.00 mm. The high-strength steels used were bake-hardenable steels,

Examples:

Shown above — chromium steel alloy with about 1%
chromium and 0.20% (0.002 parts or 20 points) carbon

1015 — plain carbon steel with 0.15% (0.0015 parts
or 15 points) carbon

E52100 — chromium steel alloy with about 20%
chromium and 1 point carbon produced in an
electric arc furnace

Figure 5-33 (a) SAE-AISI steel designation.

Figure 5-33 (b) The many alloys of steel. The Boeing Commercial Airplane assembly plant has the capacity to assemble seven of the widebody 777 planes per month. This plant is the world's largest building by volume, covering
98.36 acres (39.84 hectares). Many ferrous metal alloys (mostly steels) are seen in this picture, including the skele-
tal structure of a variety of structural steel shapes that frame the building. Various steels are also used in the elaborate fixtures used for the assembly. These fixtures are designed by tool engineers whose responsibility is to provide
tooling to assemble the many parts and components that make up this large commercial aircraft. Various steel alloys
are also chosen for the many tools used in assembling and maintaining the aircraft. (The Boeing Company)

isotropic steels, high-strength low-alloy steels, high-strength interstitial free steels, and dual-
phase steels. Ultrahigh-strength steel was used for most of the lower cross members. Up until
this time mild steel was the principal choice for fabricators. This meant that with such an array of new steel materials, designers and engineers had to become familiar with a steel's capability to spring back when stamped. These materials and the advanced processes such as hydroforming and tailored blanking enabled the design engineers to consolidate functions in
fewer parts, resulting in mass savings and improved performance.

TABLE 5-2 SAE-AISI SYSTEMS OF STEEL CLASSIFICATION

Digit designation[a]	Types of steel
10xx	Plain carbon
11xx	Sulfurized (free-cutting)
12xx	Phosphorized
13xx	High manganese
2xxx	Nickel alloys
30xx	Nickel (0.70%), chromium (0.70%)
31xx	Nickel (1.25%), chromium (0.60%)
32xx	Nickel (1.75%), chromium (1.00%)
33xx	Nickel (3.5%), chromium (1.50%)
34xx	Nickel (3.00%), chromium (0.80%)
30xxx	Corrosion and heat resistant
4xxx	Molybdenum
41xx	Chromium–molybdenum
43xx	Nickel–chromium–molybdenum
46xx	Nickel (0.75 to 2.00%), molybdenum (0.15 to 0.30%)
48xx	Nickel (3.25 to 3.75%), molybdenum (0.20 to 0.30%)
5xxx	Chromium alloys
6xxx	Chromium–vanadium alloys
81xx	Nickel (0.30%), chromium (0.30%), molybdenum (0.12%)
86xx	Nickel (0.30%), chromium (0.50%), molybdenum (0.20%)
87xx	Nickel (0.55%), chromium (0.50%), molybdenum (0.25%)
88xx	Nickel (0.55%), chromium (0.50%), molybdenum (0.35%)
93xx	Nickel (3.25%), chromium (1.20%), molybdenum (0.11%)
98xx	Nickel (1.10%), chromium (0.80%), molybdenum (0.25%)
9xxx	Silicon–manganese alloys

[a]x's indicate that numerals vary with the percentage of carbon in the alloy.

Steel Sandwich

In their quest to create a lightweight, steel autobody, automakers have developed a ***steel sandwich*** material consisting of a thermoplastic core sandwiched between two thin steel skins. Its main property is bending stiffness. The core made of polypropylene acts as a spacer between two outer sheets separating the outer surfaces from the neutral axis when a bending load is applied. This material can be up to 50% lighter than a comparable sheet of homogeneous sheet without compromising performance. By definition this new material could be classified as a true composite material. Processing of this material is the same as for sheet steel with the exception that it cannot be welded. The consortium of 35 sheet steel producers from 18 countries came together in 1994 to design and validate a ULSAB that would reduce the weight of steel autobody structures while maintaining their performance and affordability. In 1998 they published the results which culminated in a weight savings of up to 36% and substantially improved performance. In their designed vehicle the steel sandwich material was used for the spare tire tub and the dash panel. The yield strength for the steel used for the spare tire tub is 240 MPa with a width of 1050 mm and a thickness of 0.14 mm. The core has a thickness of 0.65 mm. Note that the steel used for the spare tire tub is considered to be a high-strength steel having a yield strength of 240 MPa. ***Adhesive bonding*** is used to bond steel sandwich material. It provides structure as well as a seal. The bonding material is a two-component, nonconductive, high modulus, high viscous, chemically curing polyurethane adhesive/sealant that cures virtually independent of temperature and moisture. The parts to be joined must be securely held in place while the adhesive bonding cures. Processing temperature is 10° to 35°C and working time is about 10 minutes at 23°C and 50% relative humidity.

5.8.2.2 High-strength low-alloy (HSLA) steel.

HSLA steel is a product of recent technology aimed at producing strong, lightweight steel at a price competitive with that of carbon steels. Although the price per pound of HSLA steel is greater than carbon steel, thickness is reduced due to a higher strength of 414 kPa (60,000 psi) versus 276 kPa (40,000 psi) for carbon steel; consequently, overall cost may be better for the HSLA, and significant weight savings are realized. The transportation industry, especially the automotive section, has employed

HSLA steel in numerous structural applications. Not as malleable as carbon steel, sheet HSLA steel could not be used in auto bodies, but a modification resulted in a dual-phase steel acceptable for the small bending radii required on auto bodies.

5.8.2.3 Alloy steel.   The classification of alloy steel is applied when one or more of the following maximum limits are exceeded: Mg, 1.65%; Si, 0.60%; Cu, 0.60%; or through the addition of specified amounts of aluminum (Al), boron (B), chromium (Cr, up to 3.99%), cobalt (Co), niobium (Nb), molybdenum (Mo), nickel (Ni), titanium (Ti), tungsten (W), vanadium (V), zirconium (Zr), or others. Alloy steels are grouped as low-, medium-, or high-alloy steel, with high-alloy steels encompassing the stainless steel group. Table 5-2 shows the SAE–AISI classification systems used for certain alloy steels. Elements added to steel can dissolve in iron to strengthen ferrites or α-iron (bcc) and form with carbon in the austenite or γ-iron phase (fcc) to produce carbides to improve hardness. High-temperature (heat resistance) alloys must maintain corrosion resistance, mechanical strength, creep resistance, stress–rupture strength, and toughness at temperatures greater than 425°C. Chromium is most important for oxidation resistance; cobalt, aluminum, silicon, and the rare earth elements [the 14 elements in the sixth row of the periodic table called the lanthanides, such as cerium (Ce), atomic number 58] also contribute to producing a stable oxide surface layer. Nickel provides strength, stability, and toughness; and tungsten and molybdenum increase high-temperature strength. The new high-temperature alloys generally contain high levels of chromium and cobalt, which are found outside the United States. Consequently, newer alloys are being developed that eliminate completely or reduce the amounts of these two elements.

Chromium is effective in increasing strength, hardness, and corrosion resistance. Copper forms in austenite to reduce rusting. Manganese is an austenite former that, much like carbon, increases hardness and strength. Vanadium forms with ferrite to improve hardness, toughness, and strength. Molybdenum combines in carbide to improve high-temperature tensile strength and high hardness. Silicon dissolves in ferrite to improve electromagnetic properties, plus toughness and ductility. Nickel is an austenite former that both improves high-temperature toughness and ductility and provides rust resistance. Aluminum is effective as a ferrite former in reducing grain size, thus giving improved mechanical properties. Cobalt dissolves in austenite to improve magnetic properties and high-temperature hardness. Tungsten dissolves in ferrite to increase both hardness and toughness at elevated temperatures. Table 10-11 provides a comparison of selected alloys.

A recent example of the advances in the technology of alloy steels is AerMet 100, a nickel–cobalt alloy steel strengthened by carbon, chromium, and molybdenum from Carpenter Technology Corp. The patented alloy, with a designation AMS (Aerospace Materials Specification) 6532, has an nominal analysis of 13.4 Co, 11.1 Ni, 3.1 Cr, 1.2 Mo, 0.23 C, with the balance Fe. This alloy steel has the highest fracture toughness of any commercially available steel. Through heat treating it can obtain 1930–2700 MPa (280–300 ksi) tensile strength and exceed a fracture toughness of 110 MPa · $\sqrt{m}$ at 1930 (100 ksi · in. at 280 ksi). Stress-corrosion cracking and fatigue resistance are two other attributes. The superb combination of high strength and hardness coupled with high fracture toughness and ductility make it a superior alloy steel for applications outside the aerospace industry, for which it was developed. Figure 5-34a shows one aircraft application of this steel's unique combination of properties that make it lighter and tougher and reduce the size without sacrificing strength.

5.8.2.4 Stainless steel.   This group of high-alloy steels contains at least 10.5% chromium; it is more correctly called ***corrosion-resistant steel (CRES).*** The 10.5% chromium does not ensure that the steel will not rust; a sufficiently high content of carbon or other alloys may negate the passivity of the chromium (see Figure 5-34b). As with other steels, stainless may be wrought or cast. Wrought stainless is grouped by its structure as ferritic, martensitic, austenitic, or precipitation hardening (PH). Cast stainless may be classified as heat resistant or corrosion resistant.

EMT WEBLINK

Stainless Steel Dendrites—*http://www.lucent.com/microscapes/13.html*

(a)

(b)

Figure 5-34   (a) A superior alloy steel is the AMS 6532, seen used for the landing gear of the carrier-based McDonnell Douglas F/A-18 because of the alloy's high strength and stress–corrosion cracking resistance. (b) Stainless steel. Cookware of nickel–chrome stainless. (International Nickel Company)

A new alloy named ***Liquidmetal*** made of nickel, zirconium, titanium, copper, and beryllium claiming to be twice as strong as titanium is appearing in the construction of golf clubs. A putter made of Liquidmetal cost up to $400. This metal alloy absorbs less energy than other alloys when the club strikes the ball allowing more energy transfer from the club to the ball. Its density lies between titanium and steel. A set of irons costs up to $1998. Another advanced material, ***Alpha Managing Metal,*** a stainless steel with high nickel and chrome content, is being used in fairway woods in the popular TriMetal clubs produced by Orlimer Golf (see Figure 5-36). Other clubs are being made with stainless steel combined with strong, shallow faces and copper tungsten weights on the bottom to give the clubs a low center of gravity which

Figure 5-35 New Orlimer irons and woods with face inserts (club head faces) of Carpenter Steel Custom 465™ stainless steel. The club head face is key to flight of a golf ball as its surface is subjected to the all-important impact with the ball. (Carpenter Technology Corp., Reading, PA)

Figure 5-36 The design of the dragonfly wing is unique with wings that move independently of one another. It can hover like a helicopter, fly slowly backwards, shoot forward, dive suddenly, or dart to one side. As the wing flutters, it also turns to give optimum effect, and it does all that about 30 times a second. The wing pictured was produced by the Strip Division of Sandvik made of precision strip steel. (Courtesy of Sandvik Steel)

allows the golfer more control without sacrificing distance. The golfing industry also has two next-generation alloys from Carpenter Technology Corp. Both of these alloys were originally developed for the aerospace industry. Carpenter AerMet™ alloy was a candidate for the landing gear of U.S. Navy carrier-based F-18E/F jet fighters (see Figure 5-34a). It offers a unique combination of high strength and hardness, fracture toughness, exceptional ductility, and resistance to both fatigue and stress-corrosion cracking. Table 5-3 contains additional mechanical properties of four different alloy steels contending for the lead material in the design of club heads. Among the many interesting properties is the elongation percentage of the various steels. The higher this percentage the greater the ductility of the steel; and the more ductility the greater the steel's toughness in withstanding the impact of, in this instance, a golf ball. Toughness is discussed in several modules of the text.

TABLE 5-3 ALLOY PROPERTIES

	AerMet™ Alloy (875°F Aged)*	Custom 465™ Stainless (900°F Aged)*	17Cr-4Ni Stainless (900°F Aged)*	Ti-6Al-4V Alloy (Annealed)	Ti-6 Al-4V Alloy (900°F Aged)*
U.T.S. MPa	2083	1795	1365	895	1140
(ksi)	(302)	(260)	(198)	(130)	(165)
Y.S. MPa	1780	1655	1260	830	1035
(ksi)	(258)	(240)	(183)	(120)	(150)
Elong. %	14	13	15	10	10
R.A. %	64	58	52	25	20
Fracture Toughness					
MPa ($M^{1/2}$)	109	83	61	77	39
[ksi (in.$^{1/2}$)	(99)	(75)	(55)	(70)	(35)
Density kg/m^3	7889	7833	7806	4429	4429
(lb./in.^{3})	(0.285)	(0.283)	(0.282)	(0.160)	(0.160)
Modulus of Elasticity					
Gpa	192.5	198.6	196.5	110.3	113.8
(10^3 ksi)	(27.9×10^3)	(28.8×10^3)	(28.5×10^3)	(16.0×10^3)	(16.5×10^3)
Strength-to-Density Ratio					
km	26.9	23.3	17.8	20.7	26.2
(10^3 in.)	(1060)	(919)	(702)	(813)	(1031)
Hardness HRC	54	51	44	30	40

*Except for AerMet alloy aged for 5 hrs. at 468°C (875°F), all aging done at 482°C (900°F). Custom 465 stainless aged 4 hrs., 17Cr-4Ni stainless aged 1 hr., and Ti-6Al-4V aged 6–8 hrs. Ref: Carpenter internal documents and Aerospace Structural Metals Handbook.

Carpenter Technology Corp., Reading, PA.

Carpenter *Custom 465*™ stainless steel alloy is a martensitic, age-hardened stainless steel. It provides an ultimate tensile strength of nearly 1800 MPa retaining high fracture toughness and yield strength. Note in Table 5-3 that titanium with its low density no longer occupies the highest position in specific strength. This means that these three alloy steels that possess high specific strength will absorb less energy at the point of impact between a club and a ball than titanium with the result that more energy is transferred to the ball. Testing has determined that the impact between the club head and the golf ball can produce a stress of up to 20 MPa With a high modulus of elasticity and high specific strength it is possible to design a thinner face that resists damage, allows for a greater sweet spot with no increase in weight, and absorbs less energy at the point of impact (transfers more energy to the ball). The result is that a golfer can drive a ball farther and straighter without swinging harder.

Titanium is also being used in the production of golf balls. One innovative design uses titanium compound as a core material (price for one dozen is $20), because the harder the core is the more transfer of energy and the greater the range will be. Still another manufacturer uses titanium on the cover for durability and greater distance. The computer is now analyzing the golfer's swing and showing the golfer where he or she is deviating from the perfect swing which may reduce the dollar input of the titanium balls.

The ASTM and SAE, along with other groups, have developed the Unified Numbering System (UNS), a five-digit designation with an S prefix to replace the AISI designations for stainless steels. Stainless steels in the S30000 (AISI 300) series are nickel–chromium steels, while the S40000 (AISI 400) series have chromium as the major alloy. Series S20000 (AISI 200) are austenitic alloys, with manganese and nitrogen replacing some of the nickel—far less expensive alternatives when high formability and good machinability are not required. *Austenitic stainless steel* is a single-phase solid solution that has good corrosion resistance.

Martensitic stainless steels in the S40000 (AISI 400) series have high carbon content, up to 1.2%, with 12–18% chromium. The higher carbon content allows formation of more g iron, which quenches to a hard martensitic (up to 100%) steel, but the high carbon content reduces some of the corrosion resistance. If an austenitic stainless steel is heated sufficiently so that carbon precipitates out of solid solution as chromium carbide (which leaves less than 12% chromium in some segments of the alloy), it promotes intergranular corrosion.

The *ferritic stainless steels* have low carbon (0.12% or less) content and high chromium content (14–27%) in a solid solution and do not harden by heat treatment. They are in the S40000 (AISI 400) series and, unlike martensitic and austenitic, are magnetic. The ferritic grades have good formability, machinability, and corrosion resistance above martensitics. Specific properties of each type of stainless steel are found in standard references on steels. Table 10-11 shows selected stainless steels and their properties.

Sandvik Steel provides a cold-rolled strip steel principally as austenitic stainless steel, which offers high corrosion resistance combined with excellent spring properties. Using the latest rolling technique this company supplies super-thin strip steel to extremely close thickness tolerances and excellent flatness. Thinnest strip is available in thicknesses down to approximately 0.015 mm. Figure 5-36 shows one wing of a dragon fly made of such steel.

5.8.2.5 Other steel alloys.   Beyond and including high-alloy steels and stainless steels previously discussed, there are many specialty alloys. The ASM publishes several volumes of *Metals Handbook; Volume I, Properties and Selection of Metals,* gives in-depth coverage of most metals. Included in the steel alloys is a range of tool steels, high-yield-strength (HY) steels, magnetic and electrical steels, ultrahigh-strength steel (see Table 10-12 for properties), maraging steels, low-expansion alloys, ferrous powdered metals (PMs), and ferrous amorphous metals.

SELF-ASSESSMENT

5-32. Carbon steel is an alloy of carbon and iron. Is stainless steel an alloy of carbon steel? Is it an alloy steel?

5-33. How does the composition of ferrite differ from pearlite?

5-34. Write the chemical symbol for cementite and attach another name to this ingredient of steel.

5-35. Wrought iron is mainly pure iron mixed with a slag composed mostly of iron silicate. Specify an excellent application for wrought iron. Could wrought iron be alloyed with another metal? If so, which metal would you choose, and why?

5-36. What are some barriers to the introduction of advanced materials?

5-37. Plain-carbon steels (i.e., low-, medium-, and high-carbon steels) contain up to a maximum of about 1% carbon. These steels provide much of the steel that goes into welded parts, bolted and riveted structures, springs, and free machining parts. Why is it necessary to develop more sophisticated specialty steels containing alloying elements other than carbon?

5-38. What crystalline phase of steel is referred to as a very fine platelike (lamellar) mixture of ferrite and cementite?

5-39. What is the maximum solubility of carbon (in percent) in the interstitial solid solution of carbon dissolved in gamma iron?

5-40. Modern steels are sometimes referred to as examples of human-made composite materials. Explain the significance of this statement.

5-41. The production of metals and the casting of metal parts take place under the influences of the earth's atmospheric pressure and gravitational force. Speculate about how the crystalline structures of metals, and hence their properties, would change if these processes could be accomplished with the absence of gravity.

5-42. What is the base metal in all ferrous metals, and what raw materials are used in extracting this metal from ore?

5-43. **(a)** What is the carbon content of cast iron? **(b)** Explain the designations ASTM 48 class 35 and 32510.

5-44. List one application, the key structural difference, and the best quality for each of the following cast irons: **(a)** ductile cast iron, **(b)** gray cast iron, **(c)** white iron, **(d)** malleable iron, **(e)** wrought iron.

5-45. **(a)** How is steel different from cast iron? **(b)** How is CRS superior to HRS?

5-46. Explain the following designations: **(a)** SAE 1020, **(b)** ASTM S 40000, **(c)** AISI 400, **(d)** SAE E 52100, **(e)** 6016 T 2, **(f)** CRES.

5-47. **(a)** What drawback does HSLA present to auto body makers? **(b)** What steel is normally used for auto bodies?

5-48. What is the major function of the following in alloy steel: **(a)** copper, **(b)** vanadium, **(c)** chromium, **(d)** aluminum, **(e)** lead, **(f)** nickel, and **(g)** tungsten?

5-49. Describe the major advantage of each of the following stainless steels: **(a)** martensitic, **(b)** austenitic, and **(c)** ferritic.

5-50. What is the name of the 14 elements in Period 6 of the periodic table that contribute to a stable oxide layer on alloy steels?

5-51. Name two physical properties needed in a naval fighter aircraft's landing gear.

5-52. From Tables 10-11 and 10-12, name the two highest hardness steel alloys.

OBJECTIVE QUESTIONS

5-23. The carbon content in a 1020 steel.

    a. 20%   **b.** 0.20%

    c. 2%   **d.** 0.20

5-24. Steel with alloying elements to resist corrosion.

    a. CRS   **b.** HRS

    c. CRES   **d.** TTT

5-25. Metal with no iron.

    a. Alloy   **b.** Ferrous

    c. Nonferrous   **d.** Magnetic

5-26. As the steelmaking flowline shows, _____ is a major source of raw material.

    a. Scrap   **b.** Lead

    c. Aluminum   **d.** Silica

5-27. Carbon content of cast iron.

    a. 0.6% to 1.0%   **b.** 2% to 4%   **c.** 4% to 6%

5-28. Graphite in a metal promotes _____ .

    a. lubricity   **b.** malleability

    c. toughness   **d.** conductivity

5-29. Silicon added to gray cast iron helps produce _____ .

    a. strength   **b.** malleability

    c. ductility   **d.** magnetism

5-30. The least expensive of all metallic materials.

    a. Carbon steel   **b.** Gray cast iron

    c. Wrought iron   **d.** CRES

5-31. Structural steel shapes are produced from _____ .

    a. billets   **b.** slabs

    c. ingots   **d.** blooms

5-32. Another name for nodular cast iron.

    a. Malleable   **b.** Gray

    c. Ductile   **d.** Brittle

5-33. What ferrous material was used to construct large buildings prior to the development of steel in the late 1890s?

    a. Cast iron   **b.** Wrought iron

    c. Ductile iron   **d.** Stainless

5-34. Slag is found in _____ .

    a. cast iron   **b.** wrought iron

    c. low carbon steel   **d.** CRES

5-35. Another name for plain carbon steel.

    a. Low-carbon steel   **b.** Mild steel

    c. Wrought steel   **d.** HSLA

5-36. To compete with lighter metal and plastics the steel industry offers _____ .

    a. low-carbon steel   **b.** mild steel

    c. wrought steel   **d.** HSLA

5-37. Copper forms in austenite to reduce _____ .

    a. rust   **b.** spheres

    c. graphite   **d.** strength

5-38. In addition to corrosion resistance, chromium also gives _____ to steel.

    a. magnetism   **b.** formability

    c. castability   **d.** strength

5-39. A steel alloy with nickel-cobalt and rated with the highest fracture toughness.

    a. AISI-SAE 2430   **b.** AerMet 100

    c. HSLA   **d.** AISI 400

5-40. Metal designation system used to indicate the alloy and percentage of alloy in steel.

    a. AA   **b.** ASTM D2000

    c. AMS   **d.** AISI-SAE

5-41. SAE 4000 steel contains the major alloying element _____ .

    a. manganese   **b.** molybdenum

    c. nickel   **d.** lead

5-42. Aluminum, as an alloying element, improves the steel's _____ .

    a. magnetic properties   **b.** high-temperature properties

    c. mechanical properties   **d.** processing properties

5-43. A steel with alloying elements to resist corrosion.

    a. CRS   **b.** HRS   **c.** CRES   **d.** TTT

5-44. Two major alloying elements needed for high-temperature steel that are not mined in the United States.

    a. Lead and tin   **b.** Nickel and silicon

    c. Chromium and cobalt   **d.** Iron and aluminum

5-45. Martensitic stainless steels have at least _____ of Cr.

    a. 0.8%   **b.** 4%

    c. 12%   **d.** 21%

5-46. From Table 10-11, what is the hardness capability of type 302 stainless?

    a. 85 HRB to 40 HRC   **b.** 75 HRB

    c. 325 HB   **d.** 180 HB

5-47. From Table 10-12, what ultrahigh-strength steel would provide these properties: tensile strength, 2140 MPa; yield strength, 1650 MPa; hardness, 54.5 HRC?

    a. UNS G41300   **b.** UNS K44220

    c. UNS G61500   **d.** UNS T20811 Mod

777 Uses Advanced Materials
for Reliable and Economic Operation

Advanced alloy applications
- Advanced alloys improve durability
- Advanced alloys save 3,200 lb of
  structural weight

Advanced 2000-series aluminum alloys

2000-series aluminum alloys

Advanced 7000-series

7000-series aluminum alloys

Advanced titanium alloys (main landing gear
fittings not shown)

Advanced nonferrous alloys for the Boeing 777. Various aluminum and titanium alloys (shown with symbols) used on the 777 improve durability while saving on structural weight. (Courtesy of the Boeing Company.)*

*The Public Broadcasting System (PBS) series "Twenty-first Century Jet" provides an excellent in-depth case study of the product development process, from design through testing, manufacturing, and delivery. The series includes many examples of engineering materials technology. Call PBS at 1-800-344-3337 to obtain further information on the video series.

TABLE 5-4 PROCESSABILITY OF COMMON METALS

	Steel	Iron	Aluminum	Copper	Nickel	Magnesium	Zinc	Titanium	Tin
Castability									
Centrifugal	■	■	■	■	■				
Continuous	■	■	■	■					
Investment	■		■	■	■				
Permanent mold	■	■	■	■	■	■	■	■	■
Die casting			■	■		■	■		■
Formability									
Cold	■		■	■	■	■	■		
Hot	■		■	■	■	■		■	
Machinability		■	■	■	■	■	■		
Powder metal Compacting	■	■	■	■	■				
Weldability									
Gas	■	■	■	■	◤			■	
Inert arc	■	■	■	■	◤	■		■	
Electrical Resistance	■	■	■	■	◤			■	

■ – Common on most alloys

◤ – Used on some alloys

Source: Adapted from *Machine Design,* March 15, 1976.

5.9 NONFERROUS METALS

Although the ferrous alloys are used much more than other metals, nonferrous metals comprise three-fourths of the known elements. However, the problem of extracting certain metals from the earth or ocean and the practical commercial value limits the number of metals used in significant quantity. The most commonly used nonferrous metals include aluminum, copper, zinc, nickel, chromium, tin, magnesium, beryllium, tungsten, lead, molybdenum, titanium, tantalum, and the noble (or precious) metals such as gold, platinum, silver, and rhodium. With demands for superior properties through advances in technology, each metal receives careful consideration for its potential value in unique circumstances. Catalytic converters used in automobile exhaust systems contain platinum to aid in the removal of pollutants in our environment. Ruthenium is an example of a metal that has little practical application alone, but as an alloy with platinum, it is valued in thin-film circuitry (ceramic substrates with printed circuits) for solid-state electronics. In addition to the properties of a material, selection for design bears heavily on the processibility of the material. Table 5-4 provides a simplified comparison of metals to common processes.

5.9.1 Aluminum

The most abundant of metals, aluminum is locked up with other elements in the form of bauxite ore. In 1852, this metal was only rarely available in usable form and was prized by Emperor Napoleon as dinnerware; it sold for $545 a pound. The development by Charles Hall in Ohio and by Paul Heroult in Paris, France, of an electrolytic process for economical extraction of

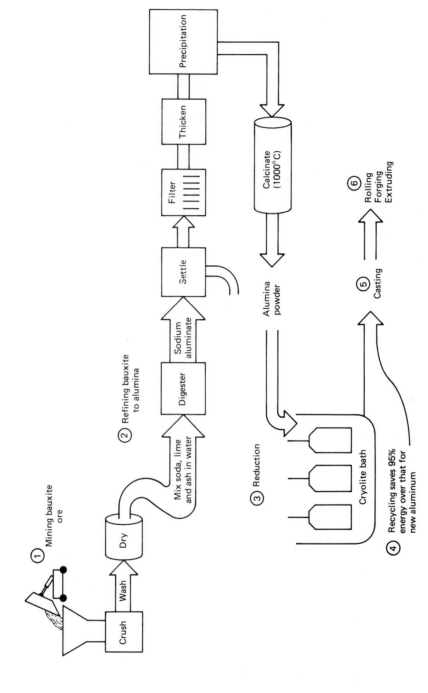

Figure 5-37 Making aluminum. (Adapted from the Aluminum Association flowcharts)

Figure 5-38 (a) Roof structures for arenas and gymnasiums are usually 6063 or 6061 extruded tube, covered with 5xxx alloy sheet. Alloy 6262 is also used in brake housings and brake pistons. (b) The inner hull stiffener structure of high-speed single-hull yachts is constructed using 5083 machined plate for hulls, hull stiffeners, decking, and super-structure. (c) Automotive structures are likely to employ increasing amounts of 5754-0 formed sheet for parts such as internal door stiffeners or the entire body-in-white. (d) Aircraft internal structures include extrusions and plate of 2xxx and 7xxx alloys such as 2024, 2124, and 2618. If used as external sheet skin they may be cladded with higher purity cladding to provide greater corrosion protection to the Al–Cu alloys that will otherwise darken with age. (*Aluminum Association*)

bauxite occurred simultaneously in 1886. Even though the price had dropped to 24 cents per pound by 1964, the cost of electrical energy keeps up the price of virgin aluminum. About 4% of our metal-processing energy is consumed in the production of aluminum. However, recycling of aluminum drastically reduces the cost of the metal.

Figure 5-37 shows the steps in the production of aluminum from bauxite. Aluminum has many favorable properties, including excellent thermal conductivity, good strength, corrosion resistance, and light weight (one-third that of steel). Figures 5-38 and 5-39 show products that stem from aluminum properties. Table 5-4 shows that it is processible by the major methods. Wrought aluminum and cast aluminum are designated by two different numbering systems. The Aluminum Association uses four digits to specify wrought alloy groups. The meaning of each digit is shown in Figure 5-40. Wrought and cast designations are similar; the last digit in cast aluminum indicates cast (0) or ingot (1 or 2). Temper designations follow the four digits; they indicate thermal treatment (T through T10) and solution heat treatment (W), as-fabricated (F), annealed (0), or strain hardened (H).

Aluminum of high purity (1000 series), or at least 99% purity, finds applications as thin foils for electronics applications; as plate and bar stock for chemical apparatus, fuel tanks, and cooking utensils; and as paint pigments and in gasoline production. Aluminum alloyed with

 EMT WEBLINK

Aluminum—*http://me.edu/2.01/Taxonomy/Characteristics/Aluminum.html*

(a)

(b)

Figure 5-39   (a) The Ford Contour concept car uses an extruded-aluminum spaceframe that reduces weight by 50% over a steel frame, while providing savings in time and materials over steel stamped frames. A pound of aluminum is twice as expensive as a pound of steel, but CAFE (corporate average fuel economy) mileage standards and virtually no scrap make Al competitive as frame material. (b) Additional cost and weight savings will come from adhesive bonding of the Al extrusions, onto which composite plastics body panels will be bolted. No welding will be involved. (Ford Motor Company)

copper (2000 series) extends the tensile strength (greater than mild steel), improves machinability, and proves good specific strength but has reduced corrosion resistance and weldability. Aluminum–manganese alloys (3000 series) provide good strength, workability, formability, and weldability, and the high corrosion resistance makes this a popular alloy for applications such as highway signs, furniture, cooking utensils, architectural shapes, and truck panels.

The aluminum and silicon alloys (4000 series) have high wear resistance and relatively low coefficients of thermal expansion; they permit good castings and forgings and find applications as welding and brazing rods, anodized architectural shapes, marine equipment, and automotive pistons. Aluminum alloyed with magnesium (5000 series) yields moderate to high strength, good corrosion resistance in marine environments, and good weldability and formability for uses as extrusions, ship and boat parts, and automotive structural parts. The 6000 series alloys of

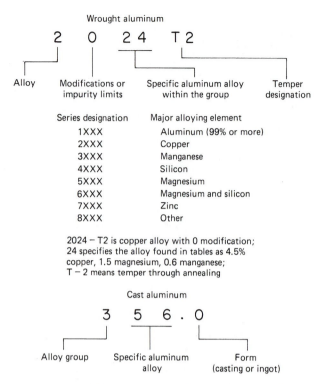

Figure 5-40 Aluminum Association number system. (a) wrought alloys, (b) casting alloys, (c) tempers designation system.

aluminum, magnesium, and silicon have excellent corrosion resistance, plus good weldability, machinability, and formability; they find uses as bridge rails, automotive sheets, piping, and extrusions. A 6000 series aluminum alloy is replacing 5000 series aluminum alloys which develop stretcher strains (elongated markings that appear on the surfaces of some sheet materials when deformed just past their yield point). Up until recently 5000 series alloys were widely used as automotive body panels. Nippon Steel Corp. and Sky Aluminum Co. of Tokyo, Japan, have developed a 6000 series alloy with slightly lower density than a 5000 series alloy that possesses the four key properties required of automotive body panels: strength, formability, bending processibility, and corrosion resistance. Nissan Motor Co. will be the first automaker to use the new alloy in the hood of its new Skyline GT-R model. The aluminum–zinc series 7000 alloy is heat treatable with tensile strength up to 606 MPa (88,000 psi) and has good corrosion resistance in rural areas but is poor in marine environments. Uses for the 7000 alloys are chiefly as aircraft structures and other equipment requiring high specific strength. The 8000 series includes alloys of combinations not included above, for example, titanium or zirconium. Table 10-11 provides comparisons of various aluminum alloys.

Corrosion resistance in aluminum is a result of the natural occurrence of an aluminum oxide film that prevents the metal from further corrosion. By making aluminum the anode— a metal is oxidized (i.e., it loses electrons forming metallic ions, at the anode of an electrolytic cell)—an electrochemical conversion of the surface of the aluminum from metallic aluminum to aluminum oxide (Al_2O_3) takes place, which is an extremely hard and porous ceramic. (Refer to Module 4 for further information on oxidation and corrosion.) For best corrosion protection the porosity is reduced by a sealing operation using hot water to immerse the anodized coating. This converts the oxide to a hydrated form which results in a reduced pore size. Dyes can be added to the sealing water which will permeate the coating and thus form an integral part of the coating. Once formed they resist chipping or peeling. Automobile trim is an example of such treatment. **_Anodizing_** (the word _anodize_ is a contraction of anode plus oxidize) is mostly applied to aluminum and magnesium alloys but it may also be used with zinc, titanium, and other less common metals. Anodized coatings are used primarily for decorative purposes, but they fill a major role in corrosive protection of metal surfaces. Most aluminum alloys can be **hard anodized,** forming thicknesses that range from 2.5 to 20 μm. Thus, anodizing is one of many surface modifications performed on metals that is helping reduce the cost of corrosion in the United States, estimated to be $3 billion a year according to Battelle and NIST estimates. Through an *anodizing* process, it is possible to build up the oxide layer and even add color to the aluminum surface. Concerns about the toxicity of materials and their effect on the environment have led to an improved process for anodizing aluminum. The older process used chromic acid and sodium dichromate, which contained hexavalent chromium, a carcinogen. The newer process uses sulfuric acid and nickel acetate, which produces thinner coats (2 to 5 μm thick) that offer better corrosion and fatigue resistance than the older method. Aluminum oxide formed in the anodizing process is quite brittle and rough, creating a poor wear surface. Adding zirconium by ion implantation after anodizing greatly improves its wear resistance (see Section 7.10). Alloying, cold working, and heat treatment can reduce the corrosion resistance of aluminum. For example, intergranular corrosion can occur in a precipitation-hardened aluminum because of the lack of homogeneity (i.e., some areas have a high copper content and become anodic to copper-depleted areas). **Spray formed aluminum alloy** billets are being produced by Osprey Ltd., UK. A high-strength aluminum–zinc alloy, a low-density aluminum–lithium alloy, and a wear-resistant aluminum–silicon alloy are the first to be spray formed, thus providing many microstructural and property advantages that result from rapid solidification technologies but at a lower cost because only one step is needed from molten metals to cylindrical billet.

5.9.1.1 Superplastic forming (SPF) and diffusion bonding (DB).

When deformed in tension at elevated temperatures, some advanced aluminum alloys are also capable of achieving large, uniform elongation (>500%) at relatively low stresses prior to localized thinning and fracture. The property of superplasticity joined with diffusion bonding provides a space-age composite. A metal alloy that is not considered very ductile at or near room temperatures but at certain higher temperatures possesses extremely high ductility is said to exhibit **superplasticity.** Titanium is such a metal with normal plastic deformation around 20%, but at certain high temperatures it can be deformed as much as 2000%. Diffusion bonding or hot pressing is a widely used technique for metal–fiber composite fabrication. Examples of such composites are aluminum and titanium alloys reinforced with boron or Borsic fibers. The matrix is usually in sheet form, with the reinforcing filaments or wires mechanically spaced and oriented between the sheets to form alternating layers of filaments and matrix. Surface coatings such as carburizing, nitriding, and bimetallic castings are also examples of diffusion bonding, but they do not use metal filament reinforcement.

The first step in diffusion bonding is to wind the fiber over a metal-foil-covered drum. The resulting mats are cut and made into tapes by diffusion bonding under high temperature and pressure. The tapes, in turn, are cut into plies, stacked in a die, and consolidated by further diffusion bonding.

Complex structural shapes can be formed in a single operation by the large forming strains produced by SPF. Superplastic forming/diffusion bonding (SPF/DB) uses several alloys to fabricate major structural components of aircraft. Figure 5-41 depicts the SPF process. The older fabrication method (Figure 5-41a), using machining and chemical milling, resulted in 80% of the starting plate machined away as scrap. The newer built-up structure approach, using an SPF beaded-web stiffener that was joined to the skin material by resistance spot-welding techniques, produced only a 15% loss in scrap material while reducing the material cost for fabrication (Figure 5-41a). SPF increases the efficiency of hat stiffeners by as much as 60% by using more complex configurations such as the beaded web, which increases stiffness. With these reductions in cost and increase in efficiencies, the use of more costly aluminum–lithium alloys will permit further reductions in weight and cost to be realized.

Compared to steel or aluminum, titanium alloys have a high strength-to-weight ratio, particularly at elevated temperatures. One main drawback to their use is difficulty in cutting, milling, and forming. This new process, with many of the details still secret, uses one or more titanium

Figure 5-41 (a) An aluminum cryotank structure for the space shuttle produced by old machining methods (left) and newer SPF/DB (right). Note alloy designations 2219, 7475. (b) Steps for superplastic forming with diffusion bonding. (NASA)

sheets laid on top of each other and held together by a jig while being heated in a furnace (Figure 5-41b). Argon gas is blown through tiny holes drilled in the sheets at desired locations. When the titanium reaches a soft and supple superplastic state, the gas assists the titanium to expand into cells, with the other sheets remaining flat. The internal structure takes on whatever shape is intended. Through this process, the outer titanium sheets flow together under pressure to form a complex expanded sandwich unit that has the strength of a single piece. A new material for automobiles is even lighter and stronger than aluminum. ***Aluminum foam,*** developed by Karmann Corp., is processed to leave the foam with a smooth aluminum interior and exterior. This new ma-

Figure 5-42 Casting alloys. Aluminum alloys with ceramic particle reinforcement can yield near-net shapes. Representative aluminum-alloy parts cast from MMC foundry ingot include, clockwise from top left, an investment-cast pressure vessel; a green-sand cast automobile-brake rotor; permanent-mold cast pistons; shell-investment cast airbus components; a dry-sand cast automobile control arm; and high-pressure, die-cast bicycle sprockets. At center is an investment-cast, aircraft-camera gimbal. (*Advanced Materials & Processes,* ASM International)

terial is reported to produce automobile bodies that are stronger, stiffer, and lighter than existing prototypes. A prototype chassis built by Volkswagen demonstrates the viability of the process.

5.9.1.2 Tube hydroforming and sheet hydroforming.

Hydroforming technology has been used by industry for many years wherever complex shapes must be formed with a high degree of accuracy using both steel and aluminum. This includes their use in the automotive industry; however it has been only recently that designers and engineers have been convinced that a tubular component can replace a stamped assembly with attendant benefits in costs, weight, structural integrity, dimensional capability, and design flexibility. Today tube hydroforming is recognized as a cutting-edge technology in the automotive industry where it is being used in spaceframes, engine cradles, roof headers, frame rails, and other subassemblies. In essence, *tube hydroformed* parts are replacing stamp and welded components resulting in lighter, stronger, and fewer required parts. The tube hydroformed part-making process (Figure 5-43a) incorporates four steps: (1) making the tube, (2) bending the tube, (3) preforming the prebent tube, and (4) hydroforming the preformed tube into final component shape. Initial steps use a CNC bender to bend the tube blank to approximately the final shape. Later steps use a die built to the dimensions of the final component along with fluid pressure (water or oil) to force the tube to flow completely into the die. In some cases punches may be mounted in the forming die to pierce holes in the forming tube; and slides may be mounted in the die to provide indented surfaces on the tube side walls. The ULSAB's roof was *sheet hydroformed* (Figure 5-43b) for weight reduction. The sheet chosen had a thinner gauge (0.70 mm) with greater strength (210 MPa) than its replacement due to the work hardening effect achieved from using fluid pressure to stretch the blanks (plastic deformation) in the opposite direction toward the punch. *Hydrostretch forming* is a related process capable of using conventional double-acting presses. Quality is improved because there is no metal-to-metal contact on the outer portion of the part surface. Continued research is needed to reduce both cycle times and costs.

5.9.1.3 Tailor-welded blanks.

Tailored blanks refer to steel blanks of various gauges and grades being welded into a single blank prior to forming in a press (see Figure 5-44). This enables manufacturing engineers to "tailor" the best properties of different steels so they are located precisely where they are needed. The result is reduced mass, reduced tooling and assembly costs, dimensional variation reduction, improved crash reduction, reduction in the number of spot welds, part integration, and improved structural integrity, quality, and manufacturing efficiency. These blanks are laser welded with its smaller heat-affected zone

Tubular hydroforming

Sheet hydroforming

Figure 5-43 Tube and sheet hydroforming. a, Rapid prototype tube hydroforming is becoming a fast growth manufacturing process for automotive exhaust and frame components as well as aerospace components. Benefits include reduced weight, fewer components and assembly operations, and improved strength and quality. b, Sheet hydroforming involves the forming of large surface body shells. For example, automobile roofs, hoods, or doors can be formed using large presses that can exert pressures up to 100,000 kN employing over a dozen programmable axes to provide greater flexibility. (*Advance Materials and Processes*)

(HAZ). Figure 5-45 is a schematic representation of a fusion-welded metal joint showing the heat-affected zone. A small portion of the metal between the HAZ and the filler metal is made up of fused base (parent) metal. The HAZ experiences changes in its microstructure and subsequent change in its properties due to the high temperatures and subsequent cooling. For steels, this material may have formed austenite which would produce brittleness. Residual stresses also may form in this region that would weaken the joint. Refer to explosive welding in Module 6 for another reference to the HAZ. Nearly half of ULSAB's mass consists of parts using tailored blanks which promote smooth load flow, reduce structural discontinuities, and allow for the combination of thicker and higher-strength materials within the same part. The database for aluminum tailor-welded blanks is not extensive but laser, electron beam, arc, and friction stir-welding techniques can be used to fabricate aluminum blanks with advantages such as less weight, reduced waste generation, and stamping tooling costs.

5.9.1.4 Aluminum in MMCs.   As explained in Module 8, metal–matrix composites (MMCs) can involve fiber reinforcing with boron, carbon, and silicon. Aluminum alloys with ceramic-particle reinforcing are prime candidates for castable MMCs. Ingots of aluminum MMCs can be used for many casting processes (Figure 5-42).

(a)

Rear quarter panel

1.3 mm (280 MPa)

0.7 mm (210 MPa)

1.5 mm
(350 MPa)

0.9 mm (280 MPa)

1.7 mm (350 MPa)

Tailored blank: panel body side outer

(b)

Figure 5-44   Tailor welded blanks. A tailor welded blank is composed of several blank flat sheets of steel each having different strengths, thicknesses, and coatings. A laser blank welding process welds these individual sheets together before they are stamped into a single automobile body panel. (*Advance Materials and Processes*)

5.9.2 Copper

Both as an alloying element in alloys such as steels and aluminum and as a base for copper alloys, copper is a valuable metal. Included in the copper-based alloys are brass (copper and zinc), bronze (copper and tin), cupronickel (copper and nickel), leaded copper, leaded brass, aluminum bronzes, nickel silver (copper, nickel, and zinc), and high copper. Figure 5-46 shows two widely different applications of copper's alloys. Table 5-4 shows that copper and its alloys are processed easily by most common methods, except certain types of welding. Several systems of designation exist for copper and its alloys but are beyond the scope of this book. Besides its value as an alloy, copper has its greatest value as an electrical conductor. High purity of over 99.9% is sought to attain the best conductivity and formability. Due to its scarcity and weight, the trend is, when possible, to move away from copper as electrical conductors. Copper and copper alloys develop a protective film that will not corrode in water or nonoxidizing acids. It will corrode if liquids, solids, or gases break down the protective film or if other conditions cause electrolysis to develop. In brass the presence of oxidizing chemicals can result in dezincification, in which the zinc dissolves, leaving a porous metal sponge. ***Spray forming*** of copper is now an established technology along with ferrous and nonferrous alloys for the production of preforms and near-net shapes. This technology will be mentioned in other

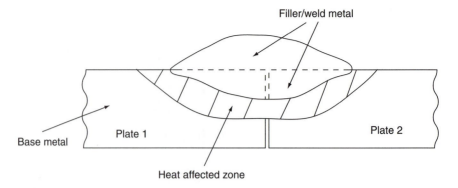

Figure 5-45 Fusion-welded metal joint showing the heat-affected zone (HAZ). High temperatures caused microstructural changes in the HAZ and thus its properties. It is in this region where welding failures occur. Even though the base metal adjacent to the HAZ has not changed in its structure it is usually left in a state of residual stress. Laser welding produces a much narrower HAZ.

(a)

(b)

Figure 5-46 Copper-based alloys such as brass and bronze offer excellent weatherability and toughness as seen with (a) American Civil War cannon dating back to the 1860s, and (b) sculpture memorializing war heroes. Note the detail and realism achieved on the statue.

Figure 5-47   An assemblage of coins.

articles such as powder metallurgy, rapid prototyping, spray forming of composites, and various laser-forming topics throughout the text.

Copper provides an excellent example of new applications of traditional engineering materials to satisfy modern needs as a consequence of the never-ending changes in the technology of materials brought about by the emphasis on industrial research and development efforts. Recent breakthroughs in brazing, external corrosion protection and laser welding are making new copper radiators for automobiles and commercial vehicles competitive with their aluminum counterparts. These new processing techniques, combined with copper's superiority in thermal conductivity, corrosion resistance, and mechanical strength, will result in radiators that are significantly lighter, more compact, and more durable than those presently in wide use throughout the world. Although copper has been known as a coinage metal along with silver and gold (see Figure 5-47), it is the only element that is now primarily used in making the world's coinage. Copper's properties that place it in demand are its durability and corrosion resistance (to air oxidation). It is used to coat (plate) other materials. Since 1982, U.S. pennies have been copper-coated zinc. Since 1948, U.S. nickels consisted of 75% copper plating over 24% nickel. The eight new euro coins (Figure 5-48), the official coinage of the 11 European Monetary countries, use copper to plate various other elements. For example, the 1, 2, and 5 euro cent pieces are made of copper-plated steel. The 2 euro bimetallic coin consists of two sections, an outer ring and an inner disk. The outer ring is 75% copper–25% nickel; the inner disk is pure nickel (99.2% min) clad on both sides with nickel brass (75% copper–20% zinc–5% nickel) with the nickel layer making up 12% of the clad's total thickness. This coin is 25.75 mm in diameter and weighs 8.5 g. All eight coins may be distinguished by form, color, weight, and thickness. Table 10-11 lists several copper alloys with corresponding designations, compositions, and properties.

5.9.3 Nickel

Nickel ranks twenty-fourth in abundance among the elements in the earth's crust. Nearly 80% of nickel is used in alloys. About 57% is used in making stainless steel corrosion resistant. Some 10% is used in making heat-resistant and low-alloy steels and 13% is used in nickel-based alloys to improve their heat-resistant and corrosion-resistant properties. Nickel-based superalloys are the most widely used for the hottest parts in aircraft gas turbine engines where they constitute nearly 50% of the weight of these advanced aircraft engines. These alloys can

Figure 5-48 On January 1, 2002 eight new euro coins will go into circulation and become the official coinage of eleven European Monetary Countries. Britain, Sweden, Denmark, and Greece did not adopt this currency in January 1999 when it became the official currency of the Europe Union (EU) and stock exchanges and large businesses began using euro notes to keep track of their finances. The top 2 euro coin down to the bottom 1 euro cent will all have a common design or face on the obverse side (front or main side) shown on the left. This common face represents Europe. On the reverse side (the national face) each country will have its own design. For the above coins the national face is that for Finland. Note the rib design on the 20 cent coin. It could be noted that these coins could be called "**Green**" coins since the producer of the coin blanks has reduced the nickel content of the coins "as far a possible to avoid the potential risk of provoking allergic reactions". (*Advance Materials and Processes*)

Inside the Power Cell

Nickel-Metal Hydride Battery

1. Positive Cap
Connected to the nickel electrode; serves as the positive terminal

2. Can
Connected to the metal-hydride electrode; serves as the negative terminal

3. Separator
Prevents direct internal short circuits

4. Cathode
Composed of nickel oxide; serves as the positive electrode of the cell

5. Anode
Composed of metallic alloy; serves as negative electrode of the cell

6. Battery Case

Figure 5-49 (a) Nickel-metal hydride battery.

withstand temperatures of up to 1110°C. Directional solidification, perfected in the 1970s, uses nickel-based superalloys (e.g., IN-6203 developed in 1981) that withstand temperatures up to 1020°C. The world mine production of nickel totaled 1.1 million tons in 1998, the highest output ever recorded. The principal producing countries were Russia (22%), Canada (19%), New Caledonia (13%), and Australia (12%). Production of refined primary nickel is dominated by two major producers. Together, Norilsk (Russia), Inco (Canada), and the UK account for approximately one-third of world output.

Table 5-4 shows that several of the common methods of processing are not recommended for nickel. While alloyed in steel, copper, and other materials, nickel is also electroplated onto objects for decorative and corrosion-resistance purposes. After development at the end of the nineteenth century, nickel–cadmium batteries have undergone tremendous refinement, to the point that these rechargeable batteries are found in hand electronic calculators, aircraft-engine starters, emergency power equipment, and portable medical equipment. Figure 5-49 shows the construction of nickel–metal hydride batteries and how a typical battery works. One type of nickel–metal hydride battery is being developed, with the involvement of the U.S. Battery Consortium (USBAC), for possible use in the next generation of electric vehicles.

Common alloys of nickel include Monel (nickel–copper), Inconel (nickel–chromium), Hastelloy (nickel–molybdenum–iron), Duranickel (nickel–aluminum), and illium (nickel–chromium–molybdenum–copper). Nickel is magnetic up to 360°C and is alloyed with iron, aluminum, cobalt, and copper in the powerful permanent alinco magnet. Figure 5-51 is a new flexible automotive exhaust coupling that is made of an alloy composed of (in weight %) 34 Ni, 2.1 Cr, 4.2 Mo, 1.8 Si, 0.6 Ti, bal Fe and designated Incoloy 864 by Inco Alloys International Inc. Two of the main effects of higher nickel content in austenitic alloys are an increase in metallurgical stability (maintaining its strength during processing) and an increase in resistance to chloride stress–corrosion cracking. Figure 5-52 is a graph comparing the 760°C hot salt corrosion resistance of stainless steels (SS) and nickel alloys. The presence of deicing salts and alternate wet and dry conditions support aqueous corrosion mechanisms such as stress corrosion cracking, pitting, and general corrosion of automotive exhaust system components such

Batteries may seem simple, but delivery of packaged power is a very complicated electrochemical process. Electric current in the form of electrons begins to flow in the external circuit when the device, in this case a light bulb, is turned on. At that time, the anode material, zinc (Zn), gives up two electrons (e^-) per atom in a process called *oxidation,* leaving unstable zinc ions (Zn^{2+}) behind. After the electrons do their work powering the light bulb, they re-enter the cell at the cathode, where they combine with the active material, manganese dioxide (MnO_2), in a process called *reduction.* The combined processes of oxidation and reduction could not occur in a power cell without an internal way to carry electrons back to the anode, balancing the external flow of current. This process is accomplished by the movement of negatively charged hydroxide ions (OH^-) present in the water solution called the electrolyte. Every electron entering the cathode reacts with the manganese dioxide to form $MnOO^-$ ion. Then, $MnOO^-$ reacts with water from the electrolyte. In that reaction, the water splits, releasing hydroxide ions into the electrolyte and hydrogen ions (H^+) that combine with $MnOO^-$ ion to form MnOOH. The internal circuit is completed when the hydroxide ions produced in this reaction at the cathode flow to the anode in the form of ionic current. There, they combine with unstable zinc ions, which were formed at the anode when the electrons were originally given up to the external circuit. This produces zinc oxide (ZnO) and water (H_2O).

Figure 5-49   (b) How a typical power cell (battery) works. (Duracell International Inc., *The Story of Packaged Power*)

as wire braid and exhaust gas recirculation tubes. Table 10-1 shows the composition and properties of certain nickel alloys.

5.9.4 Magnesium

A chief source of magnesium is seawater. Magnesium is abundant in supply and 100% recyclable. About 6 million tons of magnesium are available for processing in each cubic mile of seawater, which makes magnesium an abundant metal. But the need for electrolysis to extract the metal from mineral deposits or seawater increases its cost. The lightest of all structural metals, with a specific gravity of 1.75, magnesium weighs 1.5 times less than an equal volume

Figure 5-50 (c) Extruded magnesium seat frame represents an advanced, lightweight (3.6 kg) seat system. It is made from magnesium tubular extrusions and stampings, and it reduces seat frame weight by 40%, while retaining its strength integrity. (Findlay Industries)

Figure 5-51 Incoloy alloy 864 was developed specifically for automotive exhaust system flexible couplings and bellows. These components are attacked by a wide array of corrosion mechanisms. In addition, higher exhaust temperatures and warranties lengthened to 100,000 miles or 10 years of service life are becoming prevalent. (*Advance Materials and Processes*)

Depth of Attack Measurements For Nickel Alloys And Stainless Steels
In A Hot Salt Corrosion Test At 1400° F (760° C)

Figure 5-52 A comparison of hot salt corrosion resistance of stainless steel (SS) 321 and 316 Ti and nickel alloys 864 and 625 LCF. Depth of attack measurements for nickel alloys and stainless steels in a hot salt corrosion test at 1400°F (760°C).

of aluminum and 4 times less than zinc. It has good strength, stiffness, and dimensional stability, and in pure form and alloyed with other elements, it provides a high strength-to-weight ratio. Magnesium alloys have relatively high thermal and electrical conductivities; good energy absorption characteristics (i.e., its ability to damp vibration compares with cast iron); and nonmagnetic properties. Often used as an alloying element in engine parts, magnesium is used to produce wheels for racing cars.

A recent lightweight magnesium seat frame made from tubular extrusions and stampings (see Figure 5-50) was developed by Findlay Industries for use in automotive and/or aircraft industries. The purpose of this development was to design a frame that reduced the overall weight of an existing tubular steel frame by 40% to 50%, retained its integrity to meet federal safety standards, and met customer requirements. Other material candidates considered were HSLA steel, aluminum, carbon fiber, and Asdel (compression-molded plastic). Carbon fiber and compression-molded plastic were eliminated from consideration because of the high investment cost for prototype samples. The HSLA steel provided a 25% weight reduction, and a 20% weight reduction if the extruded tube shape was altered. Aluminum also showed substantial weight savings, but certain manufacturing problems had to be solved. For example, the aluminum tubing of the same size specified would provide a 35% weight reduction, but the tube was too brittle to bend to the required shape. Heat treating the entire frame, along with increasing the thickness of the tubing 100%, was too impractical, especially when the overall weight reduction would be only 15% to 20%. Welding problems were also encountered. Magnesium AZ61 and AZ31B alloys offered the best weight savings. These alloys are 4.51 times lighter than steel, but AZ31B won out because of its greater ductility. The seat frame weighs only 3.6 kg (7.9 lb) in bucket seat form, which is about half the weight of a steel frame. Other advantages of using magnesium for this application are that it is noncorrosive and weldable, it has greater design flexibility to accommodate customer preferences and styling concepts, it has excellent design and performance/packaging characteristics (i.e., improved comfort, better vehicle packaging, less foam required), and it is applicable to front and rear seats (bucket or bench type). Finally, the design meets all federal standards (FMVSS 201, 202, 207, and 208).

5.9.5 Titanium

As with aluminum and magnesium, titanium (Ti) is a very abundant metal in its mineral form. As a raw material element, it ranks fourth in availability of the structural metal elements. Unfortunately, the fabrication processes are costly. With a density between aluminum and stain-

INVESTMENT CASTING PROCESS

Sequence of Operations

Die Construction   **Wax Injection**   **Wax Assembly**

Slurry Coating   **Stuccoing**   **Dewaxing**   **Casting**

Shell Removal   **Cut Off**   **Finishing**   **Inspection**

(a)

Figure 5-53   (a) Investment casting. From teeth to jet engines come products made from a process that evolved from the ancient art of "lost wax casting." Investment casting serves many fields ranging from medicine for dental and bone implants, to aircraft with intricate gas turbine engine components (Figure 5-53). Constant innovations in the process came with improved nonferrous metal alloys such as titanium and Vitallium (cobalt, chromium, molybdenum, and nickel), refinements in ceramic mold materials, control of metal grain structure, and the ability to produce larger castings. (Howmet Corporation)

less steel, titanium, "the world's most glamorous metal," was discovered in 1795. It was not until 1930 that a practical method of extracting the metal from titanium oxide (TiO_2), its ore, was commercially feasible. This process produces titanium sponge, which is then further refined in an electric furnace to form ingots, which, in turn, are further remelted to form a final pure titanium. Boeing's 777 (see page 000) uses the largest amount of titanium ever in a Boeing aircraft. In fact, commercial and military aircraft now use 80% of American-made titanium. A major effort is under way to reduce the weight of such aircraft by substituting lighter-weight, advanced materials such as continuous-fiber ceramic composites (CFCCs) for titanium alloys, thus permitting larger payloads and greater hauling distances. ***Investment casting*** of the versatile titanium alloys results in very complex shapes with excellent surface finishes and unique properties often not possible with other processes (Figure 5-53a).

Titanium's alloys offer superior specific strength (up to 26.5×10^3 in.) in high temperatures (over 590°C) and low temperatures (-253°C), which makes it a popular structural metal in ultrahigh-speed aircraft and accounts for its use on the space shuttle. The *superplastic* nature of titanium allows it to deform over 2000% without nicking or cracking when it is heated to around 925°C in a process known as superplastic forming. Advances in new technologies such as *superplasticity forming* and *diffusion bonding* (illustrated in Figure 5-40) have furthered the potential of titanium. Titanium is nonmagnetic and has a lower linear coefficient of expansion and lower thermal conductivity than steel alloys or aluminum. Like zirconium and beryllium, titanium is allotropic and exists in a cph structure (alpha) at below 885°C and a bcc structure (beta) above that temperature (see Table 3-2). The composition of any titanium binary alloy consists of three phase fields in the solid state: α, ($\alpha + \beta$), and β. Various alloying elements alter the effect of the structure and stabilize the alpha or beta phases. For example, aluminum stabilizes the alpha structure, raising the temperature of the alpha–beta transformation and enlarging the α phase field. Other elements are β stabilizers and they enlarge the β field. The presence of these three distinct phases in titanium alloys produce different properties. Alpha alloys have superior creep resistance and are heat treatable. Beta phase alloys have excellent forgeability and fracture toughness. Alpha–beta alloys have good superplastic elongation and good strength characteristics. Alloys that are based on the intermetallic compound gamma titanium aluminide (γ TiAl) have good stiffness, high temperature strength, and low density. Their applications in advanced gas engines, airframes, and automotive components are quite easy to understand. These alloys are usually fabricated by casting or by powder metallurgy which makes their processing costs high. The skin of the future X-33 Space Plane or the engines of the future F-22 jet fighter will be a logical choice for applications of this alloy that can withstand temperatures over 300°C hotter than conventional titanium alloys. Table 10-11 shows some titanium alloys, compositions, and properties.

Biomaterials are materials that are compatible with human and animal systems, allowing the material to be implanted or manipulated in people and animals. Titanium is an example of a biomaterial that serves as implants for joint replacements and dental reconstruction (Figure 5-53b).

5.9.6 White Metals

The general term *white metal* includes zinc, tin, lead, antimony, cadmium, and bismuth, which all have relatively low melting temperatures (Table 5-5). *Zinc* is a readily available and inexpensive metal that can easily be applied to steel to serve as the sacrificial protection coating, as discussed previously. Zinc is also die cast for housings and decorative trim; used as additives in rubber, plastics, and paint; and used in wrought forms that can achieve high surface hardness of over 70 HRC. Tin has long been used as a coating for "tin cans" made of tin plate and also finds application as a valuable alloying element with lead as soft solder, in tin babbit, and as antimonial tin solder. Solder balls as small as 2 mils (0.002) hold semiconductors onto metal-pinned ceramics that are used for computers. *Pewter* (alloyed with lead) was a popular alloy for cooking and eating utensils, plus ornamental casting, in colonial America. The modern tin-based pewter is lead-free with 91% tin, 7% antimony, and 2% copper and is often used to replicate colonial dishes. *Lead* is not only an important alloy in solder, but when evenly dispersed in leaded steels at 0.15–0.35%, it provides built-in lubrication to ease machining and produce tightly curled chips. Lead is also important in electrical batteries as terminals and, alloyed with antimony, for grid plates (Figure 5-54). Its resistance to many corrosive chemicals, x-rays, and gamma rays and its sound-damping capacities coupled with its high density find unique applications in the medical, chemical, and nuclear industries. It is compounded with ceramic glazes. Small portions of lead built up in the human body cause lead poisoning (plumbism). Consequently, it is desirable to eliminate lead from paint on children's furniture, in gasoline, and in soldered tin cans and to reduce exposure by industrial workers. Tin is now replacing lead in steel alloys as an environmentally friendly alternative. Dubbed "*green steel,*" this new alloy is being used by Ford Motor Company. *Cadmium* is harder than tin and serves as a corrosion-resistant coating electroplated

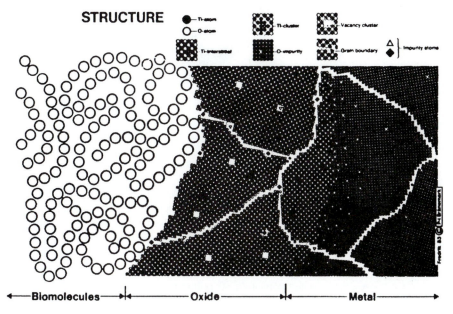

osseointegration

(b)

Figure 5-53 (b) Titanium is a biomaterial compatible with the human system for dental and joint implants. The large-scale sketch shows the interface of bone tissue with the titanium and TiO₂ boundary layer. (*Ledger-Star*)

TABLE 5-5   MELTING TEMPERATURE OF WHITE METALS

Metal	°C	Metal	°C
Antimony (Sb)	631	Lead (Pb)	327
Bismuth (Bi)	271	Tin (Sn)	232
Cadmium (Cd)	321	Zinc (Zn)	420

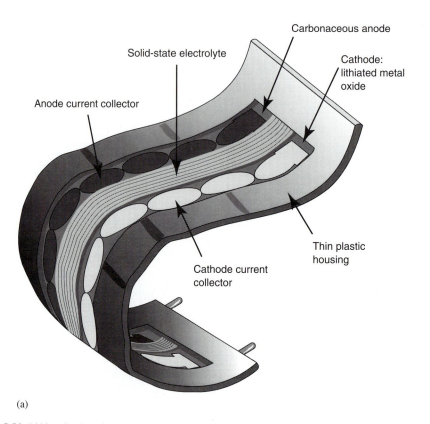

(a)

Figure 5-54 Lithium-ion batteries have replaced the nickel-based batteries for many portable notebook computers. (a) These new batteries combine lithium technology with solid polymer electrolyte for a thin, flexible, lightweight, and high-energy-density battery that can be shaped to fix most computer and portable electronics appliances.

on steels, especially fasteners. A coating of 0.0008 mm has the equal protection of a 0.025-mm zinc coat. It also serves as an alloy in copper to improve hardness.

  Battery technology, spurred on by the desire for greater mobility and long life for heart pacemakers, telephones, portable radios/CD players, space vehicles, and notebook computers with multiple functions (e.g., fax/modem, CD-ROM, and sound), continues to make strides. State legislatures gave battery technology a boost by mandating a percentage of electric vehicles for their respective state to offset the air pollution from gasoline vehicles. A variety of metals in chemical systems serve as batteries for a wide range of applications in both primary batteries and rechargeable batteries. They include the following: lithium/alkaline-manganese, manganese dioxide, zinc air, silver oxide, lithium ion (see Figure 5-54), nickel-metal hydride, and nickel–cadmium.

5.9.7 Other Metals

Beryllium (Be) is a high-cost metal that serves as an alloying element in copper for age-hardening to enhance oxidation resistance and refined grains. Beryllium is ideal for space telescopes because it is the only metal with stiffness greater than glass that has a density lower than aluminum. ***Refractory metals*** are those with melting points above 1980°C and include tungsten (W) (3370°C), tantalum (Ta) (2850°C), columbium (Cb) or niobium (Nb)

Lithium Cell

1. Positive Cap
Formed nickel-plated
steel disk which
serves as positive
terminal

2. Seal
Molded plastic part
which contains cell
materials and
provides a safety
vent

3. PTC Device
Polymeric device
which shuts off
current in a cell that
has been acciden-
tally short-circuited

4. Electrode Jelly Roll
Assembly of positive
and negative elec-
trodes which are
separated by micro-
porous polymer
separator

5. Electrolyte
Solution of complex
salts which are
dissolved in organic
solvents

(b)

Figure 5-54   (b) Construction of a standard (AA, C, D, etc.) lithium cell. (Duracell International Inc., *The Story of Packaged Power*)

(2415°C), and molybdenum (Mo) (2620°C), which are commonly alloyed with many of the metals discussed above to impart strength and hardness at high temperatures, such as type 6–6–2 molybdenum–tungsten high-speed steel. Because of the high temperatures and fabrication difficulties, refractories are often formed as powdered metal parts. Numerous nonferrous alloys known as *superalloys* have been developed to meet the requirements of high-technology innovations, such as jet and rocket engines, for high strength and for corrosion and creep resistance at extremely high temperatures (exceeding 1100°C). These superalloys use iron and nickel, nickel, or cobalt as the base metal. Examples of such alloys are Inconel (nickel base), Incoloy (iron–nickel base), and S-816 (cobalt base). Possessing such high strength, most of the alloys are cast or use powder metallurgy techniques in their manufacture. Where machining is required, chipless-type machining methods are mandated.

5.9.7.1 Ferrous amorphous metals.

Metal alloys possess crystalline atomic structures in which individual atoms are arranged in ordered, repeating patterns. *Amorphous metals* differ in that they consist of atoms arranged in near random configurations lacking any long-range order. These noncrystalline structures are common in nature being found in nonmetallic solid materials such as silicates which, in their viscous liquid state, the individual molecules have limited mobility and crystallization proceeds slowly. This allows the use of modest cooling rates to suppress crystallization completely. Liquid metal alloys, on the other hand, have low viscosity and high diffusivity because they consist of loosely bonded atoms rather than bulky clusters or molecules. This indicates that due to this free movement of atoms extraordinary cooling rates would be necessary to bypass crystallization. The discovery of amorphous metals is credited to P. Duwez in 1960, who used a pressurized gas gun to propel small droplets of molten alloy onto a polished copper plate. On contact with the plate, the molten metal droplets solidified into flake or "splat" form. Thus, the splat quenching process originated for the rapid solidification of amorphous metals. In the early 1970s the formula for all amorphous metal products was established and consisted of various percentages of three main constituents: one or more transition elements, a nonmetallic element such as boron, phosphorus, or carbon, and a metalloid such as silicon, aluminum, or germanium. *The development of amorphous metals is a classic example of the intimate linkage that exists among the three main principles in materials science, namely, the structure, the processing, and the properties of a material.* Further developments in processing techniques led to the production of continuous, amorphous metal filaments by the process called **chill block melt spinning** in which a stream of molten metal is directed at a substrate. Two later variations of this process are known as **free-jet melt spinning** and **planar flow casting.** Using the free-jet process in 1973, Allied Chemical manufactured the first commercial amorphous metal in the form of continuous ribbon 1.7 mm wide and 50 μm thick. Planar flow casting has allowed ribbon width to increase such that widths up to 210 mm are commercially available.

Initial developments with amorphous metals dealt with precious metals such as gold and palladium. Subsequent events proved that such amorphous metals could be produced with common, inexpensive metallic elements but, in addition, such metals could display ferromagnetic properties. **Ferromagnetism** is one of three forms of magnetism in materials. Each form is distinguished by different degrees of susceptibility and permeability, of which ferromagnetism has large values of both. Ferromagnetic materials such as iron, nickel, and cobalt are strongly attracted to magnets. Such magnetic metals are characterized by an electronic structure in which electronic orbits of an internal shell of bound electrons are not completely full. In iron the unfilled 3d orbits with five electrons have a positive spin and one electron has a reverse negative spin. Such a condition is the source of ferromagnetism, the tendency for the spin of these electrons in any given atom to affect the spin of the corresponding electrons in neighboring atoms. The distance between neighboring atoms is critical. For further discussion of magnetic properties see Module 9. Amorphous metals magnetize more easily than crystalline metals because they do not contain the dislocations and grain boundaries that impede the movement of domain walls. **Magnetic domains** are regions of ferromagnetic materials in which the spins of all their atomic magnets are aligned. Detectors are under development that will read the status of these domains in a material undergoing load and thus aid in determining when and if the material will undergo fatigue failure (see Module 4). Such intelligence would be available long before our present attempts to predetermine when fatigue failure occurs (i.e., when a crack is observed and is accessible for analysis). Even a demagnetized steel bar has magnetic domains that shift positions with respect to each other as the bar is crushed or bent. Such shifting produces detectable changes in the magnetic field. As defects in the bar's steel structure appear, the "flutter" in the magnetic field becomes more pronounced. As the bar begins to crack, spikes appear in the magnetic field. This development may result in instruments that will more accurately determine when a part is about to fail. Metal fatigue is difficult to detect. Once cracks have been discovered, the part undergoing fatigue has been potentially dangerous for an unknown period of time. See further treatment of domains in Section 9.3.2, "Relative Permeability."

In the United States, the Electric Power Research Institute (EPRI) is recognized as the major impetus behind the development of amorphous metal transformers. Several technologies were developed to allow amorphous metal to be used in transformer manufacturing in a manner similar to silicon steel. To simulate the thickness of silicon steel sheet, a process called **prespooling** produced multiple-layered packages of amorphous metal ribbon which subsequently formed a strip of multiple-layer thickness. The replacement of grain-oriented silicon steel in the transformer cores with amorphous metal reduced core losses by 75%. In the United States alone, the EPA estimated the energy equivalent saving at over 70 million barrels of oil while reducing the volume of harmful CO_2, NO_x, and SO_2 emissions.

Smart metals, the *shape-memory alloys (SMAs)* such as *Nitinol* (nickel–titanium alloy), have been around for many years and are beginning to find wide application as designers learn more about them. One application is in eyeglass frames. The high-quality, shape-memory frames made of this "smart material" return to their original shape when bent. SMAs are also implanted in "smart structures," and are used as controls for robot fingers and dental braces.

SELF-ASSESSMENT

5-53. Cobalt, chromium, titanium, tungsten, manganese, and platinum are some of the strategic metals that the United States must import to produce the vital specialty alloy steels needed in our advanced technological society. Most of these metals are found in the developing nations. Stockpiling aside, what does this situation signify to the many materials scientists, engineers, and technicians working in the metals industries?

5-54. Name three noble metals and give an application of each.

5-55. Explain why aluminum, magnesium, and titanium are expensive when they are so readily available as raw materials. Give a reason for recycling aluminum.

5-56. Explain the corrosion-resistance mechanism of copper and aluminum. Which metal offers the widest range of processibility?

5-57. Give three applications of nickel and explain why nickel is used in these applications.

5-58. What are (a) brass, (b) bronze, (c) Monel, (d) Inconel, and (e) refractory metals?

5-59. Name one barrier and two advantages for advanced metals technologies such as MMCs, near-net-shape forming with SPF and SSF, and light aluminum alloys.

5-60. (a) What common characteristic do white metals possess? (b) Provide a typical application for each of these metals. (c) Why is lead eliminated from paint, plumbing, gasoline, ceramics cookware, and similar uses?

5-61. Name one refractory and one white metal.

5-62. Describe the reasons for four material substitutions in engines like that used in Ford's Contour concept car. Emphasize specific environmental and economic factors.

5-63. What are the advantages of the new anodizing process for aluminum using sulfuric acid?

5-64. List four properties of magnesium alloys. Cite one example of an application where a magnesium alloy was selected because it possessed one of these properties contained in your list.

5-65. Explain why pure titanium is expensive to produce.

5-66. List the sequences of operation for the investment casting process. Explain how a typical casting is moved from one sequence to another in a modern casting operation.

5-67. The recrystallization temperature of about 600°C for Monel metal is the approximate temperature at which a highly cold-worked Monel part completely recrystallizes in an hour. Other recrystallization temperatures are low-carbon steel, 540°C; zinc, 10°C; tin and lead, −4°C. Working of tin, lead, or zinc at room temperature (about 20°C) is considered hot working. Explain.

5-68. Describe the steps in diffusion bonding of metal-fiber composites.

5-69. Refractory metals have melting points above 1980°C. With the present demand for materials with high strength-to-weight ratios, and high mechanical strength and hardness at service temperatures above 1650°C, what is the main reason why refractory metals are not used to any great extent for these types of applications. Explain how you would overcome this problem, assuming the technology was available to you.

5-70. Use Table 10-11 to select a metal alloy for the axle of a racing bike. Cite properties in explaining your selection.

OBJECTIVE QUESTIONS

5-48. Advanced alloys saved _____ pounds of structural weight in the Boeing 777.

    a. 1000   **b.** 2000   **c.** 3200   **d.** 5400

5-49. Nonferrous metals comprise _____ of the known elements.

    a. 1/2   **b.** 2/3   **c.** 3/4   **d.** 4/5

5-50. The temper designation (W) that follows the four digits specifying wrought alloy groups for the Aluminum Association stands for _____ .

    a. solution heat treatment   **b.** as fabricated   **c.** strain hardened

5-51. Recycling saves _____ energy over that for new aluminum.

    a. 60%   **b.** 20%   **c.** 80%   **d.** 95%

5-52. Titanium, at certain temperatures, can be deformed as much as _____ .

    a. 100%   **b.** 200%   **c.** 500%   **d.** 2000%

5-53. What metal is added to the surface of anodized aluminum to improve its wear resistance?

    a. Beryllium   **b.** Silicon   **c.** Zirconium

5-54. A pound of aluminum is _____ as expensive as a pound of steel.

    a. two times   **b.** three times   **c.** four times

5-55. _____ is a metal and an example of a biomaterial that serves as implants for joint replacements in humans.

    a. Titanium   **b.** Steel   **c.** Copper   **d.** Lead

5-56. The lightest of all structural metals, with a specific gravity of 1.75, is _____ .

    a. lithium   **b.** manganese   **c.** molybdenum   **d.** magnesium

5-57. _____ is added to copper to improve its oxidation resistance.

    a. Aluminum   **b.** Brass   **c.** Niobium   **d.** Beryllium

5-58. Brass is a metal alloy made from two elements: _____ and _____ .

    a. tin, lead   **b.** nickel, copper   **c.** zinc, copper

5-59. Monel is a metal alloy consisting of two elements: _____ and _____ .

    a. tin, lead   **b.** nickel, copper   **c.** zinc, copper

5-60. Inconel is a metal alloy consisting of two elements: _____ and _____ .

    a. nickel, copper   **b.** nickel, chromium   **c.** nickel, aluminum

5-61. Which metal listed below is capable of being superplastically deformed 2000%?

    a. Nickel   **b.** Aluminum   **c.** Titanium   **d.** Magnesium

5-62. Which metal listed below is a white metal?

    a. Zinc   **b.** Nickel   **c.** Aluminum   **d.** Magnesium

5-63. Superalloys are made from a particular base metal, such as one of the following:

    a. Cobalt   **b.** Iron   **c.** Aluminum   **d.** Beryllium

Scanning electron photomicrograph of a nonsagging AKS tungsten (W) filament after several hundred hours of operation at 2500°C in a common 60-watt light bulb. (a) The "coiled coil" geometry of the filament is possible because of the availability of ductile, Coolidge-process P/M, tungsten wire. (b) A closeup reveals some of the tungsten evaporated, thus exposing atomic facets of the individual W grains and characteristic interlocking grain boundaries (chevron features). This grain boundary structure prohibits creep deformation (sagging) through grain boundary sliding, which results in longer lasting filaments. (Jerry P. Wittenauer, Lockheed Missiles and Space Co., and *Advanced Materials and Processes*)

(a)   (b)

Figure 5-55 (a) Two examples of porous PM parts, a brass gear with carbon on top which will be heated and infiltrated into the porous brass to provide lubrication. The porous brass filter on the right threads onto the oil nozzle to filter out impurities. (b) Enlargement of the sheet in left foreground shows this stainless steel (AISI 316) sheet filter is made of fine metal fibers that are sintered at fiber's intersections to bind them together. (c) Sintered 2050° F.

(c)

5.10 METAL POWDERS

Through a variety of techniques, powders of selected materials are produced for ***powder metallurgy (P/M)*** applications. While there is evidence that P/M technology existed at least 5000 years ago, only recently has the technology made significant strides. Developments from the eighteenth century through World War I resulted primarily from the inability of metallurgists to fuse metals such as platinum and tungsten, so they turned to powder compacted and heated to allow diffusion bonding. This last step is known as ***sintering,*** and P/M parts are often referred to as *sintered metals.* (See Figure 5-55) Current powder metallurgy deals with most metals and a host of processes. Figure 5-56 shows the flow of the raw powder through compacting ***(briquetting)*** to finished P/M parts. Some parts undergo secondary operations to improve tolerances or finishes, to apply lubricants or coatings, or for heat treatment. A major advantage of the P/M part is the reduction or elimination of scrap and machining, so if too many secondary operations are required, this advantage can be reduced. Once the part is compacted and sintered, it may undergo secondary operations such as *coining* to squeeze the part to closer tolerances; ***infiltration,*** in which a lower-melting-point metal is added to close pores and increase density; ***impregnation***

RAW
MATERIALS

Elemental or Alloy
METAL POWDERS

OTHER ADDITIONS
(die lubricants, graphite)

MIX

MIXING

FORM

PRESSURELESS MOLDING
SLIP CASTING

DIE COMPACTING
(mechanical or hydraulic)

ISOSTATIC COMPACTING
ROLL COMPACTING
HIGH ENERGY FORMING

SOLID
STATE
FUSION
OR
BONDING

(SINTERING)

EXTRUSION

GAS PRESSURE
BONDING

SPARK
SINTERING

SINTERING

PRESINTERING
FOR MACHINING

SINTRATING

HOT
PRESSING

HOT
FORGING

OPTIONAL
OR
SECONDARY
OPERATIONS

OPTIONAL FINISHING STEPS
(heat treat, plate, tumble, machine,
steam treat, oil or plastic impregnate)

OPTIONAL MANUFACTURING STEPS
(metal infiltration, size, coin,
forge or reform, repress, resinter)

FINISHED
PRODUCT

FINISHED P/M PRODUCTS

Figure 5-56 Schematic flow diagram of P/M process. (Hoeanaer)

Open pores —
for oil or grease
impregnation
or filters

Densely packed —
for structured
parts

Figure 5-57 Impregnation.

(Figure 5-57), which adds a lubricant for self-lubricating bearings; or *machining* to achieve a profile not practical through the pressing operation. Not all metal powder is compacted and sintered. Iron powder is added to human and livestock food as a dietary supplement. Iron powders are also used in nondestructive testing (NDT) and magnetic-particle testing, for welding electrodes, and in oxyacetylene cutting and scarfing (exothermic reaction increases the heat of flame to melt oxides formed on high alloys or refractories). Platinum powder is a catalyst used in making gasoline, copper powder goes into marine paints, carbon powder is used in automotive ignition wires and in making xerographic paper copies, and aluminum powder aids in paper manufacture.

In addition to the advantages given, P/M allows the manipulation of a part's density through control of particle size and degree of compaction. Particles range from 0.2 to 2000 μm in size. Smaller particles can pack tightly for increased density; or if open pores are required, as in filters or oil-impregnated, self-lubricating bearings, larger particles would be compacted together (Figure 5-58). P/M structural parts of steel powder have tensile strengths from 310 MPa

Figure 5-58 (a) Iron-carbon sintering involves four steps: 1) establishing particle bonds at their area of contact and growth of these bonds, 2) grain growth and migration of grain bounderies formed at bonds, 3) spheroidisation of pores, and possible growth of larger pores. (b) A pressed. (c) Sintered 2050°F. (Hoeganaes)

(45,000 psi) to over 1200 MPa (175,000 psi), which makes them competitive or superior in terms of strength with cast iron, steel, and nonferrous alloys in certain applications. Although conventional P/M processing can produce parts up to 35 lb, most are smaller and weigh less than 5 lb. The most common P/M materials are iron, steel, copper and copper alloys, stainless and other alloy steels, nickel alloys, aluminum alloys, and refractories such as tungsten carbides, but most other metals find certain unique P/M applications. Tungsten light bulb filaments are made with a P/M process (see page 301). Powder metal parts are now found in automobile transmissions and engines. Ford 194R70W and E40D transmissions each contain 11 to 127 kg of P/M parts, whereas DaimlerChrysler's 2.7-L engine contains 11 kg.

EMT WEBLINKS

Powder Metals—*http://www.mpif.org*
http://www.mpif.org/chart.html
http://mpif.org/design/start.html

**Gasoline Engine
Lightweight Components**

Valve spring retainers
Valve springs

Valves

Pistons

Connecting
rods

Materials and Applications
- Titanium Alloys
  (Valve, retainers, connecting rods spring)
- Rapidly Solidified Aluminum Alloys
  (Intake valves, connecting rods, retainers)
- Metal Matrix Composites
  (Piston, connecting rods, retainers)
- Ceramic, Aluminides
  (Valves)

Figure 5-59 Gasoline engine: Lightweight components and advanced materials applications. (Ford Motor Company)

Figure 5-60 Part at left is the original, designed in cast nodular iron. Part at right shows SSF aluminum design. Note the forged-in-place wear pad and pulley mounting stud. The only machining area on the SSF part is the large hole where the pivot bolt mounts. (*Advanced Materials & Processes*)

5.11 MATERIALS SELECTION

Throughout this book and in many technical publications, advanced materials receive attention because of favorable properties and processing possibilities. The cross-sectional view in Figure 5-59 of a concept gasoline engine that uses lightweight components brings together several of the advanced-materials components illustrated in this book and provides support for the market forecast for advanced materials that encompass advanced metals alloys and metal-matrix composites (MMCs), engineering plastics (especially thermoplastics that are easier to recycle), advanced ceramics, and advanced composites. Requests for advanced-materials technology come from many areas, including demand for increases in fuel efficiency with lighter, hotter-running engines; lighter, faster, safer transportation vehicles; more powerful, smaller, friendlier computers; cleaner, cheaper, more abundant energy; and on and on. Advanced material, AlBe-Met, a composite of aluminum beryllium metal matrix composite (MMC) is a lightweight, high-strength, oxidation-resistant material. This material combines the low density and high stiffness found in beryllium with the ductility and low cost of aluminum. It reduces weight, removes heat, and increases the lifetime of electrical systems. Used now in rotary actuators for disc drives, it improves operations by having a lighter and stiffer arm which allows the heads to move faster.

A number of examples of materials substitutions for engine parts are in other parts of the book and show how materials selection and engineered materials can make more efficient engines. Various cast iron alloys have been widely used for a wide variety of engines since the beginning of automobiles' manufacture. Figure 5-60 compares two versions of a timing-belt tensioner-pulley pivot bracket; the earlier one made of nodular cast iron and the redesign of *semisolid forged (SSF)* aluminum. SSF uses the principles of thixotropic metal processing (refer to the discussion for Figures 3-1 through 3-3) to make alloys for forging. The SSF aluminum bracket was redesigned using CAD with finite element analysis. Weight was reduced from 0.31 kg (0.69 lb) for nodular iron to 0.16 kg (0.36 lb) for the SSF aluminum. The SSF process provides near-net shape since it enables the wear pad and the pulley mounting bolt to be integrally formed into the pivot brackets. This eliminated the machining operation required for the wear pad and pulley mounting bolt areas on the nodular iron version.

High cost offers the main barrier to advanced materials. Usually, they cost several times to tens of times more per pound than traditional engineering materials. Any forecast for the twenty-first century will contain the expectation that materials costs will decline as materials producers and product fabricators gain more experience with the new materials, thereby increasing the demand for greater production rates of the materials. Because advanced technology brings with it lower *total-systems cost* and improved systems performance, these forecasts can be made with confidence. However, the changeover to advanced-materials technology will evolve with traditional engineering materials technology coexisting. Advanced materials are the driving force behind the auto materials market as they fill the need for reduced weight and, in turn, reduce emissions and fuel consumption. Automakers agreed to produce cleaner automobiles, minivans, and small trucks (less than 6000 lb) beginning with the 1999 models sold in the Northeast states and nationwide two years later. These new vehicles emit nearly 70% less smog-causing tailpipe pollution (nitrogen oxide and hydrocarbons) than earlier model vehicles. For example, the gasoline engine that uses advanced-materials components (Figure 5-59) is still expected to have an engine block of cast iron or aluminum for some time. As a materials user, designer, manufacturer, tester, or in whatever role you interact with materials, expect to continue to use traditional metals for many years. Table 10-11 offers only a tiny sampling of the wide variety of traditional metals available. Many handbooks and volumes of references, some listed throughout this book, will enable you to make wise metals selections.

Failure to consult reliable references can meet with tragic consequences. Twelve people lost their lives when a Swiss swimming-pool-building roof collapsed. Failure analysis revealed that 207 stainless-steel hooks had corroded and could no longer support the concrete roof, which weighed 166 tons. The hooks corroded over 13 years from the pool's aggressive chloride atmosphere, which caused intergranular stress-corrosion cracking.

APPLICATIONS & ALTERNATIVES

LIFE CYCLE OF AUTOMOTIVE MATERIALS AND GREEN MANUFACTURING OF THE AUTOMOBILE

The materials cycle in Module 1 depicted in Figure 1-7 traces the *life cycle of materials* from "cradle to the grave." Figure 1-10 expanded on a phase of the materials cycle called the manufacturing engineered materials. Within this phase is a vital activity called engineering design, which concerns itself among other areas with a unified life-cycle approach to manufacturing. *Engineering* or *product design* can now be called *"green design"* because it, too, takes into account the life cycle of a material whether it can be recycled, disposed of as waste, or reused in some other form. In this assessment of a product the effect of ecology (that branch of biology that studies the relations between living organisms and their environment), the health of human beings, and the world's dwindling resources are concerns not only for the local environment but the global environment as well. Figure 1-10 deals with the manufacture of a product which has taken into consideration all these phases in the life of a material. In the caption to this figure note the inclusion in the list of advantages this new material has as applied to an automobile component, such as "molded in color which is preferred by processors because of the potential to eliminate painting, reduced emission of VOCs (volatile organic compounds), and enhanced recycling." This is an example of a life-cycle analysis or assessment of a product in order to improve its environmental performance. This type of manufacturing, which is epitomized by U.S. auto manufacturers, can truly be characterized by the term *"green manufacturing."* In 1992, the U.S. Council for Automotive Research (USCAR), an umbrella organization of DaimlerChrysler, Ford, and General Motors, was formed to further strengthen the technology base of the domestic auto industry through cooperative, precompetitive research. The UltraLight Steel Auto Body (ULSAB) Consortium was formed in 1994 with a goal to reduce the mass of steel in autobody structures while maintaining their performance and affordability. The consortium consisted of 35 prominent sheet steel producers from 18 countries worldwide. This weight-reduction program produced a plethora of new alloyed high- (210 to 550 MPa) and ultrahigh-strength steels (greater than 550 MPa) which comprise over 90% of the body structure and thus saved mass. Advanced forming and joining techniques enabled engineers to consolidate functions into fewer parts. Tailor-welded blanks removed mass of metal that did not contribute to performance, and hydroforming both tubular and sheet helped reduce part count to 108 parts as compared with more than 200 total parts for an existing typical body structure. Other features were the use of fewer spot welds (about one-third less) and significantly more laser welding. Steel is the most recycled material in the world and consequently the whole ULSAB structure is recyclable. With less mass the consumption of gasoline will be lessened along with its contribution of pollutants to the atmosphere.

Figure 5-61 (a) 6xxx alloys utilizing Mg and Si are heat treatable. They possess great formability and corrosion resistance with high strength. Alloy 6061, a versatile alloy, is used to produce extruded autobody components such as brake cylinders and bumper reinforcements. 6022 is used in outer and inner body panels. These alloys provide flexibility that can be described as "putting the metal where you want it."

Automotive Aluminum Recycling Loop

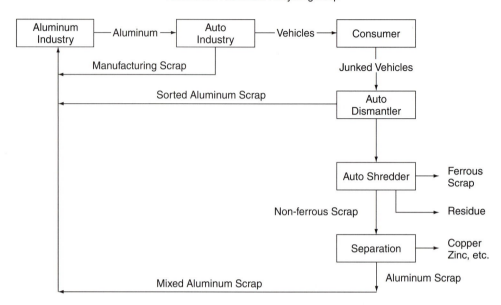

Figure 5-62   Automotive aluminum recycling loop. Aluminum recovery generally follows one of two paths. The first is at the dismantler where large aluminum parts that are easily separated from an auto hulk are sold to a scrap metal dealer. Such parts as radiators, wheels, and bumpers are examples. The other path is through the post shredder recovery. After the hulk passes through the shredder the nonferrous portion is sent through heavy media separation where the remaining aluminum is recovered. (Courtesy of the Aluminum Association)

Aluminum, because of its corrosion resistance, is nonbiodegradable. In fact, aluminum and all of its alloys are the most recycled nonferrous metals. The production of aluminum is itself a large consumer of energy and the process of extracting aluminum from its ore (bauxite) also contributes pollution to the environment. Hence, the low energy required to recycle aluminum makes its recycling an attractive process. In joining the U.S. Department of Energy's "Industries of the Future" partnership for supporting national objectives for clean and energy-efficient production, the aluminum industry has engaged in research leading to the development of an advanced aluminum cell to replace the energy-intensive Hall-Heroult process for primary production of aluminum. This century-old process produces significant emissions of greenhouse gases, particularly carbon dioxide and perfluorocarbons.

The Auto Aluminum Alliance, an interindustry alliance, was created in June 1999 to promote collaborative research efforts aimed at accelerating the use of new and improved aluminum technologies for motor vehicles. Together with USCAR this alliance has an aim to reduce the mass of today's typical midsize family sedan by 40%. This goal was established by the Partnership for a New Generation of Vehicles (PNGV), consisting of the Department of Energy and USCAR, which calls for a new generation of fuel-efficient (up to 80 mpg) vehicles by the year 2004. Over the past 10 years the use of high-strength, lightweight automotive aluminum has doubled (see Figure 5-60). Aluminum alloys (5xxx-Al-Mg) are finding wide usage such as 5754 for autobody panel and frame applications. Aluminum alloys allow the designer to create special shapes that place the metal where it can carry the required load most efficiently, thus eliminating the need for several separate parts joined together into one specially shaped part. Aluminum is flexible in that it has numerous alloys (see Figure 5-61). The unique attributes of aluminum plus new developments in its production, use, and reuse are helping the United States achieve its energy conservation, environmental, and economic goals. One of the highest-priority projects is the development of low-cost continuously cast aluminum alloy sheet for wider use by auto manufacturers. By lightweighting the nation's vehicles, fuel consumption and associated emissions will be reduced dramatically.

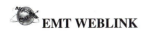 **EMT WEBLINK**

Thixotropic Injection Molding Processes with image Gallery—*http://www.thixomat.com*

One phase of the materials life cycle is the recycling/disposing of used products and systems. In Section 1.2.6 the subject of ***recycling/disposal/reuse*** included, among other things, the fact that recycling has been "accomplished by requiring deposits on beverage containers to provide financial incentives to people to return cans and bottles." A more recent example is from the Aluminum Association partnering with Habit for Humanity International with a goal to recycle aluminum cans to raise funds to build decent, affordable housing for low-income families nationwide. Included in this goal is to fund the construction of at least 250 Habitat homes every year. Similar activities have been started throughout the nation. Aluminum has been recycled since it first appeared on the hood of Henry Ford's Model T. In 1996 an average of 252 pounds of aluminum per passenger car was recycled. Figure 5-62 shows the automotive aluminum recycling loop. Aluminum has proven sustained recyclability; that is, it can be recycled many times with no loss of performance or quality. Some automotive parts have as high as 100% recycled content. Aluminum engines, manifolds, motor housings, and other parts are often made from 100% recycled aluminum. The phrase "recycled content" implies that it is from postconsumer (its origin is in the marketplace) or postmanufacturing scrap. Secondary aluminum producers make aluminum alloys using a high percentage of scrap whereas primary producers make aluminum from basic raw materials. In 1999 recycled aluminum amounted to 63% of the total aluminum used in vehicles.

SELF-ASSESSMENT

5-71. **(a)** Provide three reasons for the use of powder metal. **(b)** List three applications of P/M.

5-72. Discuss the possibilities, if any, of using P/M with high-melting-point materials.

5-73. Explain why it is more economical to use P/M for parts that require very close dimensional tolerances.

5-74. Discuss the relationship between compacting and sintering. Are they the same operation? State the objectives of each, and discuss where diffusion bonding actually takes place with the ultimate result of the final elimination of remaining voids between particles in a structure.

5-75. Discuss the reasons behind the continual drive to produce advanced materials.

5-76. List reasons why the production of near-net-shape parts by P/M is economical.

5-77. Cite an example in which the poor selection of a material resulted in tragic circumstances, and explain how better procedures in selection can reduce or eliminate such failures in the future.

5-78. Suggest a remedy for the problem of graphitizing of gray cast iron water pipes in underground freshwater service systems.

OBJECTIVE QUESTIONS

5-64. Another name for sintered metals is _____ .
    a. fused   **b.** compacted   **c.** P/M

5-65. In P/M the powder is compacted and heated to allow for _____ .
    a. diffusion bonding   **b.** reduced grain growth   **c.** lower density

5-66. Name another use for iron powders (other than P/M).
    a. Making steel   **b.** Destructive testing   **c.** Incendiaries   **d.** Welding electrodes

5-67. The primary purpose of briquetting is to _____ .
    a. facilitate handling   **b.** diffusion bond   **c.** sinter

5-68. Which of the following is the main barrier to the adoption of advanced materials?

   a. Lack of database   **b.** High cost   **c.** Lack of experience with material

5-69. Generally speaking, lightweight materials are selected for what particular area of application?

   a. Building construction   **b.** Machine tools   **c.** Transportation

5-70. Name one source of information about materials that serves as a valuable aid in their selection.

   a. Handbooks   **b.** CD-ROM   **c.** Internet   **d.** Textbooks

5-71. The following operation is not an example of a secondary operation in the P/M process.

   a. Mixing   **b.** Coining   **c.** Impregnation

5-72. A P/M part designed as a self-lubricating bearing requires _____ .

   a. extremely close-packed particles   **b.** loose-packed particles

   c. high diffusion bonded particles

5-73 Sintering implies _____ .

   a. high pressure   **b.** high temperature   **c.** both high pressure and high temperature

REFERENCES & RELATED READINGS

ALUMINUM ALLOYS. SELECTION AND APPLICATIONS. Aluminum Association, Washington, DC, 1998.

ALUMINUM ASSOCIATION. *The Story of Aluminum.* New York: AA, 1987.

ALUMINUM INDUSTRY ROADMAP FOR THE AUTOMOTIVE MARKET, May 1999, Aluminum Association, Washington, DC 20006.

ALUMINUM NOW, Vol. 1 No. 1, Summer 1999, Aluminum Association, Washington, DC 20006.

AMERICAN SOCIETY FOR TESTING AND MATERIALS. *1993 Annual Book of ASTM Standards.* Philadelphia: ASTM, 1993.

AMERICAN SOCIETY FOR TESTING AND MATERIALS. *2001 Annual Book of ASTM Standards.* Philadelphia: ASTM, 2001.

ASM INTERNATIONAL. *Alloy Finder* CD-ROM software. Materials Park, OH: ASM International, 1996.

ASM INTERNATIONAL. *ASM Handbook,* Vol. 1, *Properties and Selection: Irons, Steels and High Performance Alloys;* Vol. 2, *Properties and Selection: Nonferrous Alloys and Special Purpose Materials;* Vol. 3, *Alloy Phase Diagrams;* Vol. 4, *Heat Treating.* 1990, 1991, 1993, 1994.

ASM INTERNATIONAL. *Heat Treating Library* CD-ROM software. Materials Park, OH: ASM International, 1999.

ASM INTERNATIONAL. *SteCal,* software for heat treating. Materials Park, OH: ASM International, 1992.

AUTOMOTIVE ALUMINUM RECYCLING. Aluminum Association, Washington, DC 20006.

AVEDESIAN, MICHAEL, and HUGH BAKER. *Magnesium and Magnesium Alloys.* Materials Park, OH: ASM International, 1999.

CALLISTER, WILLIAM D., JR. *Materials Science and Engineering. An Introduction,* 2nd ed. New York: John Wiley & Sons, 1999.

CAPUS, JOSEPH M. "Advances in Powder Metallurgy Processing," *Advanced Materials & Processes,* September 1999, pp. 33–36.

COHEN, ARTHUR. "Copper and Copper Alloys," *ASTM Standardization News,* October 1992, pp. 62–71.

CUBBERLY, WILLIAM H., and RAMON BAKERIJIAN. *Tool and Manufacturing Engineers' Handbook,* desk edition. Dearborn, MI: Society of Manufacturing Engineers, 1989.

DANIELSON, PAUL. "Micro-images of Metals," *Advanced Materials & Processes,* July 1999, p. 35.

DAVIS, J. R. (ed.) *Metals Handbook Desk Edition,* 2nd ed. Materials Park, OH: ASM International, 1998.

DEEVI, SEETHARAMA. "Intermetallics for the Third Millennium," *Advanced Materials & Processes,* September 1999, pp. 44–47.

DURACELL INTERNATIONAL INC. *The Story of Packaged Power.* Bethel, CT: Duracell International Inc., 1995. To request free copy, call 1-800-551-2355.

EMMONS, JOHN R., and K. J. KALLENBORN. *Anodizing and Selecting Aluminum in Nonchromated Solutions.* Marshall Space Flight Center. NASA Tech Briefs, May 1995, pp. 87–88.

FELLERS, WILLIAM O. *Materials Science, Testing and Properties for Technicians.* Englewood Cliffs, NJ: Prentice-Hall, 1990.

FIELD, FRANK R., and JOEL P. CLARK. "Lightweight Cars," *Technology Review,* January 1997, pp. 20–39.

GIANCOLI, DOUGLAS C. *Physics. Principles with Applications,* 4th ed. Englewood Cliffs, NJ: Prentice-Hall, 1995.

GILLESPIE, RONALD J., DONALD R. EATON, DAVID A. HUMPHREYS, and EDWARD A. ROBINSON. *Atoms, Molecules, and Reactions. An Introduction to Chemistry.* Englewood Cliffs, NJ: Prentice-Hall, 1999.

GROOVER, MIKELL P. *Fundamentals of Modern Manufacturing. Materials, Processes, and Systems.* Englewood Cliffs, NJ: Prentice-Hall, 1992.

High-Strength Steel (HSS) Bulletin, 18th ed., Spring 1998. ULSAB*: A Lightweight, Affordable, Structurally Superior Steel Auto Body, Auto Steel Partnership* (A/SP), Southfield, MI 48075-1137.

Inco. *A Quick Refresher on Stainless Steel.* New York: INCO, 1999.

KALPAKJIAN, SEROPE. *Manufacturing Engineering and Technology,* 2nd ed. Reading, MA: Addison-Wesley, 1992.

KALPAKJIAN, SEROPE. *Manufacturing Engineering and Technology,* 3rd ed. Reading, MA: Addison-Wesley, 1995.

KUHN, HOWARD A. *Metalworking Technology Update.* National Center for Excellence in Metalworking Technology, 3rd quarter, 1992.

LAMB, STEPHEN (ed.). *Practical Handbook of Stainless Steels and Nickel Alloys.* ASM International, 1999.

LEFFLER THOMAS J. Manager of Advertising and Projects, Findlay Industries, Crooks Road, Troy, MI 48084. Letter dated June 7, 1995.

Light Truck Structure (LTS) Study. Final Report, Automotive Applications Committee, AISI, AAC-2686 0498 4M HM, Southfield, MI 48075-1137.

Mallardi, Joseph L. *From Teeth to Jet Engines.* Howmet Corporation, 1992.

MANGONON, PAT L. *The Principles of Materials Selection for Engineering Design.* Englewood Cliffs, NJ: Prentice-Hall, 1999.

METAL POWDER INDUSTRIES. *Powder Metallurgy.*

MIELNIK, EDWARD M. *Metalworking Science and Engineering.* New York: McGraw-Hill, 1991.

NEELY, JOHN E., and THOMAS J. BERTONE. *Practical Metallurgy and Materials of Industry.* Upper Saddle River, NJ: Prentice-Hall, 2000.

NOEVER, DAVID A. *Improved Tennis Racquets Have Tapered Strings.* Marshall Space Flight Center. NASA Tech Briefs, March 1995, pp. 88–89.

OSTWALD, PHILLIP F. *Engineering Cost Estimating,* 3rd ed. Englewood Cliffs, NJ: Prentice-Hall, 1992.

PARTNERSHIP FOR EXCELLENCE—AUTO/STEEL PARTNERSHIP ACCOMPLISHMENTS, REVISED JULY 1999, SOUTHFIELD, MI.

Powder Metal Industries Federation, web site Process Diagram, *http://www.mpif.org/chart.html,* November 19, 1999.

PRACTICES FOR THE REPAIR OF AUTOMOTIVE SHEET ALUMINUM. Aluminum Association, Pub. AT4, January 1998.

PRISM. *AMERICAN SOCIETY FOR ENGINEERING EDUCATION,* November 1999.

SAMUELS, L. E. *Metals Engineering: A Technical Guide.* Materials Park, OH: ASM International, 1988.

SCHAFFER, JAMES P., ASHOK SAXENA, STEPHEN D. ANTOLOVICH, THOMAS H. SANDERS, JR., and STEVEN B. WARNER. *The Science and Design of Engineering Materials.* Dubuque, IA: WCB/McGraw-Hill, 1999.

SIMANAITIS, D., and KIM REYNOLDS. "Ecostar Electric." *Road & Track,* August 1994 reprint.

SIMON, JOHN G. "Advanced Materials for the 21st Century," talk to Hampton Roads Chapter of ASM International, Hampton, VA, April 21, 1992.

"THE FEATHERWEIGHT TITAN." *COMPRESSED AIR MAGAZINE,* Vol. 99, No. 1, Jan./Feb. 1994, p. 40.

ULTRALIGHT STEEL AUTO BODY (ULSAB). FINAL REPORT, 2ND ED. AMERICAN IRON & STEEL INSTITUTE (AISI), MAY 1998.

WITTENAUER, JERRY P., and JEFFERY WADSWORTH. "Refractory Metals Forum: Tungsten and Its Alloys," *Advanced Materials and Processes,* September 1992, pp. 28–37.

Periodicals

Advanced Materials and Processes   *Journal of Materials Education*
Compressed Air Magazine   *Machine Design*

EXPERIMENTS & DEMONSTRATIONS IN METALS FROM NATIONAL EDUCATORS' WORKSHOP

EMSET CD-ROM 2nd Edition—Over 300 experiments and demonstrations available in easy to adapt
Acrobat Reader format; department licensed.
ISBN 0-13-648486-7 ???????? 2nd ed. ISBN) Call 1 8000-922-0579

Module

6

Polymeric Materials

After studying this module, you should be able to do the following:

6–1. With the aid of diagrams, explain synthesis of polymers in terms of raw materials, bonding, and by-products, and explain how specific types develop their properties.

6–2. Analyze applications of polymers to determine the effects on structure as a result of branching, degree of crystallinity, molecular weight, and networking on the required properties needed for the finished product.

6–3. Calculate molecular weight for a given compound and explain its significance in polymers.

6–4. Explain the techniques to improve or change polymer properties such as optical, mechanical, chemical, electrical, color, thermal, and density.

6–5. Develop a rationale for emphasizing green design and green manufacturing with regard to practices involving polymers. Describe how Life Cycle Analysis should promote green manufacturing.

6–6. Analyze the properties of various plastics to determine how specific plastics would compete with other plastics, wood, glass and other ceramics, and metals for specific design applications and cost factors. Use tables and handbooks to select plastics, elastomers, and wood to meet specified environments and stresses. Calculate the specific strength of materials to make comparisons.

6–7. Recall the characteristics and applications for general-purpose, engineering, and cellular plastics, and determine selection criteria for plastics.

6–8. Recall the types of natural and manufactured fibers; describe their properties and typical uses.

6–9. Diagram thermoplastic and cross-linked rubber; describe the effects of rubber bonds (lengthening and straightening); recall types, properties, and comparative cost of elastomers; and describe rubber additives and hardness.

6–10. Use Marra's adhesive joint model to explain the principles of bonding, name the conditions and events required for an adhesively bonded joint, and describe main methods by which adhesives work. Explain why interfacial technology is gaining wider study and application.

6–11. Recognize the value and uses for protective coatings, sealants, smart polymers, and other important polymers.

6–12. Analyze wood structure and explain how its physical and mechanical properties develop. Cite typical uses for hardwood, softwood, composite wood, and wood-by-products. Give reasons why wood is being rediscovered as a valuable engineering material.

6–13. State some popular myths and rules regarding municipal solid waste (MSW). Describe recycling of scrap automobiles and compare that to the discussion of the use of plastics in vehicles in terms of "greening" the materials cycle. Explain the savings gained in recycling metals, plastics, paper, and glass. Describe trends in materials technology with regard to improved design, materials selection, and the environment.

6–14. Use EMT Weblinks to explore sites for new developments and evolving concepts related to polymeric materials.

PAUSE & PONDER

The Right Stuff?

As a result of the many advances in materials science and technology, there seems to be no limit to the variety of materials for use in today's products. How do we choose the right "stuff" or material for a given application? Figure 6-1 shows several examples of materials applications. In Figure 6-1a we see containers made of steel, aluminum, glass, and a variety of plastics. On the left is a Coke bottle made of polyethylene terephthalate (PET), next to it is a cleanser bottle made of high-density polyethylene (HDPE), and the syrup bottle on the right is polypropylene (PP). Have you seen these same types of products in glass containers? In the middle of the picture is a "barrier" plastic catsup bottle, and to the right a glass catsup bottle. Is the one made of glass better than the one made of plastic? Shown on the right front are two more barrier containers—the one in front contains yogurt and the box contains fruit juice. How do barrier containers keep milk and juice products from spoiling in the absence of refrigeration? This module will explore those issues and relate them back to the materials cycle—product life-cycle analysis (Figure 1-7).

Figure 6-1b shows a variety of materials used in a kitchen. Note the oak cabinets, maple cutting board, acrylic countertop, polypropylene bowl, and ceramic bowl. What advantages does wood or ceramic offer? For cooking you see a combination glass top cooking surface unit

(a)   (b)

(c)   (d)

Figure 6-1 (a) Containers made of steel, aluminum, glass, and a variety of plastics. Left—PET, HDPE cleanser bottle; right—PP (polypropylene) syrup bottle; middle—composite "barrier" plastic catsup bottles; right front—two more barrier containers for yogurt and fruit juice. (b) Materials used in a kitchen; oak cabinets, maple cutting board, acrylic countertop, polypropylene bowl and ceramic bowl. (c) Acrylic photoelastic model of a jet turbine engine component under load reveals stress concentrations (closely spaced lines). (d) Camera with magnesium case and internal parts mostly of various plastics, glass lenses; film cartridge mostly of plastic including the film.

and to its right a stainless steel unit with metal cooking coils; a glass pot is used with the mostly plastic coffeemaker. Why were steel and glass used? Could plastic be used instead?

Acrylic plastics also serve as photoelastic materials (Figure 6-1c) for use in analyzing stresses. With polarized lighting, an acrylic model of a jet turbine engine component under load reveals stress concentrations (closely spaced lines) as diffracted light breaks into color spectra. Such photoelastic analysis permits engineers to test their design in the form of plastic prototypes prior to committing to costly parts production.

The high-quality camera in Figure 6-1d has a case made of magnesium with most of the internal parts made of various plastics. Finely polished glass lenses compose the optical system. The Advanced Photo System film cartridge is also made mostly of plastic, including the film. Why was the case made of metal and the balance of the camera made mostly of plastics? Contrast the materials of construction for this camera with the disposable camera later in this module (Figure 6-24b). How does consumer appeal influence decisions regarding the designer's choice of materials? What other concerns do designers have in materials selection? Do single-use products help achieve the goals of green manufacturing?

How much thought do you give to the materials used on products you purchase? Do you consider plastics as cheap materials and not very durable? What advantages can you give for selection of plastics over metals or ceramics or the natural polymer wood? What recycling issues come into play with polymers? Can plastic be part of green manufacturing? By now you should realize that achieving the goals of materials and processes which have limited environmental impact and low cost requires the use of complex problem solving. In the websites listed in this module, note how corporations and technical societies tout their efforts at being environmentally friendly.

This module will provide you with many concepts related to polymers, thus enabling you to develop competencies related to their synthesis/processing, structure, properties, and applications. The information and activities will allow you to be a wiser materials consumer and deal with polymers in your career.

 EMT WEBLINKS

Introduction to polymers with more links—*http://www.umr.edu/~wlf.*

MIT's World of Materials (Polymers)—*http://www-dmse.mit.edu/wom/#polymers*

(*EMT WEBLINKS listed throughout the book will link you to Internet sites related to topics being covered. The dynamic nature of the Web brings frequent changes, so some of these sites, while available at the time of writing, may not be up now.)

6.1 POLYMERICS

Section 2.2.3 provided an introduction to polymeric materials (often classified simply as polymers). The emphasis in this module is on the main polymeric materials used in engineering, manufacturing, and construction, and excludes natural polymers such as skin; hide (leather); bone; or natural fibers such as cotton, silk, and jute. Plastics, woods, elastomers, adhesives and many protective coatings are all polymers. While plastics, adhesives, and synthetic rubber are people-made, wood and natural rubber are naturally occurring materials that are used in industry with various alterations to their natural state. Polymers can include many types of industrial materials, as seen in Figure 6-2.

6.2 NATURE OF POLYMERS

The basic makeup of polymers consists of smaller units (**mers**) joined together either naturally or synthetically to produce polymers or macromolecules. This section on structure deals with the structures common to most polymers, and most of the examples cited are plastics, due to their abundant use in industry. The discussion will concentrate on the synthetic methods of

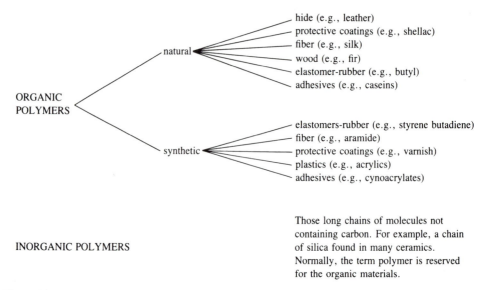

Figure 6-2 tree diagram (organic polymers / inorganic polymers):

ORGANIC POLYMERS

- natural
  - hide (e.g., leather)
  - protective coatings (e.g., shellac)
  - fiber (e.g., silk)
  - wood (e.g., fir)
  - elastomer-rubber (e.g., butyl)
  - adhesives (e.g., caseins)
- synthetic
  - elastomers-rubber (e.g., styrene butadiene)
  - fiber (e.g., aramide)
  - protective coatings (e.g., varnish)
  - plastics (e.g., acrylics)
  - adhesives (e.g., cynoacrylates)

INORGANIC POLYMERS

Those long chains of molecules not containing carbon. For example, a chain of silica found in many ceramics. Normally, the term polymer is reserved for the organic materials.

Figure 6-2   Industrial polymers.

producing polymers known as *polymerization* and will not deal with the biological process by which the natural or biological polymers are grown.

The initial step in polymer production is achieved by petrochemists or petrochemical engineers, engineering technicians and engineering technologists, or others with chemistry in their education and training. This initial step involves breaking down raw materials such as crude oil, coal, limestone, salt, natural gas, and air. With cellulose plastics, the raw material is cotton or wood; however, they are not broken down, as are most other polymer ingredients. Whether it be the cracking of natural gas, distillation of coal or crude oil, or the esterification of cotton or wood, processing the basic organic (carbon base) raw materials yields the element carbon together with hydrogen. Other basic elements are also obtained in the initial stages, including oxygen, nitrogen, silicon, fluorine, chlorine, and sulfur. Basic ingredients of polymers such as ethylene and naphtha are but two of the many products obtained in the cracking and distilling processes. An ***ester*** is an organic compound that is a colorless liquid, insoluble in water, with a very pleasant aroma. Esters find use in perfumes and in the manufacture of flavoring agents for the soft drink industry. When an acid and an alcohol chemically react (***esterification***), the product is an ester, such as the polyester resin. From 100 gallons of crude oil, at a market value of around $25, it is possible through distillation and further processing to produce products such as gasoline, motor oil, propylene, and a large variety of polymers—styrene–butadiene rubber, polyethylene plastics, polypropylene fibers for fabrics, and others—that combine to make products with a market value of over $3500. While oil is the major raw material for plastics, plastics take only a very small percentage (about 1.3%) of the world's petroleum production. Yet shortages in oil cause problems in the plastics and polymers industry, so alternative raw materials continue to receive study.

6.2.1 Polymerization

Scientists observed nature's methods of joining elements into chains and duplicated that natural process to produce macromolecules or polymers. Module 2 covered the principles of bonding, which provide a basic understanding of polymerization. ***Polymerization*** is the linking together of smaller units (monomers) into long chains. The repeating units (mers) of some polymer chains are identical, as in polyethylene (Figure 6-3a), polystyrene, and poly(vinyl chloride); these are labeled ***homo***polymers (Figure 6-3b). *Co*polymers contain two different types of monomers, such as poly(vinyl chloride) mixed with vinyl acetate to produce poly(vinyl acetate) (Figure 6-3c), and *ter*polymers such as ABS (acrylonitrile–butadiene–styrene) contain three types of monomers. ***Isomers*** (Figure 6-3d) are variations in the molecular structure of the same compo-

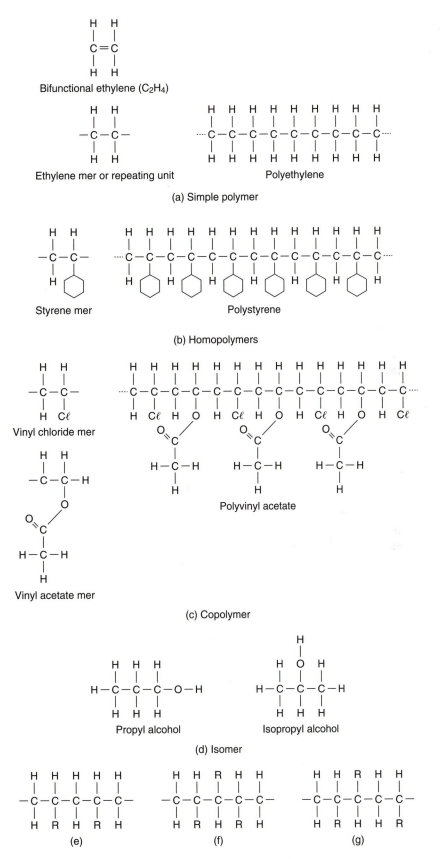

Figure 6-3 Types of polymers: (a) Simple polymer. (b) Homopolymers. (c) Copolymers. (d) Isomers. (e) Isotactic stereoisomer. (f) Syndiotactic stereoisomer. (g) Atactic stereoisomer.

sition. Isomers are found not only in hydrocarbons (HCs) but also in polymeric molecules. One class of isomers is known as ***stereoisomers,*** which in turn is divided into three categories. The first category is called ***isotactic*** stereoisomerism. Using a single carbon (C) chain of atoms with some hydrogen (H) atoms being replaced by another atom or group of atoms, denoted by the symbol **R,** one can demonstrate that the **R** group atoms are situated on the same side of the carbon chain. In a ***syndiotactic*** stereoisometric arrangement, the **R** group atoms are located on opposite sides of the carbon chain in an alternating fashion. The third category is called ***atactic*** stereoisomerism, in which the **R** groups are in a random position. Refer to Figure 6-3e, f, and g. It is important to note that copolymers and terpolymers consist of units from each contributing mer and are not an alloy of mers. If polymerization permitted only the production of homopolymers, the properties of polymers would be severely limited.

   Most polymers are produced by ***unsaturated hydrocarbons,*** which means that they have one or more multiple covalent bonds, such as ethylene:

$$
\begin{array}{ccc}
\text{H} & & \text{H} \\
| & & | \\
\text{C} & = & \text{C} \\
| & & | \\
\text{H} & & \text{H}
\end{array}
$$

or adipic acid, which is used in nylon synthesis and has two hydroxyl monomers:

$$
\text{HO} - \overset{\text{O}}{\underset{\parallel}{\text{C}}} - (\text{CH}_2)_4 - \overset{\text{O}}{\underset{\parallel}{\text{C}}} - \text{OH}
$$

Saturated hydrocarbons have all single bonds. To achieve the polymerization process, monomers must be capable of reacting with at least two neighboring monomers or be ***bifunctional.*** Because of copolymerization, terpolymerization, or other multicomponent polymerizations, a large variety of polymers is available. One example is styrene, a brittle polymer that has limited toughness and poor chemical resistance. But through copolymerization of acrylonitrile with styrene, it is possible to obtain a chemically resistant, more rigid, and stronger plastic. Copolymerization of butadiene with styrene yields an elastomer; and terpolymerization of acrylonitrile, butadiene, and styrene produces ABS plastics that are tough and elastomeric, and have good chemical resistance (see Table 10-9).

   Catalysts begin the polymerization process. A catalyst is a substance that increases the rate of a chemical reaction but is not used up in the reaction. In addition, these chemicals serve a variety of purposes in chemical processing, ranging from serving as molecular sieves to agents for extracting heating oil, gasoline, and ethane from crude oil, to serving as a molecular matchmaker that holds chemicals together for a reaction to occur such as forming long-chain polyethylene polymers from molecules of ethane. ***Zeolites*** are catalysts composed of inorganic grains of alumina (Al_2O_3) and silica (SiO_2) through which molecules of crude oil are strained to gain oil by-products. Activated carbon, platinum, and nickel are also used to catalyze synthetics. ***Aerogels*** are open-cell, often silica, foams that have ultrafine cell or pore size less than 50 nm, high surface areas (500 to 1000 m^2/g), and density approaching that of air. They are the lightest, most transparent human-made solids. Known as the world's best insulators, they do not conduct heat. With a density of 15 mg/cm^3 and a melting point of 1552°F, the properties of these materials approach those of glass. Aerogels of organic and inorganic systems of fibrous chains provide catalyst support. To improve fuel cell efficiency materials scientists tailor silica aerogels by adding minuscule particles of elements, such as carbon and platinum, for a degree of electrical conductivity and catalytic reactions. The development of scanning tunneling microscopes (STMs) has allowed nanoscale examination of catalysts. Thereby, new designer catalysts can catalyze products more efficiently without creating polluting by-products.

   The main polymerization processes are ***addition (chain reaction) polymerization*** and ***condensation (step reaction) polymerization.*** The addition process is the simpler of the two. By use of heat and pressure in an ***autoclave*** or reactor, double bonds of unsaturated monomers (Figure 6-4) break loose and then link up into a chain. These addition reactions (in addition

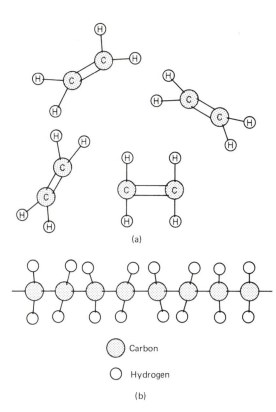

Figure 6-4   (a) Monomer of ethylene (C_2H_4). Carbon atoms with four valence electrons have four arms (shared electrons) for covalent bonding. Two shared electrons hold another carbon atom in a double bond. Two other shared electrons, each-covalently bonded, hold one shared electron of hydrogen, which has only one valence electron for single bonds. Double bonds are not stable and form ***reactive sites.*** (b) Polyethylene, chain of ethylene monomers. During polymerization, heat and pressure break the hold of one shared electron on each of the double-bonded (reactive sites) carbon atoms, thereby leaving each carbon atom of the monomers free to grasp another carbon atom (covalent bond) from other ethylene monomers, which form into a chain of thousands of ethylene monomers (polyethylene). The single bonds satisfy the carbon bond arrangement and produce the most stable saturated polymers because they have no reactive sites.

⬤ Carbon

◯ Hydrogen

(b)

polymerization for unsaturated HCs) are atoms or groups of atoms that attach themselves to the carbon atoms at the sites of multiple bonds. No products other than the polymer are formed. Saturated hydrocarbons undergo substitutional reactions in which hydrogen atoms are replaced by other atoms or groups of atoms. The products of addition polymerization, also referred to as chain reaction polymerization, include polypropylene (PP), polyethylene (PE), poly(vinyl acetate) (PVA), poly(vinyl chloride) (PVC), acrylonitrile–butadiene–styrene (ABS), and poly-tetrafluoroethylene (PTFE).* *Chain polymers* generally fit into the ***thermoplastic*** (soften when heated) group. Most plastics that cannot be resoftened when heated (***thermosetting polymers***) come from condensation polymerization, also known as ***step reaction polymerization.*** This group gets its name from the by-product (***condensate***) of the polymerization, which is often water, but it may be a gas. Phenolic (PF), polyester (PET), silicon (SI), and urethanes (PUR) are typical thermosets from the step reaction synthesis, while nylon (PA) and polycarbonates (PC)* are thermoplastic resins synthesized through step polymerization. Figure 6-5a diagrams the addition process for the production of some typical polymers, and Figure 6-5b shows a modern polymer processing plant with its complex equipment used to produce resins. Figure 6-6 shows the flow for production of nylon through condensation polymerization.

The term ***resin*** covers both solid and semisolid organic polymers. Resins are often considered as the uncompounded ingredients or monomers that are mixed but not yet polymerized. For example, thermosetting resins or pellets are molded into thermosetting plastic or elastomeric parts. Sometimes the term *resin* is used synonymously with plastics (e.g., acetal resins of acetal plastics, or thermoplastic resin instead of thermoplastic plastics).

6.2.1.1 Polymer chain lengths.

Considerable variation in polymer chains results during polymerization. The ***chain lengths*** play an important role in processing polymers and their final properties. While we cannot see the length of the polymers, the molecular weight provides a good indication. Refer to Module 2 for an introduction to molecular structure and bonding.

Molecular weight provides a weight for a single unit or molecule in a compound. To determine the ***molecular weight of the compound,*** it is necessary to multiply the weight of a molecule

* ASTM abbreviations.

(b)

Figure 6-5   (a) Typical polymers produced through addition (chain) polymerization. (b) A portion of a modern polymer processing plant showing some of the complex apparatus and piping neccessary to produce resins.

by the number of molecules or monomers. We said polyethylene typically has about 2000 monomers, so we multiply 2000 times the molecular weight of one monomer to obtain 56,000:

Illustrative Problem

Molecule or monomer	Molecular weight		Number of monomers		Molecular weight of compound
C_2H_4	28	×	2000	=	56,000

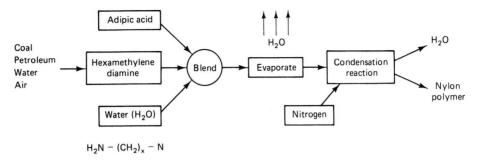

Figure 6-6   Graphical illustration of condensation (step) polymerization of a polyamide (nylon).

Involved chemical analysis is required to determine polymer chain lengths, and such analysis reveals that chains are not uniform in length. Consequently, the value given for a particular resin would be either its *average molecular weight* or *degree of polymerization (DP),* which indicates the number of repeating mers in the material. The degree of polymerization is the ratio of the molecular weight of the molecule to its mer's weight. Long-chained polymers have a greater molecular weight, which causes more entanglement of the chains and thereby increases *viscosity* (resistance to flow). The degree of viscosity affects the processing of the polymer, so manufacturers must take it into account.

EMT WEBLINKS

Addition polymerization—*http://www-dmse.mit.edu/wom/additionzation.html*
Condensation polymerization—*http://www-dmse.mit.edu/wom/Polycondensation.html*

6.2.2 Crystalline Structures

In addition, polymerized neighboring chains are held together by weak secondary (intermolecular) bonds known as van der Waals forces, as described in Module 2. As the chains grow during polymerization, they intermingle in a random pattern that will lead to an *amorphous* (disorderly) *structure.* Bear in mind that the growth of chains is three-dimensional and not simply flat, as normally diagrammed. Visualize a bowl of cooked spaghetti with its uneven strands entangled; that gives a good idea of the arrangement of an amorphous polymer structure.

Actually, polymers are semicrystalline in varying degrees, with the amorphous structured polymers having only slight regularity, while other polymers may have a high degree of crystallinity (Figure 6-7). Metals achieve crystallinity due to the uniform nature of the unit cells of their space lattices. Crystallinity in polymers can alter strength and toughness. The twisting, coiling, and branching of polymer chains causes disorder, and amorphous regions develop in polymers as they solidify. Chain configuration and rate of cooling affect the orderliness. Slow cooling allows chains to move into alignment. While metals are normally fully crystalline, the degree of crystallinity in polymers can range from mostly amorphous to mostly crystalline. Mostly amorphous polymers will be less dense than mostly crystalline polymers because the chains pack closely together.

Molecular motion of short polymer chain segments and polymer branches cause amorphous polymers to be stiff, hard, and brittle at room temperatures. Increases in temperature cause greater molecular motion through thermal mixing of the atoms and molecules. This causes an increase in the volume of the material. The thermal mixing from higher temperatures also increases the spacing between molecular segments and permits increased flow of the materials. A discussion of glass transition later in this module deals further with the concept of molecular motion and flow in polymers.

The degree of crystallinity is determined by structural regularity, compactness, and amount of flexibility, which allows packing of chains. Reduction of random chain lengths provides regularity. Stronger secondary forces allow greater compacting. *Liquid-crystal polymers* (LCPs) develop highly oriented (rodlike) molecules for increased directional strength (Figure 6-7d) and LCP

(a)   (b)   (c)

(d)

(e)

Figure 6-7 Degrees of crystalline structure of polymers. (a) Mostly amorphous (b) Semicrystalline (crystallites encircled) (c) Mostly crystalline (d) Liquid crystal polymer (e) Ultra-thin section of transmitted, cross-polarized, light circles, polymer matrix composite—carbon fiber (cut at 90° and seen as black circles) in a matrix of PEEK thermoplastic with liquid-crystalline shperoids at 2000x/100x obj. (Luther Gammon, Boeing Materials Technology) spheroids (Figure 6-7e). Other chemical factors affect crystallinity, including configuration and tacticity (see Figure 6-3e–g). See the references at the end of the module for more information.

Polyethylene serves as a good example of a polymer capable of a high degree of crystallinity because (1) the linear structure of the chains is conducive to packing, and (2) its molecular pattern is flexible, which provides easy packing even though it has weaker secondary bonding. Secondary bonds (van der Waals forces) normally promote crystallinity.

Polyethylene (PE) has the potential for a wide range of properties because of technology's ability to control its molecular weight and crystallinity. Low-density polyethylene (LDPE) has a molecular weight below 10,000, while ultrahigh-density polyethylene (UHMWPE) has molecular weights much above 1.0×10^6 (see Figure 6-8a).

Crystallinity of polymers is also achieved through processes such as extrusion and drawing. *Oriented polymer fibers* are obtained by drawing an amorphous polymer through a die, which improves their strengths or *orientation* in the direction of drawing (see Figure 6-8b). This concept can be illustrated by stretching a polyethylene sandwich bag. The method of manufacturing orients the polymers in one direction. By pulling on the bag in perpendicular directions it is possible to notice the greater resistance from the oriented direction; also, as the bag is stretched, it becomes stronger due to further orientation.

The cooling rate and processing during cooling affect the crystal patterns. *Spherulites* form, in which nuclei generate a spherical pattern that grows until several spherulites melt at boundaries, as a result of supercooling. When large spherulites are allowed to grow, weakness results. Heat treatment of polymers can change their crystal structure; for example, the

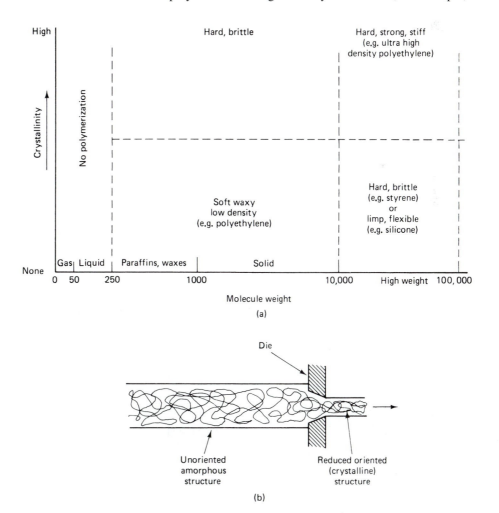

Figure 6-8   Polymers and crystallinity. (a) Properties: relation to molecular weight and crystallinity. (b) Effect of pulling an amorphous polymer through a die.

annealing processes reheat the polymer for a specified time to permit crystal thickening. In Figure 6-9, electron micrographs show polyethylene in the amorphous state, then as a single linear crystal, and then as an annealed crystal.

6.2.3 Branched and Network Structures

As seen in Figure 6-10c, some polymers do not simply grow linear chains (Figure 6-10a and b); rather, the chains develop branches much like those of a tree. As a result, the branched polymers are not as symmetrical and consequently will not achieve the degree of crystallinity obtained in the more regular linear polymers. Numerous techniques are employed by polymer chemists and materials technologists to vary both the length of linear chains and the degree and uniformity of branching.

Networks of polymers develop, and primary bonds form between chains. As contrasted to the weaker secondary forces in linear polymers, ***cross-linking*** is the development of covalent bonds between chains, as seen in Figure 6-10d and e. The cross-linked network polymers do not soften when heated, as do the ***thermoplastic*** polymers that are held together by van der Waals forces (Figure 6-10f). Heating of thermoplastics causes these weaker bonds to lose their hold, and thus movement of the mers and softening of the polymer. Thermosetting polymers are like hard-boiled eggs. Once cooked, the egg cannot be softened to flow into another shape. But a thermoplastic material such as a candle can continually be reheated and molded into new shapes. Normally, polyethylene is a thermoplastic material, but through electron irradiation cross-links are developed. Epoxy can exist as a thermoplastic polymer, or through the connections of end groups, it will become a thermosetting polymer. Sulfur is used in rubber to form numerous cross-links in the ***vulcanization*** process. The more cross-links present, the harder the rubber is. A reference to cross-linking is the ***netting index,*** which designates the number of cross-links per 100 linear bonds.

The number of double bonds in a polymer provides more sites for reaction than those with fewer double bonds. Natural rubber has hundreds more double bonds than butyl rubber, which has more tightly placed, single-bonded molecules. A hard rubber has a netting index of 10 to 20, while a hard thermosetting plastic such as phenolic has an index of about 50.

EMT WEBLINKS

Introduction to polymers—*http://www.umr.edu/~wlf*

Polymers and liquid crystals—*http://abalone.phys.cwru.edu/tutorial/enhanced/main.htm*

Introduction to plastics primer—*http://www.polymers.com/dotcom/polycon/primer.html*

6.3 GREEN DESIGN AND GREEN MANUFACTURING

Although polymers offer a wide range of benefits to society, they also pose potential burdens for the environment. These burdens present challenges at all stages of the materials cycle be-

(a)   (b)   (c)

Figure 6-9 Crystallinity in polymers. (a) Electron micrograph of amorphous polyethylene. (b) Single crystal of polyethylene. (c) Single crystal of linear polyethylene annealed at 120°C for 30 minutes. (Dr. Phillip H. Geil, University of Illinois)

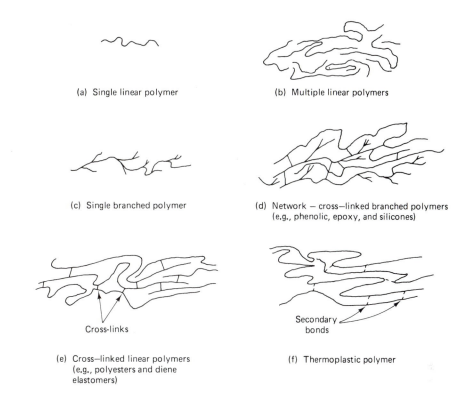

(a) Single linear polymer

(b) Multiple linear polymers

(c) Single branched polymer

(d) Network — cross—linked branched polymers
(e.g., phenolic, epoxy, and silicones)

Cross-links

Secondary
bonds

(e) Cross—linked linear polymers
(e.g., polyesters and diene
elastomers)

(f) Thermoplastic polymer

Figure 6-10   Various polymer structures.

ginning with the synthesis/processes through disposal/recycling. At the end of this module, we provide a perspective on recycling plastics. To promote *green design* (design aimed at reducing burdens on the environment), the recycling stage can be the beginning of the materials cycle for certain polymers. When we introduced the materials cycle we also tied it to life-cycle analysis, which also may be referred to as *life-cycle assessment (LCA)* which involves life-cycle inventory (LCI). LCA seeks to quantify environmental burdens of a product over its entire life cycle. The United States Automotive Materials Partnership (USAMP) provides the following objectives and intended results of LCI.

Objectives of LCI—Life-Cycle Inventory*

1. Assess the cradle-to-grave environmental performances of vehicles
2. Conduct an LCI for an entire vehicle, in this case, the generic family car (often referred to as PNGV's Super Car)
3. Quantify energy and material flows over a product's life cycle

Use of LCI Results*

1. Serve as a benchmark for future vehicles (e.g., PNGVs)
2. Communicate with public and private sector
3. Not intended for materials comparisons
4. Stress that cost and performance must also be considered

*Adapted from Susan G. Yester, "Life Cycle Inventory," NEW: Update 99, November 1, 1999, Auburn Hills, MI.

To illustrate how LCI works for a product with many environmental impacts, Figure 6-11 depicts a materials cycle for an automobile part as developed by the Automotive Composites Alliance, another group working with USAMP and PNGV. This diagram helps to see the points for application of LCI. Emissions from synthetic polymers occur during the stages of materials acquisition & processing and component manufacture & assembly. Those polymers and hydrocarbon fuels used for powering vehicles, with the lubricating, hydraulic, and cooling fluids, have potential to pollute water and the atmosphere; contribute to global warming; and affect our ozone layer. More dangers occur at the recovery, reuse & disposal stage as discussed later. During normal lifetime use & maintenance, synthetic polymers provide many advantages that tend to offset the harmful effects. For example, lightweight, noncorroding plastics and plastic-based composites that provide good design flexibility can yield more fuel-efficient vehicles. Table 6-1 shows data categories used for LCI including types of energy, water consumption, air emissions, water effluents, solid wastes, and raw materials involved in LCI for vehicles. During a life cycle of the generic family sedan it (1) consumes 961 GJ of energy, (2) consumes 46,000 pounds of hydrocarbon and (3) generates 130,000 pounds of carbon dioxide. LCI allows automakers and suppliers to monitor environmental impacts and set goals for reducing these burdens.

Figure 6-11b, shows more detail of material flow for the Generic Vehicle Product System, excluding the Energy Flow shown in 6-11a. This figure follows the description outlined for the Materials Cycle Figure 1-7. Beginning on the left, see how Primary Materials (steel, aluminum and plastics) along with other materials (glass, rubber, paint, etc) enter the manufacturing process. Parts are assembled into sub-components. The sub-components join with glass parts, tires, paints, motor oil, etc. for the component and vehicle assembly within the Vehicle System Boundary. After assembly, operation of the vehicle carries the product through its useful life. After the car is taken out of service the last stage of the Materials Cycle begins and couples with the first stage of the Cycle; i.e. Disposal/Recycling—Extracting Raw Materials. In this case the raw materials of Secondary Steel, Secondary Aluminum and other recycled materials rather than Primary Materials. A, B, C and D labels help to connect Primary Materials with Secondary Materials for Primary Manufacturing 1. Note how scrap moves in the cycle and where materials disposal occurs.

Cost will always influence materials selection. Even though a lighter, more fuel-efficient car will save the owner money over the life of the vehicle, purchase price drives the decision. The following list of broad costs for raw materials shows why steel and iron have such a strong hold on the average car for body materials, even though plastics and other polymer-based materials are growing in usage (to be discussed in the next section). With the automakers collaborating through such efforts as PNGV and USAMP using LCA coupled with legislation and consumer interest, green manufacturing will steadily improve.

Raw Materials by Dollar per Pound*

Steel	$0.40	Glass Fiber Composites	$1.20
Iron	$0.50	Magnesium	$1.50
Aluminum	$0.90	Carbon Fiber Composites	$6.00
Plastics	$1.00	Titanium	$10.00

Green architecture connects actions to consequences. It looks downstream to sources of materials and upstream to how the building will impact the environment. Buildings can have life spans up to hundreds of years. Green architecture, defined in terms of a building,

> . . . serves the needs of the people who inhabit it. It supports and nurtures their health, satisfaction, productivity, and spirit. It requires the careful application of the acknowledged strategies of sustainable architecture—non-toxic construction, the use of durable, natural, resource efficient materials, reliance on the sun for daylighting, thermal and electric power, and recycling of wastes into nutrients. An elegant architectural integration of these strategies produces a building which honors the aspirations of those who use it and engages the natural world. . . . aim for buildings which im-

*Adapted from John Fillion, "Trends for Automobile Materials," talk presented to NEW: Update 98, November 4, 1998, Brookhaven National Laboratory, Long Island, NY.

Major Lifecycle Stages for
Automotive Part

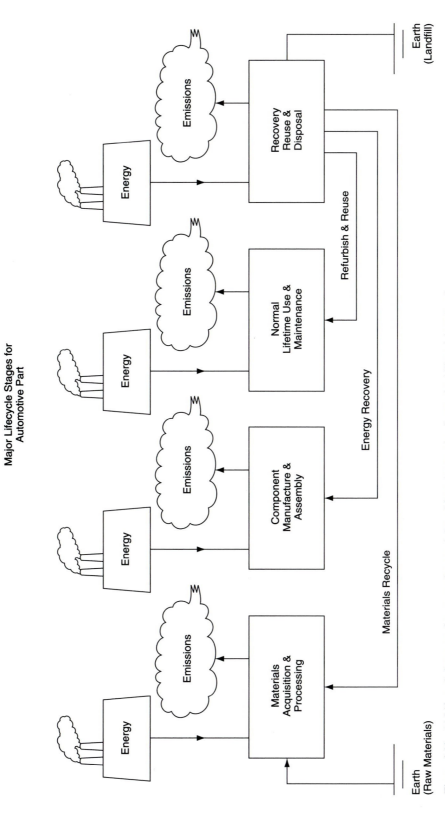

Figure 6-11 (a) Life cycle inventory for automotive part (adapted from Automotive Composite Alliance) (*continued*)

Figure 6-11 (b) Material flow for the Family Sedan (Generic Vehicle Product System).

328

TABLE 6-1   DATA CATEGORIES FOR LCI*

Energy	Air Emissions	Water Effluents
Fossil and Nonfossil	CO_2, CO, SO_x	Dissolved Solids
Process	NO_X, CH_4, HCl	Suspended Solids
Transportation	HF, NMHC, Pb	Heavy Metals
Feedstock	Dust and Particles	Oils and Greases
		PO_4 and NH_3
Solid Wastes	Water Consumption	
Sanitary and Municipal	Ground	
Total Solid Waste	Surface	
Raw Materials		

*Adapted from Susan G. Yester, "Life Cycle Inventory," NEW: Update 99, November 1, 1999, Auburn Hills, MI.

prove the quality of the air and water, produce surplus power and food, and convert waste into nutrients and useful products. (*http://www.coldhamarch.com/green/what.html,* January 9, 2000)

Also regarded as ***green*** or ***sustainable design,*** green architecture faces issues of change as explained in Module 1. Applying new design and materials, construction techniques depart from the norm. Using principles of concurrent engineering, design must bring together clients, architects, suppliers, builders, and others to integrate new building technologies, to meet the clients' needs while having a positive impact on people, and to help sustain a renewable, wholesome environment. Keeping in mind the challenges of securing the ideal material, the ideal green building should:

- impact positively on the earth,
- provide a wholesome environment for the occupants,
- utilize sunlight and produce energy,
- collect and cleanse water, and
- produce food and turn wastes into nutrients.

In attempts at building technology that aim to minimize energy consumption and use synthetic polymers, sometimes the effort results in "sick buildings" that do not provide a wholesome habitat. Carpets, furniture, paints, and other finishes may pollute internal environments. Volatile organic compounds (VOCs) released from such synthetic materials as polyvinyl chloride plastics, dyes, and adhesives combine with particulates such as crystalline silicates in paint and spackle, pollen, and tobacco smoke to make people sick.

In the next section you will learn about the living environments and see the potential offered by plastics in building technology. See if you can detect efforts for green design with this concept house.

EMT WEBLINKS

Green Design Network—*http://Greendesign.net/*

Green architecture—*http://www.coldhamarch.com/green/*

Greening the Government—EPA's Environmentally Preferable Purchasing Program— *http://www.epa.gov/opptintr/epp*

Green design materials—*http://data.oikos.com/products/*

Environment Show—*http://www.enn.com/envshow/*

SELF-ASSESSMENT

6-1. What is the name for a chemical that serves as a molecular matchmaker, holding chemicals together for a reaction? Provide examples of these chemicals and some of the specific reactions in which they are used.

6-2.   Define the term *polymer*. List the major groups of polymers, and cite examples of materials in each group.

6-3.   Discuss the sources of raw materials used in the production of synthetic polymers. Explain how the depletion of fossil fuels can affect synthetic polymers. What alternatives can meet the needs served by present-day synthetic polymers?

6-4.   Explain the necessity of unsaturated hydrocarbons and bifunctional monomers for polymerization. Describe the two main polymerization processes and list the typical products of each.

6-5.   What type of polymer develops highly oriented (rodlike) molecules for increased directional strength?

6-6.   Discuss the effect of crystallinity on polymers and explain methods of achieving crystallinity.

6-7.   Calculate the molecular weight for poly (vinyl chloride), polystyrene, and nylon 6.

6-8.   Calculate the average molecular weight of a polyethylene compound with 10,000 monomers. Would this be classified as a low-, medium-, or high-molecular-weight polymer?

6-9.   What mechanism produces network polymers? Do thermoplastic polymers form networks?

6-10.   Does condensation polymerization usually produce thermosetting or thermoplastic polymers? What are the exceptions? What terms describe cross-linking in rubber? What chemical produces this cross-linking?

6-11.   What provides for greater cross-linking in polymers? What is the nature of a polymer with a netting index of 50?

6-12.   Justify the extra effort and cost for green manufacturing practices in terms of LCA and burdens on the environment.

6-13.   Explain the benefits of LCI on estimating the impacts of a product on the environment, include some categories of data for evaluating.

OBJECTIVE QUESTIONS

6-1. Chemical that serves as "molecular matchmaker" necessary to begin polymerization reaction.

    a. Monomer   **b.** Isotope   **c.** Catalyst   **d.** Filler

6-2. Polymer with one or more multiple covalent bonds, such as ethylene.

    a. Unsaturated hydrocarbons   **b.** Saturated hydrocarbons

    c. Aerogels   **d.** Zeolites

6-3. Adding sulfur to rubber to strengthen it.

    a. Condensation   **b.** Vulcanizing   **c.** Wetting   **d.** Addition

6-4. Multiple covalent bonds that provide active sites for cross-linking.

    a. Isotopes   **b.** Single bonds   **c.** Double bonds   **d.** Ablation

6-5. Highly oriented (rodlike) molecules that increase directional strength in polymers.

    a. Liquid-crystal polymers   **b.** Linear polymers

    c. Amorphous polymers   **d.** Branched polymers

6-6. What does the *netting index* indicate about a polymer?

    a. Degree of hardness   **b.** Ability to weave fiber

    c. Degree of branching   **d.** Level of adhesion

6-7. Allowing large crystals to grow in polymers will usually do what?

    a. Strengthen   **b.** Weaken

6-8. What effect will drawing polymers through dies yield?

    a. Increases amorphous structure   **b.** Increases crystallinity

    c. Develops cross-link   **d.** Develops branches

6-9. A method to quantify environmental burdens of a product over its entire life cycle.

    a. LCA   **b.** green design   **c.** algorithm   **d.** recycling

6-10. Plastics compete best with steel on which factor?

    a. recycling infrastructure   **b.** raw materials availability

    c. cost per pound   **d.** design flexibility

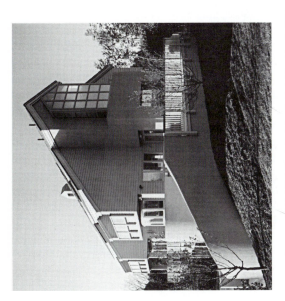

(a) Smart house. This prototype house uses smart plastics, intelligent structures, and smart electronic systems to provide the homeowner with energy-efficient, high-tech comfort and convenience. Teams of specialists in computer technology, materials engineering, industrial design, and architecture want to become major players in the home construction industry by utilizing advances in technologies. (General Electric Plastics) (b) Cutaway model of "living environments" showing unique features. On the right is interior Technopolymer Structures (TPS) of high-strength composites (glass fiber mat and Azdel resin). TPS is used for (1) the distribution floor made of Azdel skin and low-density foam and (2) radiant wall panels. Azdel is a glass-fiber-reinforced thermoplastic resin. The hollow floor encases plumbing and electrical components, as well as storage bins. (3) Total Environmental Control (TEC) are modules made of Noryl (modified polyphenylene oxide resin that has excellent mechanical and electrical properties) and Alloy (composite of amorphous polymers or semicrystalline alloys and blends, reinforced with glass fiber) for heating, cooling, hot water, air purification, and energy recovery; the number of modules depends on house size and climate. (4) Heating and air-conditioning ducts made of TEC Altum (polyetherimide resin, structural foam, sheets, and film). Altum is exceptionally strong and flame and heat resistant. If burned, it produces a low amount of smoke; it comes in clear or opaque. (5) Floor panels of Azdel and Alloy floor joist. (6) Doors of Noryl low-density foam. (7) Plastic extrusions form lighting tracks that fit into wall panels. Left cutaway of exterior; (8) precast concrete foundation wall panels with integral low-density plastic foam wall inside; plastic forms for casting concrete can be left in place. (9) Vinyl siding. (10) Windows Lexan (polycarbonate resin) with liquid crystals (LCD) are energy-efficient and photochromic (change from transparent to translucent). (11) Roof of low-density plastic foam can look like cedar shakes or ceramic tile, and can also make use of photovoltaic solar panels to generate electricity. Silicone rubber sealing creates an air and moisture barrier. Modular components such as roof systems could be fabricated in a factory to assemble solar panels, skylights, gutters, and roofing, which would then be transported to the building site and placed over roof trusses. (General Electric Plastics)

6.4 PLASTICS

Shelter is one of our basic needs, and for most people, their homes will account for the largest outlay of money over their lifetime, whether they buy or rent. Affordable shelter, including homes, is a major goal of all societies. So it is wise to give careful consideration to your home, especially if you plan to purchase a house. The materials selected for a house are very important in terms of their initial cost; the materials also will affect the ease and long-term cost to live in and maintain a house.

House construction technology has evolved over the history of civilization. Currently, the average house in the United States is a custom home, framed on its site (stick construction) with wood, using plastic or masonry siding, glass windows, and wood or metal doors. House designers and materials manufacturers believe that it is possible to make tremendous improvements in house building to achieve more cost-efficient and enjoyable homes. The "living environments" *smart house* shown on the facing page reflects the desire of teams of specialists in building technology, computer technology, materials engineering, industrial design, and architecture to utilize advances in technologies for improvements in the home construction industry.

Prototype homes are designed to provide the homeowner with energy-efficient, high-tech comfort and convenience. This "living environments" concept house was built to convince the house construction industry that plastics can have a beneficial impact on houses in a manner similar to that for automobiles. The concept house, built by General Electric (GE) and other business partners, serves as a research tool and a showplace for design, materials engineering, manufacturing and construction processes, and home systems (e.g., heating, air conditioning, security, entertainment, and appliances).

Approximately 30% of the "living environments" concept house is plastic, including the roof, windows, siding, plumbing system, foundation, and electrical and mechanical systems. Plastics are used in harmony with the traditional materials (wood, metal, and ceramics). Many of the plastics are engineering thermoplastics. Because plastics generally do not biodegrade and their processing may lead to pollution, they may present environmental concerns. The GE project also focuses on demonstrating that plastics can be environmentally friendly. Recycled engineered plastics should keep many plastics out of municipal landfills. For instance, it is possible to reuse discarded plastic packaging (bottles or boxes) to produce automotive parts. Once the automotive parts are discarded, they can be reclaimed for materials in house construction.

Another potential of the concept house comes from building factory-manufactured houses rather than traditional site-built houses. Factory-built houses can be more affordable, of higher quality, and less expensive to run and maintain, and they should have features that make for more enjoyable living. Among the unique features in the house shown in the cutaway (b) are the modular wall and flooring systems. Walls include foam-polymer sections that radiate heat into the room and serve in a system for heating, air conditioning, humidity control, and air filtration, with the capacity to recover wasted heat. Hollow composite plastic sandwich floors provide easy access, and they encase plumbing and electrical components and provide storage bins. Roofing is done with shingles of lightweight fiber-reinforced plastic that eliminate underlayment.

Plastics provide low emissivity and therefore conserve energy that might otherwise escape through roofs and walls. Instead of ceramic tiles, a modified polystyrene foam serves as floor tiling that is held in place with Velcro. Other components, such as doors, drawers, and cabinets, snap together, which allows for easy assembly and disassembly. Acrylic windows can be changed from clear to white because of embedded liquid crystals. This affords privacy and blocks the sun to avoid bleaching of colors. Replacing traditional baseboards are hollow raceways that carry all wiring: telephone, coaxial cable, and low-voltage electrical.

Other green architectural aspects of "living environments" include special features in the kitchen: a breakfast counter's polished surface with the look of marble and engineered from recycled computer housings and plastic automotive parts, and a built-in bottle and food package recycling unit with a bar code reader to segregate engineering plastics from general-purpose or commodity plastics. Engineering plastics can be ground and reused whereas commodity plastics can be incinerated to reclaim their Btu value.

 EMT WEBLINK

> Tour GE's living environments house with movies—*http://www.crd.ge.com/esl/cgsp/ projects/le/endtour.html*

The Society of Plastics Industry (SPI) defines **plastics** as

> any one of a large and varied group of materials consisting wholly or in part of combinations of carbon with oxygen, hydrogen, nitrogen, and other organic or inorganic elements which while solid in its finished state, at some stage in manufacture is made liquid, and thus capable of being formed into various shapes, most usually through the application, either singly or together, of heat and pressure.

This definition is very broad and rather awkward, but it points out that plastics come from such wide ranges of raw materials and processes, have so many varied properties, and take such diverse forms that they almost defy definition. Even natural and synthetic rubber fit this description.

6.4.1 Classification of Plastics

Several methods are used in the classification of plastics. The older and perhaps obsolete system of thermosetting plastics and thermoplastic plastics groups is still commonly used; however, many plastics, and someday possibly all plastics, can fit into either grouping. Using a grouping of thermosetting or thermoplastic is useful as a **processing classification** because it indicates what types of processes a certain plastic can undergo. To illustrate, a thermoplastic sheet can be used as a finished product or heated to be reshaped. Once a thermoset part has been molded, it can no longer be reheated for further shaping. Classifying plastics as either **linear** or **cross-linked network** provides a more descriptive system that is useful to the designer in determining general properties. For example, linear plastics can continually be remolded and usually have low heat resistance, whereas the cross-linked network plastics have greater heat and chemical resistance. Another grouping system involves the *nature of the material:* rigid, flexible, or elastic. Closely connected with the nature of the material classification are the *uses:* general purpose or engineering. **General purpose** would include the bulk of plastics that we encounter daily, such as cellulosic, acrylics, and vinyls. **Engineering plastics** are those that substitute for traditional engineering materials such as steel and wood. Some engineering plastics are considered **structural plastics** capable of bearing supporting loads. Much overlap exists for plastics in these classification systems.

    Cellular (foam) plastics are listed by the SPI as a plastics group. Cellular plastics are those that have been foamed or had gas and/or air entrapped in the polymer resins to reduce the density of the finished product (Figure 6-12a). Foaming agents include gases (nitrogen, carbon dioxide, or air) and liquids (chlorinated aliphatic hydrocarbons, alcohols, or ethers). Hollow spheres of glass or resin are also used to create voids. Whipping action introduces air, carbon dioxide is generated from chemical reactions in the resin, and volatile liquids vaporize into gas through **endothermic** reactions (absorbed heat when resin bonds break). Water is produced through **exothermic** chemical reactions (heat given off) and it volatilizes to produce cells. Cellular polymers may be open-cell or closed-cell polymers. Closed-cells are discrete, separate cells whereas open-cells are interconnected with openings between cells. Aerogels, which include both polymer and ceramic open-celled structures, have nanoscale matrices composed of interconnecting colloidal-like fibrous chains with diameters of about 10 nm. They serve as refrigerator insulators and solar windows, with the potential for many similar uses due to their density (a few times that of air) and low thermal conductivity (1% of full-density glass). Practically all thermosetting and thermoplastic resins can be foamed.

(a)

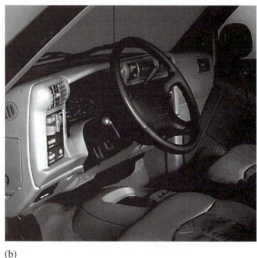

(b)

Figure 6-12   High-density, low-density, and energy-absorbing cellular foam polymer. (a) Photomicrograph enlargement of foamed polymer reveals cells created by gas. (b) Cellular plastics and elastomers are found on many
vehicles, as seen here for the seats and steering wheel. The heat-resistant, high-impact, dimensionally stable thermoplastic polycarbonate plastics are also seen on the instrument panels for GM pickup trucks and sports utility vehicles. The ML6143H grade of Lexan™ provides ease of processing plus the use of NVOCs (nonvolatile organic
compounds) and water-based painting. (GE Plastics)

Advantages offered by cellular plastics include weight reduction, increased bulk, improved thermal insulation values, greater shock absorption, increased strength-to-weight ratio,
and ability to duplicate wood in terms of texture, density, and feel. Cost savings come with
foaming. The addition of glass microspheres or bubbles to *sheet molding compounds (SMC)**
can reduce weight while improving impact strength because the hollow spheres dissipate energy through shock-absorbing action (dampening). The tiny bubbles, which are from 10 to 200
μm in diameter, also improve mechanical and chemical properties while facilitating processing. Cellular plastics are available in many densities (1.6 to 96 kg/m³) as rigid, semirigid, or
flexible compounds, and with or without color. The auto industry has accepted many cellular
plastics for their numerous favorable properties, especially their weight-saving value (Figure
6-12b). The polyurethane (PUR)* foams produced through reaction injection molding (RIM)*
composited with reinforcing fibers (FR)* serve as hoods, seat frames, or doors.

Cellular plastics are widespread and include polystyrene foams (Styrofoam™) for building construction, packaging, and appliance insulation; cellular polyethylene for electrical wire
insulation; polyamide foams with high heat resistance for aircraft; epoxy foams for flotation
devices; cellular silicone for electrical and electronic encapsulation, chemical-resistant fillers,
and structural insulation; flexible and rigid poly(vinyl chloride) (PVC)* for cushioning, upholstery, carpet backing, and toys; and foamed acrylics for decorative materials. Syntactic
glass-filled foams of polyester are used in sports equipment for composite tennis rackets,
bowling balls, floating golf balls, helmets, skateboards, and skis.

Manufactured fibers are polymer strands classified as either regenerated cellulosics or
noncellulosics (i.e., synthetics). *Natural fibers* such as jute, hemp, wool, and cotton have been
used since early times and still are important commercial materials. Fibers are woven into fabrics and used for line, string, and rope (e.g., nylon-monofilament fishing line and tennis string
or polypropylene rope). In Module 8, we will further define fibers and related terms and develop their importance in composites.

*ASTM abbreviation.

(a)

(b)

Figure 6-13   (a) Expanded PTFE Gore-Tex® fiber. Photomicrograph reveals porous microstructure. (b) PTFE fibers for synthetic knee ligament. (W. L. Gore & Associates, Inc.)

Because cellulose does not melt or dissolve, it must be processed (regenerated) into a syrup; and from there it is formed into fibers such as rayon and acetate. Synthetic fibers created by chemical synthesis methods (condensation polymerization and addition polymerization) include polyamide, polyester, and polyurethane fibers. Blending of natural fibers such as cotton and wool will provide desired properties, as may be sought in woven fabrics for clothing and upholstery. Engineering of fibers allows desired properties to develop and is accomplished by varying molecular weight, varying cross-section, and orienting molecules through drawing and stretching. The ability of fibers to *wet* (absorb or *wick* moisture) is important in coloring them or in using them in clothing. Olefins wick well and make good sports clothing; acrylic fibers do not absorb moisture well. Fibers treated with poly(ethylene glycol) (PEG), a polytherm coating, are *enthalpic* (absorb, store, and release large amounts of heat) and *hygroscopic* (release moisture as vapor, not droplets), so as woven fabric they make excellent sporting and cold-weather clothing. *Microfibers* represent another polymer advancement which transforms synthetic fibers such as polyester to behave like the natural fibers cotton and wool.

Newer formulations of synthetic fibers have produced the superstrong, lightweight aramid fibers that are woven into bulletproof vests and serve as reinforcers for tires, firefighting and racing helmets, and clothing. DuPont's aramid fiber, **Kevlar,** is a liquid-crystal polymer (LCP) fiber. Poly-*p*-benzamide can be produced only as fibers; it cannot be molded into solid bulk shapes. Even superior to aramid, the polyolefin fiber **Spectra** consists of ultrahigh-molecular-weight polyethylene polymers processed to yield highly oriented fibers. This polyolefin fiber is woven into sails and space suit/boot/glove structures because of its superior abrasion resistance, high flexural endurance, chemical resistance, low moisture absorption, and superior specific strength.

Typical of advanced materials, the engineered Gore-Tex fiber possesses a string of properties that makes it suitable for a wide range of applications. The expanded polytetrafluoroethylene (PTFE) (Figure 6-13a) is chemically inert (not affected by any common chemical), has a low friction coefficient, can function in a wide temperature range, does not age, and is weather durable. Such properties have resulted in its selection for wide-ranging applications such as clothing, joint sealants, nonshedding filters to provide microcontamination control for computer disk drives and manufacturing clean rooms, and membranes that offer microfiltration of bacteria and other particulates from air or liquids. Because Gore-Tex materials have superior strength and are biocompatible, hydrophobic, porous, and air permeable, they have been selected for clothing for sports, outdoors, and walking in outer space. They are used in surgery for threadlike sutures to sew together tissue, as heart patches, for tubes to replace diseased arteries, and as synthetic knee ligaments that provide stability for replacing the natural anterior cruciate ligaments (Figure 6-13b).

Polybenzimidazole (PBI) is woven into fabrics that will not burn in air. Sulfar resists melting until 285°C and acts as an excellent filter that, through reverse osmosis, can separate toxins and heavy metals from water. *Osmosis* is the passage of a liquid from a dilute to a more concentrated solution through a semipermeable membrane. This type of membrane allows passage of the liquid but not the dissolved solids in the concentrated solution. *Reversed osmosis* is the forced reversal of this natural phenomenon by the application of enough pressure to the concentrated solution to overcome the pressure (osmotic) of the less concentrated solution.

Fiber-reinforced plastics (FRP), along with many laminated and filled plastics, comprise a large group of plastics that fit diverse engineering and general-purpose applications. For both woven fabrics and as reinforcing fibers in FRP, the *aramid fiber* is a plastic of superior qualities. Under tensile loads, this polymer equals steel when compared to similar sizes, but has a superior strength-to-weight ratio. The superior strength develops as a result of the stringing out of the carbon atoms, which line up rather than coiling like a spring in most polymers. The built-in stiffness of the C—C bond greatly resists stretching under tensile loads. See Module 8 for additional discussion of this remarkable fiber with the trade name Kevlar.

Table 10-9 provides comparisons of various plastic classification systems. When using the table it is important to realize that these are general data, and any given plastic may have considerable grade variations due to fillers, additives, and structure. The table also provides the basic molecular structure, chemical formula, American Society for Testing and Materials (ASTM) abbreviations, some selected trade names, and typical uses of the plastics listed. Note that carbon (C) is the backbone for all those plastics shown except silicone, which has an inorganic silicon (Si) and oxygen (O) backbone. In the polyolefins group, ionomers come from ethylene gas combined with inorganic compounds consisting of sodium (Na), zinc (Zn), magnesium (Mg), or potassium (K) salts that bond both covalently and ionically. Ionomers are resilient, oil resistant, and tough, while having very high transparency.

Although the large number of plastics generated from materials technology provide a wide array of properties, they create problems for green manufacturing in terms of recycling. The auto industry seeks to reduce the types of plastics in a move to save costs and improve the recyclability of automotive plastics.

Numerous websites provide current information on plastics. For example, the American Plastics Council (see EMT Weblink) offers an extensive site, including a glossary of terms used with plastics. Plastic suppliers sites are a good place to compare types and properties to supplement the information provided in Appendix-Table 10-9 and the properties tables in this module.

 EMT WEBLINKS

American Plastics Council glossary—*www.plastics.org*

Figure 6-14   Engineering thermoplastics. The complex, integrated, plastic molded fuel system: the tank houses components for a fuel system consisting of six layers of coextruded high-density polyethylene (HDPE) plus two layers of ethylene vinyl alcohol (EVOH). The EVOH acts to provide on-board fueling vapor recovery (ORVR) that emits less than 0.1 gram of hydrocarbons from the gasoline per day. Advantages offered by these systems include reduced hydrocarbon emissions, design flexibility, light weight, and lower cost. (Solvay Automotive, Inc.)

> Comparison of polymers with tables and other graphics—*http://www.dupont.com/ enggpolymers/europe/techlit/TopTip/toptip/html/e1tip.html*
>
> Teflon—*http://www.dupont.com/teflon/*
>
> Gore-Tex—*http://www.gore.com/corp/divisions.html*

6.4.2 Properties of Plastics

In earlier discussion, the internal structure of polymers was shown to dictate properties. The nearly infinite variations of structure in plastics provide a wide range of properties that continue to grow annually as the demands of technology prompt new discoveries in the manipulation of polymeric structures. These properties fall under the categories of mechanical, chemical, thermal, electrical, and general. Additives and fillers so significantly affect the properties of plastics that one must turn to a handbook or manufacturers' specifications for a given plastic to ascertain its properties (see Figure 6-14).

6.4.2.1 Additives (modifiers).   Most plastics are truly composites because of the many additives compounded with the resin to enhance its properties. Beyond the catalysts (also known as accelerators or initiators) that are added in minute quantities to start polymerization, additives serve to increase processibility, reduce oxidation (corrosion), add color, reduce molecular weight, increase flexibility, reinforce, retard flammability, or increase electrical conductivity. Most polymers will absorb foreign elements that can cause degradation. *Stabilizers* are added to prevent this. Ultraviolet radiation and oxygen (ozone) can alter the chemical bonds of plastics so that *free radicals* develop within chains. A *free radical* is an atom, ion, or molecule with an unpaired electron. The Pauli exclusion principle states that no two electrons in an atom can have the same four quantum numbers. For that reason, only electrons with opposite spins can pair, thus creating a covalent bond. The free radicals are segmentations within the polymer that can easily combine with other elements to cause a breakdown of the plastics structure and severely affect properties. Cross-links may form to produce a more brittle plastic or cause the disruption of the chain structure, which will limit tensile strength. *Carbon black,* produced through the burning of *carbonaceous* materials such as gas or oil, will block out ultraviolet light while strengthening the plastic. Metals such as barium, cadmium, and lead in compounds are used as stabilizers. Amine and phenol chemicals serve as sunscreens as they interact with photons.

Colorants.   Both organic *dyes* and inorganic (metal based) pigments add color to plastics, and some serve dual roles such as stabilizing. Generally, the inorganic pigments will withstand high temperatures without charring or fading. *Pigments* disperse rather than dissolve in plastics and reduce the transparency of the material. The pigments also hide flaws such as air bubbles, making it difficult to judge the quality. While nearly an infinite range of color, transparency, translucency, and opacity is possible, the Food and Drug Administration (FDA) allows only certain types of colorants for plastics that make contact with food and drugs.

Plasticizers.   Plasticizers are additives that increase flexibility, while cross-linking agents such as organic peroxides cause hardening to produce free radicals. Plasticizers reduce the attraction (secondary valence bonds) between polymer chains and are normally solvents such as alcohol. Poly(vinyl chloride) (PVC) can be very hard and rigid for water and sewage piping. The addition of a plasticizer such as carbon alphates alcohol produces a resin of low volatility that is used in the slush or dip molding of very flexible products such as tool handle coatings, rain boots, or doll parts. *Flame retardants* have become more important in the United States as our search for greater product safety broadens. Elements such as boron, nitrogen, chlorine, antimony, and phosphorus reduce the flammability of plastics by preventing oxygen reactions and improving charring. *Charring* is seen when wood burns and leaves a protective residue or ash that slows burning; this is known as *ablation.* Flame retardants can cause problems by reducing flexibility, tear resistance, tensile strength, and heat deflection. Fluorocarbons, polyvinyls, and polyimide plastics can offer low- or nonburning properties without the use of flame retardants.

Fillers.   Fillers improve plastics by increasing bulk, tensile strength, hardness, abrasion resistance, and rigidity; they improve electrical and thermal properties, appearance, and chemical resistance while either increasing or decreasing specific gravity. Common fillers are woodflour, quartz, glass spheres, talc, wollastonite ($CaSiO_3$), calcium carbonate ($CaCO_3$), carbon black, clay, and alumina trihydrate (ATH). For example, glass spheres, clay, or calcium carbonate are added to sheet-molded compounds (SMC) or bulk-molded compounds (BMC) to decrease cost while increasing the rigidity of polyester resins.

Other additives include *antistatic agents* and *coupling agents* to aid bonding between plastics and inorganic materials in composites, *foaming agents* to produce cellular plastics, *heat stabilizers* for processing and end product durability, *lubricants* to decrease friction and resin melt during processing, *mold release agents, preservatives* to retain physical and chemical properties and in some cases prevent growth of bacteria and algae, and *viscosity depressants* to reduce viscosity during processing.

6.4.2.2 Mechanical properties.

As discussed previously (see Figure 6-8a), variations in properties are achieved through variations of molecular weight and crystallinity. With mechanical properties, an increase in crystallinity or density, plus an increase in molecular weight, produces corresponding increases in tensile strength, hardness, creep resistance, and flexural strength, but decreases the impact resistance and percentage of elongation. Additives, fillers, and reinforcers also change mechanical properties to a great extent.

Table 6-2 lists a few selected plastics that reveal the range of *tensile strength* for plastics and also includes other engineering materials for comparison. Because many different varieties and grades are available within a certain plastics group, there is a wide range of tensile strength for plastics of the same name. For example, some flexible cast epoxies have a tensile strength lower than 10 MPa, while a cycloaliphatic epoxy casting resin exceeds 100 MPa. The values in Table 6-2 are for specimens tested at *standard temperature and pressure (STP).* Many plastics lose strength rapidly at relatively low temperatures (20°C), while others, such as polyimides, can maintain full strength near 500°C over short periods while also withstanding cryogenic temperatures around –300°C. Figure 6-14 shows an application of engineering thermoplastics for automotive use. Note the integration of various plastics that form a materials system for the fuel tank

TABLE 6-2   RANGES[A] OF TENSILES STRENGTH AT ROOM TEMPERATURE FOR SELECTED PLASTICS COMPARED TO OTHER MATERIALS

The tensile strength scale [MPa (ksi)] across the top reads:
0, 10 (1.5), 20 (2.9), 30 (4.4), 40 (5.8), 50 (7.3), 60 (8.7), 70 (10.2), 80 (11.6), 90 (13.1), 100 (14.5), 150 (21.8), 175 (25.4), 200 (29.0), 225 (32.6), 250 (36.3), 500 (72.5), 750 (108.8), 1000 (145.0)

Plastics[b]	Tensile strength range (X marks)
ABS, e	XXXXXX
Acetals, e	XXXX
Acrylics (PMMA)	XXXXX
Alkyds	XXXXXXXXX
Amino	XXXXX
Cellulose acetate (CA)	XXXXXX
Cellulose butyrate (CB)	XXXXXXX
Diethylene glycol bisallyl carbonate	XXXX
Epoxy (EP)	XXXXXXXXXXXXXXXXXX
Fluorocarbons (TFE), e	XXXXX
Nylon, e	XXXXXXXX
Phenolic (PF)	XXXXXX
Phenylene oxide (PPO)	XXX
Poly (amide/imide), e	XXXXX
Polycarbonate (PC), e	XXX
Polyester, TS	XXXXXXXXXXX
Polyester, TP	X
Polyethylene (LDPE)	XX
Polyethylene (HDPE), e	XX
Polyethylene (UHMWPE), e	XXXXX
Polyimide, e	XXXXXXXX
Polypropylene (PP), e	XXXX
Polystyrene (PS)	XXXXXXXXXX
Polysulfone (PES), e	XXX
Polyurethane (PUR), e	XXXXXXXXXXXXX
Silicones (SI)	XXXXXXXXX
Vinyl (PVC)	XXXXX
Vinyl (PVDC)	XXXXXX[c] XXXXXXXXXXXXX[d]

Comparison Materials

Material	Tensile strength range (X marks)
Alumina	XXXXX
Aluminum alloys	XXXXXXXXXXXXXXXXXXXXXXXX
Brass	XXXX
Cast iron	XXXXXXXXXXXXXXXXX→
Fiber-reinforced plastic	XXXXXXXXXXXXXXXXXXXXXXXXXXXXXX→
Glass	XXXXXXXXXXXXXXXXXXX E & S fibers →
Stainless steel	XXXX →
Steel—plain carbon	XXXXX →
Wood	X

[a]Ranges reflect the varieties of the named plastic. Does include certain filled varieties but not fiber-reinforced plastics (FRP). Ranges for other materials are general.

[b]e, Engineering plastic; TS, thermoset; TP, thermoplastic.

[c]Unoriented.

[d]Oriented.

in Figure 6-14. Figure 6-15 shows applications of engineering thermoplastics for building construction. Figure 6-16 provides another example of an engineering thermoplastic, polyethersulfone (PES), seen in this case used for fiber-optic (FO) connectors. Radel®, a PES blend from Amoco, was selected for use as FO connectors because of the following properties: allows precise injection moldings; facilitates concurrent engineering due to the ease of design that does not require metal inserts or glass reinforcers; makes tough, chemical resistant, and dimensionally sta-

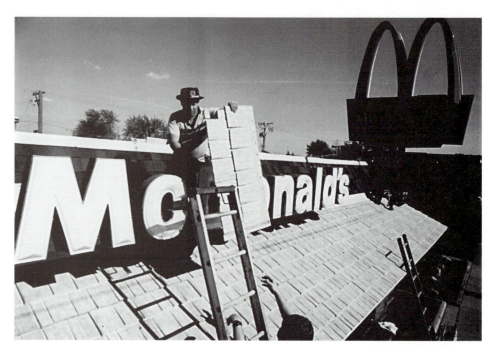

Figure 6-15 Polyphenylene oxide (PPO), Noryl®, offers many advantages over traditional roofing materials, including light weight, easy installation, very good impact resistance, weatherability, walkability, and fire performance. Because of the easy moldability, it can be molded to resemble cedar shake, slate, and barrel tile for the construction industry. (GE Plastics)

Figure 6-16 Polyethersulfone engineering thermoplastic allows for a wide range of injection-molded fiber-optic connectors and components. Intricate parts can be molded to extremely close tolerances to meet the critical alignment requirements for FO cables. (*Advanced Materials and Processes*)

ble components in temperature ranges from $-40°$ to $+85°C$ ($-40°$ to $+185°F$); and permits use of organic colorant systems to meet industry standard color codes for FO connectors. Table 6-2 shows that other engineering materials have far greater tensile strengths than the engineering plastics, but the low density of plastics often yields a competitive or higher strength-to-weight ratio. Using *specific strength* as the comparison factor, we have

$$\text{Specific strength (meters)} = \frac{\text{Tensile strength}}{\text{density (weight)}}$$

Polyester pultrusion, fiber-reinforced plastics (FRP) have specific strengths around 35,000 m, while stainless steel has a specific strength around 25,000 m. Other FRPs, dealt with in Module 8, have even higher specific strength ratios (e.g., up to 90,000 m for epoxy reinforced with carbon fibers). Table 6-7 can be used with the tensile strengths given in Table 6-2 to calculate other specific strengths for comparisons.

Illustrative Problem

Use Table 6-2 and Table 6-7 to determine the mid-value in the range of tensile strengths and densities of plain-carbon steel and polyimide. Substitute the values into the formula above and determine which has the higher specific strength. Use Table 10-7 for conversion of units.

Solution

$$\text{Specific tensile strength} = \frac{\text{tensile strength}}{\text{density (weight)}}$$

From Table 6-2,

Plain-carbon steel	680 MPa
Polyimide	76 MPa

From Table 6-7,

Plain-carbon steel	8200 kgf/m^3
Polyimide	1260 kgf/m^3

Substituting in the formula above, we find that

$$\text{Specific tensile strength} = \frac{680 \text{ MPa}}{8200 \text{ kgf/m}^3}$$

Reduce units to meters (m) by using the following unit conversions from Table 10-7; 1 kgf = 9.807 N, 1 Pa = 1 N/m^2, and M = 10^6. The specific tensile strength for plain-carbon steel is:

$$\text{Specific tensile strength} = \frac{680 \text{ MPa}}{8200 \text{ kgf/m}^3}\left(\frac{1 \text{ N/m}^2}{1 \text{ Pa}}\right)\left(\frac{10^6}{M}\right)\left(\frac{1 \text{ kgf}}{9.807 \text{ N}}\right) = 8456 \text{ m}$$

The specific tensile strength for polyimide using the same conversion units is

$$\text{specific tensile strength} = \frac{76 \text{ MPa}}{1260 \text{ kgf/m}^3}\left(\frac{1 \text{ N/m}^2}{1 \text{ Pa}}\right)\left(\frac{10^6}{M}\right)\left(\frac{1 \text{ kgf}}{9.807 \text{ N}}\right) = 6150 \text{ m}$$

Plastics compared to steel and most other engineering materials are much softer. Table 4-1 showed various hardness scales in an approximate comparison for matching hardness of selected materials. Hardness comparisons of most common plastics are made with the Shore Durometer A and D method and Rockwell M (HRM) tests. The problem with hardness tests of plastics is that they do not closely correlate to wear or abrasion resistance as do most other materials: polystyrene has a Rockwell M value of 72 but scratches easily; diethylene glycol bisallyl carbonate, marketed as CR39® (allyl diglycol carbonate) by PPG Industries, has a Rockwell M value of 95 to 100 but has abrasion resistance approximating that of glass. This makes it a competitor with glass because it has optical properties approaching those of glass. Methyl methacrylate, under trade names such as Plexiglas and Lucite, has an average hardness of 93 to 98 HRM but abrasion resistance 30 to 40 times less than CR39 using a modified Taber test. *Abrasion resistance* and *wear resistance* are measured by Taber and other tests that determine the *percentage of haze* (loss of clarity) due to marking, or *percentage of material lost* through rubbing with abrasives such as aluminum oxide. Table 6-3 shows the results of Taber tests on several plastics and their hardnesses.

Toughness or *impact resistance* is generally better in thermoplastics than thermosets. As seen in Table 6-4, polycarbonate is the toughest of the transparent rigid plastics at 12 to 14

TABLE 6-3   TABER TEST RESULTS

Plastic	Hardness	Abrasion resistance (mg loss per 1000 cycles)
Nylon 6/6	114 HRR	5
Acetals	95 HRM	16
ABS	100 HRR	84
Polysulfone	120 HRR	20
Polyimides	98 HRM	0.08

TABLE 6-4   COMPARATIVE IMPACT (IZOD) STRENGTH OF SELECTED PLASTICS

Plastic[a]	Notched impact strength (ft·lb/in.)										
	0	2	4	6	8	10	12	14	16	18	20
ABS, e		XXXXXX									
Acetals, e		X									
Acrylics (PMMA)	XX										
Alkyds		XXX									
Aminos—melamines	X										
Cellulose acetate (CA)		XXXXXXX									
Epoxy (EP)	X										
Fluorocarbon (TFE), e		XXX									
Nylon 6/6, e		XX									
Phenolic (PF)	X										
Phenylene oxide (PPO), e				X							
Poly (amide/amide), e		XXX									
Polycarbonate (PC)								XXXXXX			
Polyester (PBT)		X									
Polyester (PET)		X									
Polyethylene (LPDE)										XXX→	
Polyethylene (HDPE)		XXXXXXXXXXXXXXXXXXXXXXXXXXXXXXXXXXXXXXXXX									
Polyimide (GRP), e										XXX	
Polypropylene (PP), e		XXX									
Polystyrene (PS)	X										
Polysulfone (PES), e		XX									
Polyurethane (PUR), e				XXXXXXXX							
Silicones (SI)	X										
Vinyl (PVC)		XXXXXXXXXXXXXXXXXXXXXXXXXXXXXXXXXXXXXXX									

[a]e, Engineering plastic.

pounds per inch and far exceeds polystyrene (10.4 ft·lb/in.) and acrylics (0.8 to 1.6 ft·lb/in.). Polyurethane, with an impact strength of 5 to 8 ft·lb/in., has gained acceptance as automobile bumpers and hoods. Both polyethylene and poly(vinyl chloride) (PVC) have very broad ranges of impact strength, as you may know from the thin-walled tough polyethylene milk jugs and PVC piping that have begun to replace much of the copper and cast-iron pipe used for water and sewage. The values given in Table 6-4 are notched specimens. Notching of most plastics severely limits their toughness, so designs should avoid sharp corners or conditions in which parts subjected to impact might develop scratches or cuts. Instead of the notched Izod test, which may not give the most reliable results, some designers prefer data obtained from impacts made with a falling weight on sheet plastic (ASTM D3029), film (ASTM D1709), and pipes and fittings (ASTM D2444). Figure 6-17 reveals that certain plastics compete well with metals when toughness is required.

6.4.2.3 Viscoelasticity.   Viscoelasticity is a property unique to polymers (plastics, elastomers, adhesives, and wood) and not found in metals or ceramics. Because of the viscoelastic property of plastics, engineering technologists must become involved in *rheology* (study of flow and deformation of matter) to determine why plastics will react to mechanical acetal gears   polycarbonate/ABS

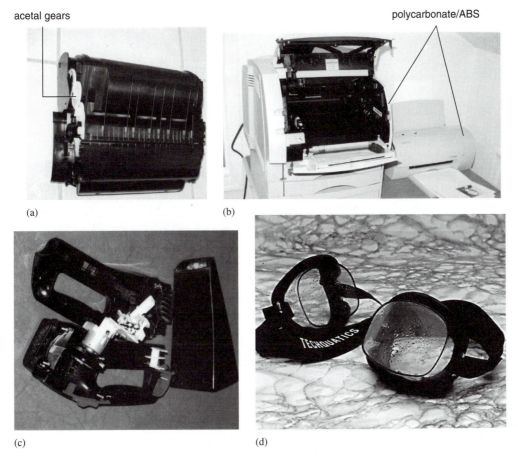

(a)   (b)

(c)   (d)

Figure 6-17   Engineering plastics provide toughness for computer applications and appliances. (a) Tough gears used on laser printer toner cartridge made of duPont's Derlin (acetal) also provide excellent tribological properties (resist wear and friction), light weight, corrosion resistance, and quiet operation. (b) Examination of internal and external parts of a computer and their peripherals reveal very few nonplastic structural parts; figure shows cases made of polycarbonate/ABS copolymers. (c) Nylons 6/6 and acetals, and other engineering plastics, in disassembled vacuum cleaner, make up most of the parts as with many appliances; fan/impeller, electrical harness, motor and batter holder, and housings, connectors, and switches. (d) Swim mask of tough polycarbonate has optical clarity and antifogging (due to coating). Neoprene rubber shirt conforms to the face for swimmers ages 8 and up, and Velcro tabs provide precise adjustability (G.E. Plastics).

stresses in a different manner than metals. The property *viscoelasticity* incorporates two properties: viscosity and elasticity. *Viscosity* refers to the nature of a liquid's resistance to flow. Motor oil used in an automobile engine is rated by its viscosity; heavy oil, such as 40 SAE (Society of Automotive Engineers), is thick syrup and is used in hot weather, while lighter oil (10 SAE) that flows much like water is used in temperatures below zero. As with oil, plastics become less viscous or flow more easily with increased temperatures. This makes them easier to process but also means that they lose strength with heat.

Ultimately *Elasticity* (discussed in Module 4) refers to the ability of a material to return to its original size and shape once a load is removed. With plastics, the two factors come into play so that a sudden impact on the plastic will not result in immediate strain or immediate and full recovery when the stress is removed. Rather, there is a viscous flow internally as the force is absorbed through shearing of molecular bonds, much the way that a shock absorber works in an automobile. The shock absorber is a combination of *dashpot* (force gradually released through a slow bleeding of fluid) and spring. Even after all load is removed, the deformation is not fully recovered as with a metal that experiences elastic loading. The effect of heat on viscoelastic properties is discussed under thermal properties. Figure 6-18 illustrates the mechanical models (Maxwell and Voigt) that simulate viscoelasticity.

Viscoelastic properties are determined by measuring the stress relaxation (Section 6.4.2.4) of a material subjected to a constant strain at different temperatures as specified by

(a) Shock absorber

(b) Maxwell series model   (c) Voigt parallel model   **Figure 6-18**   Viscoelasticity.

ASTM D2991. The reduction in stress is measured until it reaches equilibrium. Materials that offer the best strength retention and have good viscoelastic properties are needed for many applications where, for example, a force is required to be maintained for contact purposes or for clip strength. Figure 8-42 (in Module 8) exemplifies such an application for the electronics industry.

6.4.2.4 Thermal properties.

Many plastics lose their strength at relatively low temperatures. Continuous-service-temperature comparisons of plastics reveal most common plastics can endure temperatures no more than 150°C when under low or no stress. Slight increases in stress would reduce the continuous-service temperature even more. Newer plastics, such as the polyimide thermosetting resins, resist intermittent heat up to 500°C with low stress and up to 250°C for thousands of hours. Graphite-reinforced polyimides can withstand flexural stress to nearly 69,000 Pa at 250°C. Plastic composites have replaced aluminum pistons in some race cars, because the aluminum with high thermal conductivity acts as a heat sink and begins to soften at around 150°C. Plastic composites of polyimide and epoxy with glass and graphite reinforcement withstand temperatures approaching 290°C and tremendous mechanical stress, yet they offer weight savings over metals.

Glass-transition temperatures or ***glass point*** (T_g) is the point at which polymers act as glass or become viscous liquids (see Figure 6-19). The glass transition is a reversible change that occurs when a resin polymer is heated to a certain temperature (T_g), resulting in a sudden change or transition from a rigid polymer to a flexible, rubbery material or a viscous liquid. Such a material suffers a major loss in strength when subjected to a temperature near its T_g value, leading to catastrophic failure even if the service (use) temperature exceeded the T_g value for only a short time.

In cooling, the thermal mixing of the atoms and molecular chains slows, and the volume of the plastic decreases because they pack closer together. In packing, many polymers do not form crystals as do metals; instead, they freeze into an amorphous glassy structure. Of course, highly crystalline polymers such as high-density polyethylene (HDPE) would not follow this pattern, and the degree of crystallinity of plastics would control T_g. Cooling to the glass point results in noncrystalline polymers becoming brittle. This occurs at or above room temperature for certain plastics, for example, polystyrene and acrylic, at about 100°C. In crystalline plastics with low T_g, such as polypropylene at −10°C, flexibility and impact strength are maintained. In reheating an amorphous thermoplastic, it passes from the brittle glassy structure

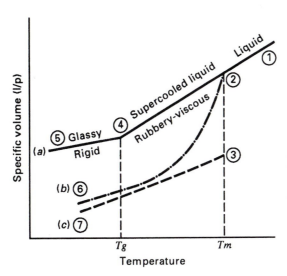

Figure 6-19 Specific volume versus temperature for polymers. A) Linear amorphous polymer [path (a) 1,2,4,5]. B) Partially crystalline polymer [path (b) 1,2,6]. C) Crystalline polymer [path (c) 1,2,3,7]. D) Path through points 1,2,3,7 is a theoretical path approaching a crystalline solid such as a metal. E) T_g (glass temperature) and T_m (melting temperature) for a crystalline polymer.

through its T_g point, and then turns into a tough rubbery form, progressing to a softer and more pliable stage into a viscous liquid upon reaching the melting point (T_M).

Exact melting points are determined optically in crystalline and semicrystalline plastics when the **birefringence** (double refraction) is lost due to the change of structure from the crystalline to the glassy state. The earlier discussion on mechanical properties covered viscoelasticity. One can see how T_g will affect this property. A linear amorphous plastic will be quite brittle and not exhibit viscous flow below its T_g, but it will be quite rigid and elastic. As the degree of crystallinity increases in plastics, they become more viscous and less capable of recovering from deformation under load as they possess less elasticity.

Time becomes an important factor in the loading of plastics. A stressed linear amorphous polymer at temperatures above T_g over sufficient time will begin to deform without additional loading or heat as the molecular chains become untangled through viscous flow; this is **stress relaxation.** A network thermoset at relatively low temperatures, under constant stress over a long time, will begin to deform and continue deforming with progressively less stress; this is **creep.** The softening temperature of polymers can also be increased through an increase in average molecular weight or polymer chain lengths. Thermoplastics are far more prone to creep than thermosets.

Thermal conductivity is low in most plastics, which makes them valuable as thermal insulators. In the cellular or foamed state, air cavities, which do not carry heat or cold, improve even further the insulating properties. However, the rate of *thermal expansion* is quite high for most plastics, generally 10 times as much as metals. Table 6-5 provides a comparison of the rate of thermal conductivity and coefficient of thermal expansion for selected plastics and compares them with other materials.

Plastic films made of select plastics for packaging can withstand heat from conventional cooking ovens; most are not affected by microwaves but some melt from the heat of the food as it is cooked. A polyvinylidene chloride (PVDC) product such as **Saran Wrap** has oxygen protection like glass and will not melt with **microwave cooking;** crystallized polyester (CPET) will withstand conventional oven temperatures, to 425°F, and microwaves. Polyethylene (PE) products such as Handiwrap and Glad Wrap do not provide the oxygen protection of PVDC or CPET. They can be used in microwave cooking if the fat and sugar content of the food is low enough so that they do not melt the plastic. Polystyrene (PS) foam, often used to package food, is safe for heating liquids such as water, tea, and coffee but will melt if used to heat food such as hamburgers or french fries.

The range of *flammability* for plastics is quite wide. For plastics that sustain combustion, a burning rate test (ASTM D635), which measures the propagation of flame (in./min) on a given size specimen, permits a measure of comparability. For instance, phenolics, polyimides, and fluorocarbons without fire-retardant additives are considered nonburning, whereas the cellulosics

TABLE 6-5 COMPARISONS OF THERMAL PROPERTIES FOR SELECTED PLASTICS AND OTHER MATERIALS[a]

Coefficient of thermal expansion	10^{-6} m/m/K				
	60	120	180	240	300
Plastics					
ABS,e	X				
Acetals (POM), e		X			
Acrylics (PMMA)	XXXXX	XXXX			
Cellulose (CA)		XXXX	XXXXXXXX		
Epoxies (EP)	XXXXXX				
Fluorocarbons (PTFE), e		X			
Nylon		X			
Phenolic (PF)	XXX				
Polycarbonate (PC)	XXXXX	XXXX			
Polyester (PBT)		XX			
Polyethylene (LPDE)			XXXXXX		
Polyethylene (HDPE), e			XXXXXXXXX	XXXXXXXXXX	XXXXX
Polyimide	X				
Polypropylene (PP)		XXX			
Polystyrene (SAN)		X			
Polysulfone, e		X			
Polyurethane (PUR)	X				
Silicone (SI), e	XXXXX	XX			
Vinyls (PVC)	XX				
Alumina	X	. . .			
Aluminum	XXX		. . . .		
Cast iron	XX	. .			
Copper	XX				. . . .
Glass	XX				
Rubber	X				
Silver	X				
Steel	X				
Stainless steel	X				
Wood			. . . .		
Zinc	X	. . . .			
	W/(m · K)				
Thermal conductivity	0 . . . . . . 10 . . . . . . 50 . . . . . . 100 . . . . . . 150 . . . . . . 200 . . . . . . 300 400				

[a]e, Engineering plastic.

are highly flammable. Certain plastics such as polyurethane and vinyls give off highly toxic fumes when burned. The vinyl siding used on housing ignites above 370°C, whereas wood siding ignites at about 270°C. Generally, thermoplastics are more flammable than thermosets. Thermoplastics can be ground into chips and melted for recycling. Although thermosets have the potential for recycling through the use of chemical solvents, the cost may not be justified; however, thermosets can be ground and used as a composite filler. The high thermal energy possible from thermosets, nearly 3 MJ/kg (megajoules per kilogram), has value as an energy source in which scraps are burned to generate heat for steam-turbine generators in the production of electrical energy. As with wood, plastics develop a char layer that serves as an *ablative* or protective shield, which insulates the unexposed area. Nylon and phenolic resins were used on earlier NASA spacecraft as ablative heat shields for reentry into the earth's atmosphere. Although the polymers did burn when exposed to temperatures over 6500°C for several seconds, the charred ablative shield protected the spacecraft because of its thermal insulating properties.

6.4.2.5 Chemical properties.
Plastics do not corrode like metals do, but they are subject to deterioration and chemical attack. Whereas the corrosion of metals is determined by weight loss, a thermoplastic's deterioration is measured by weight gain as the attacking chem-

777 composite materials

• Toughened material for improved damage resistance and damage tolerance, and parts are designed for simple, bolted or bonded repairs
• Corrosion and fatigue resistant
• Weight savings

Toughened carbon fiber reinforced plastic (CFRP)

CFRP   Hybrid   Fiberglass

Nose radome

Outboard aileron

Outboard flap

Flaperon

Inboard flap

Floor beams

Wing fixed leading edge

Trailing edge panels

Strut fwd and aft fairings

Main landing gear doors

Nose gear doors

Rudder

Elevator

Fin torque box

Stabilizer torque box

Leading and trailing edge panels

Wing-to-body fairing

Inboard and outboard spoilers

Flap track fairing

Engine cowlings

Figure 6-20 Boeing's 777 lightweight composite structure makes extensive use of polymer matrix composites (PMCs). After thorough testing of PMCs in military and research aircraft, polymers proved reliable for commercial aircraft. The "paperless" project utilized concurrent engineering (CE), in which design engineers and manufacturing engineers worked in design-build teams. In addition to advanced metals, the 777 structure used graphite, toughened graphite, fiberglass, and hybrid PMC composites, which were selected for corrosion and fatigue resistance among the other properties listed above. (Courtesy of the Boeing Company.)*

*The Public Broadcasting System (PBS) series, *Twenty-First Century Jet,* provides an excellent in-depth case study of the product development process, from design through testing, manufacturing, and delivery. The series includes many examples of engineering materials technology. Call PBS at 1-800-344-3337 to obtain further information on the video series.

347

TABLE 6-6 COMPARATIVE CHEMICAL PROPERTIES OF SELECTED PLASTICS AND OTHER MATERIALS[a]

Plastics[b]	Weak acid resistance (Little / Fair / Great)	Organic solvent resistance (Little / Fair / Great)	Water absorption resistance (Little / Fair / Great)	Weatherability (Little / Fair / Great)
ABS, e	XXXXXXX	XXXXXX	XXXX	XXXXX
Acetals (POM), e	XXXX	XXXXXXXXX	XXXX	XXXXXXX
Acrylics (PMMA)	XXXXXXXXX	X	XXXXXX	XXXXXXXXX
Alkyds	X	X	XXXXXXXX	XXXXXXX
Aminos—melamides	XXXX	XXXXXXXXX	XXXXXXXXX	
Cellulose acetate (CA)	XXXXXX	XXXXXX	X	XXXXXX
Cellulose butyrate (CAB)	XXXX	XXXXX	XX	XX
Epoxy (EP)	XXXXXXXXX	XXXXX	XXXXXXXX	XXXXXXXXX
Fluorocarbons (PTFE), e	XXXXXXXXX	XXXXXXXXX	XXXXXXXXX	XXXXXXXXX
Nylon 6/6, e	XXXX	XXXX	X	XXXX
Phenolic (PF)	XXX	XXX	XXXX	XXXX
Phenylene oxide (PPO), e	XXXXXXXXX	XXXXX	XXXXXXXXX	XXXXXXXXX
Poly(amide/amide), e	XXXXXXX	XXXXXXX	XXXX	XXXXX
Polycarbonate (PC)	X	X	XXXXXXXXX	XXXXX
Polyester, TP	XXXXXXXX	XXXX	XXXXXXXXX	XXXXXXXXX
Polyester, TS	XXXX	XXXX	XXXXXX	XXXXXXXXX
Polyethylene (LDPE)	XXXXXXXXX	XXXXX	XXXXXXXX	XXXXXXXXX
Polyethylene (HDPE), e	XXXXXXXXX	XXXXX	XXXXXXXX	XXXXXXXXX
Polyethylene (UHMWPE), e	XXXXXXXXX	XXXXX	XXXXXXXX	XXXXXXXXX
Polyimide, e	XXXXX	XXXXX	XXXXXXXXX	XXXXXXXXX
Polypropylene (PP), e	XXXXXXXXX	XXXXXXX	XXXXXXXXX	XXXXX
Polystyrene (PS)	XXXXXXX		XXXXXXXX	
Polysulfone, e	XXXXXXXXX	XXXXXXXXX	XXXXXXX	XXXXX
Polyurethane (PUR), e	XXXX	XXXXXXXXX	XXXXXXX	XXXXXX
Silicones (SI)	XXXX	X	XXXXXX	XXXXXXXXX
Vinyl (PVC)	XXXXXXXXX	XXXXXX	XXXXXXXXX	XXXXXXXXX
Vinyl (PVDC)	XXXXXXXXX	XXXXXX	XXXXXXXX	XXXXXXXXX

Comparison materials

	Weak acid resistance	Organic solvent resistance	Water absorption resistance	Weatherability
Alumina	XXXXXXXX	XXXXXXXXX	XXXXXXXXX	XXXXXXXXX
Aluminum	XXXXXXXX	XXXXX	XXXXXXXX	XXXXXXXX
Brass	XXX	XXXXXXXXX	XXXXXXXX	XXXXXXXXX
Cast iron	X	X	XXXXXXXX	XXXXXXX
Fiber-reinforced plastics vary with plastic matrix from little to great.				
Glass	XXXXXXXXX	XXXXXXXXX	XXXXXXXX	XXXXXXX
Stainless steel	XXXXXXXXX	XXXXX	XXXXXXX	XXXXXXXXX
Steel—plain carbon	X	X	X	
Wood	XXXXXXXXX	XXXXXXXXX	X	XXXX

[a]Exposures at room temperatures except weatherability.

[b]e, Engineering plastic; TS, thermoset; TP, thermoplastic.

ical combines with the plastic. The result is usually discoloration, swelling, or crazing (fine cracks) with corresponding loss of tensile, impact, and flexural strengths. Polymer-matrix composites (PMCs) have undergone considerable development to arrive at the point where they can be used with confidence in the aerospace industry. Wide use of PMCs on the Boeing 777 (Figure 6-20) attests to the corrosion resistance and many other favorable properties of these polymer-based materials.

As with mechanical properties, *chemical resistance* will decrease in plastics as temperatures increase. Table 6-6 provides data for plastics tested at room temperature for their degree of resistance to weak acid, organic solvents, and water absorption. Some common strong acids include hydrochloric and nitric, and weaker acids are lactic, boric, and citric. Common organic solvents include gasoline, alcohol, and acetone. The *weatherability* com-

(a)   (b)

Figure 6-21   (a) Chem suit used when entering chemical tanks relies on optical grade CR39® monomer which resists solvents, acids, scratching, and impact. (PPG Industries) (b) Plastics under the hood of a car such as battery cases, cowling, ducts, and reservoirs are subjected to heat and chemicals. Hotter running engines require newer formulations of thermoplastics such as nylon electrical connectors.

parison involves several chemical stresses, including heat, moisture, ultraviolet light, and chemicals in the air such as ozone (allotropic form of oxygen) and hydrochloric acid (see Figure 6-21). Notice that the fluorocarbons, with their inertness, have great resistance to weak acids, organic solvents, water absorption, and weathering. Acetal homopolymer resins have great resistance to acetone at room temperature, but at elevated temperatures (65°C) they are unacceptable for service and do not perform well when in environments with strong acids. The copolymerization of acetal resins greatly improves their resistance to inorganic liquids at elevated temperatures, so they are utilized as plumbing products such as valves and pumps. High-density polyethylene (HDPE) has great resistance to acids, water absorption, and weathering, which, coupled with its low cost, makes it a good candidate for numerous packing applications such as blow-molded bottles used for the home and industry; it also serves as electrical wire insulation, oil storage tanks, and recreational body vehicles. As seen in Table 6-6, acrylics (Lucite and Plexiglas) have great resistance to *weathering,* which provides good service as exterior windows because their impact resistance is better than that of soda-lime glass and they transmit light (92%) as well as fine optical glass. Certain plastics with little *water-absorption resistance,* such as nylon 6/6, are dried before processing so that the absorbed moisture will not cause corrosion of machinery. Moisture absorption causes swelling in plastics, which creates problems in holding dimensional accuracy. When high accuracy must be maintained, plastic of low moisture absorption is required. Polyester used as the resin in fiberglass has great water-absorption resistance and serves well in marine applications such as boat hulls and surfboards.

Aseptic packages use polymer (e.g., paper and polyethylelene) and metal (usually aluminum) layers or laminates to protect the package contents from contamination. Milk and juices do not require refrigeration when stored in aseptic packages as shown in Figure 6-1a. *Barrier containers* such as the catsup bottle seen on the left in Figure 6-22a employ polymers to prolong shelf life of the contents. The five layers of polymers, diagrammed in Figure 6-22b, consist of a copolymer of polyethylene/vinyl alcohol to form an oxygen barrier, polypropylene for a moisture barrier, polycarbonate to give strength as the hot contents fill the bottle, plus two adhesive layers to bond the other three layers. Toothpaste tubes use a similar construction and include an aluminum layer to enable the tube to be rolled. Earlier tubes used lead which required a material substitution due to its toxicity.

(a)   (b)

Figure 6-22   Plastic food containers—(a) tough and lightweight properties of plastics make them very competitive with glass and steel. The barrier plastic ketchup bottle on the left, made of layered composite plastics [see detail in b], weighs 80% less than the glass catsup bottle on the right. The plastic container is less likely to break than the glass bottle and saves transportation costs. Why would one company choose glass and the other company plastic? (b) Five layers of barrier plastic container.

6.4.2.6 Density.   The density (kilograms per cubic meter) of plastic is generally lower than that of other engineering and general-purpose materials. Table 6-7 shows that only magnesium and some wood have as low mass per unit volume as plastics. Even glass-reinforced plastics (GRPs) are lighter than aluminum and steel, while their strength approaches or, in some instances, equals that of these popular engineering metals. CR39® (allyl diglycol carbonate) (Figure 6-21), developed by Pittsburgh Plate Glass for competition with optical glass, has a density about one-half that of glass. This plastic is a practical substitute for glass in larger stylish eyeglasses and for industrial safety glasses because it combines light weight with impact resistance and is superior to acrylics in wear resistance.

6.4.2.7 Specific gravity.   Specific gravity is the ratio of the mass of a measured volume of any material to the mass of an equal volume of water at a standard temperature. This ratio is often used for weight comparisons. Specific gravity ratios for plastics range from below 0.06 for foams to over 2.0 for fluorocarbons, compared to 0.5 for softwoods, 0.7 for hardwoods, 2.2 to nearly 4 for glasses and ceramics, nearly 3 for aluminum, and around 8 for steels.

6.4.2.8 Optical properties.   A number of plastics have optical properties comparable to glass and offer impact strength superior to most glasses. In addition, they process more easily and, when broken, do not produce sharp splinters or chards. Refer to Section 9.5 for the definition of optical properties used here. Opacity develops in plastics with an increase in crystallinity, whereas an amorphous structure produces transparency. Low-density polyethylene film, such as sandwich bags, is clear, while HDPE for detergent bottles is opaque. Acrylics are available in an unlimited range of colors and are as transparent as the finest optical glass (light transmission equals 92%), can be made opaque, or can be produced through a full range of translucencies. Acrylics have a *refractive index* of 1.49 compared to 1.52 for soda-lime glass, 1.47 for borosilicate glass, and 1.46 for 96% silica glass. (*Refractive index* indicates the ratio

TABLE 6-7   MASS WEIGHT DENSITY COMPARISONS OF SELECTED PLASTICS AND OTHER MATERIALS

Kilograms per cubic meter (pounds per cubic inch)

Plastics[a]	500 (.018)	1000 (.036)	1500 (.054)	2000 (.072)	2500 (.090)	3000 (.108)	3500 (.126)	4000 (.145)	4500 (.163)	5000 (.180)	5500 (.199)	6000 (.217)	6500 (.235)	6500 (.235)	7000 (.253)	7500 (.270)	8000 (2.89)	8500 (.307)	9000 (.325)
ABS, e		|																	
Acetals (POM), e			|																
Acrylics (PMMA)			|																
Amino (MF)			|																
Cellulose (CA)			|																
Epoxy (EP)																			
Fluorocarbons (PTFE), e								|											
Nylon, e			|																
Phenolic (PF)								|											
Polycarbonate (PC), e			|																
Polyesters			■																
Polyethylene (LDPE)			|																
Polyethylene (HMWPE), e			|																
Poly(ethylene terephthalate) (PET)			|																
Polyimide			|																
Polyphenylene oxide (PPO)			|																
Polypropylene (PP), e			|																
Polystyrene (SAN)			|																
Polysulfone, e				|															
Polyurethane (PUR), e				|															
Silicones (SI), e				|															
Vinyls (PVC)				|															
Plastic foams		■	■																
Glass-reinforced plastics			■	■	■	■													

Comparison materials

	500 (.018)	1000 (.036)	1500 (.054)	2000 (.072)	2500 (.090)	3000 (.108)	3500 (.126)	4000 (.145)	4500 (.163)	5000 (.180)	5500 (.199)	6000 (.217)	6500 (.235)	6500 (.235)	7000 (.253)	7500 (.270)	8000 (2.89)	8500 (.307)	9000 (.325)
Alumina										|									
Aluminum					■	■													
Brass																			|
Cast iron																|			
Glass					■														
Magnesium				|															
Steel																		■	
Wood	■	■	■																

[a]e, Engineering plastic.

of the speed of light through the materials compared to light traveling through a vacuum. Higher values indicate greater bending of the light.) The limits of transparency of some other plastics follow:

Acrylics	92%	Polyethylene	80%
Cellulose	88%	Polypropylene	90%
Ionomer	92%	Polystyrene	92%
Polycarbonates	90%		
Amino	29%[a]	ABS	33%[a]

[a]Translucent.

Translucent plastics transmit light but objects cannot be clearly seen through them.

Certain plastics, such as acrylics, polyesters, cellulosics, and polystyrene, show colorful stress concentrations when viewed with a polarized-light filter. This photoelastic effect is used as a design tool. ***Photoelastic plastic*** is shaped to a specific design and stressed under polarized light to determine what areas will receive the greatest stress, thus allowing for added material

(a)

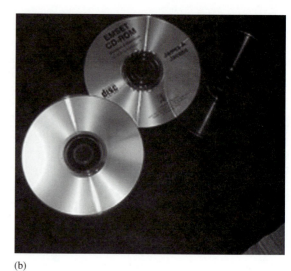
(b)

Figure 6-23(a) Photoelastic plastics allow stress analysis. (Measurement Group, Inc.) (b) Polycarbonates provide substrates for CD-ROMs; on right is CD blank, left shows read side, middle shows label side. See Figure 9-2 for construction of CD.

or redesign to handle expected loads (Figure 6-23). As with glass, many plastics can effectively bend light and serve as *light pipes* for optical fibers used in medical applications, signs, telecommunications, and plastic art. Polycarbonates (PC) possess excellent optical properties as well as harness and toughness. In Figure 23b PC provides the substrate for CD-ROMs that can store up to 660 megabytes of information as compared to less than 2 megabytes on floppy disks.

6.4.2.9 Electrical properties.
Advances in materials have fostered many technological developments. The electrical and electronic developments over the past 75 years owe much to the continuing breakthroughs in plastic materials. Superior insulating properties coupled with good heat resistance of silicones and fluorocarbons lead to large reductions in the weight of electrical motors. Epoxies serve to encapsulate electronic components subjected to temperatures as high as 150°C, corrosive chemical environments, and high vibrations and shocks. Phenolics have the oldest history as electrical insulators and find wide use as housings for automotive-ignition parts (coils and distributors), switches, receptacles, terminal blocks, and bulb bases.

The *dielectric strength* (maximum voltage that a dielectric can withstand without rupture) of most plastics makes them the logical choice as insulators. They range from 79×10^6 volts per meter (V/m) for fluorocarbons to 12×10^6 V/m for certain phenolics. These values compare to 19×10^6 V/m for porcelain and 12×10^6 V/m for alumina. The *dielectric constant* (ability to store electrical energy) of certain plastics puts them into the condenser category. Condensers in electrical circuits maintain voltage with less fluctuation. At 60 hertz, comparative dielectric constants (K) compare as follows: poly(vinyl fluoride), 8.5; cellulose acetate, 4.0 to 5.0; polyethylene (LDPE), 2.25 to 2.35; and polystyrene, 2.45 to 4.75; as compared to alumina, 8 to 10; mica, 5.4 to 8.7; and glass, 3.8 to 4.6. Sections 9.1 and 9.2 describe electrical properties discussed in this section.

Arc resistance is the ability of a material to withstand the arcing effect of an electric current. Insulators, terminals, and switches are subjected to arcing effects, which can burn through or damage the material's surface. The arc test (ASTM D495) measures the total elapsed time in seconds that an electrical current must arc to cause failure. The plastic may *carbonize* and become a conductor; burst into flames; produce thin, wiry lines or *tracks* (surface imperfections in the form of wiry lines) between electrodes; or become an incandescent (glowing hot) conductor. Arc resistance for unfilled polyimide is 125 s; melamines, 110 to 150 s; FEP fluorocarbon, over 300 s; macerated fabric and cord-filled phenolic show tracking but do not conduct; thermosetting cast polyester, 100 to 125 s; and polyethylene (LDPE), 135 to 160 s.

6.4.3 Plastic Selection

In the selection of a material, cost is of paramount concern. Assuming that properties are equivalent and other features such as appearance are similar when comparing materials, then cost is usually the overriding factor in the selection of one material over the other. Often, a trade-off is made when costs are higher for one material than another. For example, epoxy reinforced with graphite fibers offers tensile strength nearly three times that of stainless steel; however, the cost for stainless steel may be 10 or 20 times less than that of the epoxy composites. If weight is important, as in aircraft, the fact that the stainless steel is over five times as dense as the graphite composite can cause the selection of the reinforced plastic. These comparisons are for exotic materials. More common materials such as plain-carbon steel, which is used more frequently, find strong competition from low-cost plastics such as polyethylene, polypropylene, and vinyls.

High demand for a particular material may result in price drops if suppliers increase their production capabilities and competition results in improved processing methods. In the mid-1970s, graphite fiber composites cost about $24 to $29 a pound. To purchase a fishing rod or tennis racket of this new composite, you would have to pay $150 to $250. Now it is possible to buy even better designed rods and rackets of graphite composites for as low as $35 to $60. The processing technology, driven by demand and competition, yields benefits to the consumer. In turn, that demand should cause even better composite processing technology, which will drive prices down while improving composite quality. Looking at materials from more mature plastics processing technology reflects how low prices may go. Plastics in high demand, such as polyethylene and polyvinyls, sell for prices around $1 to $3 a pound.

Green manufacturing practices affect cost. Demands by the public and legislation for green design and manufacturing, in the short term, increase cost as new technologies must be applied to lessen burdens of synthesis and processes on the environment. As improved greening technologies evolve and more raw materials come from recycled plastics, the industry should reap a decline in costs just as recycled aluminum and steel save energy and overall costs.

Much of our materials innovation results from market pressures and legislation. Once a material is available, the next problem is informing designers of its availability. Some companies specialize in developing databases for a variety of materials by many producers. Producing companies also provide databases to customers. The PC–based *Plastics Materials DATA Digest* gives information on over 13,000 TP and TS resins, including properties, ASTM test descriptions, and manufacturers and suppliers.

Corporate average fuel economy (CAFE) is a federal legislative concept that mandates that automakers achieve set average fuel efficiency for the entire line of cars they produce. As the miles per gallon (mpg) values continually rise to meet CAFE, plastics come into greater use due to good strength-to-weight ratios. To encourage auto and truck designers to use its plastics, General Electric Plastics created the EDD-PC engineering-design database. It is coupled with an Alpha I multimaterial, multiprocessing machine for development of large, integrated thermoplastic parts such as bumper systems and automotive hoods. Customers can perform material evaluation 24 hours a day using the latest test and data-acquisition hardware and software. Figure 6-24a shows an example of a large, integrated component that reflects the results of CAFE and industrial design/materials applications of new technology.

Materials selection takes many paths as manufacturers seek profitable products. **Disposal** or **single-use products,** although convenient and often low cost, usually do not follow the principles of reduce, reuse, and recycle products and materials. Still, business opportunities drive manufacturers to market this kind of product. Figure 6-24b details how Kodak's materials selections and methods make its single-use camera more acceptable. How do the tenets of green manufacturing jibe with single-use products?

6.4.4 Processing Plastics

The plastics industry has been very successful in developing a large variety of resins and numerous processes, and plastics have become so competitive with other materials that plastics

Figure 6-24 (a) The Super Plug is a patented, "next-generation" door hardware module developed as a result of a special, five-year program between GE Plastics and Delphi Interior & Lighting Systems. The module, which can consolidate up to 61 separate door parts into one system, provides original equipment manufacturers (OEMs) with a reduction in system costs and weight and vehicle assembly time. GE Plastics custom compounded a new XENOY® PC/Polyester, 30% glass-fiber blend material for proprietary use in the Super Plug. (GE Plastic) Also see Figure 1-16 for more detailed illustrations. (*continued*)

are now found in most markets. Designers continue to seek lightweight, strong materials that allow improved forming capabilities, and plastic-matrix composites, reinforced with fibers and particles, and as laminates, are often favored as the engineering material of choice. Reduced costs are often realized through near-net and net-shape processes inherent in composite processes. Module 8 provides thorough coverage of plastic matrix composites.

Intelligent processing of materials (IPM) involves placing nondestructive evaluation (NDE) sensors in the processing equipment to allow computer monitoring and *adaptive control* (mixtures, temperatures, pressures, etc., adjustment in "real time" during process) to control the complex variables matched against the complex process model or expert system. IPM lends itself to the trend for *total quality management (TQM)* and *statistical process control (SPC),* techniques associated with improved quality in manufacturing products and reduced waste and overall costs. World competitiveness is driving the move for these processing improvements in ceramics and metals as well.

Manufacturing engineers working with designers, materials engineers, and technicians engage in *concurrent* or *simultaneous engineering,* which brings together team members from materials and manufacturing and design during design stages. The designers then provide ongoing engineering throughout production stages. The demand for shorter cycles from design to production, often coupled with low-volume products, favors plastics. Materials scientists and engineers are continually developing new resins to meet performance demands.

Thermosetting (TS) plastics are typically processed by casting, spraying, compression molding, transfer molding, and reaction injection molding (RIM) (Figure 6-25). Because thermosets cannot be resoftened, they are not suitable for recycling except as regrinds used for fillers. They are burned for energy production. Other processes are discussed in Module 8.

Thermoplastics (TP) plastics are much easier to recycle and are therefore favored as the plastics of choice in more and more applications. *Liquid-crystal polymers* (LCPs) are a newer class of engineering thermoplastics that include aromatic polyesters and phenylene polyamide.

KODAK FUN SAVER 35 Cameras
A Recycling Success Story

KODAK FUN SAVER Pocket Flash Camera

Figure 6-24   (b) Plastics in the materials cycle. Kodak's Fun Saver 35 "single-use" cameras seem to present disposal problems. However, Kodak has set up a "closed-loop" recycling system. The cameras are returned to the photofinisher, who has the option of sending them back to Kodak. Kodak reimburses the photofinisher for each camera, plus shipping costs. The components of the camera shown here are recycled in the following manner: A: Film—processed by photofinisher. B: Covers—polystyrene ground up and processed for new covers; paperboard outershell made of recycled paper; polycarbonate plastic shells sold to recyclers for nonphotographic materials. C: Label—ground up for recycling with covers. D: Camera mechanism and electronic flash system are tested, inspected, and reused. E: Viewfinder—polycarbonate plastic is reground and recycled into new internal camera parts. F: Lens—to ensure optical purity, new lenses are installed on each camera; acrylic lenses are ground up and sold to an outside company for raw materials on other products. G: Battery—donated to charity, or photorefinisher may use it. (Courtesy © Eastman Kodak Company)

LCPs lend themselves to easier melt processing that develops highly oriented (rodlike) molecules (Figure 6-7d). The resulting molded LCP parts possess anisotropic properties with chemical and thermal stability at service temperatures up to 190°C. In a manner similar to wood, which is composed of cellulose fibers that generate multidirectional strength, the rod-shaped LCP molecules provide self-reinforcement. Amoco's Xydar, an LCP copolyester resin, represents the newer engineering thermoplastic, which can be molded into complex shapes in injection-molding processes. Xydar has a heat-deflecting temperature of 350°C and a tensile strength of 115 MPa, making it suitable for automotive engine parts. The LCPs can replace ceramics for chemical-processing equipment and find application as engineering plastics, including tennis rackets, electric-motor insulators, and printed circuit boards onto which components are surface mounted. The LCP can withstand temperatures about 40°C hotter than the molten solder.

Thermoplastics are typically processed by casting, extrusion, injection molding, blow molding, thermoforming, drawing, rolling, calendaring, rotational molding, resin transfer molding (RTM), foaming, and spinning (Figure 6-26). For FRP, other processes are discussed in Module 8. Because of the resoftening abilities, scrap thermoplastics can often be put back into the process. After service life, recycling works well.

Cyclic thermoplastic polymers, improved in the late 1980s by General Electric (GE), provided a breakthrough for the automobile industry, which was searching for a low-viscosity engineering thermoplastic that could easily wet fibers for high-production structural composites.

Casting

Spraying

Compression Molding

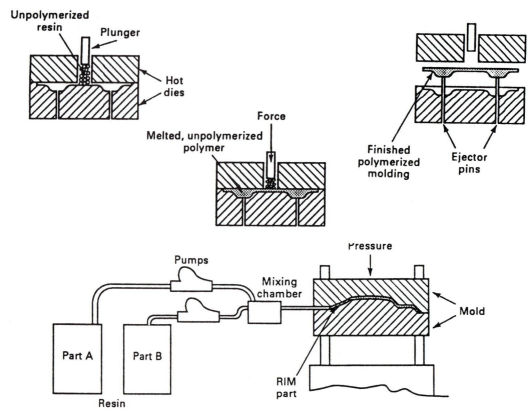

Reaction Injection Molding (RIM)

Figure 6-25 Typical thermosetting processes.

Figure 6-26   (a) Typical thermoplastic plastic processing. (*continued*)

The resulting FRP provided strength comparable to steel with about a third of the weight to provide significant gains in automobile fuel efficiency. Demonstration of the feasibility of this material, for the *Advanced Technology Program,* fell to a joint GE, Ford Motor, NIST venture as part of the big-three automakers' consortium, the *Automobile Composite Consortium* (ACC), which sought to improve U.S. automakers' world competitiveness. For the first time, this advance in cyclic polymer technology made it possible to use engineering thermoplastics in high-speed, low-cost, resin-transfer molding (RTM) composite-manufacturing processes.

To form a structural thermoplastic matrix of bisphenol-A polycarbonate, which is used to impregnate fibers, two stages of polymerization occur. First, an ***oligomer*** (*olig* is a Greek word for "few") of 2 to 20 mers forms into rings (cyclic) of low-viscosity resin that easily flows

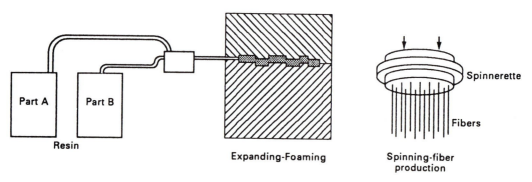

Figure 6-26   (b) (*continued*)

into the fiber within the mold. Next, a special "initiator" is added to cause the linear chains to form. The processes for making cyclic bisphenol-A polycarbonates (BPAPC) involve production of an oligomer [low-molecular-weight (2 to 20) polymer], which has a consistency similar to water rather than the viscosity of cold molasses, as is normal in polycarbonates. The polymer ring (*cyclic polymer*) shown in Figure 6-27a, with its low melt viscosity below temperatures of 300°C, flows like water around the reinforcing polymer preform for complete impregnation. Once that is accomplished, a special "initiator" triggers a polymerization reaction that rapidly transforms the rings into long, linear (Figure 6-27b), high-molecular-weight molecules (M_W 50,000 to 300,000) that give thermoplastics their characteristic strength.

    In addition to excellent damage resistance and favorable processes characteristics, the cyclic ester material offers excellent resistance to common automobile solvents such as gaso-

Figure 6-27   Steps in making bisphenol-A polycarbonates.

line and brake fluids. This BPA polycarbonate is also environmentally friendly because it does not give off volatile or nonvolatile by-products in the condensation polymerization; it offers recycling opportunities because it is thermoplastic.

As an example of applying RTM in the auto industry (see Figure 6-28a and b), Chrysler chose resin-transfer molding for its Viper sports car because it offered a 40% weight savings over sheet steel, with reduced tooling time and tooling costs. Compared to sheet-molding-compound (SMC) composites, RTM gives lighter, stronger panels at lower molding pressure with less labor. With only minor hand finishing, the RTM provides a class A finish. Some problems of holding tolerances on the large complex hood slowed the production rate.

Applying new engineering materials technology that couples new analysis technology such as *magnetic resonance imaging (MRI)* with the ability to engineer materials can result in remarkable design and production innovations. For example, Toyota engineers produced a Super Olefin Polymer as a result of a desire to change the conventional bumper fascia material *elastomer modified polypropylene (EMPP)*. Figure 6-28c provides a property comparison of the two modified polymers. The inner octagon is the EMPP and the outer polygon depicts the improved properties of the Super Olefin Polymer. EMPP is a blend of polypropylene (PP) and *ethylene propylene rubber (EPR)*. Toyota's newly engineered polymer met a material goal for engineering a Super Olefin Polymer with 50% minimum improvement in flexural modulus, melt flow rate, and hardness over the conventional EMPP, while retaining a low coefficient of thermal expansion (CTE) and high impact resistance at $-30°C$ ($-20°F$) and below. The reengineered polymer composite used an EPR elastomeric matrix to provide the following basic performance characteristics required of a bumper fascia:

a) Maintained low-temperature impact resistance and a low CTE.
b) Introduced hard segments into the EPR phase to increase hardness, toughness, and stiffness. The segments form a microscale interpenetrating network (IPN) structure discussed below.
c) Reduced molecular weight and maximized the crystallization of the PP component to increase melt flow rate and hardness.
d) Enhanced the formation of the IPN-like structure by controlling its crystallization rate as well as the crystalline component of the PP.
e) Achieved a recyclable polymer component.

The following comparison reveals the dramatic improvements.

	Super Olefin Polymer	EMPP
Wall thickness, mm	3.8	4.8
Fascia weight, kg/car	6.8	8.7
Molding cycle time, s	85	115
Relative production cost	89	100
Molding temperature, °C	210	230

(a)

(b)

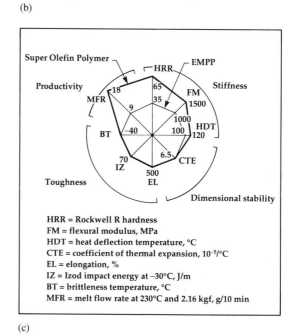

(c)

Figure 6-28 (a) The Dodge Viper was an experiment in new technologies, including concurrent engineering, low-volume and low-investment production, unique design, and numerous materials and processes innovations. (b) An exploded view of the Viper showing RTM plastic panels as the majority of outer and inner body panels; only the lower front body enclosure used SMC. (Chrysler) (c) Major advantages of Super Olefin Polymer over conventional EMPP are dramatized in this property comparison. Note the toughness values of brittleness, temperature, impact strength, and elongation were maintained. Productivity values increased with improved flow rate, as did stiffness values (flexural modulus and heat deflection temperature). There was a slight improvement with dimensional stability, which is a combination of coefficient of thermal expansion (COT or CTE) and stiffness. (*Advanced Materials and Processes*)

Solid Imaging Concept

Figure 6-29 Photopolymers permit a solids imaging technique, known as rapid prototyping, that uses laser beams to process resin into solid models and prototypes.

Because plastics offer so much to the designer, new processes coupled with improved resins will continue to emerge. Another recent development in plastics processing, *solid imaging,* is the use of lasers to polymerize plastic for net-shape processing (Figure 6-29). *Photopolymer* plastics are based on acrylic monomers and oligomers and are generally inferior to typical engineering plastics. Photoactive polymers have been used for decades in applications such as printing plates and inks, ultraviolet (UV) and electron-beam (EB) curable coatings, and photoresists, and for production of holograms.

As one of several *rapid phototyping* techniques, UV laser beams follow paths generated by computer-aided design (CAD) data to create solid parts layer by layer as the lasers polymerize resins. As each layer is cured, a very accurate, solid plastic model emerges on an elevator from the vat of resin. Various techniques of rapid prototyping have been developed, allowing designers to shorten the time required to go from drawing to a solid model without the need for numerous molding, sculpting, cutting, joining, and similar processes traditionally involved in model making. Although photopolymers are suitable for model making, fully functional parts await new polymer development. Rapid prototyping is now being integrated with solidification modeling and computer tomography to produce investment-cast rapid tooling for the die-casting, permanent-mold, and investment-casting industries. Existing parts can be scanned with computer tomography to produce a CAD file. The CAD data can be used with solidification modeling software to determine the best method for casting a tool. Rapid prototyping equipment can be used to produce a pattern for investment casting, while computed tomography can be used again to characterize both the tool and the part it produces. Still another promising technique for processing plastics is the free electron laser, discussed in Module 9.

Dendrimers, another new class of oligomers (Figure 6-30), are spherical or rod-shaped nanostructures with hollow interiors that lead to easier engineering of desired properties. They offer the potential for superior protective coatings, nonlinear fiber optics, adhesives, conductive polymers, lubricants, and reinforcers for composite matrices.

Interpenetrating polyimide-network (IPN) plastics represent research aimed at having the best of both worlds: in this instance, thermosets (TS) and thermoplastics (TP). NASA research has developed a tough, high-temperature polyimide that resists microcracking. The IPN *neat resin* (no additives or reinforcers) makes a valuable composite matrix for graphite

Figure 6-30 Computer model of carbon-based dendrimers, which resembles the structure of atoms, but at the molecular level. The symmetry provides special physical and chemical properties. (George R. Newkome, University of South Florida)

reinforcement suitable for aircraft engines and aerospace structural applications. By combining cross-linked (TS) polyimide with linear (TP) polyimide (Figure 6-31), the ease of processing TS is retained while the toughness of TP is achieved.

Block copolymers (identical mers clustered in blocks along the backbone chain polymer) are also valuable as composite matrices because of their ability to combine the stiffness of a plastic with the toughness of an elastomer. One example of this synergy is with plastic-resin polystyrene, which provides stiffness in a block copolymer with the elastomer polyisoprene that possesses toughness. This is illustrated in the TP rubber (refer to Figure 6-39).

6.4.5 Smart Polymeric Materials

Because polymer chemists can synthesize a wide range of organic polymers, many opportunities for smart polymers wait to be exploited. Current examples of smart polymers include multigrade and synthetic lubrication oils. A 10W40 SAE motor oil will become more viscous with increased heat and less viscous as the temperature drops because of polymers that expand and contract with heat variations.

Electrorheological (ER) and *magnetorheological (MR) fluids* also change viscosity, plasticity, or elasticity with variations in the application of either electrical current or magnetic fields. ERs and MRs go from liquids to gels in milliseconds with the application of external forces. These *smart fluids* consist of polymers (oils and additives) and ceramic particles suspended in liquids. The liquid carrier must be a good insulator and the particles either good conductors or ferromagnetic. When the field is applied, the particles link together to form a gel. The particles assume their random pattern with removal of the external field (Figure 6-32). The smart fluids offer potential in applications such as robotics, hydraulic valves, power transmissions, and other automated equipment that can react to computerized controls.

Shape-memory polymers, much like the shape-memory alloy Nitinol, offer control after polymerization. Polynorbornene, a shape-memory polymer, is molded to a desired shape to which it will return when subjected to loads. Other smart polymers, including liquid crystals, change colors, opacity, and other characteristics with variations in external forces. These polymers find applications in diverse fields, such as medicine (color thermometers), sports (smart tennis rackets), and building construction (LCD heating display and smart windows).

Polyvinylidene chloride (PVDC) film serves as a smart polymer, is piezoelectric (see Module 9), and senses motion. PVDC sensors have applications for automatic lights and auto alarm systems. Liquid-crystal polymers (LCPs) change colors with changes in temperature for thermometers (Figure 6-33a) as well as novelties such as drinking cups, cereal bowls and straws (Figure 6-33b and c). Figure 6-33d shows another example of novelty memory polymers.

Addition-type
thermosetting polyimide
• Easy to process
• Brittle

+

Condensation-type
thermoplastic polyimide
• Difficult to process
• Tough

Semi-IPN
• Easy to process
• Tough

Figure 6-31   Formation of an IPN plastic matrix.

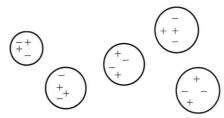

No electrical field present, charges are random, no particle attraction.

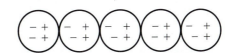

Probe energizing the fluid, particles linked together.

Figure 6-32   Smart polymers—electrorheological (ER) fluids. (From John A. Marshall, "Liquids That Take Only Milliseconds to Turn into Solids," National Educators' Workshop: *Update 93—Standard Experiments in Engineering, Materials Science, and Technology.* NASA Conference Publication 3259, April 1994, pp. 315–322.)

EMT WEBLINK

Biosphere 2—*http://www.bio2.edu/*

DuPont homepage—*www.dupont.com*

Society of Plastics Engineering—*http://4spe.org/*

Introduction to plastics—*http://www.polymers.com/dotcom/polycon/primer.html*

GE Plastics Education through cartoons on history, properties and terminology, design
 and processing—*http://www.ge.com/plastics/resins/education/cartoon.html*

Nylon—*http://me.mit.edu/2.01/Taxonomy/Characteristics/Polyamides.html*

Dow Chemical—*http://www.dow.com*

(a) Note F°76 display

(b)

(c)

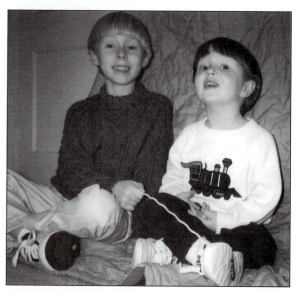

(d)

Figure 6-33 (a) LCP serve as thermometers. (b) Child's cereal bowl made of a smart LCP changes from green to yellow with hot meal and (c) drinking straw turns from white to red upon contact with cold liquids. (d) Smart shoe strings—Luke straightens Sean's polymer shoe laces; note the one not under tension stays curled. The laces have a memory and stay coiled thus not requiring tying. What a relief for the child's parents!

SELF-ASSESSMENT

6-15. What does "organic" refer to in the SPI definition of plastics? What are the other major ingredients in plastics?

6-16. What advantages do most plastics offer compared to linear plastics?

6-17. Describe the following plastic classifications: **(a)** general purpose, **(b)** engineering, **(c)** cellular.

6-18. What term means that a material is capable of bearing supporting loads?

6-19. Name some natural fibers. Name some manufactured (synthetic) fibers and give typical applications.

6-20. Why are cellular plastics increasing in use? List typical applications of foam plastics.

6-21. What is FRP?

6-22. What element serves as the backbone for most plastics? Name one plastic that has an inorganic backbone.

6-23. What does ASTM stand for?

6-24. Describe the benefits of smart polymers. Explain the advantage of shape-memory polymers.

6-25. What additive would be valuable in a plastic that is exposed frequently to sunlight? How does it alter the plastic chemically?

6-26. List additives for plastics that moderate the following service conditions: **(a)** aircraft storage compartments subject to fire; **(b)** coating for a wrench handle; **(c)** vegetable-storage drawers in refrigerated bins; **(d)** flexible gloves; **(e)** stiff, lightweight automobile interior door panels.

6-27. Determine whether thermosetting or thermoplastic plastics are *generally* best for the following type of service conditions: **(a)** high impact, **(b)** creep, **(c)** chemical exposure.

6-28. Although a certain plastic might have lower specific gravity, better toughness, and equal optical properties, why might glass still be preferred for a service condition such as storm doors, eyeglasses, or lenses on instrument panels? What plastic competes strongly with optical glasses? List some of the competing factors.

6-29. What electrical property is of primary consideration when selecting an electrical insulator? How do plastics compare as electrical insulators?

6-30. When would it be undesirable to have a plastic with low water-absorption resistance? Name three plastics with good water-absorption resistance.

6-31. What are the combined stresses that act on a material subjected to weathering? Select the three plastics most capable of service conditions that involve weathering. How would the three compare to **(a)** plain-carbon steel, **(b)** alumina, and **(c)** glass? Give the coefficient of thermal expansion and the rate of thermal conductivity for each of the three materials.

6-32. List five properties essential for the rope used to pull water skiers. Select two plastics suitable for this service and explain the selections. Use the ASTM abbreviations.

6-33. Given the choice among PS, PMMA, ABS, and PF as possibilities for motorcycle helmets, select one and explain your choice.

6-34. Use Tables 6-2 and 6-7 to calculate the specific strength of aluminum, and compare it to the specific strength of an acetal plastic.

6-35. Select two plastics that would withstand the service conditions under the hood of an automobile. Use the ASTM abbreviation. Explain your selection.

6-36. Give some reasons why viscoelasticity offers a designer certain advantages and disadvantages when using polymers in product design.

6-37. What technique involves placing NDE sensors in processing equipment to allow computer monitoring and adaptive control (adjustment in mixtures, temperatures, pressures, etc., in real time during the process) to control the complex variables that are matched against the complex process model? What factors cause a demand for simultaneous engineering?

6-38. What two stages do cyclic thermoplastic polymers go through as they are processed into matrices for engineering composites?

6-39. From the tables in this module and current prices listed in the reference issues of periodicals such as *Machine Design* and *Modern Plastics,* calculate the specific strength and cost per unit volume for the following materials: **(a)** acrylic, **(b)** polycarbonate, **(c)** plain carbon, **(d)** steel, **(e)** wood, **(f)** epoxy, **(g)** aluminum.

OBJECTIVE QUESTIONS

6-12.   What main advantage do cellular plastics offer?

    a. Reduced density   **b.** Improved crystallinity

    c. Greater density   **d.** Improved strength

6-13.   The objective of adding microspheres to plastics is which of the following?

    a. Increase weight   **b.** Improve impact strength

    c. Improve processibility   **d.** Reduce crystallinity

6-14.   Which group of polymer fibers provides superior strength and resistance to biological attack?

    a. Manufactured   **b.** Cotton   **c.** Jute   **d.** Hemp

6-15.   In transportation vehicles, structural plastics are often not used because they lack good specific strength.

    T (true)

    F (false)

6-16.   A property unique to polymers that gives them the ability to return to their original size and shape once a load is removed, much like a shock absorber handles bumps.

    a. Viscoelasticity   **b.** Voigt   **c.** Specific strength   **d.** Branching

6-17.   A problem with many plastics is that they often creep at room temperature over time with fairly small loads.

    T (true)

    F (false)

6-18.   Paper is a better packaging material because, in the modern landfill, it will biodegrade.

    T (true)

    F (false)

6-19.   Thermosetting plastics are not good for recycling because they cannot be softened for reprocessing.

    T (true)

    F (false)

6-20.   Common processes for thermoplastic plastics are casting, extrusion, blow molding, and thermoforming.

    T (true)

    F (false)

6-21.   The first stage of a cyclic thermoplastic plastic that makes processing easier.

    a. Oligomer   **b.** Linear   **c.** Hevea   **d.** RIM

6-22.   Property found in wood that results from grain patterns.

    a. Anisotropy   **b.** Isotropy   **c.** Homogenous   **d.** Interface

6-23.   A ratio of strength to weight.

    a. Isotropic   **b.** Tensile   **c.** Specific strength   **d.** Density

6-24.   Which advantage do smart polymeric materials offer designers?

    a. Eliminate selection problems   **b.** Reduce raw materials costs

    c. Adapt to environmental changes   **d.** Reduce processing

6-25.   Which application would benefit most from the properties of electrorheological fluids?

    a. Multigrade engine oil   **b.** Robotic power transmission

    c. Superconductors   **d.** Telephone switching

Colorful & Cool Stuff

Through the ages, objects made from available materials gained value for both their utility and as objects of art. Objects from early times shown here now have value as art and are collected for their beauty and value. In earlier times the very wealthy could afford the colorful objects. Today's objects, composed of an infinite combination of materials, provide colorful and cool "stuff." Due to improvements in materials technology, today and into the future, the average citizen may possess colorful, cool stuff at very reasonable prices. The images shown here represent a very small sample of the materials available, both natural and man-made, for art and utility. Both macro and micro scale images reveal colors and textures that tell a lot about the properties, structures, and processes.

Figure C-1 Ancient Art Cast glass *Groove-Decorated Bowl* circa 200 BC. [Chrysler Museum of Art http://www.chrysler.org]

Figure C-4 Color Polymer Stuff: products made of various plastics and elastomers demonstrate the wide ranges of colors, forms, and properties possible with synthetic polymers. Note the white and blue and white and black toothbrushes in the middle with flexible heads, contoured shape, and soft rubber dots and strips for gripping. What types of plastics and elastomers are shown here?

Figure C-2 Timex Race Car Road racing has grown to a year-round sport. Racing car owners recognize and welcome the introduction of newly developed materials and manufacturing processes in helping them maintain their competitive edge in racing. Sports Car Club of America's Circle 66 car is representative of this trend with its use of lightweight steel engine parts, carbon/Kevlar brake pads, polyurethane bushings, HD flame-resistant foam, and ThermoTec header wrap. Cryogenic processing of various engine and brake parts reduces wear and costs of operation. Refer to Pause & Ponder in Module 4. [Wally and Liz Hunt]

Figure C-3 Modern Art A structural vessel form made of glass, silver, and copper, 26" high x 6-11/16" diameter, by American artist Hiroshi Yamano in 1991 [Chrysler Museum of Art http://www.chrysler.org]

Figure C-5 Photoelastic plastic and polarized light show stress distributions of a dovetail opening on a turbine disk of a jet engine turbine blade, thus allowing for evaluation of creep properties. [Measurement Group Inc.]

Figure C-6 Magnification (about 130X) allows characterization of the delamination of layered oxide $K_4Nb_6O_{17}$ due to hydration. Details of the hydration-induced delamination of surface of micaceous oxide $K_4Nb_6O_{17}$ are revealed by interference-contrast microscopy. The method provides quantifiable information concerning the morphology of the crenulated outer surface layer oxide following hydration after exposure to air. Substrate interplanar separations resulting from hydration-induced delamination of micaceous layers of this compound. [L. Boatner, J. Ramey, T. Geer and *Advanced Materials & Processes*]

Metals

Figure C-7 Totem Artists use a wide full range of materials and processes to express themselves, as in this example of a painted steel sculpture that measures 105" x 109". Created by Alexander Calder in the 20th century. [Chrysler Museum of Art http://www.chrysler.org]

Figure C-8 Dragonfly Lamp A combination of stained glass bonded with copper foil on a bronze base, this 1910 lamp from the Tiffany Studio shows how mixing materials can provide beauty and utility. 1991 [Chrysler Museum of Art http://www.chrysler.org]

Figure C-9 Aluminum Engine Alloys of the 3000 series (Al-Si+Cu or Mg alloys) are some of the most widely used due to the flexibility provided by the high silicon content and its contribution to fluidity–plus their response to heat treatment, which provides a variety of high-strength options. Furthermore, the 3xx.x series may be cast by a variety of techniques ranging from relatively simple sand or die casting to very intricate permanent mold, lost foam/lost wax type castings, and the newer thixocasting and squeeze casting technologies. [Aluminum Association]

Figure C-10 Zinc alloy die castings, both natural and plated, reflect the intricate shapes possible when using this low melting point metal and casting process. Finishes include nickel, brass, tin, copper, and silver.

Figure C-11 Semi-solid forming produced this as cast ingot of A356 alloy with anodized microstructure in polarized light at 250X. Refer to Pause & Ponder in Module 3. [A. Kazakov and *Advanced Materials & Processes*]

Figure C-13 As cast Al-3.12 wt%Cu mounted in epoxy, manually polished, and etched with Kellar's reagent and photographed in light microscopy under Normarski differential interference contrast. [P. Crawford and *Advanced Materials & Processes*]

Figure C-12a
HAZ (Heat Affected Zone) of a weld. Longitudinal view (15X, bright field) of weld joining C103 and Ta-10W alloys in an anodized nozzle. Green indicates the C103 (NbHfTi) alloy; blue, the weld metal; and yellow, the Ta-10W alloy. Red indicates where hafnium came out of the C103 alloy because of overheating during welding.

Figure C-12b
HAZ (Heat Affected Zone) of a weld. Transverse view of a weld joining Ta-10W and C 103 (NbHfTi) anodized sheet alloys. Blue shows C103 alloys, yellow shows Ta-10W, and red shows weld metal (15X). [P. Danielson and *Advanced Materials & Processes*]

Figure C-14 Saint John sculpture from the 13th century made of wood and painted; 15 1/2" tall [Chrysler Museum of Art http://www.chrysler.org]

Figure C-16 White oak and red oak showing: top is end-grained; middle is edgegrain, quarter-sawed; and bottom is flat-grained, plainsawed. [Forest Products Laboratory]

White oak   Red oak

Figure C-15 Hardwoods and softwoods 1) Birch, 2) Hickory, 3) Eastern Red Cedar, 4) Southern Yellow Pine, 5) White Oak Group, 6) Western Red Cedar, 7) Douglas Fir, 8) White Pine, 9) Black Walnut, 10) Red Gum, 11) Redwood, 12) Hard Maple, 13) Red Oak Group, 14) White Ash [Heritage Workshop]

Figure C-17 The look of wood without some of its maintenance problems. Using virgin or recycled thermoplastic plastic, Extrudawood™ has the warm colors and rich texture of wood, but offers weather and insect resistance. It is possible to extrude with simulated wood colors or it can be painted. http://www.cpiplastics.com

Figure C-18 Many colors, forms, and textures of plastics: 1. Polystyrene foams (gas-filled beads, preformed expanded beads), 2. Polystyrene float, 3. Foamed PS sheet 4. Laminate insulation foam, 5. Polycarbonate pellets, 6. Acetal sheet, 8. Rigid PVC powder, 9. Rigid PVC pellets, 10. Rigid PVC (window frame, jointed tubing sitting on top of wood grain laminate), 11. Flexible PVC pellets, 12. Elasticized PVC (tubing, stain-edging, coated wire, and electric plug).

Figure C-19 Architectural acrylic (Plexiglas™) samples exhibit the wide range of colors from clear as glass to translucent and opaque.

Ceramics

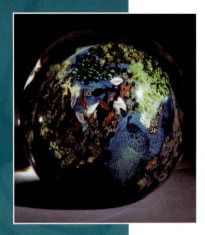

Figure C-20

Megaworld—a 1992 American glass: cane, furnace-gathered, solid and flame worked glass by Josh Simpson. 10 1/4" diameter.
[Chrysler Museum of Art http://www.chrysler.org]

Figure C-21 *Intruders*—an English glass: blown, acid etched, cameo-carved by George and Thomas Woodall circa 1893. [Chrysler Museum of Art http://www.chrysler.org]

Figure C-22 Ceramic industrial wear parts are used as mechanical seals, erosion-resistant liners, metal processing dies, slurry pump rotors, and more. Materials include alumina, silicon nitride, zirconia, silicon carbide, boron carbide, and boron nitride. [American Ceramic Society]

Figure C-23 Advanced ceramics of silicon nitride and silicon carbide parts such as turbo rotors, valves, and inserts provide superior heat resistance in automobile engines. [American Ceramic Society]

Figure C-24

Glass optical fibers have provided a technological breakthrough in the area of telecommunications. Information that was once carried electrically through hundreds of copper wires can now be carried through just one high-quality transparent silica (glass) fiber. Using this technology has increased the speed and volume of information that can be carried by orders of magnitude over that which is possible using copper cable. The reliability of the transmitted information is also greatly improved with fiber optic fibers. [American Ceramic Society]

Figure C-25

Masonry (brick and concrete) supports aluminum dome. The 6003 or 6061 (Al-Mg-Si) alloys are extrudable, making them the first choice for architectural and structural members where unusual or particularly strength- or stiffness-criticality is important. Room structures for arenas and gymnasiums are usually 6003 or 6061 extruded tube, covered with 5XXX alloy sheet. [Aluminum Association]

Figure C-26 Polymer matrix composites and reinforcers: 1. S glass fibers, 2. Carbon fibers, 3. Kevlar fibers, 4. High-temperature performance polymer powder, 5. LARC TP1 resin, 6. Graphite epoxy prepreg, 7. Epoxy fiber (orientation 0/90/90/0 4ply), 8. Nomex honeycomb, 9. Fiberglass pipe, 10. Aluminum/polysulfone pultrusion, 11. Pultrusion hat section, 12. Fiberglass

Figure C-27 Fiberglass reinforced polymer composite (PPS) thermoplastic, liquid crystalline spheroids in row-nucleation along the fiber interface. [Luther M. Gammon, The Boeing Company]

Figure C-28
Carbon fiber reinforced polymer composite cohesive fracture of adhesive. [Luther M. Gammon, The Boeing Company]

Figure C-29 Carbon fiber reinforced polymer composite Toray thermoset, produced with heat ramp of 5 dec./min. [Luther M. Gammon, The Boeing Company]

Figure C-30 Defects seen in glass fiber-filled polyethermide composite. With 500X magnification and ultra-thin section transmitted differential interference contrast, voids are revealed within the matrix filled with short chopped glass fibers. [Luther M. Gammon, The Boeing Company]

Materials Systems

Figure C-31 NASA's contribution to advanced materials systems. In order to travel into outer space, many new high-strength lightweight materials were developed by NASA and its contractors. From improved aluminum alloys to complex "designed" polymer matrix composites to optoelectronic material systems we now benefit on earth with "spin-offs" from aerospace materials research. [Aluminum Association]

Figure C-32a
Materials systems in cars. The earliest automobiles used many natural materials, including wood, leather, rubber, and common iron and steel alloys.

Figure C-32b
Efforts to make use of new materials and materials systems are often seen in concept cars such as this 1956 Firebird II, which used a titanium body and acrylic canopy and was powered by a gas tubine engine, thereby helping designers to try new technology.

Figure C-33
Material systems in the home. User-friendly appliances rely on powered metal parts to run quietly, and never need lubrication. The acrylic Corian countertop has the durability of ceramic, but greater toughness and ease of forming. Tempered glass makes an improved cutting board over wood because it reduces bacteria formation. Thin, glass-ceramic, Corelle dinnerware is lightweight, microwavable, and break-resistant.

Figure C-34
Sporting equipment has also taken advantage of the *Materials Age*. The persimmon club head (seen on the bottom driver) from an earlier time has been replaced with metals such as aluminum and titanium and exotic trimetals. Graphite shafts also compete with steel. However, persimmon seems to be making a comeback on finer golf clubs.

Figure C-35
Abrasives have evolved into many forms and are made in a wide variety of ceramics, adhesives, and cloth and paper backings to provide the finishes necessary for high quality and high performance demanded in today's products.

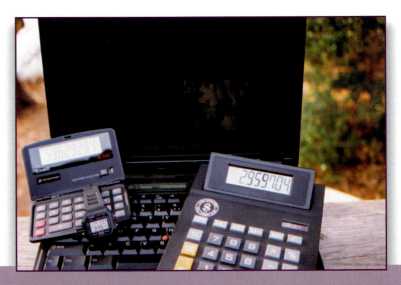

Figure C-36

Computers and all of the hardware of the *Information Age* make advances through improvement of materials systems such as semiconductors, batteries, and liquid crystal displays.

Figure C-37

Materials technologies develop techniques to grow synthetic jewels (rubies and sapphires) and machine these materials to tolerances of .000030" diameter, .000007" roundness, and .000040" concentricity to provide the necessary lenses, fiber optic guides, substrates, and rods for communication devices.

Figure C-38

Spaced Out Composite. The rugged appearance of the surface of this graphite/polyimide specimen is typical for the graphite reinforced polymer matrix composites which flew on the leading edge of NASA's Long Duration Exposure Facility or LDEF. Specimen at 2000X magnification, 3 micron ultra thin section for light, differential interference contrast, 420 nanometer wave shift.

The material specimens and systems on board this 60-ton experiment-laden satellite were exposed to low earth orbit for 5.8 years before a last-minute retrieval by the space shuttle *Columbia*. The jagged peaks and valleys seen in this photomicrograph were created by the extremely erosive nature of atomic oxygen, the primary atmospheric constituent at LDEF mission altitudes. In addition to atomic oxygen, LDEF was bombarded by ultraviolet radiation, interstellar dust, micrometeroids, and a variety of subatomic particles. Although the damage seen here is a few mils deep, space hardware designers now have a better understanding of what materials can be used where and for how long thanks to LDEF. [Luther M. Gammon, The Boeing Company]

Polymers in cars. How many types of polymers can you think of in the typical car? This diagram depicts applications of rubber. The rubber reduces vibration and isolates against noise, as seen in (1) engine mounts, (2) strut mounts, (3) bushings, (4) couplings, (5) air sleeves, (6) insulation against impact, (7) insulating foams, (8) gaskets, (9) seat padding, (10) trim, (11) cushions for headrests, (12) door cushions, and (13) roof trim. It's easy to notice the rubber tires, which involve blending of various natural and synthetic rubber with reinforcement wire and fibers. Rubber is also used to make cooling and brakes hoses, floor mats, and fan belts. Each of the applications has special requirements that result in many rubber compounds. Other important polymers include sealants, lubricants, oils, natural leather, and numerous adhesives. Of course, many plastics are also used in the modern automobile. (Bridgestone Firestone)

6.5 ELASTOMERS: NATURAL AND SYNTHETIC RUBBERS

Elastomers refer to the group of polymers that exhibit rubbery or elastic behavior, including *natural rubber* and *synthetic rubber.* The American Society for Testing and Materials (ASTM) defines **elastomers** in ASTM D1566-66T as "macromolecular material that returns rapidly to approximately the initial dimension and shape after substantial deformation by a weak stress and release of the stress." Elastomers will stretch from 200% to over 900% of their original length at room temperature and return to their original state when the tensile stress is released. Natural rubber, known as **hevea rubber,** comes from the rubber tree *Hevea brasiliensis.* Columbus brought this bouncy material back to Europe with other curiosities from the New World, but nearly 300 years passed before it was given serious attention. The name **rubber** resulted from the material's ability to erase black pencil marks from paper by rubbing. Synthetic rubber, as with plastic, is a hydrocarbon with petroleum as its major source. Development of the processing of synthetic rubber into a method suitable for high-volume production did not fully materialize until the beginning of World War II, when the United States was cut off from the natural rubber supply in India and was forced to use large quantities of synthetic rubber.

6.5.1 Natural Rubber and Vulcanization

Natural rubber (NR), in the form of liquid resin secreted from the inner bark of the **Hevea brasiliensis** tree, is known as latex. Latex is not tree sap. It consists of isoprene molecules (Figure 6-34a). Polyisoprene, or rubber, is formed through a natural polymerization in the tree (Figure 6-34b). The liquid latex dries into a thermoplastic material. The structure is an amorphous mass of coiled and kinked chains with constant, thermally induced motion of the atoms and chains. Rather than stretch when heated, rubber shrinks because the thermal agitation causes the chains to entangle and draw up. The structure provides the *resilience* for rubber to spring back into shape when compressed or stretched (Figure 6-34c and d). When a tensile load is applied, the structure changes as the chains straighten (bond straightening) and stretch (bond lengthening) (Figure 6-35a and b), and crystallinity is achieved in varying degrees depending on the amount of stress. With increased crystallinity comes a greater strength, increased rigidity, and increased hardness. The structural changes in rubber make it good for tires because the portion of the tire under high stress is crystalline and provides support for the vehicle it is carrying, while at the same time the portions of the tire not under high stress are still resilient and absorb shock from bumps in the road.

The problems with natural rubber are that it is too soft and has too many reactive sites (double bonds), which cause rapid oxidation and *dry rot.* It is also somewhat plastic and will not recover from high stress. In 1839, Charles Goodyear discovered, through the addition of sulfur to natural rubber compounds, that it was possible to increase hardness and reduce susceptibility to oxidation and reaction with other chemicals. The sulfur *vulcanizes* the rubber, or changes it into a thermosetting polymer, by linking together the molecular chains at their double bonds (Figure 6-36). Two sulfur atoms per pair of isoprene mers are needed, and with 5% of the mer pairs crosslinked, a flexible, resilient rubber exists. Further cross-linking increases hardness. Vulcanization of only sulfur to rubber requires several hours and temperatures around 145°C. However, accelerators and activators added to the compound will result in vulcanization within minutes.

In addition to natural rubber are sulfur cross-links polybutadiene (BR), polyisoprene (IR), and acrylate (ACM) elastomers. Along with sulfur, other cross-linking agents vulcanize certain synthetic elastomers. Butyl rubber (IIR) and ethylene–propylene copolymers (EPM) rely on sulfur and phenolic agents; zinc oxides form cross-links in polysulfide (PTR) and polychlorapene (CSM). Vulcanized natural rubber has excellent flexural strength or deformability, tensile strength, and abrasion resistance, but it is attacked by petroleum oil, greases, and gasoline. Its value in automobile tires comes from low heat buildup, discussed later. Natural rubber has superior overall engineering capabilities compared to synthetics. The choice of a synthetic rubber over natural rubber or of one synthetic over another boils down to specific properties required, price, and processibility.

○ Carbon atom

○ Hydrogen atom

══ Double-covalent bond
(active site)

── Single-covalent bond

(a) Isoprene

(b) Rubber (natural polyisoprene)

(c)

Figure 6-34   (a) and (b) Natural rubber. (c) tennis, running and other sport shoes possess variations in density and stiffness as a result of blending rubber compounds. (*continued*)

Because natural rubber comes from a renewable source and because it has excellent properties, natural rubber science and technology continue to make strides. Deproteinized natural rubber (DPNR) is an example of developments in improving this natural elastomer. By removal of ingredients with an affinity for water, such as protein, polyols, and inorganic salts, fatigue resistance and other mechanical properties are enhanced. Research has been renewed in the United States into the use of guayule shrubs to produce rubber; they were grown in California during World War II. Only 5% to 10% yield was possible then; a 20% yield is required to justify the use of this raw material.

6.5.2 Synthetic Rubber

Even with the improved properties obtained through vulcanization, natural rubber has poor resistance to aging. It is attacked by ultraviolet light, oxygen, and heat because it still has many reactive sites. Due to these shortcomings and because the rubber tree grows only in special environments that were vulnerable to political sanctions, the search for a substitute produced several synthetic rubbers in the early to middle part of the twentieth century. The number of synthetic rubbers has grown to include many special-purpose elastomers. Improvements in synthetic elastomers have also brought improvement in the additives and fillers for natural rubber (NR). Almost identical to NR is synthetic polyisoprene (IR), except that it has greater stretching ability. Synthetic rubbers have the same raw materials used in plastics. Starting with crude oil and natural gas, many formulations are possible. Styrene, the monomer for

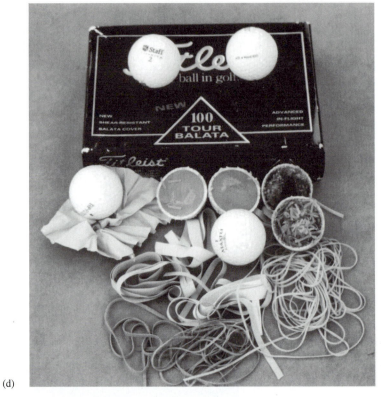

(d)

Figure 6-34   (d) Golf balls exemplify the variety of properties achieved with synthetic and natural rubbers. Note the variety on core rubber, which ranges from solid core, to thin band winding and liquid center. Ball covers of synthetic balata rubber provide a soft feel as the ball is struck, yet are tough and scuff resistant. Considerable effort goes into the dimple design on the covers in order to achieve optimum aerodynamics.

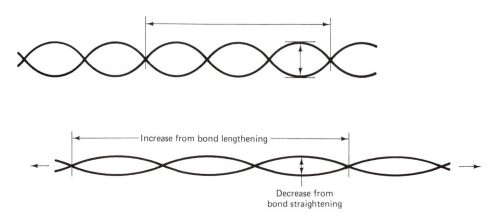

Figure 6-35   Effect of tension on bonding: (a) no load, (b) tensile load.

styrene–butadiene rubber (SBR), is derived from coal as vinyl benzene, $CH = CH_2$, a product of benzene and ethylene. Butadiene is derived from petroleum. The hydrocarbon is obtained in fractionation of cracking petroleum used for olefins, polymers, or gasoline. The 1-butene is separated and catalytically dehydrogenated in the vapor phase to produce butadiene.

Figure 6-37a shows the steps for modern automobile tire production. During the final processing step, heat and pressure are used to form the final tire shape, then vulcanize the rubber that was in a thermoplastic condition during forming. Note, at the beginning stage, the mix-

Figure 6-36   Vulcanized rubber. (a) Natural rubber—thermoplastic. (b) Vulcanized (cross-linked) rubber—thermoset.

ing of natural and synthetic rubbers with carbon black, sulfur, and other additives. A runflat tire and wheel system is shown in Figure 6-37b.

    6.5.2.1 Additives (modifiers).   Additives (modifiers) to synthetic rubber are introduced at the initial processing stage and also during compounding. Soap and water are added to styrene and butadiene in the making of raw styrene–butadiene rubber (SBR) to produce an emulsion that keeps droplets of the monomers from separating out. *Accelerators* and *activators* speed sulfur vulcanizing, while *retarders* prevent vulcanizing before it is required. For those saturated synthetic elastomers that have no double bonds, *peroxides* promote sulfur vulcanizing. *Peptizers* soften raw rubber, *pigmenters* add color, *abrasives* provide abrasing action for products such as erasers, *hardeners* increase rigidity, and *antioxidants* and *antiozonants* prevent aging or dry rot from the sunlight and ozone.

    Fillers help to increase bulk to the compound while holding down cost. Inert fillers include talc, chalk, and clay. Reinforced rubbers use many of the same *fillers* as plastics, including metals, glass, and polymer fibers. *Carbon black* also serves as a strengthening filler and imparts hardness.

Mixing process

Natural/synthetic rubber

Carbon black

Sulfur and other chemical agents

Form into sheet strips

Cool

Tread extruding process

Apply heat to make rubber elastic

Cool

Extrude rubber

Cut to tire length

Cord manufacturing process

Twist into tire cord of nylon, polyester or steel

Coat fabric with rubber

Weave into cord fabric

Cutting process

Cut at proper angle into specific length and width

Bead-making process

Bead wire

Align bead wires

Coat with rubber

Coil to form bead rings

(a)

Figure 6-37 (a) Steps in tire production. (Bridgestone Firestone) (*continued*)

■ Building process

Tread rubber, sidetread rubber

Bead rings

Body ply, steel belts

Begin with body ply

Attach bead rings

Apply sidetread rubber, steel belts and tread rubber

"Green" tire

■ Vulcanizing process

Apply heat and pressure to green tire

■ Inspection process

Trim and perform appearance inspection and balance/uniformity check

(b)

Figure 6-37 (b) Steps in tire production (*continued*).

LOW PRESSURE SUPPORT INSERT

BEAD RETENTION WEDGE HUMP

LOW PRESSURE WARNING SENSOR

Figure 6-37   (c) Runflat tire and wheel system. (Bridgestone Firestone)

(c)

6.5.2.2 Thermoplastic rubbers or elastoplastics.   Thermoplastic rubbers (TR) or elastoplastics do not cross-link as do most elastomers, even though their properties are similar to many rubbers. A copolymer of styrene and butadiene was introduced in 1965 as a thermoplastic rubber. The elastoplastics have the ability to soften when heated for processing and, upon cooling, become solid yet maintain their elastic behavior. Vulcanization is eliminated. The thermoplastic behavior develops as a result of the structure, in which the chains of one monomer form links in the center of block molecules of other monomers. Figure 6-38 shows that the domains in the middle have a hard linking effect, while the other end of the segments has elastomeric behavior. Upon heating through T_g, the hard centers soften to plasticize the rubber for processing. When cooled, the hard blocks re-form to again link the copolymers. This linking is especially effective at low temperatures.

Two-block, copolymer, thermoplastic elastomers manufactured under the trade name Kraton include styrene–butadiene styrene (SBS) and styrene–isoprene styrene (SIS). The styrene (S) occurs in two thermoplastic blocks, as shown in Figure 6-39a, on either end of a rubber block of butadiene (B) or isoprene (I). Rather than having randomly distributed monomers, as do most copolymers, block copolymers form uniformly distributed sections as blocks. Figure 6-39b shows that a network of block chains has formed. It consists of a highly uniform distribution of *polystyrene domains* and the separate butadiene rubber chains. The *rubber network,* which is physically joined, links the domains and cannot move because the domains are immobile, thus providing a physical rather than a chemical cross-linking. The physical cross-linking can be broken when heated, which makes the compound a thermoplastic. Kraton is therefore capable of continuous molding and remolding. The polystyrene domain consists of actual particles that become hard and glasslike at room temperature; they take on cylindrical, spheroidal, and platal configurations.

The major elastoplastics are polyester copolymers, olefinics, polyurethanes, and styrene copolymers. The wide service temperature ranges, low-temperature flexibility, and ease of painting place thermoplastic rubbers in contention as a substitute for thermosetting rubbers. Their ease of processibility has given them wide acceptance in the auto industry as flexible bumpers and other exterior panels, steering wheels, hose covers, housings for seat

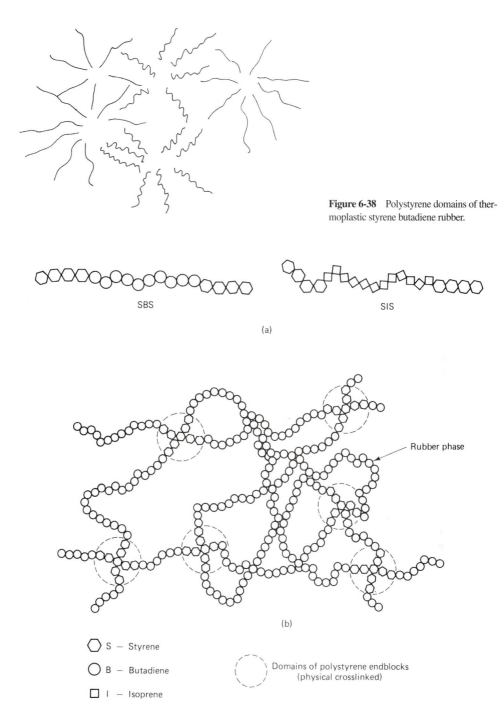

Figure 6-38   Polystyrene domains of thermoplastic styrene butadiene rubber.

SBS

SIS

(a)

Rubber phase

(b)

⬡ S — Styrene

◯ B — Butadiene

▢ I — Isoprene

Domains of polystyrene endblocks (physical crosslinked)

Figure 6-39   Thermoplastic rubber "network" structure. (a) Block copolymers. (b) SBS.

belts and horns, wire covering, and oil seals. Other applications include housewares, toys, sporting equipment, adhesives, rainwear, footwear, and skate wheels. As with plastics, scrap thermoplastic elastomers can be ground and remolded. Because of the complex nature of the modern automotive tire, their "after-life" has been a real problem. Many methods have been employed to recycle the millions of discarded tires, including grinding them into materials for road paving , as tiles for patios and tennis courts, as fence posts, and for burning to produce electricity.

Type	Test temp (°C)	Class	Maximum swell (%)
A	70	A	No requirement
B	100	B	140
C	125	C	120
D	150	D	100
E	175	E	80
F	200	F	60
G	225	G	40
H	250	H	30
J	275	J	20
		K	10

(a) ASTM D 2000/SAE J200 type and class

(b) ASTM designation example

Figure 6-40   Elastomer designation. (a) ASTM D2000/SAE J200. (b) ASTM designation example.

Shape-memory polymers are a class of "smart materials" that return to their original fabricated shape under the influence of temperature. *Bioelastics* are synthetic fibers that behave much like muscles. Based on the natural protein, elastin, these biomaterials offer medical applications for surgical repairs as well as industrial uses such as for robotic hands and automatic valve controls to replace metal springs and hydraulic and pneumatic devices. Bioelastics are nontoxic, biocompatible with human tissue, and biodegradable. Because bioelastics can convert chemical energy into mechanical energy, they offer potential as sensors or transducers.

6.5.3 Classification and Properties of Elastomers

Designations of elastomers by the American Society for Testing and Materials (ASTM D2000 standard) and the Society of Automotive Engineers (SAE) group them by type and class (Figure 6-40). *Type* reveals the maximum service temperature, which ranges from A (70°C) to J (275°C). *Class* indicates the maximum percentage of swell when immersed in oil and ranges from A, with no requirement, to K, at 10% swell. Table 10-10 shows the class and type for selected elastomers. ASTM D1418 establishes abbreviations for elastomers, also shown in Table 10-10. To illustrate, PTR, AK designates a polysulfide elastomer with a maximum service temperature of 70°C that can swell as much as 10%. Thiokol is the trade name for polysulfide. Additional systems of designation indicate hardness and minimum tensile strength (Figure 6-40). ASTM AA625 designates (AA) natural rubber with Shore durometer hardness (6), meaning 60 plus or minus 5 points on the scale and (25) 2500 as the minimum tensile strength.

Table 10-10 compares properties for common elastomers. NR is the standard for comparison of the synthetics, as can be seen by its overall superiority, with excellent resistance to abrasion, tear, impact, electrical current, and water absorption, plus a tensile strength of 3100 Pa and a wide hardness range of 30 to 100 Shore A. Through designed formulation, the synthetic rubbers can surpass NR for special properties. NBR, ACM, FPM, and PTR offer excellent resistance to oil, kerosene, and gasoline, while NR is not recommended for service when exposure to these chemicals is expected. For high-temperature service, SL and VE are excellent. Neoprene resists deterioration from weathering, oxygen, ozone, oil, gasoline, and greases and has good tear and abrasion resistance. It finds applications as soles and heels for work shoes and liners for chemical tanks and pipelines, serves as an adhesive, and is applied like paint as a protective coating. Fluoroelastomers such as FPM are space-age elastomers that meet the demands of the aircraft and aerospace industry because of excellent resistance to heat and fluids, but they carry a high price tag. NR has only fair *gas permeability,* which means that it will allow some gas to pass through it. Butyl rubber (IIR) resists gas permeating and serves well as inner tubes, gas hoses, tubing, and diaphragms.

Rubber plays an important role in health care. Silicone elastomers, the most stable group of all elastomers, have polymer chains composed of silicon and oxygen backbones. Silicone rubber is used for various body replacement and enhancement components. Silicone is often used for plastic, reconstructive surgery of body tissue; recent problems with breast implants have caused concern for silicone as an implant. Silicone is a derivative of the second most abundant material on earth—silicon. Silicon composes about 28% of the earth's crust and is usually found in combination with other elements. One combination is silica, a naturally occurring silicon–oxygen compound known as silicon dioxide. Sand is probably the best known silica. Silicones are a family of extraordinary, synthetic materials made from polymers. Of high molecular weight, their molecular structure consists of spines of oxygen alternating with silicon. By attaching various organic, carbon-containing groups of elements to the spines, liquids, pastes, and rubberlike materials can be produced. The liquids and pastes can be cured into solids with application of heat or exposure to air to make adhesives and sealants. In other forms they are used as elastomers, coatings, oils, lubricants, and resins. If the attachment of these compounds to the spines of the SiO_2 polymer is short, the result is a volatile liquid that has the appearance of water and evaporates in air. If the attachment is long and ropelike, the physical properties change and the materials are more viscous, ranging from syrupy to a "solid" that flows if left unattended. You can stretch it, pound it, roll it into a ball and bounce it, and hit it with a hammer so it shatters. This material has the trade name Silly Putty.

Uses for silicone include baby bottle nipples, heat-resistant gaskets, cable insulation, spark-plug boots, encapsulations, bumper gels, brake fluid, adhesives for exterior tile on space shuttles, sealants, and weatherproof caulks. Theoretically, millions of silicone materials can be made just by altering the structure of the silicone chain. Only a few thousand have been produced and studied so far.

Natural-rubber latex has gained considerable attention recently with the concern for the transmission of AIDS and other sexually transmitted diseases. Latex surgical and examination gloves and condoms must meet the highest standards of manufacturing and testing to ensure the highest reliability. ASTM standards D3577, D2578, and D3492 address these procedures.

Hysteresis is an energy loss through heating in elastomers and creates a problem with many synthetic elastomers. If you stretch a rubber band and hold it against your cheek, you can feel the release of heat. The constant flexing of automobile tires generates enough heat in elastomers such as SBR that, if a tire were made wholly of this compound, it would quickly deteriorate. NR, on the other hand, converts the stressing into elasticity and quickly rebounds with less heating. The high hysteresis of SBR has an advantage in preventing slippage on wet surfaces and providing better abrasion resistance than NR, so it is used for the tread of tires while the NR goes into the sidewalls.

Elastomers possess extremely diverse properties, as do plastics. Thousands of additives and fillers are available and as many as 30 may be mixed into a single elastomeric compound. The range of hardness includes soft foam rubbers used for pillows to hard-rubber (ebonite) battery cases. Figure 6-41a provides a comparison of the hardness of selected elastomers and plastics. *Foam rubbers* have been formulated to respond to body temperature and pressure. Developed for astronauts, these foams now serve as orthopedic support for bed mattresses and chairs because of their ability to provide even support, which reduces fatigue and soreness caused by prolonged body contact. Upon impact the foam firms up, and when heated, it softens. Joggers and other athletes use foam inserts in their shoes to reduce shock from the impact of running and jumping.

EPDM is a class of 13 types of materials that are a miracle of rubber technology because of their low cost and ability to be painted. Through compounding of carbon black, clay, processing oils, and cross-linking agents, the elastomer compounds can be molded or extruded as

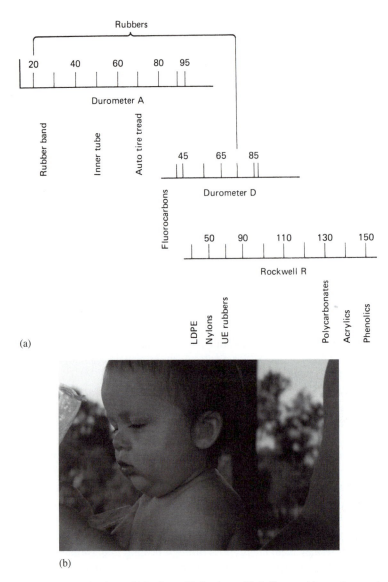

(a)

(b)

Figure 6-41　(a) Comparisons of polymer hardness. (b) Easily molded and remolded silicone rubber makes good ear plugs.

thermoplastics or thermosets (also see Figure 6-41b). Temperature insensitivity, impact resistance, high deformation recovery, dimensional stability, and distortion resistance make them good choices for flexible exterior parts on automobiles, including bumpers and fascia.

Heat resistance is an important property in certain elastomer applications. Fluoroelastomers (FPM) can withstand temperatures over 340°C and continuous service temperatures up to nearly 290°C in exposure to oils, steam, and certain fluids. They have been improved to handle low temperatures down to around −35°C, at which point they become brittle. Among the silicon elastomers, room-temperature vulcanizing (RTV) compounds have stable properties through a range of temperatures, from nearly −53° to 250°C. Ethylene acrylics used for belts, seals, rollers, hoses, and gaskets can withstand hydraulic fluids, engine coolants, and hot oils for over a year at temperatures over 200°C. Chlorinated polyethylene elastomer, which is a random compounding of chlorine with high-density polyethylene, has good weather, hydrocarbon fuel, and oil resistance down to −15°C and up to nearly 150°C of continuous service. Grades of thermoplastic elastomers (TP) such as polyurethane provide good weathering as low as −15°C and as high as 120°C. Appendix Table 10-10 provides general ranges of maximum service temperatures for the groups of elastomers listed.

6.5.4 Selection of Elastomers

Whether natural or synthetic, rubber offers distinguishing properties. It is flexible, airtight, tough and resilient, and waterproof, and it resists corrosion caused by most common chemicals. Its ability to stretch in any direction and return to its original shape make it valuable in a variety of applications, from tires, to ball bladders, to hoses and gaskets. Reinforcing fibers can make it a nonstretch composite that retains the other valuable attributes of elastomers. Some high-performance elastomers, such as polyurethane (PUR), can be synthesized as thermosets or thermoplastics (TPUs); TS PUR will elongate 650% and has a tensile strength up to 100,000 kPa; TPU can elongate 600% and has a tensile strength of about 40,000 kPa.

The *cost of elastomers* varies widely and has a tremendous influence on selection of one over another. The fluoroelastomer mentioned previously that is capable of temperatures over 340°C sells for over $4500 per kilogram on finished parts, whereas the price for SBR would range from $0.60 to $0.80 per kilogram. Natural rubber sells for about $0.90 per kilogram, SI for $3.75 to $11 per kilogram, and heat-resistant ethylene acrylics sell for $3 to $4 per kilogram. While all fluoroelastomers do not cost as much as $4500 per kilogram, they are generally the most expensive elastomers, selling at around $28 per kilogram, and as copolymers with silicones, most fluorosilicones reach prices of $65 to over $90 per kilogram. Of the thermoplastics, the following indicate general prices per kilogram: copolyesters, $3.20 to $3.65; olefins, $1.37 to $1.75; and styrenes, down to $0.90 to $1.37.

Environmental concerns regarding rubber center on disposal of reinforced rubber, especially vehicle tires. Many approaches have been applied to the disposal issue as can be seen in the categories listed with links to *Recycler's World* website *http://www.recycle.net/recycle/Rubber/index.html,* January 15, 2000, as follows:

Scrap Tire Disposal and Recycling, Used Tire, Retread Tires and Casings, Tire Parts Recycling, Non Tire Rubber Recycling, Granulated Rubber (Tire Crumb), Tubes and Flaps Recycling, Tire Pyrolysis & Gasification

These recycling approaches range from (1) burning the tires as fuel for heat and power production, to (2) using them as filler and encasing whole tires in very large concrete blocks used as bulkheads to prevent soil erosion, to (3) shredding them for filler for roads and gardening mulch as seen in Module 1.

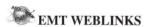 **EMT WEBLINKS**

Making rubber—*http://www.negromex.com.mx/english/Learn.htm*
Firestone rubber products—*http://www.firesyn.com/*
Enterplast rubber products—*http://www.enterplast.com/rubber.html*

6.6 ADHESIVES

Until the mid-1940s, the selection of adhesives for joining metallic materials together normally included either mechanical fasteners, such as bolts, rivets, and pins, or a thermal bonding method, such as welding, brazing, or soldering. Today, the knowledgeable designer learns that adhesive bonding must receive equal consideration over the traditional methods of joining parts together. The technician and crafter involved in maintenance and repairs can choose adhesives as an alternative technique for fastening broken or unjoined components. Generally, welding, riveting, and adhesive joining are classified as permanent joining methods, while mechanical fasteners such as bolts, screws, nuts, and pins allow disassembly.

The ideal material described in Module 1 does not exist, and neither does the ***ideal adhesive.*** If it did exist, it would (a) stick to any material, (b) need no surface preparation, and (c) maintain its strength and other properties forever under all types of conditions. An adhe-

Figure 6-42 Adhesives in construction. Low-cost, lightweight panel resulting from strong honeycomb paper core bonded with adhesives between plywood sheeting. (U.S. Forest Products Laboratory)

sive bond has a great advantage over a mechanical joint, such as a nailed, screwed, bolted, or riveted joint. The strength of these mechanical joints depends on the combined strengths of the materials joined together and the fasteners involved. This interaction takes place where these fasteners come in contact with the material to be joined, causing a concentration of stress that may cause one of the materials to exceed its ultimate strength and cause the connection to fail. Not so with an adhesive joint, which permits an even distribution of stress to be applied over the complete bond area.

Adhesives provide the benefit of joining dissimilar materials. Adhesives also provide light weight; joint sealing; sound and vibration dampening; thermal and electrical insulation; uniform strength; and low-cost, low-skill techniques. These advantages mean that adhesives find many uses in joining for building construction, including the replacement of nails with mastic adhesives to bond plywood or particleboard to flooring joists and drywall panels to wall studs. *Elastomeric adhesives* are good for joining wooden furring strips to concrete. Various other adhesives are used to fasten vinyl and ceramic tiles (see Figure 6-42).

The wide use of adhesives by the aerospace industry results from its need to join thin sheets of aluminum, plastics, or composites to framework and still maintain smooth aerodynamic surfaces, light weight, uniform strength, and good fatigue resistance. *Epoxy* adhesives stick together structural members formerly joined by rivets, which caused problems due to their nonuniform stress distribution that created stress concentrations; also, rivets are heavy and prone to corrosion.

To help the automotive industry achieve more fuel-efficient vehicles through reduction of weight, adhesive-joining technology provides many unique joining methods (see Figure 5-39). Certain *structural acrylic* and *plastisol* adhesives offer the auto industry and other such assembly lines an advantage because of their compatibility with oily metal surfaces, and they bond many dissimilar metals and plastics in addition to wood, glass, hardboard, and asbestos board. These acrylics resist moderate temperatures, have high impact and peel strength, and cure rapidly with minimal shrinkage. *Urethane adhesives* bond together the steel and glass sunroofs. Elastomeric and plastic bumper systems and plastic body side molding use adhesives that dissolve the contaminating oils and films found on production lines, thus providing a dependable joint. The successful use of *cyanoacrylates,* or *"superglues,"* on electronic, electrical, appliance, and instrument assembly lines represents the widespread acceptance of these instant bonding adhesives (Figure 6-43a). Figure 43b shows a display of adhesives and substrate cleaners types and forms.

Among the major reasons for rejecting adhesive bonding over mechanical or thermal joining is heat sensitivity, with a normal service temperature ranging from 160°C to over 500°C. However, most common adhesives withstand a lower range, from −90° to 290°C. The space shuttle vehicle employs an adhesive bonding system for its ceramic heat shield tiles, with design temperatures up to 1260°C. Epoxy adhesives in the form of microcapsules serve as

Figure 6-43   (a) Cyanoacrylics set rapidly for electronic-component assembly. (General Motors) (b) This array of adhesives and substrate cleaners shows just a few of the hundreds of adhesive types and forms available. Included here are two part epoxies, contact cement, water-based vinyl adhesives, acrylic and PVC solvents, elastomeric cement, cyanoacrylates, loss of solvent and anaerobic adhesives.

locking systems for mechanical fasteners. With the shearing force of a nut on an epoxy-coated bolt, the adhesive mixes and cures quickly.

6.6.1 Adhesive Systems and the Bond Joint

An adhesively bonded joint results from the adhesive's ability to flow, wet, and set. ASTM defines **adhesive** as a substance capable of holding materials together by surface

Figure 6-44 Wetting of substrate. Poor wetting: liquid beads up on surface. Good wetting: liquid flows into micro hills and valleys on surface, and porous materials absorb liquid.

attachment. The materials held together are *adherends* or *substrates*. **Substrate** is the broader term that defines a material upon the surface of which an adhesive-containing substance is spread for any purpose such as bonding or coating. As liquids of varying viscosity ranging from low viscosities below that of water to a semisolid (viscoelastic) or syrup consistency, including hot-melted solids, adhesives must wet the surfaces to which they bond. **Wetting** involves flowing (1) into large openings such as pores of woods or cells of foam plastics or (2) into the microscopic hills and valleys that exist on the smoothest of surfaces, such as polished metals or glass. A key factor in adhesive selection is the ability of the adhesive to wet all or most of the available bonding sites of the **substrates** (surfaces making the bond joint) (Figure 6-44). Refer to Figure 6-48 for a microscopic view of wood to contrast with smoother surfaces. The free spreading of the adhesive on the substrate surface may be impaired by conditions such as air entrapment, moisture or oxide buildups, and contaminants such as oil and dirt. Figure 6-44 illustrates a nonwetting condition, such as water beading up on a waxed surface, and a wetting condition, such as water absorbed in tissue paper. To achieve wetting of an adherend, the adhesive must not have *cohesive forces* (attraction of molecules inside the adhesive) that provide a very strong *surface tension* (force contracting the liquid into a droplet). The surface tension must be less than the adhesive forces between the adherend and adhesive. A high-surface-energy liquid such as epoxy adhesive will not wet a low-energy polyethylene solid; but the polyethylene liquid with lower surface energy will wet and form good adhesion to an epoxy solid.

   Contact angle, or wetting angle, is a measure of how a liquid will spread on a solid. Many liquids spread easily on solids and have zero or very small contact angles. If the contact angle is greater than 90°, it is considered nonwetting. Water on clean glass would have a contact angle of 0°; on polyethylene the water would form an angle of over 100°. The analysis of contact angle has many applications in interface technology, ranging from the spread of adhesives, lubricants, printing inks, and protective coatings, to achieving maximum coverage with insecticides and herbicides. Once the wetting has occurred, the adhesive must *cure* into a solid (occasionally a viscoelastic) or *set* so that it develops adhesion, thus bonding materials (*adherends*) together into an adhesive joint.

   Four theories attempt to explain the forces involved in the adhesion of an adhesive to its substrates. Both *chemical* and *physical bonds* are involved in adhesive bonding, but the exact nature of all forces is not fully known. Covalent bonds and physical attraction through van der Waals forces seem to operate in adhesives as they do in other polymers, but scientists differ on how the adhesion develops. On porous materials such as wood, *mechanical interlock* (penetration of adhesives into openings) provides added surface area for the adhesive to attach, but again some scientists believe that mechanical interlock has a minor role in adhesion. Testing of the adhesive bonding systems must deal with the *interfacial adhesion* (attraction between adhesives and substrates) and *practical adhesion* (breaking strength of joint).

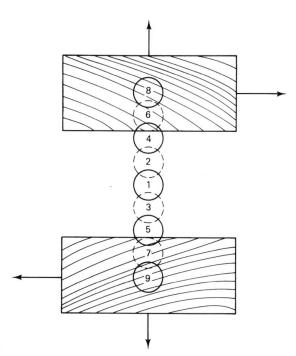

Figure 6-45   Marra's adhesive-joint model. Links of an adhesive bond: *link 1*—the adhesive film; *links 2 and 3*—intra-adhesive boundary layer, strongly influenced by the adherend; *links 4 and 5*—adhesive-adherend interface, site of adhesion forces; *links 6 and 7*—adherend subsurface, partially fractured in preparation; *links 8 and 9*—adherend proper. (Dr. Alan A. Marra, University of Massachusetts)

Synthetic resins and natural adhesives **cure** through the various polymerization processes covered earlier in this chapter. Addition and condensation polymerization develop in the thermosetting adhesives that form permanent bonds; thermoplastic adhesives can be softened to reposition parts or for disassembly.

6.6.2 Adhesive-Bond Chain Links

To grasp the complexities of adhesive bonding, see the model developed by Alan A. Marra (Figure 6-45), which illustrates the nine links in the adhesive-bond chain. The chain-link model reveals the interrelations of factors in the bonded product that include (1) adhesive composition, (2) adhesive application, (3) the bonded materials' properties, (4) their preparation, and (5) the stress and environment that the product will encounter in actual service. The physical and chemical theories and principles associated with the chain links become quite involved and require adhesive specialists to deal with them. The user of adhesives must recognize that proper adhesive bonding depends on the choice of the proper adhesive for the intended environment and on application according to specifications. The best adhesives will fail if not used properly. Link 1: The adhesive film must have proper **cohesion** to stick together when in service. Links 2 and 3: The intra-adhesive boundary layers are strongly influenced by the chemical and physical properties of the adherend; for example, extractives in wood may have an undesirable chemical interaction with the adhesive and change the composition and/or action of the adhesive as it is curing. Links 4 and 5 are critical links in the systems because this is where the adherend and adhesive engage. Problems such as surface contamination can prevent the anchoring of the adhesives at the necessary bonding sites. The compatibility of adhesive to adherend is required here so that proper wetting takes place; several theories attempt to explain the actual forces operating in these two links. Links 6 and 7: The adherend/substrate is critical in a material such as wood or ceramic, where a damaged surface must be repaired through penetration of the adhesive into the pores. Mechanical interlocking may help with the repair. Finally, the adherend material's links 8 and 9 must possess sufficient cohesive strength and compatibility with the bonded joint. For example, a porous material such as a cellular plastic or elastomer may absorb and transport harmful liquids to the bonded joint, where the chemical could degrade the adhesive.

These problems lead to four **main factors in adhesive selection:**

1. Materials to be bonded and type of joint
2. Conditions under which the bonding must be done
3. Durability and strength levels required during service of the product
4. Cost of the bonding process, including joint preparation, material cleaning, cost of adhesives, and so on

These four factors will vary considerably and require close attention to specifications by those making adhesive selections. Technical sales representatives, handbooks, and other sources of technical information guide the user, but as with any industrial production system, the manufacturer of the product must evaluate the adhesively bonded product throughout its development and maintain quality control over the adhesives and the application of those adhesives to products.

Advances in microprocessors and computers resulted from improvements in semiconductors and ceramic substrates. Many of those advances were possible because of continually improved **interface technology,** which focused on joining materials. Multilayered ceramics (MLCs) are examples of improved integration, one product of interface technology. Very-large-scale integration (VLSI) of semiconductors also presents interface challenges, as solder balls of only 0.002 in. join microthin metal circuitry deposited on ceramics. Printed-circuit boards use **surface-mount technology (SMT)** of electronic components (see Figure 8-42).

6.6.3 Types of Adhesives and Their Properties

Adhesives fit into various classifications, depending on factors such as the **substrate,** physical form (solid, liquid, viscoelastic), and bonding characteristics. Among the characteristics is the determination of whether the adhesive is structural or nonstructural. **Structural adhesives,** by ASTM definition, are those used to transfer required loads between adherents exposed to service environments typical for the structure involved. Some authorities classify structural adhesives as those whose failure could result in a threat to public safety, loss of life, or substantial property damage. These adhesives were formulated to replace nuts, bolts, rivets, and welds. A U.S. government study called the primary aircraft bond structure test (PABST) demonstrated that adhesive bonding is feasible and economical for replacing mechanical fasteners and welds for the main structural components on aircraft. Adhesives must possess **serviceability** (the ability to withstand design conditions such as heat, cold, stress, moisture, or chemical attack) and **durability** (the ability to maintain serviceability of the bond joint over its expected life).

Structural adhesives for wood include **phenol–resorcinol–formaldehyde** polymers, which are thermosetting, all-weather exterior adhesives used on laminated structures. **Casein** adhesives are structural water-resistant adhesives (moisture content of wood less than 16%) commonly used on softwood laminated beams and doors. **Nonstructural adhesives** include **urea–formaldehyde,** which is used on particleboard and furniture; **poly(vinyl acetate)** (Elmer's) for furniture and interior woodwork; and *starch* and *dextrins* for paper products. **Protein** adhesives include animal glues made from bone, hide, and fish (they are used for woodworking and gummed labels); casein from skimmed milk; and **blood** and **soybean flours** formulated with phenolic and formaldehyde resins, which cross-link to serve as water-resistant adhesives on interior plywood.

Thermoplastic adhesives are generally serviceable at temperatures below the thermosets. Among the thermoplastics, **cyanoacrylates** (Superglue) have a service range of $-5°$ to $80°C$; **polyamides,** $15°$ to $100°C$; **poly(vinyl acetate),** up to $45°C$; and **butadiene styrene,** $5°$ to $70°C$. These compare with thermosets such as **epoxy,** $-10°$ to $140°C$; **phenolic epoxy,** $-25°$ to $260°C$; and **phenolic neoprene,** $-20°$ to $95°C$. Some synthetic adhesives come as thermosets and thermoplastics, such as **polyimides** (TS up to $370°C$ and TP $-160°$ to $350°C$), **silicones** (TS $-25°$ to $250°C$ and TP $-75°C$), and polyurethanes (TS and TP $-60°$ to $175°C$).

Elastomeric adhesives are formulated with natural or synthetic rubber or blended with other resins to provide impact resistance and to withstand wide temperature variations. These adhesives include *neoprene, polysulfides, butyls,* and *nitriles.* A phenolic–rubber blend adhesive bonds brake linings that withstand severe shock and temperature and saltwater exposure.

Except for pressure-sensitive tapes that remain tacky, adhesives change from the liquid state to the solid state upon curing. Tapes such as adhesive tape, duct tape, and transparent tape do not develop the strength found in solid adhesives; however, double-coated pressure-sensitive tapes serve as hold-down tapes for use on machine tables of milling machines and surface grinders.

Curing of adhesives involves the cooling of hot melts, loss of solvents, anaerobic environments, and two-part mixtures. ***Hot-melt adhesives*** usually begin as a solid that becomes liquid in heat guns or furnaces, or with other heat sources that develop around 150°C temperatures; most cool and set within seconds after removal of the heat source. ***Loss-of-solvent adhesives*** employ volatile liquids, including water and organic solvents, that dissolve the adhesive base material to form the fluid; upon application to the joint, the adhesives harden as the solvents evaporate or penetrate porous adherends. With solvent adhesive, especially, and other adhesives, one must determine whether the adhesive will attack the substrate so severely that it may produce weaknesses in the material being joined. ***Anaerobic adhesives*** retain their fluid state when exposed to oxygen, but when squeezed into thin joints that block the oxygen, they set up into hard, strong adhesives. ***Two-part mix adhesives*** are stored with the resin and catalyst (or hardener) separated; when mixed the catalyst causes cross-linking of the resin.

Loss-of-solvent adhesives often have long setting times; and except for water-solvent types, the solvents present health and fire safety problems. Also, they rely on our short supply of petroleum, which is causing a trend toward water-based adhesives. Anaerobics come in one part, cure quickly, and develop high shear and impact strength, and some have gap-filling abilities. Other adhesives such as cyanoacrylates cure as a result of exposure to the moist alkali environment found on most surfaces. Contact cements are usually rubber-based structural adhesives that form an instant bond when substrates coated with the cured adhesives are brought into contact, thus eliminating the need for clamping. Most other adhesives require pressure long enough for the adhesive joint to cure.

Caulks and *sealants* work like adhesives and protective coatings to fill gaps and seal against weather, liquids, and many commercial applications. ***Polymer caulks*** are classified by their expected service life, known as *range.* Oil-based caulks are low range and low priced. Acrylics, butyls, and latexes are medium-range caulks and serve many construction needs. High-range caulks are the most costly; they include silicone (for considerable joint movement, ultraviolet-light exposure, waterproofing, and wide temperature variations), various copolymers, and urethane. Some high-range caulks such as urethane are also available as foam to fill large gaps. Industrial suppliers and home improvement centers have hundreds of caulks and sealants; some even allow patching and joining under water. An advanced material, Gore-Tex joint sealant, is 100% PTFE in expanded, continuous cord form that allows it to be used as an "asbestos-free" sealant. It can serve as gaskets for engine blocks (Figure 6-46a) that will never leak or need replacing and can be used on irregular or damaged surfaces because it fills in and conforms to surface imperfections to provide an ideal seal against gas or liquid. ***Sealants,*** applied as liquids or pastes, cure to form seals against gases or liquids. Often the sealants also act as adhesives, especially in building construction. See the silicone sealants in Figure 6-46b.

6.6.4 Adhesive Selection

The wide range of available adhesives and sealants is partially the result of the difficulty of finding suitable matches of adhesive to adherend. Many general guides exist, but they should be used with caution. For important production jobs, it is best to consult suppliers to obtain the best choice of adhesive. For general-purpose jobs, some broad guidelines help.

As discussed earlier, plastics present the most difficulty because of wetting problems. Table 6-8 lists some common adhesives and their typical applications and provides an overview of selected adhesives, adherents, and applications. The following list provides a basic guide for adhesive selection in selected materials groups. Read the technical data for each adhesive to learn more specific details on compatibility. For important jobs, conduct a small

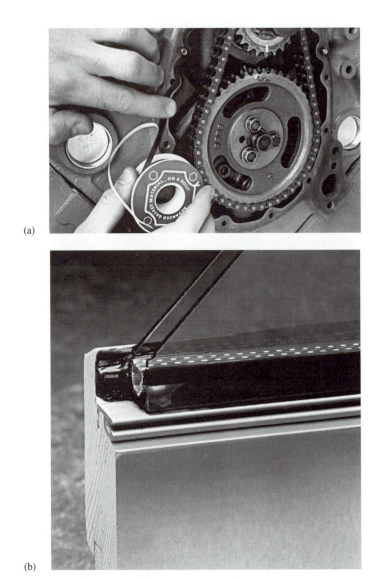

(a)

(b)

Figure 6-46   (a) Gore-Tex® gaskets of PTFE provide excellent seals for rough surfaces where oil, fluids, and temperatures may damage other sealing materials. (W. L. Gore & Associates, Inc.) (b) One part insulating glass, silicone sealant offers exceptional hydrolytic stability, fast cure, high strength, high modulus, and primerless adhesion to glass and many substrates. (G.E. Plastics)

sample test before doing the complete bonding. If adhesive tapes will not stick to an adherent, you must abrade the surface to remove coatings or apply another coating that is compatible with the adhesive.

General Material Group	General Adhesive Groups
Metals	Epoxy, cyanoacrylates, urethane
Wood	Vinyl emulsion, resin, epoxy
Ceramic	Cyanoacrylates, epoxy, urethane
Plastic	Numerous options: read packaging information

Adhesives have been developed to meet the most varied of circumstances. In medicine, methylmethacrylate resins join artificial hip sockets to natural hip bones and repair broken, damaged bones. Cyanoacrylates bond eye tissue and other skin tissue. Thermoplastic polyester resins bond copper foil to flexible backing on printed-circuit boards and,

TABLE 6-8 COMMON ADHESIVES

Adhesive	Typical Adherents and Applications
Acrylics (two-part)	Plastics (ABS, phenolic-melamine, polyester, polystyrene, polysulfone, PVC) to metal, rubber to metal, and plastics to plastics
Casein	Interior wood components
Cyanoacrylates (anaerobic)	Thread locking, gear and bushing mounting, aluminum, copper, steel, ceramics, glass, rubber, polycarbonate, phenolic, acrylics
Epoxy	Most adherents except some plastics such as acetals, polypropylenes, polyurethanes, and silicones; many modifications including nylon epoxies and epoxy phenolics
Melamine formaldehyde	Paper, textiles, hardwood, plywood for interior use
Natural rubber	Cork, foam rubber, leather, paper-building materials, pressure-sensitive tapes
Neoprene rubber	Many plastics (ABS, fluorocarbons, polystyrene), aluminum, ceramics, copper gasketing and laminating
Nitrile rubber and phenolic	Many plastics (ABS, cellulose, fluorocarbons, PVC, polyester), aluminum, ceramics, glass, magnesium, microwave isolators
Phenol–formaldehyde	Cork, cardboard, softwood, plywood for exterior use
Poly (vinyl acetate)	Textiles, polystyrene, wood, other porous and semiporous materials, interior furniture and construction
Resorcinol–formaldehyde	Asbestos, cork, paper, rubber, wood furniture, laminated exterior wooden beams, wooden boats
Silicone	Aluminum, ceramics, glass, magnesium, phenolics, polyester, acrylics, rubber, steel, textiles, integrated circuits, sealants, gasketing
Urethane elastomer	Many plastics (ABS, polyester, acrylics, PVC), aluminum, ceramics, copper, glass, magnesium, steel-bond solid propellants, cryogenic applications, insulation

with epoxy and phenolic adhesives, withstand etchants and the heat of soldering operations. **Weld bonding** combines spot welding and adhesives. The weld immobilizes the adherends to allow the adhesives to cure. The increased use of resin-based composites promises even wider use of adhesives.

Adhesives technology extends to many diverse applications. For example, the duPont website page on adhesives, *http://www.dupont.com/industrial-polymers/adhesives/index.html,* January 15, 2000, lists the following groups of adhesives and sealants:

- Hot Melt and Wax Blend Adhesives
- Extrusion and Laminating Adhesives and Sealants
- Water-Based Adhesives
- Binders and Film Formers

By visiting the duPont site you will see an array of polymers to meet those applications, with data sheets on properties, uses, and procedures.

Websites listed in EMT Weblinks provide similar data on adhesives, and some also provide activities.

 EMT WEBLINKS

DuPont industrial polymers—*http://www.dupont.com/industrial-polymers/index.html*
Devcon adhesives—*http://www.devcon.com/*
Elmers products—*http://elmers.com/*
Loctite Corporation—*http://www.loctite.com/*

6.7 OTHER IMPORTANT POLYMERICS

Polymer science and technology continue to generate innovative materials. In the medical field, research and development efforts encompass replicating human tissue such as skin and producing a polymer that will stimulate cartilage regeneration. Polymers that will not burn yet are easy to process, lightweight polymer batteries, water-soluble polymers, and controlled-release polymers are some materials of continuing interest. The following groups of polymers are other significant engineering polymers.

6.7.1 Protective Coatings and Preservatives

Because polymers include synthetics, an endless variety of them can be synthesized. Protective coatings cover another category of polymers not discussed in this module. Protective coatings include natural polymers such as shellac, oil-based paint, and many synthetic polymers, including the same general types used for plastics, elastomers, and adhesives (e.g., epoxies, silicones, latex, urethanes, and acrylics). These coatings develop their protective skins in a number of ways including polymerization, cooling of hot melts, and evaporation of solvents or water. Evaporation leaves behind a film of the polymer that was dissolved in solvent or suspended in water.

Environmental concerns about ozone layers, water quality, and overall air quality have required that the petrochemical industry put considerable effort into eliminating the traditional materials-related uses of chlorofluorocarbons (CFCs) and *volatile organic compounds (VOCs)*. VOCs create air-quality problems at the ground level (troposphere) by forming ozone and photochemical oxidants, the primary ingredients in smog. CFCs create problems in the upper atmosphere (stratosphere) by reacting with the delicately balanced ozone layer that protects the earth from the sun's ultraviolet rays. CFCs deplete the "good" ozone. Much of the current development in protective coatings focuses on eliminating VOC solvents because they create health hazards to workers and harm the atmosphere. By replacing VOCs with water as the vehicle to carry the polymer dispersions, they evaporate with few harmful effects. Much of the paint used today is water based.

As a part of the *green movement,* paints have been developed based on natural substances (plants and minerals), without pesticides, preservatives, and antimildew agents. Some paint manufacturers separate the solvents from the solids in excess paint, leaving pigment and resin that is recycled as primer or utility paint.

A polymerization process known as *group-transfer polymerization (GTP)* permits unlimited polymer-chain growth in acrylics, with the potential for high-performance auto finishes that will cure at lower temperatures (180°F versus 250°F to 300°F). The process also uses lower concentrations of polluting solvents.

Although wood offers natural beauty and warmth, it is susceptible to weathering, moisture, insect attack, and stress on weaker sections that have suffered during the life of the tree. Many protective coatings and polymer enhancements have been developed, such as diffusion of *poly(ethylene glycol) (PEG)* to preserve archeologically significant wooden pieces. Cross-linking chemicals can permanently bond cell walls of wood for stability and resistance to fungi, but their use results in decreased strength. Enhancement of mechanical properties of wood is achieved with vacuum impregnation of liquid monomers such as epoxy and methyl methacrylate into openings that result from sap leaving the tree during curing. The resins are polymerized after impregnation to make harder, stronger modified wood for boat hulls, billiard cues, knife handles, and parquet flooring. Large structural timber and lumber is both pressure and nonpressure treated (brushing, soaking, dipping, and heating) using oil-type preservatives: creosote, pentachlorophenol (penta), copper naphthenate, and waterborne inorganic arsenical compounds.

 EMT WEBLINK

Frequently Asked Questions about finishes—*http://www.fpl.fs.fed.us/faqs.htm#finishing*

6.7.2 Oils, Lubricants, and Fluids

Polymers play an important role in many types of lubricating, cutting, and hydraulic applications as solids and as additives or suspensions in liquids. Engine-oil viscosity for multigrade oils, such as SAE 10W-40, changes when the ambient and engine temperatures change because polymers control the oil thickness. New long-lasting engine oils are the result of synthetic polymers.

Electrorheological (ER) fluids (introduced in Section 6.4.5 as smart polymers) congeal from a liquid into a solid mass in a few milliseconds when an electrical current is passed through the liquid. When the current is removed, the solids return to liquids. ER fluid has promise as a hydraulic fluid for robots, automotive clutches, and artificial human limbs. Experimentation with combinations of selected oils with particles (microscopic solids of polymethacrylic acid and water in long, tangled chains) will continue in the quest for the ideal ER fluid. What possibilities might this type of solid lubricant hold for problems encountered with high-temperature ceramic engines or automotive brakes?

6.7.3 Conductive Polymers

A major disadvantage of lead-acid and other metallic batteries is their weight. The development of conductive polymers such as polyacetylene electrodes could open the way to long-sought electric automobiles. Remember that polymers do not normally serve as conductors because their electrons are locked in covalent bonding rather than as free electrons in metals. Oxidation, resulting from doping polyacetylene with iodine, allows free electron movement. Reduction in the same polymer, achieved by reaction with the metal sodium, turns polyacetylene into an anion while the sodium becomes a cation. The polymeric form of sulfur nitride becomes conductive at low temperatures without doping. Poly (*p*-phenylene), or PPP, is made up of single and double bonds that can achieve electron mobility through doping. Polypyrrole also has conductor potential because of bonding similar to that of PPP. Both can be switched back and forth between conductors and insulators. Each of these polymers offer potential as conductors, but major problems exist in converting them to reliable engineering materials. Disulfide polymers, long chains of sulfur–sulfur bonds, provide battery cathodes that act with lithium anodes to produce a depolymerization–polymerization process that yields electrochemical energy.

6.7.4 Geosynthetics

ASTM defines *geosynthetic* as:

> a planar [lying in one plane] product manufactured from polymeric material used on soil, rock, earth, or other geotechnical engineering related material as an integral part of a man-made project, structure, or system.

Geotextiles, geogrids, geofoams, geofilters, and geocomposite drains serve many purposes in building roads, dams, landfills, and similar applications where earth erosion, pollution, or containment of liquids is a factor. Civil, building, and *geotechnical* (related to the earth's crust) *engineering* often call on impermeable membranes to contain water in dams, to keep oil and other liquid chemicals in reservoirs, or to line sanitary landfills to prevent leached liquids from leaving the landfill and polluting groundwater. Porous woven polymer fabrics hold soil on river or lake banks until rocks and vegetation stabilize the banks. Polymer fabrics go into road construction to develop laminar composites that prolong the life of macadam (asphalt-and-rock composite) roads.

A major concern of the geotechnical community is how long a geosynthetic will last. A landfill has many potential pollutants. Once the geomembrane liner is in place, will it prevent pollution of the groundwater indefinitely? *Geomembranes* of PVC, PE, and ethylene copolymers undergo extensive testing based on ASTM Committee D35 standards. Geotechnical engineers also use expanded polystyrene (EPS) blocks, also known as *geofoam,* to rebuild highway embankments. Increased awareness of the fragile nature of Earth has prompted use of innovative geosynthetics; continued improvements in these synthetic polymers is inevitable.

SELF-ASSESSMENT

6-40. What is the major source of natural rubber? What are other possible sources?

6-41. Define the term *elastomer* and diagram a polymer of natural thermoplastic rubber and a polymer of vulcanized rubber.

6-42. Diagram the polymer-chain structure of rubber at rest and again with tensile stress applied.

6-43. What are the major advantages and disadvantages of natural rubber (NR)? Give examples of applications of NR to show how its advantages cause selection and its disadvantages prevent selection.

6-44. Use Table 10-9 to choose a suitable elastomer for the following containers: **(a)** hydraulic hose, **(b)** hot water bottle, **(c)** diver's wet suit, **(d)** bicycle-tire inner tube.

6-45. What is the major raw material for synthetic rubber?

6-46. At what stage in tire manufacture does vulcanizing occur?

6-47. Why is natural rubber used in automobile tires? Why are synthetic rubbers also used?

6-48. What additives to synthetic polymers accomplish the following: **(a)** retard aging from ultraviolet rays, **(b)** soften raw rubber, **(c)** increase rigidity, **(d)** strengthen?

6-49. Explain TR in terms of structure and applications.

6-50. Explain the following elastomer designation: ASTM BH 530.

6-51. What is the importance of hardness in rubber? Compare the hardness of elastomers to that of other polymers.

6-52. What is the term to describe loss of energy through heating in elastomers? When is it an advantage for this property to be **(a)** high or **(b)** low?

6-53. Cite examples of an application where the cost of an elastomer may be so high that another material would be selected.

6-54. Select elastomers and specify their ASTM abbreviation and their favorable properties for the following applications: **(a)** fire hose, **(b)** lawn mower wheels, **(c)** stopper on chemical test tubes, **(d)** golf-ball winding, **(e)** racket ball, **(f)** tennis-shoe soles, **(g)** rubber band, **(h)** seal on auto-brake master cylinder, **(i)** carpet backing, **(j)** work-shoe soles, **(k)** pad for bicycle caliper brakes, **(l)** wrestling mat.

6-55. What trends in engineering make adhesives so desirable?

6-56. List three advantages of adhesive bonding over other joining methods. Name three applications.

6-57. **(a)** In the adhesive system, what three events must occur to obtain a bonded joint? **(b)** What types of forces are involved in adhesion?

6-58. What is the difference between adhesion and cohesion?

6-59. **(a)** What is the difference between practical adhesion and interfacial adhesion? **(b)** What can prevent wetting?

6-60. Describe structural adhesives and list two examples.

6-61. **(a)** Name the five integrated factors that can affect a bonded product. **(b)** What must be considered in adhesive selection? **(c)** What contributions has interface technology made to new computers?

6-62. **(a)** Why would a thermoplastic adhesive be selected over a thermosetting adhesive? **(b)** Name two protein adhesives and give an application of each.

6-63. What adhesive can be used to bond structural components that will meet service conditions in southern deserts or in arctic regions?

6-64. Describe three methods in which adhesives develop their cured attraction between the substrate and the adherent.

6-65. Name one method of classifying caulking.

6-66. Select an adhesive for the following joining applications: **(a)** aluminum electronic component to fiberglass printed-circuit board; **(b)** PUR-foam bumper fascia to steel bumper, **(c)** ABS control mount to porcelain-enamel dishwasher panel, **(d)** plywood veneers, **(e)** aluminum identification label to PC plastic electric-drill housing; **(f)** aluminum towel rack to ceramic tile wall, **(g)** glass walls to chrome-plated steel aquarium, **(h)** PS-foam pad to wooden seat, **(i)** exterior laminated beam, **(j)** copper foil to flexible printed-circuit board, **(k)** brake liners to steel brake shoe.

6-67. What is a major thrust in the development of protective coatings? List three broad groups of protective coatings.

6-68. Name the term for synthetic polymers used on soil, rock, and earth in construction and cite three examples of these products.

6-69. Why has geotechnical engineering begun to seek greater use of synthetic polymers?

OBJECTIVE QUESTIONS

6-26. The objective of vulcanizing rubber.

    a. To improve processibility   **b.** To increase hardness and chemical resistance

    c. To reduce materials cost   **d.** To increase amorphousness

6-27. Vulcanization of rubber uses sulfur to transform it into a thermosetting elastomer.

    T (true)   **F** (false)

6-28. Before vulcanization, natural rubber has many double covalent bonds known as _____ .

    a. reactive sites   **b.** benzene rings   **c.** isoprenes   **d.** copolymers

6-29. The major source of raw materials for synthetic rubber.

    a. Hevea   **b.** Oil   **c.** Trees   **d.** Corn

6-30. The purpose of an antioxidant additive.

    a. Prevents aging   **b.** Speeds processing

    c. Accelerates curing   **d.** Strengthens

6-31. The purpose of carbon black as an additive.

    a. Prevents aging   **b.** Speeds processing

    c. Accelerates curing   **d.** Strengthens

6-32. Thermoplastic rubbers process easier than most thermosetting rubbers.

    T (true)   **F** (false)

6-33. Application for shape-memory polymers.

    a. Robotic hands   **b.** ATMs   **c.** Tire reinforcement   **d.** Netting

6-34. Energy loss through heating in elastomers, which creates problems in applications such as car tires.

    a. Ablation   **b.** Oligomer   **c.** Cohesion   **d.** Hysteresis

6-35. Elastomer used most often to help prevent the transmission of AIDS.

    a. Neoprene   **b.** Fluoroelastomers   **c.** Latex   **d.** IIR

6-36. A most stable elastomer used for human implants.

    a. Latex   **b.** NR   **c.** Fluoroelastomers   **d.** Silicone

6-37. Elastomers offer a wide range of properties from very soft to quite hard, they resist most chemicals, they provide good traction and abrasion resistance, and they are quite flexible.

    T (true)   **F** (false)

6-38. Which advantage(s) would adhesive bonding have over other joining methods?

    a. Ease of assembly and disassembly

    b. Light weight, load spreading, joint sealing

    c. Withstands very high temperatures normally exceeding the base material

    d. Easy selection for joining any type of material

6-39. A key factor in adhesive selection is the ability of the adhesive to _____ the substrate material.

    a. wet   **b.** cover   **c.** cure   **d.** entrap

6-40. Attraction of molecules within an adhesive or substrate.

    a. Adhesion   **b.** Contact   **c.** Cohesion   **d.** Interface

6-41. Penetration of adhesives into the openings of porous materials.

    a. Interface   **b.** Mechanical interlock

    c. Cross-linking   **d.** Surface tension

6-42. Which group of materials presents the greatest challenge for adhesive selection?

    a. Plastics   **b.** Wood   **c.** Metal   **d.** Ceramics

6-43. How does Marra's adhesive joint model aid in dealing with bonding?

    a. Illustrates the relationships between the adherents and adhesives

    b. Shows how to design the substrate

    c. Diagrams methods to select plastic adhesives

    d. Provides data on curing times

6-44. Loss-of-solvent curing of adhesives usually has shorter curing times than cooling of hot melts.

    T (true)   **F** (false)

6-45. Polymer strands classified as either regenerated cellulosics or noncellulosics.

    a. Natural fibers   **b.** Jute   **c.** Manufactured fibers   **d.** MLC

6-46. Materials used to work like adhesives, provide protective coatings, and keep out liquids and gases.

    a. Adhesives   **b.** Caulks   **c.** Sealants   **d.** Stains

6-47. Considerable emphasis has been placed on eliminating _____ in protective coatings to improve air quality and other aspects of our natural environment.

    a. Silicone   **b.** water-based coatings   **c.** ER   **d.** VOCs

6-48. The term for synthetic polymers used on soil, rock, and earth in construction.

    a. Aramides   **b.** Geosynthetics   **c.** Styrofoams   **d.** Earth screens

Millennium Wooden Monument. (a) Artist rendering of the Tour de la Terre, a 656-foot-high tower made mostly of wood. France's recognition of a new millennium couples with global awareness of the need to preserve our natural environment.

200	Fleur: acier inoxydable ou titane	
195	Metal "flower" structure	
190	(with five stainless steel or titanium petals)	
185		
180		
175		
170		
165		Type of wood used in tower core:
160		
155	diamètre du fût 8m	ature des bois du tût
150	core diameter = 8m	
145		
140	contreventements métal gris	pin sylvestre
135	Gray metal cross (wind) braces	ou douglas GL30
130	coques en chêne	
125	oak shells	
120		pin sylvestre armé
115	aile toile blanche	
110	white canvas wing	Reinforced Scotch fir
105		
100	terrasses 600m²	
095	niveau +95.50 700m²	pin sylvestre GL30+
090	niveau +90.10 700m²	
085	arrivée +84.70 650m²	chêne classe 1
080	arrivée +82.00 500m²	
075	diamètre du fût 12m	
070		
065	aile toile blanche white canvas wing	
060	contreventements métal gris stair	pin sylvestre GL30+ ou douglas GL30+
055	escaliers métalliques gris gray metal stairs	Scotch or
050	coques en chêne	Douglas fir
045		
040		chêne classe 2
035	ascenseurs d'urgence	Class 2 oak
030	ascenseurs panoramiques emergency	
025	and observation	
020	section du fût elevators	
015	core sectional view	
010		chêne classe 1
005		Class 1 oak
000		

(b) Drawing depicts 509-foot-high wooden pillars made of oak and fir that support four floors of office and observation spaces, as well as stainless steel and titanium metal flower sculptures. The designer sought to demonstrate the value of wood as a high-tech, modern building material, and the only renewable building material. Wooden pillars are made of lumber veneer laminate and plywood reinforced with steel or carbon rods and protected with oak shells. Tower is compared to another Paris tower, the Eiffel Tower. Tour de la Terre has a slender 56-foot diameter compared with the 410-foot base and 984-foot-high Eiffel Tower. (French Technology Press)

6.8 WOOD AND RELATED PRODUCTS

Wood, one of nature's unique, natural composites, and rock were humanity's first materials. As a ready weapon or fuel for the fire, the stick required no major processing. Since those early beginnings, wood has been the target of much technology and is still researched for methods not only of improving the yield of forests, but also for ways of better using the tree and the many by-products of wood.

As with other living organisms, trees are subject to environmental conditions. Acid rain, lack of biodiversity in tree farms, overharvesting, fires, and similar problems present constant challenges to maintaining the world's fragile balance with its forests. Many underdeveloped nations have harvested their trees too heavily to sell the timber for needed income and/or for cooking and heating because of the scarcity of other fuels. Where forests are lost, regions turn to deserts, causing disasters to all living things. As with most other materials in our family of materials, all aspects of wood must be closely monitored in the materials cycle.

Wood must be considered a valuable engineering material and should be exploited to its fullest extent, because it is environmentally friendly. Fossil fuels and minerals required an eon of time to develop, so civilization faces shortages of oil, gas, iron, chromium, and other raw materials used to produce many of our engineering materials. Trees, on the other hand, are renewable resources also crucial for the air we breathe, and they stabilize soil, provide shelter and nutrition for many species, and add to the natural beauty of our landscape. It is possible to grow wood in a variety of ways that yield a full range of construction and consumer products. If grown in plantations using new planting and harvesting techniques, then processed with new technology to maximize all of the plant, and finally recycled efficiently, wood will serve humanity well. In our modern era, wood has been replaced by other materials, such as steel, plastic, and concrete; however, experience with these alternative materials has not always been completely positive.

Many plants possess woody or cellulosic substances, but the tree is our major source of wood. *Extractives* from the tree that contribute to properties of wood such as decay resistance, color, odor, and density are also useful in a number of industries. These extractives include tannin (tannic acid), used in processing leather, and polyphenolics, used to make phenolic plastics, plus coloring agents, resins, waxes, gums, starch, and oils. In addition to the fruit and nuts that are harvested, trees serve as the raw material for paper, cellulosic plastics, explosives, rayon fibers, films, lacquers and drilling muds, ethyl alcohol, food flavoring, concrete additives, and rubber additives. Wood was our first fuel, but with the technological development of coal, oil, and gases, wood became only a romantic fuel for most of the technologically advanced people of our world. However, tree products once again are being given serious consideration as a fuel. The ability to renew wood supplies through forestry increases humanity's interest in this material as nonrenewable resources such as oil, coal, and minerals rapidly become depleted.

6.8.1 Structure

Wood develops through *photosynthesis*—the chlorophyll in trees uses carbon, hydrogen, and oxygen to produce sugar, starches, and cellulose. The dry wood used as an engineering material has an approximate composition of the following percentages by weight: cellulose (50%), lignin (16–33%), hemicelluloses (15–30%), extractives (5–30%), and ash-forming mineral (0.1–3%). *Cellulose* ($C_6H_{10}O_5$) (Figure 6-47a), a high-molecular-weight linear polymer, forms fibers that make up cell walls of vessels and ducts. *Lignin* (Figure 6-47b), an amorphous polymer, forms a matrix around the cellulose, much like the plastic matrix in fiberglass. Through removal of lignin, wood can be broken down into fibers that are used in making paper and other synthetics. The cellulose forms cellular networks of ducts, vessels, fiber rays, and pits, which transport and store the extractives and minerals throughout the living tree. The network differs, depending on whether it is a hardwood or a softwood tree. Softwoods are not always softer than hardwoods; Douglas fir, a softwood, is about twice as hard as basswood, a hardwood. *Softwood*

Figure 6-47   (a) Cellulose formula. (b) Lignin formula.

species (Figure 6-48) include firs, pines, and spruces; they bear cones and have needles or scale-like leaves. ***Hardwoods*** are broadleaved and not normally evergreen; they include maples, oaks, birches, and mahogany. ***Hemicellulose*** is another polymer closely akin to cellulose; like cellulose it breaks down into sugars when chemically treated. The ash-forming minerals include calcium, potassium, phosphate, and silica. Extractives are removed by heating the wood in water, alcohol, or other chemicals; they represent a significant commercial value.

Most common engineering materials, such as steel and concrete, are ***isotropic,*** which means that generally the strength is the same in all directions because the materials are homogeneous. Wood is ***anisotropic,*** meaning that it has greater strength in some directions than others. The anisotropy develops as a result of the way a tree grows, with various factors influencing the wood structure: cellular structure, branches, and bending by prevailing wind. A full discussion of all these factors is beyond the limits of this book, but the resulting tensile strength and stiffness of wood are greater in the radial direction than in the tangential direction (Figures 6-49 and 6-50). Rays that radiate from the pith outward on the tree tend to tie together the layers of cells (tracheids) growing longitudinally. This ***directional strength*** results partly from the complex structure of cell walls, with long polymer chains of cellulose that form layers running in varying patterns to reinforce each other. Primary covalent bonding holds together these ***microfibrils,*** which are almost parallel to the cell axis. Weaker secondary bonds operate in the perpendicular axis.

Lumber consists entirely of dead cells. Figure 6-51 illustrates a cross-section of a white oak tree trunk. The ***cambium layer*** produces all the growth of cells (wood and bark). ***Sapwood*** is a transitional wood of both living and dead cells. The sapwood's main role is to conduct sap and store food. ***Heartwood*** evolves from sapwood as the cells die and become inactive. The heartwood cells no longer transport sap but rather become storage cells. Sapwood of all species is highly susceptible to rotting, while heartwood of some species is protected from decay because it has taken on infiltrating materials that serve as preservatives. Additional information on wood structure, plus other information on wood, is available in the excellent source, *Wood Handbook: Wood as an Engineering Material.* The handbook defines ***lumber*** as the product of the saw and planing mill and is not manufactured further; ***timber*** is a classification of lumber that is nominally 5 or more inches in its least dimension.

Figure 6-48   Softwood block showing three complete and part of two other growth rings in the cross-sectional view (X). Individual cells can be detected easily in the earlywood (EW), whereas the smaller latewood cells are difficult to distinguish in the latewood (LW). Note the abrupt change from earlywood to latewood. The two longitudinal surfaces (R—radial; T—tangential) are illustrated. Rays, which consist of food-storing cells, are evident on all three surfaces. 70×. (Dr. Wilfred A. Côté, SUNY College of Environmental Science and Forestry)

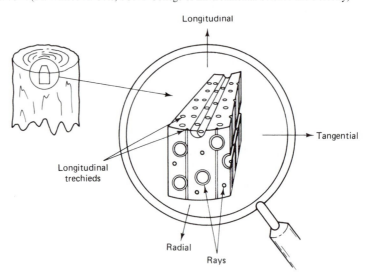

Figure 6-49   Section cut from tree shows patterns from Figure 6-48.

6.8.2 Wood Products

6.8.2.1 Commercial lumber.   Lumber is graded to provide the buyer with an indication of the strength and appearance of the lumber. Unlike most engineering materials, wood is not homogeneous. The grading thus provides an averaging of the quality of boards in a given grade. Visual grading spots defects such as knots, stains, bark pockets, decay, shakes (cracks), pitch pockets (resin buildup), and checks (splits). *Softwood lumber* is a vital material for construction and manufacturing and is graded as construction and remanufactured lumber based

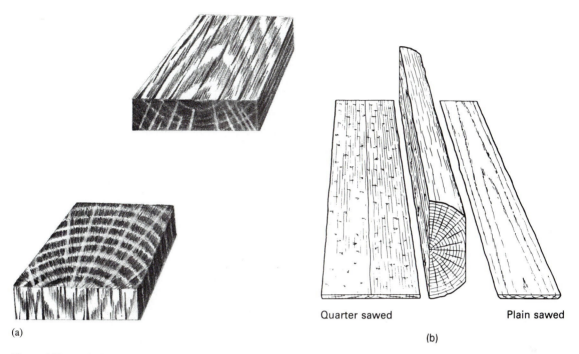

(a)

Quarter sawed   Plain sawed

(b)

Figure 6-50   (a) Grain direction. (U.S. Forest Products Laboratory, Madison, Wisconsin) (b) Sawed boards.

Figure 6-51   Cross-section of white oak tree trunk: (A) Cambium layer is inside inner bark and forms wood and bark cells. (B) Inner bark is moist and soft, and contains living tissue; carries prepared food from leaves to all growing parts of tree. (C) Outer bark containing corky layers is composed of dry dead tissue; gives protection against external injury. (D) Sapwood, which contains both living and dead tissue, is light-colored wood beneath the bark; carries sap from roots to leaves. (E) Heartwood (inactive) is formed by a gradual change in sapwood. (F) Pith is the soft tissue about which the new growth takes place in newly formed twigs. (G) Wood rays connect the various layers from pith to bark for storage and transfer of food. (U.S. Forest Products Laboratory, Madison, Wisconsin)

on the American Lumber Standard. ***Construction grade*** includes stress-graded, nonstress-graded, and appearance lumber. ***Stress-graded*** provides an indication of engineering properties and covers pieces of 2 to 4 in. nominal size, larger timbers (5 or more in. nominal thickness), posts, beams, decking, stringers, and boards (less than 2 in. nominal thickness). ***Nonstress-graded*** lumber includes a combination of allowable properties and visual defects that is used for boards, lath, batten, and planks. ***Appearance lumber*** that is nonstress graded is cut to patterns such as trim, siding, flooring, and finished boards. ***Remanufactured*** grade is lumber that will receive further processing or secondary manufacturing, such as cabinet stock, molding, and pencil stock. Grading specifies conformity of the lumber to its end use. ***Hardwood lumber*** is graded as factory lumber (amount of usable lumber in a piece), dimensional parts, and finished market products (maple flooring, stair treads, trim, and molding). The furniture industry is the prime user of most hardwood grading.

A system for designating the degree of dressing or finish ***surfacing*** specifies one side surfaced (S1S), two sides surfaced (S2S), one edge surfaced (S1E), two edges surfaced (S2E), and various combinations, such as one side and one edge surfaced (S1S1E) or all sides and edges surfaced (S4S).

6.8.2.2 Composite wood.

To gain the maximum use of trees and to achieve properties not possible from solid wood, composite woods have been developed for use in construction and manufacture. Among composite woods are laminated timber, impreg-wood, wood and recycled plastics, plywoods and veneer, particle board, hardboard, sandwiched materials, and insulation board.

Laminated timber. Laminated timber is a product of the adhesive-joining technology. Through adhesive bonding of pieces of lumber so that the grain of all pieces is parallel to the length of the timber, it is possible to produce straight to curved structural-wood members of large size and outstanding strength. ***Glulam*** bonds sawn lumber laminations with waterproof structural adhesives, allowing nearly unlimited depth, width, and length in a variety of shapes. One glulam product, Parallam, achieves higher stress ratings and uniform strength than those of conventional materials through a patented process in which thin strands of veneer are glued using waterproof adhesives, then cured with microwaves.

Laminated beams offer the architect flowing lines and the warmth and beauty of wood. Custom-ordered laminated beams have the capability to span more than 100 m of unobstructed space for sports arenas, convention centers, entertainment halls, and churches (Figure 6-52). Seasoning of lumber for laminated beams prior to gluing provides improved strength, dimensional stability, and elimination of unsightly surface cracks and shrinkage. The ablative nature of the large beams provides an advantage in protection from the hazards of fire, since the thin layer of char protects the interior of the beam and the wood does not become plastic at high temperatures, as do metal and plastic members.

Impreg-wood. ***Impreg-wood*** achieves a very stable lumber by the bonding of phenolic resins to the cell wall microstructure. This process is accomplished through saturation of thin veneers with phenolic resin that is polymerized before the veneers are stacked into thick laminates. Applications of impreg-wood include sculptured models for the huge metal dies used to stamp automobile sheet metal parts. ***Compreg-wood*** employs impregnation of veneers with phenolic resin, but the polymerization is accomplished when the veneers are stacked and compressed into the desired final shape. The dimensionally stable, high-density, hard compreg laminate yields a glossy finish throughout the material. Knife handles, bowls, jigs for manufacturing, and parts for textile looms use the compreg laminates.

Wood and Recycled Plastics. The progress in municipal recycling has resulted in a dependable flow of recycled plastics. Plastic lumber made of recycled plastics offers many benefits such as resistance to biological attack (rotting and insects) and no need for protective finishes. For example, the Smartdeck®, a synthetic lumber system, includes rails, trim, fascia, and posts. As a total system, wood can be eliminated allowing simple deck installation due to such features as molded channels in the posts and rails. These allow wires and cables to be run for electricity and sound equipment without drilling.

(a)

(b)

(c)

(d)

(e)

Figure 6-52 Wood structures built to weather the ocean. (a) U.S. Coast Guard lifesaving station, built at Virginia Beach in 1903, and lifesaving boat reflect early twentieth-century wood construction. (b) Oak stern piece from shipwreck in 1876, recovered off Cape Henry, Virginia Beach, in 1983, attests to durability of the wood. (c) Modern wood structure also built on the Atlantic Ocean has an imposing roof support made of laminated beams and wood trusses using southern pine. (d) Solid teak table and chairs, and teak plywood cabinet and birch plywood door, reveal the warmth and beauty. (e) Wood has served arts and crafts since early times and continues as an important medium, such as this decorative Black Duck decoy, carved of basswood by Ed Morrison.

Combinations of recycled plastics and wood offer the benefit of wood's beauty, and usually lower cost, which couples with plastics durability. Durawood EX®, a wood-polymer composite (approximately 2/3 oak fiber combined with recycled plastic resins), combines the look of wood with highly detailed design, durability, and low maintenance. Like other decking material, Durawood will not rot, splinter, or show other effects of moisture typical of natural wood. After installation, it turns a light "driftwood" gray. Durawood EX® is backed by a 10-year limited warranty. Eaglebrook also makes a 100% recycled plastic lumber product called Durawood PE®, which comes in several durable colors such as redwood, cedar, gray, and oak. Durawood PE® carries a 20-year limited warranty, absorbs no moisture, and therefore provides high durability and low maintenance.

EMT WEBLINK

Recycled plastics for wood—*http://oikos.com/products/wood-plastics/eaglebrook/index.html*

Plywood. Plywood is another form of laminate that is most common in building construction. It is produced by stacking veneer with the grain direction in each layer at right angles to the next, beginning with the grain running the length of the panel (Figure 6-53a). This cross-laminating takes advantage of the fact that wood is much stronger in the direction of the grain than across the grain. Thus, plywood yields equalized strength in the plane directions, while also being shear, puncture, and split resistant. Numerous grades of plywood are available; Figure 6-53b shows the marking applied to three grades of plywood using the American Plywood Association standards. Letters assigned to veneer grades specify the quality of exposed plys (Figure 6-53c). Typing of plywood divides it into **interior** plywood, which has its laminates glued with moisture-resistant glue, including phenolformaldehyde, urea, and melamine. Veneers in interior plywood may be of lower quality than the exterior type. In contrast, **exterior** plywood is glued only with phenolformaldehyde to provide a waterproof bond. The waterproof glue can withstand aging and boiling water. Marine plywood is the superior-quality exterior plywood. The exterior and interior typing are applied to both appearance- and engineering-grade plywood.

Thin appearance plywood, in the form of paneling, is very popular for walls in houses, mobile homes, recreational vehicles, and boats. There is a nearly unlimited variety of hardwood-faced paneling. An inexpensive grade of Philippine mahogany known as **lauan** receives many types of coating and textures for decorative paneling. Thicker appearance-grade plywood is used for furniture and cabinets. Engineering-grade plywood serves as forms for casting concrete, underlayment for flooring and carpets, structural panels, roofing, decking, walls, and cabinets. Plywood siding has gained wide acceptance in contemporary-style buildings. Siding offers warmth, low maintenance, and good thermal and acoustical insulation, among other properties, and is available in various species with textures including smooth, deep grooves, brushed, rough sawn, and overlays. **Veneer** in thin sheet form is used in single thickness for baskets, for boxes, and as ornamental inlays.

Wood-Based Fiber and Particle Panel Materials. Wood-based fiber and particle panel materials include a variety of panels and boards used in building construction, packaging, furniture, and other manufactured products. ***Particleboard*** panels, also known as reconstituted panels, are produced through the use of thermosetting resins, such as urea–formaldehyde and phenolformaldehyde, which serve as a matrix to bind together wood residues or shavings in the form of small wood flakes, wood flour, and additives (Figure 6-54). Water resistance is improved through the addition of wax. The resins account for 5% of the dry panel's weight and polymerize as the particles are pressed in flat presses or extruded through thin die presses. Many particle sizes and shapes are manufactured to yield panels with specific properties. Particleboard serves as backing for plastic and wood laminates or as a decorated panel that is normally painted. ***Hardboard*** is produced by processing wood chips with steam and/or pressure, and the lignin bonds the fibers together. Medium-density, high-density,

3 ply construction (3 layers of 1 ply each)

4 ply construction (3 layers: Plies 2 and 3 have grain parallel)

5 ply construction (5 layers of 1 ply each)

(a)

Typical APA Registered Trademarks

APA
Panel grade ——— RATED STURD-I-FLOOR
Span Rating ——— **20 OC** 19/32 INCH ——— Thickness
                      SIZED FOR SPACING
Tongue-and-groove- T&G NET WIDTH 47-1/2
Exposure durability — EXPOSURE 1
classification ————— 000 ————— Mill number
                 ┌——— NRB-108
National Research Board
report number

APA
Panel grade ——— RATED SHEATHING
Span Rating ——— **32/16** 15/32 INCH —— Thickness
                      SIZED FOR SPACING
Exposure durability — EXPOSURE 1
classification ————— 000 ————— Mill number
               ┌——— NRB-108
National Research Board
report number

APA
Panel grade ——— 303 SIDING   6-S/W ——— Siding face grade
Span Rating ——— **16 OC** 19/32 INCH ——— Thickness
                           GROUP 1 ——— Species
Exposure durability — EXTERIOR   group number
classification ————— 000 ————— Mill number
          ┌——— PS 1-74 FHA-UM-64 ———┐
Product Standard   FHA recognition

(b)

Figure 6-53 (a) Plywood. (b) Marking used by American Plywood Association. (*continued*)

or special densified hardboard offers high strength, wear resistance, moisture resistance, and resistance to cracking and splitting and has good working qualities. ***Tempered hardboard*** has oils added to increase water resistance, hardness, and strength. Hardboard finds use as painted house siding, diestock, and electrical panels; with plastic and other laminates; and as components in furniture such as drawer bottoms, mirror backs, and television sides and backs.

    Sandwiched Materials.   Sandwiched materials include treated paper honeycomb faced with laminating veneers. Doors and panels employ these materials. Fiberboards made of wood fibers bonded with rosin, asphalt, alum, paraffin, oils, fire-resistant chemicals, and plastic resins

N	Smooth surface "natural finish" veneer. Select, all heartwood or all sapwood. Free of open defects. Allows not more than 6 repairs, wood only, per 4 × 8 panel, made parallel to grain and well matched for grain and color.
A	Smooth, paintable. Not more than 18 neatly made repairs, boat, sled, or router type, and parallel to grain, permitted. May be used for natural finish in less demanding applications.
B	Solid surface. Shims, circular repair plugs and tight knots to 1 inch across grain permitted. Some minor splits permitted.
C Plugged	Improved C veneer with splits limited to 1/8-inch width and knotholes and borer holes limited to 1/4 × 1/2 inch. Admits some broken grain. Synthetic repairs permitted.
C	Tight knots to 1-1/2 inch. Knotholes to 1 inch across grain and some to 1-1/2 inches if total width of knots and knotholes is within specified limits. Synthetic or wood repairs. Discoloration and sanding defects that do not impair strength permitted. Limited splits allowed. Stitching permitted.
D	Knots and knotholes to 2-1/2 inch width across grain and 1/2 inch larger within specified limits. Limited splits are permitted. Stitching permitted. Limited to interior (exposure 1 or 2) panels.

Figure 6-53   (c) Veneer grades. (American Plywood Association).

serve as insulation panels on walls and roofing, onto which some other material is added, such as siding or tar and gravel.

6.8.3 Wood By-Products

The most common by-product of lumber and plywood processing is *chips* for the production of pulp used in various paper products. Pulp is obtained through chemically defibrating the chips by dissolving the lignin to obtain fibers. Other by-products include naval stores (turpentine and rosin), bark for mulching plants, charcoal briquettes, and various resins from the tree.

6.8.4 Physical Properties

6.8.4.1 Appearance.   When selected for furniture, house trim, doors, floors, wood turning, or other applications in which wood is exposed, appearance is paramount because it lends warmth and beauty. Color, figure (patterns produced by grain, texture, and machining), luster, and the manner in which finishes affect wood all become important factors in selection.

6.8.4.2 Sawing effects.   The Forest Products Laboratory of the U.S. Department of Agriculture defines grain and texture as follows:
>   *Grain:* direction, size, arrangement, appearance, or quality of fibers in wood or lumber
>   *Texture:* finer structure of wood rather than annual rings

Twenty grain specifications cited in the *Wood Handbook* provide precise meaning to various grain patterns. The angle at which a log is cut is one factor that determines grain pattern. Figure 6-55 shows the difference in quarter-sawed and plain-sawed boards.

While lumber is used in many stages, from logs with the bark removed or rough-sawn timbers, finished lumber is the most common form encountered. Standard dimensions include 2 × 4, 1 × 12, and 4 × 4, which are nominal measurements in inches. Nominal measurements indicate the approximate size of the lumber when it is in the rough state. A finished 2 × 4 is actually 1 ½ × 3 ½ in. Following rough sawing, lumber is *seasoned* to remove the moisture, reduce warping and cracking, reduce weight, and increase strength and many other properties. There are a variety of seasoning methods, including air drying and kiln drying. Air drying is a slow process in which rough timber is stacked in a specified manner and allowed to dry naturally. Kiln drying employs ovens, in which the rough lumber is stacked and hot air is forced over the wood. Research into solar kilns shows promise for low cost and quick seasoning.

Figure 6-54   Particleboard (also called reconstituted panel) products consist of a variety of compositions and organizations. Flakeboard uses relatively square chips; oriented strandboard uses long, narrow chips that can be aligned for maximum directional strength. (U.S. Forest Products Laboratory)

Beyond seasoning, many methods are employed to improve the physical, chemical, and mechanical properties of wood. *Chemical modification* involves a chemical reaction (covalent bonding) between the cellulose, lignin, and hemicellulose and a chemical that acts as a preservative. Woods are also impregnated with polymers, oils, salts, and other solutions. A wide variety of coatings, from natural clear finishes to stains and paints, provide protective and decorative value.

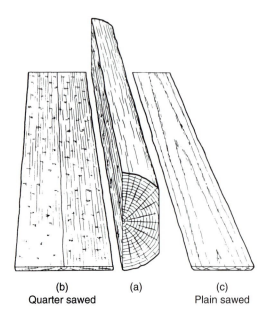

(b)   (a)
Quarter sawed

(c)
Plain sawed

BLACK CHERRY

Figure 6-55   (a) Top segment illustrates end grain. (b) Middle segment is quarter sawed. (c) Bottom segment is plain sawed. These match the 3-D sketch above.

6.8.4.3 Moisture content.

Moisture content is a measure of the water in wood. Normally expressed as a percentage of the amount of water weight to the weight of oven-dry wood, moisture content is an extremely important factor in the mechanical and other physical properties of wood. Trees have moisture content ranging from 30 to 200%. Drying of freshly cut wood removes moisture, but the hygroscopicity of wood substance allows it to pick up moisture from the surrounding atmosphere. For this reason, seasonal and environmental changes in relative humidity cause a constant shift in moisture content. You may recall that, during humid and rainy seasons, doors and drawers become hard to open, while during dry days the same doors or drawers become easy to use. Treatment of wood must include steps to minimize moisture content changes, weathering, chemicals, and biological attack. Expanding and shrinkage or improperly finished wood cause *warping, splitting, checking* (lengthwise separation across the annual rings), and other problems. Damp wood also provides opportunity for bacteria and fungi to *decay* or *rot* wood. Seasoned lumber has been dried in varying degrees. To ship lumber and avoid molding, moisture is reduced to 25% or below. Thoroughly dried lumber has from 12 to 15% moisture. The key to seasoning lumber is to attain a moisture content corresponding to the average atmospheric conditions in which the lumber will serve.

6.8.4.4 Weatherability and decay resistance.

Similar to the reaction of plastics, woods change color and structure due to heat, light, moisture, wind, and chemicals in the environment. Ultraviolet (UV) light causes chemical degradation and graying of color, while metal fasteners (screws, nails, etc.) and hardware (hinges, etc.) also cause color changes. The UV light breaks down cellulose in the fibers. However, the degradation of the surface fibers, which is rapid when wood is first exposed, slows to a negligible rate of about 1 mm per 100 years. Other effects of weathering include warping, checking, abrasion, and surface roughening. While the even gray color is attractive, the appearance of dark gray and blotchy colors indicates attack by biological organisms and probability of rotting. See Table 6-9 for the varying resistances of woods to decaying. The cut of the boards, such as vertical grain rather than flat grain, aids weatherability, but special treatment with preservatives is normally required.

Generally, heartwood provides more resistance to decay than does sapwood. The naturally occurring wood extractives determine the resistance of a particular species of heartwood to attacking fungi. The southern and eastern pines formerly were slower growing and older and thus had more heartwood, which made them more resistant to decay. New forestry methods produce faster growth but more sapwood, which reduces these species' decay resistance. Bald cypress has outstanding natural durability.

6.8.4.5 Density.

The variation in a wood's moisture content depends on the environment and preparation of the wood and will greatly affect the density, weight, and specific gravity. The range of density for most woods falls between 320 and 720 kg/m^3. Calculations of *specific gravity* are determined on either kiln-dried wood or wood with a specific percentage of moisture. Twelve percent is frequently used. At a 12% moisture level, most species of trees yield specific gravities in the range of 0.32 to 0.67, as shown in Table 6-9.

6.8.4.6 Working qualities.

In hand working, normally the lower-specific-gravity wood works best. Machinability of wood depends on characteristics such as *interlocked grain, hard deposits* of minerals such as calcium carbonates and silica, *tension wood,* and *compression wood,* in addition to density. High density and hard deposits will dull tools. Reaction wood is tension wood and compression wood. Reaction wood and interlocked grain cause binding of boards as they are fed through saws and planers. Other considerations in the workability of wood include nail splitting, screw splitting, and surface results of sanding.

6.8.4.7 Ablation.

An advantage offered by the use of wood timbers and boards in building is their ability to char when burned. This ablation property allows woods to retain much of their strength because *charring* slows the burning. Conversely, metals act as good thermal conductors and uniformly transmit the heat of a fire throughout the metal. This can cause metal supports to lose much of their strength and bend under their loads. Figure 6-56 shows the ablative value of wood; the steel beams failed (softened) under the heat, while the wood structural members held some of the load.

TABLE 6-9   PHYSICAL AND MECHANICAL PROPERTIES OF IMPORTANT COMMERCIAL WOODS
GROWN IN THE UNITED STATES (AVERAGE VALUES OF SMALL, CLEAR, STRAIGHT-
GRAINED SPECIMENS)

Specific gravity Common species	Hardness[a] (12% moisture)	Modulus of elasticity[b] (N)	Decay (MPa)	Resistance[c]
Hardwoods				
Ash, white	0.60	5,900	12,000	S
Basswood	0.37	1,800	10,000	S
Beech, American	0.64	5,800	11,900	S
Birch, yellow	0.62	5,600	13,900	S
Cherry, black	0.50	4,200	10,300	V
Elm, American	0.50	3,700	9,200	S
Maple, sugar	0.63	6,400	12,600	S
Oak, northern red	0.63	5,700	12,500	S
Oak, white	0.68	6,000	12,300	V
Walnut, black	0.55	4,500	11,600	V
Yellow, poplar	0.42	2,400	10,900	S
Softwoods				
Bald cypress	0.46	2,300	9,900	V
Cedar				
Eastern red cedar	0.47	4,000	6,100	V
Northern white cedar	0.31	1,400	5,500	V
Western red cedar	0.32	1,600	7,700	V
Douglas fir				
Coast	0.48	2,200	13,400	M
Interior west	0.50	2,300	10,300	M
Fir				
Eastern species	0.38	2,200	10,300	S
Western species	0.36	1,800	8,500	S
Pine				
Eastern white	0.35	1,700	8,500	M
Sugar	0.36	1,700	8,200	S
Western white	0.38	1,900	10,100	S
Redwood				
New-growth	0.35	1,900	7,600	V
Old-growth	0.40	2,100	9,200	V

Source: Wood Handbook: Wood as an Engineering Material, by Forest Service, U.S. Department of Agriculture.

[a]Newtons required to embed an 11.26-mm ball one-half its diameter in a direction perpendicular to the grain.

[b]Measured from a simple supported, center-loaded beam, on a span/depth ratio of 14:1.

[c]V, very resistant or resistant; M, moderately resistant; S, slight or nonresistant.

6.8.5 Mechanical Properties

Mechanical properties of woods can vary widely because of the lack of homogeneity in wood structure. Knots, cross-grain, checks, and growth rings provide varying properties within boards and between boards from the same tree. The properties shown in Table 6-9 reflect values obtained from small, clear, straight-grained specimens. They provide average values for comparing common woods and for comparing wood with other materials on the basis of modulus of elasticity.

The wooden bridges shown and described in Figure 6-57 illustrate how materials may go out of favor but then may return. Just as the Millennium Tower in France, discussed earlier in this unit on wood, demonstrates the value of wood as a high-tech construction material, so too does the rebirth of wooden bridges.

 EMT WEBLINKS

Wood products—*http://www.weyerhaeuser.com/woodprod*

Sustainable Forestry Initiative™ (SFI)—*http://www.afandpa.org/forestry/forestry.html*

Figure 6-56 Ablative nature of wood: Steel beams became plastic and draped over wooden beams in a factory fire. (U.S. Forest Products Laboratory)

SELF-ASSESSMENT

6-70. In light of diminishing supplies of many raw materials, what makes wood such a promising engineering material for future generations?

6-71. What is the relation of lignin to cellulose in a tree? When is lignin removed?

6-72. Explain how anisotropy develops in wood at both the microstructure (cellular) and macrostructure levels.

6-73. How do hardwoods differ from softwoods? Cite examples to illustrate these differences.

6-74. Describe how the various layers in the cross-section of a tree affect the properties of the lumber produced from the tree.

6-75. Name typical applications for the following grades of lumber: **(a)** nonstressed graded, **(b)** construction appearance, **(c)** remanufactured, **(d)** stress graded.

6-76. Name four wood products that are not solid wood, and cite typical uses for each.

6-77. Explain the following APA designation: B-D, Group 3, Interior.

6-78. What properties make wood a desirable material for furniture and paneling?

6-79. What properties can limit the use of wood as an engineering material?

6-80. What effects can moisture have on wood? How can they be altered?

6-81. Name a wood to fit into the following categories: **(a)** high hardness, high decay resistance, high specific gravity; **(b)** very low hardness, low specific gravity, good modulus of elasticity.

OBJECTIVE QUESTIONS

6-49. Which polymer serves as a matrix in natural wood to bond the cellulose fibers?

    a. Sulfur   **b.** Hemicellulose   **c.** Lignin   **d.** Isoprene

6-50. As a natural composite, wood's grain structure gives it _____ properties.

    a. uniform   **b.** anisotropic   **c.** isotropic   **d.** poor

Type A : Stressed Timber Deck   Type B : Stressed Timber Box Section

Type C : Stressed Timber Tee Section

Figure 6-57 Rebirth of wooden bridges. For decades, concrete and steel replaced the old timber bridges that had served humanity since early civilization. Recently, we have learned that the acid rain, road salt treatments, and other environmental conditions have rendered nearly half of the 600,000 short-span bridges in the United States unsafe and/or unusable. New technology applied to wood has resulted in timber bridges that should provide much longer service life than steel and concrete bridges. Timber treated with wood preservatives and laminated into stress timber decks, T-sections, and box sections for bridge construction offer the following advantages: pleasing aesthetics, good shock resistance, high strength-to-weight ratios, sound and thermal insulation, and weather and chemical resistance (see photo). Using wood, a renewable resource, localities should be able to harvest regional materials and labor for economical alternatives to those hundreds of thousands of defective concrete and steel bridges. (Barry Dickson, Construction Facilities Center, West Virginia University)

6-51. The main difference in hardwood and softwood is _____ .

    a. polymer structure   **b.** hardness

    c. economics   **d.** growing environment

6-52. The reason for grading lumber is its lack of _____ .

    a. value   **b.** processing   **c.** homogeneity   **d.** visual defects

6-53. A means of overcoming the problem in question 6-52 to ensure uniform strength.

    a. Seasoning   **b.** Curing   **c.** Heating   **d.** Laminated timber

6-54. The key difference between interior and exterior plywood.

    a. Number of plys   **b.** Direction of grain   **c.** Glue   **d.** Appearance

6-55. A material developed to maximize the yield of wood from forest products.

    a. Plywood   **b.** Paper   **c.** Glulam   **d.** Particleboard

6-56. A major reason wood is prized as an engineering material.

    a. Natural beauty   **b.** Low cost

    c. Plentiful supply   **d.** Uniform properties

6-57. Wood should be considered as a valuable engineering material because the raw materials are _____ , whereas those for most other materials are not.

    a. recycled   **b.** cheap   **c.** renewable   **d.** limited

6-58. To prevent wood from warping, splitting, and checking, it is important to control _____ .

    a. growth   **b.** weather   **c.** moisture   **d.** lignin

6-59. A problem with wood not found in most other engineering materials.

    a. Rotting   **b.** Weight   **c.** Processing   **d.** Supply

6-60. Because of the way that wood grows, most wood of the same species has identical properties.

    T (true)   **F** (false)

6-61. Wood timber may offer better protection in a fire than steel beams.

    T (true)   **F** (false)

6-62. Wood's ability to char when burned.

    a. Ablation   **b.** Isotropy   **c.** Homogeneous   **d.** Check

6-63. Most wood has very good workability because of the uniformity of its structure.

    T (true)   **F** (false)

Advances in polymer materials development, much like those discussed here, occur daily. If you understand the basic concepts covered in this module, you will possess the ability to understand these developments. To stay current with polymer innovations, make it a regular habit to read some of the periodicals listed at the end of this module.

6.9 LABORATORY ACTIVITIES AND CLASSROOM DEMONSTRATIONS WITH POLYMERS

The *Experiment in Materials Science, Engineering and Technology (EMSET)* CD ROM provides numerous experiments and demonstrations on polymers as a result of a cooperative effort by scores of educators and materials specialist from industry. See Prentice-Hall's website for ordering information on the latest edition of *Experiments in Materials Science* (CD-ROM), at *www.prenhall.com.*

APPLICATIONS & ALTERNATIVES

Green Manufacturing—Myths and Promises of Recycling   If green manufacturing is to be successful, it requires cooperation of both citizens and manufacturers. A headline in the December 23, 1999, *Virginia Beach Beacon,* read, "State association names Va. Beach tops in recycling." The article went on to report that 75% of the city's residents (about 440,000 in the thirty-fourth largest city in the United States) participate in a commingle recycling program, keeping more than 40,000 tons of trash out of the city's landfill. The program allows newspaper, plastic bottles (such as HDPE, PETE, and vinyl), clear, green, and brown glass bottles and jars, steel and aluminum cans, and mail to be mixed in a large container supplied by the city. In many localities, residents must separate their recyclables and place them in separate pickup containers. The high percentage of recycling participation found in Virginia Beach does not occur everywhere.

Information from National Polymers, Inc., and listed at the Environmental Factoids website—January 4, 2000, *http://envirosystemsinc.com/resrec.html*—points out some recycling benefits.

Each month, with maximum recycling participation, an average 100-unit apartment or condominium will have the following positive impact on our environment:

Save 21.93 30-foot trees
Save 8,389 kilowatts of electricity
Save 26.86 cubic yards of landfill space
Reduce 77.4 lb of air pollution

The world is running out of places to bury its garbage. There are concerns about the effect of burning rubbish: Some experts believe that it is safe to turn waste into energy by burning some garbage to generate electricity; others argue against incineration. We cannot expect simple solutions to environmental concerns. Some issues about the environment involve myths that might have sprung from facts coupled with assumptions on how those facts should (but may not) apply to a new situation. It is important for all citizens to develop a base of knowledge about technology to separate fact from myth.

There is general agreement that the best manner to handle *municipal solid waste (MSW)* is through the following procedures, listed in order of priority:

1. *Source reduction.* Keeping the amount of waste, especially toxic waste that may result from processing of the original product, to a minimum. One example of reducing waste would be the use of ceramic dishes, cups, and glasses instead of disposable items (plastic or paper) when there is no problem with the spread of disease.

2. *Removing recyclable or reusable materials.* MSW has already been reduced dramatically by recycling programs prompted by legislation. Not only does recycling reduce MSW, it saves natural resources. For recycling to be effective, there must be regional centers that can make use of the materials economically feasible.

3. *Burning for volume reduction and/or energy.* Although many localities have developed waste-incineration programs, there is still concern about global warming as a result of burning hydrocarbons. One example is rubber tires, which have created tremendous disposal problems, yet are a potential source of hydrocarbon energy. Utility companies have built tire-to-energy plants to rid regions of tires while generating electricity. Many planners feel that the existing technology is adequate to safeguard against harmful emission from incinerators. This reasoning follows the facts that gases given off by landfills are composed of about 50% methane, a much more dangerous gas in terms of the *"greenhouse effect"* (causes global warming) than the CO_2 (carbon dioxide) given off through incineration. The *NIMBY (not in my backyard)* syndrome also presents problems in terms of finding communities willing to permit incinerators for waste-to-energy processes.

4. *Landfilling.* This least desirable method of dealing with MSW is still the most common. Figure 6-58 shows the composition of the average landfill in both weight and volume. Volume became the major concern when we recognized the lack of space to bury trash. Weight is a concern in transporting waste.

In Module 1 we said that any product should be considered in terms of its total life cycle. What are the results of the processing methods? Are polluting gases released into the environment, and do toxic metals get flushed into our rivers and streams? During use, does the product safeguard our health? At the end of the product's useful life, how can it be disposed of safely? MSW is a hidden cost of a product that we pay in the form of taxes and quality of life. Recycling depends on existing technology. If the technology does not exist, recycling cannot occur. Recycling technology and availability of the technology to localities is needed to ensure a proper stream of materials flow to recycling centers.

Note in the graph in Figure 6-58 that plastics account for 18% of the volume in our landfills, while paper consumes 38% of the space. By weight, paper and cardboard comprise 40%

Landfill Composition

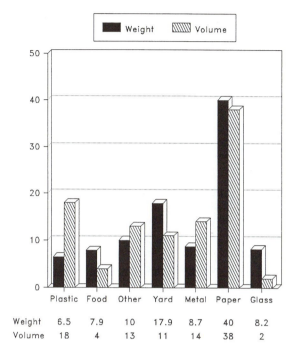

Figure 6-58   Weight and volume compositions of typical landfill. (Dow Chemical.)

	Plastic	Food	Other	Yard	Metal	Paper	Glass
Weight	6.5	7.9	10	17.9	8.7	40	8.2
Volume	18	4	13	11	14	38	2

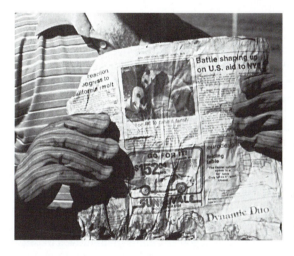

Figure 6-59   The newspaper shown here was discovered by an archaeological team digging through a landfill. It was buried 12 years. (Dow Chemical)

and plastics 6.5%. These two materials clearly occupy the greatest amount of space compared with metal, glass, and food, while yard waste (grass, limbs, shrubs, etc.) uses almost 11% of the volume. Yard waste can be reduced dramatically by using mulching lawn mowers and ways to turn limbs, leaves, and so on into mulch.

Biodegradable materials break down through the natural action of the environment. Sun, water, heat, and microorganisms will break down waste materials and reduce their effect on the MSW. However, recent digging through garbage dumps unearthed food (corn on the cob, chicken, hot dogs, etc.) that looked like it had been buried yesterday but, in fact, had been in the landfill 10 years. Newspapers buried 60 years ago can be read easily (Figure 6-59). Other items unearthed that show little sign of biodegrading after a decade or more are paper containers and wrappers, steel cans, clothing, grass clippings, and a variety of plastic products. For biodegrading to occur, there must be 65% moisture, but landfills typically are held to 20–25% moisture to prevent washing out chemicals into the groundwater below.

The materials that were dug up did not biodegrade because landfills protect much of the waste from the natural elements. On the other hand, some materials do break down because the proper conditions are present for biodegrading. Rainwater "leaches" out chemicals that enter the natural water reservoirs underneath the landfills, thus introducing toxic chemicals to water supplies that people drink.

If newspaper, packaging, and clothing do biodegrade, they release ink, dye, and paint that can pollute. This is the concern with *biodegradable plastics.* These hydrocarbons contain many elements and additives such as chlorine, formaldehyde, dyes, and metals. Like newspapers, some of these materials should not degrade. Newer landfills with liners and gas-collection systems are designed to contain the pollutants to the dump.

Photodegradable plastics are susceptible to the ultraviolet rays of the sun, whereby polymer chains break, causing the material to lose strength and crumble into small bits. These plastics are intended for products that often become litter that might harm wildlife and degrade the landscape. An example of a use for photodegradable plastics is six-pack-beverage carrying rings, which should decompose after several months of exposure to the sun. It should be pointed out that these products do not break down completely, as does rusting steel or most paper products.

Although modern landfills should be constructed to contain toxic chemicals, problems may be avoided by keeping toxic chemicals out. As with newspapers, food, and other materials, the landfill will not have the conditions necessary for biodegrading of plastics, so biodegradable plastics can present more problems than they solve. Photodegradable plastic might encourage people to litter. Also, lacking sunlight, they will not degrade and might harm the marine environment. The development of polymers that will degrade in the absence of light, leaving only carbon dioxide, water, and a small amount of biological matter (e.g., Bispol resin), should make plastics more friendly to the environment and more acceptable as packaging material.

Effective methods for recycling plastics and paper products would address 56% of the volume of MSW. To encourage recycling, a profit must be realized. Aluminum is a good example. It requires 95% more energy to produce virgin aluminum from bauxite ore than it does to melt down and use recycled aluminum. This fact has promoted a concerted effort by the aluminum industry and retailers to promote aluminum recycling. The Aluminum Association reports that, at present, 50.5% of aluminum cans are recycled.

A number of other products, such as iron, steel, and glass, are also in a healthy recycling stream. Other items, such as scrap tires and used engine oil, are a challenge to dispose of in a manner safe to the environment. Having engineering designers consider secondary uses of a product once its useful life is at an end, and designing for disassembly, should help with MSW and environmental problems.

The plastics industry has attempted to make plastic more environmentally compatible. One type of plastic, polystyrene (PS), which is widely used but also criticized because of its effect on the environment, serves as a good example for a study of plastic recycling issues. The National Polystyrene Recycling Council (NPRC) reports that PS products use an average of one-third less energy to produce than is used by comparable paper products. Some communities banned certain PS packaging based on a questionable belief that paper packaging was preferable in terms of the total materials cycle.

Polyethylene terephthalate (PET), used for soft-drink bottles, is a good recycling plastic. Some reclaimed PET goes back into food packaging, while another significant amount finds use in recycled applications such as for carpet backing, fiberfill for sleeping bags and ski jackets, and transformation into polyol, a chemical ingredient used to make rigid urethane foam.

High-density polyethylene (HDPE) is another widely recycled plastic. You encounter virgin HDPE in milk jugs and the base caps of large PET soft-drink bottles. Recycled HDPE finds use in trash cans, flowerpots, piping, and traffic cones. "Plastic lumber" made of recycled HDPE is finding uses as railroad ties, decking for boat docks, and fencing. It will not rot or splinter, nor does it need paint; the color is added during production.

In keeping with the idea of waste reduction, polycarbonate-resin returnable milk cartons provide recycling without a need to scrap the product for reprocessing. The bottles are produced to permit refilling (after sterilization) up to 100 times and should last 8 to 10 years. This keeps them out of the MSW stream for an extended period. Once their life cycle is up, they can be reground and reprocessed for uses such as engineering TP for automobiles and building components.

The National Resources Defense Council (NRDC) presents views on plastic use and recycling from a nonindustry-aligned position. They support a ban on the use of polystyrene (PS) based on several factors:

1. There is skepticism as to the cost-effectiveness of recycling.
2. The goal of 25% recycling leaves 75% as MSW.
3. During the manufacture of foam-PS, nonfood products, chlorofluorocarbons (CFCs), and gases used as blowing agents contribute to depletion of the upper atmospheric ozone layer.
4. As single-use products they create MSW problems because they are composed of small, air-filled beads that do not compress.
5. Pentane and butane, foaming gas for PS food containers, contribute to lower-atmospheric ozone (smog) problems.

The NRDC supports the use of plastics when they can be recycled. PET and HDPE have good potential here. To reduce waste, single-use (disposable) items are discouraged. We can, for example, follow the European custom of shopping with string bags instead of using single-use plastic and paper bags.

PLASTIC-CONTAINER CODING SYSTEM

Symbols on packaging containers aid in sorting plastics. These symbols, usually found on the bottom of packages and other products, are there as part of the Society of Plastic Industry, Inc. (SPI) voluntary *plastic container coding system.* Figure 6-60 shows the seven symbols used

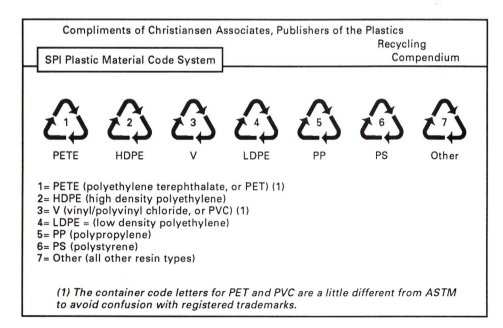

Figure 6-60   SPI's voluntary plastic container coding system. (Dow Chemical)

in this coding system and the legend of the code. Note that "1—PETE" represents polyethylene terephthalate, which has an American Society for Testing and Materials (ASTM) abbreviation of PET; the ASTM acronym for poly(vinyl chloride) or vinyl is PVC, whereas the SPI code is V. These departures from ASTM standards are made to avoid confusion with registered trademarks.

The coding system required for packaging by most U.S. states is scheduled for replacement due to the confusion it has caused. Much of the labeling on packages results in the erroneous impression that the plastic will be recycled, when in fact it will not. Only about 3% of plastic garbage actually makes it back into new products. Recycling centers can use only select materials, such as HDPE for blow-molded milk bottles or PETE for soda bottles. HDPE used for injection medicine bottles has a different composition that should not be mixed with the milk bottle resin.

In addition to SPI's container code, other organizations developed coding systems to aid plastics recycling, including the Society of Automotive Engineers (SAE), the American Society for Testing and Materials (ASTM), and the International Standards Organization (ISO). Laws may dictate the use of a particular code; otherwise, manufacturers use the coding system most appropriate for their product.

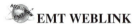 **EMT WEBLINK**

SPI Container Coding System—*http://www.plastic.org/outreach/recycling/2124.htm*

POSITIVE TREND WITH RECYCLING

Public awareness, legislative mandates, and improved technology have all helped to move society away from some of the MSW dilemma. At the beginning of the move toward recycling, municipalities established programs to collect waste, but the ability to process it was limited. Many localities had to pay for the removal of good-quality paper, plastic, and glass. However, the potential for profit in recycling spurred improved processing technology and increases in recycling processing centers. Web sites provide updates on the progress with recycling.

The auto industry also reports large gains in the plastic recycling of junked cars. Because of the savings realized from using recycled rather than virgin plastics, more new cars carry parts with recycled plastics. For example, PET from soda bottles recycles into parts such as luggage rack siding, headliners, and reinforcement panels. Ford estimates that nearly 95% of scrapped cars make it to recycling centers, where more than 75% of the contents are recycled. Research reveals that many plastics can be scrapped, shredded, pelletized, and remolded to new parts up to a dozen times.

Achieving these positive results of recycling required considerable education, discussion, and cooperation. Change is often difficult! Many people complained about the costs and inconvenience of recycling. The effort seems to be paying off, however, with fewer problems with MSW, better use of limited resources, and new opportunities for jobs and business growth.

EMT WEBLINKS

Several websites deal with municipal recycling. The following sites offer insights into waste recycling and provide links to other sites.

Institute of Scrap Recycling Industries—*http://www.recycle.net/recycle/assn/rs000103.html*

Pennsylvania's Recycling—*http://www.dep.state.pa.us/dep/deputate/airwaste/wm/recycle/recycle.html*

Macon Iron—*http://www.thescrapmarket.com/municipal_recycling.htm*

Morris County Solid Waste Management Plan—*http://www.mcmua.com/recycling/MunicipalResponsibilities.htm*

Broward County—*http://www.fll.net/oiwm.htm*

Internet Guide to Recycling—*http://www.libsci.sc.edu/bob/RECYCLE.HTM*

Center for Plastics Recycling Research (CPRR)—*http://www.recycle.net/recycle/assn/cprr.html*

For a Canadian perspective, Association of Municipal Recycling Coordinators—*http://www.amrc.guelph.org/*

REFERENCES & RELATED READINGS

Plastics

AMERICAN FIBER MANUFACTURERS ASSOCIATION. "Manufactured Fibers," video. New York: AFMA, 1989.

AMERICAN SOCIETY FOR TESTING AND MATERIALS. *Annual Book of ASTM Standards,* Vol. 08.01, *Plastics* (D-20). Philadelphia: ASTM, 1989.

ASM INTERNATIONAL. *Engineering Materials Handbook,* Vol. 2, *Engineering Plastics.* Materials Park, OH: ASM International, 1988.

BRADY, GEORGE S., and HENRY R. CLAUSER. *Materials Handbook.* New York: McGraw-Hill, 1977.

COALITION ON RESOURCE RECOVERY AND THE ENVIRONMENT. *The United States Conference of Mayors.* "Incineration of Municipal Solid Waste: Scientific and Technical Evaluation of State of the Art." Washington, D.C., February 1, 1990.

CONSTANCE, J. "Dendrimers: Materials by Design." *Compressed Air Magazine,* June 1992, pp. 12–17.

CONSUMERS UNION. "The Right Glues." *Consumer Reports,* July 1995, pp. 470–473.

CRANE, F. A. A., and J. A. CHARLES. *Selection and Use of Engineering Materials,* 2nd ed. Woburn, MA: Butterworth, 1989.

DATA BUSINESS PUBLISHING. *Plastics Materials DATA Digest.* Englewood, CO: DATA, 1996.

FINE, LEONARD W., and HERBERT BEALL. *Chemistry for Engineers and Scientists.* Saunders College Publishing, 1990, pp. 698–702.

FISCHER, DAVID. "Turning Trash into Cash." *U.S. News & World Report,* July 17, 1995, p. 43.

GOBSTEIN, SAUL. "Introduction to Plastics and How to Choose Them." ASM International continuing education short course materials. Materials Park, OH: ASM International, 1993.

HOCKING, MARTIN B. "Paper Versus Polystyrene: A Complex Choice." *Science,* June 7, 1991, pp. 504, 506.

INSTITUTE OF SCRAP RECYCLING INDUSTRIES. *http://www.recycle.net/recycle/assn/rs000103.html,* January 4, 2000.

JACOBS, JAMES. "Recycling Plastics." *The Technology Teacher,* September/October 1990, pp. 15–22.

LABANA, S. S. "Recycling of Automobiles—An Overview." Paper presented to *NEW: Update 93,* NASA-LaRC, November 4, 1993.

MARSHALL, JOHN A. "Liquids That Take Only Milliseconds to Turn into Solids." National Educators' Workshop: *Update 93—Standard Experiments in Engineering, Materials Science, and Technology.* NASA Conference Publication 3259, April 1994, pp. 315–322.

MITTAL, K. L. *Polyimides: Synthesis, Characteristics, and Applications,* Vols. 1 and 2. New York: Plenum Press, 1984.

PATER, RUTH H. "Interpenetrating Polymer Network Approach to Tough and Microcrack Resistant High Temperature Polymers." *Technical Support Package for Tech Brief,* Part 2, LAR-14338, LaRC-RP41.

PEIFFER, ROBERT W. "Solid Imaging: New Opportunities for Photopolymers." *Radtech Europe '91 Conference,* Edinburgh, September 29–October 2, 1991.

PETERSON, I. "Geometry for Segregating Polymers," *Science News,* Vol. 134, no. 19, p. 151.

ROGERS, CRAIG A. "Intelligent Materials Systems: The Dawn of a New Materials Age." Center for Intelligent Materials and Structures, Virginia Polytechnic Institute and State University, 1992.

SALEM, A. J., et al. "Fabrication of Thermoplastic Matrix Structural Composites by Resin Transfer Molding of Cyclic Bisphenol—A Polycarbonate Oligomer," *SAMPE Journal,* January/February 1991, pp. 17–22.

SEYMOUR, RAYMOND B. *Polymers for Engineering Applications.* Materials Park, OH: ASM International, 1987.

SHEER, ROBERT J. "Incorporating Intelligent Materials into Science Education." National Educators' Workshop: *Update 94—Standard Experiments in Engineering, Materials Science, and Technology.* Gaithersburg, MD, Nov. 7–9, 1994.

SOCIETY OF PLASTICS INDUSTRY. *Plastics Today—SPI's Fax-on-Demand Service.* Phone 800-774-4614 (U.S. only), January 10, 2000.

SPROW, EUGENE E. "Chrysler's Concurrent Engineering Challenge," *Manufacturing Engineering,* April 1992, pp. 35–42.

WIRKA, GENE. *Wrapped in Plastic.* Environmental Action Federation, 1990.

YESTER, SUSAN G. "Life Cycle Inventory," *NEW:Update 99,* November 1, 1999, Auburn Hills, MI.

ZAJONC, AUGUST. Green building website, *http://www.coldhamarch.com/green/,* January 17, 2000.

Elastomers

AMERICAN SOCIETY FOR TESTING AND MATERIALS. *Annual Book of ASTM Standards,* Vol. 09.01, *Rubber, Natural and Synthetic.* Philadelphia: ASTM, 1991.

BHOWMICK, A. K., and H. L. STEPHENS. *Handbook of Elastomers.* Materials Park, OH: ASM International, 1988.

BRADY, GEORGE, and H. R. CLAUSER. *Materials Handbook,* 12th ed. New York: McGraw-Hill, 1986.

HARPER, CHARLES A. (ed.). *Handbook of Plastics and Elastomers.* New York: McGraw-Hill, 1975.

INTERNATIONAL INSTITUTE OF SYNTHETIC RUBBER PRODUCTS. *Synthetic Rubber: The Story of an Industry.* New York: IISRP, 1973.

PUGH, BRADLEY. "The Role of Standards in the Development of the Condom," *ASTM Standardization News,* August 1992, pp. 23–29.

SHELL. *Kraton Thermoplastic Rubber.* Shell Chemical Company, Polymer Division, Houston, 1972.

Wood and Other Polymers

AMERICAN SOCIETY FOR TESTING AND MATERIALS. *Annual Handbook of ASTM Standards,* Vol. 04.09, *Wood* (D-7); Vol. 06.03, *Paints, Related Coatings and Aromatics.* Philadelphia: ASTM, 1992, 1991.

AMERICAN SOCIETY FOR TESTING AND MATERIALS. *Standardization News,* "Geosynthetics," special issue, September 1992.

CASSENS, DANIEL L., and WILLIAM C. FEIST. *Exterior Wood in the South: Selections, Applications and Finishes,* FPL-GTR-69. Forest Products Laboratory, 1991.

FOREST PRODUCTS LABORATORY. *Wood Handbook: Work as an Engineering Material.* Washington, D.C.: U.S. Department of Agriculture, 1974.

INDUSTRIAL FABRICS ASSOCIATION INTERNATIONAL. *A Design Primer: Geotextiles and Related Materials.* St. Paul, MN: IFAI, 1991.

MATERIALS EDUCATION COUNCIL. Vol. I, *Wood: Its Structure and Properties;* Vol. II, *Wood as a Structural Material;* Vol. III, *Adhesive Bonding of Wood and Other Structural Materials;* Vol. IV, *Wood: Design of Structural Elements.* 1983. MEC, 1988.

MITTAL, K. L. (ed.). *Adhesive Aspects of Polymeric Coatings.* New York: Plenum Press, 1983.

MITTAL, K. L. (ed.). *Adhesive Joint Formation, Characteristics, and Testing.* New York: Plenum Press, 1984.

RITTER, MICHAEL A. *Timber Bridges: Design, Construction, Inspection and Maintenance.* EM 7700-8. Washington, DC: U.S. Forest Service, 1990.

ROWELL, R. M., and P. KONKOL. *Treatments That Enhance Physical Properties of Wood.* FPL-GTR-55. Forest Products Laboratory, 1987.

SARIKAYA, MEHMET, and ILHAN A. AKSAY (eds.). *Biomimetrics: Design and Processing of Materials,* New York AIP Press, 1995.

Periodicals

Adhesive Age

Advanced Materials & Processes

Automotive Engineering

Journal of Adhesion Science and Technology

Materials Engineering

Modern Plastics

Plastics Technology

Plastics World

Popular Science

Scientific American

SME Composites in Manufacturing

Module

7

Ceramic Materials

After studying this module, you should be able to do the following:

7–1. Name some goals of industry and society that caused a push for greater use of ceramics. Describe attributes of ceramics that make them competitive with plastics and metals. Contrast traditional and advanced ceramics.

7–2. Relate bonding, structure, and processing of ceramics to their properties (electrical, thermal, optical, and mechanical). Contrast ceramics to ceramics and ceramics to metals in terms of structure and properties. Sketch common ceramic microstructures.

7–3. Recall the major processing and raw materials used to produce ceramics. Differentiate among empirical, deterministic, and probabilistic design. Describe some advanced processes and their benefits in terms of high and low technology.

7–4. List and describe the favorable major properties of ceramics (e.g., wear and heat resistance and low thermal expansion) and the units to express them. Name the common tests to measure the properties.

7–5. Use diagrams and explanations to show stress concentrations plus methods to toughen ceramics, including composite matrix ceramics. Define work of fracture and compare work of fracture versus tensile strength in glass, cements, and biomaterials. Make sketches to show methods of improved ceramic structure for greater toughness.

7–6. Express fracture toughness in MPa and psi, and density in SI and U.S. customary units. Compare typical properties such as wear, density, hardness, flexural strength, fracture toughness, and thermal expansion and thermal shock resistance of CMCs with other ceramics, metals, and polymers. Calculate the thermal shock resistance of given materials.

7–7. Explain, with the aid of sketches, methods of strengthening glass and four types of glass, and typical applications and properties allowing those methods.

7–8. Explain the difference between cement and concrete and between setting and curing. Discuss the advantages offered by cement and concrete for developing nations. Explain the advantages accrued from improved cements.

7–9. Explain the process of making polycrystalline glass and applications of this material. Draw the heating curve for glass ceramics.

7–10. Discuss some benefits of ceramics with regard to the goals of green manufacturing.

7–11. Use EMT WEBLINKs to explore sites for new developments and evolving concepts related to traditional and advanced ceramic materials.*

(*EMT WEBLINKS listed throughout the book will link you to Internet sites related to topics being covered. The dynamic nature of the Web brings frequent changes, so some of these sites, while available at the time of writing, may not be up now.)

PAUSE & PONDER

Early people learned to convert natural materials into useful products mostly by trial and error. Adobe bricks were made from wet mud, shaped, and heated in the sun to form them into a building material. Primitives used sticks, stones, and mud for their shelters. Eventually they progressed from the hunter-gatherer stage to living in a community with domesticated animals and plants which supplied their needs for food and clothing. This advancement in living style allowed them greater time to develop better and more efficient materials. With this great change came the discovery that fire could be used to dry the adobe bricks much more efficiently than the sun. Firing clay and bricks produced stronger objects that made cooking and storing of food much easier. Molds made of these natural materials also could withstand higher temperatures and, with the discovery of metals, these molds could be used to shape the metals. This led to the making of bronze tools and, later, tools and utensils made from iron. Pottery and porcelain made by heating aluminosilicate clays to high temperatures at which the particles partially melt and fuse together are examples of ***traditional silicate*** ceramics (see Figure 7-1).

Traditional ceramics are derived and processed from clay or nonclay minerals. The earth's crust or outer shell consists of rocks and minerals. Each of the crystalline substances in a rock is called a ***mineral.*** Minerals that have a commercial value are called ***ores.*** Ore called bauxite is a mineral from which the metal aluminum and alumina are produced. Limestone is a common rock made of ***calcium carbonate*** ($CaCo_3$). The word ***carbonates*** indicates compounds made up of a metal combined with C and O_2. The word ***silicates*** refers to compounds formed of metals joined with Si and O_2, and the word ***oxides*** refers to compounds formed of metals combined with oxygen. $CaCO_3$ or limestone is found in many different forms, includ-

(a)

(b)

(c)

Figure 7-1 Traditional Ceramics—(a) From the Stone Age to the present, traditional ceramics served and serve as key building materials. Restored Monticello reveals use of brick from Virginia clay, masonry mortar, and glass windows. Thomas Jefferson designed, and his slaves built and rebuilt the constantly evolving home over a period from 1768 beyond 1809. (b) Modern construction ranging from modest homes to sky scrapers to dams and bridges relies on traditional ceramics as seen in the recently constructed sea wall in Virginia Beach. Note the decorative brick walkway, concrete sculptures, and reflective glass glazing in the building. In the background, and (c) the sea wall is made of reinforced concrete with integrated huge concrete piping that connects to pumps to carry storm surged sea water back out to sea; thereby, reducing flood damage.

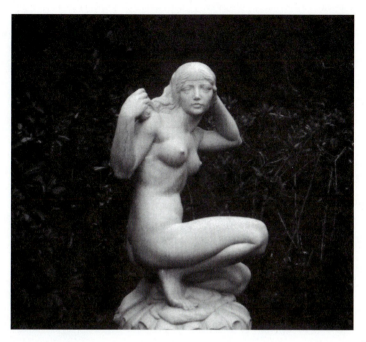

Figure 7-2 Statuary sculpted from marble, a ceramic material that has withstood the ravages of time and weather, even modern-day air pollution. Marble is a hard, crystalline form of limestone ($CaCo_3$). Sulfuric acid (H_2SO_4) in acid rain causes marble buildings and statuary to disintegrate, forming a crumbly compound of calcium sulfate ($CaSO_4$).

ing chalk (the White Cliffs of Dover in England), coral, marble (see Figure 7-2), and egg shells. In natural formations this limestone rock forms, among many others, the Rock of Gibraltar in Spain and Mammoth Cave with its stalagmites and stalactites in Kentucky. Humans have used limestone over the centuries to build the Roman Coliseum and the Taj Mahal (India). These traditional products include refractories, white wares, cement, porcelain, and structural clay ceramics. Included in the definition of traditional ceramics are the products produced by the use of silicate glass. Concrete is included in this category with two of its ingredients being ceramic materials–sand/gravel and cement. Traditional ceramics are usually heavy, fragile, and not considered for use as structural or load-bearing materials. In this modern age the demand is for lighter materials that have great strength at high temperatures and maintain their strength at these high temperatures while subjected to impact forces.

At this point it would be wise to review Figure 2-1, which traces the relative importance of the main materials groups back to the Stone Age, noting in particular the role that ceramic materials have played in supplying the needs of the various peoples that occupied the planet of the sun called Earth. A good exercise to ensure greater learning is to make a list of the traditional and advanced ceramics found in the average home. This task might be repeated later as you progress through this module. Figure 7-3a shows samples of what you may find in a typical home, and Figure 7-3b shows a typical street scene with numerous building materials that are ceramics. A third location for observing the use of ceramics would be to sit in your automobile and identify which parts are made of ceramics. Would you be able to explain why a ceramic material was selected for a particular application?

As in olden times, our civilization is developing new ceramic materials with which to meet ever-demanding expectations of our society. Brick and tile were important to the ancient civilizations both for building and decoration. Tiles were originally fabricated from wet, moldable clay that took a considerable amount of time to prepare, mold, and dry. In the late 1800s, decorated and undecorated wall and floor tiles were in universal use. Figure 7-4 shows a collage of building materials including colorful tiles—traditional ceramic materials—now fabricated using more advanced processing techniques that have greatly reduced the time and labor

(a)

(b)

Figure 7-3   (a) A mix of typical applications of ceramic materials that illustrate the vast time over which ceramics have benefited the human race—from the Stone Age to the Space Age. Shown is one example of the use of soda-lime glass; two examples of Pyrex™, borosilicate glass by Corning that are oven and microwave proof with thermal-shock resistance; two examples of clay pottery with a glazing into which has been incorporated colors and other chemicals to produce a colored, glassy, impermeable coating; and two pieces of Corelle tableware by Corning which can be placed in a preheated oven or microwave directly from the freezer. (b) Study in the use of modern building materials. A good learning experience is listing the different materials used and reasons for their selection. Note evidence of wind damage to the roof.

involved. A big difference is that today we can use our expanding scientific knowledge and technologies to help us develop new materials with structures that account for the desired properties to meet the new applications demanded by society. Figure 7-5 illustrates this by showing a mix of traditional and advanced ceramics now in our homes.

The greatest demand for advanced ceramic applications is evidenced in land, sea, air, and space industries. These industries require construction materials that are lightweight so that the energy requirements are lessened for the movement of vehicles. A challenge arises that brings to the fore the realm of advanced ceramics. Our definition of traditional ceramics does not fully describe ***new, technical, engineering,*** or ***advanced ceramics*** (see Figure 7-6). Are advanced ceramics inorganic, nonmetallic, nonmolecular solids? Yes, some are. At one time the word ***organic,*** when used with terms such as *compounds* or *chemistry,* described materials that contained carbon. Today there are a few exceptions (materials that contain CO_2 and $CaCO_3$) that

Figure 7-4 Modern ceramic building materials find use in a storefront at a popular beach resort with its multitude of colored tiles, large glassed-in front, and surrounded with a mixture of cast-cemented outdoor eating areas.

Figure 7-5 A mix of modern materials comprise a typical kitchen cooking area with stainless steel utensils, a Corning Ware casserole dish with a tempered glass cover, wood cabinets and cutting board, and Corian ceramic countertop and backsplash molding.

Figure 7-6 Examples of newer (advanced ceramics) for engines. Top, left to right: silicon nitride (Si_3N_4) turbocharger, Si_3N_4 valve, cast steel diesel engine rocker arm with partially stabilized zirconia (PSZ) cam follower and wear button, Si_3N_4 cam follower, and valve. Bottom row, left to right: valve; silicon carbide (SiC) water pump seal; piston pin, valve spring retainer, and valve guide of Si_3N_4; and PSZ diesel head plate with integrated valve seats. (ORNL.)

are not classified as organic materials. All other compounds are now defined as ***inorganic compounds. Ceramic compounds*** can be defined as inorganic compounds made by heating clay or other mineral matter to a high temperature at which they partially melt and bond together. Adobe bricks are ceramics that are made of mud which has been dried in the sun. They include both crystalline and amorphous materials. New ceramics are synthetically developed from rather simple chemical compounds, and advances in processing which have provided greater control over their structures have resulted in vast new improvements, particularly in electrical, magnetic, and optical properties. One characteristic these new ceramics possess that traditional ceramics do not is their high purity, which is the basis for their better mechanical, electrical, magnetic, and optical properties. Some examples of advanced ceramics include ***oxide ceramics*** such as alumina, beryllia, and zirconia and ***nonoxide ceramics*** which can be represented by carbides and nitrides such as boron carbide, silicon carbide, and silicon nitride.

With this discussion of advanced ceramics, could you expand on your list of ceramic materials found in the home? In summary, ceramics encompass a wide variety of materials for which humans have found applications over the past centuries. Table 7-1 compares ceramics with metals by listing some of their properties. This choice of properties can easily be broadened as the reader proceeds through this module. A good learning exercise is to construct, upon completion of the module, a listing of new properties such as chemical inertness or thermal shock resistance.

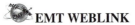

EMT WEBLINK

Aluminas—*http://me.mit.edu/2.01/Taxonomy/Characteristics/Ceramics/MainCeramics. html#Aluminas*

Ceramic is a unique material, developed in the Stone Age but finding numerous uses as a space-age material in the third millennium. Pictured in Figure 7-6 are a few of the newer ceramics that you will learn about in this module. The silicon nitride (Si_3N_4) turbocharger rotor makes possible more efficient engines that have smaller displacement and operate at higher temperatures. Si_3N_4 valves are 50% lighter than steel valves and therefore reduce friction in the valve train; they also run faster and hotter. Si_3N_4 roller bearings replace the complex steel needle bearings and cage while saving about 40% in cost. They also operate at higher temperatures, which means reduced cooling needs. Partially stabilized zirconia (PSZ) has also found applications in heat engines, as shown in Figure 7-6 in the diesel engine rocker arm, head plate, piston pins, and valve guides.

TABLE 7-1   COMPARISON OF METALS WITH CERAMIC MATERIALS

Metals	Ceramics
Crystal structure	Crystal structure
Large number of free electrons	Captive electrons
Metallic bond	Ionic/covalent bonds
Good electrical conductivity	Poor conductivity
Opaque	Transparent (in thin sections)
Uniform atoms	Different-size atoms
High tensile strength	Poor tensile strength[a]
Low shear strength	High shear strength
Good ductility	Poor ductility (brittle)
Plastic flow	None
Impact strength	Poor impact strength
Relatively high weight	Lower weight
Moderate hardness	Extreme hardness
Nonporous	Initial high porosity
High density	Initial low density

[a]Small flaws; pores act as stress concentrators that are not reduced by ductility/plastic deformation.

EMT WEBLINKS

ORNL Advanced Ceramic Technology Center—*http://www.ornl.gov/orcmt/specific/
advcer.html*

Materials Science and Technology Gateway—*http://MST-online.nsu.edu*

Most of these new advanced ceramics were developed within the past 60 years, which is quite recent in materials technology. Much of that development resulted from aerospace *research and development (R&D),* but gradually, the costs associated with advanced ceramics have made them feasible for more earthly applications, such as in the automobile, sports, and machine-tool industries. Most are actually synthetic ceramics that are produced from fine, relatively poor powders using new technology that includes microwaves, electron beams, and polymer chemistry. Figure 7-7 lists some differences between traditional and advanced ceramics, including some applications of each category of ceramic. Advanced ceramics are receiving attention for wide-ranging scitech development because they offer the designer properties such as light weight, good strength at elevated temperatures, and wear resistance. Many of these newer ceramics are composites reinforced with whiskers and fibers to improve their fracture toughness, so they do not break catastrophically as we generally expect dishes and coffee cups to do when we drop them. The wear resistance of ceramics makes them valuable as coatings for cutting tools, surgical instruments, punches, and dies. For a modest cost, it is possible to apply a TiN coating only a few micrometers thick to tools, which can extend their life 7 to 15 times. You may have seen golden TiN coatings on twist drills in the hardware store.

In the ongoing search to produce superior materials with a minimum of energy (low temperature, low pressure, and minimum number of steps), researchers have turned to nature to replicate its processing techniques. *Biomimickry* or *biomimetics* is the attempt to mimic or replicate natural processes. For example, seashells are intricate, lamellar ceramic composites made with low energy yet at a nanometric scale, yielding complex patterns that hold promise for ceramic electronic devices and similar advanced applications.

Figure 7-8 represents a taxonomy of ceramic materials based on their applications. Note that other classifications of ceramics can be developed based on such differences as their processing. New or advanced ceramics do not possess the plasticity and formability of clay when mixed with water, which necessitates new techniques in processing to be discussed later in this module. This module should help you to gain a perspective on the structure, properties, selection,

Figure 7-7 Broad classifications of ceramics. Traditional ceramics are contrasted with the newer advanced ceramics, which are also referred to as *fine ceramics, engineered ceramics, new ceramics,* and *value-added ceramics.* (ORNL)

Traditional Ceramic Materials

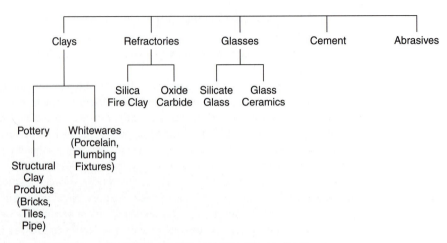

Figure 7-8   (a) Taxonomy of traditional ceramic materials based on applications.

Advanced (New) Ceramic Materials

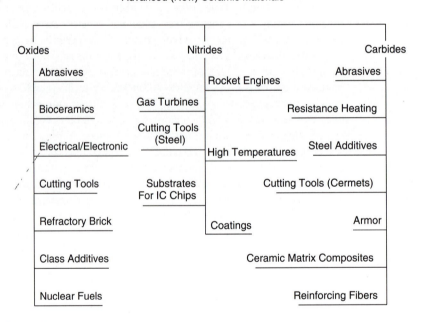

Figure 7-8   (b) Taxonomy of advanced (new) ceramics materials based on applications.

and basic processes in ceramics, both traditional and advanced. In your study of ceramics, ponder their nature and their ability to compete with metals, polymers, and composites. Be alert to news articles and advertisements that describe emerging applications of advanced ceramics. Also, imagine where there may be new opportunities for the use of ceramics. This information will serve you well in our advancing age of materials and prepare you for the dramatic changes in this era.

 EMT WEBLINKS

http://www~dmse.mitedu/wom/OxideCeramics.html

7.1 NATURE OF CERAMICS

Ceramics are loosely defined as crystalline solids composed of metallic and nonmetallic materials. They are inorganic materials; that is, they do not contain carbon and are derived from mineral sources. Ceramics are crystal structures made of metallic ions. Bonding is either partially or completely ionic. Variables involved are (1) the magnitude of the electrical charge on the ions, and (2) the relative size of the ions. The crystal must be neutral; that is, all the cation positive charges must be balanced by an equal number of anion negative charges. The *oxidation state* or *oxidation number* designates the number of electrons an atom loses or gains or otherwise uses in joining with other atoms in compounds. The total of the oxidation states of all atoms in a molecule or formula unit is zero. Na has an oxidation state of $+1$; Cl has an oxidation state of -1. When they join to form the compound expressed by the formula unit NaCl, their oxidation states must equal zero.

Some of these ceramics are intermetallic compounds (see Section 3.5.2.1). Originally, ceramics were clay-based materials. The word *ceramic* comes from the Greek *keramos,* meaning "burnt earth." Glass, with an amorphous structure (a rigid liquid), can be classified either as a traditional ceramic which includes the glasses made from silicates or in a separate category due to its amorphous structure. When glass is transformed into a largely crystalline structure by heat treatment to form glass ceramics it can be classified as an advanced ceramic. *Silicon carbide* does not occur as a mineral and is produced synthetically. Its crystalline structure is diamondlike with alternating Si and C atoms covalently bonded together to produce a material that is almost as hard as diamond. In addition, its crystals fracture in a way that leaves them with sharp edges. These properties make SiC an important industrial abrasive. Ceramics are inexpensive compared with competing materials. Consisting primarily of forms of silicon, aluminum, and oxygen—the most abundant elements in the earth's crust—ceramics can be produced at less cost than that of competing metal-alloy components. Further, many, if not all, of the strategic minerals, including cobalt, that are needed to produce sophisticated metal alloys, such as the superalloys presently used in high-temperature applications, are not found in the United States. AerMet 100, one such recently developed superalloy steel, was shown in Figure 5-34a and described in Section 5.8.2.3. What sets ceramics apart is their set of properties common to nearly all forms: (1) extreme hardness, (2) heat resistance, (3) corrosion resistance, (4) low electrical and thermal conductivity, and (5) low ductility, or *brittleness.* Table 7-1 is a comparison of these and other properties of ceramics with those of metals. Table 7-2 provides an additional comparison of the properties of ceramics and those of competing materials. Granite is a mixture of quartz and the silicate mica and feldspar. The main source of quartz is sandstone which is mixed with other minerals and clay to improve the final product. Crystalline quartz is a three-dimensional array of silicon dioxide (SiO_2), also known as *silica* (see Figure 7-59a for a sketch of the crystalline pattern formed by the Si and O atoms in silica). Crystals of pure quartz are colorless, but various impurities produce colored quartz; for example, traces of Fe (an impurity) in the crystal structure produce amethyst with its purple color.

EMT WEBLINKS

Characteristics—*http://me.mit.edu/2.01/Taxonomy/Characteristics/Ceramics.html*

Yahoo's Materials Science Site and Search Engine—*http://dir.yahoo.com/science/engineering material_science/*

The several examples that follow add emphasis to the properties of ceramics listed above. The Knoop hardness cubic boron nitride (CBN) is 5000 kg/mm^2; that of diamond, the hardest known material, is about 7500 kg/mm^2. Partially stabilized zirconia (PSZ), to be discussed in more detail later, can withstand a pressure of 78,000 psi for about 2000 hours

TABLE 7-2 TYPICAL PROPERTIES OF SELECTED ENGINEERING AND TECHNICAL CERAMICS

Properties	Ceramics				Comparison materials		
	Alumina	Beryllia	Silicon carbide	Zirconia	Mild steel	Aluminum	Nylon
Melting point (approximately °C)	2050	2550	2800	2660	1370	660.2	215
Coefficient of thermal expansion (m/m/K) $\times 10^{-6}$	8.1	10.4	4.3	6.6	14.9	24	90
Specific gravity	3.8	—	3.2	—	7.9	2.7	1.15
Density (kg/m^3)	3875	2989	3210	9965	7833	2923	1163
Dielectric strength (V/m) $\times 10^6$	11.8	—	—	9.8	—	—	18.5
Modulus of elasticity (MPa) $\times 10^4$	34.5	39.9	65.5	24.1	17.2	6.9	0.33
Hardness (Mohs)	9	9	9	8	5	3	2
Maximum service temperature (K)	2222	2672	2589	2672	—	—	422

without any appreciable effect. Its ***coefficient of friction*** with steel is only 0.17 at room temperature, and its coefficient of thermal expansion has very nearly the same value between 0° and 1000°C. ***Electronic package*** casings are a good example in which ceramic composites are finding new applications. Casings need to have superior heat transfer characteristics with low weight, high thermal conductivity, and matching coefficients of thermal expansion. Silicon ***integrated-circuit (IC) chips*** must be joined out of necessity to metal. Si has a coefficient of thermal expansion of 2.6 $\times$ 10 in/in/C and specific heat (c_p) of 702 J/kg-K. Si$_3$N$_4$ comes closest to Si with a range of thermal expansion between 2.3 and 2.9 $\times$ 10 in/in/C. Aluminum is used to fabricate metallized circuits on the surface of chips to provide circuits between devices and to bring circuits to contact pads where outside contact is established. Aluminum and its alloys have an expansion coefficient of about 25 which is much higher than Si. The IC chip itself must be packaged so that it may be bonded to a printed-circuit board. All materials mentioned must be compatible—they all need to be equal in thermal conductivity to help dissipate excess heat generated by the passage of electricity in such a small area and such extremely small circuits. Second, these materials should expand and contract together as much as possible when subjected to changes in temperature. The choice must be to use a material with matching coefficients and thermal conductivities. Some nonferrous alloys have been designed to fulfill this need, such as ***Kovar,*** a registered trademark of Carpenter Technology Corp. for its nominal 29Ni–17Co–54Fe glass-sealing alloy UNS K94610. ***Caver*** is a ternary alloy of Fe–N–Co with a coefficient of thermal expansion of 5 $\times 10^{-6}$ in/in/C. Its specific heat is 460 J/kg-K. It is important to note that materials that have excellent thermal conductivities are in demand for such applications as previously described. Designing a metal composite such as silicon carbide aluminum composite (a CMX—ceramic matrix composite) with a matching coefficient of thermal expansion with lighter weight than Kovar would be a worthwhile project. A 10°C rise in temperature reduces the life of an integrated-circuit chip by a factor of 3.

Applications for advanced ceramics have received major media attention in recent years, particularly for use as parts in a future ceramic heat engine. However, corrosion resistance, chemical inertness, thermal shock resistance, and other properties that materials scientists and engineers can design into ceramic materials make both traditional and advanced ceramics highly attractive in a large number of applications. The combination of properties mentioned make ceramics good candidates for wear-resistance applications. Electrical properties place ceramics in great demand as solid electrolytes in experimental batteries and fuel cells. Other uses include automotive sensors, packaging for integrated circuits, electronic/optical devices, fiber optics, microchips, and magnetic heads. In the marriage of the computer and communications technologies, ceramics play a major role. Two additional outstanding characteristics of ceramics, which could be added to the list under ceramics in Table 7-1, are wear resistance and chemical inertness. Wear resistance was

Figure 7-9 Ceramics are increasingly being used to repair the human body. This porous ceramic material is used as a bone void fiber and for optical implants. It is produced by converting natural coral into hydroxyapatite—the main component of human bone. (Produced by Interpore Cross International, Irvine, CA)

mentioned briefly. Replacements for human bone, particularly where bone moves in close contact with other bone components, must be wear resistant for many obvious reasons. Figure 7-9 is one example of a ceramic used in biomedicine. One not so obvious reason is that if excessive wear were to occur, the residue from such would be carried off by body fluids to cause serious pollution problems within the human system. Chemical inertness of ceramics is finding many applications in the medical field where contact with body fluids is more of a problem than with other materials. Finally, ceramics play a big role in the machine-tool industry. Their thermal and mechanical stability allows them to retain their smooth, accurate cutting surfaces longer than metals do. Coated cutting tools and inserts, some with as many as 12 extremely thin coatings, each designed to serve a special function, can run productively at faster cutting speeds and at faster feed rates than can any metal-alloy tool in the machining of hard steels, superalloys, and ceramics. What is remarkable about these multicoated carbide inserts is that, with slightly over a dozen thin ceramic layers, the total thickness of the coatings is only 10 to 12 μm. Hewitt Refractories of London, England, recently unveiled a flooring material, Manta Heat™, which is a ceramic mat that can be rolled and shipped like carpeting. It heats up when an electric current is impressed on it to act as a radiant heating element. A spin-off of an electromagnetic shield for NATO computers, it is a blend of cementitious materials and doped carbon semiconductors. Flexible or rigid and of any thickness, it can be altered during manufacture to respond to voltages from 12 to 230 volts. One disadvantage is it must be covered to prevent the possibility of electric shock when wet.

 EMT WEBLINK

High Tech Companies for Materials Science and Engineering Ceramics—*http://do1l. eng.sunysb.edu/other4.html#ceramics*

7.1.1 Porosity and Density

When referring to a solid material such as a part made from copper or stainless steel, the word *density* takes into consideration a microstructure that contains no porosity. We are not speaking about voids or vacancies in the atomic structure when we use the term *porosity.* Density in such a situation can be termed the material's theoretical density. *Mass density,* which uses the mass of a material divided by its volume, as in Section 3.4.6, refers to this theoretical density. Such a density may be expressed in terms of its weight instead of its mass, as discussed in this section. Atomic weight is a major factor in determining the density of a material. Low-atomic-weight elements have low densities. A second factor is how the atoms/ions are stacked in the microstructure. Close-packed metals are more dense than open-structured materials. Our study of crystalline materials in Module 3 revealed that a bcc unit cell with a maximum coordination number of 8 is only 68% of full density, whereas fcc and cph unit cells with a coordination number of 12 can have a density of 74% (of a maximum of 100%). Ceramics are composed primarily of different phases, each with its own density. As an example, a ceramic that has a porosity of 20% is described as being 80% of theoretical density. *Green density* is the bulk density of a compact prior to its densification. In preparing particles for the production of powdered metals or ceramics, the packing of spherical shaped, monosized (0.1 μm) particles, even when vibrated to settle the particles, does not achieve a density level much higher than 60%. Despite tapping or vibrating, these particles do not rearrange themselves completely to eliminate pores. The optimum density of spherical particles can be reached by varying the size distribution of particles to permit smaller particles (50 nm or less) to locate in the interstices of the larger ones. One disadvantage with this procedure is that the larger particles tend to grow excessively during high-temperature sintering, which takes place later in the process. After particle sizing is complete, the pores remaining must be reduced by additional steps, such as mixing and kneading of the initial green mixture prior to the making of a green compact.

Ceramics are, by nature, generally porous materials with varying degrees of porosity. The term *bulk density* is used in this instance to refer to a ceramic's density, and it includes the material's porosity and the fact that most ceramics contain both a crystalline and a noncrystalline phase. *Open porosity* refers to the network of pores in a material that is open to the surface and into which a liquid such as water can penetrate if the part were submerged in it. *Closed porosity* refers to those pores that have become sealed within the grain structure. New technologies are under development to reduce, if not eliminate, pores in ceramic materials. Pores affect the strength of ceramics in two ways. First, they produce *stress concentrations.* Once the stress reaches a critical level (see Section 7.1.5), a *crack* will form and propagate. Because ceramics possess no plastic-deformation attributes to absorb any energy transferred to these materials once a crack is initiated (started), it propagates (grows) until fracture occurs. Second, pores (i.e., their size, shape, and amount) reduce the strength of ceramics because they reduce the cross-sectional areas over which a load can be applied and, consequently, lower the stress that these materials can support. The amount, size, distribution, and the continuity of voids in a ceramic are the principal factors that determine its properties. In general, porosity is detrimental to ceramics properties. Grain boundary sliding, creep rate, and chemical reactions that attack the ceramic or reduce its melting point are more adverse conditions produced by impurities and porosity. Modern processes have been developed to minimize porosity to a great extent. One exception to this goal of minimum porosity is ceramic insulators. Porosity increases the insulating capacity of ceramics, and at the same time, increased porosity makes ceramics lighter in weight. These two characteristics, when combined, make an ideal ceramic for use in such applications where the component or object is part of a moving system propelled by the consumption of energy such as an automobile or spacecraft. The bulk density of complex geometric parts as well as their various porosities can be determined by using ASTM procedure C373, which is based on Archimedes' principle.* This principle states that the weight of an object in a fluid equals its dry weight minus the buoyant force (or the weight of the fluid displaced). Taking only three weight measurements—the object's dry weight, its saturated weight in a fluid, and its net saturated suspended weight—several physical properties relating to the density and porosity of an object can be calculated. For example, the following five relationships expressed in terms of these three weights can be calculated:

$$1. \quad d_B = \frac{W_D}{V_B} = \frac{W_D \cdot d_L}{W_S - W_{SS}}$$

$$2. \quad V_B = \frac{W_S - W_{SS}}{d_L}$$

$$3. \quad V_A = \frac{W_D - W_{SS}}{d_L}$$

$$4. \quad d_A = \frac{W_D}{V_A} = \frac{W_D \cdot d_L}{W_D - W_{SS}}$$

$$5. \quad \%P_A = \frac{W_S - W_D}{W_S - W_{SS}} \times 100\%$$

where

W_D = dry weight
W_S = saturated weight
W_{SS} = saturated suspended weight
d_L = density of saturating liquid
d_B = bulk density
d_A = apparent density
V_A and V_B = apparent volume (bulk volume)
P_A = apparent porosity

Notes:

1. If water is the saturating liquid, its density may be assumed to be 1 g/cm^3; the ratio W/d results in units of volume (cm^3).
2. The apparent density is its dry weight divided by the difference between its bulk volume and its open-pore volume.

Illustrative Problem

A specimen of sintered silicon carbide has a dry weight of 3.1 g, a saturated weight of 3.5 g, and a saturated suspended weight of 2.3 g after being suspended and soaked in water. Calculate the bulk volume, bulk density, apparent density, and apparent porosity.

Solution

$$V_B = \frac{3.5 - 2.3}{1} = 1.2 \text{ cm}^3$$

$$d_B = \frac{3.5}{1.2} = 2.6 \text{g/cm}^3$$

---

*Refer to the experiment by Gail Jordan at the end of this module. *EMSET CDROM 90.*

$$d_A = \frac{3.1 \cdot 1}{3.1 - 2.3} = 3.9 \text{g/cm}^3$$

$$\%P_A = \frac{3.5 - 3.1}{3.5 - 2.3} \times 100\% = 33.3\%$$

Note: The saturated weight is the weight of the object after it is removed from the water.

 EMT WEBLINK

American Ceramic Society—*http://www.acers.org*

7.1.2 Structure, Bonding, and Properties

The structure of most ceramics varies from relatively simple to highly complex. Being compounds, ceramics are made of different types of atoms of varying sizes. *Clay* is a complex mixture of silicates. *Silicates* are complex ionic compounds that are exceedingly numerous. Approximately 90% of the earth's crust consists of silicates, the starting material for a wide variety of products including tile, bricks, cement, glass, and china. The basic unit of the silicate structure is SiO_4 tetrahedron (see Figure 7-10). Oxygen (O) and silicon (Si) together account for about 75% of the earth's crust. Si is one of seven semimetals in the periodic table. It is a hard, gray semiconductive solid that melts at 1410°C. In nature it is generally found combined with O in *silica* (SiO_2) and in various silicate minerals. More than 1000 different silicates occur naturally depending on the different ways the SiO_4 tetrahedra can be linked. Sand consists of fine particles of quartz. Other common forms of silica are *mica* and soapstone (a form of talc). Particles of mica produce the small flecks in metallic paint used on some automobiles. Clays and mica have sheets of linked SiO_4 tetrahedra which slide over each other giving their minerals a characteristic soft feel. This layered structure can be penetrated by water molecules which act as a lubricant producing a material that is pliable and slippery. When clay is heated to over 1100°C in a kiln, the water is removed and the layered sheets bond or lock into a rigid structure. This solid structure is called a ceramic. *Silicates* include sand, clay, feldspar ($K \cdot Al \cdot Si_3O_8$), quartz (SiO_2), and the semiprecious stone, garnet. Pure silica (SiO_2) has three common polymorphs: cristobolite, tridymite (high temperature), and quartz (low temperature). The silica structure is the basic structure for glasses and many ceramics. It has an internal arrangement consisting of pyramid (tetrahedral or four sided) units, as shown in Figure 7-10a. Four large oxygen (O) atoms surround each smaller silicon (Si) atom. The silicon atoms occupy the openings (interstitials) between the oxygen atoms and share four valence electrons with the oxygen atoms through covalent bonding. One of the strongest single bonds is the Si—O bond. It is considerably stronger than the C—C bond. The simplest and most numerous silicate compound of Si is SiO_2 (silicon dioxide), or silica. Recall that these two ele-

Figure 7-10   Silica structure: (a) silicate tetrahedral, (b) single chain of silicate.

ments are the most abundant in the earth's crust. Each atom of Si is bonded to four oxygen atoms which are situated at the corners of a tetrahedron with the Si atom positioned at the center. These SiO_4 tetrahedra are joined to one another through the O_2 atoms, each of which is bonded to two Si atoms. These silica tetrahedra can be arranged in a linear manner to form fibers, rings, chains, planar sheets, or in complex three-dimensional arrays. An example of the simple anion SiO_4 is the mineral *zircon* ($ZrSiO_4$) which is an inexpensive gemstone. More common are larger anions in which two or more O atoms bridge between Si atoms to produce these various structures. Two-dimensional layer anions $(Si_3O_{10})^{5-}$ are found in clay minerals, micas, and talc. Micas can be easily cleaved into thin, transparent sheets and were used as panels for lanterns. The bonds between the layers made up of cations, principally Al^{3+}, are relatively weak and easily broken.

Asbestos, a magnesium silicate, is one of many forms of fibrous silicates that have double chains of tetrahedra. Kaolinite $[Al_2 (Si_2 O_5)(OH)_4]$ is a type of clay valuable for pottery making because its layers structure can be easily penetrated by water. When Al ions are partially substituted for Si ions, the aluminosilicates called *feldspars* are formed which are the most abundant of all minerals. *Zeolites* are aluminosilicates with three-dimensional network structures containing tunnels. They are being used as catalysts.

The silicate structures (Figure 7-10b) can link by sharing the atoms in two corners of the SiO_4 tetrahedrons, forming chain or ring structures. By sharing three corner atoms, they produce layered silicates (talc, kaolinite clay, mica); or by sharing four corner atoms, they produce framework silicates (quartz, tridymite). The alternating Si and O covalent bonding is most common in ceramics and compares to the carbon to carbon (C—C) bonding in organic materials. The unique atomic structures of ceramics leads to the properties possessed by most ceramics. Figures 7-11a and b indicate some similarities and differences between ceramic and metal unit cells. See Section 3.4.1 for metal unit cells.

As mentioned in Module 6, silica, in the form of inorganic aerogels, offers the potential for improvements in materials structures with many microstructure and nanocomposite applications. An aerogel was produced in the 1980s to become the worlds lowest density solid material with a density of 0.003 g/cm3, only three times that of air. The unique surface chemistry, pore structure, optical, and thermal properties have propelled aerogel technology. The EMT weblink below offers a good insight into this unique silica material.

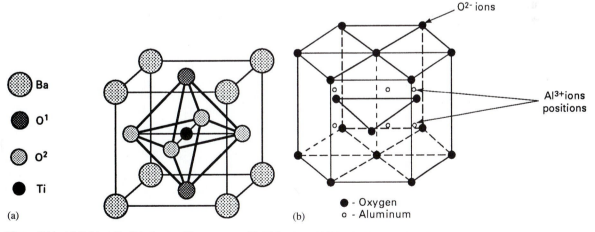

(a)   (b)

Figure 7-11 (a) Cubic cell of ideal perovskite structure of $BaTiO_3$ above 130°C with the origin on a barium ion. The ideal cubic perovskite structure, abbreviated ABX_3, has an "A," a large cation; "B," a smaller cation; and "X," an ion. Various arrangements or views of this structure might show the "B" cation in the center of the cube or the "A" cation in the center. This figure depicts this structure using $BaTiO_3$. The "A" ions are barium (Ba); the "B" ions are two oxygen ions, 01 and 02; and the "X" ion is titanium (Ti). Superconductivity occurs in oxides with this same structure, and the new high-temperature, superconducting copper oxides closely resemble it also. Such a ceramic possesses large values of dielectric constant as well as ferroelectric properties. [*Journal of Materials Education* (JME 13 (3))]. (b) Crystal structures of Al_2O_3 showing O^{2-} ions occupying the hexagonal close-packed (hcp) unit cell sites and Al^{3+} ions occupying about ⅔ of the interstitial sites to maintain electrical neutrality.

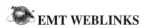

EMT WEBLINKS

Silicate Ceramics—*http://me.mit.edu/2.01/Taxonomy/Characteristics/Ceramics/Main Ceramics.html#SiliconNitrides*
Silica Aerogels—*Http://eande.lbl.gov/ECS/aerogels/satoc.htm*

The bonding mechanisms for ceramics produce the primary bonds: ionic and covalent. The crystal structure of bulk ceramic compounds is determined by the amount and type of bonds. The percentage of ionic bonds can be estimated by using electronegativity determinations (refer to Figure 2-8).

Being compounds, ceramics have different types of atoms. Hence, their resistance to shear and high-energy slip is extremely high. Because the atoms are held (bonded) so strongly compared to metals, there are fewer ways for atoms to move or slip in relation to each other. Thus, the ductility of ceramic compounds is very low and these materials act in a brittle fashion. Fracture stresses that initiate a crack build up before there is any plastic deformation and, once started, a crack will grow spontaneously. The combination of high shear stresses and reduced ductility produces high compressive strength but low tensile strength. The maximum bending stress in tension at failure is called the ***bend strength*** for ceramic materials and is often referred to as the ***modulus of rupture (MOR).*** At room temperature, metals and ceramics are often competitive, but at temperatures above 1500°F, metals weaken while ceramics retain much of their strength. Ceramics are noted for their heat resistance. The maximum service temperature for alumina is 3450°F and for silicon carbide, it is 3000°F. Heat-resistant nickel alloys are considered unserviceable above 1500°F. (Table 7-1 contrasts some characteristics of metals with those of ceramic materials.)

7.1.3 Ceramic Phase Diagrams

The statement, "the properties of materials depend on their structure," is emphasized more than once in this text. Questions need to be raised such as where the different microstructures originate, and how do we obtain the optimum microstructure that will render the properties we are seeking in a material. Microstructures of a material depend on their compositions, as well as the variables of temperature and pressure. As the composition becomes more complex, the more the number of phases present themselves. The desired microstructure comes from the ***processing.*** In processing a material we melt, refine, alloy, shape, heat treat, and finally fabricate the structure within a component which involves forming, machining, or welding. These processing steps have an effect on the microstructure; and before we can start on processing we need to know the details of the melting, solidification, and the various phases that the material goes through as it is heated and cooled. Without this information we are unable to control the processing to obtain a material with the desired characteristics. The phase diagram, equilibrium, or constitutional diagram is the map that shows us the different compositions and the temperatures under equilibrium conditions. Phase diagrams having two components were discussed in relation to metallic systems in Module 4. With ceramics, materials have at least two elements, but more than two is the norm. Figure 7-12 is a phase diagram for a binary ceramic system that shows the development of mullite ($3Al_2O_3$–$2SiO_2$), an intermetallic compound much in demand for high-temperature applications. Ceramicists dealing with three component or ternary phase diagrams need to construct three-dimensional diagrams which are difficult to not only construct but also to interpret. The concepts developed in the study of simple binary, two component systems can be directly applied to more complex ternary systems found in ceramic systems.

7.1.4 Wear Resistance

Wear can be defined as the removal of surface material. Wear is a facet of tribology that was introduced in Module 4. Several mechanisms are involved in wear. One is adhesion, or the

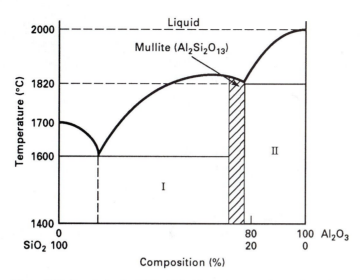

Figure 7-12   Rough sketch of SiO_2-Al_2O_2 binary system used to produce many refractories. Mullite (3 Al_2O_3 2SiO_2), an intermetallic compound with a small range of compositions, is a ceramic with many high-temperature applications.

bonding of two surfaces with subsequent removal of material. Inertial welding is an example of how this mechanism is used in the welding of metals. The roughness of the surface results in removal of part of the protrusions and depressions. When there is a large difference between the hardness of two materials in contact, gouging can occur. Gouging is how an abrasive grinding stone removes material from a workpiece. **Impact,** also known as **erosion,** occurs when abrasive particles in a fluid impinge on a surface. Other variables at play are abrasion, oxidation, and contact stress. Wear is responsible for many equipment breakdowns and attendant downtime. With wear comes friction, and with friction comes a large expenditure of financial resources to overcome friction. Without considering the loss of material from wear, the dollar value of energy lost each year due to friction is in the billions of dollars. Ceramics are likely candidates to help reduce this exorbitant expenditure of energy and funds. A quick perusal of their properties should convince people that greater selection of ceramic materials for use in their area is indeed warranted. **Laser glazing** is a steel treatment process that reduces a railroad wheel's rail-to-wheel friction by up to 40%. Friction is necessary on the rail tops for traction, but rubbing between the wheels and the sides of the rail causes problems. The wheel of a railroad train must have a flange to overlap the track to keep the train on track. The rubbing between this flange and the sides of the rail causes friction. A laser beam passed along the side rapidly melts a thin layer of steel, solidifying to form a hard glossy surface while still possessing enough elasticity to distribute the load evenly to then underlying layers of the rail. Estimates of $40 billion annually in fuel costs and $16 million in rail replacement plus reduction in derailments and the attendant loss of human lives all lend credence to this treatment.

Materials can be ranked according to their wear resistance by using a variety of wear tests. Sapphire and ruby spheres are ideal for wear resistance. Figure 7-13 shows an assortment of these miniature pieces of synthetic alumina from which are manufactured precision parts for hundreds of uses. In addition to hardness, these materials feature low coefficients of friction, shock resistance, high index of refraction, low specific gravity, and nonmagnetic and high dielectric strength. For example, ferrules with inside diameters (ID) that range from 125 to 750 μm can be manufactured with great accuracy. Many are single-purpose tests with a particular application in mind. Some concern themselves with only one type of wear factor. A recently developed test method uses flat rectangular specimens (100 × 25 × 6 mm) mounted on a cylinder that rotates inside a container filled with

(a)   (b)

Figure 7-13 (a) Ruby (red) color and sapphire (white) jewel bearings and related precision products such as orifices, nozzles, pins, valve seats, bushings, balls, ferrules, and split sleeves are manufactured from raw silica materials with tolerances such as roundness within 7×10^{-6} in., or concentrically within 4×10^{-5} in. Ruby and sapphire are both forms of alumina (Al_2O_3). The presence of chromium ions (Cr^{+3}) in alumina gives the materials its red color and small amounts of Fe and Ti give a white color to sapphire. Corundum is the name given to alumina that occurs in nature. (Microlap Technologies, Inc. (b) Ceramic industrial wear parts are used as mechanical seals, erosion-resistant liners, metal processing dies, slurry pump rotors, and more. Materials include alumina, silicon nitride, zirconia, silicon carbide, boron carbide, and boron nitride. (American Ceramic Society)

an abrasive medium tailored to simulate different types of wear. This test, called the rotating-cylinder wear test, is supposed to be a better indicator of a material's performance under service conditions. Specimens are weighed before and after the test, and the percent weight difference is an indicator of the material's wear resistance. In reporting, the test conditions must accompany the results. Figure 7-14 illustrates the properties of a ceramic metal composite produced by Alanx using a new process called Dimox, discussed later in this module. This composite is designed specifically to provide maximum wear resistance. Alanx CG 896 consists of densely packed silicon-carbide particles within a matrix of aluminum oxide and a metal alloy. The particles and the alumina resist wear, while the metal alloy provides the bonding between the two ceramic phases. The accompanying graph relates this wear-resistance composite with other materials by use of a wear-life factor. This factor compares the service lives of competitive materials: The larger the number, the longer the wear life. A slurry erosion test is used to measure the wear resistance of a test specimen. The material loss is compared to a standard aluminum test specimen and the wear-life factor is determined. Refer back to Figure 4-44 for more wear comparisons. NIST along with four manufacturers of prosthetic devices have developed an ***accelerated wear device*** that can evaluate a diverse combination of materials through imitating the wear that implants go through in the human body. Reportedly the machine produces debris and changes in surface texture that are similar to effects seen in the body. The screening takes about 1 week as opposed to the 6 months required heretofore for conventional equipment to simulate wear.

7.1.5 Stress Concentrations and Fracture Toughness

When selecting a material that might be subjected to ***cyclic loading*** or where the load cannot be distributed uniformly over an area, the subjects of stress concentration and fracture toughness come into play. For example, when a hole is drilled in a flat plate, such as that sketched in Figure 7-15a, and is loaded with a force *P,* the stress in the plate is assumed to be uniformly distributed (Figure 7-15b) across the cross-section of the plate everywhere but at the cross-section containing the hole. Figure 7-15c shows the distribution of stress as it finds its way around the

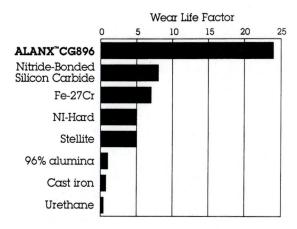

TYPICAL PROPERTIES OF ALANX™ CERAMIC/METAL COMPOSITES

	ALANX™ CG896	ALANX™ CG273	ALANX™ FGS	Developmental Metal Matrix Composite (MMC)
Wear Life Factor	24	24	12	9.2
pH Range	5-8	1-14	5-8	5-8
Density (g/ml)	3.32	3.42	3.32	2.95
Hardness (R_A)	80	80	80	57
Flexural Strength (MPa)	137	100	216	216
Fracture Toughness K_{1c} (MPa · m$^{1/2}$)	6.0	4.4	5.5	11.3
Shear Modulus (GPa)	131	144	144	79
Young's Modulus (GPa)	313	357	366	204
Poisson Ratio	0.19	0.23	0.27	0.29
Thermal Expansion (10^{-6}/K)	5.43	5.20	5.89	10.24
Thermal Conductivity (W/m · K)	160	160	160	160

Figure 7-14 Typical properties of ALANX ceramic/metal composite.

hole, thus producing a concentrated stress at the edges of the hole, as shown. Such a buildup of stress may cause the material to crack. It is important to point out that stress concentrations depend on the geometry of the flaw (hole, notch, or crack) and the geometry of the specimen or component (shape and thickness). *They have nothing to do with the properties of the material.* In materials that may contain cracks due to flaws, voids, inclusions, or oxides in their microstructure, the stress concentration is greatest at the tip of the crack. Flaws that cause fracture of ceramic materials usually are smaller than 50 μm. This size of flaw is not readily detectable by most nondestructive testing (NDT) methods.

   Fracture toughness, often perceived as the limiting factor for ceramics applications, is being improved to make ceramics competitive as engineering materials. A stress–strain diagram (Figure 7-16a) illustrates the effects of ceramic brittleness compared to those of relatively ductile metals. Figure 7-16b contains a graph contrasting the fracture strength of metals

Figure 7-15 A stress concentration resulting from a through hole in a flat plate. (a) Sketch of the plate loaded with a tensile force *P*, showing a side view of plate. (b) Cross-section of the plate at some distance from the hole, with uniform distribution of tensile stress indicated by small vectors. (c) Cross-section through the center of the hole, with the largest vectors portraying the maximum stress concentrations at either edge (side) of the hole.

with that of ceramics; the graph also shows the increase in this property using the toughening techniques discussed in this module, to the point where advanced ceramics are approaching metals in their fracture toughness.

Using fracture mechanics techniques, the critical stress–intensity factor (K_{1c}), or fracture toughness of the material, is determined independent of the nature of the material. The "1" in the symbol K_{1c} refers to the mode of loading of the specimen. In this instance, the load is parallel to the "1" or *x* material axis, producing a normal stress. When used with the symbol for fracture toughness, the numerals "2" and "3" refer to the shear loads acting on the *y* or *z* axes, respectively. Figure 7-17 is a sketch of a fiber-reinforced part being stressed parallel to the "2" axis. There are several mechanical tests, including indentation methods, used to measure the fracture toughness of ceramics. Which one is chosen depends in part on the type of information needed. One popular test used for testing ceramics is the single-edge, notched beam (SENB) test (Figure 7-18a). It is similar to a four-point bend test except that an artificial crack is placed in the specimen before testing. As Figure 7-18b illustrates, the chevron-beam test specimen is similar in loading to the SENB test but differs because it has a chevron notch rather than a straight-through notch machined or cut into the surface. The reason for this type of notch is to force the crack to propagate slowly through the specimen. It does so because it is always passing into a zone of greater load-bearing area. Both tests give a direct measure of the fracture toughness of a material. The SENB test uses the following equation:

$$K_{1c} = \frac{3P(S_1 - S_2)a^{\frac{1}{2}}Y}{2BW}$$

where

P = applied load
a = crack length
S_1, S_2 = dimensions between the outer and inner roller spans
B = specimen thickness
W = specimen height
Y = dimensionless calibration factor

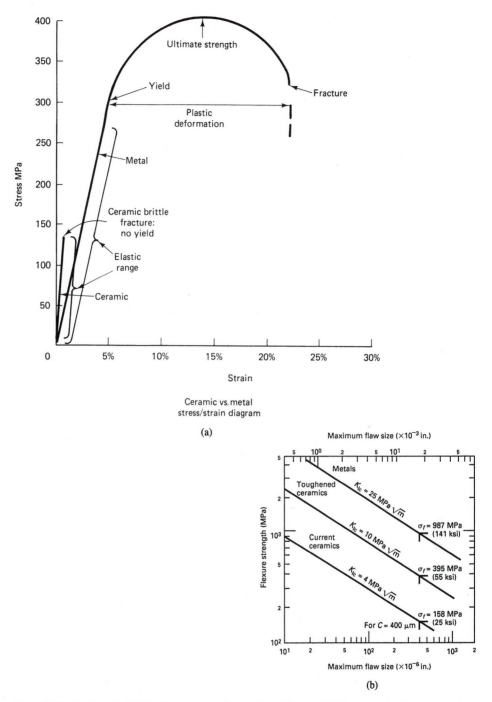

Figure 7-16 (a) Ceramics' brittle fracture results from their inability to yield like metals do. This diagram shows a large percentage of plastic deformation in metals that occurs because localized stresses are relieved by localized plastic deformation. No yielding point shows on the ceramic curve, so localized stress eventually causes catastrophic failure when the elastic range is exceeded. Percentage of strain before failure on ceramics can be measured in hundredths of a percent compared to tenths of a percent for metals. (b) A plot of flexural strength versus maximum flaw size for metals, toughened ceramics, and traditional ceramics, which translates into (represents) toughness values ranging into the 20s. (ORNL)

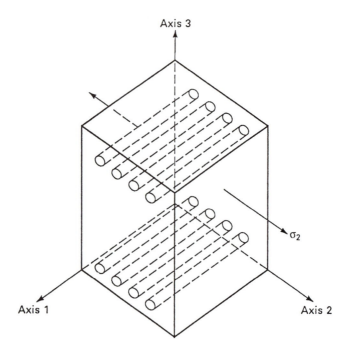

Figure 7-17 A composite material with fiber reinforcement. The axes (1, 2, and 3) are shown with a normal tensile stress acting in the direction of axis 2.

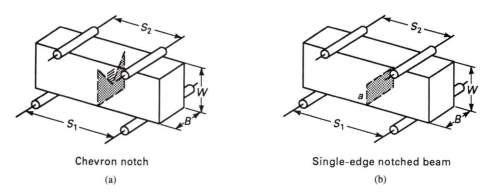

Figure 7-18 Fracture toughness is determined using single-edge notched and chevron notch beam specimens. An artificial cut is placed in both specimens. (a) A straight-through notch in the SENB is cut with a diamond-impregnated blade and (b) a chevron is machined into the chevron notch (CN) specimen.

The units for expressing fracture toughness are either psi $\times \sqrt{\text{inches}}$ or MPa $\times \sqrt{\text{meters}}$. It is of interest to note that once this critical stress–intensity factor is exceeded, the crack will propagate and the material will fracture, producing a catastrophic failure. The failure can be caused by an increase in the applied stress (load) or in the crack length. Once the fracture toughness is known, the fracture stress or the stress at which the crack will propagate can be calculated. Also, the stress at failure may be less than the yield strength. It is also noted that flaws in almost all silicate and oxide-based ceramics and glasses will grow at stress levels lower than the calculated stress that produces a catastrophic or fast fracture. This environmentally assisted crack growth is produced by a reaction between the strained bonds at the tip of the crack with water or other molecules. Figure 7-19 is a sketch of a crack zone toughened as a result of the transformation of ZrO_2, and a micrograph of the area. Transformation toughening is covered in greater detail in Section 7.4.

Figure 7-19 A diagram showing the formation of a transformation zone surrounding the tip of a crack due to the polymorphic transformation of ZrO_2, including a micrograph of the same area. (ORNL)

Flaws in ceramic materials that produce failure are usually smaller than 50 μm. Most NDT procedures do not detect flaws smaller than 50 μm. In summary, the fracture toughness, a property of a material, is an indicator of the ability of a stress concentrator (crack or flaw) to produce a catastrophic failure. To predict the ability of a material to resist fracture (fracture strength), knowledge of both fracture toughness and the presence of ***stress concentrators*** in a material is vital.

EMT WEBLINK

MSEL of NIST—*http://www.nist.gov/item/About_NIST_Materials_Science_and_ Engineering_ Laboratory.html*

7.1.6 Thermal Shock Resistance

Thermal shock is a mechanism that can produce mechanical failure in materials. It involves the buildup of thermal stress in a material as a result of exposure to a temperature difference between the surface and the interior of a material. If the temperature difference occurs rapidly, the possibility of failure increases. Testing of ceramic turbine vanes for thermal shock subjects them to repeated recycling from 340° to 1230°C in less than 2 s. At 1230°C, these vanes glow as a result of the heating. Future automobile gas-turbine engines will require materials that can withstand a rapid temperature change, from room temperature to 1200°C or more in a few seconds, and can be exposed to stresses in excess of 207 MPa. Some engine designers are contemplating future engines capable of operating for 3500 h with internal operating temperatures as high as 1745°C. Some of the material properties involved in producing thermal shock are the elastic modulus *(E)*, linear coefficient of thermal expansion (α), thermal conductivity *(k)*, tensile strength (σ), and fracture toughness (K_{1c}). Also, if a material is allotropic and undergoes a phase transformation as a consequence of the temperature change, this factor must be considered. In the selection of materials for good resistance to thermal shock, each of the variables noted above must be analyzed to determine how it affects the material's ability to limit and/or reduce the thermal shock produced. For example, a material with high thermal conductivity, high strength, a low coefficient of thermal expansion, and a low elastic modulus would be a good candidate for resisting thermal shock. Lithium aluminum silicate (LAS) has an extremely low coefficient of expansion along its hexagonal structure ($-2.0 \times 10^{-6}K^{-1}$) and, despite its low tensile strength, low thermal conductivity, and fracture toughness, is a good candidate for thermal-shock-resistance applications. A new ceramic, titanium silicon carbide (Ti_3SiC_2) combines the heat resistance of a ceramic with the machinability and ductility of a metal. Heated to 1400°C (2550°F) then plunged into cold water, it withstands a thermal shock that would shatter any other ceramic. Even the best engineering ceramics such as silicon nitride and silicon carbide are brittle at room temperature. The new ceramic can be deformed as much as 50% at room temperature before failure. It also machines like ductile iron or graphite.

Due to their ability to deform both elastically and plastically, most metals have few problems with thermal shock. The presence of free electrons, which conduct heat as well as electricity in metals, results in good thermal conductivities. Organic materials have low thermal conductivities, due to their covalent bonding, which ties up their electrons. Ceramic materials exhibit a diverse range of thermal conductivities due to their complex microstructures. Those ceramics with the least complex structures, such as silicon carbide (SiC), composed of atoms of similar size and shape, achieve the highest thermal conductivities. Although not a ceramic material, diamond has the simple crystal structure of pure carbon, with a thermal conductivity at room temperature (RT) of 900 W/m · K. Copper's conductivity is half that of diamond's. Graphite, another polymorph of carbon, has a layered microstructure, which makes it isotropic. In addition, graphite undergoes different fabrication techniques that result in different orientations of the crystal structure, which only compounds the problem of assessing its ability to conduct heat or electricity. Finally, porosity, dispersions, noncrystallinity, grain size, and grain boundaries all affect the thermal shock resistance of materials. Pores and micro-cracks are probably the most important of these.

To assist in the selection of likely candidate materials for thermal shock resistance, several parameters have been developed that take into consideration the factors mentioned above. One such parameter is the ***thermal shock index (TSI),*** also known as thermal fatigue resistance, expressed in equation form as follows:

$$TSI = \frac{\sigma \times k}{\alpha \times E}$$

where

σ = tensile strength
k = thermal conductivity
α = linear coefficient of thermal expansion
E = modulus of elasticity (Young's modulus)

TSI is a ratio of the products of four variables. High values for the variables in the numerator and low values for the variables in the denominator render large or increased values for TSI. When comparing different materials based on the value of TSI, it is essential to use the same

TABLE 7-3   THERMAL SHOCK INDEX (TSI) FOR SELECTED MATERIALS

Material	σ(MPa)	k(W/cm · °C)	α(°C^{-1} × 10^{-6})	E(GPa)	TSI^a (W/cm)
Fused SiO$_2$	68	6 × 10^{-2}	0.6	72	94
Al$_2$O$_3$	204	3 × 10^{-1}	5.4	344	33
Graphite	8.7	1.4	3.8	7.7	416
Soda-lime silica glass	69	2 × 10^{-2}	9.2	68	2.1

^{a}TSI units in the U.S. customary system are Btu · in./hr · ft^2 · F°; in the SI system, they are W/cm.

units of measure for each of the variables. For example, if you use MPa to express tensile strength for one material, you must also use MPa (not psi) for the material you are comparing it with. Table 7-3 lists the TSI and some representative property values for selected materials.

Illustrative Problem

Determine the TSI for aramid fiber and E-glass fiber, and indicate which has the best thermal shock resistance based on these calculations. Selected properties for each of these materials are provided in the table that follows:

Aramid fiber	E-glass fiber
$E = 18 \times 10^6$ psi	$E = 10.5 \times 10^6$ psi
$\sigma = 400 \times 10^3$ psi	$\sigma = 500 \times 10^3$ psi
$k = 3.5$ Btu · in./hr · ft^2 · °F	$k = 7.9$ Btu · in./hr · ft^2 · °F
$\alpha = 0.8 \times 10^{-6}$ °F^{-1} (long.)	$\alpha = 1.6 \times 10^{-6}$ °F^{-1}

Solution

$$TSI_{aramid} = \frac{4 \times 10^5 \times 3.5}{0.8 \times 10^{-6} \times 18 \times 10^6} = 9.7 \times 10^4 \text{ Btu} \cdot \text{in./hr} \cdot \text{ft}^2$$

$$TSI_{E-glass} = \frac{5 \times 10^5 \times 7}{1.6 \times 10^{-6} \times 10.5 \times 10^6} = 2.1 \times 10^5 \text{ Btu} \cdot \text{in./hr} \cdot \text{ft}^2$$

E-glass fiber was the greater and therefore has the best shock resistance.

 EMT WEBLINK

WWW Virtual Library: Technical Ceramics—*http://www.ikts.fhq.de/vl/ceramics.html*

7.1.7 Emissivity

Emissivity is an important property, particularly in high-temperature ceramics. When a surface is exposed to radiation, it may absorb part or all of the incident radiation (energy). The fraction of the energy absorbed is the emissivity *(e)* of the surface. A blackbody, the perfect absorber, has a value of $e = 1$; for the perfect reflector, $e = 0$.

Thermal barrier coatings (TBCs) on the metal components of gas turbines have low thermal conductivities and low emissivity. Having a coefficient of thermal expansion similar to that of the base metal allows the coating to adhere to the metal as the metal undergoes changes in temperature. The newer, energy-efficient window glass, which is more costly to the customer, has low emissivity, which means that a much greater percentage of the sun's radiation (heat) will be reflected than from older types of window glass. Another technique to reduce solar heat gain in glazing applications is shown in Figure 7-20a. These new windows are another example of **green manufacturing.** The benefits that accrue to all of us are staggering. With more efficient windows the most obvious benefits are as follows:

less burning of fossil fuels for heating and cooling

less greenhouse gases produced

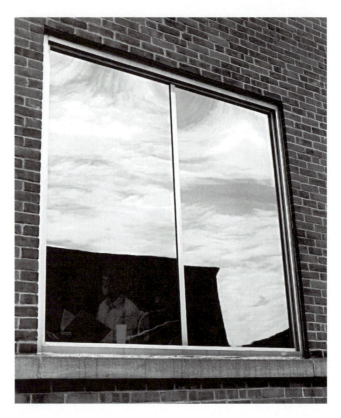

Figure 7-20 Reducing heat gain through its reflective nature. Lexan® polycarbonate sheet material (either MR® or XL-1®) is laminated with a proprietary, metallized reflective film for commercial glazing applications. The plastic also provides natural light and the high impact strength of polycarbonate sheet. It is a competitor to glass glazing. As a materials selection issue, what are the pros and cons for each type of glazing, beyond their emissivity properties? (G.E. Plastics)

less dependence on foreign oil

less money spent on utility bills

less discomfort from drafts near windows

less medical bills due to better living conditions

less outside noise

less fading of fabrics

7.2 DESIGNING CERAMICS

Because of their properties—their inherent ***brittleness;*** inability to yield, change shape, or plastically deform; sensitivity to defects; tendency to break at any point of high stress concentration; microstructures, which possess flaws that can cause failure without being detected beforehand—ceramics require a design model that calls for greater degrees of accuracy in the design and manufacture of ceramic parts. Anyone who has experienced breaking a piece of glass will know about the sensitivity of brittle materials to defects. A slight scratch on the surface of the glass and a small bending force to produce a tensile stress are all that is needed to break a relatively thick piece of glass. This flaw sensitivity means that a ceramic cannot be considered as having only one strength but rather, a distribution of strengths, depending on the flaws in the material that were added during processing, final machining, or testing, or while in service. The three general design models used by engineers are the empirical, the deterministic, and the probabilistic. ***Empirical design*** can be defined as the trial-and-error method of selecting a material for a particular application. ***Deterministic design*** is based on mathematical analysis of average properties of ductile materials, together with a factor of safety to accommodate any errors. ***Probabilistic design*** is required for designing ceramics, particularly when the parts are complex in form and when there is a complex set of applied stresses to contend with. Using the tools of fracture-mechanics theory, highly refined stress-analysis techniques, and statistical representations of material properties, this probabilistic design model allows the engineer, aided by numerous computer programs, to determine the stresses statistically, and parts are then designed not to exceed a specified probability of failure. The density, size, and severity of flaws are also determined statistically. Because ceramic parts have slight variations in their structure, which can have a severe effect on their properties, each part must be designed individually. The mathematical means for representing the statistical nature of brittle failure is the Weibull equation, which relates the cumulative probability of failure to the failure stresses of all the elements of the sample. The graphical plot of the data along with a best-fit curve is known as a ***Weibull plot*** of strength for a material. The slope of the best-fit curve, known as the ***Weibull modulus,*** indicates the degree of variance in the strength data. A high value for the modulus *(m)* is desired because it indicates low variance or a narrow range of failure stresses. Many critical parts are ***proof tested*** at a predetermined proof-stress level. Those parts with large flaws will fail the test. Several assumptions are made at this point. One is that parts that pass the test will perform satisfactorily in service. It is also assumed that the testing did not initiate cracks or cause cracks to propagate that will cause failure later in service.

7.3 CERAMIC PROCESSING

A major problem in ceramic processing is achieving the same properties every time a process is repeated. ***Intelligent processing*** of materials is an attempt to solve the problem of making complex materials more uniform in their structure and properties. As materials become more complex in composition, structure, and processing, it becomes more difficult to make a uniform product. The National Institute of Standards and Technology (NIST), in cooperation with industry, has an ongoing program to automate ***quality control*** that allows quality to be built in during processing rather than attempting to obtain it by inspection after the fact. Sensors analyze a material as it is processed to warn computers controlling the process to correct errors beginning to occur. ***Adaptive control*** is the name of the method to improve the quality of manufacturing processes. The key to tougher, stronger, more reliable ceramics lies in ceramic processing. By improving existing processes and developing more efficient ones that reduce the size and number of flaws and even out the distribution and size of particles, attaining such ceramics is being realized rapidly. In the initial preparation of ceramic materials for shaping or molding into some useful product, the raw ceramic materials must first be reduced to a powder form. The term ***comminution*** describes the operations of crushing and grinding that accomplish this task. ***Crushing*** reduces the raw material to particle size, whereas ***grinding*** takes the particle-size ceramic and further reduces it to a powder form. Grinding is accomplished by confining the ceramic powder along with extremely hard balls, pebbles, or rods in a revolving, cylindrical drum called a ball mill. This grinding action causes the reduction to a fine powder by a combination of impact and attrition. Often water is added to the powder which changes into a slurry. The hard balls and pebbles are known as ***media.*** Figure 7-57 in Section 7-9 of this module shows two examples of media. Yttria-doped tetragonal zirconia polycrystal (Y-TZP) ceramics have high strength and high wear resistance and are widely used as milling media for making powders for more advanced ceramics.

7.3.1 Thermal Processing-Sintering/Densification/Firing

Densification, also referred to as ***firing,*** is a process by which a particulate compact is transformed into a ceramic part that has adequate properties to satisfy the needs of a particular application. The powder compact, with its relatively small powder size, is heated at a temperature below the melting temperature of its components. The melting temperatures of ceramics make it impractical to actually melt the materials and thus bond the grains of powder together. The bonding between the grains during sintering is due to the action of many complex mechanisms, including diffusion. In summation, ***sintering*** is the removal, by a variety of mechanisms, of the initial pores, combined with the growth of grains and strong bonding forces to provide strength to a powder compact. The main driving force involved is the reduction in surface area (i.e., the surface energy of the item). In general, this results in a part with the same shape as the original or ***"green"*** shape but reduced in size. The density of the part has increased as well as the particle size. The increase in particle size must be controlled because the final strength of the part is determined partly by the size of the final grains. Relatively large particles have a greater number of interstices. Interstices are filled with air, which itself is a good insulating material. This may be satisfactory if the ceramic is designed for its heat resistance. Usually, a ceramic requires strength, and strength comes with a homogeneous microstructure that has a fine grain structure. The problem is that the initial pores between the original particles have not been completely eliminated. Further, the final density reaches a limit of only some 95% of the maximum *theoretical density* (see Section 7.1). Figure 7-21 shows graphs of the sintering of alumina and zirconia using microwave sintering versus standard furnace-heating temperatures. *Sintering furnaces* usually heat ceramic compacts externally by radiant and convection thermal transfer. Ceramics that have rather high dielectric-loss factors (a measure of electric energy lost as heat energy by a capacitor in an ac circuit) can take advantage of this property, as Al_2O_3 and ZrO_2 illustrate in Figure 7-21. The significant benefits of, as well as a few applications for, these new ceramic materials also accompany the graphs, which also show the benefit of using microwaves to lower temperatures for sintering and to achieve finer microstructures and better mechanical properties. In addition to the conventional sintering process as discussed, other sintering processes, such as liquid-phase and solid-state sintering, are used to densify compacted powders.

Microwave Sintering has Significant Implications for New Materials Applications

- Accelerated Kinetics:
  - Sintering occurs at lower temperatures

- Significant Potential Benefits
  - Lower temperature processing
  - Finer microstructures
  - Better mechanical properties

- Advanced Materials Applications:
  - Composites from incompatible materials
  - Self-lubricating high-temperature bearings
  - Electrical materials
  - Engine components

Figure 7-21   Plots of density percentage versus temperature for Al_2O_3 and ZrO_2, showing the vivid effects of sintering these materials with microwaves versus standard heating procedures. Note the significant increases in attained density as a result of microwave heating. (ORNL)

Many processes employed by industry improve on the pressureless sintering process as described. Some that involve the application of pressure and heat simultaneously are (1) hot pressing, (2) overpressure sintering, and (3) hot isostatic pressing (HIP). In *hot pressing,* also known as *pressure sintering,* powders are compacted using a die within an evacuated chamber. Temperatures are about one-half of the absolute melting temperature ($T_m + 273°$)* of the powdered material. Due to the uniaxial techniques used to apply the pressure in the die, hot-pressed materials may have a preferred orientation of their grain structure, which produces anisotrophy. *Hot isostatic pressing* (HIP) combines compounding (compressing and densifying loose powders into a desired shape) and sintering using temperatures as high as 2500°C and pressures as high as 200 MPa in a furnace within a pressure vessel. The process relies on plastic deformation to achieve densification. Using argon or helium gas to exert pressure from all sides, HIP results in near-shape forming of ceramic parts, particularly with cemented-carbide particulates. As with the other mentioned densification processes, HIP can be used in combination with other processes to achieve highly densified parts.

Sintering is a critical factor in the production of advanced ceramics. A green, or unsintered, ceramic material is softer than chalk, consisting of millions of powder particles. When sintered at high temperatures, usually between 1000° to 2000°C, the particles fuse to become a solid material with properties superior to many metals. *Microwave sintering* produces uniformly heated ceramic parts of large volume and irregular shape rapidly in excess of 1600°C in a vacuum or under atmospheric pressure. Results produced improved mechanical properties and homogeneous microstructures with very little porosity. Enhanced densification rates have brought on reduced sintering times and temperatures, which means that this innovative technology produces a higher quality product, and it does so at a lower cost. Microwave sintering causes the molecules throughout the workpiece to vibrate, which produces uniform heating without gradients in temperature. The key to this method is the specialized, insulated sintering chambers which trap the heat in the greenware and allow the temperature to rise rapidly. In contrast, in radiant or convection heating, energy is absorbed only at the surface of the workpiece and must be transferred into the bulk of the material by conduction.

In another novel process, graphite–carbon particles with controlled electrical properties are used as both the pressure-transmitting medium and as an electrical resistor for simultaneous heating and consolidation of the powder preforms. Cyclic times are very short compared to other methods, such as HIP (minutes versus hours), and both high temperatures (about 1000°C) and relatively low pressures are involved. This process has the ability to densify complex shapes without "canning." The term *canning* refers to the need to encapsulate preforms for HIP with glass to seal the preform in a gas-impermeable envelope. With such rapid densification, this new process has the potential for inhibiting grain growth, resulting in fine-grained microstructures with improved mechanical properties.

EMT WEBLINK

CERAM Research Ltd.—*http://www.ceram.co.uk*

7.3.2 Traditional Processing

The first ceramic materials produced by processing were bricks and lime plaster. Limestone was roasted to produce lime that was mixed with water. To this mixture was added clay, straw, or fibers to provide building materials that were used widely and are still in use today. When added to the clay materials, water provides the clay with plasticity or workability, in addition to providing the bonding mechanism that gives the material its strength. Traditional ceramics are still produced with basically the same materials (Figure 7-22). *Slip casting* (Figure 7-23), an age-old process for making vases and other hollowware, is still used as a production process for many ceramics. Many other traditional ceramics are produced by blending, pressing, and

*T_m in °C; absolute ($T_m + 273°$) is in kelvin.

(a)   (b)

Figure 7-22 (a) Examples of traditional ceramics from Colonial Williamsburg Museum Exhibit. Left to right: glazed clay pitcher, split mold made of plaster of paris for slip casting, samples of slip cast/glazed pots, potter's wheel with sample bowls and platters. Ancient ceramic processes seen here for traditional ceramics are still used today with some automated aspects. (b) Examples of modern ceramic artifacts—pottery, glass cup, flatware, and an incandescent light bulb produced by an automatic glass blowing machine that can produce over 1000 bulbs per minute. Fluorescent lights use a phosphor coating on the inside of the sealed glass tubes, a ceramic that reemits white, soft, cool light.

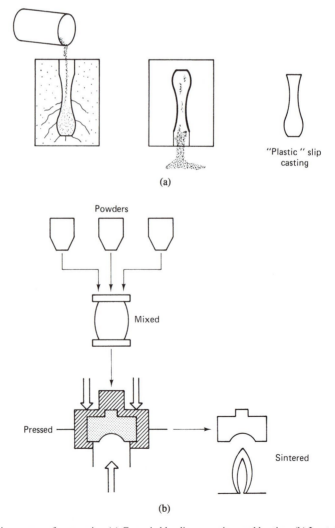

Figure 7-23 Basic processes for ceramics. (a) Ceramic blending, pressing, and heating. (b) Isostatic pressing mold in cross-section showing insulator blank. (American Ceramic Society)

(c)

(d)

Figure 7-23 (d) Stages of insulator body development (left to right): pressed "green" blank, turned "green" insulator, bisque-fired insulator, glass glazed and decorated refired to finished insulator. (American Ceramic Society)

Figure 7-24 Stages of sintering (firing) clay.

firing (Figure 7-23b–d). Often, traditional ceramics develop a vitreous or glassy phase upon cooling after firing, which accomplishes densification (Figure 7-24). Glass phases must be avoided in advanced ceramics produced from ceramic particulates. Ceramic materials processed to produce refractories, *structural clay* products (brick, pipe), or *whiteware* (dinnerware, tile pottery) can be considered traditional ceramics. In early times whiteware, such as pots or vases, was formed by pressing one's hands against material supported on a rotating wheel. This very old process, called *jiggering,* is still practiced by people engaged in making hollowware with the aid of a potter's wheel. In industry the same process is highly automated in the fabrication of hollowware.

7.3.3 Advanced Ceramic Processing

More modern ceramics with unique and more sophisticated properties needed in this highly technological age are referred to using a variety of terms. *High tech, high performance, fine,*

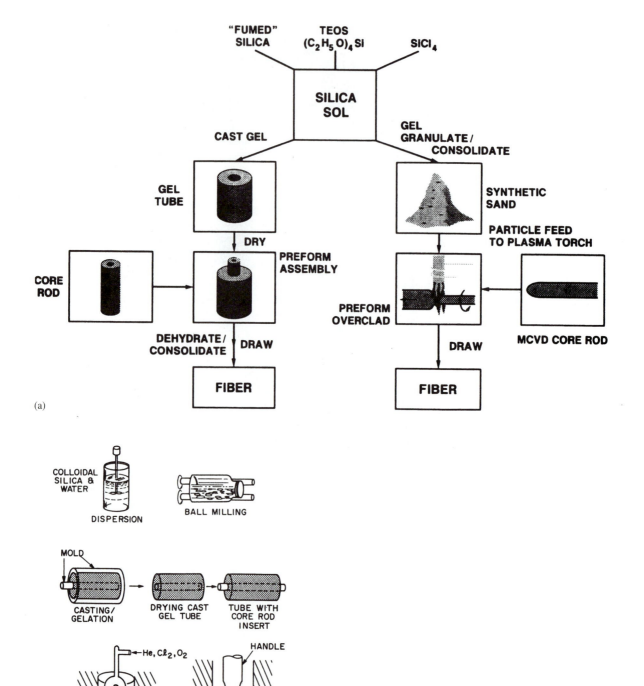

Figure 7-25 (a) A hybrid *sol-gel* strategy in which gel is cast as a tube to overclad a core rod. As an alternative, gel is granulated, then fusion-sprayed on a preform, to accomplish overcladding.
(b) Sequence of steps to prepare fiber from core-rod and cast gel body.

new, value added, engineered, engineering, and *advanced* are some of the terms seen in the literature. In this book we use the word *advanced* to describe such ceramics. Sources for advanced ceramics are not new or revolutionary. What is new are the many processes developed in recent years to produce them. For example, Figure 7-25 shows a hybrid sol-gel process for fiber optics.

Figure 7-26 Three differently designed hand scissors show the cutting surfaces made of a ceramic called zirconia which never needs sharpening. Note the differences in the way this extremely hard material (white color) is supported to augment its flexural strength.

The multistage fabrication process for ceramics includes powder production, powder conditioning, shaping (forming), drying, and densification (also called *firing* or *sintering*). Note in Figure 5-63 the similarity of ceramic processing to the processing of powder metals. Tungsten carbide (WC), a synthetic compound with extremely brittle characteristics, was developed in the late 1890s. In the 1920s, WC powders (particulate) were combined with a metal cobalt (Co), acting as a binder and sintered, producing the first modern **cemented carbide** or **cermet** (ceramic plus a metal). A true composite material, it had the trade name of Carboloy from General Electric Co. Cermets can be classified as **large-particle composites.** Relatively large-volume fractions of the particulate often exceed 90%. (See rule of mixtures in Module 8). Carboloy was used primarily as a cutting tool for cutting cast iron and nonferrous metals. Later, other cemented carbides using titanium carbide (TiC), tantalum carbide (TaC) and chromium carbide (Cr_3C_2) were developed with their powders (particulate) added to either Co or Ni as metallic binders to produce cemented carbides. Carbon black provided the carbon for these carbides. The carbide composed 80–95% of the total weight of these composite materials. Another class of cermets are oxide based, made from Al_2O_3 or MgO using Cr as a metallic binder. Some applications for these materials that possess great hardness and wear resistance are bits, dies, mining tools, valve liners, spray nozzles, gage blocks, and indenters for hardness testers. These carbides are finding increased uses for **ceramic matric composites (CMCs)** which are discussed in Module 8. These advanced materials use particulates, fibers, or whiskers embedded in a matrix of another ceramic. The source minerals for these advanced ceramics are metallic elements combined chemically to form oxides, carbides, nitrides, or borides. It is common practice to refer to oxides by a special name ending in *-ia* or *-a*. For example, Al_2O_3 is alumin*a*, MgO is magnes*ia*, Cr_2O_3 is chrom*ia*, ThO_2 is thor*ia*, and ZrO_2 is zircon*ia*. Figure 7-26 shows three different pairs of scissors whose cutting edges are made of zirconia.

Ceramic processing can be described simply as consisting of two steps: (1) a cold step, in which the ceramic part is formed or shaped into a "green" part or preform, and (2) a hot step, in which the green compact is first subjected to heat to dry up any liquid phase formed during the processing and then subjected to higher heating, known as firing, sintering, or densification. The firing consolidates the particles and bonds the ceramic particles together permanently to give the ceramic its strength and other final properties. Some processes do not require drying, nor do some require the final step in most processing—that of machining to final specifications.

Some but not all engineering-ceramics fabrication processes begin with minerals ground into fine powders before they are transformed into useful products. The processing of the powder is itself a major undertaking, involving steps such as milling and sizing to produce a powder with the desired particle size, shape, and particle-size distribution. Either physical flaws (i.e., porosity or inclusions) or chemical flaws (i.e., unwanted or unknown chemical elements that combine to produce second phases such as liquids) can occur at any of these stages and may be difficult, if not impossible, to correct or eliminate. It must be stressed that the microstructure of the final ceramic product is dependent on the range of particle sizes and the inclusion of impurities in the starting materials. Clean rooms are evidence of the importance of cleanliness in the processing of advanced ceramics. We also know that the microstructure of any material determines the material's properties. Thus, much importance has been placed on developing low-technology techniques that make the conversion of mineral ores into industrial ceramic products more efficient with the expenditure of less energy and with ever-lower temperatures. Some advanced ceramics require that the powders used in the process be produced by chemical means, which allows greater control over both impurities and the final grain structure.

7.3.3.1 Chemical processes.

Sol-gel processing of ceramics and glass is a chemical process for producing powders for use in advanced ceramics. This process and chemical-vapor-deposition techniques (discussed later) provide the means for designing and controlling the composition and structure of a ceramic at the molecular level. Their benefits include increased purity, lower processing temperatures, finer grain size, and more homogeneous microstructures. To gain further understanding of these techniques, a brief description of some terms is mandated. A *sol* is a stable dispersion in a liquid of particles less than 0.1 μm in diameter. In more general terms, it refers to a mixture of solid colloidal particles in a liquid. A *colloidal* substance consists of very small particles that remain in suspension in a liquid for some time, without settling. When a sol loses liquid, it becomes a gel. *Gels* are noncrystalline solids formed by chemical reaction rather than by melting. Described in other terms, gels are a jellylike substance formed by the coagulation of a sol (causing a liquid to become a soft, semisolid mass). Through a chemical reaction, the sol-gel process turns the solution of organometallic compounds (sol) into a gel by the formation of polymerlike bonds. The final step is raising the temperature to convert the dehydrated gel into a ceramic material. Such temperatures are much lower (less than 900°C) than those used in the conventional processes of pressing, casting, or plastic forming, which involve high-temperature sintering. Applications for sol-gel processing include preparation of glasses for coatings, optical fibers, fabrication of glass ceramics, forming of oxide matrices for fiber composites, coatings on fibers to lessen the interfacial bonding between the fibers and the matrix to increase the toughness of reinforced fiber composites, and monolithic ceramics. Sol-gel and reaction-bonded ceramic processes, **green manufacturing,** are low-temperature, relatively high-energy processes that produce many advanced ceramic materials without the many difficulties (i.e., chemical interactions and residual stresses) associated with the use of high-temperature processes. Figure 7-27 contrasts them with chemically bonded ceramics (CBCs) such as cement and concrete, discussed later in this module.

As mentioned, *gelcasting* is a new ceramic-forming process in which a slurry (a fluid suspension) of ceramic powders in a solution of organic monomers is cast in a mold. The monomer mixture is polymerized in the mold to form gelled parts. Both monolithic and composite ceramic parts, and complex-shaped and near-net-shaped parts, that require minimal final finishing are produced. Figure 7-28a shows a gelcasting process flowchart, and Figure 7-28b shows an application as well as a list of some materials that are gelcast.

The formation of the microstructures of ceramic materials is the result of chemical changes, interactions, or reactions. Chemically bonded ceramics (CBCs) can be distinguished by the manner in which consolidation of the components is brought about. In one process, consolidation results from sintering (densification) or fusion. In another, consolidation results from cementation by chemical reaction and bond formation. Figure 7-29 is a graphical representation of these two consolidation processes. Traditional sintering, depicted in Figure 7-29a, is a solid-state diffusion process that results in shrinkage; Figure 7-29b illustrates that the vol-

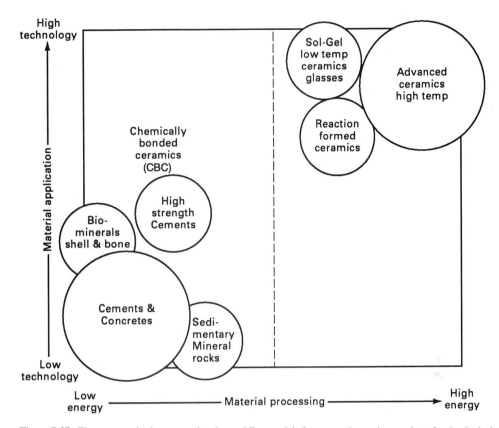

Figure 7-27   The compromise between using the world's materials for an ever-increasing number of technological materials applications and the need to reduce the energy required to process them. The higher the need for complex technology in the processing, the more energy is consumed. An underlying theme in this graphic is humankind's never-ending search to find the methods that nature uses to accomplish many of the technological goals that humans attempt to duplicate. (JME)

ume between reactants and products can be conserved, resulting in net- or near-net-shape forming. Generally, these chemical processes take place when the compacted powders are heated to their sintering temperature. On other occasions the compacted powders react with a gas or liquid. ***Reaction-bonded silicon carbide (SiC),*** or reaction-sintered SiC, is one such process. A powder mixture of Si and C is formed into the desired shape and exposed at high temperature to Si vapor or molten Si. The reaction with Si produces SiC, which bonds to the original SiC compact, and the excess Si infiltrates the compact's pores at a controlled rate. The result is a nonporous composite (SiC/C) with varying degrees of strength and elasticity. By using carbon fibers to form laminae of fibers woven within the molten Si, a range of fiber composites can be developed having the characteristics desired.

A similar process is the ***reaction-bonded Si_3N_4 (RBSN)*** process, which starts with a compacted shape of Si in a furnace under an atmosphere of N, N/He, or N/H gas. Heated initially to about 1250°C, the N permeates the compact, forming Si_3N_4. The temperature is then raised to just below the melting temperature of Si to increase the reaction rate. After about one week in the furnace, the process is completed, producing a ceramic that has good dimensional stability, thermal-shock resistance, and thermal conductivity. Its one disadvantage is its low oxidation resistance due to its interconnected porosity (open pores).

Oxidation describes the chemical reaction of oxygen when it combines with another element. If sufficient oxygen is present, it will react with the surface of the silicon nitride, producing silicon dioxide (SiO_2), which forms a protective layer at the surface that prevents further reaction. This is known as ***passive oxidation.*** If not enough oxygen is present, the gas silicon monoxide (SiO) is formed. Known as ***active oxidation,*** this type of oxidation can be continuous, resulting in complete destruction of the ceramic. Any oxygen-poor environment,

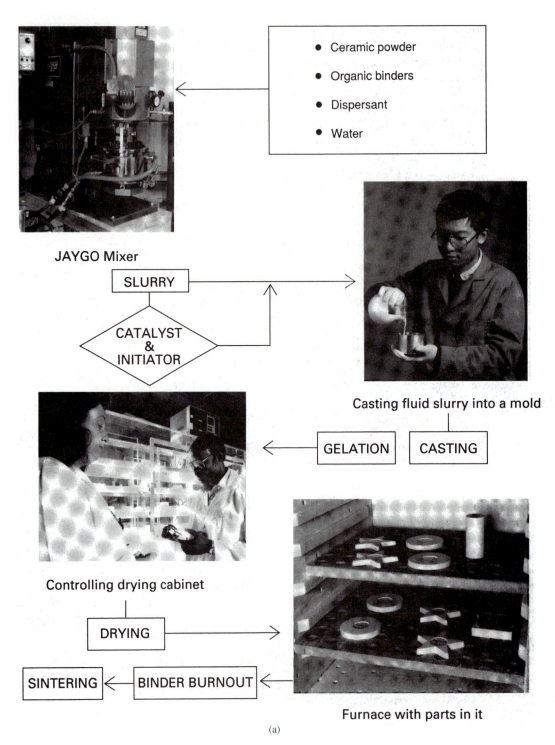

JAYGO Mixer

- Ceramic powder
- Organic binders
- Dispersant
- Water

SLURRY

CATALYST & INITIATOR

Casting fluid slurry into a mold

GELATION   CASTING

Controlling drying cabinet

DRYING

SINTERING ← BINDER BURNOUT ←

Furnace with parts in it

(a)

Figure 7-28 Gelcasting. (a) is a graphical flow chart for the Gelcasting process.

such as outer space, would be conducive to such attack. Other factors, such as temperature and porosity, influence the degree of reaction. By using oxidation initially, the surface may be coated with a layer of SiO_2, which will protect the interior from further oxidation attack. This is known as *flash oxidation,* which is useful for protecting $MoSi_2$ heating elements.

7.3.3.2 Melt processing. Most glasses and many ceramics are melted before being processed. Abrasives, glass shapes, and bricks of Al_2O_3 are made by casting from a melt.

Gelcasting Process Applications

Ceramic gears (Alumina)

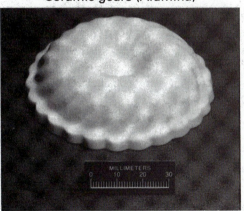

Compound curves with varied thickness

List of Materials gelcast

Monoliths

- Alumina
- Silicon
- Silicon carbide
- Zirconia
- Fused silica
- Sialon

Composites

- Nicalon fiber-reinforced reaction-bonded silicon nitride
- Alumina-zirconia

Figure 7-17 Gelcasting. (a) is a graphical flow chart for the Gelcasting process and (b) contains pictures of some spplications for the gelcasting process along with a partial listing of materials that could be selected for their manufacture. (ORNL)

Tapered two-step doughnut

Ceramic rotor fired (Alumina)

(b)

Figure 7-28 (b) contains pictures of some applications for the gelcasting process along with a partial listing of materials that could be selected for their manufacture. (ORNL)

Glass fibers for fiberglass composites are drawn or spun from a molten state. Molten droplets made from ceramic particles are sprayed onto surfaces through a variety of spray techniques that provide corrosion, wear, and thermal protection to a wide variety of metal, plastic, and ceramic products. Figure 7-30 shows photographs of various parts made of silicon nitride.

Figure 7-29 How the densification of particles is accomplished under conventional sintering (a) and reaction bonding sintering (b) processes. Conventional sintering depends, in the main, on high temperature (over 2000°C) to cause solid-state diffusion, whereas reaction bonding sintering requires temperatures around 1400°C. (JME)

(a)   (b)

Figure 7-30 (a) Silicon nitride (Si₃O₄) rolling bearings have many advantages over high-performance steel bearings: lower wear, lower heat generation, reduced noise and vibration (RNV), and excellent performance under marginal lubrication. Applications of ceramic hybrid bearings include machine-tool spindles, pump mainshaft bearings, dental drill shaft bearings, in-line skate and mountain bike hub bearings, and the space shuttle main engine oxygen fuel pump. (Produced by Norton, East Grandy, CT) (b) Turbocharger rotors and valves used in automobile engines. The development of silicon nitride gas turbine components in 1988 in Japan permitted the use of turbocharger rotors in passenger car engines because they weighed less than metal rotors, permitting the driver to accelerate more quickly without feeling a lag in power. Ceramic valves have provided consistent excellent performance. Their relative cost of manufacture prevents them from widespread use. (American Ceramic Society)

Coatings such as barrier, inhibitive, sacrificial, organic, inorganic, and metallic are used to control corrosion.

Thermal spray is a generic term for a group of coating processes used to apply metallic or nonmetallic coatings. *Plasma-arc spray, flame spray,* and *electric-arc spray* all heat the coating material (in powder, wire, or rod form) to a molten or semimolten state. The heated particles are accelerated and propelled toward a prepared surface by process gases or by atomizing jets. On impact, a bond forms with the surface and subsequent particles cause a controlled thickness buildup. Plasma, a high-temperature, ionized, electrically conducting gas, is used for spraying ceramic coatings. Flame and plasma-arc spray guns melt ceramic particles and spray the molten droplets at high velocities. One disadvantage of such a process is that even ceramics that are to be coated often require preheating to avoid thermal shock due to the high temperatures involved. Numerous techniques are used to melt ceramic particles and spray the melt onto a surface. Newer methods are used to produce uniform coatings, with "all porosity effectively sealed," that can withstand continuous exposure to temperatures of 540°C. Such coatings are referred to as ***thermal-barrier coatings*** (TBCs) (see Section 7.10). TBCs on metal components of gas turbines have low thermal conductivity and low emissivity. Materials such as ZrO_2 applied as a spray are used for such coatings because, in addition to possessing the properties noted above, such materials have a coefficient of thermal expansion similar to that of the substrate metal, which allows the coating to adhere to the metal as the metal undergoes changes in temperature. Designers working on ways to increase gas-turbine operating temperatures have concentrated their efforts on ways to improve superalloy-turbine-blade cooling with TBCs. Their goal is to develop an oxide coating that has thermal stability at temperatures up to 1650°C. Competing developmental efforts are devoted to perfecting a ceramic-matrix-reinforced composite that will be capable of replacing superalloys in such turbine applications. In the meantime, the metal components must be protected from higher and higher operating temperatures by application of various thermal coatings using several coating processes, which are themselves under constant refinement. A ***high-velocity oxyfuel (HVOF)*** system can be used to apply denser and thicker coatings than is possible using a plasma spray. The United States Navy expects to see a wide range of coatings on ships, aircraft, and land vehicles using HVOF, which produces a pore-free coating reportedly over 50% harder than conventional carbide coatings. It resists cracking and spalling even when subjected to mechanical peening. Cored wires made by extruded powder-filled metal tubes are used with arc-spray applications to deposit coatings of alloys and material combinations heretofore considered unsprayable. Computer-controlled spray systems now permit greater control over the process and provide greater safety to personnel in an inherently hazardous environment. The major thermal-spray coating processes are *plasma spray, wire flame spray, electric-arc spray, powder flame spray,* and *high-velocity oxyfuel (HVOF)*. Without devoting time to a description of their operation, a few advantages of each process will be noted. The ***plasma spray*** provides a coating that is dense and bonds strongly to the substrate. The ***wire flame*** process is possibly the best overall for applying a coating rapidly at lowest cost. ***Wire-arc spraying*** is used to deposit corrosion-inhibiting zinc on metal structures and aluminum coating on computer components to suppress electromagnetic interference. A new application is the spraying of metal on foam without damaging the foam. The metal to be sprayed is placed in an electric arc drawn between the tips of two wires of the metal to be deposited. The metal is melted and carried toward the substrate by a high-pressure flow of inert gas, such as argon. Thermal spray is used not only to spray long-lasting, corrosion-resistant coatings on nuclear weapons storage containers but also has entered the art world. Figure 7-31 represents a sculpture that more than likely was cast in bronze. The wire-arc (or electric-arc spray) process uses two electrically charged wires made of the coating materials that are fed through a spray gun. When their tips touch the gun head, an arc is created that heats and melts the wires. The molten metal is atomized and propelled onto the object to be coated by a stream of pressurized gas. This method allows artists to make a sculpture in a relatively easy-to-cast metal and then finish it with a coating of a more aesthetically pleasing but harder-to-cast material. Artists can then selectively coat with different materials. As an example, a statue could be cast in an aluminum-silicon alloy, sprayed all over with 99% nickel to a thickness of 0.8 to 1 mm, then selectively coated in certain places with a modified silicon bronze. This selective coating replaces the need to separately cast a part and then weld it to the main sculpture. In the past, thermally insulating foams were electroplated with metallic coats that protected them from damage by handling, seepage of air, or other reasons. ***Electroplating*** is more expensive. It is slow and involves toxic and polluting chemicals. Wire spraying, on the other hand, exposes the substrate to less severe temperatures, ranging from about 38° to 149°C, than do other spray techniques. The ***electric-arc*** process provides flexible tailoring of the coating characteristics and is a good choice for coating large areas. ***Powder flame*** allows for the widest choice of coating materials and provides a good combination of wear and thermal properties. ***HVOF*** provides the highest bonding strength and the lowest porosity. Thick layers and a high-quality surface finish are characteristic of this method.

Figure 7-31 Thermal spray has found a home in the art world. (Scintilla Artworks and Arizona Bronze, Tempe, Arizona)

It is well suited for the production of good-quality tungsten- and chromium-carbide coatings. The weight and inertia of brake disks are reduced in half when aluminum alloys such as AlSi-20%SiC are used instead of ferrous alloys. However, reinforced aluminum metal matrix composite (MMC) brake disks do not provide the necessary high coefficient of friction. A thermal spray coating of a powder-based Al_2O_3 optimizes thermal conductivity and friction problems such that these brake materials can be used in motorcycles and lightweight passenger cars.

7.3.3.3 Producing single-crystal materials.   *Single-crystal materials* solidified from a melt are developed to obtain materials with special properties not possessed by polycrystalline materials. Such materials are processed by several different techniques, of which two processes for forming such crystals are mentioned here. Figure 3-74 showed a typical application of this technique as it relates to metals. In the semiconductor industry, great advances have been made in growing silicon single crystals, which are now produced with few, if any, structure dislocations. An *edged-defined, film-fed growth (EFG) method* uses alumina with a molybdenum die that is immersed in molten alumina. The molten alumina is wicked to the top of the die, where it comes in contact with a seed crystal that is slowly pulled away to start the growth process. The growing crystal assumes the shape of the cross-section of the die. Plates, tubes, and various-size filaments are produced using this method. For example, some continuous filaments are produced at a rate of 62 m/h. A second technique is the *heat exchanger* or *gradient furnace,* which produces larger single-crystal components. A seed crystal is kept cool by helium flow in a crucible filled with molten alumina. The temperature of both the melt and the seed crystal is controlled such that the contents of the crucible grow into a single crystal.

7.3.3.4 Vapor processing.   *Vapor processing* involves heating a solid substance to a temperature that transforms the solid into a vapor, which is then deposited onto a surface. When the vapor contacts a cold substrate or when a gas reacts on contact with a hot substrate, the solid produced on that surface solidifies so quickly that the atoms fail to form a crystalline structure. Such noncrystalline, fine-grained, nonporous coatings are difficult, if not impossible, to produce by other means. ***Physical vapor deposition (PVD), chemical vapor deposition (CVD), chemical vapor infiltration (CVI), metalloorganic chemical vapor deposition, molecular beam epitaxy (MBE), liquid-phase epitaxy (LPE), sputtering,*** and ***ion plating*** are some of the methods that use this technology. As a group they represent a wide range of ***surface modification*** techniques to achieve surface properties in a controlled manner for ***engineered materials*** needed by today's advanced technologies. In addition to providing high-temperature materials for turbine blades or metallic coatings for plastic parts, these techniques are now involved in producing high-quality optical glass for fiber-optics communications as well as bulk forms in single shapes with structures developed on an atomic scale. These new materials that have designed structures are known as ***artificially structured materials.*** Until

recently most developments in this field were somewhat restricted to use in the semiconductor industry, but as each day passes, new applications are found for such materials.

PVD vaporizes a target material by using a high-energy source such as ions or electrons. The vapor is deposited atom by atom on a substrate, building up the required layers of atoms. *Laser ablation and deposition* uses a pulsed laser process to produce high-temperature-superconducting thin films. This new process is also based on PVD principles. Using multiple target pellets, each containing the required chemical composition, a thin film is reproduced with the exact composition. Wear problems are constantly being solved by using thin-film coatings made of zirconium, chromium, and carbon. Before a particular coating is used, a complete tribological profile of the conditions in which the part is to operate must be determined. Recently, titanium carbonitride (TiCN) has been replacing titanium nitride (TiN) and chromium nitride (CrN) coatings on *cutting* and *forming tools.* Such coatings are applied in thicknesses of 1 to 4 μm by a low-temperature, reactive ion plating PVD process. Figure 7-32a shows a schematic of the process and Figure 7-32b shows a photomicrograph of the edge view of the coating and substrate. Such coatings (Figure 7-32c) combat abrasive wear and adhesive welding when metals such as stainless steel, cast iron, brass, titanium, and aluminum alloys are machined. The TiCN coating, for example, is stable to temperatures of 400°C with a hardness of 3000 HV (25 g load). Another PVD-based process is the electron-beam vacuum evaporation process, developed primarily to coat turbine blades. This method generally produces a higher quality coating than that produced by its competitor—the plasma spray technique—with PVD's cost advantage. Two other versions of the electron-beam evaporation method—the arc method and the plasma-assisted evaporation method—are finding many applications throughout history.

Chemical vapor deposition (CVD) is a process in which a mixture of gases is passed across a heated surface. The temperature causes the gases to react or decompose to form a solid

(a)

(b)

(c)

Figure 7-32 (a) Physical vapor deposition (PVD) process used for thin film ceramic coatings. (b) Photomicrograph showing edge view with uniformity of 1×10^{-4} in. thin coating of titanium nitride (TiN) on high-speed steel substrate. (c) TiN and other thin film coatings, sometimes as many as 13 layers, are used on cutting tools and tool inserts.

Figure 7-33 A micrograph of a layer of boron nitride (BN) deposited on a substrate of magnesium oxide (MgO)—magnification 7,200,000×. (ORNL)

on contacting the surface. In an example taken from the literature, a TiN coating that protects a surface from wear is produced by CVD with a mixture of gases, $TiCl_4$–N_2–H_2, heated to a deposition temperature between 900° and 1000°C. CVD can be used to manufacture parts such as a boron-nitride (BN) crucible with wall thicknesses of several millimeters. Figure 7-33 is a micrograph of a boron-nitride layer deposited on a substrate of magnesium oxide (MgO). In the composites field, CVD is used to coat fibers that will be embedded in a metal or ceramic matrix to increase the toughness of the fiber composite. This toughening mechanism is discussed in Section 7.4. For example, silicon fibers are coated with carbon before they are used in a matrix of silicon carbide. This coating increases the toughness of the composite by reducing the interfacial bonding strength between the fibers and the matrix material. Doing so allows the fibers to debond from the matrix materials and pull out before breaking. All of these displacements require energy, decreasing the probability of catastrophic failure from a single mechanical overload. Figure 7-34 shows a toughness model sketch (a) and a micrograph (b) of fiber bridging and pullout as fibers resist the propagation of a crack. A ***near frictionless carbon (NFC)*** coating many times slicker than PTFE (see Gore Tex) developed by Argonne National Laboratory has a coefficient of friction less than 0.001 when measured in a dry nitrogen atmosphere. This is 20 times lower than the previous record holder, molybdenum disulfide. Thin NFC films can be deposited on virtually any substrate by vacuum deposition without risking damage to heat-sensitive materials. The ultrahard coatings have a host of ideal properties that make them suitable for a variety of applications such as exceptional wear resistance and durability, fast processing with large amounts being produced in only a few hours, adherence to many surfaces including polymers, finished products requiring no secondary processing, lower cost, and improved environmental compliance.

   Synthetic diamonds were first made in 1950 by General Electric using extremely high pressures and temperatures. The same technology is in use today to make diamonds. A more recent development is the making of ***synthetic diamond films*** and coatings using low-pressure processes. One of many vapor deposition techniques is CVD, which deposits the film from a gaseous mixture (CH_4–H_2) onto a substrate. Numerous substrates, such as molybdenum, sili-

Toughness Model

(a)

(b)

(c)

Figure 7-34   (a) A crack zone showing the effects of whiskers (or fibers) as they bridge the crack and pull out of the surrounding matrix material before fracturing. (b) Photomicrograph of a fractured, fiber-reinforced Lanxide ceramic composite showing the fibers spanning the fracture to resist crack propagation. By designing an interfacial bond with the correct strength between the fibers and their matrix, the fibers can expend energy overcoming friction and pulling out of the matrix, thus bridging the fracture instead of fracturing and assisting in crack propagation. (c) Ceramic composite fracture surface. (ORNL and Lanxide)

con, ceramics, and natural diamond, have been used to grow the films. The largest obstacle to successful film growth is the need to heat any substrate to temperatures above 600°C. Hydrocarbon gases, of which methane is most common, are used as the precursor gas. Various methods for forming the vapor are in use, such as microwave activation and thermal-activation and plasma-jet processes. Figure 7-35 shows a simple sketch of the process, a photograph of a typical reactor, and several grades of diamond film. Raman spectroscopy is used for detection of diamond; it consists of looking for a single, well-defined peak that is distinctive for the diamond bond. ***Natural diamond*** has a set of unique properties. It has the highest hardness reading on the Mohs scale. The elastic modulus is about five times as large as most steels, at 160×10^6 psi. Its tensile strength is higher than that of steel but lower than that of ceramic fibers, which can go as high as 10×10^6 psi. Natural diamond's 0.1 coefficient of friction ranks about the lowest for a solid material. The coefficient of thermal expansion for synthetic diamond is only about 0.8×10^{-6} °C^{-1}, a consequence of its strong covalent bonds. Electrical resistivity is about 10^{14} $\Omega \cdot$ cm. Diamond, one of the two forms of carbon, possesses another great characteristic in addition to those mentioned—its high thermal conductivity. Its thermal conductivity

Hydrogen + Methane

   Activated by hot filament

Diamond is deposited on heated
substrate positioned nearby.

(a)

(b)

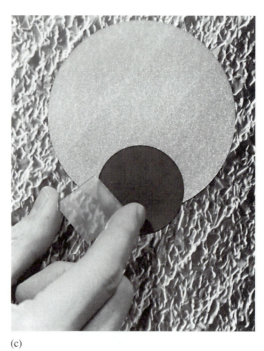

(c)

Figure 7-35 (a) Sketch of CVD process for making diamond film. A mixture of hydrogen and a hydrocarbon gas, usually methane (CH_4), is heated to about 2000°C using an electrical filament or other heat source. At this temperature, the gases disassociate, freeing the hydrogen and carbon atoms. The carbon atoms are deposited on a substrate heated to around 1000°C. Most other CVD processes are variations of this technique. (b) The chamber used in the CVD process. (c) A diamond film micrograph magnified more than 100× serves as background for several grades of thick film CVD diamond, including optical (white and clear) and cutting tool quality (black). (ORNL and Norton Diamond Film)

is 5 to 10 times greater than metals. To obtain a better picture of its hardness, consider the following statement: It takes 500,000 psi of pressure to indent metals; 0.75 million psi will indent the softest ceramic, whereas it requires more than 10 million psi to indent diamond. Diamond **cutting tools** are produced via P/M by hot pressing to full density a mixture of diamond grit and metal powders in a graphite mold. The metals serve as a binder to bond and hold the diamond in the tool during the cutting process. The latest metal alloys of copper, iron, and cobalt (not exceeding 25%) have extremely fine microstructures after hot pressing at lower temperatures than with cobalt alone, with abrasion resistance twice that of cobalt and plasticity of one-fourth that of cobalt. In addition, these new alloys have shown an increased tool (blade) life of over 30% compared to the traditional higher priced cobalt bonding materials. If synthetic diamond could be produced in sufficient quantities at an affordable price, it would replace silicon as a semiconductor material.

EMT WEBLINK

Diamond film—*http://www.nortondiamondfilm.com/*

A comparison of the various forms of diamond is interesting. Natural diamond is a single crystal made up of 99% carbon 12, with 1% carbon 13 isotope. General Electric's synthetic diamond contains more atoms per cubic centimeter than does natural diamond and can be made to any desired C_{12}/C_{13} ratio. CVD polycrystalline natural diamond has no binder, whereas polycrystalline sintered diamond contains up to 10% binder. ***Polycrystalline diamond (PCD)*** and ***polycrystalline cubic boron-nitride (PCBN),*** referred to as ***superabrasives,*** are finding more applications in the turning of hard materials, once the sole province of the grinding field. Using tipped turning inserts from the materials mentioned, material removal rates were 4 to 10 times greater than rates for grinding with cleaner surfaces. As with all materials, diamond-film properties can vary significantly, not only with deposition techniques but with substrate temperature, precursor-gas composition, and other factors that can or cannot be controlled in the process. Current research is focused on producing diamond films at lower cost and lower deposition temperatures, with larger film areas (present limits are 100 mm square and about 1 mm thick) and better adhesion between the film and the substrate.* Applications for diamond films are as cutting-tool inserts, wear surfaces for dies, heat sinks for electronics, and infrared (IR) and microwave windows (diamond is transparent to both radiations).

Chemical vapor infiltration (CVI) is somewhat related to CVD, the major difference being that the deposition occurs inside a porous preform rather than only on a surface. CVI originated in efforts to densify graphite bodies by infiltration with carbon. Infiltration is the densification of a porous preform by filling the pores with liquid, vapor, polymer, or a sol. The CVD-type coating grows with continued deposition to form the composite matrix. A wide variety of matrix materials are used, including borides, carbides, nitrides, and oxides. A great advantage of this process is that it allows the formation of high-melting-point materials such as TiB_2 ($T_M = 3200°C$) at relatively low temperatures, which reduces the risk of damage to the fiber reinforcement. Also, CVI does not use high pressures, which translates into less residual stress in the final composite materials. CVI is now used to produce carbon–carbon composites as well as ceramic–ceramic composites (see Module 8 for additional details). When fabricating fiber-reinforced ceramic composites, such as SiC high-modulus ceramic fibers with diameters of only 10 μm or less in a matrix of dense alumina, the fragile fibers cannot withstand the high stresses and handling of more conventional processes. With its low stress and gaseous precursors, CVI can infiltrate around the bundles of continuous fibers and deposit the matrix material around the fibers, filling in all pores and thus encapsulating the fibers with the matrix material and knitting them together in a strong interfacial bond. The coating of the fibers is a slow molecular process that builds up molecule by molecule. Figure 7-36 shows (a) a sketch of the CVI furnace, (b) a micrograph of SiC fibers reinforcing a SiC matrix composite material, and (c) a dramatic demonstration of the increased toughness of reinforced CVI ceramic materials. (Liquid-phase epitaxy and molecular-beam epitaxy methods are discussed in Section 9.9.)

Reactive melt infiltration is a new process that produces silicon-carbide–based ceramics and composites faster and more economically than do processes such as CVI or hot isostatic pressing (HIP). In comparison to CVI, this process takes minutes instead of days and costs about half as much. In the reactive melt infiltration process, a microporous carbon preform is infiltrated with molten silicon or a molten silicon alloy. The liquid and solid react to form a solid ceramic or composite. If the end product is to be a composite, the preform contains fibers. Products can be formed into complex shapes with full density, controlled microstructures, and tailored thermal and mechanical properties. A composite material consisting of a silicon-carbide matrix reinforced by silicon-carbide fibers has been produced. This ability to produce a composite material of complex shape and nearly final dimensions makes way for the production of silicon-carbide cutting tools.

Ion plating and *sputtering* involve plasmas. A ***plasma*** is a cloud of gas at high temperature made up of ions and electrons. Extremely high temperatures in the plasma can be reached because the cloud of gas is electrically conductive. This allows electrical heating (i.e.,

*See 7.15 Laboratory Activities & Classroom Demonstrations in Ceramics at the end of this module.

FIBER-REINFORCED CERAMIC COMPOSITES HAVE BEEN FABRICATED USING A FORCED CHEMICAL VAPOR INFILTRATION PROCESS

(a)

(b)

(c)

Figure 7-36 (a) Forced chemical vapor infiltration (CVI) process. (b) Photomicrograph of SiC ceramic produced by CVI. (c) Dramatic demonstration of the toughness obtained in ceramics as a result of the reinforcement achieved in CVI. (ORNL)

radio-frequency induction or direct contact with a metal electrode that supplies electrical current). The various factors involved in maintaining a plasma are somewhat complex, but in this brief treatment of the subject, they may be said to involve the temperatures of the ions in the plasma, energy input (heat), and the gas pressure. At high pressures, the plasma temperature can reach several thousands of degrees Celsius. Under these conditions, a powdered ceramic can be fed in the plasma, producing tiny droplets that can splatter onto a substrate as in plasma-arc spraying. At lower pressures, the gas atoms in the plasma remain cold and permit high electron temperatures to produce chemical reactions in the cold gas, the products of which can be deposited on a substrate. **Sputtering** is a basic coating process that uses a plasma, a target positioned above the substrate that is to be coated, and a sputtering gas. The target acts as the negative electrode, and the substrate acts as the positive electrode. The positive ions (usually argon ions) in the plasma are accelerated toward the target at high speed,

Figure 7-37 DIMOX™ process, demonstrating preform infiltration technology with unidirectional growth. A filler preform coated with a growth barrier is placed on top of a solid metal ingot in a refractory container. Temperatures (850°–1300°C) initiate growth, which proceeds through the filler until either the molten metal is consumed or the growth reaches the barrier. No shrinkage occurs.

resulting in the knocking off of desired atoms from the target, which are subsequently deposited on the substrate. This is known as the sputtering action. There are numerous variations to this basic process that depend, in part, on the type of substrates involved, how the coating atoms are introduced, the manner in which the plasma is maintained, or the differences in the applied voltages between the target and the substrate. One final comment is a reminder that the temperature of the substrate is extremely important to the final coating achieved. A recent innovative process for depositing diamondlike surface films on a substrate is the *microwave-assisted chemical vapor deposition (MACVD)* process. It can handle wafers up to 4 in. in diameter with a microprocessor-based programmable process control system.

7.3.3.5 Directed metal oxygen process.   The reaction of a molten metal with a gas to form near-net-shape metal and ceramic-matrix composite parts is exemplified by a Lanxide Corporation process. In the directed metal oxidation process, DIMOX™, ceramic matrices are grown around preplaced fibers (reinforcements) to produce a ceramic–matrix composite (CMC) to net or near-net shape without limitations on size or shape. Figure 7-37 is a schematic of the DIMOX™ process showing an example of unidirectional matrix growth. The preform is shaped like a filler preform of the same size and shape as the desired part. Preforms are made of particulate, which can be produced by any of the ceramic green-body-forming methods, such as pressing, isostatic pressing, slip casting, extrusion, or injection molding. The preform is covered with a special barrier material that limits matrix growth, ensuring that the part shape matches that of the preform. The parent alloy and preform are then heated to the growth temperature, at which the metal begins to oxidize rapidly in the presence of a gas oxidant. Oxidation occurs outward from the metal surface and into the preform such that the reaction product becomes the matrix surrounding the reinforcement in the preform.

    Oxidation of a metal generally forms a solid ceramic layer that limits further reaction. However, in one method, additives such as Mg or Si promote wetting of the ceramic by the molten aluminum alloy and reduce the stability of grain boundaries in the ceramic, which allows sustained reaction of the metal with the gas oxidant (oxygen). The metal is continuously made available to react with the gas oxidant, providing a growth rate that is independent of the

Figure 7-38   A micrograph of a typical microstructure of CMC and MMCs made by Lanxide's DIMOX process. Al_2O_3 reinforced by woven SiC fibers are shown with some unreacted metal also visible. Impurities in the uncoated fibers reacted with the molten metal to form a reaction layer on the fibers. In practice, fibers are given a protective coating prior to matrix formation, as found in continuous fiber ceramic composites (CFCC). (Lanxide)

Figure 7-39   A photograph of an alumina composite containing CVD-coated Nicalon® fibers produced by the Lanxide DIMOX™ process undergoing a four-point bend test. Despite a deflection greater than 1 mm, sufficient to fail an unreinforced ceramic catastrophically, the bar continues to support the load. (Lanxide)

thickness. In other words, the molten metal is drawn through its own oxidation product. The growth of the matrix into the preform occurs with little, if any, change in dimensions. This characteristic sets this process apart from traditional ceramic processing, with its inherent densification shrinkage.

Figure 7-38 shows a photomicrograph of the microstructure of a CMC made by the DIMOX™ process, composed of a matrix of Al_2O_3 reinforced by woven SiC fibers with some unreacted metal still visible. In low-temperature applications, the residual metal provides added toughness. For high-temperature use, that amount of metal can be removed, leaving a small amount of porosity, or treated to prevent melting or further oxidation during service. Figure 7-39 is a photograph of a Lanxide CMC specimen bar, consisting of an alumina matrix reinforced with CVD-coated SiC fibers, undergoing a flexure test. Table 7-4 lists some mechanical properties of SiC_p/Al_2O_3 CMC for use at high temperatures. The subscript "p" in the notation for the composite material refers to platelets or particles. Other abbreviations, "W" for whiskers and "f" for fibers, are used here.

Shape forming can be done by metal-shape replication and/or infiltration into a preform of the desired filler (reinforcement). Metal-shape replication used in making screw threads is shown in Figure 7-40. The shaped parent-metal cast or machined part is inserted into the cen-

TABLE 7-4   MECHANICAL PROPERTIES OF SIC_p/AL_2O_3 COMPOSITES PRODUCED VIA DIRECTED METAL OXIDATION (DIMOX™) PROCESS

Property	Degrees Celsius	90-X-008
Flex strength MPa (ksi)	25	179(26)
	1000	141(21)
	1400	155(22)
	1550	130(19)
Toughness (MPa · m$^{1/2}$)	25	5.5
	1000	4.0
	1400	3.0
	1550	3.8
Young's modulus (GPa)	25	391
Shear modulus (GPa)	25	153
Poisson's ratio		0.28
CTE (ppm/·C)	25–1400	6.0
Thermal conductivity (W/m-K)	25	116
	500	54
	1000	31
Bulk density (g/cc)		3.30
Carbide loading (v/o)		73
Residual metal (v/o)		6

Figure 7-40   Using the Lanxide DIMOX™ process to demonstrate the making of screw threads with a ceramic composite material through metal shape replication. (ALANX)

tral portion of the figure surrounded by the silicon-carbide bedding material. The growth will extend outward from the metal-shape preform toward the barrier bedding.

In the DIMOX™ process, the silicon-carbide bedding materials are incorporated into the ceramic matrix by the directed oxidation of molten aluminum alloy. The matrix reaction product occupies the spaces between the filler without displacing it, so that densification is achieved without shrinkage. Of course, for growth to occur, the filler used, be it fiber, whisker, particle, or platelet, must be compatible with the gas atmosphere and the parent metal. Coating of the filler materials is used to improve compatibility. Lanxide's PRIMEX™ process, a pressureless

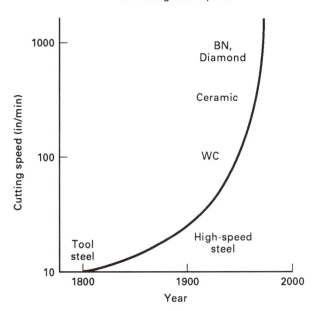

Figure 7-41 Plot of *cutting speed* versus time (years), showing the exponential rise in speed as a result of the advances in materials development.

molten-metal-infiltration technique, is discussed in the composite module, Module 8. Both of these Lanxide processes employ bulk processing techniques that have been traditional in foundries and ceramic processing for many years. This is in contrast with the expensive equipment and materials encountered in the production of many other advanced ceramics.

7.3.3.6 Finishing (machining).   In some instances, the shaped compact (preform) has not acquired the final shape or met the tolerances specified. Machining of the surfaces of ceramics is done to meet dimensional tolerances, improve surface finish, or remove surface flaws. It is to be assumed that all machining will introduce flaws of some size on the surface of ceramic parts. The machining of fine-grained ceramics such as Si_3N_4 significantly reduces their strength—more than that of ceramics of large grain size. After forming, the compact can be "green" machined (i.e., machined before densification). In other instances, final machining is required after densification.

Regardless of the type of machining (i.e., turning, milling, or wheel or profile grinding) or when the machining is used, tool materials must be made from hard, wear-resistant materials to avoid damaging or lessening the toughness of the relatively fragile ceramic workpiece. Internal flaws of less than 100 μm or surface flaws of less than 50 μm will produce fractures due to stresses in heat-engine ceramic parts. Although more costly initially, *diamond inserts* in machining tools usually result in significant cost savings in terms of reduced risk of damage to the workpiece, reduced time expended in tool changes, greater cutting speeds, and increased tool life. Figure 7-41 illustrates the phenomenal rise in cutting-tool speed over the past 200 years as a result of the research and development of metals and ceramics. Figure 7-42 is provided to emphasize the advantage of synthetic diamond as a cutting tool, particularly for ceramics.

7.4 IMPROVING TOUGHNESS

Toughness is an indication of how much resistance a material offers to a crack starting to grow through it. Most materials either contain cracks already or have structural flaws that, under stress, can readily turn into cracks. The work to be done in propagating a crack, also known as

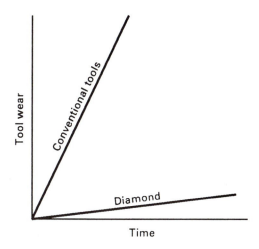

Figure 7-42 Rough plot of tool wear versus time, emphasizing the drastic reduction in tool wear as a result of the introduction of diamond-coated tools and inserts.

Figure 7-43 Plot of the specific strength versus the work of fracture for some typical ceramic materials—glass, cements, and biominerals. (JME)

the ***work of fracture,*** is measured by the energy absorbed in creating a unit area of new crack ahead of an existing crack tip. This work is a measure of toughness (G_C) and is expressed in joules/m^2. Figure 7-43 shows a graph of specific strength versus work of fracture for some typical ceramics. Work of fracture is the preferred measurement of toughness for fiber-reinforced ceramics.

Another procedure for assessing toughness is to measure the area under the stress–strain curve up to the point of fracture. This procedure is more suitable for a brittle material due to the difficulty in identifying the area under the curve that represents the energy attributable to plastic deformation (see Figure 7-44). The units for expressing toughness by this procedure are joule/m^2. ***Impact*** is defined as the sudden application of a load confined to a localized area of a material. Exemplified by the striking of a material with a hammer, this relatively quick application of a force (as opposed to a slow or static loading of a material) can cause considerable damage to a material that cannot adequately redistribute the stresses caused by the impact. Figures 4-11 and 4-13 showed a picture of an impact machine and some fractured specimens.

Ductile materials usually survive impact due to their microstructure, which allows slip to take place. Metals are thus considered tough, or having good impact resistance. Due to their inherent nature as compounds of metals and nonmetals, ceramics do not possess the ability to redistribute stresses and deform plastically. Consequently, they have poor toughness, impact resistance, and fracture toughness. To improve the impact resistance of ceramics, measures are taken to increase their fracture toughness. One such procedure is *fiber and particulate reinforcement.* Common examples are the use of glass fibers to reinforce plastics, carbon fibers to reinforce carbon-matrix composites, zirconium-diboride platelets to reinforce a zirconium-carbide matrix, and Al_2O_3 fibers in an aluminum matrix. Figure 7-44a is a micrograph of precoated SiC fibers reinforcing a SiC matrix by using the forced CVI process illustrated in Figure 7-36. Figure 7-44b is an accompanying stress–strain diagram, showing the effect on toughness of the composite by modifying the interfacial bond between fibers and the matrix material. In ceramic-matrix composites, the bond between the fibers and the matrix must be weakened to allow the fiber to absorb more of the energy that is trying to initiate or propagate a crack. This is the opposite strategy to fiber bonding with a polymeric matrix, which calls for an increase in the interfacial bonding. For example, silicon fibers are coated with carbon before they are used in a matrix of silicon carbide (Figure 7-44a). This coating increases the toughness of the composite by reducing the interfacial bonding (friction) between the fibers and the matrix material (Figure 7-44b). Doing so allows the fibers to debond from the matrix material and pull out before breaking. All of these displacements require energy expenditures and thus reduce the energy that can propagate as a crack until there is a failure of the composite. Figure 7-34a is a sketch of fiber bridging (fibers span across the crack) and pullout (fibers pull out of the matrix material) as the fibers resist the propagation of a crack; whereas Figure 7-34b is a micrograph of this same phenomenon. Figure 7-45 is a stress–strain plot showing the point of fiber pullout and matrix cracking as well as the progressive fracturing of the fibers. Each of these individual events takes energy away from that required to propagate a crack, leading to total failure of the ceramic. In addition to precoating the fibers, the toughness of the materials is increased partly because a crack that tries to propagate in the matrix encounters fibers that block its movement. The crack is diverted to run along the fiber-matrix interface. The stress applied in this direction is much lower, so the crack might stop propagating. Figure 7-46 is a micrograph of the crack pattern surrounding a diamond-hardness indentation in a polished surface of a Lanxide™ zirconium-diboride platelet, zirconium-carbide/zirconium composite. The matrix material of the composite, written or abbreviated $ZrB_{2\ p}/ZrC/Zr$, contains up to 20% metallic zirconium in addition to ZrC. These composites have fracture toughness as

(a)   (b)

Figure 7-44   (a) Micrograph of a SiC fiber precoated with carbon reinforcing a SiC matrix in a *continuous fiber ceramic composite* (CFCC). (b) Stress–strain curve showing the dramatic results of such a coating on the toughness of the composite. Note the differing results of reducing the friction or interfacial bonding between the fiber and its matrix material. (ORNL)

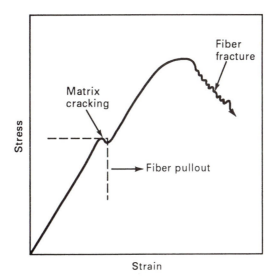

Figure 7-45 A fiber-reinforced ceramic composite has increased toughness due to the added strain energy of the fibers and the work of one in debonding the fibers from the matrix.

Figure 7-46 A micrograph of a platelet-reinforced Lanxide® ceramic composite showing the crack pattern produced by a diamond stylus indentation that effectively stops the propagation of a critical crack, which could have led to failure. (Lanxide)

high as 23 MPa · m$^{1/2}$ and flexural strength exceeding 1 GPa. These properties are the result of a combination of crack deflection and pullout of the platelets, as shown in the micrograph. Additionally, the excess zirconium provides some ductile yielding (plastic deformation) in the crack zone. These reinforcing platelets have successfully localized the damage, as shown by the numerous small cracks surrounding the indentation.

Another method for toughening materials is the use of ***second-phase reinforcement.*** Examples are the addition of small quantities (about 10%) of cobalt (Co) to tungsten carbide (WC), a ceramic (a hard refractory material), while both are in the particle stage. After additional steps in processing (pressing and sintering), a cermet is produced that has the toughness needed by cutting tools. A third method for toughening materials is called ***transformation toughening.*** In the study of steels, the formation of martensite was discussed in Section 5.6. The martensite reaction was once believed to be confined only to steels, but now a number of other alloy systems, such as iron–nickel and copper–aluminum, as well as titanium produce similar phase transformations. In the field of ceramics, ***zirconia*** or ***zirconium oxide (ZrO$_2$)*** has been found to possess some unique properties (see Figure 7-26). Zirconia possesses the lowest thermal conductivity of any ceramic material. Pure ZrO$_2$ has a stable, tetragonal crystal

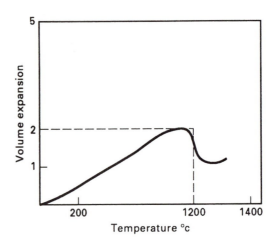

Figure 7-47   Volume expansion curve for ZrO_2 showing the effects of the displacing polymorphic transformation.

structure of temperatures above 1000°C but changes to a monoclinic crystal structure below 1000°C. This displacing transformation produces an abrupt change in the dimensions of the crystal structure, accompanied by stress buildup and possible cracking. A similar transformation occurs when silica allotropically changes from α-quartz to β-quartz, and from tridymite to cristobalite. When zirconia is cooled through 1000°C, the accompanying volume change (3.25% expansion) causes the material to fall apart. Figure 7-47 is a rough sketch of the effects of temperature on the volume of ZrO_2. However, if a small amount of yttria is added to the ZrO_2, the tetragonal phase can be stabilized down to room temperature. The stabilizer size is critical. In addition to yttria, manganese and calcium can be used. The yttria must be uniformly distributed throughout the microstructure of the ZrO_2. This is accomplished by mixing the two materials in powder form, with particle sizes varying from 0.05 to 1 μm. Note that this does not require a phase transformation. As stated, specific microstructures are crucial for optimum properties. High-temperature sintering (firing) is essential for the formation of tetragonal zirconia precipitates within a matrix of cubic stabilized zirconia. The morphology (shape) of these tetragonal precipitates can be controlled by various stabilizers. The ability of these precipitates to be transformed to monoclinic zirconia (with a 4% volumetric dilation on application of a stress or passage of a crack) imparts increased fracture toughness to these materials. The zirconia is now referred to as stabilized zirconia. By not adding enough yttria to stabilize the zirconia and with adjustments in the particle sizes and processing controls, mixtures of the stabilized cubic phase and the unstable monoclinic phase that possess high fracture toughness are obtained. This metastable mixture, known as *partially stabilized zirconia (PSZ),* will undergo transformation if it is disturbed sufficiently. If a part is made of this ceramic material and the part is subjected to impact, vibration, or another tensile stress-producing condition, the stress produced will cause the material to absorb energy sufficient to cause transformation of the monoclinic phase. The transformation takes place ahead of the tip of any crack growing in the material, resulting in a diminishing of the stress and in turn causing the crack to propagate, and hence an increase in the toughness of the ceramic material. Figure 7-19 was a sketch of the PSZ transformation zone surrounding a crack. It has been shown that Mg-PSZ, one of many PSZ systems, has the highest fracture toughness. As with other materials, it can be heat treated in a variety of ways to modify its microstructure for particular applications.

The last toughening method to be mentioned is *surface compression.* If the surface of a material is placed under compression, any applied tensile stress must exceed the compressive stress before a stress concentration can begin to build up and lead to a crack propagation. Surface compression or *prestressing* can be accomplished by ion exchange, quenching, or layering. *Ion exchange* can be brought about by exchanging ions of greater size with those of the

Figure 7-48 *Chemically strengthened glass.* Ion exchange causes relatively large potassium ions (K+) to replace smaller sodium ions (Na+), thus crowding the surface and producing compression (prestressing).

parent material (see Figure 7-48). The larger ions substitute for some of the smaller ions when the material is heated, and after the material is cooled, the larger ions near the surface cause a compressive stress. This phenomenon comes about because, at the higher temperature, the unit cells of a crystalline material or the atoms in an amorphous material have moved outward, making room for the larger ions that still find room with the smaller ions squeezed, producing the compressive stress.

When a glass material is heated sufficiently to a temperature that produces some viscous flow and then quickly cooled, the surface will cool more quickly than the interior, forming a solid, hard case around the still-somewhat-viscous interior. This process is known as *glass tempering.* Note the difference between glass tempering and tempering of martensitic steel in Module 5. Figure 7-49 is an example of tempered glass that has undergone an impact contrasted with safety glass which incorporates a laminated polymer sheet (PVB) to restrain the glass shards. Automobile windshields use a laminated, annealed glass sandwich composite with a PVB sheet between two sheets of glass. This glass sandwich composite material keeps the glass from flying about on impact yet maintains the transparency of the windshield so that the driver can maneuver, if possible. the automobile to a safe stop. By producing a surface compression with a different coefficient of thermal expansion (layering), heating to a temperature to produce some plasticity, and then cooling, the surface layers experience a compressive *prestress.* Figure 7-50 is a graphic step-by-step portrayal of this process. An example of this technique is the toughening of tempered glass and composite materials, including *prestressed concrete.* With fiber-reinforcing composites and reinforced concrete, this technique places the matrix materials under compressive stress to prevent a tensile stress from acting to cause crack propagation in the matrix. A multistep process is used to accomplish this prestressing. First, the fibers are tensioned mechanically. Second, they are encapsulated by the matrix material, which is allowed to solidify. Third, the tensioning force is released, causing the fibers to contract elastically toward their original length, placing the matrix under a compressive stress. Once an external tensile force (or load) is applied on the composite, the compressive stress in the matrix must be exceeded before the matrix feels the presence of the tensile stress.

7.5 REFRACTORIES

Refractory metals are among the few materials with significant mechanical strength above 2500°F, but they do not have sufficient strength-to-weight ratios for many sought-after

(a)

(b)

Figure 7-49   (a) *Tempered glass* is found in some automobile windows, large glass doors, and in eyeglass lenses. Note the fractured pieces of the tempered glass are small, rounded, blunt shaped, and relatively smooth and are incapable of causing injury. A piece of laminated safety glass, with an enclosed polymer adhesive made of polyvinyl butyral joins two pieces of **annealed glass** held in position by the adhesive shown in a darker color at the pointed end. The broken pieces of this glass, held in position by the adhesive made of polyvinyl butyral (PVB), shown in a darker color at the pointed end, are somewhat long, narrow, and dagger-shape shards that emanate from the origin of failure but adhere to the tough PVB interlayer. (b) Another example of *laminated safety glass* showing an extended piece of the PVB resin sheet material. Note that the polymeric membrane between the two layers of glass is relatively thick. Annealed glass when broken comparatively easy by impact or wind loads produces similar shards as the laminated glass, but an automobile windshield's transparency is retained.

applications. See refractory composites and explosive welding/bonding in Module 8 for new applications of refractories.

A *refractory material* is one with a very high melting point and other properties that make it suitable for uses such as furnace linings and kiln construction. A broad range of ceramic materials qualify as refractories. In the traditional ceramic group are products of sintered clay, consisting of crystals in a glassy matrix. These materials find use in ovens, kilns, furnaces, and melting pots; in welding and cutting; and as engine parts (see Figure 7-6). Some common refractory ceramics are alumina, alumina-silica, chromite, bauxite, zirconia, kaolin, silicon carbide, magnesite, and graphite. In addition to heat resistance, chemical properties bear heavily on the selection of refractories that come into contact with molten metals and glasses. Ceramics with weak ionic bonds have relatively low resistance to chemical attack, while those possessing strong covalent/ionic bonding are stable as linings for furnaces used for melting metal alloys and glasses at temperatures above 1200°C. The acidic or basic nature of a ceramic will determine how it reacts with a given molten material. The reaction between materials can change the properties of the metal or glass being processed. Operating temperatures of some furnaces are extremely high. For example, the basic oxygen furnace (BOF) for making steel subjects its linings for about 1 hour to a temperature of about 1700°C. Sintering furnaces may reach 2500°C. Blocks of alumina are cast after being melted in an arc furnace at temperatures well above 2000°C for use as liners in glass-melting furnaces because they are highly resistant to high-temperature corrosion.

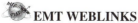

Hot

Immediately after chilling

Surfaces have shrunk and frozen-interior
is still hot and somewhat fluid-
surfaces are in tension

Room temperature

Interior has shrunk and frozen-this
action puts surfaces in compression-
interior is now in tension-interior
tension is 3 times the value of
surface compression

(e)

Figure 7-50   (a) Bending load produces tension in the lower surface of the test bar and compression in the upper surface. (b) The bar breaks when the tension on the lower surface exceeds the ultimate strength of the glass. (c) *Prestressed* bar shows compression in all surfaces. Reactive tension is buried within the bar. (d) Bending load applied to the prestressed bar must first overcome built-in compression before the surface can be put in tension. (e) Prestressing by physical tempering consists of heating glass until it begins to soften. An abrupt chill then shrinks and freezes surfaces. When the interior of the glass cools and shrinks, the surfaces are compressed. (Corning Glass Works)

EMT WEBLINKS

Coors Technical Ceramics—*http://www.coorstek.com/coorstek/home.asp*
Glass PPG Industries—*http://www.ppg.com*

These high-temperature ceramic materials must possess a varied assortment of properties. For example, the electrodes mentioned must have (1) resistance against corrosion and erosion from high-velocity gases, (2) good thermal-shock resistance, (3) high bulk density for good chemical resistance and erosion resistance, (4) good resistance against electrical-discharge arcing, and (5) good thermionic emission. Notice that strength properties were not mentioned, nor

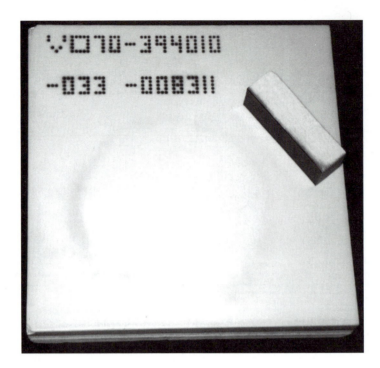

Figure 7-51 NASA's space shuttle *Orbiter* is dependent on a thermal protection system to protect it under various adverse conditions including extremes in temperatures ranging from −130°C to 1260°C. Examples of two different tiles used to cover the exterior of this space vehicle are shown. On some exterior surfaces blankets were developed consisting of quartz fiber batting sandwiched between woven quartz fabric and stitched together using quartz and glass threads. The blankets were further protected with a ceramic coating. This glassy coating protects against moisture penetration and erosion. For those surfaces exposed to higher temperatures than the above mentioned temperature range, low-density tiles were developed made of low-density silica fiber material bonded together by a sintering system (fused silica). Silica melts at about 1700°C and has good heat insulation qualities due to its large percentage of trapped porosity. Some 24,000 tiles cover about 70% of the exterior of the orbiter. (Lockheed Martin Missiles & Space) and (American Ceramic Society)

were those properties principally related to insulation against high temperatures. Many of our present engineered materials are designed for specific purposes. NASA's space shuttle *Orbiter,* a reusable cargo-carrying space vehicle, began its space missions in the early 1980s to service space vehicles in orbit. It was protected with a thermal protection system that would not only protect it in space but also during flight to and from space on a repeat basis. Though primarily a high-temperature protection system, it also was required to be, among other things, lightweight, resistant to moisture penetration, and able to withstand stresses and vibrations of flight. Figure 7-51 shows tiles that were used in two of the four different protection systems on the shuttle that would undergo different conditions of temperature and stress. Some of these tiles have densities as low as 0.14 g/cm^3. The fact that ceramic cutting tools can have more than a dozen thin coatings added to their surfaces, each with its own function, suggests that this technique may be useful for refractory ceramics as well. In the case of furnace linings, more than one liner may be installed that has special functions to perform in addition to its basic high-temperature resistance. Such functions could be structural or specialized insulating properties with low thermal conductivity in addition to high-temperature melting resistance. Finally, the possibility of using additives to these ceramics to increase their specific properties must be addressed. Fibers added to refractories serve to reduce thermal conductivities as well as to provide more open space within the material to decrease the material's ability to store up heat as a heat sink.

7.6 CEMENT AND CONCRETE

Traditional cements that were developed many years ago, and those produced in recent years, are actually a group of chemical materials in which consolidation occurs by means of chemi-

Figure 7-52 Two examples of cast concrete being used to maximize the main characteristic of concrete—weight. Tables are sitting on a cast-concrete platform. Both objects have withstood the ravages of two hurricanes without any damage.

Figure 7-53 Schematic picture of the setting and hardening process. (JME)

cal reactions at low temperatures rather than by firing or sintering at high temperatures. When mixed with water in suitable proportions, portland cement sets and hardens. Consequently, it is used in concrete to bind the sand and coarse aggregate into a solid mass. Figure 7-52 illustrates the use of cast concrete in a modern setting. Setting and hardening are caused by the chemical reactions between the cement and water. *Setting* (stiffening) occurs within hours, while with most normal varieties of cement, hardening (the subsequent development of strength) is largely complete in a month. If the water–cement mixture (paste) is kept moist, the strengthening process can continue for many years. If high strength is required in less than 1 month, alumina cements are used, which can achieve high strength in 48 hours. Figure 7-53 is a schematic of the setting and hardening process. Setting occurs when the gel coatings formed around individual grains join, and hardening results from densification of the contact areas in the gel. As seen in Figure 7-53 crystal calcite spindles (long, slender stalks or stems) grow from grains into a network strong enough to support extremely large loads. Drying times used by builders are conservative, at best, due to temperature and humidity variations during setting. However, these drying times must be complied with in determining when the concrete is ready to support the next step in construction. To speed construction, a more definitive time is required. A piezoelectric crystal device (see Module 9 for the definition of piezoelectric), which has been in service to test concrete bridges for internal cracks, is being adopted to obtain greater accuracy in calculating drying times. The operation of this device is based on the fact that the harder the concrete becomes, the faster sound travels through it. By mounting a beep-emitting piezoelectric crystal on one end of a concrete mass and a receiver on the other end to listen for the beep, the speed at which the sound travels through the drying concrete can be calculated. A more accurate prediction of drying time as to when the cement will reach a given strength is now reducing the time required in many construction projects.

7.6.1 Chemically Bonded Ceramics

More recent chemically bonded ceramics (CBCs) find uses as tooling mediums, for military armor, and in containers for storing hazardous and radioactive materials. Metal aggregates and fiber reinforcements are added to CBCs to improve their tensile strength and fracture toughness. These low-temperature CBCs are improving to the point where they have nearly 20 times the strength of concrete. As for cost, when raw materials and energy are considered, CBCs are nearly 6 times less expensive than polymer materials, 20 times less expensive than steel, and 25 times less expensive than aluminum.

Traditional hydraulic cements, as well as reaction and precipitation cements and the concretes made by combining them with various aggregates, are among the most used human-made materials. The term *cement* can include any materials that act as an adhesive to bond components together. *Cementitious* materials include pozzolans (containing natural hydrating or water-incorporating material), such as shale, lime, gypsum, asphalt, tar, synthetic plastics, and cement. *Asphalt* is the by-product of the *catalytic cracking* that separates gasoline and heating fuels from crude oil. The combination of asphalt and gravel has been used for centuries to build roads, but the amazing fact is that very little was known about this combination of materials engineering properties. Only one test, a viscosity test for high temperatures, was ever used to determine how well it would ward off rutting by vehicles in the summer; and only one grade (a certain proportion of the two ingredients) was used regardless of temperature and/or traffic conditions. This situation can be described as an *empirical* approach (based on experiment and observation without reference to any scientific principles) or, better still, it could be described as the art of road building for nearly 94% of U.S. road pavements. Do you know why there were so many pot holes and ruts in our street pavements?

The Strategic Highway Research Program was federally funded in 1987 using engineers from universities and industry to begin the 6-year job of finding a better way to build pavement. The result has been Superpave. Actually *Superpave* is not only the name for the new, superior performing asphalt product, but it includes a new set of testing equipment and protocols for mixing the materials in the proper proportions to hold up to the weather and traffic at a specific location. New tests now include both the old reliable viscosity test for high temperatures and a new viscosity test for low temperatures. In addition, a shearing test to measure deformation under a shearing type load and a load test to determine the material's ability to resist cracking are now standard practices. These tests now provide the data to determine the engineering properties of this new road-building material. The ingredients that combine with asphalt are also being investigated. Data have been collected about the correct proportion of *aggregate* (mix of rock as to its size, shape, coarseness, and fineness) such that each road is custom designed with the ideal grade of asphalt and the correct aggregation. One final factor determined by research is the optimum procedures for laying down and rolling out the Superpave. Researchers are collecting new pavement data to determine how many years a new pavement will last before rutting or cracking from temperature and fatigue. These data are from several test tracks where vehicles are run continuously to simulate many years of traffic on a particular pavement under certain weather conditions.

Ceramic cements are most familiar to us in the form of **concrete** and **mortar.** Concrete and mortar are similar because both use cement to bond together the **aggregate** (rocks and sand, which combine only through adhesion with cement), which adds bulk to the material. Concrete is more correctly classified as a composite, especially when rocks and reinforcing metal rods and wires are added to it. **Wheat concrete** is a new "green" product that uses starch—a granular, solid, complex carbohydrate found in rice, corn, wheat, and many other vegetable foods. **Masonry mortar** consists of masonry cement plus a fine aggregate sand. Masonry cement is composed of portland cement with hydrated (water-incorporating) lime, silica slag, and other additives to improve plasticity and slow setting time. Masonry mortar is applied by brick masons between cement blocks, brick, and tile.

Cements can be classified according to their bonding mechanism: hydraulic bonding, reaction bonding, and precipitation bonding. **Hydraulic cements** cure or set by interaction with water. **Portland cement,** the most common hydraulic cement, is composed primarily of anhydrous (without water) calcium silicate. An example is tricalcium silicate ($3CaO \cdot SiO_2$). Other ingredients are alumina, gypsum, and iron oxides. ASTM C150 recognizes eight types of portland cement, each with different chemical compositions and properties. Figure 7-54a depicts the raw materials used to prepare this cement. Portland cement sets by a complex reaction with water to form a hydrated composition. This reaction is **exothermic** (gives off heat), which means the heat can dry out the cement, thus stopping the reaction prematurely. Keeping the cement damp until setting is almost complete precludes this from happening. A compromise is needed here because the lower the ratio of water to cement when mixing, the higher the final strength of the mix that will be attained. When hardened, portland cement can contain a porosity of over 30%, which results in flexural strengths in the range of 4 to 10 MPa. Two factors that affect the strength of cement are the size of the pores arising from the entrapment of air and the incomplete packing of cement particles. Another hydraulic cement is gypsum cement, such as plaster of paris. These cements set using the same hydraulic reaction but recrystallize to a higher crystalline structure that has a low degree of adhesion. This property makes them good candidates for components such as wallboard and plaster molds made for slip casting of metals and ceramics. To give wallboard more strength, both sides are covered with paper, producing another example of a stressed-skin composite material.

The "greening" of concrete has been achieved with a "new age," pervious concrete that can trap contaminants as it filters runoff water in parking lots, driveways, and other large concrete pads. **Pervious concrete,** as seen in Figure 7-54b, was developed by Tarmac America, Inc. It is a porous (pervious) material that permits water, oil, sediments, and other solids and liquids to flow through the pores and slowly seep into the ground. With normal impervious concrete and asphalt pads, water and contaminants carried by the water must be directed to catch basins and storm drains, where they are piped through sewer systems to streams, rivers, lakes, and oceans. Any waterborne contaminants flow into the bodies of water and contribute to pollution. Pads made of pervious concrete provide for the slow, natural filtration of water that works its way back into groundwater reservoirs deep in the earth. To keep the pervious concrete clear, it is flushed by water.

New legislation to control storm water runoff has prompted new concrete technology. When introduced, pervious concrete cost more than double the price of impervious concrete. As with any new technology, this "green" concrete must prove its ability to withstand stresses from changing weather conditions, traffic loads, and aging. Its higher cost may be offset by savings in storm systems and improved water quality.

7.6.2 Improved Cements

Two recent advances in hydraulic cements have improved their tensile strength primarily by decreasing their porosity. In one technique, known as DSP (densified with small particles), SiO_2 particles of very fine size (ultrafine) are added. Because they are smaller than the cement particles, they fill spaces between the cement particles, decreasing the overall porosity and pore size. In addition, less water is required to produce plasticity of the mix. These factors reduce the permeability of the mix, which translates into longer service life for the many structures built with

(a)

(b)

Figure 7-54   (a) Raw materials to make portland cement. [From *Design and Control of Concrete Mixtures* (EB001.11T). Portland Cement Association, Skokie, IL.] (b) Porous concrete allows runoff water to seep through, disperses water naturally, and traps some pollutants. (John Corbitt, *The Virginian-Pilot*)

reinforced concrete that contain metal bars. However, these concrete mixes suffer corrosive attack from water seeping through the concrete.

A second technique is referred to as MDF (macrodefect-free) cement, in which additives such as small amounts of polymers and other ingredients are added to achieve better particle dispersion and packing. The absence of pores, whether due to initial particle size, particle distribution, or the elimination of entrapped air by kneading, results in the reduction of stress concentrators and a dramatic increase in tensile strength (over 200 MPa in some instances) and fracture toughness, with a corresponding decrease in permeability, all with little increase in overall cost.

High-temperature cements are produced by chemical reaction not involving water. One common reaction cement is monoaluminum phosphate, produced by the reaction between aluminum oxide and phosphoric acid. An example of these cements is dental cement based on zinc oxide. Another type of cement is precipitation cement, which is acid resistant and has good abrasion resistance. Sodium silicate, which is formed by gels that are precipitated from colloidal solutions, is a very common, inexpensive cement in this category of cements.

Improved cements, such as the CBCs, will no doubt be produced now and in the near future using techniques that call for low temperatures and low energy consumption. As a consequence, high-technology materials and processes will be called upon less often. The significance of this new, low technology is the fact that the resources needed to produce this new class of engineering materials for the twenty-first century are universally available. Third-world nations will benefit greatly, as will more developed countries that now rely on petroleum-based raw materials. As introduced in the Module 1 discussion of the materials cycle, there is much to be said for utilizing **low technology** as the world struggles to reduce consumption of limited natural resources and energy. Reduction of environmentally threatening waste (by-products of materials processing and energy consumption) and in the disposal of used products will be some of the results of the new technology. While we have studied high-technology materials and processes, we must continue to reexamine the viability of older materials such as wood (a renewable resource) and cement (from readily available resources) where it is possible to apply new processes.

7.7 CLAY

Clay is classified as either residual or sedimentary. **Residual clay** results from the wearing down of rock from the mechanical and chemical action of wind, water, earth movements, and chemicals in the soil. Such clays are found where they were formed. **Sedimentary clays** also form through mechanical and chemical erosion of rock, but they are moved by wind or water to places other than where they have formed. Common clays include kaolin ($Al_2O_3 \cdot 2SiO_2 \cdot 2H_2O$), ball clay, fire clays, stoneware clay, and slip and flint clays. In various forms, clay finds many applications: as ingredients in most traditional ceramics, structural products of brick and tile, raw materials for sculptured art, china dishes, and electronic components. Clay is an important ingredient in paper making. Silicon and aluminum are prominent in clay, comprising 28% and 8%, respectively, of the earth's upper crust. Most other elements are present in only small percentages, such as carbon, with 0.009%.

7.8 RADIOACTIVITY AND NUCLEAR FUEL

Radioactivity is the spontaneous emission of radiation from certain unstable elements. Over 1100 unstable radioisotopes are known. About 65 occur naturally and the remainder have been made in the laboratory under controlled nuclear reactions. This spontaneous emission is the disintegration or decay of the unstable nucleus. From our study of atoms and their nuclei in Module 2 we learned that the nucleus of atoms may contain the same number of protons but different numbers of neutrons. These atoms are known as isotopes. Nucleons (protons and neutrons) like electrons spin and have angular momentum. They are bound in the nucleus by a combination of forces some of which are not well understood. These nuclei are considered *stable*. Certain isotopes are not stable and they decay with the emission of some type of radiation. Unstable isotopes occur in nature and their disintegration is called *natural radioactivity*, whereas unstable isotopes produced in a laboratory by nuclear reactions have *artificial radioactivity*. When unstable nuclei undergo radiation decay, they change into other nuclei with the emission of alpha (α), beta (β), or gamma (γ) particles. Each has a different degree of penetrating power.

For example, α rays are weak and could be blocked by a single piece of paper whereas γ rays can penetrate several centimeters of lead. Gamma rays are high-energy protons. X rays also are protons with similar energy. The only difference between them is x rays are produced by an electron-atom reaction whereas a gamma ray is produced in a nuclear process. The energy released by nuclear reactions will appear as kinetic energy of the outgoing particles (radiation). This energy can be transformed into heat in a nuclear reactor which is used to boil water (to make steam) flowing through the core of a heat exchanger. Steam drives a turbine that generates electricity and the steam is then cooled in a condenser that converts the steam back to its liquid form.

Einstein's equation ($E = mc^2$) states that mass and energy are two different aspects of the same thing. During radiation, the mass of the nucleus is diminished by a tiny amount and this loss of mass is translated into radiation energy. This energy then can be used or converted to other forms. A hydrogen bomb explosion converts less than 1% of the mass to energy.

Radiation passing through materials can do considerable damage. Metals and other structural materials become brittle and lose their strength if the radiation is intense. All forms of radiation can ionize atoms by knocking out electrons. An isotope of strontium, at. no. 38, a product of radioactive fallout, concentrates in human bone causing bone cancer and destructing bone marrow (see Figure 7-9). Gamma radiation can adversely affect polymeric materials by rupturing bonds that hold molecular chains together. This degradation is called *scission.* This same radiation is used to cross-link PE molecules to increase the materials resistance to softening at high temperature.

The applications of nuclear radiation are numerous. Smoke detectors in homes and offices use a weak radiative source to ionize air molecules. When these molecules combine with smoke particles, the difference in the speed of movement is detected electronically and an alarm is triggered. Radioactive isotopes are used to detect leaks in pipes and to study corrosion and wear. In the medical field there are numerous applications of radioactivity and radiation to human beings. Radiation therapy or the treatment of disease, mainly cancer, utilizes radiation, because in large doses it will kill cancer cells. *Microspheres,* a special glass composition of S, Al, O, plus rare earth elements (lanthanum series known as the lanthanides) are being used in medicine to combat cancer (see Figure 7-55a for a listing of the lanthanide series of elements in the periodic table). By radiating this material in a nuclear reactor, the lanthanides become radioactive and give off beta, not gamma, radiation. By controlling the amount of radiation and placement of the radiative elements, the penetrating distance can be adjusted to attack cancer cells but not healthy cells in an organ or tumor. Radioactive tracers are used in medicine for diagnostic purposes such as locating the spread of disease tissues in the human body. Co, I, and the metastable state of Tc are some of the radionuclides used in medicine. (See Module 2 for a review of the notation used in designating isotopes.) Gamma radiation is used to kill bacteria and viruses and to preserve food (irradiation).

A *radioactive isotope* consists of a vast number of radioactive nuclei that do not decay all at once. They decay randomly over a period of time. The rate of decay of any isotope is often specified by giving its *half-life,* which is defined as the time it takes for half the original amount of the isotope in a given sample to decay. The longer the half-life of an isotope, the more slowly it decays. Spent fuel from nuclear reactors is extremely radioactive, so it must be kept isolated from people, animals, and the food chain. A radioactive fuel such as plutonium (Pu, at. no. 94), for example, has a half-life of about 24,000 years. Starting with a pound of plutonium, after the passage of about 24,000 years, one-half pound of this element will decay to a stable condition. After another 24,000 years, an additional one-quarter of the original pound will become stable, leaving one-quarter of the original pound still radioactive.

The *radioactive decay* of the fission products also produces heat, so cooling must be provided for the first few years that the spent fuel is stored. The usual way to provide cooling is by storing the spent fuel in water. Because of the current lack of a permanent place to dispose of spent fuel, the material is now stored on-site at power reactors. It is estimated that by the year 2020, nearly 100,000 tons of uranium-dioxide fuel will be awaiting a more permanent way of isolating it from the biosphere. Generating the electrical power to operate the average household for 3 years by burning 22.5 tons of coal would generate about 82.5 tons of carbon

39207046.1

(a)

(b)

Figure 7-55   (a) Elements typical in nuclear waste. (U.S. Dept. of Energy—Pacific Northwest Laboratory)
(b) Nuclear-waste glassmelter. (U.S. Dept. of Energy—Pacific Northwest Laboratory)

dioxide and about 2.3 tons of ash. The ash represents its own disposal problem, and the carbon dioxide would be released irretrievably into the atmosphere, where it accumulates. Carbon dioxide is one of the greenhouse effect gases, which impede the Earth from radiating energy into space. Many scientists believe that these gases could cause global warming, changing the climate worldwide.

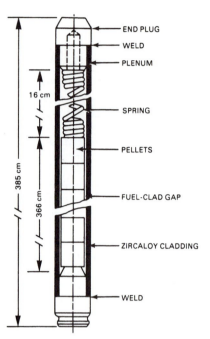

Figure 7-56 Light water reactor (LWR) fuel rod. (U.S. Dept. of Energy—Pacific Northwest Laboratory)

Uranium dioxide, a ceramic material, is used as a fuel in the current generation of nuclear reactors for the production of electric power. This oxide was chosen for its chemical stability and high melting temperature (2750°C). Figure 7-56 shows a cross-sectional view of a typical reactor fuel rod. The uranium oxide is made into pellet form by pressing and sintering; these pellets are then inserted into tubes made of a corrosion-resistant alloy of zirconia called *zircaloy,* which are then welded shut. A nuclear reactor contains thousands of these tubes, held into frameworks that allow water to flow among them. The water acts as both coolant and moderator. It slows the neutrons produced by splitting the uranium atoms so that they can then split other uranium atoms in a controlled fission process. The fuel inside each tube produces an amazingly large amount of energy, each foot of length producing about 2 kW, enough electric power to run an average household.

Because the ceramic is not a very good conductor of heat, this energy produces high temperatures, 600° to 1400°C, at the center of the oxide pellets. The water outside the fuel is at a fairly low temperature, about 300°C, so the temperature gradient is very steep. As the nuclear-fission reactions proceed, elements of lower atomic weight are produced, and they accumulate inside the solid oxide fuel. Eventually, these elements, called fission products, accumulate to the point that the fuel must be removed and new fuel inserted. The fuel produces a lot of energy before this is necessary; the spent fuel that produced all the electrical power for a household for a period of 3 years weighs about 4 oz. In a typical power reactor, a third of the fuel will be exchanged for fresh fuel every year during a planned outage. Refueling is sometimes done on a 2-year schedule, so fuel usually produces power for 3 to 6 years.

7.8.1 Disposing of Nuclear and Chemical Waste*

Nuclear materials provide an excellent example of the importance of understanding the total materials cycle. While these materials represent some of the highest levels of technological development, they also present huge hurdles in terms of their safe processing, use, maintenance, and disposal. As with other chemicals, biological organisms, and materials, materials science and engineering is often called on to immobilize the end products.

*Written for this text by L. Roy Bunnell, Battelle—Pacific Northwest Laboratory.

When in a molten state, glasses tend to be highly corrosive and dissolve other oxides readily. Glasses are nonselective in this dissolution because their noncrystalline nature permits them to incorporate other atoms without much regard for size, valence, or crystal form. One way to take advantage of this characteristic is to use particularly durable glasses as a matrix material for disposal of high-level nuclear wastes. Development of this concept started in the mid-1960s, and operating nuclear-waste glassmelters exist in several countries at present. The glass type used for nuclear-waste disposal is roughly similar in composition to Pyrex, a borosilicate glass developed by Corning Glass Company. This glass is very durable but can be processed at moderate temperatures. The various elements present in a typical nuclear waste are shown in Figure 7-55a. These elements can be dissolved by nuclear-waste glass, although solubility limits are occasionally reached for noble metals and elements such as ruthenium.

Processing of glass at temperatures of about 1100°C inside a shielded, remote-control-operated enclosure called a ***hot cell*** is a substantial engineering challenge. The most common way of heating nuclear-waste glass is to pass electricity directly through it, heating the glass by its own resistance. Glass behaves like a semiconductor, meaning that the electrical resistance drops as temperature increases, opposite to the behavior of a metal. Figure 7-55b shows a typical nuclear-waste glassmelter. Nuclear waste is combined with glassmaking ingredients and fed to the top of the melter as an aqueous slurry. The water evaporates and the ingredients combine to form a homogeneous black glass, in which the radioactive elements are dissolved at the molecular level. After a period to allow a homogeneous glass to form, the product is poured into a stainless-steel canister approximately 2 ft in diameter, where it cools and hardens to form a "log" about 10 ft long and weighing 4500 lb. After a lid is welded on and the canister is cleaned of external contamination, the waste glass is ready for transport to a permanent repository.

7.9 ABRASIVES

Abrasives are very hard particles used for grinding, sanding, and polishing. ***Abrasive materials*** include particles of flint, garnet, diamond, aluminum oxide, silicon carbide, emery, pumice (pulverized volcanic lava), rottenstone (shale rock), and corundum. Aluminum oxide and silicon carbide find the greatest industrial application, although garnet is often used as sandpaper—an abrasive paper or belt used for sanding wood. See Figure 7-57 for examples of abrasives including media used in ball mills. Adhesive papers include paper and cloth sheets, belts, disks, and drums onto which an abrasive is cemented. Wet or dry abrasive paper employs waterproof adhesives. Grinding wheels consist of abrasive particles (grains) held together by a tough bonding material such as rubber or organic resins. The particles constantly fracture on the wheel to expose new, sharp cutting edges, or they break off as the wheel becomes smaller in size. Abrasive particles are also used in loose form for sand blasting, hand rubbing, or abrasive drilling. Selection of the ***grit*** (particle size) is based on a screening method that determines the number of particles per inch, such as 60 grit or 600 grit. A 120 grit would have 120 abrasive grains per linear inch or 14,400 (120^2) per square inch. Sialon (silicon–aluminum–oxygen–nitrogen) is a hot-pressed cermet that works well on nickel-based superalloys.

Superabrasives are synthetic diamond and cubic boron nitride (CBN) in the form of grinding wheels or sintered, polycrystalline cutting tools for milling and turning. Diamond abrasives are four times as hard, and have three and one-half times the wear resistance of common abrasives such as aluminum oxide. CBN is two and one-half times as hard and has two and one-half times more wear resistance than aluminum oxide. In addition, superabrasives are excellent thermal conductors. The combined superior properties of hardness, wear resistance, and thermal conductivity provide sharper, longer-lasting tools that can operate at higher temperatures and speeds. When grinding and machining high-speed steel, cemented carbide, high-silicon aluminum alloys, cast iron, and superalloys, superabrasives offer increased productivity through high material-removal rates, achieving better quality (tolerances and surface finishes), less residual stress, reduced scrap, and lower overall manufacturing costs.

(a)

(b)

Figure 7-57 Ceramic materials are used to make abrasive products that are used to process other ceramics as well as make abrasive materials that help in the shaping, grinding, and finishing of many other types of materials. (a) A grouping of white-and-black objects represents milling or grinding media used in ball mills to reduce ceramic materials to a fine powder or slurry. (b). Note their spherical or cylindrical shapes. (b) A sampling of abrasives such as adhesive-backed sanding paper, sanding sleeves, grinding wheels, disks, disks with holes for more efficient removal of dust, sanding drum sets, diamond burr sets with assorted shanks and head shapes, a sanding flutter (flap) head, and a file.

7.10 PROTECTIVE COATINGS

Engineers have always been encouraged to look to nature to observe how certain tasks are accomplished or as an aid to stimulate the mind when involved in problem solving. One such example concerns the structure of seashells. The observations of birds dropping clams and oysters from considerable heights onto some hard surface to break the shells or sea otters smashing abalone shells with the aid of a rock all attest to the hardness of these shells. The shells are made basically from a relatively soft chalk. What makes them extremely tough is the manner in which thin alternating layers of calcium carbonate and organic polymers are intertwined much like a layered composite material. These microscopic layers could be compared to a brick wall with each brick surrounded by mortar. Cracks in an individual brick are stopped from traveling into the adjacent brick by the mortar. Scientists at Sandia National Laboratory have studied this natural process and have attempted to duplicate the process using detergent molecules similar to ordinary household detergents. An arrangement of *micelles,* a structural unit of molecules, was developed, which upon being polymerized, bonded the surfaces of the micelles together resulting in a hard surface that can be used in such applications as automobile finishes or optical lenses.

Coatings solve problems in corrosion, wear, and compatibility: Thermal spray technology processes of electric arc combustion and plasma spray coatings apply any material to the surface of another. *Surface science* is a new discipline that helps solve problems involving wear, corrosion, hardness, carburizing, nitriding, electroplating, electroless plating, phosphating, vapor deposition, and hardfacing. This new category of science will be incorporated into the broader field of materials science and engineering. An example in this field of technology and science is thin films of polymers and adhesive used to replace paint on USAF aircraft.

With the emphasis on materials that can withstand high temperatures (greater than about 320°C), whether in space applications or on earth in turbines or heat engines, new materials developments are reported almost daily. The search for such materials leads in many cases to the field of *surface engineering.* One of the first applications was the use of zirconia in the form of a surface layer applied to a metal substrate. This thermal barrier coating (TBC) is very effective in resisting high temperatures in heat engine moving parts. However, it failed to provide the necessary wear characteristics needed to protect materials in sliding contact under high temperatures. To add lubrication properties to this TBC, a new, self-lubricating material consisting of 80% silicon carbide, 10% silver, and 10% barium/fluoride eutectic has been developed that provides lubricity at high temperatures to sliding-contact-bearing surfaces of internal combustion engines made with conventional iron and aluminum. Many of these engines are now operating at such high temperatures that oil-based lubricants cannot be used. Both TBCs and self-lubricating coatings can be applied by plasma-arc spraying. This new development is another example of a true laminar composite (MMC) material consisting of an aluminum or iron substrate, with laminates consisting of a TBC and a self-lubricating wear coating superimposed on the surface of the TBC. *Nickel silicon carbide metal matrix composite* coating with a thickness of 0.08 mm for cylinder bores on the 1999 Jaguar AJ-V8 engine resists wear and corrosion and reduces the overall weight on the 4.0-liter (L), 290-HP engine by 7 kg to achieve the "best-in-class" weight of 200 kg. Since 1984, aluminum cylinder blocks experienced wear problems at the top ends and center of the bores as a result of high rubbing speeds produced by static-load reversals of the pistons. Cast iron liners solved these problems but they were too heavy. With these composite coatings, the fuel efficiency of 8.9 L per 100 kilometers or 26.4 MPG highway and 17.2 MPG city driving is attained. U.S. Chrome Corp., Stratford, Connecticut, has developed an *electroplated nickel silicon carbide coating (NiCom)*™ in which silicon carbide particles are uniformly dispersed in a nickel matrix. These composite coatings can be as thin as 0.005 to 0.015 mm and as thick as 1.5 mm. The SiC particles are co-deposited with the Ni and are uniformly disbursed throughout the coating. Coating hardness averages about 55 to 57 HRC. The company reports that nickel plating is more efficient than chromium and test results report that NiCom™ wears better than chromium in all engine bores. *Sliding wear* is its outstanding characteristic, but frictional properties occupy a high ranking. With any of these composites, adding a small amount of a laminar lubricant such as boron nitride (BN) or graphite can improve frictional properties. The *oleophyllic* (oily) nature of these coatings adds to their ability to reduce friction in internal combustion engine cylinders and hydraulic equipment. In addition, *oil retention* (the ability to maintain hydrodynamic lubrication) is better than chromium or electroless nickel. Prolonged exposure to 540°C does not result in significant loss of its physical properties.

The cost of holemaking is one of the largest machining costs in automotive engine production. Much research is devoted, therefore, to making it more efficient. Ecological factors and cost factors are both examined. Coolant disposal and recycling costs are significant. Some estimates state that the costs associated with coolant and coolant management are 16% of the machining costs in the high-production industries, compared to only 4% for the cost of the cutting tools. And of the total coolant-related costs, some 22% are charged to coolant disposal alone. The optimum objective of such research is stated in terms of eliminating all coolants. One such new process is Vari-Form's patented in-the-die hole piercing which combines hydroforming and piercing in one operation and minimizes deformations around holes. Holes can be round, hexagonal, square, rectangular, oval, or irregular shaped. This operation facilitates usage in manufacturing, access, or fastener applications. *Tube hydroforming* is a pressurized hydraulic forming process used to produce complex shapes in tubular components. It replaces traditional stamp and weld manufacturing, producing lighter, stronger, and fewer pieces. Automotive applications include radiator enclosures, space frames, dash assemblies, frame rails, and engine cradles.

Near-term objectives talk of near-dry drilling as the next step in reaching zero coolant drilling. Past attempts in this direction were the development of cermets and special carbide coatings on cutting tools. One such coating was titanium aluminum nitride (TiAlN). A combination coating of TiAlN and Al_2O_3 provides superior heat resistance in dry milling of cast iron. A more recent coating to reduce friction is a soft coating, somewhat like Teflon, that adheres well to the tool surface but provides a gliding or low-friction surface. Made of ***magnesium disulfide,*** it is called ***MOVIC.*** A second new coating is a variation of MOVIC. The soft coating is placed over a hard coating, except at the point of the tool where hardness is needed. Instead, the flutes of the drill are lubricated with the soft coating to aid chip flow. Another result of research in this area is the development of an improved titanium diboride composite material produced by the self-propagating, high-temperature synthesis (SHS) process. (See Module 8 for applications of this new composite material.)

Ceramic particles are used to coat and decorate pottery and metals when melted into a ***glassy vitreous state. Glazes*** and ***enamels*** impart color, hardness, and corrosion resistance when they are suspended in a slurry and fired to a vitreous state; as the *substrate* (base material) cools, the coating bonds tightly to the substrate surface. *Flame spraying* (see Section 7.3.3.2) uses a high-temperature heat source, which melts the ceramic oxide powder, and then sprays or blasts the coating onto the surface of a metal. The oxide material cools and bonds to increase its surface hardness while offering protection against heat and oxidation.

Mirrored sunglasses prevent the wearer's eyes from being seen. A layer of silver or other metallic compound, which is thick enough to retain its reflective properties but thin enough for the wearer to see through, provides this desired property. Some light is reflected from this "mirror," but some passes on through to the eye. In addition to this coating are up to eight other coatings that add extra properties to the glasses depending on the desires of the consumer. These coatings listed in the reverse order in which they are applied are (1) silver or metallic compound, (2) scratch resistant, (3) UV light absorbing, (4) haze resistant, (5) polarizing filter, (6) haze resistant, (7) UV light absorbing, (8) scratch resistant, and (9) antireflective. The polarizing filter absorbs over 98% of glare which is created by light bouncing off horizontal surfaces such as the ocean or a road surface. Vertically aligned light is allowed to pass through to the eyes. The multiple coatings correct, clarify, or refine the light as it passes through them. Both UV-A and UV-B light radiations are eliminated.

7.11 ELECTRONIC AND MAGNETIC CERAMICS

Typical electronic ceramics include some representative compounds and mixtures, such as ferrites (ZrFeO), silicon (SiO_2), zirconia (ZrO_3), steatite (SiO_2MgO), porcelains (Al_6SiO_{13}), and alumina (Al_2O_3). Nearly 1.2 billion pounds of ***hard ferrites*** were produced in 1997 for use in such applications as speakers, household magnets, memory cores, windshield wiper motors, and automobile power seats. ***Soft ferrites,*** which can be magnetized or demagnetized by changing the direction of the applied magnetic field, are used in television, radio, telephone, electric ignition systems, transformer cores, and tape and disk recording heads. Some 20 million pounds were produced in 1997. Module 9 deals with some of these and other electronic materials.

7.11.1 Ceramic Magnets

Ceramic magnets that contain a sufficient number of ***dipoles*** (atoms with electron spin in the same direction, discussed in Section 9.2.2) are ferromagnetic (see Section 9.3). Ceramic ferromagnetics can be either soft or hard magnets. The high electrical resistivity of ceramic magnets gives them an advantage over metal magnets for use in high-frequency devices. Their permanent magnetic behavior finds application in microwave devices. The group known as ferrites are iron oxides containing ions of elements such as zinc (Zn^{2+}), iron (Fe^{2+}), magnesium (Mg^{2+}), and nickel (Ni^{2+}) that combine with oxygen. Ferrites are commonly found in the computer industry in memory cores. Sheets of rubber and ferrite are employed for *"stealth"* coatings on some steel suspension bridges that cross navigational waterways. The coatings partially absorb radar and minimize clutter on ships' radar screens as they navigate under the bridges. Figure 7-58 shows an assortment of ceramic parts. Some depend on their ability to polarize their electric charge for their particular use whereas others are chosen because of their hardness.

7.11.2 Thick-Film and Thin-Film Ceramic Devices

Thick- and thin-film ceramic devices are of major significance to the computer industry and form a group of technical ceramic substrates (mostly alumina) onto which semiconductor integrated circuits are mounted. These substrates generally fit into the categories of single-layer or multilayer ceramic (MLC) devices that employ thick- or thin-film metal layers to serve as circuitry. A substrate is usually a base material, but in these devices it includes the composite of ceramic with the metal layer.

For *thick-film devices,* the Al_2O_3 substrate receives a layer of metal paste such as gold, platinum, palladium, or silver. The paste is printed by silk screen in a circuitry pattern (designed to carry the electron signal) onto either greenware or bisque. With the bisque substrate, the paste is sintered at about 850°C, which changes the paste structure into resistors and conductors; greenware and paste are fired together (cofiring). Computer-controlled lasers may trim resistors to exact resistance values before the chip is soldered onto the composite substrate. The layers begin as continuous slip-cast sheets of greenware, which are blanked to size, then silk screened. The ceramic substrates consist of 92% alumina and frit made up of alumina, calcia, magnesia, and silica. Frit is a form of glass that has been melted, cooled, then broken

Figure 7-58 American Ceramics Society specimen kit. Numbered items shown are: (1) Ceramic capacitors become polarized when under the flow of electrons. The electric charge distribution shifts such that one side of the ceramic becomes more positive and the other side more negative. This polarization is called the dielectric behavior of ceramics. Capacitors are required in all electrical circuits. Those that have a high degree of polarization (defined by the dielectric constant) are used to store electric charge. (AVX Corp.) (2) Cam follower roller. (Kyocera Industrial Ceramics Corp.) (3) Strip magnet-iron on elastomeric strip. Ceramic magnets are generally less dense than metallic magnets and are less costly. For high-frequency applications (e.g., microwave devices), ceramic magnets are made of rare earth ferrites, a ceramic material. (4) Armor tile. Armor to protect personnel and aircraft was developed for the Vietnam War. A ceramic material was chosen because of its hardness. Boron carbide (B_4C) was bonded to a Kevlar backing and covered with a fabric spall shield to make a composite material that defended against armor-piercing projectiles. Boron carbides hardness on the Knoop scale is 4.5×10^6 psi. (Coors Ceramics) (5) Silicon nitride ball bearings. (Cerbec, Inc.) (6) Ceramic foam filter. (Selee Corp.) (7) Photovoltaic cell on light that never requires new batteries. (8) Ruby crystal used in laser light pointer (9) Chemically toughened lens blank which will be ground to form eye glass lens. (American Ceramics Society)

into small particles. The substrates with metal paste are stacked into multilayers (5 to 35 layers), then sintered at about 1600°C.

Thin-film circuits composed of microthin layers (about 150 nm) of material, such as gold or nickel–chromium alloys, are produced through vacuum deposition onto ceramic substrates such as alumina, glass, or beryllia. Study of the references listed at the end of the module can focus on specific techniques such as sputtering, evaporation, and electron-beam deposition. Also see Section 9.7 for more detail.

7.12 GLASS

The American Society for Testing and Materials (ASTM) classifies glasses as inorganic products of fusion that have cooled to a rigid condition without crystallizing. They differ from glass ceramics in their lack of polycrystalline structure. Some authorities prefer not to consider glass as a ceramic because of its structure, and thus give it a separate classification.

7.12.1 Nature of Glass

The base raw material of glass is the very pure, white silica sand found in abundant supply in the central United States and other parts of the world. Although there are approximately 750 different glasses and glass ceramics, most can fit into six groups: soda-lime, lead-alkali, borosilicate, aluminosilicate, 96% silica, and fused silica.

The *vitreous state* of glass is mechanically rigid, like crystalline materials, yet it has an atomic space lattice that is amorphous, much like a liquid. Even though raw materials such as quartz sand (SiO_2) have a naturally crystalline structure (Figure 7-59a and Figure 7-10), when they melt, the lattice breaks up (Figure 7-59b). Slow cooling will allow some crystals to form, but normal cooling produces an amorphous vitreous or glassy structure (Figure 7-59c) such as the SiO_4^{4-} tetrahedron (silicon oxide ion with a net charge of –4).

(a)

(b)

(c)

Figure 7-59 Crystalline silica to amorphous glass. (a) Atoms of crystalline silica are held in place in a regular pattern or lattice: •—Silicon; ○—Oxygen. (b) When silica is molten, groups of atoms become detached are free to move relative to one another. (c) Atoms of silica in the glassy state are frozen in a random, or disordered, manner. (Corning Glass Works)

As discussed under phase diagrams, amorphous materials do not have clearly defined freezing or melting points, as do crystalline materials; rather, they harden or liquefy over a range of temperatures. By cooling molten glass, it becomes more viscous (Figure 7-60) and thickens to a working or softening point. Figure 7-61 plots the heating of crystalline raw materials and glass. Curve *a–b* shows an increase in volume until it becomes molten (liquidus temperature); increased temperature yields a corresponding increase in volume as the atoms expand. Glass is formed as the raw materials cool through curve *d–c–e*, but the volume remains higher at point *e* than when the raw materials were in a closely packed crystalline state. This curve represents a normally cooled glass.

Slower cooling permits some crystallization in the process known as ***devitrification,*** the conversion of glass or other noncrystalline solids into polycrystalline solids. This chemical reaction is opposite to that of *vitrification* (producing amorphous structures). Used commercially, devitrification produces high-quality, homogeneous, fine-grained, nonporous *glass ceramics* (see Section 7.12.3). Devitrification can occur only below the liquidus. To control the glass structure, it is important to know the location of the liquidus line (point), as with metals. High resistance to devitrification is desirable and prominent in commercial glasses.

A variety of raw materials yield glass compositions of assorted oxides. Oxides such as SiO_2 (silicon dioxide), B_2O_3 (boron trioxide), GeO_2 (germanium dioxide), and V_2O_5 (vanadium pentoxide) are glass formers. *Glass formers* or *network formers* are those oxides that promote the ionic linking or polymerization of oxide molecules in the glass compound. Network modifiers or fluxes such as lead, zinc, and alkali ions lower the liquidus temperature, improve

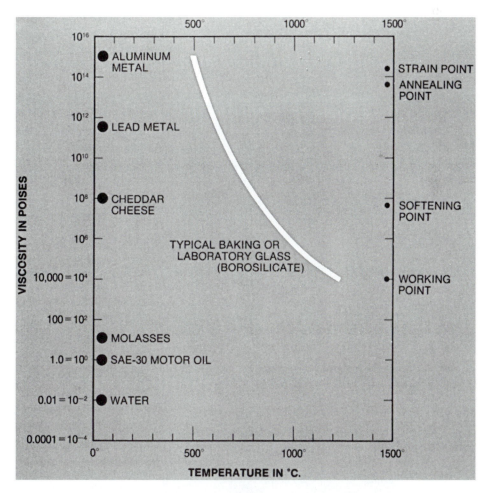

Figure 7-60 *Glass viscosity.* As glass is heated it becomes gradually less viscous. (Poise—absolute viscosity in Pascal seconds [Pa · s]. The measure of force required to overcome resistance to flow). (Corning Glass Works)

workability, and change thermal and optical properties. Other stabilizers, such as CaO (calcium oxide), improve chemical properties.

Hand *glass blowing* using a blow tube and molds dates back to around 200 B.C. Glass blowers were in evidence among the earliest settlers in Virginia. Hand blowing of glass is now used to produce glass articles for their aesthetic value primarily by hobbyists. Figure 7-62 is an example of their handiwork. The modern shaping of glass by blowing is performed by highly automated equipment. Jars, bottles, drinking glasses, incandescent light bulbs, and other pieces are produced by these automated techniques that use a mold to shape the product.

Figure 7-61   Glass formation.

Figure 7-62   A museum exhibit of hand-blown decorative glass. Note the figure of the young man for size comparison. The Romans invented glass blowing. Glass blowers were among the early settlers at Jamestown in 1607. The earliest glass factories in the United States relied on glass blowing into shaped molds. By the 1920s mold blowing was displaced by mechanical pressing into shaped tooling. Handcrafting methods are still employed for making consumer glassware items for artistic purposes.

Other shaping processes are continuous processes for making sheet and plate glasses for windows and tubing, and fiber-making processes for making fibers for such items as fiberglass composite materials, fiber optics, and insulation.

Heat treatment of glass can reduce internal stress or create high internal stress. High stresses could lead to breakage from relatively minor forces. **Annealing** of glass through slow cooling provides a homogeneous structure by reducing internal stresses to give isotropic (equal in all directions) properties. **Tempering** involves rapid cooling of the outer surface of glass while still in the plastic state. (See tempering of steel in Module 5 to note the differences in the meanings when applied to steels.) The tempering results in compressive stress on the surface and tensile stresses in the core (Figure 7-50). This condition occurs because the slower-cooling core tries to contract but is restrained by the rigid outer glass. The nonequilibrium condition causes tempered glass to fracture into small pieces rather than large splinters. Tempered glass is now required on doors in buildings and homes because of this safety feature.

Glass is a nearly perfect elastic material. At any stress below breaking stress, glass will return to its original shape when the stress is removed. Some plastic flow can be achieved, but the amount of stress must nearly equal its ultimate strength. Except when internal flaws are present, glass will fail in tension on the surface opposite the compressive stress. Whereas crystalline materials fracture along planes of slippage that may not be normal with the tensile stress, the amorphous structure of glass causes fracture to be normal to the stress. Tempering glass provides a compressive stress throughout the material, with the exception of a small amount of tensile stress in the center, which must be overcome before the tensile stresses can act on the surfaces (Figure 7-50). This is similar to prestressing concrete, which is also weak in tension.

7.12.2 Types and Properties of Glass

The six glass categories are diagrammed in Figure 7-63 and broken into soft glasses and hard glasses. The classification of hard or soft does not denote mechanical hardness but, rather, the ability to resist heat. **Soft glasses,** those that soften or fuse at relatively low temperatures, include soda-lime and lead-alkali. Soft glasses have lower heat resistance and a higher coefficient of thermal expansion than those of hard glasses. **Hard glasses** are borosilicate and aluminosilicate, and the hardest high-silica glasses are 96% silica and fused silica.

Soda-lime silica glass is the oldest glass; it dates back 4000 years and is most familiar to you in the form of window panes and bottles. It accounts for more than 90% of the glass produced. The typical composition is 74% silica (SiO_2), 15% soda (Na_2O), 10% lime (CaO), and 1% alumina (Al_2O_3). The least expensive glass, soda-lime glass has a composition of

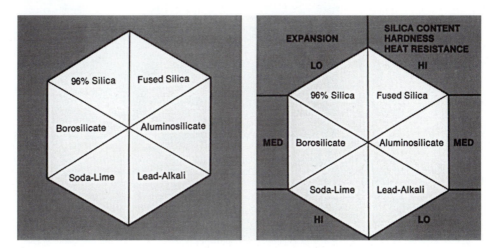

Figure 7-63 Categories of glass. *Hard glasses* are lower in thermal expansion and higher in both heat resistance and silica content than *soft glasses.* (Corning Glass Works)

oxides of silicon (from silica sand), calcium, and sodium. Although easy to form and cut into many designs, soda-lime glass has fair chemical resistance but cannot endure high temperatures or rapid thermal change (thermal shock). Because it shatters so easily into dangerous splinters, it is not recommended for doors or other glazings where persons might damage it. Thinner soda-lime glass, such as that used for small window panes, is *single-strength glass.* Increased thickness is referred to by strength: for example, *double-strength glass.*

    Lead-alkali silica glass is slightly more expensive than soda-lime, but both are considered soft glass; neither will endure high temperatures or thermal shock. ***Borosilicate glass*** is considered a hard (thermally) glass; it is the oldest type of heat-resistant glass. Pyrex is the familiar Corning trade name for this glass. Although three times as costly, ***aluminosilicate glass*** can sustain higher service temperatures than borosilicates and is similar in its ability to handle thermal shock. The ***96% silica glass,*** a valuable industrial glass, is capable of setting on a block of ice and having molten metal poured over it without breaking, attesting to its superior thermal hardness (Figure 7-64). ***Fused silica glass*** is composed of at least 99% silicon dioxide (SiO_2). This pure glass is the most costly and most resistant to heat. It is capable of service temperatures from 900° to 1200°C and has the highest corrosive resistance. It is superior for the transmission of ultraviolet (UV) rays.

    Colors are achieved in soda-lime, lead, borosilicate, and 96% silica glasses by the addition of elements that change the structure to absorb certain bands of the light spectra. For example, manganese ions (Mg^+) can produce purple glass, and iron yields green and brown bottle glass. Colorants such as iron are very undesirable in clear glasses and are kept at low levels by selecting clean sand deposits as a silica source.

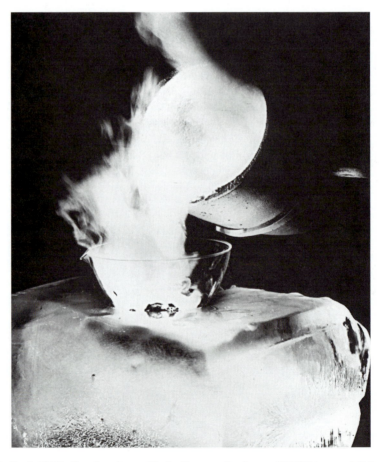

Figure 7-64 Bowl of 96% silica glass sits on a block of ice as molten bronze (2000°F) is poured. Failure to break attests to the ability of this *hard* glass to withstand thermal shock. (Corning Glass Works)

7.12.2.1 Specialty glasses. Using the six basic groups just covered, it is possible to develop glasses with unique properties to meet special needs. Some specialty glasses include tempered, optical, colored, sintered, glazing, fibrous, laminated, cellular, photosensitive, and light-sensitive glasses. Advances in optical glass technology have resulted from several societal and economic demands, including the need for greater transmittance (light transmission), safety (impact resistance), styling and convenience, and competition from the plastics industry. Consumer groups and the federal government are requiring improvements for energy efficiency in heating and cooling buildings. Specialty glass and glazing manufacturers are responding with a host of new materials and products. Alumina is a common ceramic packaging material for IC chips, because it hermetically seals the chip. Al_2O_3 is dispersed in a liquid binder, formed into sheets, and dried. The sheets are laminated by pressing and sintering to form a monolithic body. Another process sometimes chosen to seal the chip between the two ceramic sheets uses a refractory glass ($PbO–ZnO–B_2O_3$) with a melting point around 400°C. Burn-in testing may include elevated temperature testing in an oven at 125°C for 24 hours. The hard, stable nature of glass permits grinding and polishing of optical lenses for telescopes to 1/6000 of 1%. For styling and convenience, photochromic lenses provide eyeglasses that darken in bright light but lighten with subdued light. This ability to change the transmittance of light results from the use of silver ions. When photons of light strike these silver particles, they change from silver ions to metallic silver and absorb more light. When the light source is removed, they revert back to silver ions (Ag^+) (see *transmittance* in Section 9.4.5) and the lenses lighten in color. ***Photosensitive (PS)*** glass also changes its chemical characteristics when exposed to light. As mentioned, PC glass contains silver (Ag) in addition to halogens (Br and Fl). As glass cools, minute crystalline droplets (5–30 nm) of silver halide precipitate to form a dispersion so fine grained that it does not affect transparency or color of the glass. When the sun shines on the surface, droplets interact with the UV and violet wavelengths to form a tiny speck of pure Ag. The silver, in turn, absorbs a portion of the light and the glass darkens. As light disappears, the silver recombines with the halogen to form a transparent silver halide and the glass fades to colorless. Dissolving atoms in glass, exposing the glass to UV light, and heat treating produce many PC glasses with remarkable properties. The distinction between PC and PS glasses is that with PS glass the change is permanent; that is, it acts like a photographic film. Lead lanthanum zirconate titanate (PLZT) discovered by Haertling at Sandia National Laboratory in 1969 is a transparent polymeric crystalline form of an ***electrooptic ceramic.*** Used in a lens that allows light to pass through under normal lighting conditions and coupled with a photodiode sensor to detect light from a nuclear explosion, the diode transmits an electrical signal to the lens which changes the electric field in the lens making it opaque in less than 150 μs. ***Photodiode sensors*** detect light and send out an electrical signal. Gallium arsenide, a compound that can convert light directly into electricity (***photoconduction***), is used in LEDs and transistors.

7.12.2.2 Chemically toughened glasses. This type of ***tempered glass,*** also known as chemically tempered glass, is produced by using a molten salt bath in which eyeglass lenses are soaked for hours to achieve appropriate impact strength. During the long-term soaking, larger potassium ions from the salt bath replace sodium ions, as shown in Figures 7-48 and 7-50. This crowds the surface layer and causes compressive stresses. The impact strength of the tempered glass is then a product of its untempered strength and the amount of surface compression developed in chemical toughening. A chemically tempered glass is stronger than a heat-tempered glass.

7.12.2.3 Glazing. Glass is an important architectural material that adds beauty, and it can improve energy efficiency if properly used in design. Previously in this module, the word ***glaze*** or ***glazing*** described a hard, glossy finish or coating on an object for purposes of adding color and/or making the coated object impervious to moisture. The present meaning of these words refers to window glass. For example, *glazing point* or *glazing compound* refers to setting a glass in a window frame. A variety of types of glass serve as glazing, or glass windows. Most older buildings utilize single soda-lime sheet glass. Because costs for energy to supply heating and cooling have increased, improved glass technology has resulted in **insulating glasses** that are constructed of two or three panes, which are often fused together on the edges. Spaces between the sheets of glass are sometimes filled with gases to provide low thermal conductivity and improved sound dampening. Additives to glazing glass and reflective metallic/plastic films are also techniques to reduce the transmission of ultraviolet (UV) rays (see Section 9.4 and Figure 7-20).

7.12.2.4 Glass fibers.   These threads of glass are produced by various techniques, including drawing and blowing operations. The composition of glass fibers differs according to function. E-glass, used as reinforcing fiber, consists of calcium alumino-borosilicate which produces high tensile strength (3.45 GPa). Along with excellent electrical properties, dimensional stability, and low cost it is the most widely used fiber. Printed-circuit boards and reinforcing fiber for plastic materials are two large uses of this fiber. S-glass containing magnesium alumino silicate has greater tensile strength than E-glass (4.59 GPa). Combined with high thermal stability, S-glass finds primary use in the aircraft/aerospace industry. It is more expensive than E-glass. Fibers can be produced as thin, continuous filaments or discontinuous fiber segments. Starting with a 5/8-in.-diameter glass marble, you can produce 0.97 mile of filament. These thin fibers serve as reinforcers in standard, woven, and other forms for plastic resins, such as fiberglass. (Fiberglass is discussed extensively in Module 8.) Segmented or discontinuous glass fiber is also used as a very effective thermal insulation in buildings, on refrigeration and heating units, and in land and air vehicles. Its light weight and good insulation properties, coupled with low cost, make glass fiber very popular. It is used in both sheet and loose form. Glass fiber is also woven and matted into sheets for filters in heating and air-conditioning equipment.

Optical fibers, used to transmit light, are gaining wide acceptance in the communications field because they can be produced from the plentiful silica sand to replace heavier, bulkier, and more expensive copper and aluminum conductors. Optical fibers of glass and plastic have been used for many years in the medical profession and by engineers as inspection tubes. Flexible fibers on probes or endoscopes are inserted into the human body or into a motor and attached to a television system or magnifying lens to allow viewing of these otherwise inaccessible places. A single optical fiber the size of a hair has the potential to transmit several thousand voice signals, compared to the fewer than 50 voice signals that can be carried on a copper wire of the same size. For this reason, telephone, television, computer, and other communications systems are moving to the smaller coaxial fiber-optics cable to replace copper-stranded cables. Photons of electromagnetic radiation (light) replace electrons in signal transmission through these transparent optical fibers. This has increased the speed, the distance, and the amount of information (density) beyond all imagination. Figure 7-65 show some optical fibers and their connectors. Several forms of optical fibers exist, including glass cores clad with silica and coated with plastic (Figure 7-66). This high-purity glass fiber can transmit more than 95% of the light beamed into it for over 1 km and has a tensile strength of over 4137 MPa and good flexural strength.

EMT WEBLINK

Optical Properties and Electronic Structure—*http://www.lrsm.upenn.edu/~frenchrh/ opes.htm*

Cladding of fiber-optic cables with a transparent covering, whose index of refraction is lower than that of the fiber, prevents leakage of light. The transatlantic cable TAT-8 uses two pairs of glass fibers simultaneously. Because of the quality of the older fibers in transmitting signals, it is necessary to boost the signal every 50 km using repeater stations. With the use of *laser light* and thinner cable (10 μm in diameter), little energy is lost and boosting stations are seldom needed. Figure 7-67 is a sketch of an optical fiber showing the cladding and the coating of the cable, both of which consist of sophisticated glass-fiber materials designed to enhance and protect the signal transmission through the core of the fiber. Present-day technol-

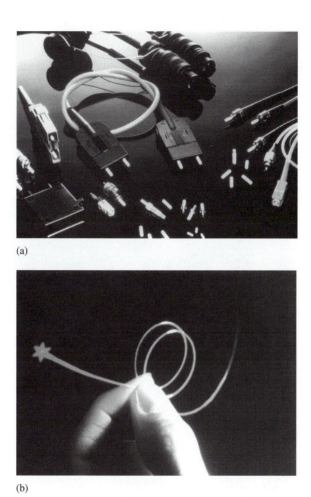

(a)

(b)

Figure 7-65 (a) Zirconia ferrules and examples of fiber-optic connectors in which they are used. (Coors Ceramics Co.) (b) Glass optical fibers have provided a technological breakthrough in the area of telecommunications. Information that was once carried electrically through hundreds of copper wires can now be carried through just one high-quality transparent silica (glass) fiber. Using this technology has increased the speed and volume of information that can be carried by orders of magnitude over which is possible using copper cable. The reliability of the transmitted information is also greatly improved with fiber-optic cables. (American Ceramic Society)

ogy limits the laser transmission of data over fiber-optic cable to some 50 mi using numerous lasers to keep the signals stable. This represents a data capacity of about 40 gigabits per second (40 Gbits/s or 40×10^9 bits/s) with the cable's capacity of 10 Tbits/s (10×10^{12} bits/s). Silkroad of San Diego, California, has developed a single-laser transmission that is now capable of transmitting 200 Gbits/s over 200 mi. This new technique is called *refraction synchronization communications,* which is bidirectional or capable of being used interactively.

Table 7-5 provides a comparison of the electrical properties of various insulating materials. The last material in the table, *steatite,* is an alumina-based ceramic used in spark plug insulators. Most of the glasses and ceramics have higher dielectric constants than those of plastics and rubber. Except for soda-lime glass, they offer greater volume resistivity than that of organic polymers. Table 10-13 provides a comparison of the mechanical, physical, and chemical properties of selected glasses. With petroleum getting scarcer, it is conceivable that glass and other ceramics will replace plastics in many applications.

7.12.3 Glass-Ceramics

The glass-ceramics group includes β-cordierite ($2MgO \cdot 2Al_2O_3 \cdot 5SO_2$) (Figure 7-68) and β-spodumene ($Li_2O \cdot Al_2O_3 \cdot 4SiO$), which are specific phases of glass that possess very low expansion coefficients and superior resistance to oxidation when subjected to high heat. As

Figure 7-66 Optical fibers. The fibermaking process uses highly purified glass layers that form as chemical vapors react within the silica tube; the tube and its contents are heated until they collapse into a solid glass rod, from which nearly 10 miles of fiber are drawn in the modified chemical vapor deposition process. Single-strand, optical wave-guide has the potential, through fiber-optic technology, to transmit 10,000 simultaneous telephone conversations when used in pairs. (Corning Glass Works)

C_o – Core
C_d – Cladding
C_l – Coating

Figure 7-67 The core consists of extremely high purity (flow free) silica glass with diameters that range from 5 to 100 μm. The containment of the light rays within the core is the job of the cladding while the outer coating protects both the core and the cladding. Each of these components are designed to accomplish certain objectives, and each is the product of sophisticated and advanced processing techniques. This rendering is not to scale.

discussed earlier (Section 7.12.1), glass will crystallize under slower cooling processes than are normally used in glass manufacture. Crystals form on the surfaces of glass with impurities. These surface crystals will cause failure of the glass part. However, controlled heat treatment can produce a fully crystallized material known as glass-ceramic.

Glass-ceramics are polycrystalline glass that has four to five times the strength of glass, with mechanical hardness about equal to that of tool steel. This polycrystalline glass is capable of developing structures that resist extreme thermal shock (rapid change from cold to hot, or vice versa). Figure 7-68 is an application of glass-ceramics that was selected and designed for an array of sophisticated properties. Cookware is a familiar form of glass ceramic. The fine crystalline structure of glass ceramics is achieved by the introduction of a nucleating agent into the ceramic compound (Figure 7-69). Typical nucleating agents include phosphorus oxide (P_2O_5), fluorides, titanium oxide (TiO_2), platinum, and zirconium oxide (ZrO_2). These nucleating agents are barely soluble in glass and remain in solution at high temperatures. At lower temperatures the agents precipitate out of solution to become seeds or nuclei around which crystals grow. Therefore, the process shown in Figure 7-70a involves allowing a ceramic composition to cool down as a normal glass *(b–c),* raise slightly to nucleation temperature and hold *(d–e–f),* and then raise and hold the temperature for crystal growth *(f–g–h).* Figure 7-70b shows the formation of a glass crystal from an amorphous opal glass with fluorine-rich droplets dispersed throughout. Through heat treatment, it is transformed to a fully crystalline structure. Crystallization of glass ceramics exceeds 95% and forms grains between 0.1 and 1 μm, which are smaller than in typical ceramics. For-

TABLE 7-5 COMPARISON OF ELECTRICAL PROPERTIES OF INSULATING MATERIALS AT ROOM TEMPERATURE

Thickness material	Intrinsic dielectric strength[a]				Dielectric constant	Volume resistivity ($\Omega \cdot cm$)
	mm	in. $\times 10^{-3}$	kV/cm	kV/in.		
Cellulose acetate	0.025–0.12	0.98–4.7	2,300[b]	5,840	5.5	10^{12}
Glass						
Borosilicate code 7740	0.10	3.9	4,800[c]	12,200	4.8	10^{16}
Soda-lime	0.10	3.9	4,500[c]	11,400	7.0	10^{12}
Soda-lead	0.10	3.9	3,100[c]	7,880	8.2	10^{14}
Mica, muscovite clear	0.020	0.79	3,000	7,620	7.3	10^{17}
ruby	0.10	3.9	8,200[b]	20,850		
Phenolic resin	0.012	0.47	2,600	6,600	7.5	10^{11}
	0.04	1.6	3,300[b]	8,380		
Porcelain, electrical			380[b]	965	4.4–6.8	10^{14}
Silica, fused			5,000[c]	12,700	3.5	10^{18}
Rubber, hard	0.10	3.9	2,150[b]	5,460	2.8	10^{13}
	0.30	11.8				
Porcelain, steatite—low loss			500[b]	1,270	6.0–6.5	10^{15}

Source: C. J. Phillips, *Glass, the Miracle Maker,* Pitman Publishing Co., New York, 1987.

[a]Intrinsic dielectric strength can be realized only under special test conditions and is very much higher than the working dielectric strength attainable in ordinary service. These data are listed for purposes of comparison.

[b]Values of S. Whitehead, *World Power,* p. 72, Sept. 1936.

[c]Values of P. H. Moon and A. S. Norcross, *Trans. AIEE* 49, 755 (1930)

Figure 7-68 Automobile internal engines contribute to photochemical smog by exhausting unburned and partially burned hydrocarbons (HC), compounds containing carbon and hydrogen. Most (65%) HC vapor in the atmosphere comes from exhaust gas. Catalytic converters reduce HC and carbon monoxide (CO) emissions in auto exhaust by coating a substrate with a catalyst such as platinum (Pt) and allowing oxidation of the exhaust products to take place producing hardless water, nitrogen (N), and carbon dioxide (CO_2). Cordierite, a glass-ceramic and one of over 1000 different silcates a magnesium aluminum silicate (MAS), is the main ceramic material used as a substrate for this application. One reason for its selection is its high temperature capability. (Corning Inc., Corning, NY)

mation of glass ceramic parts follows the processes used for glassmaking, including pressing, blowing, drawing, and centrifugal casting. One polycrystalline glass ceramic consists of mica crystals in an opal-glass matrix. Known by the trade name *Macor,* it can be machined with standard metalworking tools (Figure 7-71). Randomly oriented crystals about

Figure 7-69   A radome is a protective covering or fairing as well as a window for electronic guidance and protection equipment in aircraft, missiles, and spacecraft. It provides protection against high-velocity impacts of various particles in the atmosphere as well as rain. It functions as a window to allow the transmission and receipt of electromagnetic radiation from and to the craft (high transparency to radar). Most radomes are made of Pyroceram 9606 ™ glass-ceramic produced by Corning Glass Works.

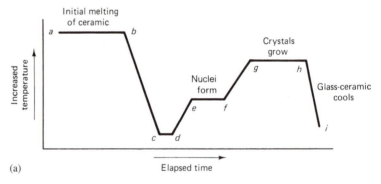

Figure 7-70   (a) Heating curve for crystal formation in glass ceramics.

20 μm in length and width keep fracturing locally instead of propagating (spreading), as would a crack in an amorphous glass or plastic.

7.12.4 Glass and Green Manufacturing

The rules for *green manufacturing* of 1) reduce, 2) reuse and 3) recycle apply to glass. Because glass is very hard and usually easily cleaned, it has many secondary uses. How often have you washed a mayonnaise or pickle jar and reused it to store another food in the refrigerator? The auto scrap industry finds uses for the glass in junked cars.

Recycling of glass, much like that of steel, involves a well established infrastructure. Most commercial glass, such as bottles and jars, requires recycled glass for its production. About 25% of these products consist of recycled glass. Like steel, glass will last forever in the *materials cycle.*

Resources saved by recycling glass include about 28% of the processing energy as a result of not having to extract glass from raw materials. Also, we save the raw materials for glass making including sand, soda ash, limestone and feldspar. About a ton of raw materials are saved for every ton of recycled glass used.

At the same time recycling reduces air pollution that would result from processing the raw materials. This amounts to a reduction of about 28 pounds of particulate from the air for

(b1)

(b2)

(b3)

(b4)

Figure 7-70   (b) Formation of glass ceramics. (b1) Droplet-imbedded parent-glass material before heat treatment. (b2) First intermediate crystal phase seen after heating to 750°C. (b3) Second intermediate crystal phase seen after heating to 825°C. (b4) Beginning of mica-crystal formation seen after heating to 850°C. (b5) Fully crystallized Macor glass ceramic seen after heating to 950°C. (Five micrographs from Corning Glass Works)

(b5)

every ton of glass made or a savings of about 17%. Mineral waste from extracting the raw materials is also saved. Recycling glass also keeps it out of municipal landfills.

  EMT WEBLINK

Recycling Glass—<*http://envirosystemsinc.com/glassfac.html*>

Figure 7-71 Machinable glass ceramic. (Corning Glass Works)

7.13 SMART CERAMICS MATERIALS

Among the smart ceramics materials, *electrorheological (ER)* and *magnetorheological (MR)* fluid use inorganic particles and liquid crystal displays that change their structure with the application of an electrical current. They are covered in other parts of this book. *Artificial nerves* and *sensors* are constructed of optical fibers, and are part of fiber-optic technology. The glass optical fibers can be imbedded in a host of materials or bonded to surfaces. The fibers are connected to instruments to receive signals. As a smart materials system, movement of embedded fibers breaks a light path to reveal material movement or rupture due to various external forces, such as earthquakes shaking a building or bridge.

Optical fibers embedded in composites with "smart skins" employ various light-sensing systems to indicate material damage due to variation in pressure, such as may be experienced by the surface of a model aircraft in a wind tunnel. A similar approach can be used to test designs of automobiles. *Piezoelectric materials,* explained in Module 9, include lead zirconate titanate (PZT), which is the most popular piezoceramic. PZT acts as a control device in intelligent materials systems for active acoustic attenuation (reducing sound), active structural damping (reducing vibration), and active damage control (see Figure 7-72).

Another of the smart ceramics, *magnetostrictive actuator materials,* respond to magnetic fields in a manner different than PZT, which changes its microstructure from the introduction of electrical currents. This group of intelligent materials includes Terfenol-D, a compound with the rare earth element terbium at. no. 65. Terfenol-D is controlled with magnetic fields that bring on an alignment of magnetic domains, thereby causing controlled expansion of the magnetostrictive material. It can be used in vibration damping systems; high-torque, low-speed rotary motors; and hydraulic actuators.

7.14 MATERIALS SELECTION APPLICATION

A material(s) is needed to provide good thermal-shock resistance to serve as a liner for a high-temperature vacuum furnace. The liner will be in contact with molten aluminum. Strength re-

Figure 7-72   "Smart Skis." All structures are subject to vibrations which move or deform them. Snow skis are structures similar to a cantilever beam. Unwanted vibrations cause the edges of skis to lift off the snow resulting in reduced control and responsiveness. Piezoelectric ceramics convert mechanical vibration energy into electrical energy which through a shunt, a tuned circuit dissipates the energy in the form of light and heat, reducing the vibrations. Active Control eXperts, Inc. (ACX) has designed and developed a line of piezo control modules for damping vibrations for military applications and for the commercial market such as mountain bikes and Formula One race cars. (American Ceramic Society)

 EMT WEBLINK

Smart skis—*http://www.acx.com./cool_smartski.html*

quirements for the liner are minimal, but resistance to crack propagation is desired. The requirements are as follows:

1. Based on your knowledge of structures, bonding, and properties of ceramics, make a list of properties that the selected material(s) should have, based on the information given above.
2. For each property listed in (1), explain why each property was chosen (i.e., justify your choice in terms of how the property adds to the overall characteristics of the material selected).
3. List suitable candidate materials that possess all or some of the properties listed in (2).
4. Select the material(s) and justify your selection. Consult the textbook tables of data or outside sources for appropriate data to include with your justification.

Hint: Multicoatings and additives such as fibers can be considered in making your selection.

7.15 LABORATORY ACTIVITIES AND CLASSROOM DEMONSTRATIONS WITH CERAMICS

The *Experiment in Materials Science, Engineering and Technology (EMSET) CD ROM* provides numerous experiments and demonstrations of ceramics as a result of a cooperative effort by scores of educators and materials specialists from industry. See Prentice-Hall's website for ordering information on the latest edition of ***Experiments in Materials Science (CD-ROM),*** at *www.prenhall.com*

APPLICATIONS & ALTERNATIVES

The Use of Glass-Ceramics in Dentistry*

Glass-ceramics are highly crystalline ceramics with some residual glass matrix prepared by controlled crystallization of glasses.[1] The unique properties of glass-ceramics allows them to be formed into complex shapes in the glassy state and to retain their finished dimensions during firing. Along with their aesthetic qualities, their high strength and toughness, insensitivity to abrasion damage, chemical durability, thermal-shock resistance, and polishability make them appropriate materials for use in dental restorations.

The use of glass-ceramics in dentistry was first proposed by W. T. MacCulloch in 1968 to provide an alternative to the dental porcelains used at that time. He produced denture teeth using a continuous glassmolding process. He also proposed the fabrication of dental crowns and inlays by centrifugal casting of molten glass.

Glass-ceramics are used in the fabrication of crowns, inlays, onlays, veneers, and inserts for composite restorations. The glass-ceramics used in these applications are based on three different crystalline phases: tetrasilicic fluoromica, β-quartz, and leucite.[2,3,4]

2mm above gum line

*This section is a portion of a paper entitled "Dental Applications of Ceramics," written by Laurie A. George, ADA, Health Foundation, and published in the book entitled *Bioceramics: Materials and Applications Ceramic Transactions,* Vol. 48, Westerville, OH: American Ceramic Society, 1995.

[1]P. W. McMillan, Chapter 1, 2nd ed., Academic Press, London, 1979.

[2]Piddock and A. J. E. Qualtrough, *Glass Ceramics—An Update,"* J. Dent., 18, 5, 227–235, 1990.

[3]L. A. George, F. C. Eichmiller, and R. L. Bowen, "An intrinsically colored microcrystalline glass-ceramic for use in dental restorations," *Ceramic Bulletin,* 71, 7, 1073–1076, 1992.

[4]J. K. Dong, H. Lüthy, A. Wohlwend, and P. Schärer, "Heat-Pressed Ceramics: Technology and Strength," *Int. J. Prosth.,* 5, 1, 9–16, 1992.

One of the first commercially available castable glass-ceramic materials used in dental restorations was a fluoromica containing glass-ceramic. This glass-ceramic is used in crowns, inlays, onlays, and veneers. The microstructure of this glass-ceramic consists of small, interlocking fluoromica crystals that have a platelike morphology. The interlocking of these crystals results in greater strength and reinforcement. These crystals also help to divert fractures on a microscopic level and permit the glass ceramic to be tooled with rotary instruments without physical deterioration. This material is supplied in the form of silica-based glass ingots containing magnesium fluoride, which serves as a nucleating agent for devitrification. The refractive index of the crystals is similar to that of glass surrounding them; this condition helps to maintain translucency in the glass-ceramic casting.[5] Several important properties of dental restorative materials, including density, hardness, wear resistance, thermal conductivity, and radiographic density, are similar for this glass-ceramic and tooth enamel.

Restorations are formed from this glass-ceramic by using laboratory techniques similar to the lost-wax casting process used in cast-metal restorations. A wax pattern is invested in a phosphate-bonded refractory. The glass ingot is melted in a crucible in the muffle of the specially designed casting machine. The resultant molten glass, which is heated to 1357°C, is centrifugally thrown into the prepared mold. Following formation, the casting is **cerammed** (heat treatment where structure changes from amorphous, to part crystalline and part amorphous) at a temperature of 1075°C for 6 hours.[6] The aesthetics of the restoration is enhanced with thin layers of tinted porcelain fused to the surface through sintering.

[5]D. G. Grossman, "Cast Glass Ceramics," *Dent. Clin. North Am.,* 29, 4, 725–739, 1985.
[6]DICOR Technique Manual, 14, Dentsply International, York, 1988.

A second glass-ceramic used for the fabrication of crowns, inlays, and veneers is a leucite-containing glass-ceramic (i.e., the crystalline portion of this glass-ceramic is composed of leucite crystals). This glass-ceramic is formed from a glass composed of SiO_2, Al_2O_3, K_2O, Na_2O, B_2O_3, CeO_2, CaO, BaO, and TiO_2.[7] This material is supplied in preformed ceramic tablets.

The initial steps in the fabrication of restorations from this glass-ceramic are similar to the techniques used to produce porcelain fused to metal restorations. These steps include fabrication of the impression, waxing of the object and spruing it with a wax cylinder, and investing with a phosphate-bonded investment. The set investment, preformed ceramic tablets, and a small alumina press cylinder are placed in a burn-out oven heated to 850°C at 3–6°C/min and held for a minimum of 90 minutes. The investment ring, ceramic tablets, and the heat-saturated plunger are placed in the specially designed furnace at 700°C. Automatic pressing, vacuum, and heating processes are activated. The furnace is heated to 1100°C at 60°C/min and held at this temperature for 20 minutes. The pressing process then begins and continues until the press cylinder movement is less than 0.3 mm in 3 minutes. Once the pressing process is complete, the investment is bench-cooled to room temperature and the object is removed from the mold. Shading of the restoration is achieved with a surface staining technique or a layer technique.[8]

Improvements in the physical properties of dental composite restorations, including reduced microleakage and increased stiffness and durability, have resulted from the incorporation of a glass-ceramic insert into the restoration. ***Dental composites*** are a two-phase system that uses polymer as a matrix for the dispersion of ceramic particles.

The glass-ceramic used to form these inserts is largely composed of beta-quartz solid solution. This glass-ceramic is formed from a lithium aluminosilicate glass system. The inserts are formed by using standard pressing techniques. After formation, the inserts are heat-treated to convert the glass to a glass-ceramic. Following conversion to a glass-ceramic, the inserts are coated with an organofunctional silane that provides for chemical coupling between the insert and the dental composite used for the restoration. The inserts are available in three shapes and sizes, corresponding to the most frequently encountered cavity configurations. These shapes include a round cylinder, a tapered cylinder, and an L-shaped inlay. The

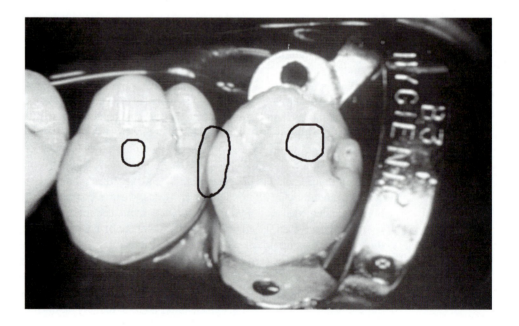

[7]J. K. Dong, H. Lüthy, A. Wohlwend, and P. Schärer, "Heat-Pressed Ceramics: Technology and Strength," *Int. J. Prosth.*, 5, 1, 9–16, 1992.

[8]G. Beham, "IPS-Empress: A New Ceramic Technology," Ivoclar-Vivadent Report, 6, 1, 1990.

insert fills approximately 50% to 75% of the cavity; thus the overall restoration takes on more of the advantageous properties of the insert. The inserts are intended primarily for use in posterior dental restorations.

Using standard composite filling techniques, the dentist places the inserts in a cavity preparation prefilled with composite. The excess composite extruded from the preparation as a result of insert placement is removed, and the restoration is cured with a dental curing light. The handle portion (sprue) of the insert is removed, and the restoration is finished with a series of diamond dental burs and polishing paste.

The refractive index of this glass-ceramic is similar to that of the composite resin, thus providing restorations with good aesthetic properties. The x-ray capacity of the glass-ceramic is nearly equal to that of enamel, providing for accurate interpretation in dental diagnosis. The inherent color of the inserts is similar to that of composite to allow for appropriate aesthetics.

SELF-ASSESSMENT

7-1. From the Pause & Ponder, cite examples of traditional ceramics both in early American and modern society. Contrast them to advanced ceramics while citing examples of types, applications and the rational for using the newer ceramics.

7-2. How does biomimetics represent the kind of appeal offered by ceramics for materials selection? What other attributes make ceramics competitive with metals and polymers?

7-3. Ceramics are bonded by what types of bonding mechanisms?

7-4. A material's resistance to the removal of its surface material is known by what terms?

7-5. Name one test used to measure the wear resistance of a material.

7-6. What is a suitable set of units to express thermal conductivity?

7-7. At what location does stress concentrate when a material cracks under load?

7-8. Name one test used to measure or test a material's toughness.

7-9. Explain why the machining of fine-grained ceramic parts reduces their strength and toughness more than that of large-grained ceramic parts.

7-10. What material is added to brittle ceramics to increase their resistance to crack propagation?

7-11. List two properties that a material should have to be a candidate for resisting thermal shock.

7-12. Describe one technique for reducing the emissivity of a roof covering on a two-story house located along a treeless section of an oceanfront property.

7-13. The trial-and-error method of selecting a material for a particular application is also known by what other name?

7-14. In ceramic processing the term *green* is used to describe a particular part. What condition does this term describe?

7-15. At least three ceramic processes involve the application of pressure as well as heat in the sintering of ceramic materials. Name one such process.

7-16. What symbol is used to denote absolute temperatures based on the Celsius scale?

7-17. What are the chemical name and chemical formula for alumina?

7-18. Explain the difference between cement and concrete.

7-19. Surface modification techniques are developed and employed to accomplish what particular goals?

7-20. What is the main difference between CVD and CVI processes?

7-21. Explain how a material is affected by machining with either dull or improper tools.

7-22. Why do ductile materials withstand impact better than brittle materials?

7-23. There are many reasons why fibers add toughness to a composite material. Discuss one such reason.

7-24. The word *cermet* stands for what two words?

7-25. What happens to the volume of ZrO_2 as it is heated past 1200°C?

7-26. Hydraulic cements require what addition for them to set or cure?

7-27. A radioactive material has a half-life of 10,000 years. Using the pound as a unit of mass, how much of the material is stable after 20,000 years?

7-28. Superabrasives possess high hardness and wear resistance. What other property do they have? Name a material from which they are made.

7-29. In terms of microstructure, why do crystalline ceramics have different properties from metals that are also crystalline?

7-30. Match the name of the ceramic to its chemical formula or composition.

(a) Portland cement	(1) Al_2O_3
(b) Soda-lime silica glass	(2) SiO_2
(c) Alumina	(3) 99% SiO_2
(d) Quartz	(4) $Al_2(Si_2O_5)(OH)_4$
(e) Fused silica glass	(5) 74% SiO_2, 15% Na_2O, 10% CaO, 10% Al_2O_3
(f) Kaolinite	(6) C_3S, C_2S, C_3A, C_4AF

7-31. Explain **(a)** how crystals are formed in glass, **(b)** when crystals are undesirable, and **(c)** when crystals are desirable.

7-32. **(a)** What effect does tempering have on glass?

  (b) How does light cause photochromic lenses to darken?

  (c) How does toughened glass obtain improved impact strength?

7-33. Match the ceramic to the application for which it is best suited:

(a) Tempered glass	(1) Grinding and polishing
(b) Fused silica	(2) Furnace windows
(c) Ferrite	(3) Protect metal from oxidation
(d) Glaze (porcelain)	(4) Computer memory discs
(e) 96% silica glass	(5) Storm door glazing
(f) Diamond	(6) Contain hot acids

7-34. Consider the characteristics of the general classes of ceramics, plastics, and metals. List the class that is generally superior in the following characteristics; then write down a specific material and value to illustrate its superiority: **(a)** stiffness, **(b)** chemical resistance, **(c)** abundance of raw materials, **(d)** tensile strength, **(e)** lowest coefficient of thermal expansion, **(f)** thermal insulation, **(g)** hardness, **(h)** impact strength, **(i)** flexural strength, **(j)** density.

7-35. Express a cutting tool speed of 1000 in./min in units of meters per second.

7-36. List at least 10 applications for ceramic materials within a home.

7-37. In prestressing a composite material by the layering technique, explain how the surface or coating material's coefficient of thermal expansion should differ, if at all, from that of the substrate or matrix material.

7-38. Using the data from Table 7-3, determine the SI units of TSI for graphite.

7-39. The terms *soft* and *hard* used to describe glass refer to what property of glass?

7-40. What must be done to introduce color into glass? Does the structure of glass affect its color?

7-41. What is the purpose of soaking eyeglass lenses in a molten salt bath?

7-42. Explain the difference between a glass-ceramic material and the ceramic glass.

7-43. In studying the relationship of a finished product in service with its properties and the processes used to produce the product from its basic material(s), one interesting observation is that the final properties desired in a product are usually the cause of the problems that need to be overcome in its processing. An example taken from ceramics deals with the desired properties of high hardness and temperature resistance. These are the very properties that must be overcome in processing the product. Using metals and/or polymers, cite a general statement that supports the foregoing observation.

7-44. When selecting a ceramic material to serve as a heat insulator, would a material with a high density qualify? Discuss your answer.

7-45. Name one property that is most desired of a ceramic material for use as an abrasive for a grinding wheel.

7-46. When choosing a ceramic material for a cutting-tool insert, would a ceramic-composite material using SiC fibers be suitable? List the properties required of this material.

7-47. Ceramic armor, initially developed for the Vietnam War and also used in the Gulf War, was a composite made up of boron carbide (B_4C) and fiberglass plus aramid fiber matting. The ceramic has a density of 2.4 g/cm^3. Is this a low- or a high-density ceramic? Explain why you think this ceramic was chosen for this application.

7-48. Match the correct ceramic to its favorable characteristics. Some will have the same characteristics.

(a) Boron nitride	(1) Low cost
(b) Graphite	(2) Machinable
(c) Al$_2$O$_3$	(3) Best lubricity
(d) Glass ceramic	(4) Maximum hardness

7-49. Express a fracture toughness of 186 Mpa $\cdot \sqrt{m}$ in terms of psi $\cdot \sqrt{in}$.

7-50. Write an abbreviated notation for a CMC that consists of SiC whiskers in an alumina matrix.

7-51. Explain the difference between setting and hardening of cement.

7-52. In terms of their atomic bonding, explain why metals are good conductors and most ceramics are good insulators.

7-53. Explain some of the goals sought when designing advanced, high-tech, high-performance, or engineering ceramics.

7-54. Discuss the advantages of improved low-tech cements, such as CBCs, now and in the near future for developing nations and a greener materials cycle.

7-55. Name two advantages of metals over ceramics.

7-56. The text shows several photomicrographs and sketches of methods used to toughen ceramics against cracks and impact. Sketch one of those techniques.

7-57. Discuss the possibilities of zero coolant machining. Mention P/M, surface engineering, protective coatings, and environmental concerns.

7-58. Describe a smart ceramic including examples of materials and applications.

7-59. Support a position for using glass rather than plastics in terms of meeting the goals of green manufacturing.

OBJECTIVE QUESTIONS

7-1. A broad classification of ceramic materials that use artificial raw materials, exhibit specialized properties, and require sophisticated processing.

    a. Traditional ceramics   **b.** Advanced ceramics   **c.** Cements   **d.** Glass

7-2. Ceramics based primarily on natural raw materials of clay and silicates.

    a. Traditional ceramics   **b.** Advanced ceramics   **c.** Cements   **d.** Glass

7-3. A major objective in designing new ceramics is to overcome _____ .

    a. hardness   **b.** brittleness   **c.** reactivity   **d.** conductivity

7-4. Generally, ceramics compete well with metals when there is a need for lower weight, lower density, high shear strength, and good tensile strength.

    T (true)

    F (false)

7-5. Metals are usually superior to ceramics in areas such as impact strength, ductility, and plastic flow.

    T (true)

    F (false)

7-6. The term for ceramic's density that includes its porosity and both crystalline and noncrystalline phases.

    a. Bulk density   **b.** Open density   **c.** Closed porosity   **d.** Green density

7-7. A polymer with a backbone of SiO; the ASTM designation is SI.

    a. Silicon   **b.** Silica   **c.** Silicone   **d.** Silicates

7-8. The element Si.

    a. Silicon   **b.** Silica   **c.** Silicone   **d.** Silicates

7-9. The ceramic SiO.

    a. Silicon   **b.** Silica   **c.** Silicone   **d.** Silicates

7-10. Because of their bonding, different atoms, and unit cell structure, ceramics offer high resistance to shear and are good insulators.

T (true) **F** (false)

7-11. Using Figure 7-14, place in order from good to poor the following materials in terms of wear resistance.

   a. Cast iron   **b.** Silicon carbide   **c.** Urethane   **d.** 96% alumina

7-12. Transformation zones such as ZrO_2 serve what purpose in ceramics?

   a. Thermal shock resistance   **b.** Fracture toughness

   c. Improved conductivity   **d.** Tensile strength

7-13. Ceramic that has been molded but is still in a plastic condition.

   a. Frit   **b.** Bisque   **c.** Glaze   **d.** Greenware

7-14. Ceramic that has been fired but is porous and will absorb liquids.

   a. Frit   **b.** Bisque   **c.** Glaze   **d.** Greenware

7-15. The glassy finish such as porcelain fired onto porous ceramic.

   a. Frit   **b.** Bisque   **c.** Glaze   **d.** Greenware

7-16. An environmentally friendly concrete known as pervious concrete offers which major advantage?

   a. Uses cementitious materials   **b.** Porous structure allows natural seepage of water

   c. Offers superior toughness   **d.** Relies on glass-fiber reinforcement

7-17. A very hard glass that withstands extremes of thermal shock.

   a. Soda-lime glass   **b.** Polycrystalline glass

   c. Lead glass   **d.** 96% silica glass

7-18. A ceramic-structured material produced by controlled heat-treating of glass that has nucleating agents added to the mixture.

   a. Soda-lime glass   **b.** Polycrystalline glass

   c. Lead glass   **d.** 96% silica glass

7-19. Glass that was rapidly cooled while in the plastic state to place the surface in compression and the core in tension.

   a. Annealed glass   **b.** Tempered glass

   c. Soft glass   **d.** Chemically toughened glass

7-20. A method to immobilize hazardous waste by encasing it in glass at the site of the waste.

   a. Biomimetics   **b.** Densification   **c.** In situ vitrification   **d.** CVD

7-21. Bonding in ceramic materials is _____ .

   a. ionic   **b.** covalent   **c.** secondary

7-22. Oxidation numbers designate the number of _____ taking part in the bonding of elements during a chemical reaction.

   a. Protons   **b.** Atoms   **c.** Electrons   **d.** Ions

7-23. Mullite, an intermetallic compound, is a ceramic with many applications that take advantage of its properties dealing with _____ .

   a. high strength   **b.** high conductivity

   c. high temperature   **d.** porosity

7-24. Normal sintering depends on convection heating as well as heating by _____ .

   a. contact   **b.** radiance   **c.** conduction   **d.** microwave

7-25. Spraying a protective metal on foam uses _____ .

   a. plasma arc   **b.** HVOF   **c.** wire arc

7-26. The ceramic with the lowest conductivity.

   a. Thoria   **b.** Al_2O_3 **c.** ZrO_2

7-27. The limit on the number of coatings that can be applied to a substrate.

   a. Two   **b.** One   **c.** No limit

7-28. In machining, the tool costs are _____ .

   a. greater than the coolant-related costs

      b. less than the coolant-related costs

      c. the same as the coolant-related costs

7-29. The process in which high temperatures cause some ceramics to melt into a glassy matrix.

      a. Sintering   **b.** Atomization   **c.** Vitrification   **d.** Annealing

7-30. The process in which clay is heated to change a soft material to a hard thermoset.

      a. Sintering   **b.** Atomization   **c.** Vitrification   **d.** Annealing

7-31. Most often the structure of this material is amorphous.

      a. Diamond   **b.** Cement   **c.** Ceramic   **d.** Glass

7-32. The measure of energy absorbed in creating a unit area of a new crack ahead of an existing crack.

      a. Impact   **b.** SENB   **c.** Flaw   **d.** Work of fracture

7-33. A ceramic metal-matrix composite (densely packed SiC particles in Al_2O_3) produced by the DIMOX™ process that gives maximum wear resistance.

      a. CVI   **b.** RHM   **c.** ALANX   **d.** CFCC

7-34. Extremely hard metals such as tungsten carbide or molybdenum that are used for cutting tools.

      a. Cement   **b.** RHM   **c.** Cermet   **d.** Abrasives

7-35. A ceramic, such as masonry mortar, used to bond bricks together.

      a. Cement   **b.** RHM   **c.** Cermet   **d.** Abrasives

7-36. Ceramic particles floating in a liquid such as water.

      a. Slurry   **b.** Cement   **c.** Amorphous   **d.** Composition

7-37. In terms of green manufacturing, which IS NOT an advantage that glass containers would have over plastics in containers?

      a. Save weight in transportation

      b. 25% of waste glass used to make glass

      c. 1 ton of raw materials saved for every ton of recycled glass used to make glass

      d. Glass will last forever in the materials cycle

7-38. A category of ceramics such as piezoelectric materials act as sensors and for acoustic attenuation.

      a. glass-ceramic   **b.** hard ceramic   **c.** traditional ceramic   **d.** smart ceramic

7-39. From the Applications & Alternatives section discussing glass-ceramics for dental applications, what would be one of their major advantages?

      a. complex shape that retain finish   **b.** low tech processing

      c. lower costs   **d.** brittle nature of insert

REFERENCES & RELATED READINGS

AMERICAN SOCIETY FOR TESTING AND MATERIALS. *Annual Book of ASTM Standards* Part 17. *Refractories, Glass and Other Ceramic Materials; Carbon and Graphite Products.* Philadelphia: ASTM, 1991, or current edition.

ANDERSON, RICHARD M. "Testing Advanced Ceramics," *Advanced Materials and Processes,* Vol. 135, No. 3, 1989.

ASM International. *Engineering Materials Handbook,* Desk Edition. Materials Park, OH. 2000.

BEHRENDT, DONALD R., and MRITYUNJAY SINGH. *Melt-Infiltration Process for SiC Ceramics and Composites.* Lewis Research Center (LEW-15767), *NASA Tech Briefs,* August 1994, p. 59.

BESMANN T. M., B. W. SHELDON, R. A. LOWDEN, AND D. P. STINTON. "Vapor-Phase Fabrication and Properties of Continuous-Filament Ceramic Composites," *Science,* Vol. 253, September 6, 1991.

BONDS, JAMES W. et al. "Wire-Arc Spraying of Metal onto Insulating Foam." Rockwell International Corp. for Marshall Space Flight Center, *NASA Tech Briefs,* May 1995, p. 84.

BUNNELL, L. ROY. "Tempered Glass and Thermal Shock of Ceramic Materials," *NEW: Update 91,* Oak Ridge, TN: Oak Ridge National Laboratory, November 12–14, 1991.

CORNING GLASS WORKS. *All about Glass.* Corning, NY: CGW, 1968.

CORNING GLASS WORKS. *Properties of Glasses and Glass-Ceramics.* Corning, NY: CGW, 1973.

CRAIG, DOUGLAS F. "Structural Ceramics," *NEW:Update 91,* Oak Ridge, TN: Oak Ridge National Laboratory, November 12–14, 1991.

CSELLE, TIBOR. "New Directions in Drilling." Applications Research & Development, Guhring, Inc., Brookfield, WI, *Manufacturing Engineering,* August 1995, pp. 77–80.

DOUBLE, DAVID D. "Chemically Bonded Ceramics; Taking the Heat Out of Making Ceramics," *Journal of Materials Education,* Vol. 12, No. 5/6, 1990, p. 353.

EVANS, JAMES W., and LUTGARD C. DEJONGHE. *The Production of Inorganic Materials.* New York: Macmillan, 1991.

GEORGE, LAURIE A. "Dental Applications of Ceramics," in *Bioceramics: Materials and Applications Ceramic Transactions,* Vol. 48, Westerville, OH: American Ceramic Society, 1995.

HARPER, SCOTT. "New-Age Concrete Acts Like a Big Filter." *The Virginian-Pilot,* Sept. 7, 1995, p. B3.

JORDAN, GAIL W. "Adapting Archimedes' Method for Defining Densities and Porosities of Small Ceramic Samples," *NEW:Update 90,* Gaithersburg, MD: NIST, November 12–14, 1990.

LEHMANN, RICHARD L. "Primer on Engineering Ceramics," *Advanced Materials and Processes,* Vol. 141, No. 6, 1992.

MACCHESNEY, J. B. "The Materials Development of Optical Fiber: A Case History," *Journal of Materials Education,* Vol. 11, No. 4, 1989, p. 321.

MCCOLM, I. J. *Ceramic Hardness.* New York: Plenum Press, 1990.

MPR. "Net and Near-Net Shaped Ceramic Composites Formed by Lanxide Technology," *Materials and Processing Report,* Vol. 3, No. 3, 1988.

OMETITE, OGBENI O., MARK A. JANNAYA, AND RICHARD A. STREHLOW. "Gelcasting: A New Ceramic Forming Process," *American Ceramics Society,* Vol. 64, No. 12, 1985.

PFAENDER, HEINZ G. *Schott Guide to Glass.* New York: Van Nostrand Reinhold, 1983.

REED, JAMES S. *Introduction to the Principles of Ceramic Processing.* New York: Wiley, 1988.

RICHERSON, DAVID W. *Modern Ceramic Engineering: Properties, Processing and Use in Design,* 2nd ed. New York: Marcel Dekker, 1992.

RICHERSON, DAVID W. *The Magic of Ceramics.* Westerville, OH: The American Ceramic Society, 2000.

ROGERS, CRAIG A. "Intelligent Materials Systems: The Dawn of a New Materials Age." Center for Intelligent Materials and Structures, Virginia Polytechnic Institute and State University, 1992.

ROY, DELLA M., ed. "Instructional Module in Cement Science," *Journal of Materials Education.* Materials Education Council, 1985.

SWAIN, BERTA. "Microwave Processing of Ceramics," *Advanced Materials and Processes,* Vol. 134, No. 3, 1988, p. 76.

URQUHART, A. W., G. H. SCHIROKY, AND B. W. SORENSON, "Ceramic Composites for Gas Turbine Engines via a New Process," ASME 89-GT-316, *Gas Turbine and Aeroengine Congress and Exposition,* Toronto, Canada, June 4–8, 1989.

Periodicals

Advanced Materials and Processes   *Journal of Adhesion Science and Technology*
Ceramic Bulletin   *Manufacturing Engineering*
Ceramic Monthly   *MRS Bulletin*

Module

8

Composite Materials

After studying this module, you should be able to do the following:

8–1.   Sketch the Venn diagram that shows how the sets of three main Family of Materials groups intersect with reinforcers to form subsets of polymer matrix composites (PMCs), ceramic matrix composites (CMCs), and metal matrix composites (MMCs). Also diagram how the PMC subset expands into a wide array of composites materials.

8–2.   Select examples of composites and advanced composites, sketch their macrostructure with labels of constituents (matrix, interface zone, etc.) , and justify their use over monolithic materials. Recall reasons to use composites and cite the applications of advanced composites.

8–3.   Discuss carbon and graphite fibers in terms of their definitions; production; and important properties including coefficient of thermal expansion (CTE), specific strength, and wear (erosion) rates.

8–4.   Use diagrams to explain isostress and isostrain, and use the rule of mixtures (ROM) to express the elastic modulus in tension for fiber-reinforced plastic (FRP). Explain the value of CMCs and MMCs using examples such as molten-metal infiltration (PRIMEX) processes.

8–5.   Describe the relationship among methods of fabricating composites and the properties that develop from the processes. Explain the factors in machining and repairing advanced composites.

8–6.   Calculate the specific strength and specific stiffness of given materials and determine when these properties are major reasons for material selection.

8–7.   Explain the uses of advanced composites in automobiles. Also recall techniques for recycling composites.

8–8.   Use composite databases to determine the applications of select composites. Describe NDE techniques to evaluate composite designs and manufacture.

8–9.   Use EMT Weblinks* to explore sites for new developments and evolving concepts related to composite materials.

*EMT Weblinks listed throughout the book will link you to Internet sites related to topics being covered. The dynamic nature of the Web brings frequent changes, so some of these sites, although available at the time of writing, may not be up now.

PAUSE & PONDER

Economics, often the overriding factor in product and system design, takes into account all aspects of the materials cycle outlined in Module 1. To review some considerations in a case study of the use of advanced composites, we turn to two programs: the Advanced Composites Technology (ACT) program and the Boeing 777 program. ACT, a scitech partnership between NASA and the U.S. aeronautical industry, works with a broad objective to develop appropriate technology to design and manufacture the primary structure for commercial aircraft. Cost-effective use of lightweight advanced composites is the emphasis. The goal of ACT is to use advanced composites so that they will provide a 25% reduction in costs below aluminum structures.

Gaining enough confidence in advanced composites to use them on commercial aircraft requires considerable research and development. Among the tools of research, microscopy provides a means to examine the microstructure of failed composites in an attempt to determine the reasons for failure.

Initially, carbon-fiber-reinforced plastic composite materials had low damage resistance and fracture toughness. These properties severely limited their application particularly in the aircraft industry. One effort to rectify this problem led to a study of the relationship between microstructure and materials performance. One outcome was the development of several specimen preparation methods that revealed the size and distribution of individual resin phases and subsequently the role played by these resin phases in the improvement of damage resistance. One final result was that changes in formulations and manufacturing processing could be evaluated from a microstructural viewpoint.

The technique of ***thin-film microscopy**** was one development that provided greater morphological detail than traditional reflected-light microscopy. This ***thin-section microscopy*** uses transmitted, polarized light. Specimen sections only 1 to 5 μm thick are used since the composites materials are opaque. At these levels of thickness, the materials become optically transparent. The result is that transmitted light methods reveal more details of the morphology of polymers and similar materials than are observable using any other microscope techniques available. Ultrathin microscopy also reveals information about fracture processes in structural laminates which leads to improved composites that resist delamination. It also supplements other failure analysis techniques. Fiber-reinforced laminates with multiple plies, honeycomb-core sandwich structures with laminated face sheets, and polymer matrix composites with finely dispersed particulates are likely candidates for thin-section techniques. As discovered in the preparation of metallic specimens for observation by a metallograph, good microscopy is 99% sample preparation. Preparing a flat, scratch-free, strain-free cross-section without introducing artifacts (human-made changes) is both a science and an art. To grind/polish a material such as a fiber composite material with its relatively hard fibers (carbon, Kevlar, etc.) embedded in a soft, polymeric matrix certainly requires special care. Most composite microscopy involves looking for porosity and microcracks. Microcracking is the result of stresses which surpass the ultimate strength of the polymer, or polymer/fiber interface bond. From a careful study of the microcracking, scientists can determine the durability of composites in service and investigate the relationship of process parameters and environmental exposure to micrographs. High-temperature thermoplastic composites behave like metallic structures in that mechanical properties change when these composites so-

*A short course on "Microscopy of Fiber-reinforced Polymer Composites" containing numerous experiments and demonstrations as well as color photomicrographs, one of many valuable composite resources is available on the *CD-ROM Experiments* in Material Science, Engineering and Technology, Version 2 from Prentice-Hall at 1-800-526-0485.

Figure 8-1 Stress-Strain Before Your Eyes. Shows a thermoplastic resin with fiber reinforcement under impact loading. Ultrathin sectioning and polarized light microscopy reveal in a series of photomicrographs (a thru d) the effects of delamination and other fracture processes in this structural laminate composite material. (a) The composite material was impacted at 3000 in-lb/in with a 0.625-in. diameter spherical indentor. The specimen dimensions were 6 in. by 4 in. The characteristic fracture path propagates into the laminate like a radially expanding spiral staircase. Magnification 8.4X (b) Shear deformation along the interplies is the main energy absorption mechanism. In contrast, the intraplies absorb less energy due to fiber spacing constraints. In this photo, spreading of the damage zone in the intraply inhibits delamination. Magnification 250X (c) As shear strains in the interply reach a critical value, the resolved tensile strains initiate cracks 45° to the shear plane, becoming hackles as the crack tips link together. Hackle spacing gives qualitative information about resin toughness and strength. Magnification 500X. (d) A large interply deformation zone runs parallel to the intraply fracture. Fibers alter the local strain field by restricting polymer alignment. Magnification 500X. (e) In contrast to K3B Laminate, a toughened bismaleimide laminate exhibits a deformation region that does not span the interply. The hackles are generally smaller and more jagged in appearance due to the more brittle nature of the resin. However, the specific strength of the bismaleimide resin is higher. (Luther M. Gammon, Boeing Materials Technology)** (*continued*)

lidify at different rates. Controlling the cooling rates, and microstructural phenomenon such as crystal sizes, shape, distribution, etc., determine, to a great extent, mechanical properties of the composite. Several photomicrographs taken with thin-section microscopy were provided by Luther M. Gammon, Boeing Materials Technology, Boeing Commercial Airplane Group, Seattle, Washington, who was instrumental in the development of this technique. Figure 8-1a through d, entitled "Stress/Strain Before Your Eyes," consists of five photomicrographs

---

**A short course on "Microscopy of Fiber-reinforced Polymer Composites" containing numerous experiments and demonstrations as well as color photomicrographs, one of many valuable composite resources is available on the *CD-ROM Experiments* in Material Science, Engineering and Technology, Version 2 from Prentice-Hall at 1-800-526-0485.

777 Lightweight Composite Structure

777 composite structure:
- Toughened materials for improved damage resistance and damage tolerance
- Designed for simple, low-temperature bolted repairs
- Corrosion and fatigue resistant
- Weighs less (composite empennage saves over 1,500 lb compared with prior aluminum structure)

- Toughened graphite
- Graphite
- Hybrid
- Fiberglass

Rudder
Elevator
Leading and trailing edge panels
Fin torque box
Stabilizer torque box
Inboard and outboard spoilers
Floor beams
Wing-to-body fairing
Flaps
Flaperon
Outboard aileron
Outboard flap
Wing landing gear doors
Engine cowlings
Wing fixed leading edge
Trailing edge panels
Strut fwd and aft fairings
Main landing gear doors
Nose gear doors
Nose radome

Figure 8-1 (f) The application of composites showing the advantages of reduced-weight, corrosion-free, and durable structures. (Courtesy of The Boeing Company.) The Public Broadcasting System (PBS) series "21st Century Jet" provides an excellent in-depth case study of the product development process, from design through testing, manufacturing, and delivery. The series includes many examples of engineering materials technology. Call PBS at 1-800-344-3337 to obtain further information on the video series. Also consult the book *21st Century Jet: The Making and Marketing of the Boeing 777*, Karl Sabbagh, Scribner, 1995.

with various defects that illustrate this technique of ultrathin sections with polarized light microscopy for a composite material undergoing impact testing. Such studies among other R&D efforts led to decisions to make large use of composites in commercial aircraft.

The Boeing 777 is the first of a new family of wide-body airplanes that makes extensive use of carbon-fiber-reinforced plastics (CFRP) advanced composites (Figure 8-1f). While weight savings, and corresponding fuel savings, from the use of composite materials would provide large economic benefits over the life of the aircraft, the poor economic health of the airline industry demands that the initial cost of the finished airplane be reduced if composites are to be selected. New design and innovative manufacturing techniques using polymer matrix composites are required to make them competitive because the cost per pound of composites is higher than aluminum, and current production costs of composites is twice that of aluminum aircraft structures.

Concurrent engineering of the industrial partners involves parallel developments in materials and their processing, determination of the properties for new materials, design and manufacture of individual elements and the structural components they form, and continual analysis of elements and components. This integrated approach to advanced composite technology for primary aircraft structures will carry through the next generation of commercial transport aircraft.

Boeing's decision to use advanced materials for its totally redesigned 777 resulted in the application of various lightweight composites, which amount to 10 times more mass of composites than in their 757. These composites are approximately 9% of the new plane's structural weight. The 777 success, or lack thereof, may well predict the future of advanced composites in transport aircraft. Uses of fiberglass, as seen in Figure 8-1e, range from the wing fixed leading edge to the strut fairings. Graphite composites' use continued, as on the Boeing 767 and 757, on rudders, elevators, and flaps. Hybrid composites served as wing-to-body fairing.

The *toughened graphite* (Figure 8-1e) is a new, toughened epoxy composite used on the fin torque box, stabilizer torque box, and floor beams. Toughened graphite addresses a major concern for composites because of its outstanding resistance to impact: nearly seven times tougher than the previous CFRP.

The costs of all aspects of design, tooling, materials, and manufacturing for the 777's wing were greater than those of a competing aluminum wing. Why? For example, in design and its related testing, more costs from additional analysis spring from a lack of experience and the databases that come with the use of aluminum. Of course, more experience with CFRP should bring reduced costs. The ACT program faces a formidable challenge to reduce cost for advanced composites. The benefits of using advanced composites, as fully developed in this module, can be enjoyed by the entire aerospace industry worldwide and will provide spin-offs for the auto industry, building construction, road construction, and other fields. The 777 represents the global nature of modern manufacturing. Components of the airplane are produced around the world and shipped to the United States and on to Everett, Washington, for assembly.

8.1 INTRODUCTION

A *composite material* or *composite* is a complex solid material composed of two or more materials that, on a macroscopic scale, form a useful material. The composite is designed to exhibit the best properties or qualities of its constituents or some properties possessed by neither. The combining of the two or more existing materials is done by physical means, as opposed to the chemical bonding that takes place in the alloys of monolithic solid materials. A true composite might be considered to have a matrix material completely surrounding its reinforcing material in which the two phases act together to produce characteristics not attainable by either constituent acting alone. Note that the insoluble phases or main constituents in a composite do not lose their identity. This is not true of solids, which are metallic alloys or copolymers whose phases are lost to the naked eye because these phases are formed as a result of natural phenomena. In other words, steel is a multiphase metal alloy but is not considered a composite. A metallic alloy of Al and Cu, with the addition of Al_2O_3, is considered a dispersion-strengthened copper composite in which a dilute copper/aluminum alloy powder is exposed to an oxidizing environment and the aluminum is oxidized within the copper matrix. The aluminum oxide particles are inert and act as inhibitors of dislocation movement and thus strengthen the alloy at high temperatures. By adding niobium (10%) to this same extruded composite in the form of a uniformly distributed powder, even greater hardness and strength are produced with only a minimal reduction in electrical conductivity. Two phase alloys are not considered composites by some because their phases are not formed in separate processes but originate in a single manufacturing process.

The above discussion reveals that the definition of a composite is still somewhat arbitrary (see Section 2.2.5). A broad definition of composite materials includes the naturally occurring composites, such as wood, as well as the synthetic or human-made composites. A recent example of the development of composites in the medical field is cited. A mineral paste composed of monocalcium phosphate monohydrate, α-tricalcium phosphate, and calcium carbonate, dry mixed, to which a sodium phosphate solution is added to form a paste, is surgically implanted into acute bone fractures and hardens within ten minutes after injection. This paste holds fractured bones in place while the native bone remodeling process replaces the implant with living bone. This process causes the in-situ formation of the mineral phase of bone. Attaining a compressive strength greater than that of long bones, the paste's tensile strength of about 2.1 MPa is comparable to long bones. This new material eliminates the need for heavy and uncomfortable casts.

In answer to the question, ***Why use composites?*** one can reply, in part, as follows:

1. To increase stiffness, strength, or dimensional stability
2. To increase toughness (impact strength)
3. To increase heat-deflection temperature
4. To increase mechanical damping
5. To reduce permeability to gases and liquids
6. To modify electrical properties (e.g., increase electrical resistivity)
7. To reduce costs
8. To decrease water absorption
9. To decrease thermal expansion
10. To increase chemical wear and corrosion resistance
11. To reduce weight
12. To maintain strength/stiffness at high temperatures while under strain conditions in a corrosive environment
13. To increase secondary uses and recyclability, and to reduce any negative impact on the environment
14. To improve design flexibility

The objective of this module is to present a fundamental knowledge of composites; the terms used in this specialized field; the nature, structure, and properties of composites; and an arbitrary classification of the major types. Examples of the applications of some composites in a limited number of industries are included with a view toward providing the readers, many of whom occupy positions in industry that require a degree of expertise in the selection of materials, with a base of knowledge of composites from which to make intelligent selection decisions.

8.1.1 Development of Composites

Composite materials have been used from earliest time. Mongol bows of the thirteenth century utilized a materials system consisting of animal tendons, wood, silk, and adhesives. The ancient Israelites used straw to reinforce mud bricks. The early Egyptians fabricated a type of plywood. Medieval swords and armor are examples of our present-day laminated metal materials. Nature has provided composite materials in living things such as seaweed, bamboo, wood, and human bone.

Such materials were recognized for their strength and light weight when used in the construction of the first all-composite, radar-proof airplane—the Mosquito—circa 1940. Composite materials in the form of sandwich construction showed that primary aircraft structures could be fabricated from these materials. World War II saw the birth of glass-fiber-polyester composites for radomes and secondary aircraft structures, such as doors and fairings, which were designed and placed into production. Glass-fiber composites were recognized as valid space materials when they were picked for fabrication and production of Polaris-submarine missile casings.

In the 1950s, fiber technology identified the need for fibers that could compete in strength and stiffness when the state-of-the-art development led to high-performance glass fibers such as the S-type. In the late 1950s, research efforts focused on lightweight elements in the search for fibers of even greater strength that could compete successfully in the marketplace with aluminum and titanium. Boron fibers were the first result of this effort (1963), followed by carbon, beryllium oxide, and graphite. These filaments are surrounded by a material such as aluminum that serves as a matrix. Beryllium-oxide fiber technology developed short fibers, whereas boron research developed continuous filaments with greater strength properties. These developments, made through the collective efforts of government, NASA, industry, and universities, gave rise to **advanced composites.** Figure 8-2a, a Venn diagram, shows how ceramics, metals, and polymers are integrated into sets and subsets of advanced composites. The sets of CMC, MMC, and PMC multiply further into subsets based on the form of reinforcement and the general structural arrangement, as illustrated in Figure 8-2b for fiber-reinforced composites, a subset of PMC. **Nanocomposites,** a new subset of CMC, MMC, or PMC, are composites with a substrate that is reinforced with nanometer (10^{-9}m) size particles which enhance the substrates' physical properties. For example, a new nanocomposite, a subset of PMC, with a thermoplastic olefin (TPO) substrate has been developed that uses submicron particles of clay. An addition of only 5% of this natural clay provides significant benefits over traditional thermoplastic materials and fillers such as talc, mica, and $CaCO_3$. Stiffness characteristics are equivocal to 25–35% additions of talc-filled materials. In addition, this is accomplished without increasing weight or losing temperature properties or opacity. This new material was developed specifically for automobile applications which should improve the recyclability of automobiles in the future. The other major ingredients needed for successful production—engineering design and manufacturing—kept pace with developments in materials and resulted in the continuous introduction of new composite materials for use throughout industry.

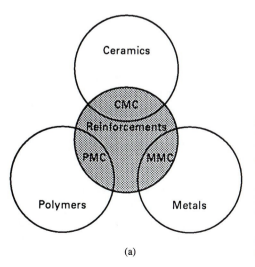

(a)

Figure 8-2   (a) Composite materials. The design process for engineering materials involves the selection of components from major members of the family of materials, which, when combined into a composite material, will possess a mix of desired material properties that will satisfy the requirements of a particular application. A ceramic selected as a matrix material combined with a reinforcement in any form and quantity produces a ceramic matrix composite (CMC). A similar combination produces a metal matrix composite (MMC) or a polymer matrix composite (PMC). (*continued*)

RRIM: reinforced reaction injection molding
SMC: sheet molding compound
TMC: thick molding compound
BMC: bulk molding compound
XMC: x-pattern molding compound (trademark of PPG ind.)
DMC: dough molding compound

(b)

Figure 8-2   (b) Composite materials—their matrices and reinforcements produce an infinite array of materials.

Figure 8-2a also represents the continual search for materials systems that can withstand even greater loads under hostile environments. These composite materials, with ceramic and metal matrices reinforced with similar materials, are being developed and tested for use in many high-temperature applications.

The Advanced Composites Technology (ACT) program, introduced in the Pause & Ponder, (see Fig. 8-3) has the broad objective of developing the technology to design and manufacture a cost-effective and structurally optimized, lightweight, composite airframe primary structure. In its first phase, the program developed three major manufacturing methods for fabricating these cost-effective composite structures, namely, ***stitched dry preform, textile preform,*** and ***automated tow placement.*** Production costs of a composite structure, once twice the costs of a corresponding aluminum airframe structure, continue to creep down as new technology evolves. It is reasonable to speculate that materials designers will eventually use materials that do not yet exist. This materials effort calls for a "synergistic advancement" of technology where a partnership between materials scientists and materials users will, for the first time, be put into practice.

Down on the ground, designers of the full range of products, which include clothing, athletic gear, appliances, furniture, office supplies, home entertainment systems, packaging, computer systems, boating, camping gear, and housing, have called on materials technology for new composite systems that make better products. Shoes are one example of a line of products making use of new composite systems. Look at inflatable basketball and tennis shoes and cut open a pair of jogging shoes to find the air chambers and wide variety of polymers in the form

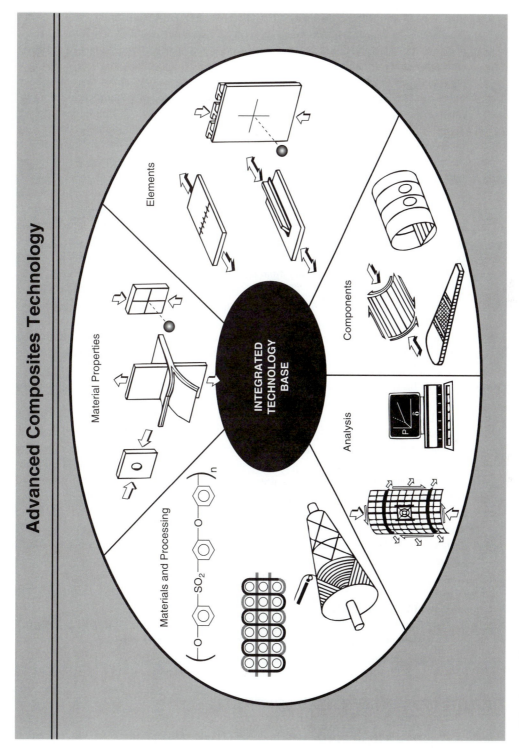

Figure 8-3 ACT program uses an integrated technology base and concurrent engineering to promote reduced costs, improved manufacturing techniques, and high-quality products. (NASA-LaRC.)

521

of rigid plastics, soft foams, dense foams, nylon mesh, and hard rubber. Designers used CAD technology coupled with materials technology to keep the marketplace humming with ever-improving products. The customer should be the winner.

A composite consisting of 80% advanced polymer and 20% aluminum has been developed for automotive applications by XCORP, Beverly Hills, California, at a cost said to be comparable to that of stamped steel. Designated "Composite X," it is reportedly resistant to abrasion and wear, and offers superior crashworthiness. The material is one of a group XCORP calls *environmental composites,* because they are easily recycled and are made using low energy and a clean manufacturing system. The company has applied for more than 50 patents covering the composition of the materials, proprietary fastening techniques, molding systems, production, and recycling methods. XCORP has combined these technologies to produce its Series 2 SuperCar, which has an all-composite driveshaft, aluminum/polymer sandwich bumpers, and a variety of other composite parts. The Partnership for a New Generation of Vehicles (PNGV) includes the Automotive Composites Alliance. Throughout this module you will see examples of PNGV efforts for improving composites.

The National Composites Center (NCC) in Kettering, Ohio, studies new methods for (a) infrastructure (building construction related to roads, bridges, etc.), (b) automobiles, and (c) aerospace. The NCC and others who bring new developments in building technology continue to look to composite materials. A building material—a fiber-reinforced polymer composite (graphite fibers embedded in an epoxy resin)—was used to build a vehicle bridge which can support loads heavier than a comparable bridge made of traditional materials. First tested in 1995 on a 26-foot span, the "Tech" (Materials Technology for the 21st Century) vehicle bridge (33 feet in length and capable of supporting more than 36 tons) opened in Butler County, Ohio, in 1997. It is the first all-composite vehicle bridge in Ohio and the first fully instrumented bridge in the United States. Its performance can be remotely monitored by computer. There are 230,000 bridges across the country that are structurally deficient or functionally obsolete. Replacing or reinforcing these bridges with new building materials would prolong the lives of these structures—tripling their life expectancies, improving highway safety, and saving billions of tax dollars.

Due to the many advances in recent years in composite systems, not only as spin-offs from the nation's space program but also as a consequence of the growing energy conservation and concern about the depletion of the world's nonrenewable materials resources by more and more citizens, the emphasis placed on the group of composite materials in this module is deemed appropriate. This is not an attempt to downgrade the importance of the many other composite materials, without which our economic well-being would suffer adverse consequences. In industries' new thrust for "design for manufacturability" and "design for automation," designers seek to reduce the number of components in a product, eliminate mechanical fasteners, and design to facilitate assembly by robots. Composites offer great promise to help meet at least some of these goals.

EMT WEBLINKS

Automotive Composites Alliance—*http://autocomposites.org/*
National Composite Center—*http://www.compositecenter.org*

8.2 CONSTITUENTS OF FIBER-REINFORCED COMPOSITES

A *fiber-reinforced composite* is a material system made primarily of varying amounts of particular fiber reinforcement embedded in a protective material called a matrix, with a coupling agent applied to the fiber to improve the adhesion of the fiber to the matrix material. Many of the terms in this modern materials technology field originated in the textile industry. A few such terms are selected for brief definitions to facilitate an understanding of these highly important engineering materials. See the glossary at the end of this module, Section 8.13, for other definitions.

High-performance fiber-reinforced composites exhibit higher structural performance than that of commercial E-glass and polyester composites. The degree of performance over a fiberglass composite depends on the fiber, its orientation and loading, and the matrix. A significant processing aspect that affects performance is the degree to which the fibers are saturated with wet matrix, leaving few or no voids. The matrix's ability to adhere to the fibers, as well as its inherent toughness, also make a contribution. As for fibers, their loading (density); continuous lengths (high aspect ratios); and orientation, including straightness, degree of knitting, and interlacing, all play a role in contributing to the properties.

In the study of atomic structure, we learned that a material's strength in a certain direction or plane is directly proportional to the type of atomic density and the atomic bonding existing in that particular direction or plane. If the atoms are tightly packed and are bonded with strong primary bonds, the material will possess a high degree of strength. We also learned that properties, particularly in brittle materials, are adversely affected by defects or flaws. The larger the flaw, the less the ability to withstand any degree of load. As an example of the application of this knowledge to the production of carbon fibers (to be discussed), we recall that carbon, an element, is polymorphic. In Figure 2-14, the structure of graphite was sketched showing two basal planes, consisting of atoms covalently bonded, which were connected in a lateral direction by weak secondary bonds. In the production of carbon fiber, for example, the original precursor material is a similar state. As part of the process, the basal planes are oriented along the axis of the fiber, thus placing these atoms with the strongest bonding in the best position to withstand a longitudinal force. Defects in the atomic structure or macrodefects on the surfaces of the fibers are dealt with by reducing the amount of cross-sectional area in the fiber. The chance of a defect remaining when the diameter of a fiber decreases is minimized considerably. This last statement explains the ongoing program to produce very fine fibers, down to the nanometer range.

8.2.1 Fiber

A *fiber* is the basic individual filament of raw material from which threads or fabrics are made. Fibers constitute one of the oldest engineering materials. Jute, hemp, flax, cotton, and animal fibers have been used from the earliest days of history. Today, fibers can be organic (plant, animal, and mineral), synthetic (human-made from a polymer or ceramic material), or metallic. ASTM further defines a fiber as having a length at least 100 times its diameter, with a minimum length of at least 5 mm. A fiber can be a filament or a staple. Manufactured polymeric fibers were discussed in Module 6. *Filaments* are long, continuous fibers, whereas *staple* fibers are less than 150 mm in length. Most natural fibers are in staple form, whereas synthetic fibers may exist in both forms.

Whiskers are formed from extremely thin single crystals of high perfection with a large length-to-diameter ratio used to reinforce composite materials. They are virtually flaw free and possess exceptionally high strength. Whiskers are, however extremely expensive to produce and difficult to handle in forming composite materials. SiC and Si_3N_4 ceramic whiskers inhibit crack propagation (fracture toughness) and strength. These two properties increase with an increase in whisker/fiber content in the composite materials such as metal matrix composites. Whiskers are examples of discontinuous reinforcements, as are chopped fibers of alumina or carbon and particulates of SiC and alumina.

Continuous fibers are essentially infinite in length and extend continuously throughout the matrix in a composite material. *Discontinuous fibers* are not less than 3 mm in length and tend to orient themselves in the direction of resin flow as a composite material is processed. Particulate composites with reinforcements of SiC and alumina are an example of discontinuous fibers. Whisker-reinforced alumina is a composite used to make inserts for cutting tools. Other examples of discontinuous reinforcements are chopped fibers of alumina and carbon. Discontinuous reinforcements in metal matrix composites (MMC) do not add any problems to the shaping of these composites by metal-forming operations such as rolling.

Other terms relating to fibers and fiber composites can be found in Section 8.13.

For fiber-reinforced composites to improve their strength, the fibers must be able to transmit their load-carrying abilities to the surrounding matrix. If a fiber ends in a matrix, then the fiber cannot transmit its stress to the matrix and thus has little strength reinforcement.

A formula has been devised to determine the *critical length of a fiber (lc).* This formula uses the smallest of either the fiber diameter, its tensile strength, the shear strength of the matrix, or the fiber-matrix bond strength. A fiber that is longer than the critical length (l > 15 lc) is termed a *continuous fiber.* Metal matrix composites (MMCs) with a matrix of alumina and continuous boron fiber reinforcement are used on the space shuttle, *Orbiter.* Carbon fibers in an alumina matrix are used on the Hubble Telescope. Other examples of continuous fibers are SiC, alumina, and refractory metals.

$$l_c = \frac{\sigma_f d}{2\tau_c}$$

σ_f = tensile strength

d = fiber diameter

τ_c = shear yield strength

8.2.2 Matrix

The functions of a *matrix,* the binder material, whether organic, ceramic, or metallic, are to support and protect the fibers, the principal load-carrying agent, and to provide a means of distributing the load among and transmitting it between the fibers without itself fracturing. When a fiber breaks, the load from one side of the broken fiber is first transferred to the matrix and subsequently to the other side of the broken fiber and to adjacent fibers. The load-transferring mechanism is the shearing stress in the matrix. Typically, the matrix has a considerably lower density, stiffness (modulus), and strength than those of the reinforcing fiber material, but the combination of the two main constituents (matrix and fiber) produces high strength and stiffness, while still possessing a relatively low density.

The matrix serves as the structure, keeping the reinforcing fibers at the same distance from the bending axis, while the filler provides the desired properties. There is no significant chemical reaction between the two phases except for the bonding action at their interface. If this were not true, any reaction between the two materials would have a varying negative effect on the inherent properties of the filled composite.

The structure or matrix of a filled composite can take a porous or spongelike structural form. This network of open pores can be impregnated with a variety of materials, from plastics to metals to ceramics, and the final shape of the filled open-pore composite is formed during the processing. Finally, the matrix can have a predetermined shape and size formed by an open honeycomb core made of metal impregnated with a ceramic filler for high-temperature applications.

Maximum operating temperatures for resins as matrices in fiber-reinforced composites are listed in Table 8-1 to give some idea of their limitations. No attempt is made to introduce

TABLE 8-1　OPERATING TEMPERATURES FOR COMMON POLYMER RESINS

Resin	Maximum temperature (°C)
Polyester	Room temperature (RT)
High-performance thermoplastic polyesters	150
Epoxies	200
Phenolics	260
Polyimides	300
Polybenimidazole	Above 300

the element of time in these figures, other than to state that the time a material is subjected to these extremes in temperature is a critical factor that must be taken into consideration in any design. An example taken from the literature illustrates how this heat-resistant property is described for a series of polyimide molding compounds, Tribolo PI-600, available from Tribol Industries. Such compounds perform at 316°C continually and 427°C intermittently. A typical molding retains 56% of its mechanical strength after 100 h at 316°C. Weight loss is 2.1% after 500 h.

Both thermosetting (TS) plastics and thermoplastic (TP) plastics are used with hybrid composites, defined later. Epoxies are the chief TS, but thermoset polyester is a strong contender, particularly for auto applications. Epoxies have a wide range of properties. Intermediate modulus epoxies are used in hybrids. They have good high-temperature properties but suffer from moisture absorption when exposed to temperatures near 180°C. Poly(ether ether ketone) (PEEK), a thermoplastic that is tougher and potentially less expensive to manufacture than conventional resins, is used as doors of F-15E fighters and T-38 trainer aircraft. Polyimides find use for extended operations near 260°C and above. TS polyester, formulated to be hard, brittle, or tough, is used with chopped glass in many applications, seen later in Figure 8-4e.

TP resins are the latest matrix materials, and an example of TP use for automotive components is seen later in Figure 8-4f. Nylon; polysufone (PSU); poly(phenylene sulfide) (PPS); and poly(phenylene terephthalate) (PET), better known as thermoplastic polyester, are used in hybrids and advanced composites. Also, bismaleimide (BMI), which is manufactured like conventional composites, is able to withstand higher operational temperatures and has been tested similar to the manner described above for PEEK. Improved resin-transfer molding (RTM) resins have a longer pot life and lower viscosity. User-friendly prepreg resins have better controlled reactivity and an extended out-time.

Resins can be divided into two main groups: structural pastes and structural film adhesives. Pastes can be stored for a year at room temperature (RT) and cure at RT. Four versions are available: one is aluminum filled for easy handling and high compressive strength, another is a thixotropic nonmetallic for composite surfaces and random structures, another is an unfilled version designed for wet-layup composite repairs, and the final version has high shear modulus at elevated temperatures and can be stored 6 months at RT. Structural film adhesives resist moisture in the bond line and are extremely resistant to severe environmental exposure. They can be cocured with the composite structure and features at 250° or 350°F. Thixotropic metals are discussed in Module 3. Figure 3-2 contains a micrograph of a semisolid, thixotropic alloy.

Polyimide molding compounds capable of withstanding continuous-use temperatures ranging from 480° to 600°F and intermittent service at 800°F are now available. A typical molding retains 56% of mechanical strength after 100 h at 600°F. Weight loss is 2.1% after 500 h. All are compatible with reinforcing fibers.

Glass ceramics, polycrystalline solids produced by the controlled crystallization of glass, have highly desirable properties for service as matrix materials. Dimensional stability, thermal stability, heat resistance, corrosion resistance, and the ability to maintain strength at high temperatures make these materials attractive despite their inherent brittleness. Their coefficient of thermal expansion is zero up to about 450°C. Glass ceramic matrices act just the opposite to polymer- and metal-matrix materials because the matrix fails first, having the lowest strain to failure. Failure occurs when the matrix viscosity becomes too low to transfer the load to the reinforcement. Thermal compatibility between the matrix and fibers must be good. If the coefficients of thermal expansion are not similar, the brittle matrix will not comply with any differences. Also, the need for high forming temperatures to bring about matrix flow among the reinforcing fibers requires thermochemical compatibility between the fibers and matrix material to prevent reactions at the fiber-matrix interface that produce unwanted interphases that degrade the fibers and reduce their strength.

Some of the most common glass-ceramic matrices are litha–alumina–silicate (LAH; $Li_2O_2–Al_2O_3–SiO_2$); silica, SiO_2; and $BaO–SiO_2–Al_2O_3–Si_3N_4$ with SiC (Nicalon), the most commonly used reinforcing fiber for glass-ceramic matrices.

(a)

(b)

(c)

(d)

Figure 8-4 (a) Cross-section of fiber-reinforcing fibers in a matrix. (b) Coated fibers create two interfaces that act like grain boundaries inhibiting crack propagation. (c) Two fibers prevented from making direct contact with each other stop crack propagation. (See also Figure 8-12 and discussions of methods to increase toughness of CMCs.) (d) Short chopped glass fibers and voids (produced by air entrapment) in polyethemide ultrathin section transmitted differential interference contrast −500×/20× obj. (Luther M. Gammon, Boeing Commercial Airplane) (*continued*)

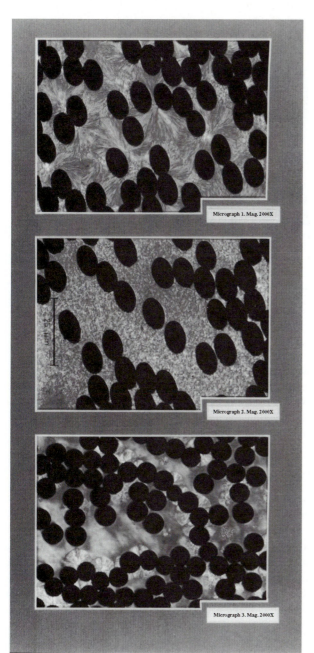

(e)

Figure 8-4 (e) (Luther M. Gammon, Boeing Commercial Airplane)
(*continued*)

CLASS 2

MICROSTRUCTURES OF THERMOPLASTIC COMPOSITES

L.M. GAMMON, M.Y. ALMASSY, C. H. SHEPPARD,
S.L. STANEK. A.M. THORSTENSON
BOEING MATERIALS LABORATORY

Objective

Establish a qualitative relation between mechanical behavior, microstructure, and solidification kinetics in high temperature thermoplastic composites, with an emphasis on Poly Ether Ketone (PEEK)

Background

It has been found that high temperature thermoplastic composites behave like metallic structures, in that mechanical properties change when these composites solidfy at different rates. A precise correlation has not been established between mechanical properties and the amount of polymer crystallinity, which is affected by solidification rates determined by X-ray diffraction, thermal analysis, or density measurements. By extension of the analogy with metals, microstructural phenomenon, such as crystal size, shape, distribution, etc., should affect mechanical properties and may be controlled by the cooling rate.

Procedure

Composite specimens made from PEEK and carbon fiber solidified at various cooling rates (low, moderate and fast [followed by annealing]) were prepared for optical microscopy using transmitted, polarized light. These specimens were 2- to 4-micron thick wafers prepared by mounting and polishing one side, then bonding the polished surface to a Lucite slide. A specially designed hand vise was used to obtain parallel polished surfaces.

Specimen preparation by polishing minimizes internal damage (cold work). Controlling specimen thickness is essential to 1. resolve spherulites that can be stacked behind each other in thick sections, and 2. avoid ghost-like images due to lack of refractive material in thin sections. Thickness may be determined geometrically (refer to sketch).

$$T = \frac{L\sin(0)-D}{\cos(0)}$$

Results

Micrographs 1 and 2 show PEEK/carbon fiber composites with comparable percentages of crystallinity. While both of these matrices are uniformly covered with spherulite, those in Micrograph 1 are larger than those in Micrograph 2 (In metallurgical terms: coarse grain structure versus fine grain structure.) The large spherulites of Micrograph 1 were obtained by cooling slowly, while the smaller spherulites of Micrograph 2 were formed by quenching (rapid cooling) and subsequent annealing.

Matrix crystallinity can be eliminated by rapid cooling which produces a thermodynamically unstable system. A more stable system can be obtained by annealing to nucleate some crystals which then incorporate the surrounding noncrystalline material into what can be viewed as spherulites. Micrograph 3 indicates that the most favorable nucleation sites for crystals tend to be located along the carbon fiber surfaces.

Conclusions

The techniques of metallography described here are proving invaluable for researching matrix crystallization. Mechanical properties of semicrystalline thermoplastic/fiber composites are related to percent of crystallinity and microstructure analogous to metals. These attributes can be controlled during processing by varying the cooling rate from the polymer melt. Material characterization of these materials should include metallography, along with thermal analysis and X-ray diffraction.

A ***coupling agent,*** also known as a bonding agent or binder, provides a flexible layer at the interface between fiber and matrix that will improve their adhesion and reduce the number of voids trapped in the material. ***Voids,*** air pockets in the matrix, are harmful because the fiber passing through the void is not supported by the matrix. The fiber under load may buckle and transfer the stresses to the weaker matrix, which could crack.

8.2.2.1 Interface and interphase.   By definition, composites contain a combination of mutually insoluble components (constituents). With fiber-reinforced plastics (FRPs), the fibers and the polymer resin are the main constituents. A wide variety of other material additives, such as coupling agents, fillers, pigments, catalysts, and fire retardants, are also used to satisfy specific needs. Consequently, there exists an ***interface***(s), a surface that forms a common

Figure 8-4   (f) TP resins serve as a matrix for glass fibers in the new thermoplastic composite used in Ford's Lincoln Continental bumper beams. AZDEL®Plus is a high-strength, lightweight, impact-resistant composite available in a layered structure of glass mat or chopped glass impregnated with thermoplastic resins. Its superior properties to polypropylene-based glass mat help meet higher bumper collision standards (5 mph versus 2.5 mph). Demonstrating up to 90% higher peak loads, these new bumpers weigh five pounds less than their steel counterparts. This joint development by PPG Industries and AZDEL, Inc. has led to a 40% improvement in the performance of molded parts. (G.E. Plastics)

area or boundary similar in many respects to grain boundaries between any two constituents in a monolithic material. An *interphase* is the region formed between two interfaces. It is therefore a distinct phase of itself with its own identity, forming a region contiguous to two phases. Figure 8-4 includes sketches showing these main constituents of an FRP. In metal matrix composite materials (MMCs) the reactions between the metal matrix and the reinforcement takes place in the interface zone outlined in Figure 8-4b.

8.2.3 Fiber/Matrix Consolidation

Fibers can be saturated with resinous material through a process called *preimpregnation.* The resinous material is subsequently used as a matrix material. The preimpregnated fibers are called *prepregs.* Unidirectional preimpregnated fibers with a removable backing that prevents the fibers from sticking together in a roll are known as *prepreg tape. Prepreg cloth* or *mats* are also available. Prepregs are ready-to-be-molded resin and fiber laminates (see Figure 8-5a). The fibers impregnated with resin are partially cured (B-staged) and ready to be softened and formed into a permanent shape with the application of heat and/or pressure. Some important advantages of using prepregs are the elimination of the handling of liquid resins, the simplification of the manufacturing of reinforced plastic forms and shapes, and the savings in cost with small production lots. A few disadvantages are the elimination of room-temperature cures, the need for refrigerated storage for the prepregs, and the additional cost of the prepregs over the dry blended or undispersed materials. The latter disadvantage must be weighed against the use of higher mold pressures and increased molding cycle times for undispersed material to effect the required dispersion of the reinforcement throughout the matrix material for the attainment of the optimum properties of the composite.

Polyimides are attractive to the aerospace industry because of their toughness, thermal and thermo-oxidative stability, resistance to solvents, and excellent mechanical and electrical properties over a wide range of temperatures. One disadvantage is that they have poor melt-flow characteristics during processing, which leads to voids in the prepreg laminates. By blending linear, high-molecular-weight polyamic acid solutions with semicrystalline thermoplastic polyimide powders, a slurry is formed that can make void-free prepregs. The slurry can be applied to film, fabric, metal, polymer, or composite surfaces. As discussed in Module 6, cyclic thermoplastic resins developed recently have also improved wet-out of reinforcing fibers.

An *intermetallic compound* or *phase* (see Section 3.5.2.1 for a discussion about intermediate phase materials), formed from two or more metals, has a mixture of metallic and ionic bonds. Cementite, Fe_3C, is a compound or alloy that serves as the backbone for strengthening steel. Titanium aluminides, high-strength, low-density intermetallics, are superseding nickel-based superalloys. The density of $Ti_2 \cdot Al \cdot Nb$ (titanium–aluminum–niobium) alloys is less

MAGNAMITE GRAPHITE PREPREG
X-AS4/1904

Magnamite graphite prepreg X-AS4/1904 is a 250°F curable epoxy resin reinforced with unidirectional graphite fibers. The reinforcement is Magnamite continuous AS4 graphite fiber that has been surface-treated to increase the composite-shear and transverse-tensile strength. The 1904 resin matrix was developed to cure under tape-wrap or vacuum-bag conditions at 250°F and to operate at temperatures up to 180°F.

	At 77°F	
Typical Composite Properties	Without Glass Scrim	With Glass Scrim
0° flexural strength, psi $\times$ 10^3	225	215
0° flexural modulus, psi $\times$ 10^6	18.4	16.1
0° tensile strength, psi $\times$ 10^3	240	—
0° tensile modulus, psi $\times$ 10^6	20.1	—
Shear strength, psi $\times$ 10^3	14.5	15.0
Cured ply thickness, mils	5.6	6.4
Fiber volume, %	58	58
Typical Prepreg Characteristics		
Tape width, in.	12	12
Resin content, %	40 ± 3	40 ± 3
Fiber content, [a] g/m²	145 ± 4	145 ± 4
Glass scrim[b]	—	104
Yield, ft/lb	19	18

(a) Various fiber contents are available between 95 and 195 g/m².

(b) 104, 108, and 120 glass scrim have been used.

(a)

(b)

(c)

Figure 8-5   (a) Hercules prepregs. (Hercules, Inc.) (b) Schematic of machines used to make SMC sheet and (c) mold SMC plaques.

than two-thirds that of nickel-based superalloys such as UNS N-07718. High specific strength at high temperatures, improved ductility, and fracture toughness are some of the improved properties of such intermetallics that are making these materials feasible for gas-turbine applications at temperatures approaching 1000°C.

Much development work is involved in using intermetallics as matrix materials for metal matrix composites. Intermetallics such as Ni_3Al, $TiAl$, and $NbAl_3$ reinforced with Al_2O_3 or ZrO_2 fibers form composites that are finding use at temperatures from 1000° to 1400°C. Room-temperature ductility of titanium aluminides containing niobium is improved for these composites by the addition of small quantities of magnesium. A special "pressure-casting" technique, in which the molten intermetallic is poured onto preheated fibers, is used to fabricate these composites. Argon gas pressure of 3 MPa ensures complete infiltration of the ceramic fibers. Titanium or yttrium is added to promote fiber wetting.

Intermetallics will certainly be the key to the future of technology development. Their high-temperature properties will find future applications in road vehicles, in aviation, and in industrial equipment. One such intermetallic compound is molybdenum–silicon–boron (Mo_5Si_3B) which is likely to replace *superalloys* in the next generation of aircraft and automobile engines.

Particulates defines a large, diverse group of materials that consist of minute particles. Many of their functions overlap and ambiguity results. Carbon black, for example, is a filler because it fills space in the matrix and reduces cost; but it also is a reinforcing agent for elastomers. Carbon black particles are essentially spherical and range in size from 20 to 50 mm. The length-to-diameter ratio (aspect ratio) for spherical particles is 1, which means that the particles can be easily moved or that they offer no resistance to being moved or realigned. Compare these particles with those that have an aspect ratio other than 1 and thus offer resistance to movement. Their characteristics may include spherical, saucer, needlelike, nodules, platelets, or short fibers. Aspect ratios determine their resistance or realignment in the matrix. Particulate density will affect the composites' final density as in the case of hollow spheres. The amount used (loading) and their distribution in the matrix greatly influence a composite's performance.

Preforms are custom-shaped, resin-bonded mats for reinforcement of molded parts with complex shapes. In other words, a preform is a bundle of fibers having the desired volume and architecture to fill a mold cavity for a one-shot molding.

A *lamina* is a flat arrangement of unidirectional fibers or woven fibers in a matrix.

A *laminate* is a stack of lamina with various orientations of principal materials directions in the lamina. Laminates can be built up with plates or plies of different materials or of the same material such as glass fibers. Shear stresses are always present between the layers of a laminate because each layer tends to deform independently of its neighboring layers due to each layer having different properties. These shear stresses, including the transverse normal stresses, are a cause of delamination.

Bulk molding compounds (BMCs) are a premixed material of short fibers (chopped-glass strands) preimpregnated with resin and various additives. A dough molding compound (DMC) is an alternate term for a BMC. Some thermoset resins are quite thick and are called molding doughs. Parts made by BMCs are limited to about 16 inches in their longest dimension due to the problem with the separation of the components of the molding compound during molding.

Sheet molding compounds (SMCs) are impregnated, continuous sheets of composite material. SMCs cut to proper size and stacked to provide the required thickness before heat curing in matched metal molds are used in applications such as automotive bodies and large structural parts (see Figure 8-5f). They come in three types, depending on the length of the glass fibers used in the reinforcement. Fibers can be as short as 25 mm in length and arranged in random fashion in the resin. Longer fibers (200 to 300 mm) can be oriented in one direction (directional fibers). Continuous fibers laid in only one direction make up another designation. Various combinations of these SMC types are used with fibrous glass reinforcements, reaching 65% by weight of the composite. A recent molding compound designated XMC can contain up to 80% glass fiber by weight and uses continuous fibers running in an X pattern (see Figure 8-16). Thick molding compounds (TMCs) are similar to SMCs but are up to 2 inches thicker.

The more common fibers used in composites are described on the following pages. As research continues, more materials (organic, polymeric, ceramic, and metallic) are becoming sources of fibers for different composites designed for ever-increasing applications in an energy-conscious age.

8.3 REINFORCING FIBERS

8.3.1 Glass Fiber

Glass fibers are made by letting molten glass drop through minute orifices and then attenuating (lengthening) them by air jet. The standard glass fiber used in glass-reinforced composite materials is E-glass, a borosilicate type of glass. The glass fibers produced, with diameters from 5 to 25 μm, are formed into strands having a tensile strength of 5 GPa. Chopped glass used as a filler material in polymeric resins for molding consists of glass fibers chopped into very short lengths.

E-glass is the first glass developed for use as continuous fibers. It is composed of 55% silica, 20% calcium oxide, 15% aluminum oxide, and 10% boron oxide. It is the standard grade of glass used in fiber glass and has a tensile strength of about 3.45 GPa and high resistivity. Fiber diameters range between 3 and 20 μm.

S-glass was developed for high-tensile-strength applications in the aerospace industry. It is about one-third stronger than E-glass and is composed of 65% silicon dioxide, 25% aluminum oxide, and 10% magnesium oxide.

Automakers are finding more uses for *fiberglass* fiber reinforcements for both thermosets and thermoplastic polymers in their automobiles. Both exterior and interior parts as well as under the hood reflect this change. Air intake manifolds, radiator caps, cables and connectors, bumpers, fenders and hoods, seats, and instrument panels increasingly contain fiberglass. PPG Industries projects additional growth of 19% by 2001. Air intake manifolds represent a dramatic growth area for glass fiber in auto applications. In 1993, one U.S. vehicle featured a nylon manifold which required, overall, 40,000 pounds of glass fiber reinforcement. Four years later, 12 models had incorporated the part, raising their fiberglass consumption by 7.3 million pounds. Composite manifolds were initially developed in the United States, but European automakers use three times more glass in manifolds than U.S. car manufacturers. The demand for this reinforcement originated from the desire of manufacturers to build cars that cost less, are visually appealing, and are easy to maintain.

8.3.2 Boron Fiber

Advanced composites with boron fibers were developed in the early 1960s by the U.S. Air Force. The year 1963 marked the birth of boron technology. Boron fibers are composites with tungsten, silica coated with graphite, or carbon filaments as a substrate upon which boron is deposited by a vapor-deposition process (CVD). The final boron fiber has a specific gravity of about 2.6, a diameter between 0.01 and 0.15 mm, a tensile strength of about 3.45 GPa, and a modulus around 413 GPa. Figure 8-6a is a representation of a boron fiber.

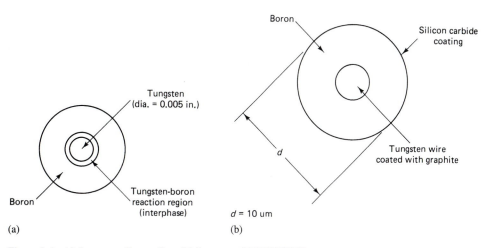

Figure 8-6   (a) Structure of boron fiber. (b) Structure of BORSIC™ fiber.

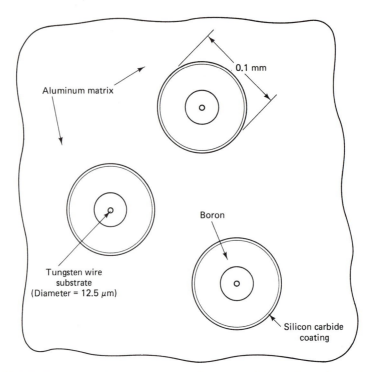

Figure 8-7 BORSIC™ fibers in an aluminum matrix would produce a MMC. The boron fiber itself is a composite, having been deposited on a Ti substrate and coated with SiC.

　　　Another boron-fiber composite with the addition of a silicon carbide coating deposited over the boron surface provides a fiber that is more compatible with metal–matrix materials, particularly at high operating temperatures. Figures 8-6b and 8-7 show a sketch of a BORSIC™ filament and an aluminum matrix in which boron is deposited on a graphite-coated tungsten wire, which makes the fiber itself a composite.

　　　Boron is more expensive than graphite and requires expensive equipment to place the fibers in a resin matrix with a high degree of precision. Problems with chemical reactions between boron and other metals continue to be a source of concern to materials technologists who are exploring ways to overcome this limitation. One approach to this problem was the coating of the BORSIC fiber with silicon carbide (see Figure 8-7).

8.3.3 Carbon and Graphite Fibers

Carbon is a nonmetallic element. Black crystalline carbon, known as graphite, has a specific gravity of 2.25. Transparent carbon, diamond, has a specific gravity of 3.5. Graphite and diamond are allotropes of carbon, as are amorphous forms of carbon such as coke and charcoal. Both graphite and diamond have very high melting points of over 3732 K, which is explained upon examining the crystal structure and seeing the nearly infinite network of carbon–carbon covalent bonds that must be broken to melt these materials.

　　　The terms *carbon* and *graphite* are often used interchangeably. However, a line of demarcation has been established in terms of modulus and carbon content. Carbon fiber usually has a modulus of less than 344 GPa and a carbon content between 80% and 95%. Graphite fibers have a modulus of over 344 GPa and a carbon content of 99% or greater. Another distinguishing feature is the pyrolyzing temperatures. For carbon, this temperature is around 1315°C, and for graphite it is around 1900° to 2480°C.

　　　One authority classifies carbon as a ceramic. Another says the pyrolized organic material is a polymer, even though it is mostly carbon. They admit that carbon itself is an element, but if you believe that the word *carbon* really means "graphite," then carbon is a ceramic. Another authority uses the word *carbon* to describe fibers because the fibers are composed of crystalline graphite regions. In general, *carbon* describes both carbon and graphite fibers.

Pyrolysis is the thermal decomposition of a polymer. This chemical decomposition by heat can vary from 1000° to 3000°C, depending on the desired fiber properties. A carbon-containing polymer can decompose to carbon in an inert or reducing atmosphere and thus produce carbon fibers and/or carbon–carbon composite materials. If elements other than carbon are present in the original polymer, these elements will form a residue as a result of pyrolysis of ceramic compounds. An example of pyrolysis is the production of SiC fibers from polycarbosilane polymers having tensile strengths of 2.7 GPa.

As with composite materials, graphite fibers are not a new development. They were first produced in small quantities in the nineteenth century for use in incandescent lamp filaments.

Several methods are used today to make carbon fibers of varying length and diameter and having the versatility of glass fibers. The oldest method of producing these fibers, used in the late 1960s, was *graphitization* of organic fibers, such as a rayon, at temperatures up to 3250 K. An acrylic fiber, polyacrylonitrile, abbreviated PAN, is also used as a source of such fibers, which the Royal Air Force (RAF) in England produced in 1961. Heated (pyrolized) under tension to stabilize the molecular structure at temperatures between 920 and 1140 K, the noncarbon elements (O_2, N_2, H_2) are driven off, leaving a fiber high in carbon content. A more recent process for producing high-modulus carbon fibers uses low-cost pitch; the pitch is converted to a liquid crystal, or mesophase, state before it is spun into fibers.

Twenty different carbon fibers with cross-sectional configurations, varying from circular to kidney shaped, in diameters of 0.008 to 0.01 mm, with strengths ranging from 1.72 to 3.1 GPa, and modulus values from 193 to 517 GPa, are now available from six manufacturers. One carbon yarn comes in plies from 2 to 30, with each ply composed of 720 continuous filaments about 8 μm in diameter. This fiber is 99.5% carbon and maintains dimensional stability to 3420 K. In addition to carbon, glass and aluminum fibers are supplied in the form of tows (rovings or strands) consisting of many individual, continuous filaments. Tows can consist of 3000 to 30,000 filaments. Although carbon fibers have a maximum-use temperature range from 250° to 2000°C, the use temperature of a composite is controlled by the use temperature of the matrix. Another grade of fiber consists of 2000 filaments gathered together into a uniform strand that provides a modulus of 379 GPa and a tensile strength of 2.0 GPa. Thornel 300 carbon fiber, produced by Union Carbide, consists of 6000 filaments in a one-ply construction. The surface of the fiber has been treated with sizing to increase the interlaminar shear strength in a polymer–matrix composite in excess of 89.6 MPa. The tensile strength is listed as 2.69 GPa, with an accompanying modulus (tensile) of 229 GPa. Carbon and graphite fibers and yarns are used to produce various fabrics to meet the ever-increasing demands of industry. See Figure 8-8.

Carbon fibers, first introduced in 1959 at a price of over $500 per pound, have steadily dropped in price. By 1984, the price had declined to $13 to $15 a pound for lower-grade graphite fiber and about $19 a pound for aerospace-grade fibers. Generally, the decline in prices of materials can be expected to continue with an increase in use, as long as there is a good availability of raw materials and improved production technology. Graphite fibers are linear elastic and, like other carbon materials, are anisotropic. Graphite possesses excellent creep and fatigue resistance, as well as dimensional stability at high temperatures. It is one of the few truly efficient structural reinforcements, with both high specific modulus and strength. Figure 8-9 compares the strength and modulus of two grades of graphite with some other fiber materials.

Thermal properties also must be considered unusual. The most important properties are low *coefficient of thermal expansion* (CTE or COT), high thermal conductivity, and dimensional stability at temperatures of 2000°C and above. The CTE of graphite is negative at 0°C, but increases with temperature (see Table 8-2). However, its generally negative CTE, coupled with the ability to tailor the orientation of the fibers, permits the designer to adjust the thermal coefficient over a wide range of values. Its thermal conductivity in the same temperature range is similar to that of many metals, including steel and nickel. Aluminum is the only metal that has a significant advantage in thermal conductivity over graphite. However, aluminum's high CTE and low-temperature capability make it useful only in low-temperature composite curing

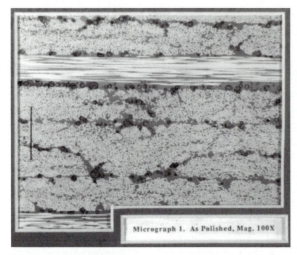

Micrograph 1. As Polished, Mag. 100X

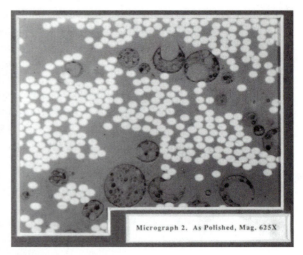

Micrograph 2. As Polished, Mag. 625X

Class 8
Microstructure of Carbon Fiber Reinforced Composites
L. M. Gammon, J. S. Chen, R. A. Grove, A. M. Thorstenson
BOEING MATERIALS TECHNOLOGY

The use of composite, *Carbon Fiber Reinforced Plastic* (CFRP), for airplane components has resulted in significant weight savings while yielding improved performance and fuel efficiency. Low damage resistance and fracture toughness have commonly limited the application of CFRP throughout the industry.

The Boeing Company, during the past several years, has actively engaged in the development and evaluation of tougher and less prone to damage CFRP composites. A promising composite candidate was researched and identified in 1987. It was the result of a breakthrough in resin technology that this new composite exhibited significant improvements in damage resistance without compromising other vital properties. The main objective within the laboratory has been to come to a full understanding of those key microstructural features which relate directly to damage resistance, fracture toughness, and process and formulation changes.

Boeing Materials Technology research has lead to the development of several specimen preparation methods, and has resulted in an accurate and broader understanding of relationships between microstructure and material performance. It was found that size and distribution of individual resin phases would directly affect damage resistance and the changes in formulation and manufacturing processing could also be evaluated from a microstructural viewpoint. The latter benefit allowed the development of a "fingerprint" of the morphology for maintaining material consistency.

This presentation reveals three successive specimen preparation techniques for CFRP laminates developed with a complex, multiphase resin system. Micrographs of the same field are shown for each of the following techniques: 1. Polishing. 2. Etching; and, 3. Epi-fluorescent dying/microscopy. For each integral specimen preparation method, the correlation between key microstructural features and material performance is being determined.

Micrographs 1 and 2
Specimen preparation
Silk wheel primary polished at 1200 rpm, high pressure and alumina/H_2O slurry. Used 5.0 u alumina followed by 0.3 u alumina. Final polished with medium nap wheel, high pressure and 0.05 u alumina/H_2O slurry.
Observation
In the "as polished" condition, a stacking sequence of fibers (white) can be seen and spheres (Phase 1) are distributed within the resin (Phase 2). Although this method provides a superior polish, this is what many microstructural investigations of composites have been limited to in the past.

Figure 8-8* *(continued)*

TABLE 8-2 DIMENSIONAL STABILITY OF SOME FIBER MATERIALS (CTE)

Material	Thermal Expansion (in./in./°F) $\times 10^{-6}$	
	0°	0°, ±45°, 90°
Graphite AS	−0.20	1.01
Graphite HTS	−0.25	0.80
Graphite HMS	−0.30	0.45
Fiber glass	3.50	6.00
Aluminum	13.0	—
Steel	7.0	—
Nylon (6/6)	45.0	—

*A short course on "Microscopy of Fiber-reinforced Polymer Composites" containing numerous experiments and demonstrations as well as color photomicrographs, one of many valuable composite resources is available on the CD-Rom Experiments in Material Science, Engineering and Technology, Version 2 from Prentice-Hall at 1-800-526-0485.

Micrograph 3. Etched, Mag. 625X

Micrograph 4. EPI - Fluorescence, Mag. 600X

Micrograph 3
Specimen preparation
Polished specimen is etched for 30 seconds using [CrO_3 (12.5 g): Conc. HNO_3 (50 ml): H_2O (50 ml)] solution.
Observation
The resin (Phase 2) gives indications of two sub-phases: a continuous phase (2a) and a discrete phase (2b). Precise detail can be seen within each phase, especially within the spheres (Phase 1). This information allowed Boeing to visually characterize the microstructure and correlate material performance with formulation and processing changes and also enabled them to understand how each of the phases contributed to damage resistance and fracture toughness.

Micrograph 4
Specimen preparation
Specimen is immersed for 30 minutes in a solution of Rhodamine B/methylene chloride dye. Ultrasonically cleaned excess dye using methylene chloride, followed by a bake at 250°F to remove solvent.
Observation
Using epi fluorescent microscopy with a 190 110 mg excitation range each individual phase becomes distinctly colored. This breakthrough allows investigation of image analysis for quantitative morphology characterization (where more contrast is required) and also more accurately allows definition of the chemical nature of each phase with finer detail.

Figure 8-8*

Figure 8-9   Comparison of specific strength and specific modulus (specific stiffness) of commonly used fiber materials. (DuPont)

applications. Finally, graphite is essentially free from residual stresses and never needs stress-relief annealing, as is common with many metals.

Graphite is commercially available in three different types: high modulus, intermediate modulus and tensile strength, and high stiffness and strength. All three have low density but also low impact resistance. The strength of graphite fiber/polycarbonate matrix composites is increased by precoating (sizing) the fibers with a thin layer of polycarbonate. Sizing can promote the wetting of fibers by matrix polymers, thereby enhancing fiber/matrix adhesion. The COT of graphite increases with temperature.

Graphite fibers can be coated with other materials to extend the usefulness of their properties to higher temperatures. Titanium carbide (TiC) is used to coat graphite or carbon/carbon composites with usual thicknesses (about 150 mm) to provide exceptional resistance to erosive wear and to corrosion by chemical agents. In addition, this coating has demonstrated very high thermal-shock resistance. Table 8-3 provides a list of a few properties of TiC-coated graphite and a graph showing the erosion rates of various materials, including TiC-coated graphite.

Electrically conducting composites, made from bromine-intercalated (insertion between something) graphite fibers in an epoxy material, are used for many applications, such as grounding planes. The electrical conductivity of graphite fibers (derived from pitch blend) has been increased from about 4,000 to about 50,000 $\Omega^{-1}\text{cm}^{-1}$ through *intercalation.* Common metals have conductivities that range from 10,000 to 550,000 $\Omega^{-1}\text{cm}^{-1}$. The process of intercalation introduces donor/acceptor atoms, such as bromine, between the layers of graphite atoms in their characteristic layer arrangement. A cloth made of graphite fibers immersed in bromine is stacked between epoxy films making up the composite, which is then pressed and cured.

One disadvantage of graphite is its poor oxygen resistance. It begins to oxidize in air at about 700 K. Unlike metals, graphite does not form a protective film. Instead, the graphite oxide is volatile. To protect graphite from oxidation, several coatings have been developed, one of which, silicon carbide, protects graphite for a limited time at temperatures approaching 1920 K. Another limitation of carbon fiber is that it is a brittle, strain-sensitive material that cannot be depended on to offer much impact resistance.

Graphite fibers are available to the fabricator in the following forms:

1. Continuous fiber
2. Unidirectional prepreg as tow or tape in widths from 3 to 36 inches to form laminations
3. Chopped fiber
4. Pultruded shapes
5. Woven fabrics as dry cloth or prepreg for laminations

8.3.3.1 Nano composites.   Future potential reinforcements for composite materials are high modulus and high elastic strain nanotubes. *Nanotubes* are carbon molecules. A carbon nanotube is constructed of a single sheet of graphite rolled into a tube with fullerine caps on the ends. (See discussion of fullerines in Module 2 and the molecule in Figure 2-14.) These single-walled elongated *fullerine molecules* contain millions of carbon atoms forming defect-free hollow structures just a few atoms in circumference. Nanotubes are many thousands of times longer than their diameters (see aspect ratio discussion in Section 8.4.1) and they possess exceptional mechanical, electrical, and thermal properties. For example, their tensile strength is at least 10 times stronger and their weight is less than half that of conventional carbon fibers with thermal conductivity as high as that of diamond. These properties can be tailored through processing to fit a multitude of applications. Development of these remarkable molecules will be, at the least, difficult.

8.3.4 Kevlar Fiber

Kevlar, an organic fiber introduced in 1972 by DuPont for use in radial tires, is an aramid, or aromatic, polyamide fiber. This organic fiber is melt-spun from a liquid polymer solution. The aromatic ring structure (see Figure 2-13) results in high thermal stability. The rodlike nature of the molecules classifies *Kevlar* as a liquid-crystalline polymer characterized by its ability to form ordered domains in which the stiff, rodlike molecules line up in parallel arrays. These domains orient and align themselves in the direction of flow during processing, which results in a

TABLE 8-3   ENGINEERING PROPERTIES OF TiC-COATED GRAPHITE

Flexural strength [MPa (ksi)]	90	(13)
Elastic modulus [GPa (Msi)]	11	(1.6)
CTE [ppm/K (ppm/°F)]	8.4	(4.7)
Thermal conductivity [W/m · K (Btu/hr · ft · °F)]	121	(70)
Density [g/cm³ (lb/ft³)]	1.81	(113)

Source: Courtesy of Lanxide.

(a)

Figure 8-10   (a) Bending stress–strain curves of some fiber materials. (DuPont) (b) An artist's concept drawing of the luxury yacht *Evviva* shows a 49-m (160-ft) custom motor yacht constructed from a plastic superstructure reinforced with a hybrid combination of Kevlar aramid fibers and E-glass. Nomex aramid structural sheet, fabricated into honeycomb core panels, is used for the infrastructure and accessories such as furniture. The advanced composites allowed the designer to engineer the 180-ton displacement yacht for luxury and also service as a high-speed craft that will cruise at 37 km/h (20 knots) over a 9260-km (5000-nautical-mile) range. (*Advanced Materials and Processes*)

high degree of alignment parallel to the fiber axis. This alignment results in anisotropy with high strength and tensile modulus in the fiber-longitudinal direction. Refer to Figure 6-7 and the accompanying text for additional discussion of polymer crystallinity. The overall properties above, plus a density of 1.44 g/cm³ (about one-half that of glass), are the key to Kevlar's use in weight-limited applications. There are three grades of Kevlar. Kevlar 29 provides high toughness with a tensile strength of about 3.4 GPa for use where resistance to stretch and penetration are important. Kevlar 49 is a high-tensile-strength modulus of 130 GPa for use with structural composites. Kevlar 149 is an ultrahigh-tensile-strength modulus of 180 GPa. Figure 8-10a is a stress–strain diagram showing the relation of Kevlar 49 to other common reinforcing fibers. The luxury yacht seen in Figure 8-10b makes use of Kevlar aramid fibers, which were selected over competing materials shown in Figure 8-10a. The craft's plastic superstructure is reinforced with Kevlar and E-glass, which provides nearly a 20% weight savings while maintaining structural integrity equivalent to more conventionally built vessels.

8.3.5 Ceramic Fibers

Both continuous and discontinuous ceramic fibers that are based on oxide, carbide, and nitride compositions comprise most of the production of these fibers. Their development was started by the need for high-temperature reinforcing fibers in composites for the aerospace industry. Most oxide fibers are compositions of Al_2O_3 and SiO_2, although a few are almost pure oxides of aluminum and silicon (alumina and silica). Average properties of continuous oxide fibers are as follows: density, 3 g/cm^3; diameter, 12 μm; tensile strength, 2 GPa; tensile modulus, 200 GPa; and use temperature, around 1300°C.

Carbide fiber properties differ significantly from those of the oxides. Representative of continuous carbide fibers is SiC. **Silicon carbide (SiC),** a ceramic fiber, is made by one of two methods. The first method uses CVD to deposit Si and C onto a pyrolytic graphite-coated carbon core. The resultant fiber designated SCS-6 is similar in size and microstructure to boron fiber and has a diameter of about 140 μm. The second method uses controlled pyrolysis of a polymeric precursor. The filament produced (trade name Nicalon) is similar to carbon filaments in size (14μm) and microstructure with a diameter of about one-tenth that of an SCS-6 fiber. It is more flexible and its fibers are arranged in rows of 250 to 500 filaments per row. SiC has a melting point of 2830°C. A new coating for SiC particulates is SiO_2 produced from rice hulls (SiC_w is the designation for this treatment). SiC is available in three different forms. Monofilaments are 140 μm in diameter. Yarns consist of 500 fibers. Each fiber has a diameter of 14.5 μm. SiC whiskers have diameters ranging from 0.1 to 1 μm. Allied Signal Inc.'s SiC yarn fabrication process is similar to those used for manufacturing fiber from PAN. SiC fibers are the reinforcement of choice for use with aluminum, magnesium, titanium and titanium aluminide, and toughened ceramics materials because such fibers offer better resistance to oxidation and galvanic corrosion than graphite fibers. Approximations of SiC's properties are as follows: density, 2.5 g/cm^3; diameter, 10 to 20 mm; tensile strength, 3.0 GPa; tensile modulus, 100 to 400 GPa; and use temperature, 1200°C. Both SiC and Si_3N_4 are available as discontinuous (whisker) forms. The principal advantage of carbide fibers over oxide fibers is the great increase in the modulus of elasticity.

8.3.6 High-Performance Manufactured Polymeric Fibers

Included in this category are fibers based on polyester, nylon, aramid, and polyolefin. As discussed in Module 6, (6.4.1) fibers such as aramid (Kevlar and Nomex), polybenzimidazole (PBI), Sulfar, and Spectra have increased the range of choices for materials engineers in designing materials with tailor-made properties. However, most do not possess the thermal properties found in ceramic- and metal-based fibers. Kevlar, a trademarked material, is discussed separately in this module. A polyolefin fiber, **Spectra** (ultrahigh-molecular-weight polyethylene) has a low specific gravity (0.9), which allows 40% more fiber per pound than aramid (Kevlar) fibers. It also has higher specific strength and specific stiffness, and has 7 to 10 times more abrasive resistance than aramid, with low moisture absorption and good chemical resistance.

8.4 FIBER PROPERTIES

8.4.1 Fiber Strength

A material's brittleness greatly affects its strength. Brittle materials have little resistance to the propagation of cracks. Pure metals possess varying degrees of ductility (the opposite of brittleness), which allows for some yielding in the face of longitudinal stress concentrations, which in turn prevents fracture. Brittle fractures, on the other hand, are usually catastrophic and come without advance warning. Alloying of metals and thermal treatments are two ways of improving the resistance of a material to plastic deformation (strength) by providing for the dispersion of harder particles within a matrix of softer material, which tends to limit the motion of dislocations through

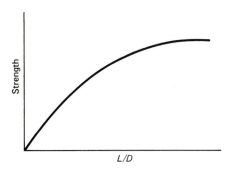

Figure 8-11 Strength versus fiber-aspect ratio (*L/D*).

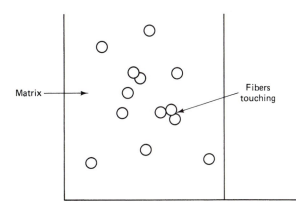

Figure 8-12 Cross-section of a fiber-reinforced material showing a poor spatial arrangement of fibers within the matrix material. See also Figure 8-4.

the matrix. Another way to increase the strength of a material is to add another material that has greater load-carrying capacity. The most prevalent form of this added material is fiber. How well fibers strengthen a material depends on the efficiency with which the relatively soft matrix material interacts with the fibers to transfer the load between them. The strength of an individual fiber is dependent on the absence of surface defects. The presence of microcracks anywhere in the small fiber causes localized stress and load concentrations and eventual failure of the individual fiber. The properties of a fiber are mainly dependent on the fiber's length, diameter, and orientation. The optimum reinforcing fiber for a composite material has a ratio of length *(L)* to diameter *(D)* of about 150. As this ratio, known as the ***fiber-aspect ratio,*** increases, the strength of the composite increases (see Figure 8-11). A single polymer chain has the highest aspect ratio of any "fiber." Incorporating molecular fibers into composite polymers is a technique with great potential for the future. With fibers of small diameter, 1 to 25 μm, aspect ratios range from 100 to 15,000.

Illustrative Problem

Given a fiber that is 25 μm in diameter and has an aspect ratio of 15,000, find the length of the fiber expressed in meters.

Solution

$$\text{Length } (L) = \text{diameter } (D) \text{ of fiber} \times \text{aspect ratio (AR)}$$
$$L = 15 \times 10^3 \times 25 \times 10^{-6}$$
$$L = 0.375 \text{ m}$$

The shape of the reinforcing phase, rodlike for fibers, plays an important role in determining the performance of a composite, as does the shape of the unit cell in a homogeneous solid. Their size and distribution control the texture of the material and determine, in part, the interfacial area between the fibers and the surrounding matrix. The topology of the fibers (i.e., their spatial relation to each other) is important because certain properties of the composite may be affected by how much the individual fibers or filaments touch each other (see Figure 8-12).

8.4.2 Specific Tensile Strength

Fiber composites find use in many weight-sensitive applications in space exploration, general aviation, and the automobile and sporting-goods industries. One indicator used to portray the effectiveness of the strength of a fiber is specific strength.

Specific tensile strength, the ratio of the tensile strength of a fiber material to its weight density, is an indicator of structural efficiency, giving the relative load-carrying ability of equal weights of material. Typical units of specific strength are millimeters (mm), inches (in.), or kilometers (km). Table 8-4 and Figures 8-9 and 8-13 list the specific strength for some typical materials from which fibers are formed for use in composite materials.

8.4.3 Specific Stiffness—Specific Modulus

Stiffness is the ability of a material to maintain its shape when acted upon by a load. Another indicator of the special properties of fiber composites, in particular, the effectiveness of a fiber, is *specific stiffness.* It is a ratio of the modulus of elasticity (or tensile stiffness) to the weight density of the fiber. Table 8-5, using the same materials as in Table 8-4, and Figures 8-9 and

TABLE 8-4   SPECIFIC STRENGTH

Materials	Weight density, ρ (kN/m^3)	Tensile strength, S (GN/m^2)	Specific strength, S/ρ (km)
S-glass	24.4	4.8	197
E-glass	25.0	3.4	137
Boron	25.2	3.4	137
Carbon and graphite	13.8	1.7	123
Beryllium	18.2	1.7	93
Steel	77	4.1	54
Titanim	46	1.9	40
Aluminum	26.2	0.62	24

Note: The specific strength of fibers can be more than 16 times that of structural metals. E-glass has a tensile strength of about 500 ksi (3.44 GPa). Its modulus of elasticity (E) is 10.5 Msi (72.3 GPa) and its density is 2.54 g/cm^3. S-glass has a higher strength-to-weight ratio and is more expensive than E-glass. Its tensile strength is about 650 ksi (4.48 GPa). Its modulus of elasticity is about 12. 4 Msi (85.4 GPa). E-glass fibers have diameters that range from 3 to 20 μm.

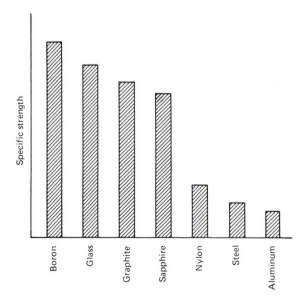

Figure 8-13   Relative specific strengths of typical materials used in composites.

TABLE 8-5   SPECIFIC STIFFNESS/MODULUS

Materials	Weight density, ρ (kN/m^3)	Tensile elasticity, E (GN/m^2)	Specific Stiffness, E/ρ (Mm)
Graphite	13.8	250	18
Beryllium	18.2	300	16
Boron	25.2	400	16
Carbon	13.8	190	14
S-glass	24.4	86	3.5
E-glass	25.0	72	3.0
Steel	77	207	2.7
Titanium	46	115	2.6
Aluminum	26.2	73	2.8

Note: The specific stiffness of fibers can be more than 13 times that of structural metals.

Figure 8-14   (a) Relative specific stiffness of typical materials used in composites. (b) A light, strong, and rigid aerodynamic composite shell surrounded the rider of the Dexter Hysol Cheetah bicycle as it traveled through a 200-meter speed trap at San Luis Valley, Colorado, going 68.73 mph (110.6 km/h) to capture the new human-powered vehicle speed record. (*Advanced Materials and Processes*)

Note: Graphite has a specific stiffness of 5000 versus steel with 25 in units of 10^{6}N 'm/kg (using mass density in kg/m^3)

(a)

(b)

8-14 show the average specific stiffness of such fiber materials, starting with materials with the highest values. Note that graphite, boron, and carbon have values almost six times that of steel. Graphite is known to have six times the strength of steel with six times less weight. These values of both strength and stiffness of materials in fiber form over the same materials in bulk form are most significant. As an example, the strength of glass in bulk form, such as plate glass, is only a few megapascal, yet in fiber form this figure rises to around 3.5 GPa. Structural steel (plain-carbon steel) in bulk form has a tensile strength around 0.5 GPa, but in fiber form the value is 4 GPa. This contrast is due to the fact that the structure of the material differs between the two forms of the material. In fiber form, the microstructure approaches the nearly perfect structure, with the crystals aligning themselves along the fiber axis in ordered fashion. As a result, there are fewer internal defects or dislocations than would be present in bulk form. As we recall from our previous discussions, the presence of dislocations explains many properties of a material. In this situation, the movement of dislocations permits the material to yield, changing the internal structure accordingly and accounting for ductility and accompanying decreases in strength and stiffness. Figure 4-5 showed the increase in stiffness gained by reinforcing materials (aluminum and titanium) with boron fibers.

Specific stiffness and specific strength can be expressed in terms of the mass density rather than the weight density. Such a practice is more compatible with the SI. Typically, units of mass density are expressed in $kg \cdot m^{-3}$. Modulus of elasticity and tensile strength can be expressed either in $N \cdot m^{-2}$ or Pa. The specific ratio will have units of $m \cdot N \cdot kg^{-1}$. For example, a wood's longitudinal modulus (E) is $1.2 \times 10^{10} \, N \cdot m^{-2}$ and its density measured at 12% moisture content is $0.68 \times 10^3 \, kg \cdot m^{-3}$. The specific modulus of this wood is

$$\frac{E}{\rho} = \frac{\dfrac{1.2 \times 10^{10} N}{m^2}}{\dfrac{0.68 \times 10^3 \, kg}{m^3}} = 17.6 \times 10^6 N \cdot m \cdot kg^{-1}$$

Values of the stiffness–mass density parameter (stiffness modulus) for some other materials are:

Graphite	$5000 \times 10^6 \, N \cdot m \cdot kg^{-1}$
Boron	$190 \times 10^6 \, N \cdot m \cdot kg^{-1}$
Steel	$25 \times 10^6 \, N \cdot m \cdot kg^{-1}$
Nylon	$3 \times 10^6 \, N \cdot m \cdot kg^{-1}$

Figure 8-14b provides a striking example of how the specific tensile strength and specific stiffness values found in composites provide designers with the ability to select and design engineered materials to form a materials system for specific applications. The human-powered, record-setting vehicle used aerospace-grade adhesive bonding to join the lightweight, rigid structure. The bicycle system used paste adhesives that bond the frame, seat mold, fairing mounts, aluminum inserts, steering assemblies, and composite parts. *Syntactic core material* (reinforced plastic foams made with microspheres instead of entrapped gas voids) was used for the bucket seat and as a laminar core stiffener. The windshield and fairing were attached by epoxy patch kits. It should be noted that these strength and stiffness properties of fibers are reduced significantly when the fibers are used with a matrix material to form a composite. Nevertheless, the significant weight savings achieved through the use of fiber composites in engineering structures is, in the majority of cases, the primary consideration. The weight of some mountain bikes from Cannondale is less than 25 pounds due to design innovations such as "an internal magnesium spine with carbon-fiber skin."

8.4.4 Fiber Loading

Fiber loading refers to the amount of reinforcement in a composite material. The strength of the composite is directly proportional to the volume of fiber *(volume fraction)* present. In addition to the fiber loading, the arrangement or orientation of the fibers plays a major role in the strength of the given product. Using the analogy of filling a shoebox with pipe cleaners, the maximum number of fibers that can be placed in a given volume (shoebox) is determined by the arrangement of the fibers. If all fibers are placed parallel to each other, a maximum number can be attained. In fiber loadings, a maximum of about 85% can be achieved. Between 50 and 75% load range can be reached if half the fibers are arranged at right angles to each other (fabrics), while a random arrangement (chopped strands) permits a range of only 15 to 50% (see Section 8.4.5). A new fiber designated *Securus,* made by Allied Signal, is termed a "smart response" fiber. This fiber was developed as an automobile seat belt fiber that increases occupant safety. Working in conjunction with the airbag in a collision, the fiber replaces the me-

Figure 8-15   SN Diagrams (a) Range of fiber-composite tensile fatigue performance. (General Motors) (b) Effect of fiber orientation on fatigue of XMC. (General Motors)

chanical load-limiting device such that the fibers stretch or relax as needed to limit the force imposed in a collision. The new fiber is made from a category of synthetic fibers known as Pelco fibers, which are based on a polyester-caprolactone block copolymer. The fiber begins to stretch at 400 pounds of force and then stiffens again to prevent the occupant from hitting the dashboard. Costing 25% less, these new seat belts provide protection for a wider range of occupant sizes. These new belts will appear in 2000 when the manufacturing capacity is sufficient to supply some 5 million cars.

8.4.5 Fiber Orientation

As the direction of the applied load moves 90° to the fiber orientation, the strength of a directionally oriented fiber composite decreases to about 20–30% of the longitudinal direction. Several techniques are employed to orient the fibers in a fiber composite. If continuous fibers are oriented such that their length is in the direction (longitudinal) of the loading, this type of arrangement is known as ***directionally oriented.*** If the applied loads and/or their directions are not known, a random oriented arrangement with continuous fibers running at various directions may be used. Alternating layers of continuous fibers at various angles provide full strength at these various directions. However, this calls for additional layup time, with its increased costs. Discontinuous fibers less than 30 cm long, which tend to orient themselves in the direction of the resin–matrix flow, provide another partial solution to providing a balanced strength in several directions. This technique is a compromise between the continuous and random orientation of continuous fibers. By sandwiching random short fibers between continuous fibers, a multiple-configuration fiber composite results that is superior to a fully randomly reinforced material. The effect of adding some continuous fibers to a random, chopped-fiber reinforced SMC is highlighted in Figure 8-15a, which compares, among other things, SMC-R50, a 50% (by weight) of 25-mm chopped E-glass fibers, 16% calcium–carbonate filler, and 34% polyester–resin–matrix formulation, to SMC-C20/R30, a hybrid of chopped (30%) and continuous (20%) E-glass fibers in a polyester-resin matrix. Figure 8-15b shows the effect of a change in the fiber orientation on the fatigue failure of XMC, an X-pattern molding compound produced by PPG Industries. All preceding arrangements are available in SMC form. Figure 8-16 contains sketches of some of these arrangements of fibers in a matrix.

8.4.6 Smart Fibers

There is an increased demand from various sources for fabrics that "react to our environment and provide greater degrees of comfort or performance." ***Antimicrobials,*** which prevent the spread of bacteria and reduced odors caused by them, are being investigated by textile manufacturers for applications particularly in the sportswear area. Fiber-based antibacterials have the advantage of

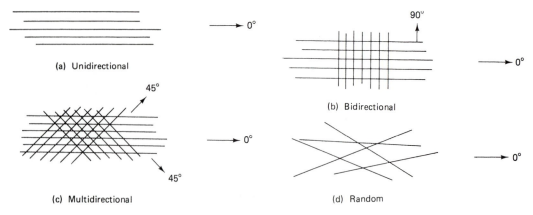

Figure 8-16 Fiber orientations.

conserving their germ-killing properties even after numerous launderings. The National Textile Center has funded research into *"smart" fibers* that can alter themselves in different conditions. Athletic wear can react to environmental conditions by warming or cooling as needed.

Electrically conductive polymers that act like metals can conduct electricity. Microcircuits within fabrics would permit the fabrics to store information. Future applications for such a fabric are extensive. Other smart fibers using *photonics* could respond to the environment by changing colors like a chameleon or, when exposed to a source of radiation such as light or gamma radiation, could respond by identifying the source of such radiation. Other fibers could be developed that would be inert to radar.

These fibers are being developed not for their strength characteristics but for a multitude of other reasons that will find future, beneficial applications for consumers, the military, and our environment.

8.5 STRUCTURE AND PROPERTIES OF COMPOSITES

Throughout this book we have stressed that the properties of a material are dependent on the material's structure. In monolithic materials such as steel or polymers, the structure commonly referred to is that of the crystalline- or molecular-material microstructure. Many times we peer beyond this atomic or molecular level to the subatomic level for further understanding of the behavior of such materials. With composite materials, we can identify with the naked eye (macroscopic) the major ingredients or constituents, such as metal particles or glass fibers and different matrix materials involved in understanding the collective performance of these components in the composite material. Just as with steel alloys, where we can change the microstructure of these multiphase materials to obtain different properties by varying the amount of the basic ingredients (elements) or by changing the number, size, and shape of the ingredients through appropriate thermal processing, we can vary the properties of composite materials by varying their composition and their structure. We know also that the final behavior of materials such as steel depends on how well the various phases interact with each other. This is vitally important with composites. As human-made materials, composites can be designed to have isotropic properties and, if so designed, they probably would be homogeneous materials with their components distributed and arranged in a uniform pattern; or the components can be distributed in a nonuniform manner and, if so, the composites would have anisotropic properties. Third, similar to forged metals, a composite material can have its components distributed and arranged in a particular orientation that would result in directional properties. These last statements help explain some of the interest in composites in recent years by materials scientists who can, with the new technologies of materials, design their own material to fit the specific requirements of a particular application that cannot be met by existing technology or conventional homogeneous materials.

The characteristic properties of individual constituents of a composite interact in various ways to produce the collective properties or behavior of the composite. Some properties obey the *rule of mixtures (ROM)* because the composite properties are the weighted sums of the values of the individual constituents. In other words, properties are a function of the amounts and the distribution of the contributing material. Density, specific heat, thermal and electrical conductivities, and some mechanical properties, such as modulus of elasticity, follow this rule (see Section 8.5.1). In some composites the properties of the components are somewhat independent and supplement each other to produce a collective performance by the composite. Other composites have resultant properties that are the net results of the interaction or interdependence of the components with each other. This later type of composite behavior is by far the most important because the end result of combining materials into a materials system is a performance or set of properties that far exceeds the individual properties possessed by the components acting alone. An example of this type of composite can be taken from the fiber-reinforced plastics (FRP) composites, such as glass fiber embedded in plastic resin. The glass fibers possess extremely high tensile strength, but being very brittle they cannot be used alone. A plastic resin, on the other hand, is relatively weak but very ductile. Once a sufficient number of glass fibers are embedded in a plastic-resin matrix in a unidirectional manner parallel to the direction of the load, the two components act together as a unit to withstand the load by deforming equal amounts and sharing the load proportionately so that the composite or materials system achieves higher tensile strengths than otherwise is possible by the individual components.

8.5.1 Rule of Mixtures

Using fiber-glass filler and a polyester-resin matrix as our composite, we can sketch the stress–strain curves for each material as shown in Figure 8-17. The modulus of elasticity *(E)* for glass is about 69 GPa and for polyester, around 69 MPa. Tensile strength for glass (single fiber) is about 3.45 GPa.

The curves show the great differences in the stiffness (slope of the curves) between the two materials. Glass is extremely brittle and shows little evidence of any elastic strain up to the point of fracture. As indicated by its values of modulus and the slope of its stress–strain curve, polyester is a very ductile material. If each material were deformed an equal amount, the stress carried by the glass would be over 1000 times greater than the stress on the polyester.

If these two materials were combined in one ideal material, the glass would act as continuous fibers completely surrounded by a matrix of polyester, which forms a bond between the fibers and the matrix. The bond would allow both to deform (elongate) the same amount under a uniaxial load.

Research shows that the modulus of a unidirectional composite material in which (1) the load is applied parallel to the continuous fiber and (2) the bond between the fiber and the matrix material is strong enough to permit both materials to deform equally is the weighted sum of the modulus of the fiber and the modulus of the matrix. This last statement is an example of

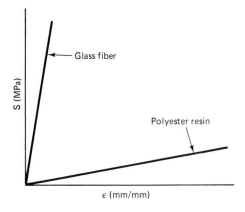

Figure 8-17   Stress–strain curves for glass fiber and polyester resin.

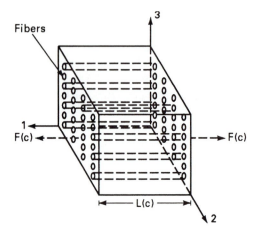

Figure 8-18 Fiber-reinforced-composite material undergoing isostrain loading conditions.

a *simple mixture law.* Mixture laws for physical properties such as specific heat, density, and conductivity for a composite material can also be developed. For such composite materials, most of the flow of thermal energy or electrical current will be through the component that is the best conductor.

The preceding discussion assumed that the chemical bond between the fibers and the matrix was greater than the strength of the matrix materials. This is not always the situation. Much research goes into determining the conditions necessary to improve the bonding so that the stress transfer between the matrix and the fibers can be made without fracture. Studies of metal–matrix composite materials have led to the development of composite materials that can withstand temperatures in excess of 670 K and retain great stiffness. Such materials are represented by tungsten, boron, graphite, or silicon carbide fibers in aluminum, cobalt, or nickel matrices.

Certain properties of a composite depend on the relative amounts and properties of the components of the composite material. Using the law or rule of mixtures (ROM), an estimate of properties such as mass, density, modulus of elasticity, thermal conductivity, or electrical conductivity can be determined provided that the reinforcing fibers are continuous, unidirectional, and have the same length as the matrix. Also, the load on the composite is a tensile load parallel to the longitudinal axis of the fibers. It is to be emphasized that not all properties of a composite follow the rule (i.e., some composite properties are in simple proportion or are weighted averages of their components).

The ROM will be developed to express the elastic modulus in tension for a fiber-reinforced matrix material conforming to the constraints stated above. Figure 8-18 is a sketch of the composite loaded with a tensile load, F(c). The following abbreviations represent the quantities involved in this development and in further discussions of the ROM to follow:

$$V = \text{volume fraction}$$

$$s = \text{stress}$$

$$e = \text{unit deformation, strain}$$

$$E = \text{modulus of elasticity in tension}$$

$$A = \text{area}$$

$$(f) = \text{refers to fibers}$$

$$(m) = \text{refers to matrix}$$

$$(c) = \text{refers to the composite material}$$

$$S = \text{tensile strength}$$

$$F = \text{load or force acting on a material}$$

$$L = \text{length}$$

$$p = \text{density}$$

$$m = \text{mass}$$

Referring to Figure 8-18, the load *(F)* is carried by both the fibers and the matrix acting to share the burden:

$$F(c) = F(f) + F(m) \tag{1}$$

Assuming that the bond between the fibers and the matrix is good, we can say that both the fibers and the matrix will stretch or deform the same amount in withstanding the tensile load.

$$e(c) = e(f) = e(m) \tag{2}$$

Using the definition of stress (Section 4.2.1), $s = F/A$, and Hooke's law (Section 4.2.3), $s = E/e$, equation (1) can be written

$$s(c)A(c) = s(f)A(f) + s(m)A(m) \tag{3}$$

Dividing equation (3) by $A(c)$ yields

$$s(c) = \frac{s(f)A(f)}{A(c)} + \frac{s(m)A(m)}{A(c)} \tag{4}$$

The ratios $A(f)/A(c)$ and $A(m)/A(c)$ are area fractions. Because the composite and its components are arranged in parallel, areas $A(f)$ and $A(m)$ are proportional to their volumes. Thus, volume fractions can be substituted for area fractions:

$$s(c) = s(f)V(f) + s(m)V(m) \tag{5}$$

Returning to equation (2) with Hooke's law and substituting for s yields

$$E(c) = E(f)V(f) + E(m)V(m) \tag{6}$$

Equation (6) expresses the ROM in terms of the elastic modulus for the ***isostrain*** situation described in Figure 8-18. Note that if one volume fraction is known, the other volume fraction can be found using the relationship $V(f) + V(m) = 1$.

Illustrative Problem

A laminar composite of epoxy resin reinforced with graphite fibers is loaded similarly to the model in Figure 8-18. Assume that there are equal volumes of graphite and epoxy sustaining the load. The modulus of elasticity for the graphite fibers is 58×10^6 psi and for epoxy it is 0.6×10^6 psi. Find the modulus for the composite.

Solution

$$E(c) = E(f)V(f) + E(m)V(m)$$

$$= 58 \times 10^6 \times \tfrac{1}{2} + 0.6 \times 10^6 \times \tfrac{1}{2}$$

$$= 29.3 \times 10^6 \text{ psi}$$

Figure 8-19 illustrates a model of the ***isostress*** loading of a composite material. The load, $F(c)$, acts transversely as opposed to longitudinally or parallel to the long axis of the fibers. The fibers and the matrix each resist the load equally. Stated in equation form, $F(c) = F(f) = F(m)$. This also means that the stress in the fibers and in the matrix is equal. However, the strain

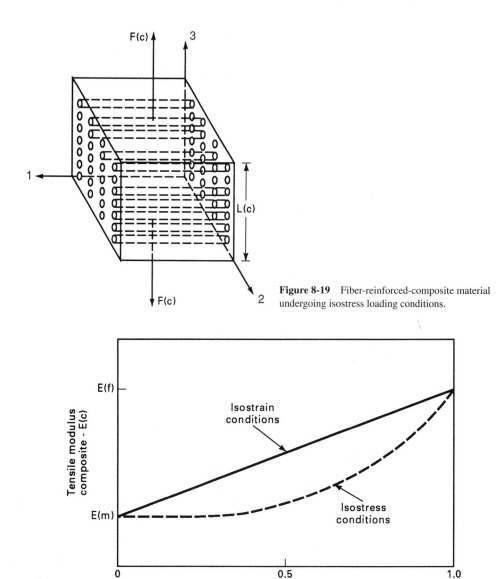

Figure 8-19 Fiber-reinforced-composite material undergoing isostress loading conditions.

Figure 8-20 Graphical plot of the tensile modulus of a unidirectional composite material loaded under isostrain and isostress conditions as a function of volume fraction of reinforcing fiber.

or deformation in each material is different. The weaker material will deform more than the stronger component. In equation form,

$$e(c) = e(f) + e(m)$$

The equation that expresses this type of loading, known as the ***inverse rule of mixtures,*** reveals that this isostress loading condition does not follow the rule of mixtures. In equation form, the inverse rule of mixtures is expressed arbitrarily in terms of modulus:

$$\frac{1}{E(c)} = \frac{V(f)}{E(f)} + \frac{V(m)}{E(m)}$$

Figure 8-20 contrasts these two types of loading and helps illustrate that, for a given volume fraction of fiber in a composite, the composite loaded under isostrain conditions is more efficient, having the higher modulus of elasticity for the composite. In fact, when looking at the curve in Figure 8-20, you can see that the modulus for the composite is not increased much over the modulus for the weaker component in the composite. On the contrary, using the same materials components and with the same volume fractions arranged in the isostress model, the composite modulus is about 10% less than that obtained in the isostrain model.

When determining how to arrange the components in a composite, it is important to recognize how a composite material is to be loaded. If the direction and approximate size of the loading are known beforehand, this simplifies the problem. If these conditions are not known, the correct components must be chosen and best arranged to support the load regardless of its magnitude (within limits) and direction.

8.5.2 Design and Fatigue

The design of a new or improved product for an industrial and/or technical application entails the detailed analyses of many factors. The design process links the scientific knowledge of materials with modern industrial methods to reach an overall objective of producing a better product at a lower cost. The interrelationship among materials, process, and design is nowhere more pronounced than in the use of composite materials to replace traditional materials such as steel in both structural and nonstructural components. Figure 8-21a is a flowchart showing the interaction of design, materials, and manufacturing processes that must exist to ensure the success of any design project.

Composite materials owe their success in large part to the fact that they provide the engineer with considerable design latitude compared to metals. This fact is in addition to their many other desirable properties. Fatigue in metals, discussed in Module 5, differs from fatigue in composites (see Figure 8-21b and Figure 8-24a). The mechanisms of crack initiation and crack propagation result in the failure of metals under cyclic, repeated loads at stress levels well below their ultimate strengths. Crack initiation is caused by stress concentrations, inclusions, or voids, which are then acted on by the mechanism of crack propagation, reducing the net-load-bearing area to the point where the fatigue stress exceeds the metal's ultimate strength. Final failure in metals usually results in a relatively clean fracture surface. In many instances the failure is of catastrophic proportions. Finally, metals exhibit an ***endurance limit,*** a stress level below which fatigue failure does not occur, regardless of the number of fatigue cycles applied.

Engineers have always been encouraged to look to nature to observe how certain tasks are accomplished or as an aid to stimulate the mind when involved in problem solving. One such example concerns the structure of seashells. The observations of birds dropping clams and oysters from considerable heights onto some hard surface to break the shells or sea otters smashing abalone shells with the aid of a rock attest to the hardness of these shells. The shells are made basically from a relatively soft chalk. What makes them extremely tough is the manner in which thin alternating layers of calcium carbonate and organic polymers are intertwined much like a layered composite material. These microscopic layers could be compared to a brick wall with each brick surrounded by mortar. Cracks in an individual brick are stopped from traveling into the adjacent brick by the mortar. Scientists at Sandia National Laboratory have studied this natural process and have attempted to duplicate it by using detergent molecules similar to ordinary household detergents. An arrangement of ***micelles,*** a structural unit made up of molecules, was developed, which upon being polymerized, bonded the surfaces of the micelles together resulting in a hard surface that can be used in such applications as automobile finishes or optical lenses.

In composites, crack initiation and propagation produce simultaneous growth of cracks that may (1) extend through the matrix, (2) be stopped at a fiber, or (3) propagate along a fiber–matrix interface. Cracks are initiated by factors such as filler or fiber debonding, voids, or fiber discontinuities. The crack propagation results in cracks joining each other to the extent that the matrix is unable to perform its basic function of transferring the load from one fiber to the next in fiber composites. The fracture surface usually shows evidence of a complex assortment of matrix failure, fiber failure, and fiber pullout.

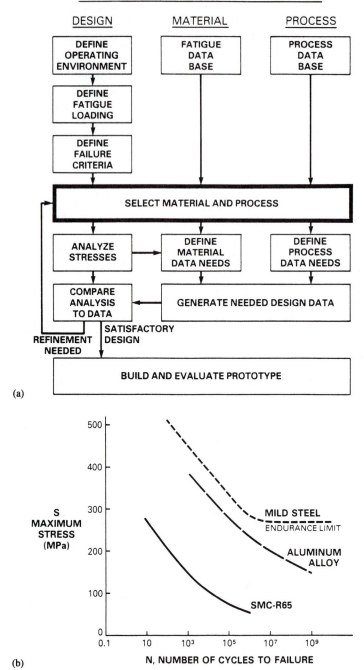

Figure 8-21 (a) Flowchart for fatigue design with fiber-reinforced PMCs showing need for databases. (General Motors) (b) Comparison of metals and PMC typical flexural fatigue sN diagram curves. (General Motors)

Other *fatigue–failure characteristics* of FRP composites are as follows:

1. A gradual decrease in the slope of the stress–strain curve (modulus of elasticity, *E*) occurs during cyclic loading.
2. Elevated temperatures normally reduce the performance of matrix-dominated composites (stress carried mainly by the matrix material). Fiber fatigue properties are better than those of the polymer matrix.

3. Moisture and chemical exposure can greatly affect the polymer matrix.

4. Notch sensitivity in high-cycle tensile-fatigue conditions is less than for metals.

5. There are considerably fewer matrix-dominated properties than fiber-dominated properties. Hence tensile, compressive, and shear stresses in the interlaminar (through-the-thickness) direction, being matrix dominated, should be kept to a minimum.

8.5.3 Toughness and Impact Strength

As measured by the ASTM D256 test on unnotched Izod specimens, the impact strength at room temperature for "tough" plastic molding materials is at least 8 J/cm (15 ft · lb/in.). A value of at least 10 J/cm^2 (50 ft · lb/in.^{2}) using the high-speed tensile impact test (ASTM D1822) also qualifies such materials as being tough. Considerable impact testing of such materials using glass- or carbon-fiber reinforcement in varying amounts with different resin materials has been done. Only recently, however, have such tests been carried out at low temperatures (down to 200 K). From our study of other materials, we know that the impact strength of a notched specimen is less than that of an unnotched specimen because notches produce stress concentrations. We also know that it takes energy not only to start a crack but also to propagate one. With notched specimens the crack is already provided, and depending on the sharpness of the notch, less energy is usually needed to propagate it. The tremendous effect of surface imperfections such as cuts, tool marks, and scratches on the toughness of materials is well known. For many reinforced thermoplastics, toughness drops off and the notch sensitivity increases at low temperatures. However, there are exceptions. Polypropylene, as well as acetal copolymer, becomes tougher, as evidenced by higher test values for unnotched specimens. Glass-reinforced polycarbonates show little change in toughness, although their notch sensitivity is greater at lower temperatures. With DuPont's "super-tough" nylons (Zytel ST resin), notch sensitivity decreases with increased glass- or carbon-fiber content, while at low temperatures, notched Izod impact-strength readings increase with fiber content. Using unnotched specimens, the toughness of these nylon resins increases with increasing fiber content at both room temperature and low temperatures (Figure 8-22).

Impact testing measures the material's resistance to fracture under certain prescribed test conditions when a standard specimen is struck at high velocity. In addition to the Izod and Charpy notched and unnotched specimens using a swinging pendulum, other impact tests, such as the falling ball, falling dart, high-velocity tensile stress, and fracture-toughness tests, are also used to measure toughness. *Fracture toughness,* in units of stress times the square root of a crack length (MPa · m$^{1/2}$), is an indicator of a material's resistance to the extension of a preexisting crack. The high-speed tensile test uses unnotched specimens and defined impact strength as the area under the stress–strain curve, or the energy required to break a material (toughness). Impact strength using this test setup is expressed in units such as kilojoule per square meter (kJ/m^2). Those tests using notched Izod or Charpy specimens express impact toughness in terms of the energy per length of notch, or kJ/m, as one example.

8.5.4 Test Values and Data

Test values and other data presented in this module and elsewhere are typical values for the material. They are offered as an aid to understanding. The properties of parts fabricated from a particular material are too contingent on many factors, such as part configuration, molding techniques, or curing times, to guarantee reproducibility of the data.

 EMT WEBLINKS

Advanced Materials Instruments and Analysis Inc.—*www.amia.com*

Detroit Testing Laboratory Inc.—*www.Dettest.com*

Materials Research & Engineering Inc.—*www.henge.com/mre*

Figure 8-22   An application that takes advantage of the low notch sensitivity of glass-reinforced Zytel nylon.

Polymer Diagnostics Inc.—*www.polymer diagnostics.com*
University of Dayton Research Institute—*www.udri.udayton.edu/*
Wyle Laboratories Inc.—*www.wylelabs.com*

8.6 TYPES OF COMPOSITES

Products resulting from different manufacturing techniques, different reinforcing components, and the specialized nature of the parts themselves result in many types of composites. We have divided this ever-increasing variety of composites into fiber, laminated sandwich, particulate, flake, and filled composites (see Figure 8-2b).

8.6.1 Fiber-Reinforced Composites

8.6.1.1 Polymer matrix composites (PMCs) and glass-reinforced plastics (GRPs).   GRPs represent the earliest and the most widely used (over three-fourths of total fiber-reinforced composite production) fiber–resin composites. With glass fibers in various forms coupled with either a thermosetting or thermoplastic resin, these composites can be produced without the need for high curing temperatures or pressures. The product contains a good balance of properties, has high corrosion resistance, and is low in cost for a multitude of uses as structural, industrial, and consumer-related products, ranging in size from minute circuit boards to boat hulls. Using 20–40% fiber loadings, the composites will, in general, double the strength and stiffness of the plastic resin used alone. Continuous fibers will increase these properties fourfold, with accompanying desirable decreases in thermal expansion and creep rate and with increases in impact strength, heat-deflection temperatures, and dimensional stability. These fibers may take the form of continuous filaments (monofilaments) or yarn. Figure 8-23 shows applications of GRPs.

(a)

(b)

Figure 8-23   (a) Fiberglass composite wind-turbine blades (layers of glass mat and rovings in a vinyl resin matrix). GRP provides high strength, light weight, low cost, extended service life, and increased weatherability. (Morrison Molded Fiberglass Co.) (b) A redesign of the Ford Taurus chassis. This prototype chassis reflects designers' keen interest in using reinforced-plastic composite technology for improved vehicles. GRP would allow 8 or 10 parts to replace over 400 stamped steel parts. A GRP chassis would save about 30% of the car weight because a lighter chassis means downsizing engines, brakes, and other components. Additional benefits come through more efficient manufacturing techniques. (Courtesy of Ford Motor Company)

The disadvantages accompanying these composites arise, in the main, from the fact that they are especially two-phase structures. This leads to a degree of environmental degradation greater than that experienced by either component material alone. Residual stresses and electrochemical effects result from the marriage of two dissimilar materials. Furthermore, the diffusion of fiber materials into the matrix materials, and vice versa, may take place at several stages. Variation in the thermal expansion of these two materials leads to thermally induced stresses that result in warping, plastic deformation, cracking, or combinations thereof.

A recently developed FRP composite rebar, called **C-bar,** is a likely candidate to replace epoxy-coated steel reinforcing bar (rebar) in many applications. C-bar is composed of unidirectional glass fibers wetted by recycled terephthalic acid resin in the center core. Unidirectional mat and ½-in. chopped fibers with a polyester surfacing coat are added to the exterior. The base resin used only on the outside is a polyurethane-modified vinyl ester. What sets C-bar apart from other forms of FRP rebar is the combined use of both mat and fibers, plus the existence of a mechanical and a chemical bond between the acid resin in the center core and the vinyl ester used on the outside. Tensile strength is reported to be more than twice that of steel rebar, with four times less weight. C-bar is nonmagnetic and noncorrosive, and has a more compatible coefficient of thermal expansion than that of concrete, attributes that can lead to many potential uses. The hybrid process combines pultrusion with compression molding (see Section 8.7), applying the fiber circumferentially to the inner core of the rebar. The result is a consistent, fully cured cross-section at a processing rate of 20 ft/min, which is unattainable with any pultrusion process.

Naturally, the advantages of such composites outweigh the disadvantages. Some of these major advantages can be summarized briefly as follows:

1. Greater matrix strength and stiffness, which permits little or no cross-plying and greater joint and stress concentration load strength
2. Increased operating temperatures and increased durations of time spent at those temperatures
3. Increased resistance to high temperatures and humidity environments
4. Better comformability with existing metal fabrication techniques

The FRP industry is competing with the manufacturers of railroad cross ties made of treated wood to produce composite cross ties. The traditional market uses 75 million ties per year with 20% of that volume sold in the United States. Seaward International, Inc. has developed **thermoplastic cross ties** reinforced with **fiberglass rebar** based on the company's successful marine timber technology. This latter technology uses a continuous, coextruded process, with the company's Duralin Plastic as a matrix and fiberglass rebar as reinforcement. A high-strength outer skin of plastic covers the inner core of matrix and rebar. Another company, TieTek, Inc., is using fiberglass as chopped strand throughout the matrix which also includes other materials such as rubber from recycled tires. Both companies feature recycled high-density polyethylene in their designs. Accelerated aging tests of TieTek's composite cross ties show that they lose only 25% of their performance properties in 15 years whereas the properties of wooden ties drop off by 50%.

The recent switch from steel to FRP composites for **dowel bars** was brought about after almost 15 years of service on heavily traveled roads in Ohio. The dowel bar serves as the transfer joint mechanism connecting adjacent slabs of concrete roadway. When vehicles travel over the joint, dowel bars restrain vertical movement between slabs by transferring loads from one slab to another. Carbon steel (mild steel, general-purpose grade) bars are susceptible to corrosion from moisture and deicing salts, limiting their life span to 7 to 15 years depending on their design and location, whereas concrete slabs can last 35 to 40 years [see Module 4 for more details on corrosion and refer to the American Concrete Paving Association (ACPA) for further details]. When a steel dowel corrodes, it loses its ability to transfer loads and can "lock" in the joint which produces excessive loads and causes the concrete to break down. This, in turn, produces a familiar "thump-thump" as wheels pass over the joints. Traditional joints use a steel dowel typically 1.5 in. diameter by 18 in. long, coated with epoxy. After 13 to 15 years of service during the testing period, the epoxy coating on the steel bars delaminated and could be removed as a thin film shell. The composite dowels were 1.00 to 1.25 in. diameter pultruded by Strongwell (Bristol, Virginia) of 78% E-glass fiber-reinforced vinyl ester. When these composite dowels were removed for testing after 13 to 15 years they "looked like new dowels that just came out of a pultrusion machine" (see Section 8.7.5 for details on pultrusion). In the past 2 years composite dowels have been installed in roads and highways of Iowa, Wisconsin, and Illinois.

Figure 8-24   (a) Metal matrix composites (MMCs): photomicrographs. (NASA) (b) Specific strength versus temperature: metal alloys and composites. (NASA) (*continued*)

8.6.1.2 Metal matrix composites (MMCs) and ceramic matrix composites (CMCs).   As with plastic resins, many fibers are used with metal matrices to form composites. The volume fraction of reinforcing materials plus interface–interphase reactions play a leading role in determining the ultimate success of any metal matrix composite in meeting the demanding requirements of a particular design. Figure 8-24a shows photomicrographs of metal reinforcing fiber embedded in a metal matrix.

The major rationale for the development of such composite materials is to satisfy the important need for tough, strong materials capable of maintaining their special properties under

(c)

Figure 8-24 (c) A titanium-matrix composite reinforced with silicon carbide continuous fibers that have been coated with a beta titanium alloy by sputtering. The composite was formed using vacuum hot pressing. Magnification 200X. (Courtesy of James Baughman, Analytical Services and Materials, NASA Langley Research Center, Langley, VA) (*continued*)

high-temperature operating conditions. Figure 8-24b is a plot showing how the specific strength of some materials varies with temperature.

Originally developed for the aerospace industry, these composite materials are now finding many uses in other industries. Metal matrix composites (MMCs) and ceramic matrix composites (CMCs) consist of a base (matrix) that is reinforced with continuous fibers or discontinuous materials in particulate form.

MMCs are metals that are reinforced with particulates or filaments of other metals or nonmetallics that provide properties not possible with metals alone. These properties include higher-temperature operating limits than the base metal and the tailorability to give improved strength, thermal conductivity, stiffness, abrasion and creep resistance, and dimensional stability. An MMC has a monolithic alloy as a continuous matrix and the reinforcement is a high-performance carbon, metallic, or ceramic addition. Being a metallic alloy, the matrix is not attacked by organic fluids such as fuels and solvents. Reinforcement takes the form of continuous or discontinuous reinforcements and may represent up to 60vol% of the MMC. Continuous fibers may be carbon, silicon carbide (SiC), boron, alumina (Al_2O_3), and refractory metals. Discontinuous reinforcements may be SiC in the form of whiskers or particulates, alumina in the form of particulates or short or chopped fibers, and titanium diboride (TiB_2) in the form of particulates.

Processing techniques are divided into two parts—primary and secondary processing. Primary processing involves the introduction of the reinforcement into the matrix in the proper amount, location, and orientation and the bonding of the reinforcement to the matrix. Secondary processing involves translating the composite into a finished hardware component.

Many reinforcing agents are not compatible with their matrix metals—either a negative reaction causes one or both components to suffer degradation or a bonding failure occurs. This situation involves the interface between them. (Refer back to Figure 8-4 showing the interface in a fiber-reinforced composite.) Solutions to this problem take the form of coatings, surface treatment of the reinforcement, or additions to the matrix to make it more compatible.

There are six primary processing methods for fabricating MMCs. Solid-state processing involves diffusion bonding of materials and powder metallurgy techniques, stir casting, squeeze casting, rheocasting (also known as compocasting), pressureless metal infiltration (see Lanxide Primex process in this module), and spray deposition.

The following types of matrix composites incorporate different materials and properties to provide numerous applications.

• ***Aluminum matrix MMCs.*** Aluminum is the most popular matrix material because of its melting point. Most commercial applications do not require too high a melting point for their matrix materials. A relatively low melting point simplifies and makes composite processing relatively less complex and less costly. Aluminum also imposes few compatibility problems with ***reinforcements.***

   • *Boron-aluminum* MMCs are continuous boron (B) fiber composites. Made by a hot press diffusion bonding or plasma spray process, boron is expensive and finds use primarily in the aerospace industry. The midfuselage structure of *Orbiter* is made of tubular truss members that employ B–Al composites.

   • *Silicon carbide–aluminum* composites can maintain their strength up to 500°F. SiC_d/Al are discontinuous silicon carbide composites materials using SiC whiskers, nodules, flakes, platelets, or short fibers in an aluminum matrix. The subscript following the abbreviation for carbon indicates the reinforcement is a particulate of some form. The letter *p* is also used to signify this type of reinforcement.

   • *Graphite-aluminum* MMCs introduced in the 1960s are an example of the negative effect of the matrix material and its reinforcement. Molten aluminum does not wet the carbon fibers; it attacks the carbon and the carbon itself will oxidize. Combined, these situations make processing difficult. Liquid metal infiltration, vapor deposition, diffusion bonding, pultrusion, and squeeze casting are some of the processes used in fabricating this composite.

   • *Alumina-aluminum* MMCs using inexpensive ceramic fibers have improved resistance to wear and thermal fatigue. Continuous fiber alumina-aluminum composites use a preform made of alumina tapes wherein the fibers are arranged in the required direction.

• ***Magnesium matrix composites.*** Magnesium composites have similar properties to aluminum composites that include low coefficient of thermal expansion (CTE), high stiffness, and light weight (Mg is nearly two-thirds as dense as Al). Continuous fiber C/Mg composites are used in the aerospace industry; Al/Mg composites using alumina short fibers are used in making automobile engine components; and SiC_d or B_4C/Mg composites find use in engine components and low expansion electronic packaging materials.

(d)

Figure 8-24   (d) Ceramic-matrix-composite (CMC) parts, such as this cam roller follower, are made from a liquid process, developed by Sullivan Mining, that does not use the conventional powder process. The process is less damaging to reinforcing fibers and capable of mass producing complex shapes at about one-tenth the cost of parts made by conventional processing. At least as durable as steel rollers, they may run more quietly. (*Advanced Materials and Processes*)

- *Titanium matrix composites.* Titanium retains its strength at moderately high temperatures and has excellent corrosion resistance. Titanium reacts negatively with many reinforcements. To overcome this problem, titanium composites are hot isostatic pressed and vacuum hot pressed rather successfully. Another method calls for plasma spraying the titanium matrix onto SiC fibers. These processes result in a titanium composite that is significantly stronger than unreinforced titanium. Figure 8-24c is a photograph of the cross-sectional area of a titanium matrix composite consisting of continuous SiC fiber reinforcement that was sputter coated with a beta titanium alloy and vacuum hot pressed forming the composite material. This coating is necessitated by the adverse reaction between the titanium matrix and the SiC fibers. One of many film deposition techniques, *sputtering* is a coating process in which ions in a plasma are used to knock off atoms of the coating material (the target) in a vacuum chamber which diffuse to the adjacent substrate, effectively coating it with the target material. Continuous fiber reinforcements are used in aerospace for structural applications and compressor blades for advanced turbine engines.
- *Copper-matrix composites.* Copper is a better conductor both electrically and thermally than aluminum. As a composite, copper has higher temperature strength properties than aluminum. W/Cu fiber composites with their high-strength properties up to about 925°C are finding more uses in rocket engine designs. C/Cu MMCs that are less dense, stiffer, and better thermal conductors than copper have been developed. One result of this effort is the making of graphite fibers coated with pure copper that are flexible enough to be made into a fabric.
- *Superalloy matrix composites.* Since the 1950s, superalloys have found use in turbine engines. In the 1960s, superalloy composites have been developed for similar applications. Using refractory metals reinforcement of which tungsten, molybdenum, tantalum, and niobium fibers received the most attention, the strongest fiber developed was tungsten with a strength of more than 2070 MPa at 1095°C. Iron, nickel, and cobalt matrix materials permit the enclosing of tungsten reinforcements without adverse effects. A W/FeCrAlY superalloy composite has been used to fabricate a hollow composite blade.

Metal matrix composites due to their improved mechanical properties over traditional alloys are finding applications as shaped charged liners, military armor, and piston engine cylinder liners. A novel method of producing MMCs is *spray atomization* and *co-deposition.* Several advantages of this new processing technique are (1) low processing temperatures can be used due to the highly efficient heat convention during atomization, (2) spray deposited materials exhibit some characteristics of rapid solidification which produces increased solid solubility and nonequilibrium phases among others, (3) the process has the potential of being utilized for near-net-shape manufacturing of difficult-to-form materials such as *refractory metals,* and (4) the process has the potential of being a cost-effective manufacturing technology because it avoids powder handling.

A second family of aluminum metal matrix composites, *ceramic reinforced aluminum MMCs,* provides a new dimension in tailorability. Structural, thermal, and abrasion-resistant properties are now adjustable to provide customized materials that are castable and extrudable into complex to near-net shapes for a variety of applications.

Figure 8-24d depicts how the Chrysler Corporation used the Sullivan process to produce a single-piece, CMC Si_3Ni_4, cam roller follower. As discussed in Module 7, this CMC component replaces 18 small steel bearings, potentially costs less to mass-produce than steel parts, and offers longer wear life. *Continuous reinforcement* is defined as fibers, usually oriented in one direction, that produce an anisotropic composite in which the fibers predominate, with the matrix serving as a vehicle for transmitting the load to the fibers. Nickel-based superalloys are slowly being replaced by composites in high-temperature applications for aircraft engines. This development began in the 1960s. SiC fiber reinforcement added to a matrix, which is itself a composite material, makes for another interesting material. The matrix consists of molybdenum disiliside ($MoSi_2$) which contains a reinforcement of 30 to 50 vol% of Si_3N_4 particles. This composite material has a higher temperature oxidation behavior with a lower density and a higher melting point temperature than super alloys. It retains excellent strength and toughness improvements at temperatures up to 1400°C. The addition of Si_3N_4 to $MoSi_2$ doubled the room temperature toughness and reduced the high-temperature creep rate by about five orders of magnitude. The addition of SiC fibers to the matrix materials increased the room temperature fracture toughness and impact resistance.

To satisfy the need for materials that are lighter, stronger, more corrosion resistant, and capable of performing at elevated temperatures, the U.S. Department of Energy established a comprehensive program to develop a relatively new family of materials known as ***continuous fiber ceramic composites (CFCCs).*** Near-term applications are in the mid-temperature (500° to 1000°C), secondary aerospace structures such as exhaust components, hot gas ducting, and leading edges. In the longer term, high-temperature applications in the automotive and industrial fields will have great potential, provided the costs of fabrication can be lowered to a competitive basis. Several methods of fabricating these composites are used. Among them is the use of preceramic polymers for the low-cost fabrication of large, complex-shaped CFCCs. After forming in the polymer condition, the composite is pyrolyzed to form the ceramic matrix and then densified through repeated reinfiltration/pyrolysis cycles. Regardless of what fabricating process is finally determined to have the most merit, the deciding factor will be the development of a cost-effective, oxidation-stable fiber interface that can withstand high strain fracture.

Continuous fiber ceramic composites (CFCCs) have been developed to overcome the brittle nature of ceramics and to capitalize on their favorable properties (see Figure 7-31). ***Discontinuous reinforcement*** refers to whiskers, short or chopped fiber, or particulates that produce a composite exhibiting a blend of the reinforcement and matrix properties. ***Discontinuous reinforced aluminum (DRA)*** can refer to a technology, a defense department program, or a specific metal matrix composite material. This technology is now capable of competing with conventional aluminum and titanium alloys as well as organic matrix composites. DRA is an aluminum metal matrix composite in which the reinforcement is a particulate, flake, whisker, or short fiber typically of a strong, stiff ceramic material such as Al_2O_3 or SiC. Highly versatile, this isotropic composite has a unique combination of strength, stiffness, and affordability as well as the ability to be processed and finished by conventional metal processing. It has good wear and corrosion resistance and can be anodized. Because its density is similar to conventional aluminum, its specific characteristics (specific compressive strength and stiffness) are superior to conventional aerospace materials. Figure 8-25a is a graph of DRA's compressive strength compared to other aerospace materials. Military applications are numerous especially in fighter aircraft. Figure 8-25b shows a U.S. Air Force F-16 fighter aircraft with ventral fins made of discontinuous reinforced aluminum MMC. These composite fins are projected to last more than 6000 flight hours as opposed to less than 400 hours obtained using the less stiffer aluminum alloy sheets bonded to honeycomb panels. The Hubble space telescope antenna wave guide mast and electronic packages for satellites are further aerospace applications for this new composite material. Automotive applications include Chevrolet Corvette and GMS/T truck drive shafts, Honda Prelude cylinder liners, Plymouth Prowler brake rotors, and GM Electronic Vehicle-1 brake drums. The last two applications take advantage of DRA's low density and high wear resistance. Table 8-6 compares the physical properties of two MMCs reinforced by particulates, emphasizing the difference in properties due primarily to the type of reinforcement. Most of these composites tend to be isotropic. MMCs and CMCs are composed of three constituents: the fibers, having high strength and stiffness; the matrix, which holds the fibers together and distributes the applied load; and the interface zone between the fibers and the matrix, which determines the wetting and bonding (coupling between the fiber and the matrix) as well as the transfer of the load between the fibers and the matrix (see Figures 8-4 and 8-7). The formation of a shallow interface zone is generally desirable to ensure a strong bond.

Most MMCs and CMCs can be synthesized using standard metal-working practices such as extrusion, hot forging, hot molding, superplastic forming, squeeze casting, powder metallurgy, liquid-melt infiltration, and rolling. A recent Mixalloy process uses turbulent eddies and jet impingement to mix two or more liquid metal streams at turbulent velocities in a chamber. The resulting mixture is then solidified fast enough to preserve the required microstructure. This process is unique because it is successful in generating nanoscale particles (e.g., 50-nm refractory boride) for use in a copper matrix. Coarser particles can also be obtained by careful

(a)

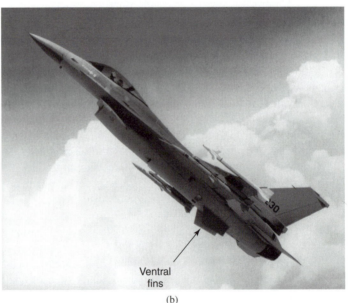

(b)

Figure 8-25 (a) DRA metal matric composite is an isotropic composite with a broad range of properties. With a density similar to conventional aluminum its specific properties are superior to conventional aerospace materials. (Courtesy of DWA Aluminum Composites) (b) The U.S. Air Force F-16 Fulcan fighter aircraft uses ventral fins located under the aft section of the fuselage to provide added stability during tight, high-speed turns. This material replacement program is expected to save the Air Force approximately $21 million in life-cycle costs. (Courtesy U.S. Air Force)

control of the mixing conditions. This new process is replacing older methods of producing a wide variety of dispersion-strengthened alloys (see Sections 8.6.4 and 8.6.4.2 for further information). In addition, brass and bronze can be strengthened by the Mixalloy process using boride particles. A coating of MMC consisting of a nickel matrix in which silicon carbide particles are uniformly disbursed has been developed for engine cylinder bores by U.S. Chrome Corp. Designated NiCom, the coating can be as thin as 0.005 to 0.0015 mm and as thick as

TABLE 8-6   PHYSICAL PROPERTIES[a] FOR SILICON-CARBIDE- AND ALUMINA PARTICULATE-REINFORCED (SiC_P/AL AND AL_2O_3/AL) METAL MATRIX COMPOSITES

	Reinforcement			
Property	Alumina (Al_2O_{3p}/Al)		Silicon carbide (SiC_p/Al)	
Filler content[b](%)	50	—	55	—
Density [g/cm^3 (lb/in.^{3})]	3.2	(0.12)	2.9	(0.10)
Tensile strength [MPa (ksi)]	530	(77)	250	(36)
Modulus [GPa (Msi)]	160	(23)	200	(29)
Elongation (%)	0.6	—	0.3	—
Compressive strength				
0.2% yield [MPa (ksi)]	520	(75)	460	(66)
Ultimate [MPa (ksi)]	900	(130)	490	(70)
Fracture toughness[c] [MPa$\sqrt{m}$ (ksi $\sqrt{in.}$]	18.5	(130)	12	(11)
Coefficient of thermal expansion [ppm/°C (ppm/°F)]	13.5	(16.8)	8.5	(4.7)
Thermal conductivity $\left[\dfrac{W}{m \cdot K}\left(\dfrac{Btu \cdot in.}{hr \cdot ft^2 \cdot °F}\right)\right]$		(7.5)	160	1100

Source: Courtesy of Lanxide.

[a]Room temperature (25°C) properties. Elevated temperature properties available.

[b]Filler content can be tailored (20–80%) to obtain desired properties. Properties as a function of filler content available.

[c]Chevron notch beam.

1.5 mm. Average micro hardness ranges from 57 to 55 HRC. The SiC particles can range from as large as 20 μm to submicron size. According to the company, nickel coating is more efficient than chromium. Test results show that NiCom wears better than chromium in all engine bores. Other properties of NiCom are excellent frictional characteristics, oil retention, and environmental friendliness.

Whisker-reinforced MMCs have distinct advantages over conventional alloys. These MMCs, classified by some as dispersion-strengthened alloys, provide the highest-strength discontinuous-reinforced MMCs. Typical fibers are SiC, Al_2O_3, and Si_3N_4, which are combined with some common metal matrices, such as Al, Mg, Ti, and Cu. Powder metal (P/M) or liquid-melt infiltration methods are used to synthesize the fine-mesh metal-alloy powders and the whiskers by thorough mixing, blending, and consolidating to produce a near-net-shape component. SiC whiskers reinforcing an Al matrix have transverse tensile properties that are nearly equal to the longitudinal properties, making them capable of outperforming fiber-reinforced composites in a situation requiring multidirectional reinforcement.

Lanxide Corporation fabricates MMCs by using a pressureless, molten-metal-infiltration process (trademarked Primex) that features excellent wetting of the reinforcement by the matrix alloy. (See Module 7 for information on Lanxide's fabrication of CMCs.) Net- and near-net-shape parts, small to large, having complex shapes can be produced economically. Figure 8-26a is a schematic of the process for making an electronics package. To produce an aluminum-alloy MMC, for example, the infiltration is carried out in a nitrogenous atmosphere at a temperature somewhat above the melting point of aluminum. No external pressure or vacuum is needed. The MMC produced is completely infiltrated with metal alloy and features high and uniform loadings of particles or fibers. Figure 8-26b is a photomicrograph of an aluminum matrix MMC filled with fused-Al_2O_3 particles, and Figure 8-27 is a similar photomicrograph showing Al_2O_3 agglomerates (clusters of particles). These figures illustrate complete infiltration of the spaces between the particles by the molten aluminum alloy. One application of this SiC/Al MMC is for support structures for electronic components. Typical requirements include a low coefficient of thermal expansion (CTE) to reduce mechanical stresses imposed on the component during attachment and operation, high thermal conductivity for heat dissipation, and low density for minimum weight. Table 8-7 compares the physical properties of SiC_p/Al MMCs to a typical aluminum alloy. In addition, the dimensional stability of this MMC can be more than 3.2 times that of unreinforced steels, aluminum alloys, and titanium alloys. Combined with high thermal conductivity and low CTE, this array of properties provides this

(a)

(b)

Figure 8-26 (a) Primex® pressureless metal infiltration process by Lanxide for producing MMC parts such as this electronics package to net- and near-net shape, ranging in size from small to large. (Lanxide Corp.) (b) A photomicrograph of Al_2O_{3p}/Al MMC showing a high and uniform particle loading. (Lanxide Corp.)

(a)   (b)

Figure 8-27 (a) A photomicrograph of an aluminum-matrix MMC filled with Al_2O_3 agglomerates. The molten metal completely infiltrates both the interagglomerate spaces and the minute openings within the agglomerates. (Lanxide Corp.) (b) The Primex process for aluminum MMCs, with 60–70% (volume change) particulate reinforcement, can be used in engine components such as connecting rods, piston pins, brake parts, brackets, and stiffeners. (Lanxide Corp.)

MMC with excellent rigidity and dimensional stability for use in space stations, automotive-brakes calipers, or optical system components. Table 8-8 lists some properties of $(Al_2O_3)_p$/Al MMC in which the particles are more coarse, resulting in a composite that has higher strength and fracture toughness than those of the parent alloy. Despite the lower ductility, these MMCs retain good fracture toughness. Table 8-9 provides an opportunity to compare properties of a

TABLE 8-7 PROPERTIES OF SiC-PARTICLE-FILLED Al MMCS FOR ELECTRONIC
APPLICATIONS

	Composite examples		Typical Al alloys (for comparison)
SiC loading (vol%)	55	75	—
Coefficient of thermal expansion	8.5	6.2	22–24
[$\times 10^{-6}$ ft^2 K ($\times 10^{-6}$ ft^2 °F)]	(4.7)	(3.4)	(12–13)
Thermal conductivity	160	170	150–180
[W/m · K (Btu/hr · ft^2 · °F)]	(93)	(99)	(87–104)
Density [g/cm^3 (lb/in.^{3})]	2.95	3.0	2.7
	(0.106)	(0.108)	(0.097)
Elastic modulus [GPa (msi)]	200	270	70
	(29)	(39)	(10)

Source: Courtesy of Lanxide.

TABLE 8-8 PROPERTIES OF AN Al$_2$O$_3$-PARTICLE-FILLED Al MMC

	Composite (51% Al$_2$O$_3$)	Base alloy (AA 520)
Tensile strength	531	331
[MPa (ksi)]	(77)	(48)
Elastic modulus	161	65
[GPa (msi)]	(23)	(9)
Fracture toughness[a]	18.5	N.A.[b]
[MPa · m$^{1/2}$ (ksi · in.$^{1/2}$)]	(16.8)	
Elongation (%)	0.63	16

[a]Chevron notch method.

[b]N.A. = not applicable.

CMC with an MMC. In particular, note the retention of strength properties at high temperatures for this CMC.

A major inhibition to the use of MMC parts is the difficulty in joining them. A newly developed fabrication process produces MMC parts with integral inserts. Parts consisting of a magnesium matrix and graphite fibers with titanium inserts are formed in final shape in a single casting step using a pressure casting machine. No finish machining is required. Molten magnesium is infiltrated into the graphite fiber preform reinforcement, which is inserted into a mold cavity along with the titanium insert. Pressures, temperatures, and times are calibrated to produce a complete infiltration of the densely packed preform and rapid solidification of the melt. Subsequent tensile testing produced fracture, at around 10,000 N, at a point well away from the insert.

The designation for aluminum MMCs, approved by the Aluminum Association Inc., uses a four-part system to identify the matrix alloy and the composite's amount and type of reinforcement. For example, 6063/Al$_2$O$_3$/15$_p$-T6, 6063 designates an alloy-6063 matrix composite reinforced by 15 vol% alumina particles and heat treated to T6 temper.

Carbon–carbon composites have been used on nose cones and leading edges of high-performance space vehicles subjected to very high temperatures (3000°C). Another high-temperature application has been as brake discs in racing cars. Such composites are stronger at high temperatures than at low temperatures, have high specific strength and good resistance to thermal shock, and are self-lubricating. Composed of a carbon fabric and an organic resin, the fabric is placed in a mold that is infiltrated (one method is CVD, described in Module 7) with a resin that is subsequently ***pyrolized*** (heated to a high temperature), driving off the noncarbon atoms in the polymer and thus forming a carbonized matrix. A coating of silicon carbide is added to reduce oxidation, which limits the high-temperature applications of these composite materials. Carbon–carbon composites are noted for their light weight and strength, and thus

TABLE 8-9 PHYSICAL PROPERTIES FOR SILICON–CARBIDE FIBER-REINFORCED ALUMINA SiC$_F$/Al$_2$)O$_3$[a]

Property	Temperature					
	25°C	(73°F)	1000°C	(1832°F)	1200°C	(2192°F)
Fiber content (eight plies) (%)	35	—	35	—	35	—
Density [g/cm^3 (lb/in.^{3})]	2.9	(90.10)	2.9	(0.10)	2.9	(0.10)
Tensile strength [MPa (ksi)]	283	(41)	—	—	—	—
Modulus [GPa (Msi)]	122	(17.8)	—	—	—	—
Poisson ratio	0.10	—	—	—	—	—
Flexural strength[b] [MPa (ksi)]	450	(65)	400	(58)	350	(51)
Modulus [GPa (Msi)]	200	(29)	—	—	—	—
Poisson ratio	0.29	—	—	—	—	—
Fracture toughness[c] [MPa $\sqrt{m}$ (ksi $\sqrt{in.}$)]	28	(26)	23	(21.5)	23	(21.5)
Interlaminar shear strength[d] [MPa (ksi)]	63	(9)	—	—	—	—
Shear modulus[e] [GPa (Msi)]	90	(13.1)	—	—	—	—
Coefficient of thermal expansion [ppm/°C (ppm/°F)]	5.8	(3.2)	—	—	—	—
Thermal conductivity $\left[\dfrac{W}{m \cdot K}\left(\dfrac{Btu \cdot in.}{hr \cdot ft^2 \cdot °F}\right)\right]$	8.7	(60)	5.7	(39)	5.5	(38)

Thermal shock			
(% retained flexural strength after quenching into RT water)	1000 to 25°C	92.7%	
	1000 to 25°C	85.2% . . . after 5 cycles	
	1200 to 25°C	83.6%	

Source: Courtesy of Lanxide.

[a]Silicon–carbide fiber (Nicalon); 12-harness satin weave, 0/90 two-dimensional preform layup.

[b]Four-point bend.

[c]Chevron notch beam.

[d]Short-beam method, ASTM D2344-84.

[e]Sonic method.

are likely candidates for applications in the aircraft and auto industries. Recent research in the field of damage detection, where sensors are either attached to or embedded in the structural material, led to the discovery of the electrical properties of carbon composite materials. Subjected to changes in temperature, the composite material produces changes in its electrical properties similar to the behavior of semiconductors. A most promising application for this technological advance is using carbon composites to act either primarily or secondarily as a sensor to eliminate the need for embedding of conventional sensors in aircraft parts.

Carbon–carbon composites also are replacing graphite electrodes which are difficult to use in fuel cells. Using a chemical vapor infiltration technique, a light carbon composite electrode just 2 mm thick with good electrical conductivity and low permeability has been developed.

8.6.2 Laminar Composites

Laminar composites (Figure 8-28) consist of layers (*lamina*) of at least two different solid materials bonded together so the fiber orientation runs at different angles (e.g., 0°, 45°, 90°). Lamination allows the designer of this composite material to use the best properties of each layer to achieve a more useful material. Properties such as wear resistance, low weight, corrosion resistance, strength, stiffness, and many more can be accented by a wise selection of different constituent layers. A fiber-reinforced composite made up of layers of tape, fabric, or mat can be considered a laminar composite. (Figure 8-28d). A well-known example of a laminar composite is plywood. Plywood has isotropic properties in the plane of each sheet due to the layers of wood bonded with a thermoset resin such that the longitudinal direction is at right angles in adjacent plies. Only their orientations differ. ***Clad metals*** such as bimetallic strip or copper-clad stainless steel are composed of two different materials. In the former, the two sheets of metal bonded together have different amounts of thermal expansion, and in the latter example the thermal conductivity property is the rationale for the design of the composite ma-

(a)

UNIDIRECTIONAL

CROSSPLIED
QUASI-ISOTROPIC

$0°$
$0°$
$0°$
$0°$
$0°$
$0°$
$0°$
$0°$

$0°$
$90°$
$+45°$
$-45°$
$-45°$
$+45°$
$90°$
$0°$

(b)

(c)

Wind
& Rain

Outer fabric

GORE-TEX®
membrane

Liner

Sweat vapor

Figure 8-28 (a) Fiber orientation within laminates. Laminates can be tailored to meet specific requirements, with savings in time and material. (Hercules, Inc.) (b) Gore-TEX® laminated membrane used by manufacturers of clothing and footwear for use outdoors in all types of weather. (W. L. Gore & Associates, Inc.) (c) Gore-TEX® lamina between liner and outer fabric allows sweat vapor to escape but keeps large water drops out of foul-weather gear. (*continued*)

(d)

Figure 8-28   (d) Figure 2 shows the traditional reflected light bright-field technique. The carbon fibers are seen as white rods, while the polymer resin is grey. It is important to note that the soft polymer matrix encapsulating the hard carbon fibers must be polished with no relief at the interface. An optical wedge created by thinning the polymer just under the surface adjacent to the fiber results in the observed prism. The multiphase resin morphology is difficult to see with reflected light microscopy.

In contrast to reflected light microscopy, thin-section micrographs of baseline interior (3a-3b), aged interior (4a-4b) and aged surface (5a-5b) reveal changes in resin morphology caused by thermal and oxidative processes. Since thin-section microscopy utilized transmitted light, carbon fibers are seen as black rods and the resin morphology is visible in high resolution.

The process of making thin-sections is unique, and specimen preparation takes about three days. The technique requires polishing two surfaces one to four microns apart, while keeping each surface flat. The specimens must be strain-free, because stresses will distort the microstructure. Micro-toming is not an acceptable thin-section preparation technique because severe distortions result from stresses during cutting. Complimentary microscopy techniques including PC, DID, and polarize light with waveplate can be implemented when the specimen preparation is complete.

Conclusion

Thin-section microscopy is useful for understanding resin morphology in multi-phase systems and changes that result from environmental factors, such as isothermal oven aging. Thin-sectioning requires time consuming and careful specimen preparation, but yields information on resin morphology that is not found using other microscopic approaches. Extensive details of the resin morphology are seen using PC, DIC, and waveplate techniques, which are required for a thorough evaluation of resin degradation. (Luther M. Gammon, Boeing Commercial Airplane)

terial. Similar treatments with paper produce isotropic properties. Laminated paper with plastic film or metal foil, or using polymers, such as nylon fabric, as a base and laminating the polymer with layers of metal, produces an endless variety of plastic-based laminates. Roofing paper, Formica, and Kevlar fibers laminated with a resin to produce a material for bulletproof vests are additional examples of these composite materials on the market today.

Safety glass is a laminated glass consisting of a layer of poly(vinyl butyral) bonded between two layers of glass. Glass alone is quite brittle and dangerous due to its proclivity to shatter into many sharp-edged pieces. Poly(vinyl butyral) is very tough but weak in scratch resistance. The glass lamination protects the plastic from scratching and gives stiffness, while the plastic contributes its toughness. By protecting the other in different ways, each material allows the other to contribute to the composite material's vastly improved properties over those of its constituents.

A layered composite known as *Graftskin* is an artificial human skin replacement tissue that replaces both the outer epidermal and inner epidermal skin layers. It blends dermal cells with collagen—the protein that forms a fibrous framework for skin cells. The collagen is taken from cows. Graftskin is used successfully for treating chronic skin diseases, burns, and deep wounds. The results of such treatment show quick wound closure, accelerated healing, and reduced scarring.

Laminates formed by high pressure and heat are referred to as *rigid laminates.* Some examples of rigid laminates are countertops, rods, tubes, and printed-circuit boards (PCBs) for the electronics industry.

Unsolved manufacturing problems associated with the processing of bulk-laminated ceramic or metal composites limit their potential use in structural applications. Design engineers, in their search for solutions to structure problems, have observed how nature produces many examples of laminated composites that possess excellent mechanical properties. Seashells would be one example of this type of structure. Ceramics, with their inherent properties of high-temperature strength, hardness, and wear and oxidation resistance, also possess low toughness. Module 7 contains much information on how materials engineers and scientists are combating this problem of brittleness in ceramics by mechanisms such as crack bridging and transformation of residual stresses, to mention but a few. The field of *surface coatings* has had great success in the application of *ceramic coatings* to provide other materials with some of ceramic's properties mentioned above. However, the application of coatings is expensive. A processing time of 500 hours is mentioned in the literature to achieve a laminated structure 1 cm thick. Metallic laminated structures made of ultrahigh-carbon steel and iron–silicon alloys using roll bonding would have limitations in producing a bulk ceramic/metal laminated composite due to concern about adhesion between the layers.

Refractory metals are among the few materials with significant mechanical strength above 2500°F, but they do not have sufficient strength-to-weight ratios for many applications. A material with a high strength-to-weight ratio and good mechanical strength at service temperatures above 3000°F is sorely needed for the many high-temperature applications such as space shields, rocket engine fuel turbines, and gas turbines. Combining a dense refractory metal such as tungsten, molybdenum, vanadium, or tantalum with a lighter-weight matrix material creates a unique material—a lightweight *refractory composite.* A fabricating process, *explosive welding,* is used to join these incompatible metals. Explosive welding or *explosive bonding* is a solid-state welding process that uses controlled explosive detonations to force two or more metals together at high pressures. The contaminant surface films are plastically jetted off the base metals as a result of the collision of the two metals. The time duration is so short that the heat-affected zone (haz) is microscopic. During the process, the first few layers of each metal become plasma due to the high velocity impact (200 to 500 m/s). The remaining thickness of each metal maintains a near-ambient temperature and acts as a large heat sink, cooling the plasma rapidly. The resultant composite system is joined with a high-quality metallurgical bond. Some explosive bonded materials in industry include niobium/copper, copper/aluminum (for high-voltage busbars), aluminum/stainless steel, titanium/stainless steel (for pacemakers), and zirconium/stainless steel.

Microinfiltrated macrolaminated composites (MIMLCs) offer a new, economically feasible approach to the production of tough composites using cermet/metallic, ceramic/metallic, and ceramic/intermetallic-compound bulk-laminated composites. Figure 8-29 is a schematic of an MIMLC illustrating the conceptual structure. Such materials attempt to duplicate a case-hardened metal with a ductile interior. Such a metal has the surface hardness and strength needed for many applications, such as a cutting tool, along with the ductility to possess great impact resistance (toughness). One problem that must be overcome is the proper adhesion between the layers of the bulk composite. Figures 8-30a and b illustrate poor and good solid-state bonding with a W–Ni–Fe/Ni MIMLC. W–Ni–Fe is a ceramic alloy that represents a hard material having high strength and modulus but low toughness. It could be compared to the fiber reinforcement in a metal- or ceramic-matrix composite. The nickel (Ni) represents a material having low strength, elastic modulus, and hardness but high toughness, and could be compared to the matrix material in a fiber-reinforced composite. Figure 8-30a shows a crack-like discontinuity between the bottom side of the Ni foil and the W–Ni–Fe heavy sheet when the composite was subjected only to a sintering temperature of 1400°C. Figure 8-30b shows proper solid-state bonding and consolidation of the layers when pressure was added to the process. Many variations of the fabrication process are necessary to accommodate the differences in solubility of the combinations of materials that can be used. Figure 8-31a illustrates the situation in which the two materials in a ceramic and a metal that have very little or no solubility in one another are combined. Porous alumina (Al_2O_3) and Ni are tape cast (to be defined) to produce an alumina/nickel MIMLC. Using a temperature just below the melting point of Ni and applying pressure, the Ni is extruded into the fine pores of the alumina structure. With further controls of variables such as time, temperature, and pressure, more adequate bonding can be achieved.

Figure 8-29 Schematic of the conceptual architecture of an MIMLC. The basic double-layer structure, consisting of a soft, ductile material (low modulus and strength and high toughness) and a hard, brittle material (high modulus and strength and low toughness), is repeated as many times as necessary to form a bulk composite material. The brittle material is also infiltrated with the ductile component, tending to increase the brittle material's toughness. (Bose & Lankford, *Advanced Materials & Processing*)

(a)   (b)

Figure 8-30 (a) W–Ni–Fe/Ni MIMLC composite that was sintered at 1400°C without pressure, showing improper bonding between layers. Note a crack along the bottom side of the nickel foil. (Bose & Lankford, *Advanced Materials & Processing*) (b) W–Ni–Fe/Ni MIMLC composite that was sintered under pressure at 1400°C, showing good bonding between the nickel foil and the heavy alloy sheet, which creates one continuous matrix that interpenetrates the tungsten grains. (Bose & Lankford, *Advanced Materials & Processing*)

Figure 8-31 (a) Photomicrograph of the microstructure of a hot-pressed alumina/nickel MIMLC where the nickel layer has infiltrated to only a small depth in the alumina sheet, which remains porous. Additional increases in time, temperature, and pressure would be necessary to achieve adequate bonding. (Bose & Lankford, *Advanced Materials & Processing*) (b) An X-ray dot map of a portion of a properly bonded W–Ni–Fe sheet/Ni sheet MIMLC, showing the tungsten spheroids imbedded in a matrix of nickel and iron. (Bose & Lankford, *Advanced Materials & Processing*)

Additional details of the fabrication of the W–Ni–Fe/Ni MIMLC may assist in greater understanding of this process. In this process, both materials are tape cast. *Tape casting* consists of mixing powders with a suitable binder to form a slurry, which is then cast as a thin tape (sheet) of material. The material is then cut in the green state when it is relatively soft or after it is sintered. The W–Ni alloy is tape cast and sintered to some 60% of theoretical density to produce an interconnected porosity. An 80:20 ratio of Ni to Fe powders is also tape cast and sintered to full density. Laid up in alternating layers and heated to about 40°C above its melting temperature, the molten Ni–Fe alloy infiltrates the porous W sheet and takes into solution a fraction of the W. In Figure 8-31b the lower white portion (Ni–Fe alloy) shows the infiltration of the porous tungsten sheet. The top, darker portion shows the W spheroids surrounded by the diffusing Ni–Fe matrix, which takes into solution a fraction of the W (darker portion).

This discussion points out the many process control variables, as well as the large number of material combinations of a hard, brittle phase interpenetrated by a soft, ductile material, that can be modified to produce a multitude of MIMLCs. With a plethora of different properties, they will find applications as armor plating; high-temperature composites; tough cutting tools; wear-resistant composites; and components with low density, high strength and high modulus, good fracture toughness, and impact resistance for use in aerospace and automotive structures and components.

8.6.3 Sandwich (Stressed-Skin) Composites

Sandwich composites can also be classified as laminar composites. Their outer surfaces or *facings* are made of some material higher in density than the inner material or *core* that supports the facings. The primary purpose of sandwich composites is the achievement of high strength with less weight, specifically a high *strength-to-weight ratio* or *specific strength*. A sandwich composite may be compared to a structural I beam. The high-density facings correspond to the I-beam flanges carrying most of the applied load, particularly the bending loads. Like the I-beam web, the low-density core allows the facings to be placed at a relatively large distance from the neutral plane to produce a large section modulus. The core carries the shear stresses. Overall sandwich composites are more efficient than I beams, because they possess a combination of high section modulus and low weight.

Typical sandwich facing materials are aluminum, wood, vinyl, paper, glass-reinforced plastics, and stainless steel. Core materials represent all families of materials, are primarily cellular in form, and take on the configurations known as honeycomb, waffle, corrugated, tubes, and cones. These rigid cores provide the greatest strength and stiffness, with honeycomb possessing isotropic properties. Metals that can be made into very thin sheets capable of being diffusion bonded are finding greater use as cellular-core materials. Solid cores made of plywood, as well as foamed cores of polystyrene or ceramics, are examples of other types of core materials. An *aluminum honeycomb material* is finding applications in automobiles to improve head support protection due to its unique shock-absorbing ability in thicknesses down to 10 mm. The nested hexagonal cell material is very light (weight savings of up to 30%), easy to form, simple to install, and is 100% recyclable. It is less sensitive to temperature changes and environmental degradation than foam materials.

Ceramic honeycomb panels withstanding temperatures to 1800°C serve as lightweight structural panels. Using CVD, a vapor-depositing ceramic is deposited on a fabric substrate woven in a honeycomb pattern and to make a honeycomb structure. The substrate is ultimately eliminated. The fabric can be made of a loosely woven polymer such as polyacrylonitrile (PAN), which is impregnated with an organic binder such as phenolic resin for stiffness. The ceramic materials are not limited to SiC—others are silicon boride, silicon nitride, or boron nitride. Heated to temperatures between 500° and 1000°C in an atmosphere of 2–5% oxygen, the fabric and binder oxidize and pass through the pores of the ceramic as a gas, leaving a microstructure of voids. The voids can be filled with ceramic by additional CVD, as can the hexagonal holes of the honeycomb.

The honeycomb structure or shape, (see Figure 8-32a) mathematically the strongest possible and the most economical for a mass of adjacent cells, dramatically demonstrates its strength (Figure 8-32b). *Nomex* honeycomb paneling provides superior specific modulus and finds many applications for structural elements in aircraft, spacecraft, racing vehicles, and watercraft. A *honeycomb composite* sandwich material used by Boeing Co. to build a one-piece satellite body is 40% lighter than a traditional aluminum structure, four times stiffer, 40% stronger, and weighs only 14 kg. Fabricating time is halved by reducing the number of parts and eliminating the number of manufacturing and assembly steps to one. The composite face sheets are made of an ultrahigh modulus graphite fiber and polycyanate ester resin composite that is about 20% lighter than conventional composites. This material releases fewer contaminants in space than conventional epoxy resin systems and has good resistance to microcracking. As good a heat conductor as aluminum and an excellent thermal conductor, the honeycomb composite permits passive thermal control which is impossible with low-conductivity composites.

8.6.4 Particulate (Particle) Composites

Particulate composite materials contain reinforcing particles of one or more materials suspended in a matrix of a different material. The particles, either metallic or nonmetallic, by definition do not chemically combine with the metallic or nonmetallic matrix material. As with nearly all materials, structure determines properties, and so it is with these composite materials. The size, shape, and spacing of particles, their volume fraction, and their distribution all contribute to the properties of these materials. How particles influence the properties of a material are explained in the discussion about dispersed-strengthened alloys (Module 8.6.4.2). Particulate composites are many times divided into subclasses by using some characteristic or combination of particular characteristics. This category of composites does not include particulates flat in shape, which possess sufficiently different properties to warrant a special classification. A reinforcing particle can be defined as having all its dimensions nearly equal.

Up to 30% titanium diboride (TiB_2) incorporated into the steel matrix by powder metallurgy has produced a *particle-reinforced steel composite.* Hot isostatic pressing consolidates the powders to 100% density. Tensile strengths are in excess of 1000 MPa and the specific modulus is double that of titanium alloys. Several matrix materials are being used such as pure iron, plain carbon steel, and gear steel. Titanium diboride is the particulate reinforcement due to its high modulus, low density, and good chemical compatibility with steel. The composite has been hot

(a)

Cross-section shows honeycomb of NOMEX is 90% air.

(b)

Figure 8-32 (a) Honeycomb . . . fly with me. Class 2 The honeycomb sandwich composite is a very efficient yet complex structure widely used in the aircraft industry. Use of ultra thin sections (1-10 μm) and transmitted light enhance our ability to characterize the morphology of the composite for later improvement of the materials and refinement of the process parameters. By imitating the natural geometric structure of a beehive, the honeycomb core imparts strength and light weight to sandwich panels, while supporting the prepreg skins. The bond between honeycomb core and the carbon/epoxy skin, shown in this cross section, is created by the adhesive film layer. A surfacing film is often co-cured with the composite to improve the appearance of sandwich panel and provide a smooth, uniform and nonporous surface. With the appropriate technique, transmitted light microscopy allows us to see differences in morphology between the prepreg, adhesive film and surfacing film. Understanding the intermingling of the resin systems enables the proper selection of materials and/or processing steps. Base Figure: Transmitted cross polarized light with 530 nanometer compensator plate, 200X, 10 micron thin cross section. The 3K-70-PW carbon fibers are visible as black rods, and the amorphous resins are first order magenta. The randomly oriented, birefringent polyester scrim fibers are observed in the 3M, AF310 surfacing film. BMS 8-245 adhesive, and in the Nomex BMS 8-124 core cell wall. A filler mineral, talc, appears as birefringent flakes and needles in the surfacing film. Insects A, B and C: Transmitted light Hoffman contrast, 200X. This technique reveals the subtle amorphous morphology of the ultra thin section. With this contrast mode, a three dimensional appearance is given, indicated by different contrast due to different refractive indices of the resins. Therefore, one can easily trace resin intermingling. Figure A shows the resin intermingling of the AF310 surfacing film and the BMS 8-256 carbon/epoxy skin. Figure B shows the Nomex core scrim and phenolic dips, and the BMS 8-245 adhesive fillet which bonds the core to the skin. Figure C shows the resin intermingling of the BMS 8-256 skins and BMS 8-245 adhesive. Inset D: Reflected light bright field, 200X. This technique shows the fibers in the carbon/epoxy skin as light grey rods, but does little to enhance the morphology shown in transmitted light. A white reflective surface placed on the back of the slide allows one to see a three dimensional view of the fiber. (b) A 10-in. square by 22-in. high column of honeycomb of Nomex™ aramid weighing 11 pounds supports a load of 27 tons. (DuPont.) (Luther M. Gammon, Boeing Commercial Airplane) forged and extruded and a gear component has been manufactured to date. This new composite should help meet the demands for higher strength for high-performance structural applications.

Advanced Ceramics Corp., Cleveland, Ohio, has developed a process in which small particles of ceramics, graphite, and metal are encapsulated with ***pure metallic coatings.*** Particle size ranges from less than 5μm to more than 5000 μm depending on the particle size and shape. Coatings include copper, tin, cobalt, zinc, nickel, or a combination of these metals. Successfully coated particles include graphite, boron nitride, silicon carbide, tungsten carbide, zirconia, titanium diboride, and stainless steel. Figure 8-33 shows photomicrographs of some of these particles. With the ability to put thick coatings on a variety of particles, applications will be easy to find in thermal spray coatings and as reinforcements in metal matrix composites.

A carbon fiber material called ***graphite fibrils,*** each piece about 1/10,000 the size of a human hair, added to a resin creates a conductive resin for electrostatic painting of automotive side mirrors made of ABS. This GE and United Technologies Automotive Inc. development may lead to a better way of painting plastic parts.

8.6.4.1 Cermets.   Cermets provide an excellent example of a particulate composite material. Cermets are produced by sintering a mixture of ceramic and metal powders to form a structure that consists of a dispersion of ceramic particles in a continuous metallic matrix. Carbides of tungsten, chromium, and titanium are widely used in combination with cobalt, nickel, and stainless-steel matrices to provide materials with very high hardness for wire-drawing dies, for very high corrosion resistance for valves, or for very high temperature applications such as turbine parts. Uranium-oxide and boron-carbide cermets suspended in stainless steel find several uses in nuclear-fuel elements and control rods.

8.6.4.2 Dispersion-strengthened alloys/composites.   Dispersion-strengthened alloys are similar to cermets and to precipitation-hardened alloys. Differences among these three categories are spelled out in terms of the constituent particles. The particles in dispersion-strengthened alloys are smaller (between 0.01 and 0.1μm) and of a lesser volume fraction (at 3% by volume) than found in cermets. Dispersion-strengthened alloys are formed by the mechanical dispersion of particles, as opposed to precipitation-hardened alloys, in which compounds are precipitated from the matrix by heat treatment. Cold solder is a metal powder suspended in a thermoset to provide a metal that is not only hard and strong, but also a good conductor of heat and electricity.

High-temperature structural silicides combine the desired qualities of ceramics with the high-temperature ductility and fracture resistance of metals. New patents have been issued in this field that describe the dispersing of silicon carbide throughout a matrix material of molybdenum disilicide that produces a molybdenum disilicide–alloy matrix composite with eight times the high-temperature strength of molybdenum disilicide by itself. Such a composite has the potential of meeting the demands of advanced high-temperature structural applications in the range of 1200° to 1600°C in oxidizing and aggressive environments. In addition to possessing specialized properties, the composite material is cost-effective in its manufacture compared to other advanced composites, such as silicon carbide ceramics materials.

At this point, it is timely to ask what the differences are between ***precipitation hardening*** and ***dispersion strengthening*** (DS). Most copper alloys achieve their properties through the addition of alloying elements and cold rolling; **beryllium copper (BeCu)** achieves its properties through alloying elements (Be 1.60–1.79% plus trace amounts of other elements), cold rolling, and a ***precipitation-hardening*** heat treatment. The Be precipitates out of a solid solution to form microscopic copper beryllides. Due to their submicron size, very fine distribution, and interaction with the copper matrix, these precipitates are responsible for the unique properties that set apart BeCu from other copper alloys. (See Module 5.5 for further details on precipitation hardening.) PH metal alloys are strengthened by the in situ precipitation of a hard second-phase particle in a soft, ductile matrix through a multistage heat treat process. The size and amount of the second-phase can be controlled minimally by adjusting the alloy composition and the aging temperature, but certainly not as much as can be produced by the DS process. The second-phase par-

Cu Coated SiC Whisker

Ni Coated SiC Particle
(SIKA-I)

Fe Coated Natural Graphite Flake

Ni Coated Synthetic Graphite Particle

Cu Coated TiB$_2$ Platelet

Ni Coated Yttria Stabilized Zirconia Hollow Sphere
(Metco 204 NS, Lot # P-3165)

Figure 8-33 Metal-coated particulates used in MMCs can improve mechanical properties by 30–100% or more. Particles can range from less than 5 micrometers to more than 5000 micrometers. Coatings can be a thick as 50% or more by volume or as thin as 0.5% by volume. Virtually any particle that can be wetted by water and does not react with water and has a specific gravity greater than 1 is a candidate for coating. (Courtesy of Advanced Ceramics Corp.)

ticles are *not* stable at high temperatures, which means that, as temperatures rise, these particles, including those under service conditions, will coalesce and agglomerate, reducing the number of particles and thus reducing the hardness and strength of the alloy. DS materials (composites) are produced by mechanically combining a hard and strong oxide particle with soft, ductile, metallic matrix particles. Various combinations of these particles can be used in varying proportions, sizes, and shapes by the powder metallurgy process, yielding a composite material with many desirable properties. The oxide particles chosen for these composite materials are chemically inert. Hence, they do not affect electrical properties. Since they do *not* grow or dissolve in the matrix at high temperatures, they contribute to excellent resistance to softening and to mechanical properties that do not change with changes in service operating temperatures.

The words *in situ* mean "in its original position." In relation to composites, the term refers to the formation of components (i.e., the reinforcements, matrices, and coatings of the interfaces) in a material that are produced during processing from the elements that made up the material. Because these reinforcements were produced within the materials, (i.e., a metal or intermetallic), they are thermodynamically stable in the matrix (they will not degrade or disappear when heated) and they form as single crystals. These reinforcing structures are different from more conventional composites that have reinforcements in the form of fibers or particles that were deliberately added to the original material. *In situ composites* are proving to be more cost effective and to have higher mechanical strength properties, particularly at high operating temperatures. The cost savings come from the reduction or elimination of the processing steps and the need to add reinforcing components. The many chemical processes that normally occur when a material system is thermally processed, such as phase changes, recrystallization, or nucleation, produce uniform distributions of the reinforcing phases much more efficiently than composites produced by more conventional means. Examples are the processing of conventional aluminum alloys or precipitation-strengthened alloys. Also, it is extremely difficult to handle very fine particles either as reinforcement or matrix contributions smaller than 1 μm. In fact, fine powders, whiskers, or fibers are extremely reactive and present toxicity/health problems. With in situ composites, reinforcements produced are well within the nanometer range, and they require no handling. With such a scale of reinforcement comes greater mechanical strength.

A good example of in situ composites is exothermic dispersion (XD) composites, under development since 1983 by Martin Marietta. These XD materials are produced by exothermic (heat given off) reactions of precursors, and they include metals and intermetallics reinforced by dispersions of intermetallic or ceramic particles or whiskers. *XD technology* deals with composite fabrication in which reinforcing components such as particulates, short fibers, or whiskers develop in situ within metal or intermetallic matrices. One such composite is a titanium aluminide (Ti_3Al) matrix composite made from three starting materials in powder form: Ti, Ti_3Al (titanium aluminide), and TiAl. To these is added titanium diboride (TiB_2), forming TiB_2 dispersions in a matrix of Ti_3Al, which in turn is formed by reactions of titanium, aluminum, and boron. This composite is replacing nickel-based superalloys in several aerospace applications. With a significantly lower density than stainless steel, titanium aluminide composites have greater strength and stiffness at elevated temperatures. For example, the yield strength at 750°C for 17-PH stainless steel is about 345 MPa; for TiAl (Ti-47Al + 7 vol% TiB_2), it is 415 MPa. The modulus at 920 K for steel is 145 GPa, compared to 162 GPa for the in situ composite.

A titanium diboride–alumina composite material with superior high-temperature wear resistance is produced by a new process called the *self-propagating high-temperature synthesis (SHS)* process developed by the Georgia Institute of Technology. The new chemically based process yields titanium diboride material with smaller (submicron) particles and allows the composite to be produced in final form in molds. The older process using solid-state diffusion left carbides as contaminants in the powder, which accumulated in the grain boundaries. The SHS process mixes powder metal, titanium oxide, and boron oxide; places them in a high-temperature crucible; and ignites them to start a self-sustaining reaction between the components. This produces titanium diboride dispersed within either magnesium oxide or alumina, depending on the choice of metal. Finished products can be hot pressed at 1500°C in less time (2 to 4 h) than the conventional materials. By varying processing conditions for the titanium diboride–alumina composite, either a dense composite material or a porous form (into which materials such as molten metals can be infiltrated) is produced. Some physical properties for this new material are a melting point of 3000°C, hardness superior to tungsten carbide, thermal conductivity better than cubic boron nitride, fracture toughness greater than silicon nitride, and a good stiffness-to-weight ratio. Applications for this new material are cutting tools, dies, and electrodes.

A "particle-dispersed" steel made by mechanical alloying is said to have a Young's modulus of 265 to 285 GPa, the highest ever achieved for steel. Sumitomo Metal Industries, Japan, reports that they have developed the high levels of stiffness through three stages of processing. The first step was mechanical alloying, in which oxides of iron, carbon, and yttrium are comminuted (pulverized or reduced) to ultrafine particles in a ball mill. The second step was the sealing of the powder in a capsule and hot extruding it. The third step was hot recrystallization, followed by heat treatments. The composition of the steel is Fe, 13–15Cr, 1–3 Al, 0.5 Y. Its properties are tensile yield strength, 600 to 900 MPa; ultimate tensile strength, 700 to 900 MPa; and elongation, 10–20%. The steel is capable of being fabricated in bar, pipe, or sheet form.

8.6.5 Flake Composites

Flakes of mica or glass in a glass or plastic matrix form a composite material that has a primarily two-dimensional geometry with corresponding strength and stiffness in two dimensions. Flakes tend to pack parallel to and overlap each other, and they provide properties such as decreasing wear, low coefficient of thermal expansion, and increased thermal and electrical conductivity, which are dependent on higher densities in materials. Aluminum in flake form suspended in paint provides good coverage. Silver flakes similarly prepared provide good electrical conductivity.

Metal particles suspended in metal matrices such as lead added to copper and steel alloys lead to increased machinability or reduced bearing wear resistance. These usually brittle metal particles are not dissolved in their metal matrix as in the alloying of metals, but instead, the metal matrix material is infiltrated around the brittle particles in a liquid-sintering process. When extremely small flakes of aluminum are disbursed into an aluminum metal matrix, a material called **sintered aluminum powder (SAP)** is produced. The flakes (0.1 to 0.2 μm thick) are coated with alumina prior to disbursement. (See metal-coated particles in Section 8.6.4.) Nickel alloys can gain high temperature strength by disbursing thoria (ThO_2) particles into a nickel metal matrix producing **thoria-disbursed (TD) nickel.**

8.6.6 Filled Composites

A *filler* is a material added to another material to alter its physical and mechanical properties significantly or to decrease its costs. Some fillers, when added to polymer materials, improve their strength by reducing the mobility of the polymer chains, much as an appropriate amount of gravel, as a filler, improves the strength of concrete. Fillers are added to polymer materials for several other reasons (e.g., to improve frictional characteristics, control shrinkage, improve moldability, reduce dielectric properties, lower resistivity, reduce the moisture absorption characteristics, or enhance wear rate). Celluloid and Bakelite are filled polymers, as are electric-circuit boards and countertops made from phenolics, a thermosetting resin. See Module 6.3.2.1 for more details.

Since the accidental discovery early in the twentieth century that natural rubber would accept large amounts of carbon black, thus improving its mechanical properties, great advances in the technology of fillers are continuing. Particles of *carbon black,* an inexpensive material, are very small (20 to 50 nm) and spherical in shape. When added to rubber, carbon black enhances its tensile strength, toughness, and wear resistance. Its reinforcing effect on such properties is the result of its particularly good adhesive bonding with the rubber molecules. Additional reinforcement effects are attributable to the uniformly distributed particles acting as barriers to plastic deformation. Automobile tires (see Sections 6.4.2 and 6.4.3) may contain up to 30% vol of carbon black. Just as with multiphase microstructures such as steel, whose structures (micro) can be changed by the relative amounts of the phases present, the phase distribution, and the size of the individual grains making up the phases, fillers are components that are purposely added to other materials to change their properties. Glass fibers are added as fillers to plastics. Not only the type of filler but its size, shape (e.g., spheres), and distribution all play an important role in determining the desired properties of the filled composite.

8.6.7 Hybrid Composites

Hybrid composites, which combine two or more different fibers in a common matrix, greatly expand the range of properties that can be achieved with advanced composites. They can cost less than materials reinforced only with graphite or boron. The common matrix can be either a thermoset or a thermoplastic. Using combinations of continuous as well as chopped fibers also qualifies the material as a hybrid.

There are four basic types of hybrids. One is *interply,* which consists of plies from two or more different unidirectional composites stacked in a specific sequence. A second is *intraply,* which consists of two or more different fibers mixed in the same ply. A third are *interply/intraply* hybrids, in which the plies of interply/intraply hybrids are stacked in a specific sequence. *Superhybrids* are resin-matrix-composite plies and metal-matrix-composite plies stacked in a specific sequence. The curing procedure must be compatible with the matrix resin, which places something of a restraint on using plies with different matrix material. In the case of superhybrids, the resin matrix and the prepreg tape must have the same curing cycle as the adhesive used to bond the entire composite. The fabrication of hybrids is the same as for other composites.

The potential number of fiber/resin combinations for hybrids is vast. With two resins and three types of fibers, an unlimited number of different hybrid combinations can be produced. One design problem is how to make the most effective use of each fiber. Graphite fibers have poor impact resistance but high tensile strength and high modulus of elasticity. Aramid fibers have good impact but low modulus. The combination is a natural where each fiber contributes its best properties. A hybrid containing 50% graphite and 5% aramid reportedly shows flex strength on the order of three times that of a straight aramid. Another reason for this good combination is that both have coefficients of thermal expansion that are similar. In fact, the CTEs are slightly negative. This minimizes internal thermal stresses. Another combination is graphite or aramid and glass. Both fibers increase the stiffness, but with aramid there is some loss of compressive strength. Another combination is graphite and glass in polyester. The more glass that is used, the less the cost. The more graphite replaces glass, the higher the mechanical properties. For example, 25% carbon/75% glass fibers gives a modulus of 6.39×10^4 MPa, while the reverse of this proportion gives a modulus (in tension) of 12.3×10^4 MPa. Using SMCs and continuous graphite fibers with a typical loading of 30% graphite fiber and 45% chopped glass, the composite's properties replicated those of the closest all-glass system.

The B-1 bomber stabilizer was made of a hybrid composite—the largest hybrid ever produced, having an area of some 45 m²—made of aluminum/graphite–epoxy/boron–epoxy hybrid. The saving in weight over the traditional aluminum/titanium structure is about 320%; the stabilizer consists of 108 parts compared to 270 in the metal version. With fewer parts needed, there are fewer holes to drill, countersink, and inspect, and fewer fasteners, all of which reduces the expense of production. Alcoa has developed a hybrid composite ARALL™ consisting of layers of aramid/epoxy laminated with layers of aluminum. Compared to an all-metal structure, this hybrid has better fatigue life and higher specific strength and stiffness. It is similarly machined as a solid metal and the outer layers of aluminum provide a moisture barrier plus lightening protection.

The auto industry will be the biggest user, with hybrid driveshafts. Made of graphite fibers for stiffening and glass fibers to carry the torsional loads, this is another example of the intelligent selection of materials to meet the requirements of the application.

The sports industry uses graphite/aramid and graphite/boron hybrids. Graphite/boron golf shafts are more rigid than an all-graphite shaft, and they are used in golf clubs, for both woods as well as irons.

In the area of high-temperature materials, where composites are beginning to replace metals, a final example illustrates the intelligent design of hybrid composites. It must be remembered that the polymer-resin matrix is the determiner of the high-temperature performance of the composite. Carbon- and glass-fiber-reinforced amorphous composites lose their flexural strength as they approach their glass transitional temperatures (T_g). Any exposure to temperatures above T_g, even for a brief moment, could result in a catastrophic failure. Despite this, amorphous thermoplastic–resin composites are stronger than crystalline-resin composites near their T_g values, while crystalline-resin composites retain measurable strength above these temperatures. By using a hybrid matrix blend of resins (where one resin has a higher T_g value), high-temperature properties can be raised enough to meet the application requirements for a particular service temperature.

8.6.8 Smart Composite Materials

As explained in Module 1, materials engineers and designers seek to mimic biological organisms and to place control and feedback in materials and structures. The infinite variety of structures possible with hybrid composites led to the opportunity for designer *smart composites* to fulfill many of the goals sought for smart materials. Such materials are currently available, with more varieties to follow.

Hybrid PMCs incorporating *shape-memory alloys (SMAs)* such as Nitinol give the designer more choice for functionality. As a load-bearing armor material, the PMC with imbedded Nitinol improves low-velocity impact and perforation resistance compared to monolithic PMC. Nitinol's superplastic nature allows it to deform up to a 7% strain with fully reversible recovery when this load is removed, thus allowing greater impacts by the host material. Hybrid composites designed for structural and armor application include various compositions of tough polyethylene, Kevlar, nylon, glass, aluminum, and graphite using the Nitinol SMA to improve strain and impact resistance.

Another smart composite involves *imbedded particles,* which allow in-service monitoring of the host composite structure. A polymer or ceramic matrix imbedded in micrometer-size particles can provide active or passive tagging. *Active tagging* requires an external source to energize the particles. One external source is a magnetic field to energize ferromagnetic particles. *Passive tagging* uses nondestructive testing instruments (e.g., ultraviolet light, X-ray eddy current, or magnetic pickup) to detect the presence of particles and defects in the material. *Particle signatures* result from their interaction with their host material and reveal various structural conditions including internal stress, voids, inclusions, degree of curing, and delaminations. The signatures aid in quality control (QC) and quality assurance (QA). Certain QC/QA issues with composites involving defects, such as hidden voids and delamination of composite layers, remain major concerns because composites are becoming more widely accepted in buildings, roads, and other construction as replacements for traditional structural materials.

The 1995 bombing of the Alfred P. Murrah Building in Oklahoma City prompted a call for legislation requiring tagging of explosive materials. An explosive tagging technique patented in 1973 involves mixing a small amount of phosphor with the explosive. Ultraviolet light can easily detect the particles of phosphor before or after the explosion. Other methods use inorganic materials and magnetic particles, including some color coding.

8.6.9 Foam Composite Materials

To make a *foam* or *cellular plastic* (both open and closed cell), a foaming agent is added to the resin stream and mixed in the impingement chamber (in-line mixer). Foams are of two types—*rigid foams* whose walls are stiff, and *flexible foams* whose walls will collapse when pressed. Hollow glass spheres (micro balloons) are used to produce *syntactic foams.* The glass spheres have a diameter of 0.03 mm. Cellular plastics are used in flotation devices, noise abatement, and where thermal insulation is required. A stained and deeply grained composite (carbon or fiberglass) door can now replace a real wood door. Unlike real wood doors, these doors are virtually maintenance free, as they will not crack, warp, swell, or chip. With rigid foam between the indoor and outdoor composite skins, the insulation level is about six higher than solid wood. They come with a 25-year warranty. Decorative glass in the door is usually triple pane and tempered. Super secure glass is an option with a lifetime never-break warranty. New stainable, carbon composite doors offer advantages even over fiberglass-skin doors. Carbon expands less than fiberglass with temperature changes so there is less thermal bow. Carbon skins also require less forming pressure during molding so the graining remains more true.

New low-density SMCs with hollow glass microspheres are producing SMC body panels for today's automobiles that have specific gravities of 1.3 and 1.5. Panel thickness is 2.5 mm. Some of these panels are being used on the GMC Silverado pickup truck box discussed later in this module (see Figure 8-36a).

Metal foams are metallic, cellular materials that have a high porosity fraction typically ranging from 40 to 90 vol%. With their high stiffness and low specific weight, they are used in construction, packaging, noise and vibration damping situations, and filtering. They are considered by many to be a new class of engineering materials. A new powder method for fabricating these foams makes it possible to build metallic foam parts that have complex geometry. Sandwich structures composed of a metallic foam core and metallic face sheets can be produced. Common aluminum alloys, 2XXX and 6XXX, Al–Si casting alloys, and pure aluminum are used for foaming. Tin, zinc, lead, bronze, and steel can be foamed. Foamed materials are capable of absorbing large amounts of energy at relatively low stress. Automotive applications for these stiff, light structures made of aluminum foam and foam sandwich panels are hood and trunk lids. A foam sandwich panel is about eight times stiffer than the original steel panel and about 25% lighter. Potential applications are bumpers, pillars, and other locations where impact protection is required. *Honeycomb structures* could be replaced by foam structures based on their reduced cost, isotropy, and fire resistance. It is contemplated that lead and nickel foams may find use in batteries. Open-cell foams would be excellent for several applications such as heat exchangers and filters. Nickel and titanium foams should readily find applications in aerospace and biomedical fields.

8.7 COMPOSITES PROCESSING

Although ordinarily not a part of the study of materials, the numerous manufacturing processes involved in the production of composites are intimately connected with materials, because design affects manufacturing and the particular process affects the properties of the final product. Because of the one-piece forming capability of composites, in which the material is made simultaneously as the end product is manufactured, a brief mention of the major manufacturing processes as a primary factor in evaluating the costs is justified. The accomplishment of near-net-shape (NNS) components with composite manufacture epitomizes this technological development of recent years.

Unlike manufacturing methods for metals, the processes related to the manufacturing of composite materials have limitations based on shape, microstructure, and materials. The basic processes for converting the many polymeric materials into solid shapes include molding, casting, thermoforming, and extrusion. Secondary processes are concerned with finishing, coating, and machining the solid shapes and will be discussed in less detail. Molding processes shape or mold the product by some force other than gravity or atmospheric pressure. Heat and pressure are normally used depending on whether a thermoplastic (TP) or thermosetting (TS) resin is being used as a matrix material. The major processes to be discussed briefly are hand layup, compression molding, injection molding, filament winding, and pultrusion.

Two stages in the processing of most fiber composites are the *layup* or combining of the reinforcement and matrix materials and the molding or curing stage. *Curing* is the drying or

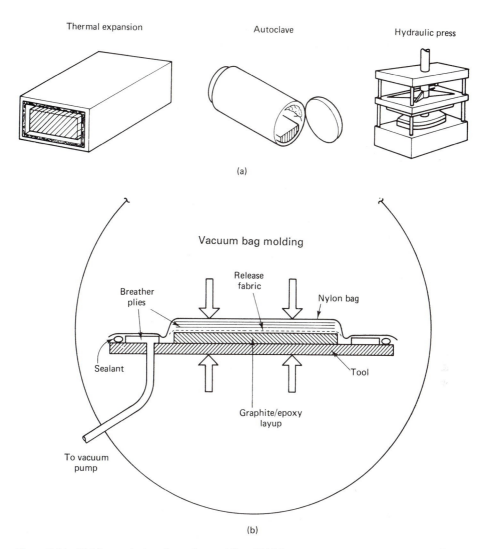

Figure 8-34   Molding methods and autoclave molding. (NASA)

polymerization of the resinous matrix to form a permanent bond between the fibers and the matrix material. It occurs unaided as with contact molding or by the application of heat and/or pressure using vacuum bags, autoclaves, pressure bags, or conventional metal stamping machinery (see Figure 8-34).

8.7.1 Contact Molding, Open Molding, and Lamination Molding Processes

Hand layup is the simplest of the processing methods. Using a single, inexpensive, open mold, reinforcement mat or fabric is placed in the mold and saturated with resin by hand. Layers are built up to the desired thickness to form a laminate that is cured at room temperature. This form of curing without the application of heat is called *contact molding.* Spray-up molding using a single mold combines short lengths of reinforcing fibers and resin in a spray gun which deposits them simultaneously on the mold surface. Contact molding follows. Hand layup is a general term to describe a number of procedures using a single mold, a gel coat, followed by the placement of various layers or reinforcements. By using prepreg materials the careful placement of the fibers in the mold can be ensured. Additional layers can be placed on top of previous layers and for that situation the prepregs are generally tacky so that the layers stick to each other. No force or pressure is used to form the composite shape. Hence, it is not considered a molding operation or technique but, rather, contact molding or open molding.

The ***open-mold processes*** for shaping FRP laminates use a single-mold surface to produce laminated FRP structures. Open-mold processes include hand layup, spray-up, bag molding, and automated tape-laying machines. These methods differ mainly in the manner in which the laminations are applied to the mold. Hand layup is the oldest and most labor-intensive method consisting of five steps common to the other methods: (1) application of the mold releasing agent, (2) application of the thin gel coating to the mold, (3) application of the dry fiber reinforcement and impregnation of the fiber with resin through the use of hand rolling, (4) curing, and (5) removal of the fully hardened part from the mold. In step (3) prepregs may be used to more closely control the placement of fibers and gain greater efficiency in adding the laminations. Molds can be made of metal, plaster, GFRP, or other materials depending on economics and other technical factors. Spray-up reduces layup time by spraying liquid resin and chopped fibers onto an open mold to build up successive laminations. The U.S. automotive aftermarket parts and accessories markets using open molding, primarily hand layup and spray-up methods, focuses on low-volume products with which vehicle owners can afford to customize their cars, trucks, or vans. Examples of such products are rear deck spoilers also known as wings, body fairing kits, fender flairs, hoods with aerodynamic air-induction scoops, fenders, bumpers, trunks, doors, running boards, and truck caps and bed covers.

Plaster casts, flexible rubber molds, epoxy resins, and composite plates that allow for varying thicknesses without adding weight have invaded the traditional-bound area of violin making. The plates are layered sandwiches of wood veneer, carbon fiber, and aeronautical foam.

8.7.2 Closed Molding Processes

Compression molding or ***matched metal die molding*** mass-produce high-strength parts limited in size by the press equipment. Pressures used by this equipment are about 1.72 MPa and temperatures are about 120°C. The materials used are in the form of SMCs or BMCs. These processes use a mold that consists of two sections that open and close during each molding cycle. Tooling cost is a determining factor in these processes due to the complex equipment required. These processes are employed when a good finish is required on all surfaces, when high production rates are involved with smaller tolerances, and where more complex three-dimensional shapes are involved.

Closed molding can be divided into (1) compression molding, also known as matched die molding, (2) injection molding, and (3) transfer molding. Compression molding or matched die molding uses TS resins placed in a cavity of a matched mold in the open position. The mold is closed, pressure is applied, the material is heated to cause cross-linking, and when complete, the mold is opened. This process is aptly described as waffle making. Parts that have hollow sections are very difficult to mold. Longer lengths of chopped fibers (approaching 100 mm) can be used in this process. If fibers are used as reinforcement and are to remain in place, a preform can be made and positioned in the mold. Then the resin is poured on top. The pressure from the mold-closing distributes the resin.

Preforms are TS granules pressed together into a particular shape for easy handling. Granules are TS, such a phenolic that liquefies when heated and then cures. A "green compact," a name given to a ceramic part before it is fired, is an example of a preform. TS materials may be removed from the mold while hot since the curing creates a dimensionally stable form. Carbon fiber reinforcement can be used in this molding operation. Compared to injection molding, compression molding uses simplified tools and can accommodate high fiber volume fractions and long fiber lengths, thus making parts with better physical and mechanical properties. Compression molding also is suitable for the production of parts with large surface areas. More common is the use of chopped fibers which have been combined with a resin and a filler to form a molding compound. This compound is then used as a charge material. Molding compounds are materials in which the fibers have been combined with the resin and a filler such as $CaCo_3$.

In addition to preforms, SMC, another molding compound, is the starting material for compression molding. Figure 8-35a shows a typical formulation for a SMC. The process begins

Material	Percent	Purpose
Calcium Carbonate	48%	Filler
Fiberglass (1 in. Lengths)	28%	Strength & Physical Properties
Polyester Resin	13%	Plastic Binder
Low Profile Additive	9%	Shrinkage Compensation & Class "A" Surface
Additives	2%	Catalyst, Mold Release, Pigment, Thickening Agent

(a)

Figure 8-35 (a) Typical SMC formulation. The presence of fiberglass as a reinforcement indicates that this is a reinforced SMC. High-strength reinforced composites have a larger percentage of fiberglass reinforcement than conventional SMC. High strength means ability to withstand high impacts (dent resistance). The low profile additive is present to ensure a smooth paintable finish for automotive panels. (*continued*)

with the making of a stack of SMC plies placed in a preheated mold which are die-cut to the required size and shape. This stack is called the charge. The top mold is lowered slowly onto the charge, and the rising temperature reduces its viscosity. As the mold continues to close, the SMC flows toward the extremities, forcing air out of the mold cavity. Both the top and bottom molds are heated. Curing times vary from 1 minute to several minutes. The top mold is opened and the part is removed with the aid of ejector pins. Cycle time is long compared to injection molding since the matrix material must be cooled in the mold below its melting temperature before removal. Compression-molded SMC composites find many uses in the automobile industry. High-strength automobile road wheels formerly made of SAE 1020 steel have been compression molded using SMC 50 polyester composite. Figure 8-35b is a graph of this composite's strength versus use temperature ranging from −50° to 125°C. Weight savings of up to 50% over standard steel were realized. Due to the composite's uniform weight distribution, the need for wheel balancing was nearly eliminated. Corrosion resistance topped that of steel wheels.

Injection molding (similar to die casting of metals), molten thermoplastic (TP), and thermosetting (TS) matrix materials are formed into a closed mold and held until the materials solidify. TS materials are forced into heated molds where they cure (cross-link) into a solid. Conversely, TP materials use a cooler mold to increase production rates. It takes longer to mold and cure a TS material than a TP material. Preheating a TS material before it is placed in the mold reduces the curing time. Injection molding is the best example of the high production of TP parts. Modifications to the injection molding equipment allowed the subsequent molding of TS parts including reinforced TS parts. Figure 8-35c is a plot of viscosity versus temperature which shows that for both types of materials, at or around room temperatures, they can easily exist as a liquid. This makes the mixing of reinforcements, fillers, colorants, and so forth much easier with simple stirring or mechanical mixing. Note also that the TP materials have a higher viscosity in the liquid form than TS materials. Filler, reinforcing agents, and other additives reduced to powder or granular form are fed into the injection machine from a hopper. RIM can be used to encapsulate an insert or core. When the core is made of fiberglass, the process is called *structural reaction injection molding (SRIM).* Two types of injection machines are the plunger type and the reciprocating screw. Injection molding of metals is discussed in module 3. Figure 3-5 is a schematic of an injection molder (screw type).

In *reaction injection molding (RIM),* the word *reaction* refers to the use of TS polymers which polymerize (cross-link or cure). During the process, two or more TS monomers are forced into an impingement chamber and the mixture is then injected into a mold cavity. Remember that to cure, these resins must be heated, so temperature must be controlled throughout this process. Polyurethane is the most commonly used resin system where the two-part

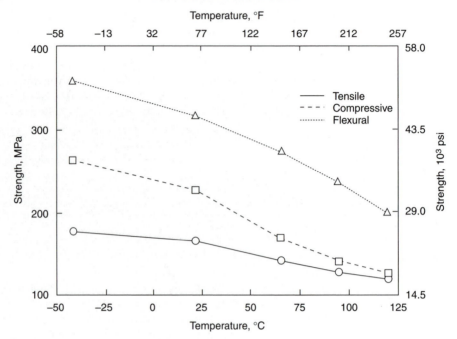

TYPICAL SMC
R50 POLYESTER COMPOSITE

Tensile, compressive and flexural strengths vs temperature.

(b)

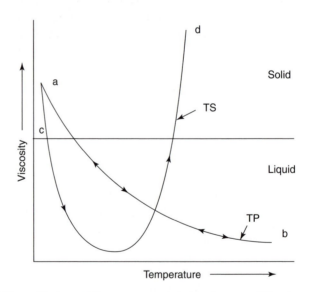

Figure 8-35 (b) SMCs are thin sheets of fiber precompounded with a thermoset (TS) resin such as polyester or vinyl ester. Low profile additives are added to compensate for any shrinkage during the cooling period outside the mold, thus assuring a fine finish as on automobile bodies that is free from surface defects. SMC R50 polyester composites maintaining their strength at high temperatures are finding many automotive applications. (Courtesy of The Budd Co.) (c) A plot of the viscosity versus temperature for thermoplastic (TP) and thermosetting (TS) resins. Point **a** on the TP curve represents a resin that is in a solid state at the coldest or room temperature. Point **b** is the same TP resin at its warmest temperature in a definite liquid form. Point **c** represent a TS resin around room temperature at which it could be in either a solid or liquid state. Point **d** represents a TS resin at the warmest temperature in a definite solid state. Note the double arrows on the TP curve which indicate that these resins can change their form of state depending on temperature while the single arrow direction on the TS curve signifies that as these resins are heated they begin to cure and solidify to a final solid state.

reaction system begins to cure and form a plastic typically in less than a minute. The two components of polyurethane are polyol and isocyanate which are metered and mixed in-line so that they begin to polymerize or cross-link. When reinforcement in the form of very short fibers (chopped fibers usually less than 100 mm in length) or particulate such as flakes are added to the mold, a more isotropic solid is produced. To accomplish this, the mold is closed and the resin is injected into it in a process called *reinforced reaction injection molding (RRIM).* Due to fiber breakage mainly, there are limits as to the strength of the fiber reinforcement using both TP and TS resins. Breakage occurs as the resins with their reinforcing phase are forced through narrow gates and openings in the mold. This restriction, in turn, limits the degree of reinforcement obtained. Another definition of aspect ratio used in this industry is the ratio of the largest to the smallest dimension. If a filler has an aspect ratio close to 1, then it would make a fairly good reinforcement for a fiber composite part. A *reinforcement* might be defined, therefore, as a filler with a ratio greater than 1. A variety of fillers are used with TS resins. Glass filler in the form of microspheres (both hollow and solid) and mineral fillers such as clay, talc mica, and metallic powders used to coat particulates such as aluminum-coated flake (see metallic-coated particles) are some examples. Some fibers used include thermoplastic, glass, aramid, and carbon fibers. Carbon and glass hybrid composites also are produced by the above mentioned processes. One application of parts made by RIM is glass-reinforced PA radiator tanks for automotive radiators. These plastic tops and bottoms of radiators must resist temperatures up to 130°C and have chemical and mechanical resistance. Cooling fans for the radiators are made of a similar material.

FRP composites of both thermosets and thermoplastics continue to find new and growing applications in sporting goods and recreation equipment. Three million bow hunters are now going to compression-molded composites to gain greater arrow speed. Rollerblade, Inc., the in-line skate market leader, has introduced an innovative children's skate that adjusts over four shoe sizes to accommodate growing feet and thus extend the skate's usable life. The Xtenblade and Xten Plus skates feature upgraded wheels and bearings and hard shell boots with a toe section that adjusts forward as needed. Of the glass-filled nylon thermoplastic composites, 28% are used for three injection-molded parts which provide torsional rigidity, stiffness, and strength. Bicycles and bicycle components constitute a $1.7 billion U.S. market. SRAM Corp. (Chicago), a specialist in mountain bike components, uses thermoplastic composites with carbon fibers and/or glass to produce shifters, front and rear derailleurs, braking systems, and related accessories.

8.7.3 Resin Transfer Molding

Resin transfer molding (RTM) differs from RIM in that a reinforcement is molded with the resin; however, the equipment is basically the same. With RTM both long and short fibers can be used as well as innovative weaving and knitting techniques that allow for greater latitude in laying out the fibers. This dry reinforcement, cut and shaped, is first placed in a mold cavity. The resin is added to wet out the reinforcement and eliminate air trapped in pores or voids in the preform. The mold is closed and the resin is cured. This technique allows the precise control of the fiber reinforcement in high concentrations. This process is capable of producing large, integrated sections with complex geometries. A variety of resins are used with reinforcements, particularly low-cost E-glass fibers and mats, and woven fabrics. Numerous variations of RTM are used. Several pertain to the application of a pressure differential on the resin flow into the mold.

Structural resin injection molding (SRIM) uses a similar preform and mold preparation to RTM. Once the mold has been closed, the SRIM resin is rapidly introduced into the mold and reacts quickly to cure fully within a few minutes. This requires the wet out of the preform to occur rapidly. Resins are required that have low viscosities that are highly reactive. The two-part resins require high-pressure impingement mixing to ensure thorough mixing prior to injection. Names of these variations include high-speed resin transfer molding (HSRTM), Seeman process, flexible resin transfer molding (FRTM), continuous resin transfer molding (CRTM), and vacuum resin transfer molding (VRTM).

(a)

Figure 8-36   (a) Composite pickup box assembly. The cargo box was fabricated by using HD-SRIM, a high-density structural injection-molding process and polyurethane as the matrix resin. Glass fiber mats and glass fiber preforms comprised the reinforcement. The wall thickness is 3 mm. This one-piece inner panel replaced four steel panels resulting in a reduction of the number of fasteners, easy disassembly, and aftermarket repair. Along with RRIM fenders and tailgate panel, overall savings in weight amounted to 50 lb over the steel box. (National Composite Center) (*continued*)

General Motors has demonstrated a pickup truck box that is lightweight, corrosion proof, and virtually indestructible. It is made from high-strength urethane molded with glass fibers. Figure 8-36a shows the composite pickup box assembly with all the major parts broken out. The box's fenders and outer tailgate panel are made of RRIM materials, and SRIM will comprise the one-piece inner panel and the tailgate liner. These two processes decrease the total weight of the truck by about 50 pounds and improve fuel economy and payload capability. Stronger than steel, the flexible body panels are dentproof. This structural composite box is being used as a test bed for future composite structures. The Silverado 1500 Series truck will be the first truck with the new composite pickup box. Figure 8-36b is a comparison of composite materials and processes for making various automotive components and a listing of applicable properties. They are compared to steel as well as to each other. It must be noted that they all do not possess the same properties to the same degree and this observation requires that with most materials and their manufacturing processes, a compromise based on experience and knowledge is needed.

The competition for selecting the best material for any vehicular component involves many considerations: Data banks are consulted, models produced, tools developed, and projected costs estimated and analyzed, to name a few. Figure 8-36c provides analysis for producing small fenders on automobiles. Four materials are under discussion with steel being the base material. The two SMC materials are low density materials containing glass microspheres (see Section 8.6.9). If weight were the only factor, then aluminum would be still under consideration. If the cost of the components is considered, including tooling and materials costs, then aluminum would lose out to SMCs. This example also shows that selection of materials is mostly a compromise, as materials possess neither the same kind of properties nor the same degree of properties.

A new process, ***vacuum-assisted custom resin injection molding (VACRIM)*** contains styrene emissions rather than releasing them into the atmosphere. This is an example of industry's efforts to reduce the amount of dangerous solvents entering the atmosphere (see Section 6.6.1 for a discussion of VOCs). ***Thixomolding*** is a new one-step process for the production of near-net-shape injection-molded magnesium parts. Current applications include

	SMC	LD-SMC	RRIM	SRIM
Weight (Specific Gravity)	1.9	1.3	1.3	1.5
Investment (Tooling)	~	~	Lowest	~
Total Program Cost	Lowest	~	~	Highest
Design Flexibility	~	~	Most	Least
Complexity Reduction	Most	Most	Least	~
Heat Resistance (degrees F)	400+	400+	375	350
Dimensional Integrity	Best	Best	~	~
Thermal Expansion	Steel	Steel	Higher	Higher
Corrosion Resistance	High	High	High	High
Dent Resistance	~	~	High	~
Structural Strength	~	~	Lowest	Highest
Stiffness (Modulus)	1.8	1.5	0.3	1.2
Recyclable	Yes	Yes	Yes	Yes
Surface Quality	Class A	Class A	Class A	Poor

(b)

Weight Comparison		
Steel	19 lb	Base
SMC (sp. gr. 1.9)	14.5 lb	−4.5 lb
SMC (sp. gr. 1.3)	10 lb	−9.0 lb
Aluminum	10 lb	−9.0 lb

Cost Comparison				
	Pc. Cost	Tooling	Amort.	Total Cost	Cost Diff.
Steel	$40.00	$8.9 Mil	$14.85	$54.85	Base
SMC (sp. gr. 1.9)	$45.00	$4.0 Mil	$ 6.65	$51.65	−$ 3.20
SMC (sp. gr. 1.3)	$50.00	$4.0 Mil	$ 6.65	$56.65	+$ 1.80
Aluminum	$60.00	$8.9 Mil	$14.85	$74.85	+$20.00

Note: The above costs are for engineering purposes only and do not represent a cost quotation. Tooling amortization assumes 2-year program at 300,000 units per year.

(c)

Figure 8-36   (b) Composite materials comparison. Compression molding uses SMC and low-density SMC as resin materials. RRIM and SRIM are molding processes in which the matrix resin and reinforcements are injected into molds. Note that all materials are high on corrosion resistance and recyclability. Both SMC materials have a coefficient of expansion the same as steel which adds to their advantage when mating composites to a steel structure. This simple comparison reveals that selection of materials requires an extensive background in materials. (c) Cost/weight comparison—small fender. The weight and cost for four candidate materials in the production of small fenders for automobiles is compared with steel acting as the base material to which the others are compared. A recent Ford Motor Co. automobile using 100 pounds of SMC components reduced vehicle mass by about 30 pounds versus steel.

computer cases, digital camera bodies, auto shift cams, and mechanical components of hand-held power tools.

8.7.4 Filament Winding and Tape Winding

Filament winding and tape winding, which produce the highest specific strength and glass content by weight of composite parts (fiber loading up to 85% by weight), is generally limited to parts with round, oval, or tapered inner surfaces. External shapes are unlimited. The continuous glass strand or filament is usually passed through a resin bath prior to winding onto a revolving mandrel. The mechanical as well as other properties of a filament-wound product can be changed by altering the *wind angle* (α), shown in Figure 8-37a. This angle is measured between the axis of the mandrel and the lay of the filaments. The tangent of this angle equals the ratio of the circumference of the mandrel and the pitch of the filaments. As the angle increases to 90°, the hoop tensile strength increases and the axial tensile strength decreases. Figure 8-37a shows the angle, the pitch, and the axial and hoop directions. The change in modulus of elasticity

(a)

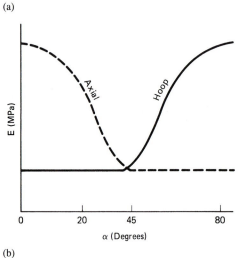

(b)

Figure 8-37 (a) Filament winding, showing the wind pitch and the wind angle (α). (b) Wind angle of a filament-wound product. (*continued*)

in the two principal directions is sketched in Figure 8-37b. Figure 8-37c shows some internal shapes possible with filament winding, along with a few special design and winding problems.

8.7.5 Pultrusion

Pultrusion is a composite-fabrication method by which an extremely long, fiber-reinforced, polymer-matrix material can be produced by pulling a fiber or a bundle of continuous filaments through a resin system for impregnation and then through a heated curing die. As the fiber/resin bundle travels through the curing die, polymerization occurs, and a composite structure emerges from the exit end of the die. Figure 8-38a shows a profile view of the pultrusion process used to produce a graphite-reinforced, epoxy composite T beam for use as a lightweight, high-strength aircraft stiffener. Figure 8-38b is a photograph of a section of the T beam, which can be produced to any desired length. Figure 8-38c is a profile view giving the dimensions of the T beam in inches, and Figure 8-38d shows the fiber orientation used in the T beam. The fiber, supplied in 12K tows, was sized with a coupling agent for epoxy-resin compatibility. The fiber layup consisted of eight plies oriented in the 0°, 90°, +45°, −45°, +45°, −45°, 90°, 0° directions (in that order). These tailored fiber orientations were obtained by locking two plies in their positions by stitching with a polyester thread; that is, the 0° (longitudinal) and 90° (transverse) fiber orientations were obtained by knitting the two plies together, thus locking them in their respective positions. The fiber volume was 54%. Preforming, preimpregnation, and layup of the plies to form the T beam were performed simultaneously in the 47-in. resin tank. Pultrusion produces solid rods, pipe, and many other structural shapes that have a constant cross-section.

THESE INTERNAL SHAPES CAN BE FILAMENT WOUND

THESE INTERNAL SHAPES CREATE SPECIAL DESIGN AND WINDING PROBLEMS

(c)

Figure 8-37   (c) Winding problems and some internal shapes of filament-wound products. (Permali, Inc.)

This new advancement in the use of pultrusion illustrates the need to minimize, if not eliminate, the labor-intensive manufacturing processes required for any advanced composites. The use of knit-locked, tailor-made reinforcement systems that produce tailor-made properties by pultrusion is contrasted with some older fabrication techniques that require the manufacture and stowage of preimpregnated materials, laborious layup and fixturing methods, vacuum bagging, autoclave curing, and debagging, which produce composite structures that are limited in size and length to the available autoclaves and drying ovens. Parts being pultruded can be inspected as they are fabricated using ultrasonic techniques to detect the presence of porosity and delaminations. This real-time feedback enables changes to be made to correct or discontinue the process before the production of a large number of defective parts that must be rejected.

PROFILE VIEW OF PULTRUSION PROCESS

Fiber creel   Resin tank   Preform dies   Cure die   Puller

(a)

(b)

T-BEAM PROFILE

FIBER ORIENTATION

3.00

.080   .170 R

2.25

.080

[0°/90°]
[+45°/-45°]
[+45°/-45°]
[0°/90°]

7 TOWS AS4-12K

(c)   (d)

Figure 8-38 (a) A profile view of the pultrusion process. (NASA.) (b) Photograph of a portion of a T beam produced by the pultrusion process. (c) T beam profile showing dimensions in inches. (d) A cross-section of the T beam showing the placement of the fibers (fiber orientation) in the T beam. (AS4 is the producer's designation for a graphite fiber.)

8.8 FASTENERS

Current fasteners for composites are not meeting all the requirements of structural designers. There is a need to develop fasteners for composites that will eliminate the problems associated with installing metallic fasteners in nonmetallic structures. Material compatibility is a necessity if problems with corrosion, lightning strikes, and low radar signatures are to be avoided.

Fiberlite fasteners (made by the Tiodize Company) have low creep, water absorption, and thermal expansion, and half the weight of aluminum with equivalent strength. Used to bolt composite structures together, these fasteners come in a variety of composite nuts and bolts with shear strengths of about 30,000 psi. They resist temperatures to 600°F. Machined from composite rod stock, the reinforcement in the rod is a three-dimensional, woven graphite-fiber structure with a polyimide-resin matrix. The cost is equivalent to titanium.

8.9 MACHINING OF COMPOSITE MATERIALS

Machining of composites demands special skill and knowledge. Composites cause severe tool wear by abrasion. Dust from the machining of graphite is a health hazard and causes fouling of the machine tools, producing maintenance problems. Graphite and carbon fibers are conductive and pose problems with numerical-controlled equipment. Coolants used with graphite can cause additional problems by combining with graphite dust to produce a slurry that is difficult to clean up. Also, some coolants tend to be absorbed and retained by the composite. Drilling bolt holes to allow joining of composites to other materials has a multitude of special problems, particularly when aligning prefabricated holes with precision tolerances. *Drilling holes* in laminated composites require special tools and techniques. The highly abrasive nature of these composites with their low interlaminar strength often results in splintering and delamination of the hole. Innovative drill designs fail to produce clean repeatable holes without some kind of backside support or frequent sharpening. A new diamond drill has been developed that is used when backing materials or backside support is not cost effective.

Solid ceramic tools boost production rates 5 to 10 times over carbide tools when machining composites. Metal-removal rates are up to 200 times higher than with diamond-coated tools. Tools of fine-grained alumina matrix reinforced with SiC fibers or whiskers are harder and more resistant than carbide. Drills, end mills, and routers can be used to cut graphite/epoxy, fiberglass, Kevlar, polyimide, graphite/graphite, and other advanced composites.

Single-layer diamond tools, made of a layer of diamond bonded to a formed metallic preform representing the negative of the shape desired, are manufactured for machining composites. Known as metal-single-layer tools, they are improving precision with longer tool life by remaining cool throughout the cut.

The Suppliers of Advanced Composite Materials Association (SACMA, 1600 Wilson Blvd., Arlington, VA 22209) has a report available that addresses the workplace hazards of advanced composite materials. Many manufacturers are abiding by the Occupational Safety and Health Administration (OSHA) guidelines for the handling of asbestos and confronting the possible hazards of composite particulates released during fabrication. The best remedy for eliminating airborne particles is adequate ventilation provided by a high-efficiency particulate air (HEPA) filtered vacuum. *Conventional machining* poses several problems, including excessive tool wear, frictional-heat generation, crazing effects when using lubricants, and localized fiber breakage and delamination due to tool impact and vibration.

Laser machining has some advantages in cutting and drilling holes in epoxy composites. The laser beam can focus on a very small spot, minimizing the heat-affected zone. Its vaporizing of material eliminates fraying. Kevlar and glass/epoxy composites present no laser-machining problems. Because of the high melting temperature and electrical/thermal conductivity of graphite, graphite/epoxy composites do not lend themselves to laser machining.

Figure 8-39   Robotic Waternife® waterjet cutting system is used here to finish-cut and trim automotive Class A surfaced rear-end panels with numerous clearance holes for taillights and a license plate mount. (General Electric.)

Waterjet machining, with or without the garnet abrasive added to the stream, is finding use in the machining of composites. Kevlar is cut with a waterjet without an abrasive. Due to safety requirements, only automated waterjet systems are used. Figure 8-39 shows a GE waterjet system used in the automobile industry. Its biggest limitation is the inability to effectively machine material thicknesses greater than 2.54 mm. Otherwise, the abrasive jets cut quickly and leave an excellent edge without tearing, fraying, or delamination. Several companies now specify that composites and other heat-sensitive parts must be cut with abrasive waterjet. Flow International's Paser II waterjet can cut composites at 15 to 30 in./min, in comparison with mechanical cutting of the same materials at 1 in./min. The same waterjet is capable of cutting 0.8-in. Kevlar at 5 in./min. Both waterjet and laser machining exhibit very little tool wear.

8.10 REPAIR OF COMPOSITE MATERIALS

A fast-repair kit produced by Ferro Corporation for fiberglass parts in the aerospace industry consists of an epoxy–resin system—a two-component thermosetting patch and repair material that, when mixed with an activator, produces a hard plastic requiring no external heat to cure. The activated resin saturates glass cloth, mat, or tape for a structurally sound composite with good dimensional stability, impact resistance, and tensile strength. Patches are abrasion and corrosion resistant and not affected by moderately high or subzero temperatures. The resin adheres to most surfaces, including steel, plastics, wood, and ceramics. Figure 8-40 is a line drawing of a typical honeycomb-core repair. Other types of repair patches for laminar composites are illustrated in Figure 8-41. Most composite-repair techniques depend on an adhesive with a certain shelf life. This requires use of the adhesive prior to its shelf-life expiration. Additional restrictions are imposed on its disposal to ensure compliance with existing environmental laws and regulations. Many adhesives require environmentally controlled storage, usually at low temperatures, and their out-times are limited, requiring close monitoring.

A functionally gradient material joint is one in which the materials gradually change in chemical composition and properties across a dissimilar joint, bridging mismatches and reducing thermally induced stress. Traditionally, the joining of dissimilar materials has been a

Figure 8-40 Honeycomb-core repair. (USAF)

(a)

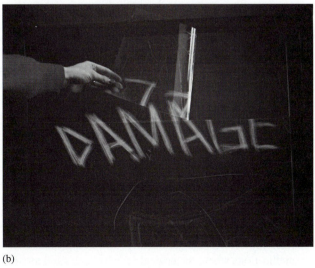

(b)

Figure 8-41 (a) Scarfed, stepped, and overlay patch repairs. (USAF.) (b) The new material, N-View, is a sheet of Lexan MR5 polycarbonate that receives an abrasive-resistant Margard surface treatment and is then laminated with a thin, abrasive-resistant film of Lexan HPH to avoid damage to plastic glazing. The thin film can be removed if damaged, leaving the MR5 thick sheet and thereby doubling the glazing life. (G.E. Plastics) difficult task. A recent exercise demonstrated the joining of silicon carbide and a nickel-base superalloy with a functional gradient material made up of elemental powders of nickel, titanium, aluminum, and carbon. Such a process may lead to the efficient and effective joining of hybrid structures consisting of mixed material systems.

8.11 NONDESTRUCTIVE EVALUATION

Composites have extended the lifetimes of helicopter blades threefold. Older blades made with aluminum sheet and extrusions bonded with adhesives were inspected by X-ray equipment, which took about 8 h. New analysis and characterization techniques, as introduced in Module 4.6.3, have drastically reduced the time and improved the reliability of nondestructive evaluation (NDE). Using computerized infrared thermography, inspecting the blades now takes about an hour. The NDE of composites during the manufacturing process will be the key to 100% reliable inspection. Composites pose a severe challenge due to the variety and complexity of the material's composition and fabricating processes. Also, there is no clear definition of defects. *Computer tomography* using X-rays provides high-resolution two- and three-dimensional images of the internal structure of advanced composites for inspecting stress-induced damage in small electronic or mechanical components. Unlike radiography, the system makes measurements from many angles, which a computer program then uses to reconstruct a slice of the object's internal structure. One challenge to adopting PMCs for use in marine structures, including surface ships, submarines, and marine platforms, is the absence of any effective test methods for electronic or mechanical components.

8.12 COMPOSITE SELECTION: DATABASE FOR COMPOSITES

In the *selection of composite materials* for high-temperature applications, particularly resin-matrix composites, it is imperative to have a detailed knowledge of a resin's glass transition temperature (T_g) (see Section 6.3.2.4). The need for an up-to-date materials *database* is also essential. To replace traditional engineering materials with composite materials requires a database that reflects varying service conditions, including combinations of conditions such as high temperatures with or without a load and/or an adverse environmental condition such as a corrosive atmosphere. P. K. Mallick, University of Michigan—Dearborn—is compiling an automobile materials database. Figure 8-21b showed a flowchart for designing an FRP composite to withstand fatigue. Note the inclusion of databases for both materials and processes. The lack of good databases inhibits the adoption of new materials by many design personnel who are thus unable to locate sufficient data about materials that would be more efficient in many applications. Using fiber-reinforced thermoplastic composites as materials, the importance of the foregoing two points can be emphasized. The amorphous resin poly(ether sulfone) (PES) and the crystalline resin poly(ether ether ketone) (PEEK), both thermoplastics, are used as matrix materials reinforced with glass or carbon fibers. These reinforcements provide increases in strength and modulus (stiffness) to the polymer resins. In general, carbon-fiber reinforcements provide a 20% improvement in tensile strength and a 30–100% improvement in flexural modulus over glass-fiber reinforcement at room temperatures, due to the ability of the matrix material to transfer the stresses produced by the load(s) from itself to the fibers. Under high-temperature conditions the matrix plays another important role—that of being the determiner of the high-temperature performance of the composite. How well it is able to maintain its strength and stiffness at high service temperatures dictates the service temperature for the composite. There are innumerable high-temperature applications where the critical properties desired are dimensional stability and stiffness. One is the need to reduce mold shrinkage. PES has a T_g of 230°C, whereas PEEK, being crystalline, has a melting temperature (T_m) of 336°C. Testing these materials at various high temperatures shows that the amorphous polymer PES has higher strength at just below its T_g than

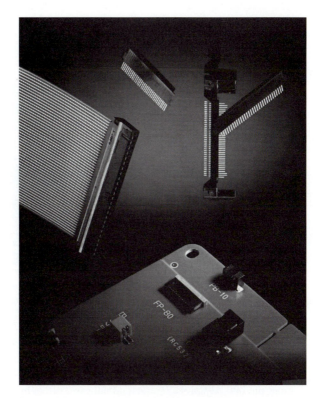

Figure 8-42   High-performance connectors used in the electronics industry's surface mount technology are made of GE Plastics SUPEC® resins with high ductility and low-flash grades. (General Electric Plastics)

Figure 8-43   Two impellers used in the chemical industry—the metal impeller on the right has an average service life of only two months in an acid environment at 230°C compared to a carbon-fiber-reinforced PEEK (the high-temperature thermoplastic composite impeller on the left) with an extended service life of 18 months. (ICI Advanced Materials, *Advanced Materials & Processing,* Vol. 140, No. 2, August 1991, Pub. by ASM Int.)

does the crystalline polymer PEEK, but the crystalline PEEK resin composite maintains its strength above the T_g of the amorphous PES composite. Figure 8-42 illustrates GE SUPEC® poly(phenylene sulfide) (PPS) resin, which has high ductility and good thermal capabilities and finds applications in ***surface-mount technology*** as high-performance connectors with good pin-insertion/retention qualities. This example illustrates that, when selecting thermoplastic composites to replace metals for high-temperature applications, a knowledge of glass transition temperatures and a database containing the results of performance testing at conditions other than ambient room temperatures are essential. Figure 8-43 illustrates a situation when a high-temperature thermoplastic composite, having an array of properties that were required not only to combat high temperatures but also a combination of stressful and corrosive environments, has substituted for a metal and performed superior to the metal.

Figure 8-44   *Spirit of Columbus,* Battelle Columbus's award-winning entry in the International Submarine Race for human-powered submersible craft, allowed their designers to gain experience with emerging composites systems technology. (*Advanced Materials and Processes.*)

The submersible craft (seen in Figure 8-44) serves as a case study to demonstrate how the many concepts of materials selection presented in this book are applied. Battelle Columbus's two-person, human-powered *Spirit of Columbus* uses varying-density PVC foam in an epoxy-bonded composite hull. Hysol 9460 structural adhesive offers high tensile lap-shear strength (24 MPa or 35,000 psi) when bonded to the aluminum frame. This aluminum-to-composite system is the same one used in Indy 500 race car bodies and reflects a movement away from mechanical fastening systems that use bolts, rivets, etc. The hull sections consist of PVC foam in different densities, sandwiched between layers of fiberglass. High-density, 0.1 g/cm^3 (6.23 lb/ft^3), foam is concentrated on the load-bearing hull bottom, to which air tanks and the drivetrain are permanently bonded. Low-density, 0.06 g/cm^3 (3.75 lb/ft^3), foam is used on the hull's top half, where no other structural support is needed. The propulsion system uses a 455-mm (18-in.) diameter drive sprocket, connected by a chain to a cable takeup pulley. Pulley-cable ends are attached to the driveshaft and connect to two hinged flappers. The flappers are made of 6061-T6 aluminum alloy with lightening holes, polystyrene foam, glass fiber, two layers of carbon fiber, and a top layer of drag-reducing riblet tape. The drag-reducing tape inhibits the transverse movement of turbulent eddies across the flapper surface. The lightweight hull must be both very rigid and very strong because the innovative, articulated linear-thrust propulsion system causes the drivetrain to move back and forth.

8.12.1 Integrated Design for Manufacturing

Textile structural composites are composite materials that are reinforced by textile structures for load-bearing applications. An *integrated design for manufacturing (IDFM)* approach has been established based on textile-fiber architecture to facilitate the communications between structural designers and composite manufacturers. Several software packages are available based on IDFM, including a cost analysis module.

Several U.S. aerospace manufacturers are engaged in a major cooperative effort to develop software to reduce the cost of composite manufacturing. The *Composite Affordability Initiative (CAI)* is a cooperative effort between government and industry to reduce the cost of aerospace structures so that engineers can take full advantage of their unique benefits. Entirely new processes for composite manufacturing are being developed.

8.12.2 Microscopy of Composite Materials

Initially, carbon-fiber-reinforced plastic composite materials had low damage resistance and fracture toughness. These properties severely limited their application particularly in the aircraft industry. One effort to rectify this problem led to a study of the relationship between microstructure and materials performance. One outcome was the development of several specimen preparation methods that revealed the size and distribution of individual resin phases and subsequently the role played by these resin phases in the improvement of damage resistance. One final result was that changes in formulations and manufacturing processing could be evaluated from a microstructural viewpoint.

The technique of ***thin-film microscopy*** was one development that provided greater morphological detail than traditional reflected-light microscopy. This ***thin-section microscopy*** uses transmitted, polarized light. Specimen sections only 1-5 μm thick are used since the composite's materials are opaque. At these levels of thickness, the materials become optically transparent. The result is that transmitted light methods reveal more details of the morphology of polymers and similar materials than are observable using any other microscope techniques available. Ultrathin microscopy also reveals information about fracture processes in structural laminates which leads to improved composites that resist delamination. It also supplements other failure analysis techniques. Fiber-reinforced laminates with multiple plies, honeycomb-core sandwich structures with laminated face sheets, and polymer-matrix composites with finely dispersed particulates are likely candidates for thin-section techniques. As discovered in the preparation of metallic specimens for observation by a metallograph, good microscopy is 99% sample preparation. Preparing a flat, scratch-free, strain-free cross-section without introducing artifacts (human-made changes) is both a science and an art. To grind/polish a material such as a fiber composite material with its relatively hard fibers (carbon, Kevlar, etc.) embedded in a soft, polymeric matrix certainly requires special care. Most composite microscopy involves looking for porosity and microcracks. Microcracking is the result of stresses which surpass the ultimate strength of the polymer, or polymer/fiber interface bond. From a careful study of the microcracking, scientists can determine the durability of composites in service and investigate the relationship of process parameters and environmental exposure to micrographs. High-temperature thermoplastic composites behave like metallic structures in that mechanical properties change when these composites solidify at different rates. Controlling the cooling rates, and microstructural phenomenon such as crystal sizes, shape, distribution, etc., determine, to a great extent, mechanical properties of the composite. Several photomicrographs taken with thin-section microscopy were provided by Luther M. Gammon, Boeing Materials Technology, Boeing Commercial Airplane Group, Seattle, Washington, who was instrumental in the development of this technique. Figure 8-1a–d, entitled **"Stress/Strain Before Your/Eyes,"** consists of five photomicrographs with various defects that illustrate this technique of ultrathin sections with polarized light microscopy for a composite material undergoing impact testing.

8.13 COMPOSITES GLOSSARY

Areal weight is the weight of fiber per unit area (width $\times$ length) of tape or fabric.

Braiding is the weaving of fibers into a tubular shape as opposed to a flat fabric.

Cavity is the location in a mold where the resin enters and fills to form a solid molded part replicating the features and size of the cavity.

Chopped strands are produced by cutting continuous strands into short lengths.

Commingled yarn is a hybrid yarn made with two types of materials intermingled into a single yarn.

Conductive composites describes electrically conductive composites using metal fibers embedded in a nonconductive matrix to provide materials for uses such as antistatic coatings; fuel-cell catalysts; solar electrical cells; or lightweight, inexpensive batteries.

Cowoven fabric is a reinforced fabric woven with two different types of fibers in individual yarns, such as thermoplastic fibers woven with carbon fibers.

Creel is a device for holding the required number of roving balls (spools) in the desired position for unwinding onto the next step in the process (i.e., weaving, braiding, or filament winding).

Curing agent describes initiators, hardeners, or any other materials that promote cross-linking. It differs from a catalyst in that it is consumed by the cross-linking reaction whereas a catalyst is recoverable once the reaction is complete.

Drape is the ability of a fabric or prepreg to conform to a contoured surface.

Fabric is made by interlacing yarn in a variety of ways, including weaving, knitting, and braiding.

Fabrics such as 2-D woven fabrics are currently used for prepregs. There are nearly a dozen types or styles of basic weave such as plain, 2/2 twill, 8HS, and triaxial wave. In 3-D woven fabrics, only z yarns interlace with the filling yarns. 3-D woven fabrics have high interlaminar strengths and find use in impact-resistant structures.

Fiber refers to filamentary materials and is synonymous with *filament,* which has an extreme length and very small diameter, usually less than 25 μm. A continuous fiber has an indefinite length, in contrast to a discontinuous fiber represented by whiskers (chopped fiber in short specified lengths) or particulates.

Filament is slender thread or fiber.

Flash is the overflow or squeeze-out of cured resin from a mold.

Harness satin is a weaving pattern that produces a satin appearance. "Eight-harness" means that the warp tow crosses over seven fill tows and under the eighth in a repeating manner.

To *impregnate* is to saturate the voids and interstices of a reinforcement with a resin.

Knitted fabrics are fabrics produced by interlooping chains of yarns.

Knitting is a process of interlooping chains of tow or yarn.

Laminate is two or more layers (plies) bonded together. Laminates are composite materials made from several layers of materials.

Macrocomposites are single parts that incorporate more than one composite. An example of the incorporation of more than one composite in a single part would be a CMC and an MMC in a composite-fabricated driveshaft.

Mat is a fibrous reinforcing material comprised of chopped filaments, along with a binder to maintain form, and is available in blankets of various lengths, widths, and weights.

Mold release agents create a barrier between the mold or tool and the part being molded primarily to prevent the part being molded from sticking in the mold when curing is completed. They also hold the part in place during cure so that the proper shape and surface finish are attained. The various types of release agents include solvent-based, water-based, paste, semipermanent, and internal (blended into the molding resin). Silicones, fluorocarbons, and beeswax serve as release agents.

Neat resin is a resin to which nothing (additives, reinforcements, etc.) has been added.

Out-time is the time a prepreg is exposed to ambient temperature (i.e., the total amount of time the prepreg is out of the freezer).

Particulate composites are materials consisting of one or more components (particles) suspended in a matrix of another material. Such particles can be either metallic or nonmetallic (see Section 8.6.4).

Pitch is the residue left from the distillation of coal and petroleum products. It is a base material for the production of some carbon fibers, as well as a matrix precursor for some carbon–carbon composites.

Ply is a single layer of tape or fabric.

Pot life or *working life* is the length of time that a catalyzed thermosetting resin retains a viscosity low enough for it to remain suitable for processing.

Prepregs are a precoating of fiber or cloth sheets with resin and mixed with a hardener to help keep the fibers in place. Fibers may be in the form of continuous roving, mat, or woven fabric. The resin is allowed to cross-link slightly until it is viscous enough to stay on the fibers. This process is called B-staging, and the result is tacky. Placed on paper to keep from sticking together, the prepreg is rolled up and stored under low temperature until used. Epoxy resins are mainly used in prepreg especially in the aerospace industry primarily due to its high strength and modulus (stiffness) that can be obtained. Rolls of prepregs have widths from 4 to 40 in. Roving can be prepreged. Prepregs are subsequently cut into required dimensions and cured into final shape by molding operations.

Ramping is the gradual, programmed increase or decrease in the temperature or pressure used to control the cure or cooling of composite parts.

Resins are mostly polymers. In reinforced plastics, the matrix is a resin that binds together the reinforcement. Resins are solids or pseudosolid organic materials, usually of high molecular weight, that will flow when subjected to stress.

Roving is a collection of bundles of continuous filaments, either as twisted yarns or as untwisted strands.

Rovings are used in continuous molding operations such as filament winding or pultrusion. They can be preimpregnated with a thin layer of polymeric resin to form prepregs.

Scrim is a low-cost reinforcing fabric made from continuous filament yarn in an open-mesh construction (see Figure 8-4).

Sizing is a compound that binds together and stiffens yarn and fibers, providing resistance to abrasion during weaving and handling. The ingredients, such as gelatin, wax, or oil, provide surface lubricity and binding action. Unlike a *finish,* which contains a *coupling agent* to improve the bond of resin to the fibers, it contains no coupling agent. The size is usually removed by heat cleaning and a finish is applied. A finish can also contain a binder.

Space-grown composites are organic composites grown in space to provide repair materials for a space station. They obviate the need to deliver repair materials from Earth at great expense.

Strand is an untwisted bundle of continuous filaments used as a unit, including tows and yarn. Sometimes a single filament or fiber is referred to as a strand.

Tape is a material in which the fibers are laid in a single direction within a resin matrix.

Tow is an untwisted bundle of continuous filaments. A tow designated as 100K has 100,000 filaments.

Warp is yarn running lengthwise in a woven fabric.

Warp refers to the 0° yarns—the 90° yarns are *fill.* The method of 2-D weaving involves interlocking two sets of yarns perpendicular to each other. The warp or the 0° direction is the continuous fabric direction. The warp refers to the long, machine, or fiber direction. The fill direction is also called the weft or woof direction.

Weave refers to the particular manner in which a fabric is formed by interlacing yarns.

Weave consists of interlacing two sets of yarn perpendicular to each other. The 0° set is called warp and the 90° set is called fill. The warp axis is the continuous fabric direction.

Weft are the transverse threads or fibers in a woven fabric. They run perpendicularly to the warp. Also called fill, yarn, or woof.

Wet layup is a method of making a reinforced product by applying the resin system as a liquid when the reinforcement is put in place.

Wet-out is the saturation of all interstices between bundles and filaments of reinforcement with resin.

Whiskers are single crystals ranging in size from about 0.5 to 10 μm in diameter, and 3 μm to 3 cm in length. They can be metallic or ceramic. Their aspect ratios are between 100 and 15,000. (See Section 8.4.1 for a definition of aspect ratio.) An example of a popular whisker in use is one made of silicon carbide. It varies in diameter from 3 to 10 μm, 10 to 100 μm in length, and has a Young's modulus of 800 GPa and an ultimate strength of 21,000 MPa. Another popular whisker made of silicon nitride varies in diameter between 0.2 and 0.5 μm, 50 to 300 μm in length, and has a modulus of elasticity in tension of 300 GPa. For example, the size of small defects in silicon carbide, such as voids, inclusions, and surface roughness, is 0.1 to 0.4 mm. This is the secret to the strength of whiskers: the smaller the fiber, the smaller the defect. Adding encapsulation to the fiber, which helps protect it from surface damage, results in fibers (whiskers) maintaining their inherent strength.

Woven roving is a heavy, coarse fabric produced by the weaving of continuous roving bundles.

Yarn is an assembly of twisted filaments, fibers, or strands, either natural or manufactured, to form a continuous length that is suitable for use in weaving or interlacing into textile materials.

APPLICATIONS & ALTERNATIVES

The composite automobile: promise and challenges.   This module has shown how the engineered materials PMC, CMC, and MMC provide many desirable properties. In transportation, fuel efficiency is of paramount importance not only because of the economics of using less fuel, but burning less fuel also means that fewer gases and particulates (hydrocarbons) are emitted into the atmosphere, hence less air pollution. Throughout the world, transportation vehicles have diminished the air quality; thus, legislation is requiring more fuel-efficient vehicles and lowered emissions. CMCs, PMCs, and ceramics offer potential for more efficient engines for air, sea, and ground vehicles. PMCs and MMCs can provide substantial weight savings for both the structures and bodies of those vehicles.

Interest in composites for automobiles dates back to Henry Ford's demonstration of the impact resistance and durability of a trunk lid made of a soybean-based composite. In 1953 the Chevrolet Corvette appeared with an all-FRP production autobody (Figure 8-45a). Since then, numerous applications of composites throughout the automobile industry have appeared, as have concept cars with plastics and composites for complete body assemblies. These developments were helped along by the 1970s fuel crisis and by the demand to increase market share among the world's car and truck manufacturers.

Each year, the pounds of steel and iron in an average car drops and the use of plastic and composites increases. An increase in the use of plastic and composite body panels, from the 8% level in the early 1990s up to a level of about 70%, would bring the following benefits:

1. Improved fuel economy and performance through weight reduction
2. Improved overall product quality and consistency in manufacturing
3. Part consolidation, resulting in lower product and manufacturing costs
4. Improved ride performance due to reduced noise, vibration, and harshness
5. Improved corrosion resistance
6. Greater impact resistance (crash integrity) with a corresponding increase in safety
7. Improved repairability and recyclability

In the United States, the Big Three automakers formed the Automobile Composite Consortium (ACC) to attack the critical area for the future use of high-performance composites—manufacturing techniques. The first two requirements are that they be *rapid* and *economic*. They must be able to combine high rates of production, precise fiber control, and a high degree of part integration. Production vehicles such as GM's Saturn (sedans and station wagons) and all-purpose vehicle (APV) vans and Dodge's Viper have used a combination of SMC and RTM (resin transfer molding) for the entire body, but most other automobile manufacturers have used selected plastic and composite panels. Prototypes of potential production vehicles were seen in concept cars such as the Ford Ghia Connecta (Figure 8-45b), with a carbon-fiber and Kevlar body, and GM's Ultralite (Figure 8-45c), with a body that sandwiched polyurethane foam between thin layers of carbon-fiber fabric. For auto frames and structural components, Ford's development of a primary PMC structural frame (see Figure 8-23) demonstrates more of the evolution of the composite automobile. AISI features the **ultralight steel autobody (ULSAB)** which destroys the current engineering paradigms that a stiffer, stronger steel autobody structure must be heavier and that a light body structure must be expensive. This model structure weighs merely 203 kg (447 lb), up to 36% less than the midsize sedans benchmarked in the study.

 EMT Weblinks

United States Council for Automotive Research (USCAR)—*http://www.uscar.org*
Automotive Composite Consortium (ACC)—*http://www.uscar.org/consortia/con-acc.htm*

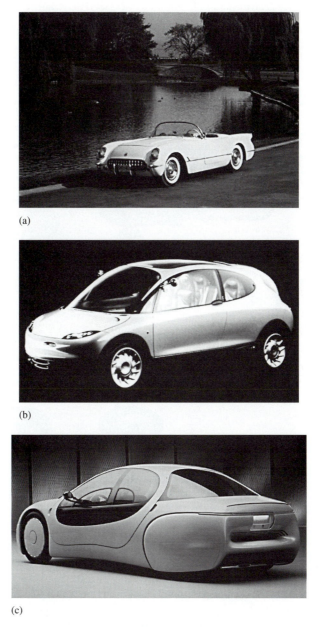

Figure 8-45   (a) Composite body: the 1953 Chevrolet Corvette was the first fiberglass production body. (Morrison Molded Fiber Glass Co.) (b) Ghia Connecta: Ford Motor Company's electrically powered concept vehicle has a carbon and Kevlar fiber composite body, double-skinned glass roof, and solar panel in the top surface powering the rear roof vent. (Ford Motor Co.) (c) GM Ultralite concept car. (General Motors)

The industry's successful three-tier (salvage dealers, shredders, and secondary-metal recoverers) recycling program has been quite effective in keeping iron, steel, aluminum, and other metals out of the ***municipal solid waste*** (MSW) stream (see Applications & Alternatives in Module 5). This tiered system recovers more than 80% of the ferrous metals and aluminum and 66% of the copper from scrap automobiles. Most batteries are salvaged to recover lead plates and the polypropylene case. Figure 6-22 showed the increased use (past and projected) of major plastics for automobile use. PMCs are also on a steady incline. There is a declining trend for iron, which is expected to continue, as discussed in Module 6.

Setting up the ***recycling infrastructure*** for PMCs is a major challenge. Currently, much of the polymer content ends up in the landfill. Not only is this unacceptable from an environmental

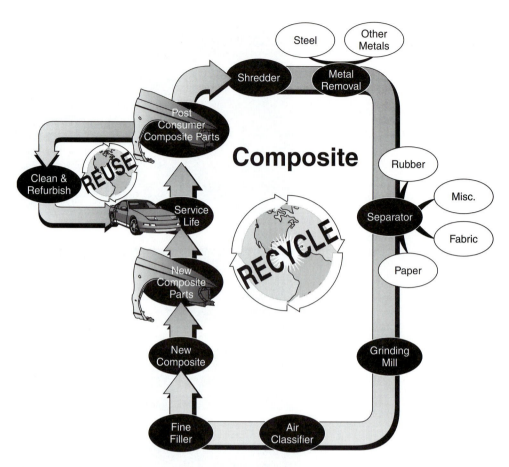

Figure 8-46   Recycling of SMCs composite parts for automobiles. Scrapped parts (end of life parts or in-plant scrap) made of thermosetting plastics (SMCs) are ground up to provide a material that can replace the filler in new formulations. New composites can contain substantial quantities of recycled materials including resin, filler, and additives. (Adapted from Automotive Composites Alliance.)

perspective, but from an economic point of view it is unwise to bury polymers at ever-increasing "tipping fees" at landfills when the materials have value as recyclable resources. In the United States, Chrysler, Ford, and GM have established a vehicle recycling partnership to promote and conduct research for improved recycling technology.

Composites Recycling/Green Manufacturing.   Composites are recyclable at the end of their own life cycle. Today's composites can contain substantial amounts of recycled materials. The *Automotive Composites Alliance (ACA)* which assists automakers in the use of composites has reported that more than 10 million cars and trucks produced annually have composite components. Reinforced thermosetting polymers, specifically SMCs, are making attractive alternatives to steel and aluminum, because their styling flexibility permits automakers to make frequent styling changes to incorporate part consolidation and weight savings at low program costs. The ACA has promoted research and development necessary to establish economically viable processes for recycling these materials. To date, many automotive parts containing recycled SMC filler are used on production vehicles. A recycled part can contain recycled content as high as 25% by weight without adverse effect on physical properties. SMC is an easily recycled plastic, more so than thermoplastics, and there is no need to remove paint and adhesive from the surface prior to processing as recycled filler. SMC is less sensitive to contamination by dissimilar materials and involves no degradation of properties. The resin portion of SMC also can contain recycled PET from pop bottles. In view of these remarks, SMC will be the material of choice for auto components when compared to other plastic materials. As this use grows, the recyclability of SMC will be one of its greatest advantages.

RECYCLING OPTIONS
Material Comparison For External Autobody Panels
(Hood, Fender, Applique, Front-End Panel, Bumper, Etc.)

	Thermosetting Composite SMC/BMC RRIM SRIM	Thermoplastic Composite Azdel GFN/GFPE	Engineering Thermoplastic PET/PBT Polycarbonate TPO	Sheet Steel	Sheet Aluminum
Recycled Content in New Part					
Post-Consumer	No (a)	No (a)	No (a)	No (a)	No (a)
Pre-Consumer	No (a)	No (a)	No (a)	Yes	Yes
Internal Reuse	No	Yes (b)	Yes (b)	Yes	Yes
Recycleable at End of Service Life					
Refurbish and Reuse	Yes (d)	No (e)	No (e)	Some (f)	Some (f)
Primary Recycling	No (a)	No (x)	No (x)	No (a)	No (a)
Secondary/Tertiary Recycling	No (a)	No (a)	No (a)	Yes	Some
Life Cycle Assessment					
Life Cycle Inventory (Energy)	Low	Low	Low	High	Med.
Environmental Impact	Med. (g)	Med. (g)	Med. (g)	Med. (g)	Med. (g)

Notes:	(a)	Feasible, but not currently done because of cost penalty
	(b)	Internal scrap used, but generally into "lower grade" parts
	(d)	Bolt-on parts commonly dismantled and reused (if not damaged)
	(e)	Generally not economical to reuse
	(f)	Bolt-on parts can be reused if free of corrosion and/or damage
	(g)	Generally no major differences between materials
	(x)	Generally not feasible or large cost penalty

Figure 8-47 Recycling options for automotive materials. Five materials used to produce external autobody panels are compared showing their potential for varying techniques of recycling. SMC automobile panels reduce usage of fuel reserves. A Lincoln Continental uses 1200 pounds of SMC components reducing vehicle mass by about 30 pounds versus steel. (Adapted from Automotive Composites Alliance.)

Figure 8-46 is a schematic of the recycling of composite thermosetting plastics. The current recycling sequence is shown in this figure via a scrap car delivered to a dismantler where useful or recyclable parts are removed. The "hulk" is passed through a shredder where ferrous materials are magnetically removed, leaving the remainder or automotive shredder residue (ASR). This residue is passed through a nonferrous separator where aluminum, copper, zinc, magnesium, and other metals are removed. The balance of the ASR is sent to a landfill. Typical recovery rates for some materials dismantled are thermoplastics, 1% dismantled of which 93% ends up in a landfill; aluminum, 15% dismantled of which 20% ends up in landfill; SMC/BMC, 15% dismantled of which 85% ends up in landfill.

The percentage of organic materials (plastics, rubber, fabrics, etc.) in an average vehicle is 15% by weight but 55% by volume. With this new recycling of SMC with its reuse by manufacturers (***green manufacturing***) this large influx of materials to landfills will be reduced considerably. Composites benefit the environment because they are energy efficient to manufacture and their inherent light weight results in fuel savings over the life of the vehicle. Energy savings are accomplished in the production of raw materials for composites, and in the processing and manufacturing of composites when compared to similar steel parts. Weight savings to the vehicle means less fuel consumed and thus a reduction in emissions.

Figure 8-47 lists the recycling options for five materials used to produce external automobile body panels. Recycled content can be used in several ways. Preconsumer content refers to postindustrial use other than in an automobile. Internal reuse refers to the manufacturing process for production of vehicles. Primary recycling refers to the reuse of material to make a similar or identical part, and secondary use is reusing material to make a lesser-value part. ***Life-cycle assessment*** is the "cradle to the grave" analysis, or ***life-cycle inventory (energy),*** as described in Module 6. This inventory of total energy consumption spans raw materials to final disposal of the part. Accompanying this analysis is an ***environmental impact assessment*** which involves natural resources depletion, toxic emissions, pollution, and other forms of negative attack on the Earth from initial manufacture to final disposal of the part.

SELF-ASSESSMENT

8-1. From the Pause & Ponder, what advantages were accrued by using engineered polymer composites to replace natural polymers?

8-2. Identify and list the composite materials present in a typical U.S. kitchen. Include furniture, equipment, and tools usually found in the room.

8-3. Reinforced concrete is a composite material. List the constituents that must act in concert with each other to produce this widely used engineering material.

8-4. Name at least three composite materials found in the human body.

8-5. A natural polymeric composite material, wood, has directional properties. Explain what is meant by this, and cite an application of wood that takes this particular property into account.

8-6. The terms *specific strength* and *specific modulus* are often used in describing composites. What does specific modulus refer to, and what does it tell us about a particular material?

8-7. Oilite bearings are an example of what type of composite material?

8-8. When speaking of interface and interphase, which term represents the larger area or volume in a composite material?

8-9. Which fiber material has the greater specific stiffness, graphite or carbon?

8-10. Name two polymorphs of carbon.

8-11. State a major disadvantage of graphite as a fiber material.

8-12. What is the trade name for aramid fibers?

8-13. Compare the tensile strength of Kevlar with nylon. Which material would make better ropes?

8-14. The abbreviation NDE stands for what?

8-15. Using Figure 8-10, which material listed would produce a bending strain of 2% with a bending load that produces a stress of 750 MPa?

8-16. Verify the units of specific strength (inches) using Figure 8-9.

8-17. Describe three tests used to measure toughness.

8-18. The properties of a reinforcing fiber are dependent on their size and length. How are these two qualities related and used in describing the strength of such a fiber?

8-19. Regardless of the name given to the processing method for producing fiber-reinforced plastic products, two stages are included in most of them. Name these two stages.

8-20. Which molding process provides the highest specific strength and glass content by weight?

8-21. Describe the effect of a decreasing wind angle on the hoop stress.

8-22. When a part that is filament wound contains a transition (concave contour), bridging is likely to occur. Explain what *transition bridging* means.

8-23. Laminations in plywood are known by what name?

8-24. List two examples of a rigid laminate.

8-25. Sandwich composites have cores of various configurations. List three such configurations.

8-26. A cermet is basically a composite made from a ceramic and a metal. Which of the two materials is softer?

8-27. What thermal-protection system protected space capsules from the intense heat of reentry into Earth's atmosphere prior to flights of the space shuttle?

8-28. Identify a possible use for a laminated composite consisting of two bonded sheets of different metals, each having a different coefficient of thermal expansion.

8-29. A popular kitchen cooking utensil contains a composite base made of copper clad to stainless steel. What is the purpose of this composite application?

8-30. Discuss the reasons why the emphasis in materials science over the past three decades has been on the research and development of fiber-composite materials as opposed to efforts to develop or refine traditional load-bearing materials such as an alloy steel.

8-31. If present materials technology could produce crystals of practical size free of defects, such high-strength whiskers could be used as reinforcing material in a composite that would outstrip the performance of existing composites. Discuss what problems could arise, if any, in combining this filler material with a suitable matrix material.

8-32. A rod made of polyester under a tensile load that is parallel to the length of the rod complies with the rule of mixtures. If another rod with the same strain under the same load conditions and composed of glass fibers in a polyester matrix is substituted for the rod of polyester, compare the two rods by finding: **(a)** the density, **(b)** the elastic modulus of the composite rod, **(c)** the difference between the cross-sectional areas of the rods, and **(d)** the diameter of the composite rod in millimeters if the area of the polyester rod is 0.7854 in.^{2}. Use the following property values:

$$E(p), \text{ modulus for polyester} = 4 \text{ GPa}$$

$$E(f), \text{ modulus for glass fibers} = 68 \text{ GPa}$$

$$p(p), \text{ density of polyester} = 1.36 \text{ g/cm}^3$$

$$p(f), \text{ density of glass fibers} = 2.54 \text{ g/cm}^3$$

$$V(f), \text{ volume fraction glass fibers in composite rod} = 0.094$$

$$V(p), \text{ volume fraction for polyester}$$

$$E(c), \text{ modulus of the composite material}$$

$$F(p), \text{ load or force on polyester rod}$$

8-33. Derive the inverse rule of mixtures in terms of modulus and volume fraction as stated in Section 8.5.1.

8-34. Solve for $E(c)$ using the inverse rule of mixtures equation in Section 8.5.1.

8-35. NASA and the Department of Defense (DOD) are both financing an experimental aircraft that will fly from a standing start on a runway into Earth orbit at speeds of mach 25. Give the name of this program and the designation for the aircraft.

8-36. Calculate the specific strength of S-glass that has a tensile strength of 4.8 GN/m^2 and a weight density of 24.4 kN/m^3.

8-37. Sketch a Venn diagram that shows how sets of the three Family of Materials groups intersect with reinforcers to form the wide array of advanced materials.

8-38. Sketch a typical fiber-reinforced composite and label the following: matrix, reinforcing fiber, interfaces, interphases.

OBJECTIVE QUESTIONS

8-1. Composite of plastic reinforced with glass.
   a. GPa   **b.** FRS   **c.** SMC   **d.** GRP

8-2. Impregnated, continuous sheets of composite materials such as those used for autobody panels.
   a. GPa   **b.** FRS   **c.** SMC   **d.** GRP

8-3. From the Pause & Ponder, list one advantage of using an engineered polymer composite to replace natural polymers.

8-4. List one type of advanced composite used on the Boeing 777, NASP, or HSCT.

8-5. The binding portion of a composite that distributes the load among fibers or particles.
   a. Matrix   **b.** Reinforcer   **c.** Coupling agent   **d.** Interface

8-6. Organic, ceramic, synthetic, or metallic materials with a length of 100 times the diameter, with a minimum length of 5 mm.
   a. Strand   **b.** Fiber   **c.** Staple   **d.** Particulate

8-7. Bonding agent of binder that provides a flexible layer between the fiber and the matrix.

    a. Matrix   **b.** Reinforcer   **c.** Coupling agent   **d.** Interface

8-8. Means of inspecting materials for flaws without adversely affecting them.

    a. NDE   **b.** SPF/DB   **c.** RRIM   **d.** RCC

8-9. Method of manufacturing structural shapes, such as channels and angles, in which prepregs are drawn through a die and heated until they set.

    a. NDE   **b.** SPF/DB   **c.** RRIM   **d.** Pultrusion

8-10. Composite primary structures in the aircraft industry now cost _____ as much as corresponding metallic structures.

    a. five times   **b.** just   **c.** two times

8-11. Fibers such as rayon that are pyrolized to drive off organic elements, leaving a very stiff, high-strength carbon fiber.

    a. E-glass   **b.** Kevlar   **c.** Borsic   **d.** Graphite

8-12. Trade name for aramid fiber, which is used for bulletproof vests and advanced composites for aerospace applications.

    a. E-glass   **b.** Kevlar   **c.** Borsic   **d.** Graphite

8-13. Directional properties in fiber composites are achieved mainly due to _____ .

    a. fiber orientation   **b.** laminate plane   **c.** interphase   **d.** prepregs

8-14. Air pockets that reduce the strength of the composite structure.

    a. Voids   **b.** Laminates   **c.** Particles   **d.** Honeycomb

8-15. A hollow core material that provides stiffening but very little weight.

    a. Voids   **b.** Laminates   **c.** Particles   **d.** Honeycomb

8-16. One advantage of composites over monolithic cast materials is _____ .

    a. low material cost   **b.** anisotropic properties   **c.** isotropic properties

8-17. An advantage of reinforced plastic composites over sheet steel is _____ .

    a. lower material cost   **b.** common fabrication methods

    c. isotropic properties   **d.** flame resistance

8-18. A molding method for producing FRP, in which parts are placed in a cylinder that is subjected to high heat and pressure.

    a. Filament winding   **b.** Injection molding   **c.** Autoclave   **d.** Pultrusion

8-19. A fiber composed of 55% silica, 20% calcium oxide, 15% aluminum oxide, and 10% boron oxide.

    a. SMC   **b.** GRP   **c.** SCAB   **d.** E-glass

8-20. The goal of the ACT program is to reduce the cost of _____ .

    a. metallic primary structures   **b.** composite primary structures   **c.** aircraft safety

8-21. What key step have automakers made to promote green manufacturing in use of thermosetting resin matrix composite parts?

    a. used ASTM symbols   **b.** use only thermoplastics resins

    c. use only CMC matrices   **d.** grind parts to use as fillers

8-22. From the table on Recycling Options and using a High, Medium, and Low rating, how would the Life Cycle Assessment of automotive materials rate the following key materials:

Life Cycle Inventory (Energy)

    a. Thermosetting Composite (e.g. SMC/BMC) _____

    b. Thermoplastic Composite (e.g. Azdel) _____

    c. Sheet Steel _____

    d. Sheet Aluminum _____

Environmental Impact

    e. Thermosetting Composite (e.g. SMC/BMC) _____

    f. Thermoplastic Composite (e.g. Azdel) _____

    g. Sheet Steel _____

    h. Sheet Aluminum _____

8-23. Environmental composites are so named because they are _____ .

    a. easily recycled   **b.** easily manufactured   **c.** cheap to fabricate

8-24. The chance of a defect remaining in a fiber as the diameter is decreased is _____ .

    a. minimized   **b.** unaffected   **c.** increased

8-25. Electrically conducting composites use components such as _____ .

    a. metal matrix   **b.** polymer fibers   **c.** graphite fibers

8-26. CFCCs for near-term applications will perform in the temperature range of _____ .

    a. 250° to 500°C   **b.** 500° to 1000°C   **c.** 1250° to 2000°C

8-27. Explosive welding is used to _____ .

    a. join incompatible metals   **b.** remove surface contaminants   **c.** weld projectiles

8-28. In precipitation-hardened alloys, the second-phase particles are _____ at high temperatures.

    a. stable   **b.** unstable   **c.** do not grow or dissolve

8-29. Carbon black is used as an inexpensive reinforcing filler in tires. Its reinforcing effect is the result of the _____ of its particles.

    a. large size   **b.** irregular shape   **c.** uniform distribution

REFERENCES & RELATED READINGS

AMERICAN SOCIETY FOR TESTING AND MATERIALS. *Annual Book of ASTM Standards,* Part 36. Philadelphia: ASTM, 1993.

AMERICAN SOCIETY FOR TESTING AND MATERIALS. *Composite Materials, Testing and Design,* ASTM STP 617. Philadelphia: ASTM, 1993.

AMERICAN SOCIETY FOR TESTING AND MATERIALS. *Fracture Mechanics of Composites,* ASTM STP 593. Philadelphia: ASTM, 1993.

ARONSON, ROBERT B., ed. "New Materials, New Challenges," *Manufacturing Engineering,* February 1995, pp. 57–59.

ASHBEE, K. H. G. *Fundamental Principals of Fiber Reinforced Composites.* Lancaster, PA: Technomic Publishing, 1989.

ASM INTERNATIONAL. *ASM Engineered Materials Reference Book.* Materials Park, OH: ASM International, 1990.

ASM INTERNATIONAL. *ASM Handbook,* Vol. 7, *Powder Metallurgy*; Vol. 15, *Casting*; Vol. 17, *NDE and QC.* Materials Park, OH: ASM International, 1987, 1988, 1989.

ASM INTERNATIONAL. *Engineered Materials Handbook,* Vol. 1, *Composites.* Materials Park, OH: ASM International, 1987.

BITTENCE, J. C., ed. *Engineering Plastics and Composites.* Materials Park, OH: ASM International, 1990.

BOSE, ANIMESH, and JAMES LANKFORD. "MIMICs: New Composite Architecture," *Advanced Materials and Processes,* Vol. 140, No. 1, 1991.

BRASHER, DAVID. "Explosive Welding: Principles and Potentials," *Outlook.* Teledyne WahChang, Vol. 15, No. 2, 2nd Quarter 1994, pp. 2–6.

BROSTOW, WITOLD. *Science of Materials.* New York: Wiley, 1979.

CARLSSON, L. A., and R. B. PIPES. *Experimental Characterization of Advanced Composite Materials.* Englewood Cliffs, NJ: Prentice-Hall, 1987.

CATLIN, CHRIS. "Aluminum/Polymer Composite Developed for Automobiles." *Advanced Materials & Processes,* Vol. 147, No. 5, May 1995, p. 8.

CHARLES, J. A., and F. A. A. CRANE. *Selection and Use of Engineering Materials,* 2nd ed. London: Butterworth, 1989.

Compressed Air Magazine. "Get Smart," Vol. 100, No. 3, April/May 1995, pp. 36–41.

EASTERLING, KEN. *Tomorrow's Materials.* Dorchester, England: Henry Ling Limited, Dorset Press, 1988.

GORDON, J. E. *The New Science of Strong Materials.* New York: Penguin, 1976.

HARRIS, C. E. "Industry-University-NASA Partnerships in Aeronautics Research and Technology Development," 4th Technical Conference: New Century Partnerships for Material Systems. Blacksburg, VA, April 23–25, 1995.

KUMAR, K. S. et al. "XD Titanium Aluminide Composites," *Advanced Materials and Processes,* Vol. 147, No. 4, April 1995, pp. 35–38.

LANGLEY RESEARCH CENTER. *Research and Technology,* NASA Technical Memorandum 4243. Hampton, VA: NASA Langley Research Center, 1993.

LEE, ARTHUR K., LUIS E. SANCHEZ-CALDERA, S. TURKER OKTAY, and NAM P. SUH. "Liquid–Metal Mixing Process Tailors MMC Microstructures," *Advanced Materials and Processes,* Vol. 142, No. 2, 1992.

LEWIS, D. III et al. "In-Situ Composites," *Advanced Materials and Processes,* Vol. 148, No. 1, July 1995, pp. 29–31.

LINDSAY, KAREN. "FRP Rebar: The Next Generation," *CI on Composites,* April/May 1995, p. 7.

LUCE, S. *Introduction to Composite Technology.* Dearborn, MI: Society of Manufacturing Engineers, 1988.

Manufacturing Engineering. "Cutting Tool Material Makeover," August 1995, p. 26.

MARC ANALYSIS RESEARCH CORP. "Creep-Forming Simulation Helps Form Al Wing Skins," *Advanced Materials and Processes,* July 1994, p. 15.

Modern Plastics Encyclopedia. New York: McGraw-Hill, 1993.

PAINE, J. S. N. "Enhanced Damage Resistant Composites Incorporating Shape Memory Alloys," 4th Technical Conference: New Century Partnerships for Material Systems. Blacksburg, VA, April 23–25, 1995.

Popular Mechanics. "Synthetic Skin," January 1944, p. 14.

POSTON, IRVIN E. "Recycling of Automotive Polymer Composites," *Society of Plastics Engineers Recycling Meeting,* Detroit, March 9, 1992.

QUINN, KEVIN R., and CARLOS A. CARRENO. "High-Temperature THERMO-PLASTIC Composites," *Advanced Materials and Processes,* August 1991.

RICHARDSON, TERRY. *Composites: A Design Guide.* New York: Industrial Press, 1987.

SCHADE J. P., and R. W. ROSS, JR. "Corrosion Control with New Nickel-Base Alloys," *Advanced Materials and Processes,* Vol. 146, No. 1, July 1994.

SCHAEFER, R. J., and M. LINZER, eds. *Hot Isostatic Pressing: Theory and Applications,* ASM Conference Book. Materials Park, OH: ASM International, 1991.

SCHWARTZ, M. *Composite Materials Handbook.* New York: McGraw-Hill, 1983.

SEYMOUR, R. B. *Engineering Polymer Sourcebook.* New York: McGraw-Hill, 1990.

STRONG, A. BRENT. *Fundamentals of Composites Manufacturing.* Dearborn, MI: Society of Manufacturing Engineers, 1992.

SUMITOMO METALS INDUSTRIES, LTD. "Particle-Dispersed Steel Has Highest Modulus," *Advanced Materials and Processes,* July 1994, p. 7.

TRACESKI, E. T. *Specifications and Standards for Plastics and Composites.* Materials Park, OH: ASM International, 1990.

URQUHART, ANDREW W. "Molten Metals Sire MMCs, CMCs," *Advanced Materials and Processes,* Vol. 140, No. 1, 1991.

WILSON, MAAYWOOD L., GARY S. JOHNSON, and ROBERT MISERENTINO. "Pultrusion Process Development for Multidirectional Graphite/Epoxy T-Beam," *SPI Paper,* NASA Langley Research Center, Hampton, VA, 1992.

ZHOU, SUWEI, ZAFFIR CHAUDRY, and CRAIG ROGERS. "Review of Particle Tagging Methods for NDE of Composite Materials and Structures," SPIE North American Conference on Smart Structures and Materials, February 26, 1995. Blacksburg, Virginia.

Periodicals

Advanced Composites.

Advanced Materials and Processes

Automotive Composites

CI on Composites

CDA-Composites Design & Application

Composites in Manufacturing

Composites Technology

Compressed Air Magazine

High Performance Composites

Journal of Composite Materials

*Journal of Composite Technology and
    Research*

Journal of Materials Research

Journal of Reinforced Plastics

Machine Design

Manufacturing Engineering

Materials Engineering

Materials Performance

NASA Tech Briefs*

Plastics Engineering

Plastics Technology

Plastics World

Popular Mechanics

Popular Science

* CD-ROM, Experiments in Materials Science, Engineering and Technology, Version 2.

Electronic and Other Important Materials and Materials Systems Technology

After studying this module, you should be able to do the following:

9–1.   Use diagrams, explanations, and calculations to determine electrical, magnetic, and optical properties of materials. Recall the nature, processing, and applications of electronic, magnetic, and optical materials.

9–2.   Describe the goals of materials science in developing smart materials and intelligent structural systems. Recall some current and potential applications of various smart materials.

9–3.   Describe the goals of materials science in research and development with biomimetic materials. Explain the properties and some applications of biomaterials.

9–4.   Name some of the techniques, explain the needs, and list some developments in nanotechnology and microtechnology.

9–5.   Gain an appreciation for the role of engineering materials technology and its value to you in the future.

9–6.   Use EMT Weblinks* to explore sites for new developments and evolving concepts related to electronic, photonic, smart materials, materials systems, and other advances.

EMT WEBLINKS listed throughout the book will link you to Internet sites related to topics being covered. The dynamic nature of the Web brings frequent changes, so some of these sites, while available at the time of writing, may not be up now.

PAUSE & PONDER

The information age gained momentum in large part from rapid innovations in materials science and technology. Key technologies involving electronic and optical materials continue to propel information technology. Figure 9-1 shows a lightweight, but remarkably powerful, notebook computer equipped with a DVD-ROM. While the DVD and CD-ROM disks may look the same, a standard digital video disk holds seven times more information [4.7 GB (gigabytes)] than a CD-ROM. More advanced DVDs, consisting of two layers for recording on each side, hold up to 17 GB of binary code.

Starting around 1986 with the adaptation of optical disks to storage media, ***CD-ROMs*** (compact disc—read-only memory) brought a revolution in the storage and retrieval of digital information. Offering a capacity to store 500 times as much data as the ordinary magnetic diskette, CD-ROM technology, with its associated laser technology, has already moved the music business away from audio tapes and now offers both computer and entertainment business arenas new variations of ***multimedia technology.*** CD-ROM provides high-density data storage, durability, and accessibility, and allows for the storage and retrieval of text, graphics (both still and motion), and sound. ***Information technology,*** the term often used to label the combination of electronics, optics, computers, telecommunications, graphics communications, entertainment, and the wide range of other forms of communication systems, owes much of its evolution to materials science and engineering developments, which spurred its progress.

Once the master disc is produced by lasers, replication of a limitless number of disks is accomplished by injection molding of polycarbonate (see Figure 9-2a). With recordings on magnetic media such as cassette tape, a recording head had to pass over the entire surface to transfer data. The 4.75-in. (121-mm) CD-ROMs contain up to 2.8 billions pits, which spiral for approximately 3 miles as measured from the hub to the outer rim. The cross-section (Figure 9-2b) shows the structure of a recordable CD. A regular CD uses aluminum instead of gold and has no dye layer.

(a)

(b)

Figure 9-1 Advances in electronic materials, especially semiconductors and optical materials, propelled the information age. (a) The notebook computer shown here represents a marriage of optical and electronic technologies among many other innovations. A powerful CPU (central processing unit), LCD (flat screen liquid-crystal display), DVD-ROM (digital video disk), internal fax modem, and high-capacity hard disk represent a few of the features that provide portability of computers. (b) DVD-ROM close up.

A recent application of amorphous metals is the new technology that is reinventing the ***compact disk (CD).*** This new CD is a medium that is capable of being recorded onto—not merely played back. The ***digital video disk (DVD),*** for video recorders and personal computers with capacities over 5 GB, uses a red laser to record digital data by heating spots on the disk's recording layer. The laser changes the structure of the heated portions of the material's surface from a shiny crystalline state to a dull, amorphous state. To erase the data, the laser heats the selected spots on the surface, again changing the structure of the heated spots on the surface back to a crystalline structure. A red laser has

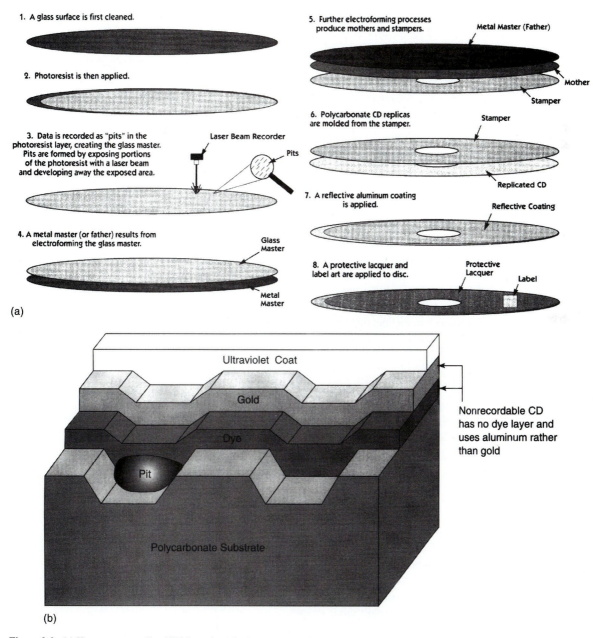

1. A glass surface is first cleaned.

2. Photoresist is then applied.

3. Data is recorded as "pits" in the photoresist layer, creating the glass master. Pits are formed by exposing portions of the photoresist with a laser beam and developing away the exposed area.

Laser Beam Recorder

Pits

4. A metal master (or father) results from electroforming the glass master.

Glass Master

Metal Master

(a)

5. Further electroforming processes produce mothers and stampers.

Metal Master (Father)

Mother

Stamper

6. Polycarbonate CD replicas are molded from the stamper.

Stamper

Replicated CD

7. A reflective aluminum coating is applied.

Reflective Coating

8. A protective lacquer and label art are applied to disc.

Protective Lacquer

Label

Ultraviolet Coat

Gold

Dye

Pit

Nonrecordable CD has no dye layer and uses aluminum rather than gold

Polycarbonate Substrate

(b)

Figure 9-2   (a) How a compact disc (CD) is made. (Nimbus Inc.) (b) Cross-sectional view of a recordable CD.

a wavelength of about 660 nanometers. A blue semiconductor laser that was recently perfected has a shorter wavelength of about 440 nanometers which increases the amount of readable data on a DVD-size disc to about 15 GB per side. In essence, this allows for the packing of pits so closely together that one disk will be able to hold four times as much information. This core technology will be the backbone of the industry producing video products such as high definition DVD that require large amounts of data storage. Sony markets a high-definition DVD player that utilizes a blue laser licensed from Nichia Chemical Industries that stores as much as 20 GB of data on a double-sided 12-cm disc—more than double that of current players. Blue lasers will be used to produce more effective blue LEDs (see Section 9.7.3.1).

A study of CD-ROM materials processing and manufacturing techniques provides insight into modern production technology. The processing of a metal master and the subsequent replication of polycarbonate CD-ROMs, shown in Figure 9-2a, involves eight basic steps:

1. Preparing the coated glass substrate
2. Applying a photoresistant coating to the glass substrate
3. Using a laser beam recorder to "cut" pits in the photoresist layer and then developing the glass master
4. Electroforming the glass master to produce a metal "master," or father
5. Additional electroforming produces the "mother" and "stamper"
6. Injection molding of polycarbonate plastic discs from the "stamper"
7. Aluminizing the disc
8. Adding a protective coating to the aluminized surface and applying a label

With multimedia computers becoming commonplace, the demand for CD-ROMs will continue. Most PC computers use these optical storage media in the form of ***OROM (optical read-only memory);*** however, the demand for recordable CD-ROMs, known as ***WORM (write once, read many),*** has already caused an increase in production and a lower price for these more versatile optical storage disks.

Pause for a minute and consider all the rapid developments in information you have witnessed in your lifetime. As with CD-ROMs, many of those developments were a direct result of advances in engineering materials technology. This module will introduce you to some of that technology.

 EMT WEBLINKS—General Coverage

Department of Energy National Laboratories—*http://home.doe.gov/people/labsmap.htm*
Thomas Jefferson National Laboratory—*http://www.jlab.org/*
MIT's Technology Review—www.techreview.com
Encyclopedia Britannica—http://www.britannica.com/

9.1 ELECTRICAL PROPERTIES

Before beginning our discussion of electrical properties, a review of some common terms used to describe electrical behavior of materials is essential. First it is important to realize that electrical behavior is affected not only by the structure of the material and how it was processed, but also by the material's environment.

Using a simple conducting wire circuit to which a source of electrical potential such as a battery or generator is applied, (see Fig. 9-3b) Ohm's law relates mathematically the electrical

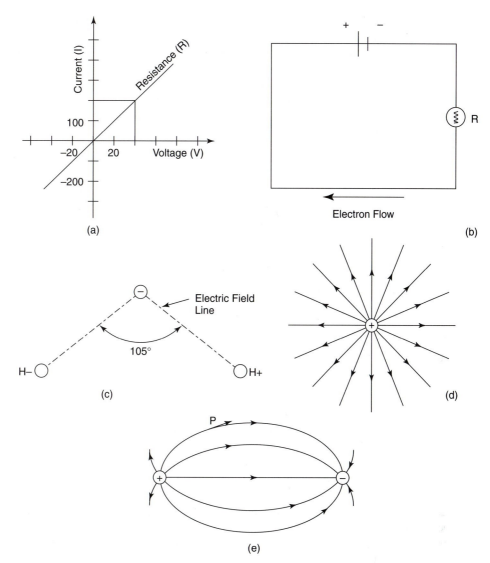

Figure 9-3 Resistance in a DC circuit-polar water molecule—electric charges and field lines (a). Ohm's law states that resistance in metal conductors is a constant in (b) direct-current circuits. The ratio of V/I is called the resistance (R). A material is ***ohmic*** whenever the resistance is constant over a range of potential differences (voltages). (c) A water molecule is a polar molecule due to the separation of charges between the hydrogen and oxygen atoms. (d) A schematic of a single positive-electric charge showing the electric field lines starting at a positive charge. The greater the number the greater the charge, and the denser the lines near the charge the stronger the field. (e) The electric field around two charges of opposite sign. The direction of the lines is from the positive to the negative charge. At point P, the direction of the field is tangent to the lines. These lines are continuous (closed curves) and they always point from the north toward the south pole of a magnet.

properties of current flow, resistance to the flow of current, and the voltage drop or potential difference across a resistance. Stated mathematically, $I = E/R$. Ohm's law states that the current flow (I), measured in amperes (A), is directly proportional to the applied voltage (E), measured in volts (V), and is inversely proportional to the resistance (R) to the current flow, measured in ohms (Ω). If the potential drop across a resistance is 1 volt when the current in it is 1 ampere, then the resistance must have unit value. This unit is called the ***ohm.*** The ohm is a resistance, across which is a potential drop of 1 volt when the current in it is 1 ampere. This mathematical relationship known as Ohm's law applies only to metal conductors. The symbol for resistance is the Greek capital letter theta (Ω) (see Appendix Table 10-3). Figure 9-3a is a plot of current versus voltage for the direct current (DC) circuit (Figure 9-3b). It shows that in this type of electrical circuit the resistance (R) is a constant; but we also know that resistance is affected by temperature changes increasing as the temperature increases. This resistance might be considered as the macroscopic definition of *electrical resistance* because it is related to the size, shape, and properties of the materials that comprise the electrical circuit. For example, a copper wire, 1000 ft in length with a diameter of 0.1 in., has the same resistance of 1 Ω as a steel rod 1000 m long with a diameter of 1 cm. The unit of current is the ampere, which is equal to a rate of flow of electric charge of 1 coulomb per second.

Illustrative Problem

The current in a circuit is 0.25 A when a potential difference (voltage) of 120 V is applied across its terminals. Find the value of the resistance.

Solution

$$I = \frac{E}{R} \text{ and } R = \frac{E}{I}$$

$$= \frac{120 \text{ V}}{0.25 \text{ A}} = 480\Omega$$

Note that in an electrical circuit the electrical current is capable of producing several effects. The principal ones are heat, magnetism, and electrolysis. A current in a lamp heats the filament to incandescence and produces illumination. A current in an electromagnet of a telegraph key magnetizes the iron core and causes a pivoted piece of iron to be attracted, thereby producing a click. An electrical current in water causes the liberation of hydrogen and oxygen by electrolysis.

Electric current is sometimes confused with *electric charge.* Using the analogy that compares an electrical generator or battery circulating charges in an electrical circuit to a pump circulating water in a pipeline, the difference between electrical current and electric charge can be distinguished. The quantity of water would be expressed in gallons while the electric charge would be in *coulombs.* The flow rate of water past any point in the pipeline would be expressed in gallons per second while the flow of electricity in the electrical circuit would be in coulombs per second or *amperes.* Electric charges can be demonstrated by rubbing an amber rod (a resinous material) with a piece of cloth, producing a charge of electricity on the rod. Amber is a material whose atoms contain loosely bound electrons. The electrons leave the amber rod by conduction and deposit themselves on the cloth. The rod now has an excess of positive charge (each proton in the nucleus can be considered as a positive charge). The cloth is now negatively charged. Both materials have acquired a charge. To demonstrate that these charges can flow, connect the amber rod to the earth (or ground the rod). The rod will lose its positive charge, because electrons flow from the earth and neutralize the amber rod. An excellent demonstration of this experiment can be done using a Teflon rod and a piece of fur. Flow of charge is further discussed when describing dielectrics later in this module. Just as two material bodies (mass m and mass M) produce forces on each other, so too do two charged bodies (charge q and charge Q).

The *law of universal gravitation* deals with the two bodies possessing a mass. It states that there is an attractive force existing between the two masses that is directly proportional to the product of their two masses and inversely proportional to the distance (r) between them ($F = mM/r^2$), whereas **Coulomb's law** concerns itself with charged bodies. Using the symbol q to represent a charge on a body, Coulomb's law states that the electric forces produced on one another by two charged bodies is directionally proportional to the product of their charges (qQ), when they are at rest, and inversely proportional to the distance (r) between them, $F = qQ/r$. This force is now referred to as the *electromagnetic force.* One major distinction between the two laws is that Coulomb's law deals with both attractive and repulsive forces. These opposite forces can be represented by use of + or − signs such as +q and −q. From our prior knowledge we remember the expression that like charges repel and unlike charges attract each other. Experiments have determined with great precision the magnitude of the electric charge between an electron and proton. This charge is the smallest found in nature. Using the SI unit of charge, the coulomb (C), the magnitude of the charge on the electron is e $= -1.602 \times 10^{-19}$ C. The mass of an electron is 9.11×10^{-31} kg, noted here for informational purposes only.

Charges give rise to an *electric field,* the space around a charged body. The concept of field is a construct that helps us understand how these separate concepts of gravity and electric charge can exert forces on objects at some distance. The earth exerts a force (your weight) on your body which can be calculated. When discussing the bonding of atoms and the polarity of some molecules in Module 2, the nomenclature contained expressions such as "attraction of electrons," "regions of positive charge," or "charge distribution." A water molecule sketched in Figure 9-3c is a permanent electric dipole, because the unequal positive charge on one side of the molecule has a strong covalent bond that holds it together. The two hydrogen atoms in the water molecule are represented as positive point charges in relation to the negative point charge of the oxygen atom. The two hydrogen atoms are separated by an angle of about 105°.

Electric fields can be visualized by using *electric field lines.* Figure 9-3d is a sketch of a single positive point charge showing the electric field lines (or lines of force) emanating radially out of the charge. These field lines extend outward into space. The closer the lines are to each other the stronger the force is in that region. A negative point charge would have its electric field pointing inward. If another point charge was placed near this first charge, it would feel a force because of the electric field from the first charge. This interaction between the two charges produces a force within this field. Figure 9-3e shows the electric field lines surrounding two charges of opposite sign. The direction of the field at any point is shown by the arrow at point P drawn tangent to the line. The field lines are directed from the positive charge to the negative charge. Any charge located at a distance from another charge will experience a force either of attraction or repulsion. Just as the earth is influenced by the presence of some other object with mass, one charge can impose a force on another even if they are separated by large distances. Electric fields spread through space with the speed of light and they carry energy. They can be calculated by measuring the magnitude and direction of the electric force acting on a small test charge or probe placed in the field. Note that this force is a vector quantity as is the electric field. The SI units of electric field are newtons per coulomb or N/C. A map of the field lines about charged bodies can be shown experimentally by sprinkling tiny bits of straw or hair on an insulating plate held between them.

9.1.1 Resistivity

Resistivity (or *volume resistivity*) is the term used to describe the relationship between electric current and the applied electric field. It is a measure of the resistance to the flow of current from a microscopic level, that is, as explained in terms of the atoms, the basic building blocks of all solid materials. The resistivity, ρ (Greek letter rho), depends on the behavior and number of free, or conduction, electrons and not on the shape of the conductor, as does resistance. Like density, this inherent property of a material will change as the structure changes, as in the alloying of metals or the doping of semiconductor materials (section 9.6.2). It depends on the movement of charge carriers, electrons in metallic conductors, or ions in ionic materials. In fact, electrical resistivity is the reciprocal of electrical conductivity (σ, the Greek letter sigma) (i.e., $\rho = 1/\sigma$). This electrical property is discussed later in this module. Microstructure again plays a large role in these properties. Any imperfections in the crystalline structure, whether they are atoms out of their normal positions, dislocations, or grain boundaries, to mention but a few, increase the collisions between electrons. This, in turn, prevents the transfer of energy in the form of electron flow to some intended user.

Figure 9-4a and b shows a sketch of the effects of an increase in temperature on metallic conductors and semiconductors, respectively. With metallic conductors, the increased vibrational energies of the atoms as a result of an increase in energy make the passage of free

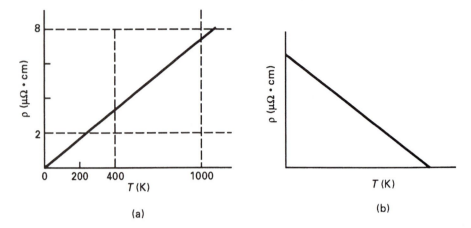

Figure 9-4   (a) Resistivity (ρ) versus temperature (T) for a metallic conductor. (b) Resistivity versus temperature for a semiconductor material.

electrons through the structure even more difficult. The ***mean free path,*** the average distance an electron can travel as a wave without hitting or deflecting off a positive-ion core (atom) in the lattice structure, is decreased. Consequently, the mobility of the electrons decreases, which produces an increase in the resistivity. For semiconductor materials, the resistivity decreases (conductivity increases) with an increase in temperature because more charge carriers become available to act as conduction electrons. Carbon's resistivity decreases with an increase in temperature. An interesting illustration of the requirements imposed on an insulating material is a silicon-based ***transistor chip.*** Some IC circuits only 5 μm thick are available but, on average, present-day ICs are about 400 μm thick. The typical insulating layer of silicon dioxide, known as the gate oxide, is the chip's smallest feature. A typical insulating layer of silicon is 24 atoms thick. It lies between the gate electrode which turns the current flow on and off and the channel through which the current flows. The gate oxide acts as an insulator to prevent a short circuit. The thickness of the gate oxide and the length of the electrode are two variables that control the transistor's switching speed. Such changes in these variables have resulted in a doubling of a transistor's switching speed every 18 to 24 months. Bell Labs recently reported the construction of an ultrathin, five-atom-thick layer of silicon dioxide. Each layer must be uniform in structure and smooth (i.e. flawless) to prevent a short circuit. Such a development will permit the further use of silicon in chip manufacturing.

The doping of semiconductor materials also lowers the resistivity by increasing the number of charge carriers. Additional information on this subject appears in Section 9.6.

In the discussion of resistance, we stated that resistance varies directly with the length *L* and indirectly with the uniform cross-sectional area *A* of a conductor. To write this as a mathematical statement, we need a constant to make the units agree on both sides of the equation. Thus,

$$R = \rho \frac{l}{A}$$

where ρ is the proportionality constant or resistivity in ohm-centimeters, assuming that *l* and *A* are expressed in centimeters. Note that a conductor, 1 cm in length with an area of 1 cm², has a resistance *R* equal to the resistivity ρ.

In selecting conducting materials for the express purpose of generating heat from the flow of electricity, such as Nichrome, a heat-resistant alloy of nickel and chromium, the material must have a carefully selected variety of properties, such as a moderate resistivity, excellent resistance to oxidation, and a capability to operate effectively at high temperatures without failure. Table 9-1 lists the electrical resistivities of some selected solids from three families of materials.

TABLE 9.1 DIELECTRIC PROPERTIES AND RESISTIVITY FOR SOME COMMON MATERIALS

Material	Resistivity ($\Omega \cdot$ m at 20°C)	Dielectric constant, κ (10^6 Hz)	Dielectric strength (10^6 V/m)	Dielectric loss factor, $\kappa \tan \delta$ (10^6Hz)
Mica	10^{11}	7	79	
Nylon 6/6	10^{13}	3.5	17	0.04
BaTiO$_3$	10^9	1600.0		
Al$_2$O$_3$	10^{12}	9	98	0.001
Steatite	10^{12}	6	8	
Phenolics—transparent	10^{10}	8	6	0.05
Paraffin	10^{15}	2.3	10	
Water	10^{12}	78		
Polyethylene	10^{15}	2.3	20	0.0001
Soda-lime-silica glass	10^{13}	7	10	0.01
Silver	16×10^{-9}			
Copper	17×10^{-9}			
Tungsten	56×10^{-9}			
Aluminum	29×10^{-9}			
Steel (301)	7×10^{-7}			
Graphite	10^{-5}			
SiC	0.1			

Illustrative Problem

(a) A metallic wire 100 cm long has a diameter of 0.05 cm. If it has a resistance of 0.08 Ω when 10 A of current are running through the circuit at standard temperature, find the wire's resistivity in $\Omega \cdot$ cm. (b) Using Table 9-1, identify the material from which the wire was fabricated.

Solution

(a) $R = \rho \, (l/A)$; therefore, $\rho = R/(A/l)$.

$$A = \frac{\pi d^2}{4} = 0.7854 \times (0.05)^2 = 1.96 \times 10^{-3} cm^2$$

$$\rho = 0.08 \, \Omega \times 1.96 \times 10^{-3} \, cm^2/10^2 = 1.7 \times 10^{-6} \, \Omega \cdot cm$$

(b) Either silver or copper would be adequate answers.

Note in our discussion that, where heat is involved in the flow of electric charge, efficiency is reduced. The incandescent light bulb is a good example. This bulb gets hot when used. A fluorescent bulb remains cool. Consequently, the fluorescent bulb is more efficient in the use of electricity (energy). As a matter of fact, the incandescent light is only about 3% efficient; that is, 97% of the energy needed to produce light is lost mainly to heat. Compare the lighting characteristics of light-emitting diodes (LEDs) in Section 9.7 to those sources of light as discussed previously.

Summarizing, the resistance R of an electrical circuit is a function of the shape, size, and nature of a solid material. Just like specific heat or density, resistivity ρ is a function of the intrinsic nature of the material itself. Instead of thinking of resistivity, one can think in terms of conductivity, σ. Low resistivity and high conductivity refer to a similar situation in different terms. To further reinforce the concept of resistivity, remember that 1 lb of aluminum has the same resistivity as 1 g of aluminum, whereas the resistance of 1 lb of aluminum is very different from that offered by 1 g. Figure 9-5 illustrates an industrial use for a human-made nonconductive material.

9.1.2 Electrical Conductivity

The reciprocal of resistivity is known as *conductivity*. It is a measure of how readily electrons flow through a material. Using the symbol σ (Greek letter sigma) to represent conductivity,

Figure 9-5 Pultruded fiberglass boom is superior to older metal booms in terms of mechanical, weathering, and electrical properties. The boom exhibits less than 100 microamp leakage over any 2-foot section after a 48-hour full immersion soak while a 100-KV (dc) test voltage is applied. (Morrison Molded Fiber Glass Co.)

this quantity is determined for a particular material by measuring the amount of electric charge that passes through a unit cube of the material per unit of time. Conductivities of common metals range from 10,000 to 550,000 $\Omega^{-1}\text{cm}^{-1}$. (See Section 8.3 for information on increasing conductivities of graphite fibers.) Consequently, σ is dependent on three factors:

1. n, the number of charge carriers in a cubic centimeter (cm^3) of material
2. q, the charge per carrier (coulombs/carrier)
3. μ, the mobility of each carrier, or the ease of movement of the charge carriers (e.g., electrons in a metal)

$$\mu = \frac{\text{velocity of the carriers (cm/s)}}{\text{voltage gradient (V/cm)}}$$

The product of these three factors, with units of $1/(\Omega \cdot \text{cm})$ or $(\Omega \cdot \text{cm})^{-1}$, is the conductivity. Expressed in mathematical terms,

$$\sigma = nq\mu$$

This calculation agrees with units obtained for conductivity if the basic equation for resistivity is used ($\rho = RA/l$).

Illustrative Problem

A tungsten wire is used in an electrical circuit at a temperature of 20°C. Find its electrical conductivity.

Solution   Conductivity is the reciprocal of resistivity. From Table 9-1, tungsten's resistivity is 56 $\times 10^{-9}\ \Omega \cdot \text{m}$, so the conductivity is $1/56 \times 10^{-9}\ \Omega \cdot \text{m} = 17.9 \times 10^6\ (\Omega \cdot \text{m})^{-1}$ or $\Omega^{-1} \cdot \text{m}^{-1}$.

Illustrative Problem

A rectangular block of aluminum has the dimensions 25 mm $\times$ 25 mm $\times$ 50 mm.
(a) Make a three-dimensional sketch of the block and place dimensions on it.
(b) What is the resistivity of the block, in $\mu\Omega \cdot \text{cm}$, measured between the two squared ends at 20°C?

(c) Determine the electrical conductivity for aluminum at 20°C.

Solution

(a)

25 mm

50 mm

25 mm

(b) The cross-sectional area is 25 mm^2 = 6.25 cm^2.

$$R = \rho\frac{l}{A} = \frac{2.9 \times 10^{-6}\,\Omega \cdot cm \times 5\,cm}{6.25\,cm^2}$$

$$R = 2.32\ \mu\Omega$$

(c) $\sigma = 1/\rho = 1/2.9\ \mu\Omega \cdot cm = 3.45 \times 10^5 (\Omega \cdot cm)^{-1}$.

9.2 DIELECTRIC PROPERTIES

9.2.1 Dielectrics

Dielectric properties describe the response of dielectric materials (poor electrical conductors) to electric fields. In other words, they are dependent on the polarization of the material. Most of these materials are ceramics or polymers. The term *dielectric* refers to nonconductive materials whose polarization is caused by an electric field. A material or medium such as a vacuum or gas is called a dielectric because it permits the passage of the lines of force of an electrostatic field but does not conduct the current. Subjecting these materials to a magnetic field also produces similar effects, although the sources are different. A dielectric has two functions to perform. One is to act as an insulator and the second is to increase the capacitance of a condenser (capacitor) beyond that which can be derived from an air gap or vacuum between the plates of the capacitor. As a covering for an electrical wire, the plastic coating is acting as an insulator. Good electrical insulators such as ceramics are referred to as dielectric materials. Even though these materials do not conduct electrical current, they are affected by the applied electric field. This effect, known as *polarization,* results in a shift in the distribution of charge within the material forming an electric dipole. *Dipoles* are atoms/ions or groups of atoms/ions that have an unbalanced or asymmetrical charge. Dipoles are atoms or molecules that have a positive pole and a negative pole; or, in other words, in a dipole the centers of positive and negative charges do not coincide. Figure 9-6a shows a sketch of a single dipole. The centers of positive and negative charges are indicated by the letter *q,* both charges being equal in magnitude. They are separated by a distance *d.* The vector labeled *p* is the *dipole moment,* defined as the product of one charge multiplied by the distance *d* or $p = qd$. When an external electric field is applied to a material containing dipoles, a force (or torque) will come to bear on a dipole to orient it with the applied field (see Figure 9-6b). The units of dipole moment are coulomb-meters (C-m). The water molecule is a permanent dipole. Its dipole moment is about 6×10^{-30} C-m. This process of dipole alignment is termed *polarization.* Polarization is affected by many structural features. The presence of permanent dipoles, ease of molecule movement, moisture, cracks, and grain boundary imperfections are only a few of these features. When subjected to an outside electric field, the dipoles become aligned with the direction of the applied field in the material, causing polarization. It should be remembered that atoms and ions move (i.e., they translate as well as rotate within their structures).

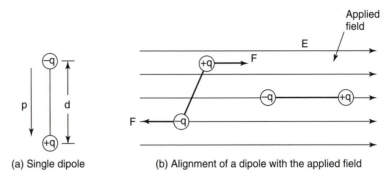

(a) Single dipole   (b) Alignment of a dipole with the applied field

Figure 9-6 (a) A single dipole; (b) a single dipole undergoing a force from an applied electric field that will orient it with the applied field.

9.2.1.1 Electronic polarization. When an external electric field (E in volts) is applied to the material, the electron clouds around atoms and molecules shift off their center of the positive nuclear charge. Electrons concentrate on one side of the nucleus near the positive end of the field. The atoms then act as a temporary or induced dipole. A temporary dipole can then induce another dipole in an adjacent atom. The two dipoles then experience a force of attraction. This occurs in molecules as well. When the field is removed this polarization disappears (see Figure 9-7a).

9.2.1.2 Ionic polarization. The applied field causes a separation of charges. The cations are displaced in one direction and the anions in the opposite direction. The cations are attracted toward the negative electrode or the negative side of the field and the anions move toward the positive electrode or the positive end of the field. Polarization may change the overall dimensions of a material (see Figure 9-7b).

9.2.1.3 Molecular, dipole, or orientation polarization. This mechanism is associated with materials that have natural dipoles. Water molecules, some glass, polar polymers, and crystalline ceramics such as barium titanate ($BaTiO_3$) exhibit this type of polarization. When the external field is removed, this polarization can be retained because of the presence of permanent electric dipoles in the material (see Figure 9-7c).

The extent of the polarization depends on the following factors: the type of atoms, the type of bonding, the frequency of the applied field, and the mass of the material involved. Each of the polarization mechanisms occurs in a limited range of frequencies. The larger the mass involved the more sluggish is the response to the external field. Electrons being less heavy than protons can respond faster to fields as they are applied and removed. Electronic polarization can respond to radiation in the visible portion of the electromagnetic spectrum. Ionic polarization will be experienced in the infrared (IR) region. Molecular polarization responds to frequencies in the subinfrared range.

When dipoles are caused to align themselves with an applied external field, the material is polarized and the magnitude of the polarization (P) can be expressed using the units of C/m^2.

Dielectrics and their properties are important in the design and fabrication of both ceramic and polymer composites. *Dielectrometry* uses electrical techniques to measure the changes in dielectric loss factor and capacitance during cure of a resin in a laminate. Ceramics are widely used as insulators and as substrates in MMCs, where the need is for high resistivities and dielectric strength. High thermal conductivity might also be needed if heat must be dissipated.

9.2.2 Dielectric Constant

The *capacitance (C)* of a condenser or capacitor is a measure of its ability to store electricity. It is determined by measuring the amount of charge (Q) that must be placed on a conductor to raise its potential by 1 V. Mathematically, $C = Q/V$, where C is expressed in farads (coulombs per second), Q in coulombs, and V in volts. Solving the expression for $Q = CV$, it can be seen

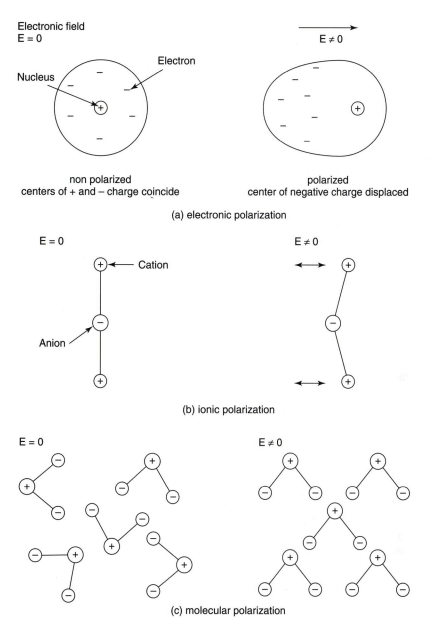

Figure 9-7 (a) Electronic polarization, a nonpolarized atom with its centers of charge coinciding at the nucleus when there is no electric filed (E = 0). When an external electric field is applied (E ≠ 0), the charge centers separate producing a polarized atom. (b) Ionic polarization showing a single ion with distorted bonds when subjected to the applied field. (c) Molecular polarization showing the natural dipoles disoriented when no electric field is applied and aligned with the field when the field is imposed.

that C is a constant of proportionality with no units. A farad (F), the unit of electric capacitance, is defined as the capacitance of a capacitor, between the plates of which there appears a potential (E) of 1 V when it is charged by a quantity of electricity of 1 coulomb. A farad is a fairly large unit, and most capacitors have small values of capacitance. So prefixes such as *pico* or *micro* are used to express smaller units of the farad (1 pF = 10^{-12} F).

Two conductors separated from each other by some insulating material (including air) form a condenser. The conductors make up the plates of the condenser, and the insulation between the plates is the dielectric. The condenser is charged by connecting its plates to a battery or a source of electrical energy. Electrons will flow and collect on one plate until equilibrium is reached and the potential difference between the two plates equals the electromotive force (emf) of the battery. The ratio of the condenser charge (Q) to the potential difference (E) defines the *capacitance* of the condenser. The capacitance of a parallel-plate condenser with free space as the dielectric is proportional to the area of the plates and inversely proportional to the distance between the plates. Expressed mathematically, we have

$$C = \epsilon_0 \frac{A}{d}$$

where A is the area of the plates, d is the thickness of the dielectric or the plate separation, and ϵ_0 (Greek letter epsilon) is the **permittivity constant of free space** with a value of 8.85×10^{-12} F/m. Note that C depends on the geometry of the arrangement.

Illustrative Problem

A condenser is made of two thin sheets of copper with a total area of 1 m². The sheets are separated by air at a distance of 0.05 mm. What is the capacitance of the condenser?

Solution

$$C = \epsilon \frac{A}{d} = \frac{8.85 \times 10^{-12} \text{ F/m} \times 1 \text{ m}^2}{5 \times 10^{-5} \text{ m}}$$

$$= 0.177 \text{ μF}$$

If a dielectric such as mica is placed between the plates of the same parallel-plate condenser using the same potential difference E, the charge Q on the plates will be greater and will produce a corresponding increase in the capacitance C of the condenser. This result can be expressed:

$$C = \kappa \epsilon_0 \frac{A}{d}$$

where κ (Greek letter kappa) is the **dielectric constant** or **relative permittivity** of the dielectric. The dielectric constant is a measure of a material's value toward making a desired capacitance when the material is placed between two conducting plates. The dielectric permits a higher potential difference to be applied than when air is used. It is a dimensionless number that indicates the reduction of the field due to the dielectric. (See Table 9-1 for dielectric constants for some common materials.)

Illustrative Problem

Using the preceding problem with only air between the plates of the condenser ($\kappa = 1$), insert paper between the plates with a dielectric constant of 3.5. Find the capacitance.

Solution

$$0.05 \text{ mm} = 5 \times 10^{-5} \text{ m}$$

$$C = \kappa \epsilon_0 \frac{A}{d}$$

$$C = 3.5 \times 8.85 \times 10^{-12} \left(\frac{\text{F}}{\text{m}} \times \frac{1 \text{ m}^2}{5 \times 10^{-5} \text{ m}} \right)$$

$$C = 6.19 \times 10^{-7} \text{ F} = 0.619 \text{ μF}$$

Note: Adding paper increased the capacitance almost fourfold.

When a dielectric material such as a polymer, glass, or ceramic is placed between the plates of a condenser, an induced charge Q appears on the plates without any change in the voltage. This naturally increases the capacitance of the condenser. The increased charge is due to the *polarization* of the entire volume of the dielectric, in turn due to the applied electric field.

Dielectrics are used in alternating current (ac) circuits with a wide range of frequencies. The higher the frequency, the faster the dipoles must switch directions to allow the device to perform satisfactorily. If the dipoles have difficulty moving, as is the situation with complex organic molecules, dipole friction results, with attendant energy losses. These effects show up in the value of the dielectric constant. Temperatures also affect the dielectric constant. In general, the dielectric constant increases with an increase in temperature. Because the dielectric constant is a ratio of similar quantities, it has no units. Some typical values for κ for some common materials taken from a variety of sources for a frequency of 10^6 Hz and 20°C are listed in Table 9-1.

High values of dielectric constant are needed for capacitors, whereas low values are used as insulating materials. *Capacitors* perform many different functions including filtering (separating) electrical signals with different frequencies, blocking direct current, permitting alternating current to pass, and serving as energy storage devices. Much research is devoted to the miniaturization of electrical circuits, including capacitors, which means that a continual search is under way to find materials such as ceramics that not only have high dielectric constants, but have such values at higher and higher frequencies and temperatures for use in, for example, dielectric heating applications. *Ferroelectrics* are a group of dielectric materials that exhibit spontaneous polarization. They contain permanent electric dipoles. When heated above their ferroelectric curie temperature (120°C), they lose their ferroelectric behavior. These materials have extremely high dielectric constants at relatively low applied field frequencies. Barium titanate ($BaTiO_3$) is an example of such a material when it exists in its tetragonal crystal structure. In this structure it is also piezoelectric (see Section 9.8 for further information on piezoelectric materials). Table 9-1 lists some additional properties of barium titanate. As a consequence, capacitors made from such materials can be significantly small. A need for materials that have low (< 3) dielectric constants for use in coatings, films, matrix resins, and fibers is in evidence, particularly in the electronics and aerospace industries. A low dielectric constant polyimide fiber with high thermal stability and good tensile properties similar to standard textile fibers has been developed for possible use in printed-circuit boards and in aircraft composites.

9.2.3 Dielectric Strength

Dielectric strength is the voltage gradient (voltage per unit thickness, such as volts per mil), which produces electrical breakdown through the dielectric. It is an expression of the maximum voltage that a dielectric can withstand before electrical discharge occurs through the dielectric. This insulating strength is dependent not only on the usual items, such as bonding type and crystalline structure, but on moisture absorption and the nature of the applied electrical energy. In the latter, the source might be a direct current (dc) or an alternating current (ac), or a combination of the two. If the magnitude of the dielectric field is sufficient, it will overcome the attraction of the electrons to their positive-charged nuclei and produce a *leakage* and eventual rupture of the material. The breakdown voltage has then been reached. With an ac source, the continual reorientation of the material's atoms and electrons, in addition to the deformation of the paths of the electrons, results in a hysteresis loss (to be explained) that produces heat in the dielectric material. Therefore, the breakdown voltage will vary inversely with the frequency of the source voltage. Some typical ranges of values of dielectric strength for some nonmetallic materials, expressed in 10^6 volts per meter (10^6 V/m), are listed in Table 9-1.

9.2.4 Dissipation Factor/Dielectric-Loss Factor

When alternating current provides the voltage (sinusoidal) needed to maintain the charge on a capacitor or any dielectric, the current leads the voltage by $90° - \delta$, where the angle δ is called the dielectric-loss angle. If the dielectric were perfect, the angle δ would be zero and no loss would occur. In this situation, the current would lead the voltage by $90°$. Most of the time, however,

(a)

Microstructure of Si₃N₄-6% Y₂O₃-2% Al₂O₃ After Annealing at 1200°C for 20 hours

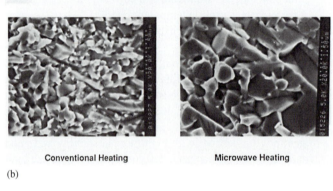

Conventional Heating   Microwave Heating

(b)

Figure 9-8   Dielectric heating using microwave assist in the drying/sintering/curing of materials as well as the bonding of composites using adhesives that have different dielectric loss factors than the materials being joined. (a) Microwave ovens are used in ceramic processing to achieve lower temperatures, reducing the possibility of inflicting damage on materials due to high-temperature processing. (b) Contrasting ceramic microstructures between conventional and microwave heating. (ORNL)

perfection is not attained and the current and voltage are therefore out of phase, resulting in a loss of electrical energy. This loss of energy, hence power, usually takes the form of heat. The tangent of the angle δ (tan δ) is known as the ***dissipation factor.*** Multiplying tan δ by the dielectric constant (κ) expresses the ***dielectric-loss factor, dielectric heat-loss factor, dielectric coefficient,*** or ***loss tangent.*** Dielectric heating is the heating of dielectric materials by subjecting them to a high-frequency, alternating electric field. This heating assists in the bonding or drying of materials (see Figure 9-8). Dielectric loss is affected by frequency and temperature changes. The loss usually increases at low frequencies and as the temperature increases. Table 9-1 lists the dielectric loss factor for a few nonmetallic materials.

9.3 MAGNETIC PROPERTIES

9.3.1 Magnets, Magnetic Field, Electromagnetic Force

A ***magnet*** is a body that has the property of attracting iron and steel and that, if suspended freely, will turn so as to point in a certain direction. Any material that a magnet attracts is known as a magnetic material. A rod of steel can be made into a magnet simply by rubbing it with another magnet or by placing it within a coil of wire carrying an electrical current. If the rod is made of hard steel it will retain its magnetism for a long time after the electrical current has ceased. It then becomes a ***permanent magnet.*** Permanent magnets made of iron can remain magnetized for long periods of time; however, if you drop a magnet you may cause it to become disoriented and thus lose some or all of its magnetism. Heating a magnet may cause the same effects. Above a characteristic temperature known as the ***Curie temperature (T_c)*** (1043 K for iron), magnets cannot be made.

Soft iron can be magnetized in the same manner, but upon interruption of the electric current it loses most of its magnetism. What magnetism remains is called ***residual magnetism.*** As long as the current continues to flow through the wire, the soft iron will be magnetized and act as an ***electromagnet.*** Magnetism and electricity are closely related. This relationship was not discovered until the nineteenth century. A magnet (named after its place of discovery) will attract iron objects such as pins or paper clips. If a magnet is suspended from a thread, its north-seeking pole will point toward the earth's geographic north. This is the principle of the magnetic compass. The other pole pointing toward the south is called the south pole. Magnetism shows its greatest effects at the ends of a magnet and practically none at its center. Any magnet has two ends called ***poles*** where the magnetic effect is the strongest. If you break a magnet in half you produce two new magnets each with a north and a south pole. There is no such thing as a magnet with only one pole. A suspended magnet will turn to a general north–south orientation with one pole always pointing in the same direction. The pole that always points toward north is called the ***north pole*** and is called its opposite pole the ***south pole.*** If two magnets are brought near one another, each exerts a force on the other. When a north pole is brought near the north pole of another magnet, this force is repulsive and can be felt before the magnets touch each other. An attractive force is felt when two unlike magnetic poles are brought near one another. Using two magnets it can be demonstrated that like poles repel when brought close to each other; and unlike poles attract each other. Besides iron, other elements such as cobalt, nickel, and gadolinium show strong magnetic effects. These materials are classified as ***ferromagnetics*** (see ferromagnetism later in this module). All other materials only show extremely small effects.

Why can a piece of iron be turned into a bar magnet (See Figure 9-9a) whereas a piece of aluminum cannot? Why do lodestones (made from the mineral magnetite) have magnetic fields? The understanding of the magnetic properties of bulk materials, that is, the magnetic behavior exhibited by these materials, is related to the atoms that make up these materials. Magnetic forces act at a distance just like gravitational and electric forces. Magnetic and electric forces have been combined into one force called ***electromagnetic force,*** which is one of the four fundamental forces in nature.

A magnet creates a magnetic field which surrounds it for a considerable distance. We can visualize the magnetic field by constructing magnetic field lines. It is this field that interacts with the magnetic field of an adjacent magnet or an electric field. (see Figure 9-9). *A magnetic field exerts a force on an electric current and electric currents produce magnetic fields.* For example, the lines of a magnetic field, due to a current in a straight wire, form circles around the wire and the field exerts a force on magnets placed adjacent to it. Figure 9-9b shows magnetic field lines that are continuous (closed curves) whereas electric field lines begin and end on the electric charges. The magnetic field is represented by a vector with the symbol B, whose magnitude is defined as the torque exerted on a compass needle when it makes a certain angle within the magnetic field. The torque applied to the needle orients the needle so that it is parallel to the magnetic field line at that point. B is sometimes referred to as *magnetic flux density* or *magnetic induction.* The force on a current-carrying wire caused by a magnetic field is determined by

$$F = I \times l \times B \times \sin\theta$$

where θ is the angle the wire makes with the magnetic field as measured from the horizontal. This angle is illustrated in Figure 9-9c. I is the value of the current in the wire and l is the length

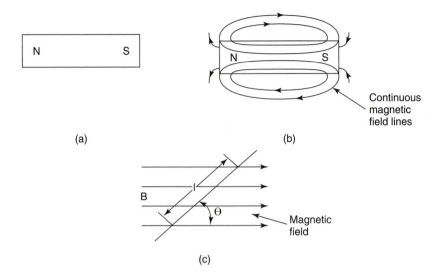

Figure 9-9   (a) Bar Magnet. (b) Magnetic field lines outside of a bar magnet. These lines are continuous (closed curves) and they always point from the north toward the south pole of a magnet. (c) A magnetic field B with a length of wire (l) and crossing it at an angle (θ) to the horizontal.

of wire in contact with the magnetic field B. The direction of the magnetic field can be defined as the direction that a north pole of a compass needle would point when placed at that point. An electric current produces a magnetic field. Place a compass needle near an electric wire carrying a current and the compass needle will deflect. Conversely, if a magnet is placed near a wire carrying a current, the magnet exerts a force on the wire. The SI unit for this force B is the tesla (T). $1\,T = 1\,N/(A - m)$. The older name for the tesla was the weber per meter squared ($1\,W_b/m^2 = 1\,T$). The earth's magnetic field at its surface is about 0.5 T. A refrigerator magnet used to hold paper notes has a magnetic field of about $10^{-2}\,T$. A strong electromagnet can produce a magnetic field strength of 3 T. Strong, super-conducting magnets have magnetic fields over 10 T. A long coil of wire consisting of many loops is a *solenoid.* Each coil of wire when carrying an electric current produces a magnetic field. An electromagnet is a solenoid with a piece of iron placed inside it. When energized by an electric current, the magnetic field produced consists of the magnetic field produced by the current plus the magnetic field produced by the alignment of the atomic domains in the soft iron core. Figure 9-10 shows a solenoid and an electromagnet; however, it does not show the densely packed continuous magnetic field lines that would be formed inside the coil in (b).

Illustrative Problem

An electric wire carries a 30 A current. It is oriented in a magnetic field whose width is 12 cm at an angle of 40°. The magnetic field (B) is 0.90 T. Find the force acting on the wire. See Figure 9-9c.

Solution   The sketch shows that if the angle θ = 90° then the sine of 90° is 1. This makes the wire perpendicular to the field and the force on the wire a maximum. If the angle is 0° then the wire would be parallel to the field and the sine of 0° is 0. For the case where the angle is 40°, and 0.9 T or $0.9\,N(A - M)^{-1}$ where N is neutrons of force and A is amperes of current flow.

$$\text{Using } F = I \times 1 \times B \times \sin\theta$$

$$F = 30\,A \times 0.12\,m \times 0.90\,T \times 0.64 = 0.1\,N$$

9.3.2 Permeability Classes of Magnetic Materials, Magnetic Domains

In a magnetic circuit, the **permeability** (μ) of a material, a measure of the ease with which a magnetic field can be set up through a material, can be compared to the conductivity in an elec-

H (external applied field strength)
B (internal field strength)

Key: n = turns per meter
       I = electric current in amperes (A)
       H = magnetic field strength (A/m)
       μ_0 = permeability of free space $(4\pi \times 10^{-7}\,\text{T·m/A})$

Note: SI unit, Tesla (T) replaces W_b/m^2.

(a)   (b)

Figure 9-10   (a) Magnetic helical coil with a vacuum core. (b) Magnetic helical coil with a solid material core.

trical circuit. The magnetic field, represented by the symbol (H) with units of amperes per meter (A/m), corresponds with the voltage or potential drop. The magnetic flux density, magnetic induction, or magnetic field (B), expressed in units of webers per square meter (Wb/m^2), corresponds with the electric current. (See SI unit in Fig 9-10a). The ratio of (B) to (H) (B/H) is known as ***permeability*** (μ) (Greek letter mu), with units of webers per ampere meter (Wb/A · m). An important distinction between electrical and magnetic circuits is that permeability is not constant. As (H) changes, so does (μ).

Figure 9-10 illustrates the permeability of a vacuum or free space in the core of an electrical coil through which an electrical current is passing. When a core of material replaces the vacuum in the coil, a change in the flux density (B) is detected. This change is reflected in a new value of (μ). If the material is made of iron, cobalt, or nickel, an exceedingly large rise in (B) is noted. The magnetic field of the earth at its surface is about 0.5×10^{-4} T.

Relative permeability (μ_r) is the ratio of two permeabilities and therefore has no units. The permeability of a material μ is thus compared mathematically with the permeability of free space (vacuum), $\mu_r = \mu/\mu_0$. This ratio may be used to classify magnetic materials into three main categories:

1. Diamagnetic material if $\mu_r < \mu_0$.
2. Paramagnetic material if $\mu_r > \mu_0$.
3. Ferromagnetic material $\mu_r \gg \mu_0$.*

Magnetic materials can be classified into several groupings depending on the interaction of the electrons (internal dipoles) and an external magnetic field. ***Diamagnetic*** materials are solids composed of atoms with completely filled electron shells so the interaction is nil. Some examples are copper (Cu), silver (Ag), gold (Au), and organic polymers. ***Paramagnetic*** materials contain unpaired electrons in their inner shells that form atomic dipoles that are free to rotate with an external field producing some interaction, rendering a relative permeability μ slightly greater than 1. Like diamagnetic materials, paramagnetic materials only exhibit magnetism when in the presence of an external magnetic field. Some examples of paramagnetism are the metals aluminum (Al), chromium (Cr), and titanium (Ti). Figure 9-11 is a sketch of the electron configuration of manganese (Mn) showing the unpaired electrons in its third inner

*The symbol $\gg$ means "much greater than."

shell. ***Ferromagnetism*** is present in some metallic materials which possess a permanent magnetic moment. Iron (Fe), cobalt (Co), nickel (Ni), and some metal alloys exhibit ferromagnetism. This magnetic behavior can be increased in magnitude by strong magnetic fields. The explanation of this type of behavior uses the word *domain*. A ***domain*** is a tiny region about 1 mm in length or width. Each domain acts like a tiny magnet with a north pole and a south pole, and is the result of the electronic structure of individual atoms. These regions are visible using powerful electron microscopes. If the magnetic material is unmagnetized ($B = 0$) the domains are arranged in a random fashion as seen in Figure 9-12. This random arrangement cancels out any magnetic effects of the domains. A magnet's domains are preferentially aligned in a single direction. These domains are capable of rotating as well as changing size. They will rotate such as to be more parallel to an external field and grow in size at the expense of their neighboring domains depending on the strength of the external field. If a ferromagnetic material is heated, the increased movement of the atoms leads to randomization of the domains and their magnetic moments. At the Curie temperature (T_c) the randomization is complete and the material is no longer ferromagnetic. The value of T_c varies with the material. For iron, T_c = 1045K. As the temperature decreases below T_c ferromagnetism returns gradually in a manner similar to a lake freezing over in the winter. Steel, even if demagnetized, has little packets of atoms that are aligned roughly in the same direction.

Using domains, one can readily understand why an iron nail placed in the vicinity of a magnet will become magnetized in a direction dictated by the location of the north-seeking end of the magnet used to magnetize the nail. Proof of the existence of domains is readily available by using a colloidal solution of iron-oxide particles spread over the polished surface of a

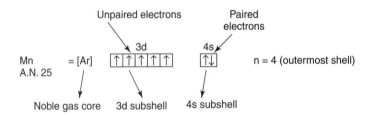

Figure 9-11 Orbital filling diagram for manganese (atomic number 25). The noble gas core consists of a shorthand notation for the orbital diagram for argon (Ar, atomic number 18). The electron configuration is written as Mm:[AR]4s^2 3d^5

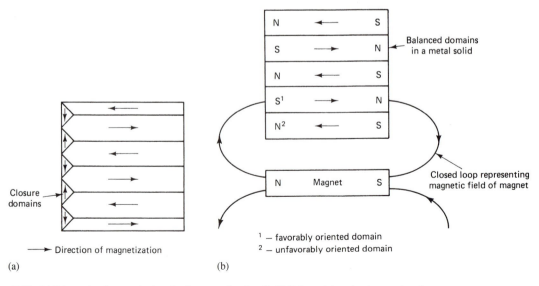

Figure 9-12 (a) Schematic of magnetic domains in a crystal at $B = 0$. (b) Balanced domains in an external magnetic field.

piece of magnetic material. When viewed under a microscope, the particles of iron oxide will be concentrated around the domain boundaries. All ferromagnetic materials begin to lose their magnetic properties as their temperatures increase. The thermal oscillations produced tend to destroy the magnetic domains. Above the *Curie temperature (T_c)*, ferromagnetism will no longer exist even when subjected to large external magnetic fields.

9.3.3 *B–H* Hysteresis (Magnetization) Curves for Ferromagnetic Materials

In our earlier discussion of permeability, we used the terms *magnetic flux density (B)* and *magnetic field (H)*. If we plot these two quantities, we arrive at a *B–H* or magnetization curve for a ferromagnetic material. Such curves are depicted in Figures 9-13 and 9-14 for soft- and hard-magnetic materials, to be discussed. Referring to Figure 9-13, an "unmagnetized" material at point *a* begins to receive an increasing *H* as the external magnetic field is built up, possibly by use of an electrical coil winding as shown in Figure 9-10. Inside the material the domains are growing, and reach point *b* where the maximum B_s (saturation) or $B_{(max)}$ is reached. At this point the material is essentially one large domain. See Figure 9-13 for a schematic of full domain remanence at point *b*. Upon decreasing the external magnetic field (*H*), the domain tends to return to a balanced domain arrangement; but because of the many structural imperfections, such as crystal defects, grain boundaries, or inclusions, the balanced domain state is not reached. Such a situation can be compared to the plastic region of a stress–strain curve. Point *c* represents the remanent or remanent magnetization with *H* = 0. As the magnetic field increases in the opposite direction, that is, *H* increases in the negative direction (−*H*), point *d* is reached, *B* = 0. The corresponding value of *H* at point *d* is known as the *coercive field* or *force*.

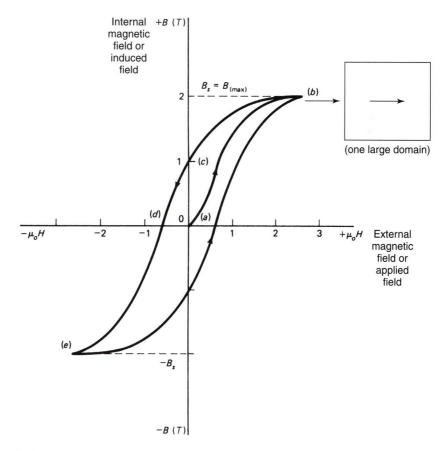

Figure 9-13   Representative *B–H,* or hysteresis, curve for soft-magnetic materials.

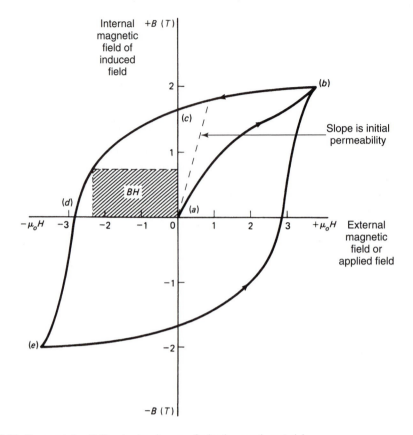

Figure 9-14 Representative *B–H,* or hysteresis, curve for hard-magnetic materials.

This value of *H* is needed to demagnetize the material (balanced domains). Most domains can be turned around at this point. Continuing to increase the field in the negative direction produces a similar domain growth, but in the opposite direction $(-B_s)$. Reversing the direction of the field a second time produces a curve that will plot out a *hysteresis loop.* Notice that the curve did not pass through the origin, point *a* in this cycle. The fact that these curves did not retrace themselves on the same path is called **hysteresis.** In such a cycle, much energy is transformed into thermal energy (friction) due to the reordering of the domains. This energy loss is proportional to the area of the hysteresis loop.

9.3.4 Soft-Magnetic Material

Using Figure 9-13 again and treating it as a representative hysteresis curve for a soft-magnetic material, note that for small values of *H,* a relatively large value of *B* is produced. Second, note that relatively little remanent magnetism needs to be canceled. Third, the area enclosed by the loop is relatively small. Such a material would find use as a transformer core. A ferromagnetic material can be demagnetized by reversing the magnetic field repeatedly while decreasing its magnitude. Iron (Fe), nickel (Ni), cobalt (Co), and certain alloys are ferromagnetic at room temperature (RT). Some alloys are ferromagnetic only at low temperatures. Soft-magnetic material is used for electromagnets since the field can be switched off, reversing the field with little loss of energy. Tape recorder heads are demagnetized in this way.

9.3.5 Hard-Magnetic Material

Figure 9-14 is a typical hysteresis curve for a hard- or permanent-magnetic material. Comparing this curve with the similar curve for a soft-magnetic material in Figure 9-13 and assuming that the scales for the axes are the same for both curves, several major differences can

TABLE 9.2   MAGNETIC PROPERTIES OF NICKEL ALLOY STEEL*

Steel[b]		Temperature						
Type	Condition	°C	°F	Initial permeability, μ_i	Maximum permeability, μ_{max}	Remanence, B (gauss)	Coercivity, H (oersteds)	Hysteresis loss (ergs/cm³/cycle)
1015 (0.10 Ni)	NT	20	68	190	1590 ($H = 4.7^c$)	11,400	4.0	2.7×10^4
		93	200	260	1680 ($H = 4.2$)	11,100	4.0	2.3×10^4
		165	392	260	1850 ($H = 3.9$)	10,250	3.8	1.9×10^4
		300	572	290	1900 ($H = 3.3$)	8,500	2.9	1.5×10^4
4340 (1.88 Ni)	NT	25	77	43	330 ($H = 30$)	13,700	22.5	11.4×10^4
		114	237	34	440 ($H = 25$)	12,700	13.9	7.6×10^4
		167	333	40	420 ($H = 28$)	12,000	18.0	7.9×10^4
		302	576	85	460 ($H = 23$)	10,250	11.0	4.9×10^4

Source: Courtesy of Inco. (Note: units not SI)

[a]These data are based on the initial magnetization curve and the magnetic hysteresis loop for applied field strengths (magnetizing forces) up to about 90 oersteds.

[b]N, normalized; NN, double normalized; Q, quenched; T, tempered.

[c]Oersted.

be observed. Hard-magnetic materials gain their magnetization in manufacture. The domains are preferentially aligned in one direction. Their essential property is their resistance to demagnetization. In other words, they must retain their magnetization when the external magnetic field is removed. The relatively large coercive field then is characteristic of this property, as well as the large hysteresis loop. Note the differences in the values of points *c* and *d* in both figures. The area of the largest rectangle that can be inscribed in the second quadrant of the hysteresis loop in Figure 9-14, called the *B–H product,* is a measure of the power of a permanent magnet. Table 9-2 is a typical listing of magnetic properties for some nickel alloy steels. Materials that make permanent magnets are said to have high *retentivity.*

Time and space limit discussion of past and ongoing research and development efforts in the field of new materials (metals, ceramics, and composites) to meet the ever-increasing needs of our technological society. Sales of permanent-magnet materials are over $3 billion and growing. The polymer-blended neodymium–iron–boron segment is growing at a 30% annual rate. NdFeB permanent magnets, also known as *rare earth iron magnets,* are produced by a sintering process called *rubber isostatic pressing (RIP).* They are being used in both the electronics and power industries. These high-performance magnets, when used in large electric motors, contribute to the world's efforts in reducing the temperature rise of the earth because they improve the efficiency of these motors substantially. These magnets could be classed as *"green magnets"* due to their effect on the earth's environment. This treatment of magnetic materials will permit a greater understanding and appreciation of the breakthroughs in this area as they are made public in the future.

9.3.6 Smart Electronic Materials: Electrorheological and Magnetorheological Materials

The word *rheology* literally means "a knowledge of flow." Today rheology is an interdisciplinary science that, in the main, deals with all aspects of deformation of materials under the influence of an external force (load, electrical or magnetic field, etc.) In fact, recent attention in the field concentrates on materials that can be classified as neither solid nor liquid. This intermediate state of matter between the solid and fluid states is represented by many rheological materials such as biological fluids, polymer melts, solutions, emulsions, and pastes. Rheology uses terms such as *elasticity, viscosity, solvents, solutes, gels,* and *suspensions,* many of which have been described previously. Two examples of such "smart" materials and their potential for future applications are *electrorheological* and *magnetorheological* materials, which were introduced in Module 6.

Electrorheological fluids change their flow characteristics when subjected to an electric field. The response, almost instantaneous, to such a field is a progressive gelling of the fluid, which is proportional to the field strength. These materials are composed of two components, a carrier fluid and the suspended particles in the fluid. The fluid does not conduct electricity. The suspended particles are typically polymers, minerals, and ceramics, which may take the form of suspended or dissolved particles or molecules. When an electrical field is applied across the fluid, the particles are polarized and link themselves together. (See Sections 2.1.1.2 and 9.2.1 for discussions of polarized molecules and polarization.) This linking produces a gelling of the carrier fluid, inhibiting its flow. Once the applied field is removed, the polarization of the particles ceases and the fluid can then flow freely (see Figure 6-30). Many future applications are envisioned in the development of automatic transmissions, brakes, clutches, hydraulic valves, and actuator devices.

A second example is ***magnetorheological materials.*** Similar in many respects to electrorheological materials, these materials use magnetic particles suspended in a nonmagnetic fluid (the host fluid). The host fluid could be a light machine oil or vegetable oil. A simple system such as one with iron filings suspended in a vegetable oil and a horseshoe magnet could amply demonstrate the principles of magnetorheological behavior. When a magnetic field is applied, the magnetically susceptible particles in the nonmagnetic fluid align themselves and remain stationary, thus restricting the movement of the suspension fluid. This situation illustrates that the viscosity of these materials is a function of the magnetic field in which they are immersed. The greater the field, the more solidified is the slurry. Applications for such materials are as varied and numerous as they are for electrorheological materials, especially for controlling fluid flow as in hydraulic braking. A recent application for such fluids is in controlling the suspension of truck seats by automatically adjusting the seat to changing driving conditions. During routine driving, a height sensor notes the downward movement of the seat and sends a signal to a microprocessor, which activates an electromagnet. The electromagnet quickly stiffens the fluid in a liquid state in a shock absorber to a jellylike near-solid. The fluid is silicon-based hydraulic oil plus an additive that permits a greater suspension of iron filings. Delphi Automotive System's new Magnaride shock absorber liquid is a combination of microscopic iron particles and conventional fluid. It changes viscosity when exposed to a magnetic field from the shock absorber piston. The change is rapid—10 times as fast as a conventional shock. It alters the suspension damping without the need for mechanical valves which allows a wide variation in shock action.

9.4 OPTICAL PROPERTIES

9.4.1 Light Characteristics

Optical properties describe a solid material's response to exposure to electromagnetic radiation and specifically to visible light, a narrow portion of this spectrum (see Figure 9-15). This response has been described using the interaction that occurs between this incident radiation and the material's atoms, ions, or electrons. These interactions and specialized terms are discussed briefly throughout the text under the headings of absorption, reflection, refraction, color, phosphorescence, electroceramics, polarization, electromagnetic spectrum, electron band theory, reflection of light, refraction, transmission, luminescence, solar cells, and lasers.

Light waves are often compared to water waves. The study of wave motion usually begins using examples such as wave motion in a body of water made by a boat, waves produced by vibrating a rope or string, or those sound waves produced by some vibrating object. Such waves are considered ***mechanical waves.*** All require some medium in order to travel. For example, without a medium, sound waves could not be heard. If one observes leaves floating on a lake when waves pass by, the leaves simply oscillate up and down and are not carried forward with the waves. A wave consists of oscillations that move without carrying matter with

Figure 9-15 Wavelengths of electromagnetic radiation characteristic of various regions of the electromagnetic spectrum. (Theodore L. Brown and H. Eugene LeMay Jr., *Chemistry,* 2nd ed., Englewood Cliffs, N.J.: Prentice-Hall, Inc., 1981.)

them. Rather, they carry energy from one place to another. Visualize ocean waves hitting a beach during a hurricane to understand the significance of this statement.

Wavelength (λ) is the distance between any two successive identical points on a wave. The frequency of a wave (f) is the number of complete cycles per second or the number of crests that pass a given point per unit time. *Frequency* is usually specified in hertz (Hz), where 1 Hz = 1 cycle per second. The period (T) is 1/f or the time required for one complete cycle. For example, if a wheel rotates at a frequency of three revolutions per second, then each revolution takes 1/3s. The wave velocity (v) is the speed at which wave crests appear to move. A wave crest that travels a distance of one wavelength in the time period T has a velocity equal to λ/T or v = λ/f.

Illustrative Problem

If a wave with a wavelength of 14 cm is traveling at a velocity of 34 m/s, what is the wave's frequency?

Solution   Convert units 14 cm = 0.14 m. Using the previous formula v = λ/f and solving for f, f = v/λ substituting values f = 34 m/s/0.14 m = 243 Hz.

Electromagnetic (EM) waves, or EM radiation, resemble the previously discussed form of mechanical waves. They are produced by electric charges that are oscillatory. Generally EM waves are produced by accelerating electrons and other charged particles. EM waves carry energy from one region of space to another. The different forms of EM radiation are represented by the EM spectrum (see Figure 9-15). They all travel in free space at the same velocity—the velocity of light—at 3.00×10^8 m/s. Visible light can be considered an electromagnetic wave with wavelengths between about 400 nm and 750 nm. Since all EM waves travel at the same velocity—that of visible light—the letter *c* is used to represent that velocity.

Illustrative Problem

Calculate the wavelength of a 60-Hz EM wave.

Solution   Using the previous formula and substituting c for v, this new relationship is c = λf. Solving for λ and substituting values:

$$\lambda = c/f = 3.00 \times 10^8 \text{ m/s}/60 \times 10^{-1\text{s}} = 5.0 \times 10^6 \text{ m}$$

(note the size of this wavelength).

When light travels from one medium into a second medium where its speed is greater, the ray bends away from the common normal. When the ray enters a medium where the speed of light is less, the ray bends toward the normal. In other words, the index of refraction depends on the wavelength. As wavelength changes with the index the frequency of a wave remains the same as the wave passes from one medium to another. The ratio of the speed of light in vacuum to the speed in a given material (v) is called the *index of refraction (n)* of that material. Because the index will vary with the wavelength of the light, a particular wavelength is specified in dealing with materials.

Illustrative Problem

> The index for diamond of 2.42, and using yellow light with a wavelength of 589 nm (see Figure 9-15), find the speed of light in diamond.
>
> **Solution**
>
> 1. n = c/v substituting and solving for v.
> 2. Solving for v,
> $$v = c/n = 3 \times 10^8 \text{ m/s}/2.42 = 1.24 \times 10^8 \text{m/s}$$

The frequency of light does not change when it enters another material, but the wavelength of light in a material differs from the wavelength of light in a vacuum. We can conclude that the index gives us a measure of the speed of light in a transparent material. It is a measure of the material's *optical density.* The greater the index the greater is the optical density of the material and the smaller the speed of light in that material. Light can be pictured as traveling in rays (light perpendicular to a wave front). When a light ray strikes a smooth surface it is reflected making an angle θ (Greek letter theta) with a line normal to the surface at the point of reflection. The angle θ^1 that the reflected ray makes with the normal obeys the equality known as the **law of reflection,** namely $\theta = \theta_i$. See Figure 9-16a.

Light is energy in the form of radiation. It is regarded as being composed of either photons in the energy range (2.5 to 5.6×10^{-19} J) or electromagnetic waves in the wavelength range of 800 nm down to 380 nm (see Figure 9-15). Light, being a form of electromagnetic radiation, interacts with the electronic structure of atoms of a material. The initial interaction is one of absorption; that is, the electrons of atoms on the surface of a material will absorb the energy of the colliding photons of light and move to the higher-energy states. The degree of absorption depends, among other things, on the number of free electrons capable of receiving this photon energy. The electrons can then do several things. They can have sufficient energy to jump into higher-energy states, which permits them to be accelerated within an electric field and thereby conduct electricity (photoconductive effect). (See Sect 9.4.4 and 9.7.1 for further discussion of photo conductivity). They can collide with atoms and release their excess energy in some form of electromagnetic radiation, or they can convert their excess energies to atoms in the form of thermal energy. When light proceeds from one medium into another (a metal, semiconductor materials, or electrically insulating materials) it may be transmitted through the materials, it may be reflected at the interface between the two media, or it may be absorbed. Figure 9-16 shows the light *intensity* (*I*) with units of watts per square meter represented by I_o, the incident light ray; I_r, the reflected light; and I_t, the transmitted light. The term *reflectivity* is a ratio of I_r/I_0 (see Figure 9-16).

Transmissivity is the ratio of I_t/I_0 and *absorptivity* is the ratio of absorbed light to the incident light. Light radiation is absorbed within solid materials by interactions between the light, a form of electromagnetic radiation (Figure 9-15), and the materials electrons. The results may be electronic polarization (see Figure 9-7) and/or electron transfers from one energy state to another (see electron band structure in Section 9.6).

An example of the effect of passing a current through a gaseous medium is the sodium lamp. A high-pressure sodium (HPS) lamp invented in the 1960s can last 6.5 years, if operated for 10 hr per day (24,000 hr). This lamp puts out 140 lumens per watt versus 15 lumens per watt for incandescent. The light is produced by passing an electric arc through vapors of sodium and mercury resulting in a temperature of greater than 2192°F in the thin-walled "arc tube" made of translucent polycrystalline Al_2O_3. The arc tube is about 5/16 in. diameter × 4 in. long. It is enclosed within a glass bulb. High-intensity lighting gives truer colors. Using metal halide vapor emits white light which is more expensive and shorter lived.

As light enters a nonmetallic material, it is absorbed (Figure 9-17) by the action of several mechanisms, such as electronic polarization (Figure 9-7) and electron transitions. The

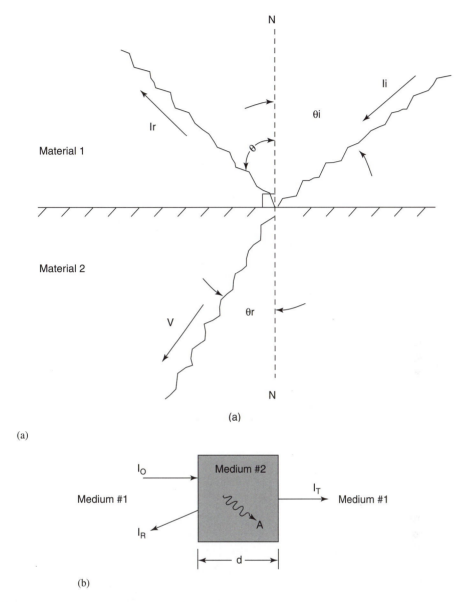

Figure 9-16   (a) Refraction of light, showing the angle of incidence (θ_i) and angle of refraction (θ_r), both measured from a common normal N. (b) The transmission of light through a transparent medium of thickness d is shown. I_0 is the incident light, I_R is the reflected light, and I_T represents the transmitted light leaving the material. The absorbed light in medium #2 is shown by a wiggly arrow labeled A. Both versions of medium #1 are the same material. Refraction is generally accompanied by reflection.

electron transitions depend on the electron energy band structure of the material. The interaction process characteristic of photons depends on their energy (E_ρ). Low-energy photons interact principally by ionization or excitation of the outer orbitals in a solid's atoms. Light is composed of low-energy photons (E_ρ is less than 10 eV) represented by infrared (IR), visible light, and ultraviolet (UV) in the electromagnetic spectrum (see Figure 9-17). UV light is being tested for curing solvent-free urethane paint on autobodies as a technique to eliminate the release of VOCs into the atmosphere from the spraying as well as the costly baking process newly painted autobodies must undergo. High-energy protons (E_ρ greater than 10^4 eV) are produced by X rays and gamma rays (Figure 9-18). The minimum energy (E_ρ) required to excite and/or ionize the component atoms of a solid is called the absorption edge or threshold.

   When crystalline materials are exposed to radiation, the change in their properties is due to a displacement of atoms from their normal locations in the crystal structure, i.e., the creation

Figure 9-17   Energy-efficient, low-emissive, double-pane glazing with a reflective film.

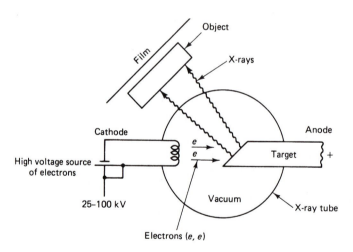

Figure 9-18   Schematic of an X-ray tube.

of crystal defects. The total energy of the incident radiation is measured in terms of dose, the amount of radiation affecting the material, or the dose rate, which is the rate of energy deposition. Irradiation conditions are often given in terms of the total number of particles incident on the sample, i.e., the ***fluence*** (fluence equals the summation of flux, or the particles per unit time deposited, over irradiation time) (see Figure 9-19). A greater understanding of this topic requires a study of the mechanisms by which energy is transferred from the radiation (in this case, light) to the atoms of a solid. In the case of metals, only the energy transferred to the nuclei by the beam of energetic, charged particles such as electrons, photons, and fission fragments is effective in creating lattice defects. The crystalline structure can also be converted to an amorphous one by sufficient exposure to radiation. Table 9-4, and Figure 9-20. In metals, the band

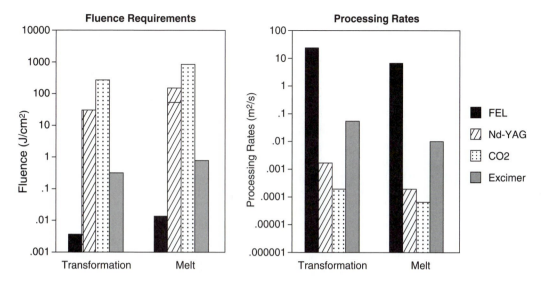

Figure 9-19 Calculated laser fluence and processing rates for metal surface processing with conventional lasers and the UV demonstration FEL. Large-scale exploitation of UV surface processing will require sources of at least a few tens of kilowatts, light that costs under a cent per kilojoule, and a full wavelength tunability. (TJNAF, Newport News, Virginia)

gap is zero and photons of all energies can be absorbed to increase the kinetic energy of the conducting electrons and holes. In polymers (covalently bonded materials), photons of suitable energy are capable of exciting vibrations directly in both molecules and crystalline structures. The chemical bonding can be altered by electronic excitation, which can also create cross-links and break primary bonds. Fig. 9-20. ***Refractive index (n)*** is determined by the chemical composition and the crystallographic properties in a material. Changes may occur which can be brought about by compositional changes, crystallographic changes, or intermingling of phases. Usually these changes cause a material's physical properties to change. Microscopic analysis of the ***morphology*** (shape and structure) of fiber-reinforced polymeric materials (composites) use the refractive index of polymeric materials since most are semitransparent to light.

When light is transmitted into the interior of a transparent solid, it experiences a decrease in its velocity. The result is the light bends as shown in Figure 9-16b. This phenomenon is known as ***refraction***. The ***index of refraction (n)*** of a solid material depends on the wavelength of the incident light. For air, the index is nearly equal to 1, whereas water has a higher index of refraction, 1.33. Other indices of refraction for some materials are as follows: silica glass, 1.46; corundum, 1.76; polyethylene, 1.51; and polypropylene, 1.49. Lucite (Plexiglas) has an index of 1.51 and diamond's is 2.42. When light passes from one medium to another having a different index of refraction, some of the light is scattered at the interface between the two media. Both of these media could be transparent materials.

9.4.2 Photoelectric Effect, Photons, Emissivity, and Lasers

Certain metallic materials will become charged positively when exposed to electromagnetic radiation. Because no electron charge can be carried by such radiation, the conclusion reached is that the radiation must interact with the electrons attached to the metallic atoms and cause the ejection of electrons from the surface of the metal. Such a phenomenon is known as the ***photoelectric effect***. The ejected electrons are called ***photoelectrons***. Furthermore, not all electromagnetic radiation produces this effect. Only when a certain frequency of radiation is reached (***threshold frequency***) does this effect occur. The threshold frequencies differ with different metallic elements. As an example, the metal sodium will produce this effect once the frequency of the incident radiation reaches 5.6×10^{14} Hz. The corresponding wavelength of 5.4×10^{-7} m places such radiation in the visible-light portion of the electromagnetic spectrum. Zinc requires a threshold frequency of 8.0×10^{14} Hz, which is in the ultraviolet region of the spectrum. The intensity of the radiation plays no part in producing the photoelectric effect.

To explain this phenomenon adequately, Einstein proposed in 1905 that this radiation be treated not as a series of waves (wave phenomenon) but as a series of particles (light photons) that interact with single electrons. Most of the questions dealing with light radiation confine themselves to large-scale bodies such as glass lenses. The films of materials taken by X rays and the electron microscope are some examples of radiation that exhibits wavelike properties. The pattern of electrons that projects itself onto the X-ray film and the highly magnified images of the internal structure of a material as viewed under an electron microscope are diffraction patterns of waves typical of those exhibited by the diffraction of light through a grating. When we begin to probe into the land of the atom with its subatomic electrons, we use the particle nature of radiation to explain the results of such interactions of radiation, not with a whole surface but with individual electrons.

A *photon* of light (a quantum of electromagnetic radiation) exchanges its energy with a single valence or conduction electron. We have learned that such electrons possess the highest kinetic energies and are the outermost electrons in an atom. When a photon of light with sufficient energy ($E = hc/\lambda$) strikes a valence electron, it causes the excited electron to leave a solid metal. Note that the photon reacts with a single electron, not with an entire atom or a solid metal electrode. The kinetic energy (KE) of the ejected electron (now called a photoelectron) can be expressed in two terms. The first term, the work function (W), is that part of a photon's energy (work) expended in digging out or removing the valence electron from the surface of the metal. The second term represents the remainder of the photon's energy appearing in the form of kinetic energy of the escaping electron. The photon disappears, having done its job of transferring its energy, and a photoelectron is emitted from the surface. The electrons in a metal that are eligible to receive energy from colliding photons may be in various energy levels of an atom. Those with the highest energies will require the least energy expenditure to break free from the surface; hence, they will have the greatest kinetic energy. Using electron volts (eV) as units, this energy transfer can be expressed mathematically as

$$\text{KE} = E - W$$

Suppose that violet light is incident on a piece of cesium (Cs). Reference books can be consulted to find the photoelectric properties of metallic elements. Referring to such sources, we find the threshold frequency for Cs to be 4.6×10^{14} Hz, the work function W is 1.9 eV, and the wavelength (λ) of violet light is 4×10^{-7} m. To find the frequency f of violet light, we use

$$f = \frac{c}{\lambda} = \frac{3 \times 10^8 \text{ m/s}}{4 \times 10^{-7} \text{ m}} = 7.5 \times 10^{14} \text{ Hz}$$

Note this frequency exceeds the threshold frequency for Cs. The photon's energy E is next calculated using

$$E = \hbar f$$

where $\hbar$ is Planck's constant (6.625×10^{-34} J · s) and 1 eV = 1.60×10^{-19} J. Thus,

$$E = \frac{(6.625 \times 10^{-34} \text{ J} \cdot \text{s})(7.5 \times 10^{14} \text{ Hz})}{1.6 \times 10^{-19} \text{ J}} = 3.1 \text{ eV}$$

The kinetic energy (KE) of the photoelectron emitted can then be determined:

$$\text{KE} = E - W$$

$$= 3.1 - 1.9 = 1.2 \text{ eV}$$

When a surface is exposed to radiation, it can absorb part or all of the incident radiation (energy). The fraction of the energy absorbed is the emissivity (e) of the surface. For a black body, the perfect absorber, $e = 1$; for the perfect reflector, a white body, $e = 0$. **Photoconductivity** describes the increase in conductivity experienced by some semiconductive materials when illuminated by a light source. This increase in conductivity occurs when light is absorbed by photon-induced electron transistors which produce additional charge carriers. Cadmium sulfide, a photoconductive material used in light meters when struck by incident light radiation, generates a photon-induced current which is a measure of the rate at which photons of light strike the photoconductive material. Gallium arsenide (GaAs) converts light directly into electricity. It is used in light-emitting diodes (LEDs) and transistors. (See Sect 9.7.1 for further discussion of photoconductive devices.)

By synchronizing the light emission from many electrons, one can produce light that is monochromatic (one frequency) and coherent (photons are in phase with each other). More and more materials are being found whose electromagnetic radiation can be controlled by technicians. These include solids, liquids, and gases. Semiconductor materials can produce laser action that emits photons with wavelengths in the range 7 to 8×10^{-7} m. A **laser** (light amplification by stimulated emission of radiation) is a special example of the emission phenomenon we have been discussing. The light produced is further amplified through stimulating the emission. To defeat lung cancer two newly developed technologies use laser light to detect and destroy cancer cells. One such technology uses certain wavelengths of laser light to illuminate the organ; cancerous tissues fluoresce in reddish hues. Another technology called **photodynamic therapy** involves injecting the patient with a photosensitive drug that concentrates in cancerous tissues; when activated by the red laser light the drug destroys these cells. A **maser** (microwave amplification by stimulated emission of radiation) differs from lasers mainly in the wavelength of the radiation produced.

The Laser Processing Consortium (LPC), a collaboration involving nine U.S. corporations and companies, seven research institutions, and the Thomas Jefferson National Accelerator Facility (TJNAF) in Newport News, Virginia, has completed its first major run demonstrating high-power infrared (IR) light with its **free-electron laser (FEL)**. The FEL is a unique high-power, high-efficiency, tunable light source for basic research, industrial processing, and military applications. "Free electrons" are electrons that are free from their respective nuclei in the atomic structure of materials and travel at the speed of light. They are not bound to their nuclei. One unique feature of the FEL is how it differs from other lasers. Conventional lasers suffer from limitations in cost, power, and choice of wavelength. Figure 9-19 compares the fluence (a measure of the irradiation on a surface) and the processing rates of commercial lasers with the FEL. Commercially available excimer lasers are limited to tenths of a kilowatt, tens of cents per kilojoule, and a few isolated specific wavelengths. Table 9-3 compares the characteristics of some commercial lasers with those demonstrated by the FEL. The FEL can be tuned over a much wider range, which allows it to lase a variety of materials with different absorbencies with a single-laser system (see **absorbtivity** in this section).

TABLE 9.3   PERFORMANCE COMPARISON OF SELECTED LASERS

Laser	Pulse Width	Repetition Rate	Output Energy (per pulse)
CO_2	100 ns	50kHz	0.1 J
Excimer	10–30 ns	~100 Hz	0.3 J
IR-DEMO FEL	~0.5–2 ps	18.7 or 37.4 MHz	25 μJ
Ti:Al_2O_3	100 fs	80 MHz	10 nJ
Nd:YAG (Q-switched)	3–20 ns	10 Hz—50 kHz	1.5 J—80 μJ
Nd:YAG (ultra-fast)	30 ps	10 Hz	50 mJ
Alexandrite	30–100 ns	20 Hz	0.5 J
HF	400 ns	100 Hz	50 mJ

TJNAF, Newport News, Virginia

TABLE 9.4 REPRESENTATIVE INDUSTRIAL APPLICATIONS OF FELS

Level of maturity	Polymer surface processing	Micromachining	Electronic materials processing	Metal surface processing
Commercializable				
Demonstrated major market	Surface texturing	Fuel injectors	Flat panel displays	Laser glazing and annealing
Cost/capacity barrier	Surface conductivity	Spinnerets	Large-area photovoltaics	Surface carburizing and nitriding
	Surface amorphization	Slitting coated films	Large-area diamond coating	
Developable				
Demonstrated	Antimicrobial nylon	Microtexturing	Pulsed laser deposition	Adhesive bond pretreatment
Development needed	Pulsed laser deposition	Subthreshold ablation		Solvent-free cleaning
				Removal of corrosives
Prospective				
Preliminary results	Surface activation of	Nanometer-scale surface	High-density CD-ROMs	Thermal barrier coatings (TBCs)
Research develop-ment needed	carbon fibers for composites	contouring for chemi-cal catalysis	Embedded layer silicon processing	Functionally gradient coatings
Enabling facilities needed				Metglas coatings
				Shape memory alloys (SMAs)

TJNAF, Newport News, Virginia

Its cost-effectiveness is increased by its ability for nearly complete recovery of energy back into the accelerating electron beam. This recycling of the electron beam extracts power that was not radiated as light. Short pulse length allows for more efficient coupling of light with the target materials. Pulse lengths are measured in pico seconds (ps). The pulse rate between 18 and 74 megahertz (MHz) can drastically increase the rate at which a material is lased; thus, speed deposition times can be reduced in laser-based deposition systems. These characteristics favor future application both in the military and for industrial applications calling for processing of large volumes of materials (***light-based manufacturing***).

Currently the FEL operates in the IR regime of the electromagnetic spectrum but the ultraviolet (UV) regime is favored by industry because UV light has higher absorbency in most relevant materials. A UV FEL could replace wet chemical methods and provide new capabilities for surface modification and other valuable procedures. Thus, FEL manufacturing would take its position as a member of "green Manufacturing," helping to reduce the pollution of the earth in which we live. Most surface modification treatments of over 40 billion pounds of fiber produced in the United States use wet chemical processes. Experiments with the surface modification of nylon and polyester to enhance certain properties such as surface friction, filtration, and visual characteristics have already begun (see Figures 9-19 and 20a and b). A NASA project involves a "***Solar Sail,***" a kilometer or larger wide sail which spacecraft would unfold in interplanetary space to harness solar winds (the winds are made up of protons). These sails could consist of a solid polyimide base material coated with aluminum. The FEL would be used to machine away most of the polyimide used as ribbing to strengthen the aluminum. Preliminary tests showed that the laser could ablate the polyimide without affecting the aluminum. The steel industry is interested in using the FEL to remove oxide layers from carbon steels that build up during the milling process. These layers are normally removed by acid etching in pickling vats. This process uses hazardous materials that call for unusual care and it raises environmental concerns.

Another process earmarked for the laser is the forming of amorphous layers on metals to produce harder, more corrosion-resistant surfaces. Figure 9-20c shows a laser forming a thin surface layer on a metal surface to eliminate grain boundaries. Other future applications for FEL are the production of particulate-free films by providing atom-by-atom vaporization of substrate materials for use in semiconductor microelectronics and optical materials. Laser-assisted chemical vapor deposition (CVD) would provide greater control over chemistry and eliminate the need to heat substrates to high temperatures. Table 9-4 sets forth the status (level of maturity) of some representative industrial applications of FELs using commercializable, developable, and prospective categories.

FEL APPLICATIONS

Tunable light can alter surface characteristics of plastics

1 Cross-linking of polymer bonds makes surface of plastic slippery, prevents adhesion of glue, paint, and other polymers

2 Light tuned to the bond frequency breaks some of the cross-linking, providing molecules on surface with available bonds to attach to.

3 Glue, paint, other polymer molecules can now attach to surface.

TYPICAL APPLICATIONS
- ▶ Anti-static coatings for carpet
- ▶ Bonding plastic auto parts
- ▶ Germicidal drug containers
- ▶ Synthetic skin, arteries, etc.

FELapptitplBjm 1/7/93

(a)

IR FEL APPLICATION

Without laser irradiation

Semicrystalline polymer surfaces bond poorly because few chains cross original plane.

WEAK ADHESION

IR FEL light

With irradiation

Chains amorphized

IR light amorphizes polymer chains, producing entanglement of polymers and good bonding.

STRONG ADHESION

dylla/FEL/mbfel/1RamorphBVG/jm 7/26/93

(b)

Figure 9-20 (a) Future light-based manufacturing will involve modifying materials' surfaces by using the UV portion of the electromagnetic spectrum. Laser processing treatments can give a polymer or fiber product new friction, filtration, wetting, or visual characteristics. (TJNAF, Newport News, Virginia) (b) Surface modification by light can add value to existing products, serve as the value basis for new products, or be incorporated into special interfaces of packaging materials or structural composites. In this application, surface amorphization (carefully tailored light energy absorption in the transformed region in the surface layers of a polymeric material) produces greater adhesion, with the additional benefit of eliminating the environmental costs of processing with wet chemistry. (TJNAF, Newport News, Virginia) (*continued*)

Common Metal Surface

• Shows grain boundaries that increase potential for corrosion and wear

Industrial Applications

• Hardening machine tools
• Long life turbine blades
• Corrosion resistant structures

(e)

Laser Treated Metal Surface

• Grain boundaries are "healed" by melting a thin surface layer
• Resulting smooth surface is highly resistant to wear and corrosion

Figure 9-20   (c) FEL Application to surface processing of metals.

Present-day understanding of the fundamental structure of matter reveals four fundamental forces in nature. Gravity and the electromagnetic (the electric) force acting over large distances are called long-range forces. The electromagnetic force binds atoms and molecules into ordinary matter. The strong force and the weak force are confined to the nucleus of the atom. The strong force is responsible for holding together the nuclei of atoms. It acts between all nucleons. The protons in the nucleus attract each other via the strong force at the same time they repel each other by the electromagnetic force. Neutrons, being electrically neutral, only attract other neutrons or protons via the strong force. The strong force operates only at short distances. The force is very strong between two nucleons if they are less than about 10^{-15} m apart and it drops to zero at distances greater than that amount.

Quarks and electrons are the basic building blocks of the world's structure. A present-day theory (QCD) as mentioned below uses an analogy to explain quarks. An ordinary atom consists of electrons bound to the nucleus by electromagnetism, the force generated by particles of light. Similarly, the proton and neutron are known to be "quark atoms" clusters of quarks bound together by the strong force—the "gluing" force generated by particles "playfully" called gluons. Now, protons and neutrons can be thought of as "atoms" made up of quarks. Two quarks, called *up* and *down,* are found in everyday matter. By using high-energy accelerators, four other quarks have been found fleetingly under these extreme conditions. Their names are *strange, charm, top,* and *bottom.* Quantum chromodynamics (QCD) theory developed in the early 1980s forms the basis for the experiments with the continuous electron beam accelerator at the Jefferson Laboratory. Experimenters seek to build understanding of everyday matter by examining nuclei in terms of quarks and the interactions the quarks experience via the strong force.

The FEL represents another development in surface modification technology. *Surface modification* techniques are referenced throughout the book, particularly in reference to heat-treated surface techniques such as case hardening and nitriding, electroplating, diamond films, titanium coatings, plating, layered composites, chemical vapor deposition (CVD), sputtering, chemical vapor infiltration (CVI), toughening, surface engineering, and the like. A recent laser technique reported by QQC, Inc. converts a surface layer into a superhard coating not only on metals, but also on ceramics and plastics. The process uses three high-powered lasers, excimer, yttrium–aluminum–garnet, and carbon dioxide, each producing a different wavelength of light whose light pulses overlap the surface. These lasers vaporize a thin layer of material creating an electrically charged **superheated plasma** of iron atoms. After the beam is removed, the atoms fall to the surface and bond together. Elements can be added to the surface by sending a stream of particles across the material. The lasers break apart the particles adding their constituent atoms to the plasma. One application describes a hot forging punch hardened with a layer of high-strength chromium–cobalt alloy which increases its wear life from 1 hour to 12 days. Specified locations can be hardened with nanometer precision. Advances in this new technology will propel many improvements in materials, manufacturing processes, and superior product quality.

Many uses for these well-controlled sources of energy will continue to be discovered for boring holes, welding delicate parts of the human eye, detecting and measuring pollutants in the atmosphere, and carrying human communications through space with little or no interference. Present applications of the optical behavior of materials, including glass and plastics that produce light through the interaction of electromagnetic radiation with the material's electrons, are numerous. Some common examples are worth mentioning. The coating of TV screens with zinc sulfide (ZnS), a semiconductor material on which electrons act to produce light, is one. A second is the phototube that "reads" the sound track on motion-picture film, ultimately producing an electrical signal that can be amplified and broadcast to an audience. Solar batteries using semiconductor materials and exposure meters carried by photographers for properly setting their cameras are additional examples of the many applications of such devices.

Luminescence is the reemission of photon energy at wavelengths in the visible spectrum as the result of the absorption of electromagnetic radiation from some outside source. If reemission occurs at the same time that the material is absorbing the radiation, the phenomenon is called **fluorescence.** In a fluorescent lamp, electrons emitted by the incandescent cathodes collide with electrons of the mercury atoms that fill the tube. The collisions cause the emission of radiation in the invisible ultraviolet range. This radiation, in turn, strikes the fluorescent or phosphor material coating the inner side of the tube and causes this material to emit radiation in the visible range of the spectrum.

Should the light emitted continue after the radiation producing it has been removed, the light is called **phosphorescence.** The color TV picture tube uses phosphors that phosphoresce. Phosphorescence is an important optical property of ceramics. Ceramic phosphors are used in many applications such as TV screens and fluorescent lights. Phosphorescent ceramic coatings are used in photocopy lamps. Various dopants are added to produce the emission of certain colors.

9.4.3 X Rays

X rays, discovered by W. Roentgen in 1895, can be produced in two ways. The first process uses a high-energy electron capable of knocking an electron from an inner energy level of an atom completely out of the atom. This removal of an electron, say, from the K level, immediately creates a hole into which an L-shell electron can fall. As an L-shell electron falls back to the lower-energy K level, the sudden decrease in potential energy is emitted as a photon of electromagnetic energy characteristic of X rays (atomic radiation). A hole is produced in the L shell, which calls for the transition of another electron from an outer level to the L shell, which produces an additional X ray. Several X rays are emitted as electrons cascade down to fill the lower-energy-level vacancies, or holes, until eventually the atom captures an electron from the surrounding region and changes from an ion to a neutral atom in its lowest equilibrium condition. The wavelength of the X ray depends on the particular energy levels involved, as do the colors in visible light emitted from a given jump. X rays lie in the wavelength range between 1×10^{-12} and 1×10^{-8} m. Another process of generating X rays (also called *cathode rays*) is through the sudden braking (deceleration) of high-energy electrons by striking them on a metal target, which results in a conversion of part of the electron's kinetic energy into a quantum of electromagnetic radiation or X ray. Such a technique is shown schematically in Figure 9-18.

9.4.4 Thermionic Emission

Another way to exchange energy with the free electrons in a solid metal is to heat the metal to such an extent that the thermal energy of the electrons is sufficient to emit them from the surface. This is known as ***thermionic emission.*** A typical low-power vacuum diode has a cathode consisting of a nickel tube coated with barium oxide, which is heated by an insulated filament contained within the tube.

9.4.5 Transmission

Optical devices possess many desirable properties, but the most important is their ability to transmit light. One measure of this ability is ***transmittance,*** defined as the percentage of an incident light ray remaining after passing through 10 mm of a material. It can also be expressed as a ratio of transmitted-light intensity to the incident-light intensity (***intensity*** is the measure of the strength of a light source). Some materials are ***opaque*** (will not let light pass through). Metals also are opaque. When white light is incident on a metal surface, most of the light is reflected, and some is refracted into the surface of the metal where the photons strike the electrons, thus transferring discrete amounts of their energy to the electrons and raising them to higher energy levels. (See Fig. 9-16). With copper, the $3d$ electrons absorb the energy of the photons corresponding to blue wavelengths through collisions, and reflect the others, which give a reddish color. The energy absorbed causes the atoms concerned to increase their energy levels, which allows them to increase their vibrations, which, in turn, results in an increase in temperature, an indicator of heat.

In covalent and ionic solids, the electrons are bound to the atoms by the very nature of the covalent and ionic bonds. It takes more energy to break such bonds and free the electrons with increased energies sufficient to move them to higher energy levels. Therefore, visible light passes through such solids without reacting with the atom's electrons. Such solids are termed ***transparent.*** If impurities are present in such a solid, the photons of the visible light could react with the electrons of the impurity atoms, which are not involved in covalent or ionic bonding. In a ruby, the chromium atoms absorb the photons of blue and green light, but allow the remaining photons to pass through, which are mostly of wavelengths that correspond to red light.

Many polymer materials are transparent. Others are ***translucent***; that is, these materials, because of their structure, allow some light to pass through, but most of the light is scattered due to the reflection and refraction of the light as it transmits across phase boundaries and interacts with pores in the structures. Research in new ceramic materials allows the production of single-phase, pore-free ceramics that are transparent.

The bending of light rays is due to the fact that the speed of light within a dense medium is less than in air. Many of us have the experience of trying to pick up an object in a shallow pool of water, only to find that the object's actual location is displaced from its apparent location.

The ratio of the speed of light in air (c) to its speed in a medium (v_m) is called the ***index of refraction*** of the medium (n). See Sect 9.4.1 for further discussion of this index.

$$n = \frac{c}{v_m}$$

Illustrative Problem

Given the index of refraction in a medium of 1.5, how fast would ordinary light travel in this medium? What material has an index with this value? (See below.)

Solution   Solving the above equation for v_m

$$v_m = c/n = (3 \times 10^8 \text{ m/s})/1.5 = 2 \times 10^8 \text{ m/s; polyethylene.}$$

The index is related to polarization in a material by the expression $n = \kappa^{1/2}$, where κ is the dielectric constant. This may vary with the change in density of materials as well as in certain crystal directions (i.e., n is anisotropic). The index also varies with frequency. Using a wavelength of 589 nm, some selected indices are: diamond, 2.4; polyethylene, 1.5; water, 1.33; and air, 1.0.

Another way in which the index may be defined is in terms of the angles of incidence (θ_i) and refraction (θ_r). Figure 9-16 is a sketch showing how these angles are measured from a common normal.

$$n = \frac{\sin \theta_i}{\sin \theta_r}$$

Illustrative Problem

Visible light traveling in air enters a medium with an index of refraction of 1.5 at an angle of incidence of 60°. Find the angle of refraction.

Solution

$$n = \frac{\sin \theta_i}{\sin \theta_r} \qquad \sin \theta_r = \frac{\sin \theta_i}{n}$$

$$\sin \theta_r = \frac{0.866}{1.5}$$

$$= 0.577$$

$$\theta_r = 35.3°$$

The refractive index n is for the medium into which the incident light penetrates. Solids with dense atomic packing generally have higher indexes of refraction. In crystals the index of refraction is different for different directions, which can result in the incident rays being split into two polarized rays.

In explaining the phenomenon of **polarized** light, we use the concept of the electromagnetic wave. All electromagnetic radiation of whatever frequency consists of two fields, one electrical and the other magnetic. Figure 9-21 shows these two fields designated by the vectors **E** and **B**. Each field is separated by 90°; that is, for each **E** there is a corresponding **B** set up in a plane at right angles to each other. For simplicity, only two fields are shown. It must be remembered that any radiation is composed of many, many waves with **E** and **B** vectors representing the electronic and magnetic fields pointing in all directions. Using just the **E** field and Figure 9-22, if you position yourself as standing out on the x axis and looking back at the origin of our x–y–z axes system, the many **E** fields could be represented by the sketch. These **E** vectors could also be represented by their respective components in the y and z directions, thus simplifying the representation of these **E** vectors. A schematic of unpolarized light would then look like the sketch in Figure 9-23a. This view of *unpolarized* light represents **E** vectors that oscillate back and forth along the y and z axes. If the radiation consists only of **E** vectors along only one axis, it is known as

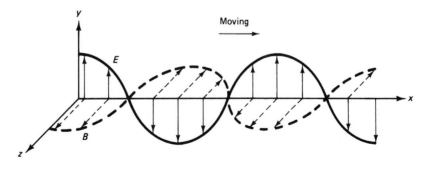

E — Electric field vector, shown in the xy plane
B — Magnetic field vector, shown in the xz plane

Figure 9-21 Electromagnetic fields in a light wave.

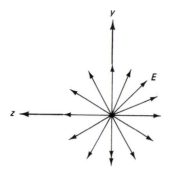

Figure 9-22 A point view of electric field vector **E** in an electromagnetic radiation of unpolarized light.

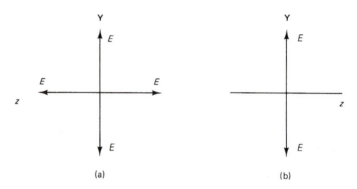

Figure 9-23 (a) Unpolarized light with **E** vector resolved into y and z axes. (b) Polarized light in *y* direction.

polarized. Polarization can be accomplished by the use of a filter that absorbs the electric field oscillations in one particular direction. A representative sketch of polarized light is shown in Figure 9-23b. The light in this example is polarized in the *y* direction.

Unpolarized and polarized light can be further understood by comparing them to vibrating a rope. If you fix a rope at one end and vibrate the other end up and down in all directions, this would be a model for ordinary or unpolarized light. If you vibrate the rope in just one direction, say, the vertical direction (up and down), this would be representative of polarized light. Furthermore, if the same rope were pulled through the slots in a picket fence, the rope could be vibrated only in this up and down manner. The picket fence then could serve as a model of an optical filter (a polarizer) that permits light vibrations in only one direction and absorbs (prevents) other vibrations from being transmitted. When light is transmitted through material, the light can become polarized due to the differences in the material structures in different directions. A material whose structure varies with the orientation of the material is not isotropic (anisotropic). As you recall, isotropy is a quality of a material that has identical properties in all directions. The term ***optical anisotropy*** refers to a material having different optical properties in different directions.

Being insulators, most polymeric materials have no free electrons. In addition, their internal structure contains boundaries and pores that interact with light and cause it to be transmitted in different directions at different speeds, hence at different refraction indexes. Polarization of light is used in many applications, such as liquid crystal displays (LCDs), Polaroid sunglasses, and photoelasticity, where polarized light is used to reveal the amount of anisotropy in a transparent-polymer (polystyrene) model of a loaded structure (see Figure 9-24). When subjected to stresses ($\sigma = P/A$), certain transparent materials become ***birefringent*** or double refracting. These substances will divide an incident ray of light into two beams that travel through the material at different speeds. In particular, these two beams are polarized at right angles to each other (see Figure 9-23).

This material property of birefringence is taken advantage of by constructing models of structures such as buildings using transparent materials having double-refraction properties.

Figure 9-24   Stress concentrations revealed in loaded transparent-polymer models. (NASA)

The models are subjected to forces (strains) and the internal stresses produced are analyzed by translating the stress fringes or isochromatics into stress magnitudes. An ***isochromatic*** appearing as a dark band (see Figure 9-24) represents a region or locus of constant difference in principal stress. A pattern of alternate bright and dark lines represents the variation in refraction produced in the model. A variation of this technique is to coat an actual structure with a photoelastic coating such as Photostress. When the part is subjected to loads, the strains are transmitted to the coating, which then becomes birefringent.

Architectural flat glass for applications such as glazing in windows and doors has good optical qualities, strength, and weathering resistance. Additional requirements imposed on flat glass mandate sophisticated industrial treatments that do not involve changing the composition of the glass. Some of these additional requirements include heat reduction, light and sound protection, and increased resistance to breakage. One example of such treatment is the atoms-thick coating, known as low-emissive coating, which provides an emissivity of approximately 0.14 (83% better than a normal glass surface). (See Sect 7.1.7 for definition of emissivity). This low-emissive coating inhibits the absorptivity of the glass surface, resulting in more of the long-wave heat energy being reflected (see Figure 9-17). All such glass must meet rigid ASTM test standards E773-83, E774-81, and E6,P-1. To determine whether a window glass is low emissive requires a special detector available from window contractors and utility companies.

9.4.6 Photoelasticity

Some materials, however, will transmit light only in a particular direction. If light waves such as those in Figure 9-22 were directed into such material, the resulting emerging beam would have vibrations (oscillations) in one plane only, as shown in Figure 9-23. Other materials rotate the plane of polarized light. If two pieces of such material are arranged so that their polarized planes are perpendicular, no light would emerge. Such polymeric materials experience this rotation as a result of the alignment of their macromolecules due to internal stress. This effect is known as *photoelasticity.*

9.5 ELECTROCERAMICS

Highly conducting electroceramics are comprised of inorganic single-crystal and polycrystalline compounds as well as glasses that can be used in a variety of electrical, optical, and magnetic applications. Oxide ceramics are the materials of primary interest, but carbides, fluorides, and sulfides are also finding many commercial applications. Additional information on similar ceramic materials can be found in Module 7.

Such ceramics are generally classified as electronic conductors, ionic conductors, mixed (electronic/ionic) conductors, and insulators (Figure 9-25a). The electronic conductors include superconductors, metallic and ionic conductors, and semiconductors. This grouping represents materials with the highest conductivities. The metallic conductors exhibit high conductivities that decrease with increasing temperature while the conductivities of oxide semiconductors increase with temperature. Ionic conductors generally exhibit conductivities in the range 10^{-8} to 100 S/m that increase exponentially with temperature. The units of conductivity used in this discussion and the accompanying figure are expressed using the special SI unit name *siemens,* with symbol S (refer to Table 10-5). A siemens is defined as the quotient of amperes divided by volts (A/V). Insulators such as high-purity alumina are at the lower extreme of the conductivity spectrum, with conductivities of 10^{-13} S/m. Figure 9-25a summarizes the electrical conductivity characteristics of these classes of ceramic materials.

9.6 MICROELECTRONICS/SEMICONDUCTOR MATERIALS/ELECTRON BAND THEORY

Our discussion of energy levels in Module 2 dealt primarily with the action of electrons in a single atom. When atoms come together to form a solid material, the electrons in one atom come under the influence of other atoms. The energy levels that may be occupied by electrons merge or broaden into bands of energy levels. The *Pauli exclusion principle* states that only two electrons in the entire solid have the same energy. According to the energy band theory there are two distinct energy bands in which electrons may exist: (1) the valence band and (2) the conduction band. Separating these two bands is an energy gap in which no electrons can normally exist. The gap is known as the *forbidden gap.* A sketch of these bands is shown in Figure 9-25b.

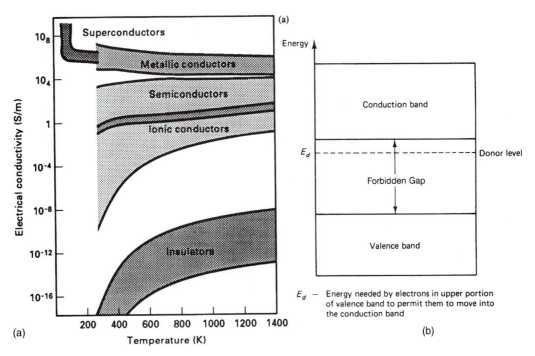

Figure 9-25   (a) Typical conductivity ranges for various classes of electroceramics as a function of temperature. (b) Electron energy band diagram showing the forbidden gap.

Electrons in the conduction band have escaped from the atomic forces that hold them to their nucleus and therefore they may move about within the material with the application of minimum energy. Electrons in the valence band, in normal orbits around their nucleus, are tightly held or restrained by the attractive forces originating in the nucleus. Much larger amounts of energy must be applied to extract an electron from or move it within this band. The basic differences among insulators, conductors, and semiconductors can be visualized using the energy band diagram sketched in Figure 9-25b. Conductors have no forbidden gap and the valence and conduction energy bands overlap. This permits large numbers of electrons to move (carry electrical current), even at extremely low temperatures. Metals have a partially filled valence band. Insulators have a wide forbidden gap with practically no electrons in the conduction band. Diamond has a large band gap. Thermal conductivity is carried out primarily by the phonon mechanism. Almost all electrons are in the valence band. Large amounts of energy (about 6 eV) are required to cause an electron to cross from the valence band to the conduction band. Finally, a semiconductor material has a narrow forbidden gap, which is more narrow than that of an insulator, with electron distribution similar to that of an insulator. However, since the forbidden gap is narrower, only small amounts of energy (0.785 eV for germanium or 1.2 eV for silicon) can raise electrons from the valence band to the conduction band. Semiconductor materials have a small band gap. Thermal conductivity is carried out by the free motion of electrons and phonons. Silicon as a semiconductive material, in addition to the above properties, allows the integration of both mechanical and electrical properties in one device. At this point it is worthwhile to review Figure 9-25a to observe any possible similarities with Figure 9-25b. The properties of semiconductor materials make possible the transistor, solar cells, tiny lasers, and the multitude of devices using integrated miniature circuits. This new technology has as its basis the development of human-made materials. Such materials permit the ***generation and precise control of electron motion within the confines of a tiny crystal of semiconductor material,*** to which has been purposely added (in most instances) a minute amount of impurity atoms forming a solid solution (Figure 9-26a). ***Transistors*** consist of a crystal of one type of doped semiconductor material sandwiched between two crystals of the

(a)

(b)

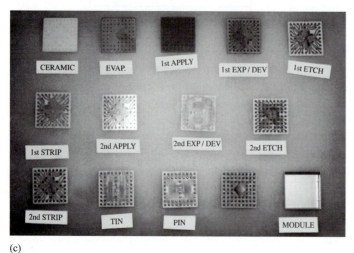

(c)

Figure 9-26 Semiconductors. (a) Silicon wafer (black disk) is cut into memory chips. Alumina (Al_2O_3) serves as the substrate to which metal circuitry is applied (left square) and memory chips are soldered to the circuitry (right). (b) Representation of the silicon-crystal lattice arrangement. (U.S. Dept. of Energy) (c) Stages for producing a memory module begin with ceramic (Al_2O_3) substrate then go through progressive stages to build up circuitry and insert connecting pins. Soldering of the memory chips is the last stage before the module is encased.

opposite type. They are the basic elements in modern electronic amplifiers, and also serve as switching devices for the processing and storage of information. Their new ***complementary metal oxide semiconductor (CMOS)*** provides more than 400 million transistors on a single chip, interconnected with up to seven layers of wiring. Semiconductors are ***microelectronic circuits.*** These circuits when interconnected are called ***integrated circuits.*** Very-large-scale integrated (VSLI) circuits, when placed on silicon, form a chip. A ***chip*** is rectangular in shape, typically 6 mm on a side (Figure 9-26c). In 1999, the maximum number of transistors on one chip was about 120 million with transistors 0.18 mm long. With knowledge of electrical properties of materials, it is somewhat easy to picture conditions that arise when transistor sizes are scaled down, interconnect wiring diameters shrink, and wiring is more closely packed. Resistance rises and capacitance increases to mention some changes with negative results for performance. TI hopes to offset these effects by replacing aluminum wiring with copper. Copper's resistance is one-third that of aluminum. If development proceeds as planned it will enable complete computer systems on a single chip with operating frequencies exceeding 1 GHz, and internal voltages as low as 1 V and below. The result will be extremely high performance combined with ultra-low power consumption and high design flexibility.

 EMT WEBLINK

Texas Instruments—*http://www.ti.com/sc/docs/news/1998/98079.html*

Pure semiconductor materials are referred to as ideal or **intrinsic** semiconducting materials. They possess a minimum of impurity atoms, certainly none added purposely. We recall that it is a very rare occurrence to find anywhere in a natural state a pure substance. Second, we have learned that impurity atoms can effect profound changes in the properties of materials. So it is with semiconducting materials. If selected impurity atoms in accurately measured amounts are added to certain pure semiconducting materials, highly significant changes in behavior will occur. These materials are now known as **extrinsic** semiconducting materials. An igniter made of a semiconductor material ignites gas appliances by heating up when electric current is passed through it. It takes only seven seconds to light an appliance. This development took nearly twenty years.

Intrinsic semiconducting materials use the elements in Group IV in the periodic table, germanium (Ge) and silicon (Si). These elements form crystal structures like that of carbon in diamond (Figure 9-26b). Each atom is bonded to its four neighboring atoms with covalent double bonds. There are no free electrons in the conduction band. How can these bound electrons be excited to act as charge carriers?

EMT WEBLINKS

Pure Element Semiconductors—*http://www-dmse.mit.edu/wom/ElementalSemiconductors.
    html*
Semiconductor Applets—*http://jas2.eng.buffalo.edu/applets/*

9.6.1 Electrically Charged Particles

Electricity is defined in terms of the flow, or movement, of electrically charged particles. This movement to other locations results in increased kinetic energy, which can be converted to some form of work for our benefit. Several forms of charged particles can satisfy these conditions, although some are not as efficient as others. The word *charged* refers to the electrical charge carried by these particles. All particles have this same charge, which is equal to 1.6×10^{-19} coulomb. This is the charge that all electrons carry. Its symbol is e^- and it is the standard unit of charge for measuring electrical charge. The magnitude of the charge in one proton is the same. However, the proton is a positive charge; its symbol is e^+. The measure of electric current (I) is the amount of charge (q) that passes a given point per unit of time (t):

$$\text{current } (I) = \frac{\text{charge } (q)}{\text{time } (t)} \qquad \text{or} \qquad I = \frac{q}{t}$$

An ampere (A) is defined as an electric current of 1 coulomb (C) of charge passing a point in 1 second (s). Knowing that each electron contains an electric charge of 1.6×10^{-19} C, and using a proportion to determine the number of electrons in 1 C of electric current, the results are 6.25×10^{18} electrons in 1 C. Therefore, 1 A of electric current is the passage of 6.25×10^{18} electrons in 1 s.

A *hole,* the absence of an electron, is considered as having the equivalent of an electron's charge only with a positive sign. As you recall, holes in the valence band are produced by supplying electrons with energy sufficient to cause them to be excited to the conduction band. Free electrons have greater mobility than holes. For comparison purposes, the mobility of electrons in pure germanium at 300 K is about 3950 cm^2 per volt second, and for holes it is about 1950.

9.6.2 Donor Doping

If a Group V element such as phosphorus (P), with its five valence electrons, is added to germanium (Ge), then only four outer electrons are needed to perfect a covalent bond with its neighboring germanium atoms to form a substitutional solid solution. The extra electron not needed for bonding will be attracted to the impurity atom (phosphorus) by a weak coulomb force of attraction of the nucleus. Because this electron is held much less tightly than the four

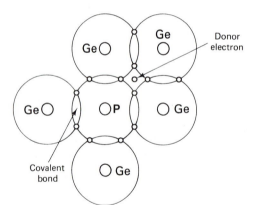

Figure 9-27 (a) Energy-band diagram of an *n*-type semiconductor. (b) Substitution of a phosphorus atom having five valence electrons for a silicon atom leaves an unbonded electron. (U.S. Dept. of Energy)

Figure 9-28 Effect of a dopant phosphorus atom substituting for a germanium atom in the crystal structure of an *n*-type semiconductor material.

bonding electrons, it and similar electrons contributed by other impurity atoms can be raised into the conduction band by the absorption of much smaller energies than those required for raising electrons from the valence band into the conduction band. This situation is depicted in Figure 9-27a and b. The extra electron is shown lying just below the conduction band at an extra energy level ((E_d) called the ***donor level***. This type of impurity atom makes the material *n*-type (possesses free electrons). The process of adding impurities is called ***doping***. Figure 9-28 shows a sketch of one phosphorus atom substituting for a germanium atom in the crystal lattice of an *n*-type germanium semiconductor material. Other donor impurities that can be used for donor doping from Group V are arsenic (As) and antimony (Sb). Note that these atoms, having five valence electrons, are referred to as *pentavalent* atoms. ***Diffusion*** was initially discussed in Section 3.4.8, and again when describing surface hardening processes such as carburizing and nitriding in Section 5.7. The diffusion of atoms and molecules across a boundary between two materials in contact is called ***mass diffusion.*** It occurs between two materials that share a common interface or boundary. In metals, the atomic motion is facilitated by vacancies and other imperfections in the crystal structure (See Sections 3.5 and 5.9.1.1). In semi-

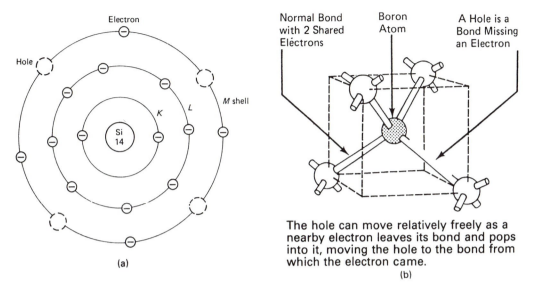

Figure 9-29 (a) Two-dimensional representation of the silicon (Si) atom. (b) Boron, a three-valence-electron atom in a silicon crystal, is normally bonded, except that one of the bonds is missing an electron, creating a hole. (U.S. Dept. of Energy)

conductor processing, diffusion is used to dope the silicon substrate with the desired amount of impurity atoms. The dopant enters the microstructure of the silicon substrate, substituting for the silicon atoms until a maximum concentration limit is reached. Temperatures around 1800°F accelerate the diffusion of dopant atoms.

9.6.3 Holes

From our study of the electronic structure of atoms, we have learned that any orbit of most atoms, other than the K shell or first shell (which is completely filled with two electrons), is stable if it contains eight electrons. Silicon (see Figure 9-29a) has only four electrons in its outermost shell (M shell). Being its outermost shell, it contains space for four additional electrons. These vacancies or absences of an electron in the valence shell are called *holes*. Holes are created in the valence band when valence electrons receive sufficient energy to move up into the conduction band. With intrinsic semiconductors, the valence electrons with the highest energies (outer-shell electrons have different energies) will be freed first when energy is added to the material as a result of an increase in temperature. The transfer of these electrons to the upper band creates holes in the lower (valence) band, which can then be occupied by electrons lower down or deeper in the valence band. When an electron receives an input of energy from some outside source, the electron will be loosened from its bonded position in an atom and be free to move about in the crystal structure (in the conduction band). If the energy received is insufficient to loosen the electron, it may just vibrate, which gives off heat. Once the electron breaks loose, it leaves behind a bond missing an electron. The bond missing an electron may also be called a hole (see Figure 9-29b). Electrons and holes freed from their positions in the crystal structure are said to be ***electron–hole pairs.*** A hole, like a free electron, is free to move about in the crystal (Figure 9-30).

9.6.4 Conduction in Solids

Electricity flow (conduction) occurs in a solid when an applied voltage (electric field) causes charge carriers within the solid to move in a desired direction. In the absence of an electric field, the movements of charge carriers are random and result in zero charge transport. If a metal is placed in an electric field (in an electrical circuit), the free electrons in the conduction band will move toward the positive terminal. As they do so, they receive additional momentum (mass × velocity) and hence more energy. Those electrons moving toward the negative terminal lose momentum and reduce their energy.

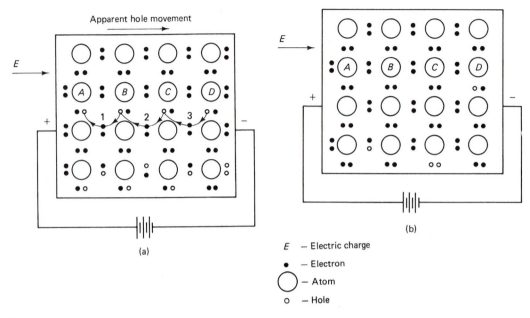

Figure 9-30 Conduction by holes in the valence band.

Hole transfer involving electrons takes place when an electron from one atom jumps to fill the hole in another atom. This electron jump leaves a hole behind it. Or we can describe this movement, not in terms of what the electron does, but in terms of the hole movement. The hole moves in the direction opposite to the electron. Therefore, the flow of electric current is brought about by the movement of free electrons in the conduction band and/or holes in the valence band. Free electrons are easier than holes to move through a semiconductor material, hence they have greater mobility (***drift velocity*** of the carrier).

With the help of Figure 9-30a, the movement of one hole can be clarified. The hole starts with atom *A*. As an electric field (*E*) is applied, a valence electron breaks free and moves to its left to fill the vacancy (hole) shown by arrow 1. The hole is now at atom *B*. A similar effect is felt by one of the valence electrons at atom *C,* which also moves to its left and fills the hole at atom *B*. Arrow 2 shows this movement. Each time an electron moves, it creates a hole. Figure 9-30b shows the final location of the hole at atom *D*. Note that the hole movement, in this instance, from left to right, is toward the negative-charged terminal of the battery, opposite to the flow of electrons as they move toward their left to fill the holes. In other words, this hole movement is a measurement of positive charge in the direction of the applied field (*E*). The action of the electrons in their moves, labeled 1, 2, and 3, in filling the just-created holes is known as *recombination,* which will be discussed later.

9.6.5 Acceptor Doping

When a Group III, or trivalent, element such as boron (B), aluminum (Al), gallium (Ga), or indium (In) is introduced as an impurity into an intrinsic semiconductor such as silicon (or other Group IV elements), a mismatch in the electronic-bonding structure occurs. Using aluminum as an example of the dopant, this element lacks one electron to satisfy the covalent bonds of the Group IV matrix element such as silicon. This mismatch is illustrated in Figures 9-31 and 9-29a. One covalent bond near each dopant atom (Al) is incomplete; that is, it is missing an electron, or it contains a hole. If an external electronic field is applied to this solid, one of the neighboring electrons from another covalent bond can acquire sufficient energy to move into the hole. This, in turn, causes the hole to move to the position formerly occupied by the electron. In this manner the hole moves through the solid as a positive-charge carrier. The electric

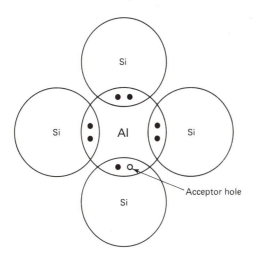

Al = Acceptor atom

Figure 9-31 Structure of silicon with aluminum atom added as an impurity (dopant).

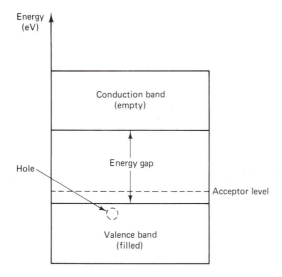

Figure 9-32 Energy-band diagram for a *p*-type semiextrinsic semiconductor material.

current produced is mainly the result of hole movement (positive-charge carrier). This structure is called *p*-type because of the presence of free positive charges (the moving holes).

In terms of the energy band diagram for this *p*-type structure (Figure 9-32), the aluminum atom has provided an energy level that is only slightly higher than the upper limit of the valence band. This puts it in the forbidden gap. Thus, a nearby valence electron could easily be excited into this intermediate level. This level is called an ***acceptor level.*** As with *n*-type extrinsic semiconductor materials, the extremely small concentration of dopant is measured in parts per million (ppm). In summary, the addition of acceptor atoms to a semiconductor material increases the number of holes in the valence band without an increase in the number of electrons. These positive-charge carriers are termed ***majority carriers,*** and the electrons are called ***minority carriers*** in *p*-type semiconductors, just the reverse of their designations in *n*-type semiconductors.

Figure 9-33 represents an *n*-type semiconductor material in a circuit with a switch in the open position. No electrons will flow in the external circuit. However, within the doped *n*-type semiconductor material, the free electrons contributed by the dopant atoms and the electrons that break away from their parent atoms in the valence band are diffusing throughout the material in a random fashion. In addition, each time a valence electron breaks free of its atom, an

Si	— Silicon atom
P	— Phosphorus donor atom
•	— Electron
○	— Hole

Figure 9-33 Random movement of electrons and holes in an *n*-type semiconductor material.

electron deficiency in the valence shell of that atom is created, which makes the atom a positive ion with an electric charge of equal magnitude to that of the free electron. The vacancy in the valence shell (energy band) of the electronic structure of the atom is called a hole. The phrase *electron–hole pair* refers to this energy transfer process. When a free electron collides with a hole, it is captured by it, which produces a balanced atom in its lowest equilibrium state. This process is known as ***recombination.*** Taking into consideration the remaining atoms that comprise the material, we recognize that overall equilibrium must be maintained, and thus a new electron–hole pair is generated with every recombination (Figure 9-34).

Next we close the switch in the external circuit in Figure 9-33. This causes electron flow to occur from the negative terminal of the source through the semiconductor material and back to the positive terminal of the source. Figure 9-33 attempts to show that only electrons are flowing in the external circuit within the material. The action of the electric field gives direction and movement (drift) to the flow of both free electrons donated by the phosphorus atoms in this instance and the holes (minority carriers) in their movement from right to left. Both the free electrons donated by the donor atoms and the electrons supplied through the external circuit combine with the holes to accomplish the conduction process through the processes of generation and recombination.

When a piece of *n*-type material is brought into contact with a piece of *p*-type material, a **pn** *junction* is formed (Figure 9-35). A junction of this type produces a ***diode,*** which is the simplest of the semiconductor-junction devices. Before observing the diode in a simple circuit, it would be helpful to study the interaction of the electrons at the *pn* junction.

The *n*-type material contains an excess of electrons, whereas the *p*-type material has an excess of holes. When the *pn* junction is formed, there is a diffusion of charges between the two materials. The electrons from the *n*-type will move across the junction to the *p*-type. This will create holes or positively charged particles in the *n*-type. The reason for this diffusion of charges is to counteract the electrical imbalance in the two materials caused by the formation of the junction. The diffusion process will stop once an equilibrium of charge is established between the two materials. This equilibrium is reached when no additional charge carriers have the energy to overcome the electric field that has been built up at the junction. It can be seen in Figure 9-36a that a boundary is formed across the *pn* junction. This boundary is usually re-

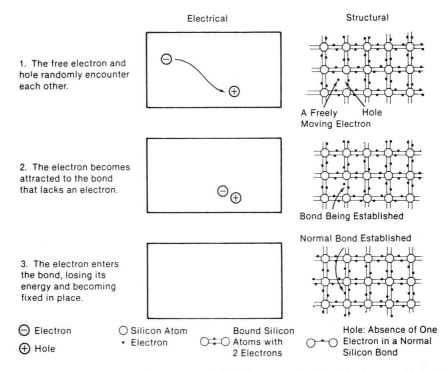

| | Electrical | Structural |

1. The free electron and hole randomly encounter each other.

2. The electron becomes attracted to the bond that lacks an electron.

3. The electron enters the bond, losing its energy and becoming fixed in place.

A Freely Moving Electron   Hole

Bond Being Established

Normal Bond Established

⊖ Electron   ○ Silicon Atom   Bound Silicon   Hole: Absence of One
⊕ Hole   • Electron   Atoms with   Electron in a Normal
                                            2 Electrons   Silicon Bond

Figure 9-34   Direct recombination of an electron and a hole. Excess energies of both the electron and hole are lost to heat. (U.S. Dept. of Energy)

ferred to as the ***depletion region*** because all mobile charge carriers have been depleted from this area.

The electric field set up by the stationary charges in the depletion region creates a difference of potential between the two materials. This electric field represents the equilibrium potential difference from *n*-type to *p*-type in direction. This voltage is called the contact potential of the junction. For silicon, it is about 0.6 V. All the previous discussion has been aimed at the *pn* junction (a diode) as a closed system. Next the junction will be observed with an externally applied voltage (Figure 9-36).

The *p*-type side of the diode is called the anode and the *n*-type side is called the cathode. Applying a voltage to the diode is called ***biasing.***

If, as in Figure 9-36, a battery is connected across the diode so that the positive lead is attached to the anode and the negative lead is attached to the cathode, a large current will flow. This is the forward-biased condition of the diode. The reason for the large current is that the junction potential (contact potential and applied potential) has been lowered due to the forward bias. Electrons from the battery that enter the anode of the device will see little resistance and can flow easily across the *narrowed* depletion region to the cathode.

In Figure 9-36, the leads of the battery have been reversed so that the positive side of the battery is across the cathode and the negative side is across the anode. This is the *reverse-biased* condition of the diode, and very little (only leakage current) will flow. The reason for the extremely small current in this condition is that the battery is acting in the same direction as the electric field across the junction; thus, the junction potential is larger. When the junction potential increases, the width of the depletion region and the amount of stationary charges will increase. This means that the charge carriers must gain an extremely large amount of energy before they can cross from cathode to anode.

The *pn* junction is the basic building block for all semiconductor-junction devices. The preceding discussion can be applied to the transistor, silicon-controlled rectifier (SCR), field-effect transistor (FET), or any other junction device.

Figure 9-35 Junction formation. Additional movement of charge carriers is stopped and equilibrium is established. (U.S. Dept. of Energy)

 EMT WEBLINK

PN Junction Diode at Equilibrium—*http://jas2.eng.buffalo.edu/applets/education/pn/ pnformation/pnformation.html*

9.6.6 COMPUTER CHIP TECHNOLOGY

Semiconductors and their properties form the basis for the design, production, and application of electronic microscopic devices. As each year passes, these devices are produced

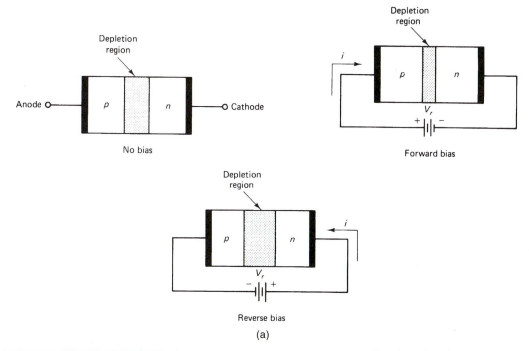

No bias

Forward bias

Reverse bias

(a)

(b)

Figure 9-36 (a) A *pn* junction (diode) showing no-bias, forward-bias, and reverse-bias conditions. (b) Time-of-flight secondary ion mass spectrometer (TOF-SIMS) is a high performance technique that uses ion beams to analyze the surface of solid materials. It is capable of shallow depth sputter depth profiling. (Applied Research Center, Newport News, Virginia) (*continued*)

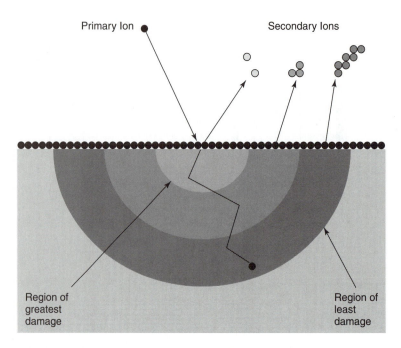

Figure 9-36   (c) The primary ion impact on the surface causes the emission of secondary ions which are extracted into the TOF analyzer. Roughly less than one in one thousand surface atoms or molecules are struck by a primary ion. (Applied Research Center, Newport News, Virginia)

smaller, such that by the year 2001 the average size will be 180 nm. Examples of such microelectronic devices are the *pn* junction diode and the *npn* bipolar junction transistor. These devices and many others are linked to fabricate microelectronic circuits known as ***integrated circuits (ICs)*** or ***chips.***

A simple IC can be an amplifier and a complex IC a complete digital computer (or microprocessor). These devices must be connected with paths that can readily conduct electric currents. In most cases, metal wires are used. These wires are made by filling tiny trenches in the surface of semiconductor wafers. Multiple layers are created by penetrating a layer to make contacts with layers above and below. The metal used for these wires has been aluminum and for interlayer connections titanium. These wires require connections to other wires using soldered joints. As the size of these devices becomes smaller, the numerous tiny, very fine, tightly packed connections become more difficult to solder. The failure of one joint only a few thousandths of an inch wide can cause failure of the entire device. A computer model has been developed that predicts when solder joints are likely to fail by calculating their precise shapes. As these devices get smaller, copper is succeeding aluminum because of its greater conductivity among other properties. A new method of laying copper wire for ICs is called the ***ion-assisted trench*** method. This technique can produce thin and uniform layers which can conform to the shape of the patterned wafer. As is often the case with other materials combinations, however, copper is not compatible with silicon: It readily diffuses in silicon but causes defects in the silicon structure. Silicon is still the favored semiconductor material. Some of its advantages are lower material cost, larger band gap permitting operations at high temperatures, integration of mechanical and electrical properties in one device, and an excellent natural oxide with a remarkable set of properties.

The fabrication of ICs begins with a thin piece of ***silicon*** (see Figure 9-26a), an intrinsic semiconductor on which well-defined surface areas are processed to contain certain amounts of electrically active dopants. These areas are created by first removing the SiO_2 surface layer. Next the dopant is added which thermally diffuses into the substrate. The SiO_2 blocks the diffusion of many types of dopants. In addition, SiO_2 is an excellent electrical insulator which is a great benefit in cases when a layer of insulating materials is required such as in the making of a capacitor (two conducting layers separated by a thin layer of insulating material). Thus, SiO_2 is the primary choice for most microelectronic materials. Gallium arsenide (GaAs) is preferred for high-speed and high-frequency applications because of its greater electron mobility. Silicon carbide (SiC), with its high-temperature stability (can withstand temperatures up to 325°C) and its resistance to chemical attack, is used in aggressive environments. Its one disadvantage involves inexpensive simple crystals, which are difficult to grow. Another new alloy is made of silicon, carbon, and germanium for use with high-temperature, high-power, and high-frequency *microelectronics* and *microelectromechanical machines (MEMs).* Germanium (Ge) permits higher speeds than Si, but it is difficult to add much Ge to Si because the Ge atoms are larger than Si atoms and consequently strain the Si lattice structure. Greater amounts of Ge can be added by the addition of 1–2% carbon. By switching from Si to SiC, additional Ge can be used (up to 4%) to produce twice the current as pure SiC. IBM's new technology, *"silicon-on-insulator (SOL)"* for chip manufacture, involving some 15 years of effort, was announced in August 1998. The new process deposits a layer of dielectric on a silicon wafer. Giving better protection to the metal circuitry allows for closer placement of transistors and more efficient flow of electrical current. Combined with another advance in chip manufacture that replaces aluminum with copper, these new advances in the research, development, and applications of materials is reported to boost performance 35%, cut power usage by one-third, and increase the service life of silicon-based chips.

The metallization process can deposit many layers of metal, each insulated by a dielectric. The layers of metal are connected by *vias.* Contacts provide access to the devices on the substrate. The width of this new circuitry is reported to be 1/400 the width of human hair. Human hair varies in diameter from 40 to 120 μm. Expressing this width in micrometers and using the smallest value for the width of a human hair, this new width would be about 0.1 μm. Knowing that the flow of electrical current, no matter how small, generates resistance which manifests itself in the form of heat, the constant challenges to electronic chip manufacturers is to decrease this resistance and thus (1) reduce the amount of heat to be dissipated, (2) reduce the amount of dielectric needed to insulate the individual layers of the circuit, (3) increase the *chip speed* (electron flow through the circuits), (4) decrease power consumption so that smaller battery-type energy sources may be used, and (5) allow for the greater placement of more smaller-size semiconductor devices such as transistors in smaller and smaller width circuits.

The new *copper-chip technology* that is replacing aluminum wiring that connects transistors increases processor speeds and decreases power consumption. Power PC 740 and 750 processors in workstations now contain this technology with PC personal computers next in line. Present cell phones use 200 to 250 chips to send and receive calls, commonly on one frequency (900, 1800, or 1900 MHZ) and one encoding format (GSM, CDMA, or TDMA). This means the traveling public would need several phones to maintain communications in different parts of the world. IBM has developed a new chip technology using silicon germanium (SiGe) that allows future cell phones to use multiple frequencies and encoding formats. This integration is brought about because SiGe is compatible with the other cellular components which act on the digital signal and reproduce the sound of the caller's voice. Future applications in personal digital assistants, watches, and jewelry will be possible with reduced power consumption, and longer operations capability with more densely packed transistor cell phones, using this new technology.

IC chips are attached to substrates (*electronic package casings*) so that they may be safely attached to structures or supported for shipment. Normally these two materials are not compatible with each other in that they differ in properties such as thermal conductivity and thermal expansion. Thermal expansion is expressed by use of the coefficient of thermal expansion (COT). Table 7-2 shows a comparison of some metals with a variety of ceramics. Silicon has a COT of 2.6×10^{-6}°C^{-1} whereas aluminum has a COT of 23.6×10^{-6} °C^{-1}. Materials with such a wide difference in COTs, if attached to one another, will experience a buildup of thermal-induced stresses as temperatures change. This condition could lead, in the case of ICs, to warping and more likely a loss of adhesion between the layers of conductive metal and dielectric ceramics or polymer materials of which most are composed. One alloy of iron–nickel–cobalt called *Kovar* has a coefficient of 5×10^{-6}°C^{-1} which is a better match for silicon chip mounting. Kovar is the registered trademark of Carpenter Technology Corp. for its nominal 29Ni–17Co–54Fe glass-sealing alloy UNS K94610.

9.7 OPTOELECTRONIC DEVICES

Optoelectronic devices are (1) operated by light (photoelectric), (2) produce or emit light, or (3) modify light. Photoelectric devices, in turn, can be categorized as photoconductive, photovoltaic, or photoemissive. A review of optical properties in Section 9.4 and of semiconductor materials in Section 9.6 is appropriate prior to studying the following treatment of these devices. **Spectroscopes,** or **spectrometers** measure accurately the wavelength of light emitted by atoms after they have been excited by the use of diffraction gratings. A **diffraction grating** is a large number of equally spaced slits. Gratings can be made by precision machining of very fine parallel lines on a glass plate. The untouched spaces between the lines serve as the slits from which light is reflected and analyzed. Another use of these devices is for the identification of atoms and molecules whose chemical analysis would be difficult. Spectroscopy is initially discussed in Section 3.2.1. **Time-of-flight Secondary ion Mass Spectrometry (TOF-SIMS)** is an analytical technique that uses a primary ion beam to probe the surface of a solid material. (Figures 9-36b) The secondary ions which desorb from the sample surface are analyzed and their mass determined with high accuracy (Figure 9-36c). As a result, the elemental and molecular chemical information can be obtained from both insulating and conducting materials. TOF-SIMS provides spectroscopy for characterization of chemical composition, imaging for determining the distribution of chemical species, and depth profiling for thin film characterization. In the spectroscopy and imaging modes, only the outermost (1-2) atomic layers of the sample are analyzed. For each primary ion pulse, a full mass spectrum is obtained by measuring the arrival times of the secondary ions at the detector and performing a simple time to mass conversion. Chemical images are generated by collecting a mass spectrum as the primary ion beam is rastered across the sample surface.

9.7.1 Photoconductive Devices

In photoconductive devices, the conductivity of the semiconductor material will vary, provided that the energy supplied by the light (visible, infrared, or ultraviolet) is sufficient to raise the electrons into the conduction band. A photomultiplier is such a device. It can produce an image that is visible when illuminated by a weak light source. A light meter is another example that operates in accordance with the same basic principle. Gallium arsenide (GaAs), a compound, can convert light directly into electricity. This example of *photoconductivity* is used in LEDs and other solid-state devices such as transistors. Photoconductivity is also discussed in Section 9.4.4.

9.7.2 Photovoltaic Devices

Photovoltaic (PV) cells are semiconductor-junction devices that convert electromagnetic energy in the form of light directly into electrical energy. The amount of electrical current (flow of electrons) is directly proportional to the amount of light that is incident on the semiconductor material. A *solar* or *photovoltaic cell* is a photodiode used to extract electrical energy directly from sunlight. This direct conversion of sunlight to electricity differs from the solar thermal conversion process used in solar heating of homes and offices, which uses panels to absorb sunlight and uses the energy to heat water or other media, including air for heating a building. Photovoltaic cells embedded in the glass of a sunroof for an automobile are a recent application for these devices. They help keep the interior cool. When temperatures inside the vehicle exceed 85°F, the cells power two fans that pull in cooler outside air and exhaust hot interior air through vents in the roof. If it starts to rain, sensors shut the vents.

A typical solar cell (Figure 9-37a) contains two extremely thin layers of silicon connected externally by a wire to the load where light energy is converted to work. When light is

Figure 9-37 (a) A typical *pn* junction, single-crystal, silicon solar cell, outlining its components. (b) Light striking a cell creates electron–hole pairs that are separated by the potential barrier, creating a voltage that drives a current through an external circuit. (c) What happens to light entering a cell? (d) Schematic of a photovoltaic system. (U.S. Dept. of Energy)

Figure 9-37 (e) Silicon solar cell on left; powers motor on right.

absorbed in silicon, it creates electron–hole pairs (Figure 9-37b). Electron–hole pairs that reach the junction are separated—holes going to the *p* side and electrons to the *n* side. This buildup of charge on either side of the junction creates a voltage that drives current through the external circuit. Reflection of sunlight from an untreated silicon cell can be as great as 30%. By chemical treatments and texturing of the surface, this value can be reduced to around 5%. Light with a certain wavelength (energy) is required to interact with specific materials before

(f)

Figure 9-37   (f) The Forever Light flashlight, powered by a solar cell needs no batteries.

the optimum electron–hole generation is achieved (Figure 9-37c). This characteristic energy is called the material's band energy. (Silicon requires 1.1 eV, gallium arsenide, 1.4 eV.) Overall, these mismatches of light with a solar cell's material waste some 55% of the energy from sunlight. Research is proceeding to find better ways to process sunlight to make it monochromatic (one wavelength) and possess an exact energy match with that required by the cell material to make the conversion of incoming sunlight more efficient. (See Figs. 9-37 e and f).

Theoretically, the silicon solar cell should be able to convert about 25% of the sun's energy into electricity. To achieve greater conversion of sunlight to electricity, two different solar cells can be used, one of aluminum–gallium–arsenide and the other of silicon. These two cells absorb a wider range of wavelengths from the solar spectrum, but various optical losses reduce the overall system efficiency to about 25%. The term *efficiency* is the percentage of energy in sunlight striking the cell that is converted into electricity. Individual solar cells have limited power. To produce electricity for most applications, they must be joined together electrically to form modules that become building blocks for larger arrays (Figure 9-37d). Superthin solar cells with a semiconductor layer thinner than a human hair and 1/100 the thickness of silicon solar cells have been developed in the laboratory. Traditional cells require relatively large amounts of semiconductor material which drives up their cost. These new cells use but a penny's worth of material. The new semiconductor material is made of copper indium diselenide which is deposited only 2 to 3 μm thick on a material such as glass. These cells can reach conversion efficiencies (sunlight to electricity) of 18% which rivals the best silicon cells. A typical problem involves the inability to manufacture the cells. The technology must be developed to pass on to industry the know-all about how to produce them.

Solar cells are very expensive primarily because of their fabrication costs. Many breakthroughs in materials technology are needed before solar cells can be developed to compete on the mass market and produce electricity costing about 50 cents per peak watt or about 6 to 8 cents per kilowatthour. Ultrapure (99.999% purity) semiconductor-grade silicon costs around $15 a pound, compared to $0.15 a pound for metallurgical silicon, with a purity of 99.5%. Approximately 30% of the cost per peak watt of solar-produced electricity represents this processing cost. *Peak watt* is a term used to express the amount of power produced by a solar cell in full sunlight at 25°C. A 4-in. cell, for example, can produce 1 peak watt at noon on a sunny day in the U.S. Southwest. This is a description of ideal conditions. A more relevant term is *average wattage,* which is about one-fifth the peak wattage.

Invented at Bell Laboratories in 1954 with a 4% efficiency, most common solar cells are made from a single crystal of pure silicon grown artificially in the form of an ingot that is sliced into wafers 0.012 in. thick, which are then polished, trimmed, and doped in an oven. The finished cells are mounted in arrays or panels containing dozens of individual cells whose diameters may range up to 4 in. One manufacturing innovation produces cells continuously in the form of thin ribbons. This thin-film technology, similar to that used in making electronic integrated circuits, could cut manufacturing costs substantially and permit the fabrication of large-area cells that would make solar cells more cost effective. (See Figs. 9-37e and f).

Today this Bell Labs process, referred to as the crystalline process, uses scrap silicon from the semiconductor industry. The silicon is melted at 1450°C and poured into 4.5-in. square ceramic molds to form a foot-long silicon crystal ingot. This ingot is sliced into 300-micrometer-wide cells which are about the width of a piece of construction paper. Thirty-six cells are soldered together into a standard 1.5-ft × 3.5-ft module, wires and electrical leads are attached to connect individual cells, and the module is placed between two pieces of glass with a plastic backing and installed into bulky frames for shipment to customers.

A **thin-film deposition** process, now called the continuous process, uses 12 silicon gas deposition chambers through which single sheets of 2 × 4-ft glass panels laser etched with conducting filaments are coated with a photovoltaic silicon layer that has a thickness of 0.5 μm. Outside layers of glass are added for protection. This process uses less materials and less labor and has wider commercial applications than the older process; however, comparative-size crystalline cells are still more efficient in converting sunlight into electricity. In spite of this loss in efficiency, the new thin-film cells cost from 10–15% less than cells produced by the crystalline process.

Bulky and less-efficient crystalline and amorphous Si thin films are becoming obsolete with the development of two new thin-film photovoltaic (PV) materials. These materials make it possible to attain efficiencies of 10% or more. Cadmium telluride and copper indium diselenide both are more efficient and less costly than current materials. The latter material is being used in the production of PV roof tiles. The highest efficiency rating for any mass-produced solar cell is held by the HIT Power 21 cell produced by Sanyo. With an efficiency of 15.2%, this cell is made of a sandwich with a single crystal layer between thin amorphous silicon layers. The combination improves the flow of electrons through the cell, cutting internal energy loss and boosting the conversion efficiency.

The U.S. government has spent some $1.5 billion of research money for the solar industry over the last 25 years and is taking steps to keep some of the benefits at home. **The Million Solar Roofs Initiative** announced in the summer of 1998 has the goal to put 1 million solar systems on the roofs of commercial and residential buildings by 2010. This goal includes both photovoltaic units and thermal systems for space heating and hot water. Another drawback to more use of PV devices is the conversion of the electricity generated from direct current to alternating current. Another technological advance is the **microinverter,** which mounts on the back of PV panels and allows the panels to generate alternating current directly at the roof. As with all innovative technological advances, with greater use comes lower prices which permits them to be more cost effective for residential use. New PV panels featuring built-in microinverters are being developed independently by Advanced Energy System and Ascention Technology. The small 250-watt microinverters allow the panels to act as their own AC generators, thus eliminating the need for DC wiring and an in-home inverter. Homeowners can now start with a small system and add on since the microinverters are built into each panel.

9.7.3 Photoemissive Devices

In a photoemissive device, the light (or photons) generated by the recombination of electron–hole pairs is emitted from the surface of the device. Electrons striking the television screen, which is coated with a semiconductor material doped with copper, raise the electrons in the coating into the conduction band, where they recombine and emit energy in the form of light (photons). This is known as **electroluminescence.**

Fluorescent lamps depend on ultraviolet light striking electrons in the inner coating of the lamps, which emit light in the visible region of the electromagnetic spectrum. To round out our coverage of these devices, a light-emitting diode is treated in more detail below. It is important to repeat at this point that, regardless of the device, the underlying mechanisms involve the mutual interaction of electrons with electromagnetic radiation.

9.7.3.1 Light-emitting diode.   A light-emitting diode (LED) is essentially a sandwich of conductive materials with electrodes attached at the top and bottom. A more technical definition would say that an LED is a device that luminesces visible light (see definition of

luminescence in Section 9.4.1). Figure 9-38 shows a cross-section of an LED. Passing an electric current through the material forces electrons and positive-charge carriers (holes) to combine in the middle of the sandwich to emit photons of light. When forward bias is placed across the junction, electrons cross from the *n* side of the junction to the *p* side and recombine with the holes giving off energy in the form of heat and light. The light will be emitted, assuming that the semiconductor materials are translucent and the gold-film cathode effectively reflects the light to the surface. Gallium phosphide (GaP) and gallium arsenide phosphide (GaAsP) are typical semiconductor materials used in LEDs to produce red, yellow, or green light. The color of the light depends on the characteristics of the materials that comprise the sandwich. Gallium nitride is the new material that makes up the thin film that produces bright blue light. The thin films are deposited by a novel metal–organic chemical vapor deposition (CVD) process. These photoconductive materials (see photoconduction in Section 9.4) must be formed into a thin film atop of a substrate. Present-day applications of LEDs can be seen as the tiny "on" lights on computers, stereos, and VCRs.

The invention of the blue laser late in 1999 will bring about further uses for LEDs. The bright blue room-temperature LED licensed from Nichia Chemical Industries is the first LED that can achieve a 10,000-hour life (deemed necessary for a commercial product). The present output is 5 milliwatts which is enough for read-only applications. For read-write applications, 30 williwatts are required. These newly developed LEDs find most of their use in digital displays. New blue and green LEDs are now the leading choice for scoreboards and stadium displays. Their next applications will be in emergency lighting, flashlights, and traffic lights. The new future goal of researchers is to produce highly efficient LEDs that will replace incandescent lighting. Consumers quickly adopt products made with optoelectronic devices which propels designers and manufacturers to create more advanced products. As a result technology advances often providing improvements at reduced prices. The educational toy seen in Figure 9-38b typifies these advances. For decades high density polyethylene, LEGO bricks offered children an inexpensive toy which had merit in teaching manipulative and problem solving skills. Now with the addition of an optical detector and microcomputer that couples with miniature motors, children can learn to assemble robotic toys and create a wide range of configurations that employ gears, belts, wheels and many new bricks.

9.7.4 Silicon Technology

Silicon, the second most abundant element in the earth's crust, exists as silica or silicon dioxide (SiO_2). The processing of quartzite, a mineral (almost 99% silica) and the source of pure silicon, raises the cost of silicon to about $70 per kilogram. Pure polycrystalline silicon is further processed into single-crystal or large-grain (5 mm) polycrystalline silicon (Figure 9-39a). The single-crystal silicon technology, the oldest and most well established, is slowly giving way to newer technologies using polycrystalline and amorphous silicon.

The latest fabrication techniques form silicon directly into usable wafers (0.5 mm thick) without intermediate steps such as sawing, thus reducing processing costs considerably. More advanced methods produce ribbons (long ultrathin, rectangular sheets). One such method produces ribbons 33 m long, 5 cm wide, and 0.5 mm thick at a linear rate of 55 cm/min, with cell efficiencies of over 10% (Figure 9-39b). *Polycrystalline silicon* is cheaper to produce and fabricate than single-crystal silicon, but at present single-crystal cells have maintained their lead in efficiency. In polycrystalline silicon, the grain boundaries impose numerous restraints on the movement of charge carriers (Figure 9-39a). However, with the constant growth of new technologies, the effect of the grain boundaries is being reduced.

A third form of silicon, *amorphous silicon* (abbreviated aSi), has been used since 1974 to produce cells with high output voltage (0.8 V), currents greater than 10 mA/cm^2, and efficiencies of 6% using *p-i-n* cells (*p*-type, intrinsic, and *n*-type layers). Figure 9-39c is a schematic of this cell, pointing out the ultrathinness of the layers of amorphous silicon. To overcome amorphous silicon's disordered atomic structure (Figure 9-39d) with its incomplete bonds, hydrogen is added (doped) to complete this bonding and increase the cells' efficiency.

(a)

(b)

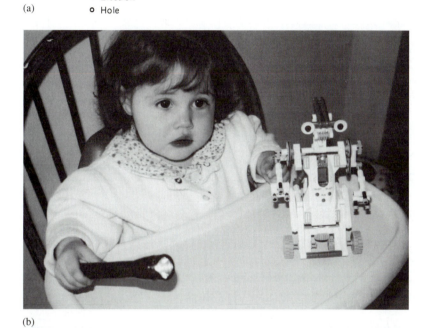

(c)

Figure 9-38 (a) Cross-section of an LED. (b) Camille is attracted to the colorful, complex Micro Scout, a smart LEGO brick. (c) The device can be programmed with push buttons on the brick then control the robot with a flashlight.

Figure 9-39   (a) Lifetime of free-charge carriers dependent on grain size and orientation. (b) Low-angle silicon sheet (LASS) growth is a fast method of drawing a silicon ribbon from a shallow trough of molten silicon. (c) A *p-i-n* device. An electric field sweeps the charge carriers to opposite ends of the cell. (d) Silicon atoms build a three-dimensional tetrahedral structure that, in amorphous silicon, is rotated randomly, producing a disordered atomic structure. (U.S. Dept. of Energy)

A distinct advantage of amorphous silicon is its ability to absorb light about 40 times more strongly than crystalline silicon. Amorphous silicon cells are used to power handheld calculators because they are more efficient and cost-effective under fluorescent light than either single-crystal or polycrystalline silicon. This characteristic of amorphous silicon allows such cells to be extremely thin, 1/40 of the thickness required to absorb the same light. The low mobility of the charge carriers provides relatively rapid recombination, which means that the charge carriers must be separated by the *p-i-n* junction in the short time during illumination. In addition, thinness translates into less material, and depositions of the material tend to be easier.

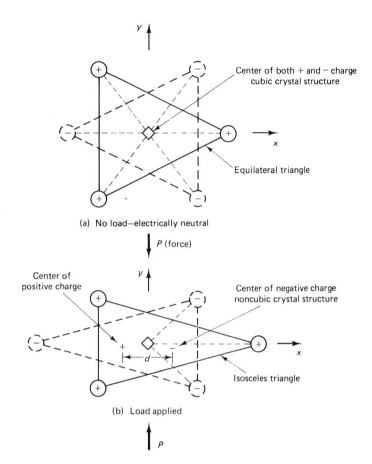

Figure 9-40   Model of piezoelectric crystal structure.

9.8 PIEZOELECTRIC MATERIALS

Piezoelectric materials, first discovered by the Curies in 1880, were used in crystal radios as slices of quartz crystal to select (receive or block) different frequencies and thus tune the radio. Now they find widespread use in such instruments as speakers, microphones, and sensors for wheel balancers, alarms, pagers, motors, transformers, ignitors, ink-jet printers, fish finders, watches, ultrasonic NDTs, and automobile airbags. By controlling the size and electrical input to the crystal, a range of audible sounds can be produced by the frequency of the vibrations. Applications are numerous. Piezoelectric noise sources are used in tops, telephones, noise alarms, seat belt buzzers, and pagers.

Piezoelectric materials can be explained most easily through a discussion of dielectrics with an emphasis on polarization. Piezoelectric crystals are physically uniform solids that are bonded together by ionic bonds. We have learned that ionic bonding is a result of the electrostatic forces of attraction between the ions of opposite charge. Normally, the number of positive-charged ions equals the number of negative-charged ions. Figure 9-40a is an attempt to show the symmetric, crystalline structure of a crystal using three representative positive ions located equidistant from each other (at the vertices of an equilateral triangle). Similarly located are three negative-charged ions. The centers of charge both occupy a common center of symmetry. Being of equal magnitude, these centers of charge not only coincide in location, but they cancel each other, leaving the electric charge of these six ions neutralized.

Figure 9-40b is a similar representation of the same structure of a crystal with the addition of a mechanical load or force (P) that produces a stress assumed to be uniformly distributed over the face of the crystal on which it acts. This force, in this instance, acting along the

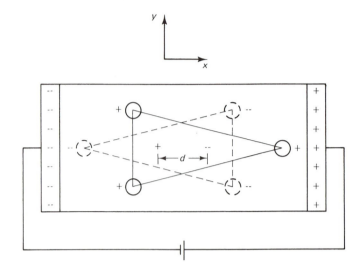

Figure 9-41   An electric field across a piezoelectric crystal.

y or mechanical axis, tends to compress the crystal, resulting in a compressive, elastic strain or unit deformation. The elastic deformation, in turn, results in the offsetting of the centers of electric charge from each other. In effect, this offsetting of centers of charge creates electric dipoles throughout the material, which combine to produce a measurable electric potential along the x or electric axis of the crystal. Conversely, if an electric potential were to be applied to the x-axis, a detectable deformation would be observed along the y-axis. The prefix *piezo-* comes from the Greek word meaning "to sit on, or press." The word ***piezoelectricity*** refers to electricity generated by exerting pressure on an ionic crystal.

Figure 9-41 is a model of such a crystal with an electric field impressed along the x-axis. This particular arrangement of electric charge acting as the input of energy would produce a mechanical deformation (contraction) similar to that of the force (P) in Figure 9-40. Note that the ions of negative charge are attracted to the right toward the positive terminal, with a similar movement of the positive ions and their center of charge to the left, thus increasing the separation of the centers of change (increasing the dipole length). If the polarity of the terminal were reversed, the deformation produced would be distorted in the opposite direction (expansion). An ac voltage, when impressed on the crystal, would cause the crystal to expand and contract (oscillate) at its driven frequency (maximum amplitude at resonance), which could be transmitted as a wave of sound energy into the surrounding medium. In other words, if this potential were produced by an alternating current, it would cause the crystal to vibrate. Each crystal has its own natural, mechanical frequency and the ac potential could be adjusted to match this frequency and produce resonance (maximum amplitude of vibration). Ultrasound applications in medicine involve high-frequency sound to create an echo image. ***Sonar*** (sound navigation and ranging) requires low-frequency sound to travel long distances in water. ***Ceramic spinning*** technology is producing ***piezo-electric composites*** in acoustic components for the U.S. Navy. These new fibers make sonar more versatile, because they can be bent to conform to curved surfaces and thus produce a clearer signal.

In summary, if a mechanical stress is applied to two opposite ends of a crystal (mechanical energy supplied), the remaining ends of the crystal are charged with electricity (output in electrical energy); if an electric voltage is supplied to one set of faces of a crystal, the other two faces of the crystal are deformed (either they contract or they expand); and if an ac potential is supplied, the output is crystal oscillation at its driven frequency, which can be transmitted into a surrounding medium as sound energy at a constant wavelength. A piezoelectronic crystal is a ***transducer*** because it not only transfers energy from one system to another, but it converts energy from one form into another.

Quartz crystals (SiO_2), possessing piezoelectric properties, are both naturally occurring and grown commercially. In its uncut state, the crystal is in the form of a hexagonal prism, as

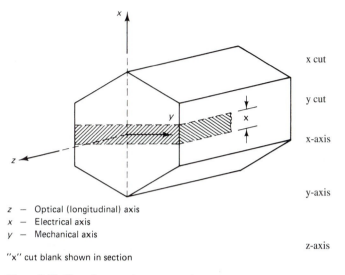

z — Optical (longitudinal) axis
x — Electrical axis
y — Mechanical axis

"x" cut blank shown in section

x cut	crystal cut with its flat sides perpendicular to the electrical axis (x-axis)
y cut	crystal cut with its flat sides perpendicular to the mechanical axis (y-axis)
x-axis	(electrical axis)—voltage appears on this axis when mechanical stress is applied to the y-axis (mechanical axis)
y-axis	(mechanical axis)—mechanical stress applied to this axis causes a voltage to be generated at the x-axis (electrical axis)
z-axis	(optical axis)—no piezoelectric effect is exhibited

Figure 9-42 Uncut hexagonal quartz crystal.

sketched in Figure 9-42. The x-axis passing through the corners of the hexagonal cross-section is the *electrical axis,* the y-axis perpendicular to the faces of the hexagonal cross-section is the *mechanical axis,* and the z-axis is the *optical axis.* The crystal is cut in a variety of ways, depending on the desired characteristics. The sections cut from the crystal are known as *blanks.* One such blank, sketched in Figure 9-42, is called an *x cut* and has a thickness that is parallel to the x-axis and a length parallel to the y-axis. A mechanical stress applied to its faces along the y-axis produces a voltage along the x-axis. The dipoles created in quartz crystals do not rotate; thus, there is a permanent orientation. This characteristic permits a quartz crystal, once cut and polished to specified dimensions, to maintain a resonance frequency with an extremely high degree of accuracy. This feature finds great use in controlling the frequency of radio broadcast signals and in chronometers (time pieces of great accuracy). A quartz crystal vibrates 32,768 times a second when electric current is passed through it. Today's digital watches contain a battery, a tiny computer, and a quartz crystal.

9.9 LIQUID-PHASE EPITAXY AND MOLECULAR-BEAM EPITAXY

Liquid-phase and molecular-beam epitaxies (MBEs) are methods developed by Bell Laboratories to produce new semiconducting materials not found in nature and having tailor-made atomic structures that provide for an array of new, built-in electronic, mechanical, and optical properties. The word *epitaxy* describes the overgrowth in layers of crystalline material deposited in a definite orientation on a base material of different crystal structure and chemical composition. In this world of human-made materials, this scientific research and development effort by Bell Laboratories demonstrates once more our ability to exert precise control over the creation of a new material by methods that ensure the exact composition and structure of the alternating atomic layers that make up a material.

Starting with a substrate of gallium arsenide, a semiconductor material, atomic beams of controlled intensity are aimed at the base material in an ultrahigh vacuum (Figure 9-43a and b). Shutters are used to turn the beams on and off as they are directed at the base from a heated oven. A layer of gallium atoms followed by layers of arsenic, aluminum, arsenic, then gallium atoms, repeated many times, produces a crystal resembling a thin, highly polished mirror. Liquid-phase epitaxy, an older technique, produces satisfactory wafers, but MBE appears to be the most promising. Using MBE, ultrathin, multilayer crystals with different structures have been grown. Quantum-well structures with up to 100 alternating layers represent one class of such crystals, with each layer as thin as several hundredths of a micrometer. Monolayer structures

Figure 9-43 (a) Atomic or molecular beams formed by heating elements in effusion ovens aimed at a base metal. (Bell Labs) (b) "Rough" surface of wafer (left) smoothed after MBE growth of one layer (right), permitting the growth of layers of uniform thickness. (Bell Labs) (c) Patterned layer. An island of gallium arsenide doped with elements with desired properties can be grown in a "sea" of semi-insulating gallium arsenide. (Bell Labs) (d) Rectangular glass tubing has made tritium lighting practical for LCDs in digital watches and calculators. LCDs are also used for notebook computers. (Corning Glassworks)

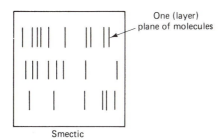

One (layer)
plane of molecules

Smectic

Figure 9-44 Model of the smectic mesophase.

with alternating layers of atoms in a stack, each layer two atom planes thick, are being produced. Multilayer structures consisting of several thousands of individual monolayers have been grown with a total thickness of 1 μm. Research has revealed that, because of the thinness of these layers, the electrons and holes normally confined to each layer will interact with their counterparts in adjacent layers and build *superlattice* structures with even more desirable properties (Figure 9-43c).

9.10 LIQUID CRYSTALS

Liquid-crystal displays (LCDs), currently used in watches, signs, and similar applications, offer a display system that does not require as much energy as LEDs (light-emitting diodes; Figure 9-43d). The image familiarly seen on the LCD produces a silver display, while the LED is the familiar red display. Images for the LCD that are generated by lasers have potential for multicolored video displays for information systems. The LCD offers many possibilities for easier information retrieval.

Most known substances can exist, given certain conditions, in a gaseous, liquid, or solid state. Very few compounds can exist in a fourth state, between the liquid and solid phases known as the ***liquid-crystal phase.*** This mesomorphic (intermediate) phase or anisotropic liquid was discovered over 100 years ago (1888) but has remained a physical oddity until recent advances in the technologies of electrooptics and thin-film components have permitted several important applications of these materials, particularly in the visual display of information.

Liquid crystals (LCs) are organic compounds that flow like a liquid while maintaining the long-range orderliness of a solid. Normally, crystals of a pure compound, when heated, show a well-defined and characteristic melting point at which the ordered crystalline lattice structure breaks down and the material becomes a liquid. In the liquid phase, the individual molecules show no preferred spatial orientation. The feature of liquid crystals is that, during the melting process, the well-ordered three-dimensional crystalline structure transforms into a one- or two-dimensional state of order. This results in a material that has some optical properties of a solid combined with the fluidity of liquids, resulting in a mix of unique properties. With both glass sheets or plates transparent, the cell is known as a ***transmitive-type cell*** (see Figure 9-47 later in this section). When only one glass plate is transparent and the other side has a reflective coating, the cell is called a ***reflective-type cell*** (see Figure 9-50a later in this section). Cholesteryl nonanoate and p-azoxyanisole are two organic compounds that exhibit these optical properties.

The molecular aggregates (collection of molecules) of an LC compound are in the form of long cigar-shaped rods. The orientation of these rodlike polar molecules forms the basis for classifying three basic types of liquid crystals: smectic, nematic, and cholesteric. The ***smectic*** phase consists of flat layers of cigar-shaped molecules with their long axes oriented perpendicular to the plane of the layer. This is the most ordered phase. The molecules within each layer remain oriented within each layer and do not move between layers. Figure 9-44 shows a structural model of this smectic mesophase. The molecules lie with their long axes parallel in layers. The molecules can move relative to each other, and consequently several types of smectic structures can be formed, depending on the inclination of the long axes of the molecules to the plane of the layers.

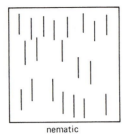

nematic

Figure 9-45 Model of the nematic mesophase.

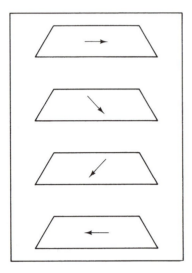

Figure 9-46 Model of the cholesteric mesophase.

The *nematic* phase also has molecules with their long axes parallel, but they are not separated into layers. Rather, their structural arrangement is similar to the ordinary packing of toothpicks in a box. Figure 9-45 depicts such a nematic mesophase structure. While maintaining their orientation, the individual molecules can move freely up and down.

The *cholesteric* mesophase (Figure 9-46) can be defined as a special type of the nematic phase in which the thin layers (one molecule thick) of mostly parallel molecules have their longitudinal axes twisted (rotated) in adjacent layers at a defined angle. Each layer is basically a nematic structure. The axes of alignment of contiguous layers differ by a small angle and produce a helix or progressive rotation of many layers in an LC material.

Because their molecules can be aligned by electric and magnetic fields to produce changes in their optical properties, the *nematic* liquid crystals find increasing use in electrooptical devices. Additionally, several mechanical influences such as pressure, impact, or temperature induce a change in the structure of nematic liquid crystals similar to those brought about by electromagnetic fields. This cholesteric structure exhibits anisotropic optical properties. When such a material is illuminated with white light at certain temperatures, iridescent (having shifting or changing colors) colors brought about by a fraction of the incident light are observed in the material. The much larger portion of light is transmitted by the LC material. This reversible color phenomenon functions over a temperature range of about −20° to 250°C. These properties of LC materials are finding many applications in heat-transfer studies (thermal mapping); nondestructive testing (NDT); toys and games; holography; medical diagnosing of vascular diseases; and flat-panel, full-color display panels.

Switchable privacy glass, produced by Marvin Corporation of Warroad, Minnesota, uses a technology similar to that found in laptop computer screens. The windows contain a thin film of liquid crystals made by 3M, sandwiched between layers of tempered glass. When a small electric charge of about 1 watt per square foot of window is applied, the crystals align, and the

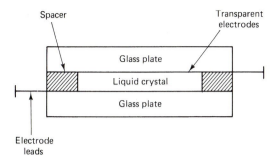

Figure 9-47   Cross-section of an LC cell.

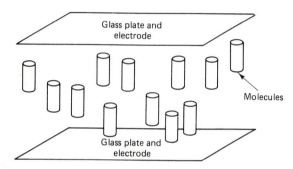

Figure 9-48   Homeotropic molecular orientation.

glass appears clear. With power off, the crystals arrange themselves in a random fashion, scattering light. The result is that the glass turns a frosty opaque white. This phenomenon is known as ***dynamic scattering,*** i.e., cells do not generate light. They transmit or reflect light from external sources. Therefore, the only energy required by the cell is that necessary to produce the dynamic scattering. A common display requires about 125 μA, whereas a corresponding-size LED display requires about 500 mA. See Figure 6-1 and the related text discussion of privacy glass used in the smart house.

9.10.1 Liquid-Crystal (LC) Cell

An LC cell consists of a layer of nematic mesophase LC material between two glass plates that are glued or fused together. The thickness of the LC material is about 10 to 25 μm. The two glass plates have transparent electrodes, made of a transparent and conducting material such as tin or indium oxide, deposited on their inside faces. Figure 9-47 shows a sketch of the construction of an LC cell.

9.10.2 Homeotropic and Homogeneous Orientations

Two preferred orientations of the LC molecules, *homeotropic* and *homogeneous,* are used within LC-display devices (LCDs). Figure 9-48 shows the homeotropic orientation with the long axes of the molecules perpendicular to the surface of the glass plates and electrodes. This orientation can be achieved by chemical doping of the nematic phase. The long, cigar-shaped rod orientation of the molecules allows the groups of molecules to act like dipoles in the presence of an electric field.

Figure 9-49 shows the orientation of the molecules to be parallel to the glass plates. This parallel orientation can be produced in the LC material by mechanically rubbing, unidirectionally, the glass plates with a leather cloth prior to assembly of the LCD. Another means of accomplishing this orientation is by the deposition of a layer of dielectric over the transparent electrodes. Each plate is then rotated 90° in relation to each other. In effect, the LC material acts now as a set of polarizers that cause the light passing through the LC cell to be rotated 90°. Another way of altering the alignment of these layers in the LC material is to apply an electric field

Figure 9-49 Homogeneous molecular orientation.

across the material. The molecules then align themselves in the direction of the electric field, which prevents the 90° twist of the molecular layers. The incident light cannot be transmitted through the cell. By judicious selection of the LC material, polarizers, glass-plate treatment, and the electric field, the incident light can be controlled through the cell; that is, it may be reflected, rotated, transmitted, or extinguished.

9.10.3 Reflective-Type LC Cell Operation

Although there is a transmittive-type LC cell, the reflective-type cell is chosen to explain the operation of a numerical-display LC device. Figure 9-50a is a sketch of the components of a reflective-type LCD. The main components of the LC cell consist of the two glass plates, electrodes, and LC material. A vertical polarizer and a horizontal polarizer, plus a reflector, are added to the cell. With the plates treated and arranged to produce the 90° twist described previously, unpolarized light enters the vertical polarizer, rotates 90° through the LC material, and passes through the horizontal polarizer and on to the reflector. The reflector reflects the light back through the same path. The face of the reflector appears light all over its surface.

If we wish to show one numeral in the display, then several LC cells are placed together in a particular configuration; each controls one segment of the digital display, as shown in Figure 9-50b. An electric field activates the selected segments, which act together to form the particular numerical character desired. The digital display can be impressed on the electrode on the front glass plate in Figure 9-50b, with the electrode on the back plate acting as the base of the common electrode. The molecules in the area of the energized segments will align with the electric field and prevent rotation of the polarized light. Consequently, the vertically polarized light passes through the cell everywhere but in the region of the energized pattern elements. The end result is that the energized display elements appear as black images against a light background. By properly choosing the correct combination of segments to be activated, any digit from 0 to 9 can be displayed.

9.11 SUPERCONDUCTORS

As the world seeks cleaner, more efficient, and readily available energy, many developments, ranging from improved solar collectors to cold fusion to high-temperature superconductivity, quickly grab the public's attention. High-temperature superconductors (Figure 9-51a) achieved just that effect back in 1986. A superconductor is a material that allows an electric current to flow without resistance. Most conductive materials, usually metals, restrict the flow of electricity to some degree; this is known as *resistance*. If metals were compared for their resistance on a scale of 1 to 10, with 1 being the most resistant and 10 being the least, iron would be close to 1 and gold and silver near 10. Copper and aluminum are somewhere in the middle. Resistance of an electric current causes energy losses due primarily to heat generated from the resistance. Superconductors have a theoretical resistance of 0 Ω. This means that there is no energy loss when an electric current passes through the superconductor. Because

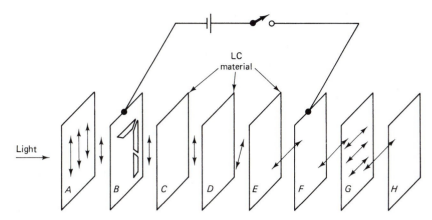

A — Vertical polarizer
*B — Front glass plate electrode
C — Liquid crystal field effect cell
D — Liquid crystal field effect cell
E — Same as C and D
*F — Back glass plate electrode
G — Horizontal polarizer
H — Reflector

* Numerical character segments
   contained in B and F.

(a)

(b)

Figure 9-50   (a) Reflective-type LCD (electric field not applied). (b) Seven-segment numeric display.

of this feature, superconductors can be of great benefit in electrical transmission, from very small millivolt signals in computers and other instrumentation to the high-powered, electrical-transmission lines that bring electricity to our homes and businesses. This brings up the question, Why aren't superconductors utilized more? This question is best answered with a little historical perspective.

Superconductivity was discovered in 1911 by a Dutch physicist, Heike Kamerlingh Onnes, while he was observing mercury at liquid-helium temperatures ($-452°F$, almost absolute zero). Because of the difficulty and expense of working with superconducting materials at such low temperatures, uses for it have been very limited. Research on superconducting materials continued, and by 1973, the temperature had risen to $-418°F$ with the discovery of superconductivity in a niobium–germanium alloy. Because metals are conductive, the most logical choice for a superconductive material would be metal. The discovery in 1986 of

(a)   (b)

(c)

Figure 9-51   (a) High-temperature superconductors resulted from the use of ceramic materials. The Meissner effect, demonstrated here and explained later in this module, shows how superconductivity permits levitation of objects. (NIST) (b) High-speed travel exceeding 300 mph over long distances with Maglev trains is a desirable alternative in some cases to air and road travel. (Transrapid) (c) Electromagnetic Maglev, depicted in this simple diagram, is one option to a suspension system whereby electromagnets, perhaps high-temperature superconductors, will levitate the train over the track. (Federal Railroad Administration)

superconducting metal oxides by K. Alex Muller and J. Georg Bednorz, physicists at the IBM Zurich Research Laboratory in Switzerland, made quite a stir in the scientific community. Metal oxides are normally nonconductive, but certain oxides, when mixed in the correct ratios, have been found to be superconducting at temperatures that are reasonably achievable.

Since the Muller and Bednorz discovery in 1986, research around the globe has boomed as people seek to find ways to commercialize a product using this so-called high-temperature superconductivity. The number of compositions has increased dramatically. One of the amazing things about this discovery is that it has been the only major discovery in the twentieth century that requires common laboratory equipment and therefore can be duplicated almost anywhere, including the high-school classroom. This means that not only are researchers experimenting with high-temperature superconductivity, but people with an interest—engineers, inventors, industrialists, and tinkers—are trying their hands at making a usable product and furthering the research.

This discovery will have a tremendous impact on society. Besides wires for conducting electric currents, we are witnessing attempts at revolutionary concepts in medical equipment, trains, batteries, machinery, computers, and so on. Figure 9-51b and c shows a magnetic-levitation (Maglev) train that uses magnets to suspend the train over, but not touching, the rails. The magnets and rails act together as a long-stator, linear induction motor to propel the high-speed supertrains. The Maglev project in the United States seeks to augment air and road travel with high-speed supertrains. Electromagnetic levitation of trains using superconducting magnets is one application of electronic materials that is sure to receive attention.

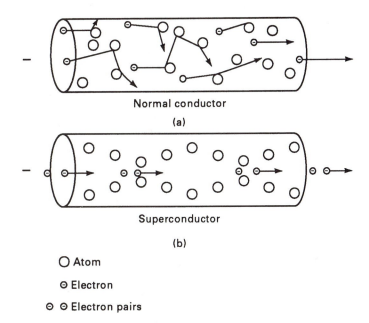

Normal conductor

(a)

Superconductor

(b)

O Atom

⊖ Electron

⊖ ⊖ Electron pairs

Figure 9-52 (a) Normal conductor. (b) Superconductor.

Superconducting metals act as normal metals and offer resistance above the ***critical temperature (T_c)*** but have zero electrical resistance below T_c. Figure 9-52a depicts electrons flowing through a normal metal conductor (Figure 9-52a) and a superconductor (Figure 9-52b) from the negative end to the positive end. In the normal conductor, electrons collide with phonons. These quanta, or packets of energy, are caused by the vibration of the atomic space lattice, in which atoms form into crystalline solids. As materials become hotter, the vibration of their atomic lattices increases, causing a corresponding increase in electrical resistivity. Silver has a resistivity of 1.6 $\mu\Omega$/cm at 20°C, but that resistance increases to 4.6 at 500°C; copper goes from 1.7 $\mu\Omega$/cm to 4.5 with the same increase in temperature. The collision of electrons with phonons is resistance, which creates its own heat in the conductor. As long as metals are maintained at moderate temperatures, they are *fairly* good conductors.

A theory known as ***BCS theory*** (after Bardeen, Cooper, and Schrieffer, 1957) holds that, in superconducting metals, pairs of electrons move with ease through the vibrating lattice (Figure 9-52b). As one of the paired electrons passes between two atoms, it causes a drawing together of the positive-charged atoms, which results from their attraction to the negative-charged electron. This is much like a speeding truck sucking cars into its draft. Once the electron passes the two atoms and they are still momentarily close together, they generate a positive attraction for the second electron in the pair and pull it into their region. This seems to set up a pulsing effect that causes masses of valence electrons to move through the superconducting material without resistance. The flow of electrons through the superconductor is perpetual, as long as the critical temperature is maintained.

High-T_c superconductors established some new lines of thought. While ceramics, with their covalent and ionic bonding, are normally insulators, a group of superconducting ceramic oxides possess superconducting properties at temperatures high enough to use the relatively cheap liquid nitrogen as a coolant to achieve superconductivity. Muller and Bednorz's now famous 1:2:3 superconductor was comprised of a ratio of 1 part yttrium to 2 parts barium and 3 parts copper. Copper oxides in these ceramics have positive- and negative-charged ends (dipole) in the same manner as a bar magnet. It is believed that valence electrons passing through the lattice cause a shifting in the polarity of the charged ends, thereby developing a strong positive charge as they pass (see Section 9.2). Again, as with the superconducting metal, the second electron in the pair is pulled in by the positive attraction. This phenomenon is repeated

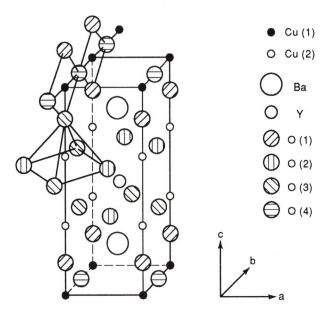

Figure 9-53 A single unit cell from the atomic space lattice of a ceramic–oxide superconductor composed of one atom of rare earth element yttrium (Y), two atoms of barium (B), 17 atoms of copper (Cu), and 27 oxygen (O) atoms or vacancies. This unit cell is repeated millions of times in the solid material. The superconductor is abbreviated YBCO.

many times to provide superconducting current. Figure 9-53 provides a diagram of a unit cell for the new $YBa_2Cu_3O_7$ (yttrium–barium–copper oxide).

The ceramic–oxide-superconductor materials provide a major challenge in processing techniques because of their brittle nature and the problems of keeping the composition in proper proportions. A *plasma-spraying technique* can coat a variety of large and small objects, including large copper panels and spherical vessels, with the ceramic oxide.

Many applications involving superconducting materials call for powerful magnets, as with magnetic-resonance-imaging (MRI) medical-diagnostic devices (Figure 9-54). *Superconducting magnets,* such as the magnets used for these devices, consist of doughnut-shaped field-coil windings made of tightly wound niobium–titanium wire. They are cooled to near 0 K and high amperage is applied. As long as the coil stays cold enough, the magnets continue to conduct the electric current, free of resistance, at a highly stable rate.

Magnetic characteristics of high-T_c superconductors result in their excluding all magnetic flux in the material. This phenomenon, known as the ***Meissner effect,*** is illustrated in Figures 9-51 and 9-55. The figures compare a superconductor to a normal ferromagnetic material. The drawings show the lines of flux (H) as arrows penetrating the normal conductor. A critical field (H_c) is reached for the superconductor when the temperature is below the critical temperature. Then no lines of flux penetrate the superconductor. When the external field is created by a permanent magnet, it can move the superconductor. Figure 9-51 is the familiar picture showing a permanent magnet levitating over a pallet of ceramic–oxide superconductor. The magnet floats in the air above the superconductor because its lines of flux are repulsed by the superconductor and the magnetic field is excluded from the magnet's interiors. In effect, the superconductor acts as a perfect magnetic mirror to the magnet. The superconductor material and magnet are firmly attracted to one another but never touch, as would a normal magnet attracting iron. The cloud beneath the pallet seen in the picture is a result of the liquid nitrogen boiling away at room temperature.

High-T_c superconductors offer many gains to society, such as cost savings in power generation. By applying superconductivity to electric-power generators, it is possible to generate electricity without the losses normally associated with the flow of current in rotor windings in conventional generators, thereby saving millions of dollars (Figure 9-56).

Figure 9-54 Two prints of a set of many slices taken through a human head by a magnetic resonance imaging (MRI) scanner to reveal bone and tissue for diagnosis of suspected cancer. (General Electric)

Normal:
$T > T_c$
or $H > H_c$

Superconducting:
$T < T_c$
or $H < H_c$

Figure 9-55 Comparison of the effect of the magnetic field (H) on normal conductors and superconductors. Critical field (H_c) is reached when the temperature (T) falls below the critical temperature.

Figure 9-56 General Electric Company has built a superconducting generator that produces 20,600 kilovolt-amperes of electricity, or enough for a community of 20,000 people. This is about twice the amount of electricity that would be generated by a conventional generator. Field windings of conventional generators are made of a copper–silver alloy. The low-temperature superconductor generator employs modular field windings consisting of hundreds of filaments of a metal composite of a niobium–titanium alloy in a copper matrix. When available, high-T_c ceramic superconductor materials will provide even more cost savings. But no one can say when those breakthroughs will occur.

Figure 9-57   Shape-memory alloys, used for the Nitinol eyeglass frame temples, "remember" their original shape and immediately return to that shape when the bending force is removed.

The demand for electricity in the United States is expected to double by the year 2030. Large industrial motors delivering 10,000 HP consume almost 15% of all U.S. electricity. If their coils could be made out of a superconducting wire, the motor could be reduced in size by one-half, with a corresponding 60% reduction in weight. The international competition to produce such a wire is fierce. To date, laboratory samples of such wire are small, consisting of tapes about $1/2$ in. wide, 2 in. long, and a fraction of an inch thick. These samples are extremely brittle and very difficult to make into wire. With a proven high-temperature, superconductor metallic oxide compound abbreviated YBCO (yttium, barium, copper, oxygen) (see Figure 9-53), researchers have used an ion-beam-assisted, deposition spray process to produce a thin film of this material. In the process, one beam of ions lays down a few hundred thousandths of an inch of zirconia on a nickel-alloy tape, while a second beam orients the crystals of YBCO as they form on a nickel substrate. The zirconia acts to prevent the YBCO from reacting with the nickel substrate, and the other ion beam forces the YBCO into a regular crystalline order. In effect, a sandwich is produced that is malleable and has retained much of its superconductivity. Much time and effort will be required to transform the lab process into a low-cost volume manufacturing process.

 EMT WEBLINK

Superconductors—*http://www.lucent.com/microscapes/16.html*

9.12 SMART MATERIALS AND INTELLIGENT STRUCTURES TECHNOLOGY

Hardly a day goes by that we don't read or hear about some "smart" technology: smart cars, smart highways, smart credit cards, smart bombs, smart explosives, smart pills, smart structures, smart batteries, smart skins, and smart fluids. With each major group of our Family of Materials, we have provided you with examples of how smart materials within each group work and explained some of their uses. This technology shows great promise because it represents the potential for materials systems that often involve electronic and optical materials in conjunction with metals, ceramics, polymers, and composites.

As Figure 9-57 shows, smart materials may find uses such as shape-memory alloys in rather simple products like eyeglass frame temples, or they may involve complex intelligent structures such as bridges and aircraft wings. An example of the use of a shape memory alloy is provided by the United States Air Force (USAF) and its use of CryoFit® permanent fittings and Cryolive® dematable end fittings in many of its aircraft hydraulic systems. These fittings are made by Advanced Metal Components, Inc., from nitinol barstock. Nitinol barstock is first machined at room temperature to an inside diameter (ID) that is smaller than the outside di-

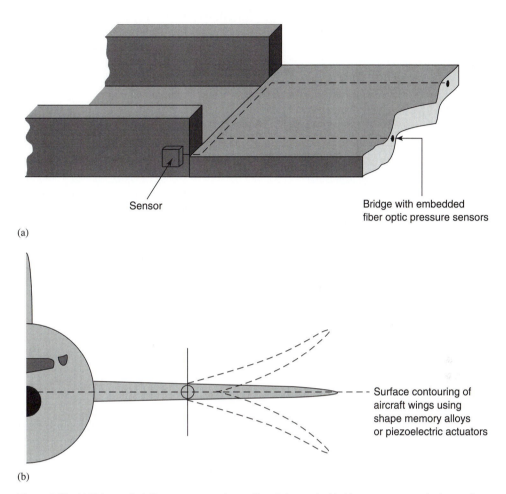

(a)

(b)

Figure 9-58 (a) Using optical fibers, sensors, and recording devices embedded in concrete or attached to steel structural members, it is possible to monitor stresses on bridges, walls, and other structures. (b) Current research may one day yield aircraft equipped with intelligent structures that can deform adaptive wings in reaction to variations in air pressure and air speed, and thus improve high-speed cruising.

ameter (OD) of the tube to be joined. The fitting is then cooled in liquid nitrogen below its martensite transformation temperature. Next it is expanded in the cold condition to an ID larger than the OD of the tube to be joined. The finished fitting is placed over the tube end, where it warms and recovers onto the substrate.

The fastest growing applications for binary NiTi alloys use this shaping memory or superelastic property. These alloys will restore up to 8% strain repeatedly in this pseudo-elastic manner. NiTi wires are used to make very flexible eyeglass frames (Figure 9-58a), which are essentially unbreakable. Highly flexible antenna wire for cellular phones is produced from these alloys. The service life of these alloys is inversely proportional to the amount of deformation. If deformed close to 8%, a NiTi alloy's life span may be only a few hundred cycles. On the other hand, if the magnitude of the deformations is less than 1%, the alloy's life may be several million cycles.

The U.S. Army is developing a smart skin for tanks. This skin uses sensors much like those seen in the bridge in Figure 9-58a. The armor skin would detect incoming projectiles, such as shells made of depleted uranium, and trigger a repulsive explosion in microseconds to counter the shell's impact. An aerospace intelligent structure is the adaptive wing shown in Figure 9-58b. It employs piezoelectric materials, which were discussed in this module. Piezoelectrics are being evaluated to see if they can deform wings in response to changes in air speed and air pressure and thus provide for more efficient cruising. Both shape memory alloys and fiber optic systems are also being evaluated by the aerospace industry.

RECTENNA

An antenna that captures
and converts RF or
microwave power to DC
power.

Advantage:

Tailored to generate either high
voltage/low current or low
voltage/high current DC power.

Piezoelectric: high V - low I
Magnetostrictive: low V - high I

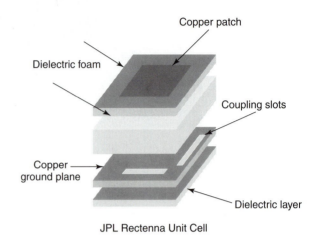

JPL Rectenna Unit Cell

Figure 9-59   Adapted from Song, Kyo D. *Development of Microwave-Driven Smart Materials Actuators.* NASA project overview, March 1, 2000.

One example of research on smart materials is being conducted at NASA Langley Research Center. It has focused on the feasibility of a smart material actuator system driven by a microwave. If such a system proves to be viable, it could dramatically reduce the cost of distributed shape-control systems and enable many of the technologically aggressive future NASA science instruments such as the Next Generation Space Telescope. The rectenna in Figure 9-59 uses piezoelectric materials that receive control signals from microwaves to move "smart strings" and "smart membranes" as control devices, such as earth-orbiting devices or micro robots on Mars. Many exotic technologies first developed for NASA space applications soon found their way into broad commercial applications. What down-to-earth applications can you think of for the smart material actuator system driven by a microwave?

In another development, the "smart battery," an intelligent materials system, integrates battery technology with semiconductor technology. The battery pack designed for laptop computers includes a microprocessor inside each lithium ion battery pack. The system keeps track of battery use and notifies the user of the time remaining on the battery. In the design of these new smart batteries, some major computer manufacturers are teaming up with major battery manufacturers to develop standard sizes and configurations. This would ease the burden on consumers. Just like with the standard AAA and D cells, it would be possible to purchase standard replacement batteries for the new laptop computers in many stores.

In Module 1, we defined *intelligent materials systems, smart materials,* or *smart structures* as materials and materials systems designed to mimic biological organisms and offer the ultimate materials system that can place control and feedback into a material structure. These materials take their cue from biological elements such as muscles, nerves, and bodily control systems, which adapt to environmental changes.

Where will smart materials and intelligent structures go from here? Much of the focus of smart materials research is based on mimicking nature. We will examine that technology next.

EMT WEBLINKS

Smart materials and structures at Sandia Lab—*http://endo.sandia.gov/9234/smas_ expand.html*

Shape memory alloys—*http://www.uni.uiuc.edu/~richlin/sma.html*

Figure 9-60   Biomimetics uses techniques that imitate nature to synthesize materials. Mollusks are examples of complex structures at nano scale, which nature produces at low pressure and low temperature, and without harmful effects on the environment. (DOE-PNL)

9.13 BIOMIMETICS AND BIOMATERIALS

9.13.1 Biomimetics

From the beginning of civilization, technology has reflected human attempts to imitate nature. In the early stages of human flight, the effort to build flying machines that imitated the structure and motion of birds met with failure. On the other hand, imitating the structure used by bees has produced advanced honeycomb composites suitable for aircraft.

Transforming raw materials into finished products usually requires large amounts of energy and many different processes. Because of the concern for the limitations of natural resources, considerable research and development is aimed at reducing the input of both raw materials and energy to process materials. Synthesis and processing determine the structure and properties of a material. How can technology conserve resources and be friendly to the environment in both synthesis and processing, and still obtain the desired structure and properties of materials?

Materials technology has turned to nature in the search for low-energy (low pressure and low temperature) and low-environmental-impact synthesis. *Biomimetics* is a technology that seeks to duplicate some of the processes that living organisms employ to construct the bones, tissues, shells, spider webs, and other materials in nature. The view of the cross-section of a seashell (Figure 9-60) reveals the intricate structure that mollusks achieve as they construct their hard, durable homes. Some doubt the value of pursuing the biomimetics route to materials synthesis for engineering materials, but others feel that it is useful to study nature to produce biomaterials. Biomaterials for dental and bone transplants have been produced using biomimetics that replicate sea coral. Biomimetic research enhances the development of the smart materials discussed previously.

Thin-film coatings offer many possibilities if they can be manipulated (1) to achieve unique structures that possess designed properties and (2) for selective application to a variety of surfaces. Organic thin films are found everywhere. Antirust primer on an automobile, Mylar film that holds the iron oxide on the audio tape, and soap-forming film around dirt are but a few of the films that make our lives simpler in the twenty-first century. The ability to orient the crystal structure of coatings offers possibilities for special optical, electronic, magnetic,

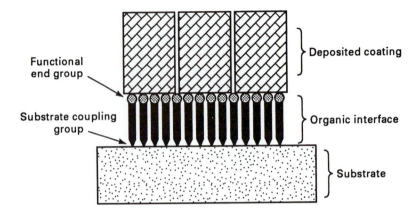

Figure 9-61   Biomimetic synthesis allows thin ceramic films (deposited coating) such as magnetic iron oxide to be deposited on substrates such as polyethylene and polystyrene plastics in a very controlled manner so that the functional end groups covalently bond (discussed in Module 2) to the substrate. (DOE-Pacific Northwest Laboratory)

and mechanical properties. Tough ceramic coatings provide increased durability for many products. The Pacific Northwest Laboratory used biomimetic synthesis to develop an organic interface that controlled crystal growth and formed desired ceramic coatings on polymer, glass, and metal surfaces of complex shapes at lower cost than current vapor deposition techniques. As depicted in Figure 9-61, the coating was deposited at temperatures below 100°C in a water solution that produced no hazardous waste.

Engineers are taught to be curious about the world around them, to observe how Mother Nature does things. Some examples that illustrate this awareness of curiosity are the following. How do honey bees construct their nests? Why doesn't a blade of grass break when stepped on by an animal or human? How does a tree withstand the force of wind? What makes the adhesive produced by barnacles so strong that sand blasting, at considerable expense due to labor and the need for drydocking of vessels, is required to remove them from underwater hulls of ships and pilings? Or how does a bat manage to capture insects as it flies in the dark of night at high speeds? Some 20 years ago NASA researchers were studying how to use networks of strings to fabricate connecting platforms in outer space. Part of their studies involved an experiment with a space-borne spider. As the spider spun a web, it used threads of nonuniform size. NASA researchers used this information recently to help design an improved tennis racket with greater power, feel, and control. Two characteristics of tennis rackets, as well as other sports equipment such as baseball bats, must first be explained. (This discussion builds on the information about testing sports equipment in Module 4.)

The point that reaches the maximum speed in swinging a racket is located at the toe of the racket or at the furthest location from the player's hand. Maximum power is generated when a ball hits that point. This point is the desired point of impact when serving a tennis ball. The other characteristic is the point of optimum vibration. Known also as the center of percussion or sweet spot, this location produces the least amount of vibration on impact of a ball. This vibration, kick, or jar is transmitted through the racket to the player's hand and, in some cases, can cause chronic arm problems. Many tennis racket designs have resulted in enlarged sweet spots to reduce this vibration. Sweet spots are usually found at the geometric center of the string area. If the center of percussion could be moved outward toward the location of the position of maximum power, improvements in the power delivered to the ball would be achieved because most of the ball impacts would now be on or close to the sweet spot. Fewer vibrations would be felt by the player, with attendant reduction in stress on the player's arm and hand. Better control would also be experienced because of the increase in the racket's stability. (The racket is less likely to twist in the player's hand when impact occurs at the center of percussion.)

Solving the problem was greatly helped by the knowledge received in the spider experiment. Using a computer and a genetic algorithm, and the laws of physics dealing with vibrating strings, an optimum design that involved distributing the required mass along each string (or tapering the individual strings) was obtained. A *genetic algorithm* is defined as one that implements a mathematical evolutionary process that resembles biological evolution in some respects and that arrives at an optimized design by following a sequence of random design changes and preserving those changes that improve performance. Any string material can be used, and the tapering can be achieved by spinning the string thicker or bonding different string materials along a central braid in the string.

An object that moves, such as a vehicle or a tennis racket, usually carries the requirement that it be light in weight. In the following case, weight is important, but *where* it is concentrated in an object is of great concern, because extra weight usually is not a positive aspect. Taylor Made's new FireSole golf clubs have weighted, metal plugs placed in them made of a combination of titanium, tungsten, and steel. The quantity, size, composition, and placement of the plugs changes the center of gravity for each club to provide weight and power where it is most needed. The company has not disclosed the details of this new innovation. Shafts are made of graphite. A graphite driver sells for around $389. Could you speculate as to why three different metals were chosen or where the locations of the plugs might be?

9.13.2 Biomaterials

A *biomaterial* is any substance other than food or drugs contained in a *therapeutic* (serving to cure or heal) system that is in contact with biological tissues or fluids. One biomaterial, dental fillings, is the most common. Dental fillings have been in use for more than 2000 years. Synthetic polymers are being developed to produce implantable (in humans) dispensing machines to deliver drugs at the proper levels, to direct the release of drugs to a particular cell type or tissue, or to release drugs at a precise rate. A conservative estimate states that up to 100,000 deaths each year in the United States can be attributed to adverse drug events such as overdose. As a consequence of new developments similar to those mentioned, the drug-delivery industry is a multibillion-dollar industry. A second avenue of development is in the area of engineered tissue in which synthetic polymers are used in the successful attempt to grow replacements of body parts. The greatest challenge to the research teams of chemical engineers, materials scientists, and surgeons is to produce a material that is not harmful to the human body, and once inside a human body, will degrade after fulfilling its task. Roger Langer, professor of chemical and biomedical engineering at MIT has been instrumental in molding materials that turn biological discoveries into medical cures. He has discovered a family of polymers called *polyanhydrides* that among other properties is degradable. These polymers are now being used to treat brain cancers. Another avenue of development is the use of "*smart systems*" that can deliver drugs in response to specific signals. For example, an implantable biosensor that detects glucose levels and tells a drug-delivery implant to release insulin when it is needed may be the next "crazy thing" that proves successful. Better still, an implantable microchip that can release specific drugs or vaccines into the body on demand by telemetry is not far off.

The field of *tissue engineering* integrates knowledge from cell biology, materials science, and process science to create new tissue. The first engineered tissue was skin, which is grown on a two-dimensional surface. Synthetic polymers are used to construct three-dimensional structures that provide a framework or template for the cells to grow and take the shape of the template. Consider the process of using cartilage cells (cells that eventually grow into bone) and degradable polymeric foam to make a human nose. The cells are placed throughout the foam and allowed to grow, producing an extracellular matrix which provides strength while the polymer dissolves. The result is pure cartilage in the shape of a human nose.

More than 50 companies have received licenses based on recent inventions using such materials as polyanhydride polymers. Tough, stretchy *collagen threads* produced by marine mussels possess extraordinary properties. They are 5 times stronger (tensile strength) and 16 times as resilient as human tendons. Scientists are looking for a material that is strong enough to withstand the strain of a human joint and be so resilient that it can stretch dramatically and still return to its original size. Such a material could be a breakthrough in helping to repair human tendons and ligaments. Inspection of the molecular structure reveals that the threads are made up of two types of collagen—a fibrous protein commonly found in bone, cartilage, and connective tissue. Further, the manner in which nature has created this tendonlike material that is seamlessly woven to achieve a gradient between stiffness and elasticity will require considerable investigation before any developmental work can be undertaken.

The term *soft materials* is used to categorize many of these new materials. Materials science is not only interested in mimicking nature but in contributing to bioengineering materials and components that work with human and animal systems. *Biomaterials* are compatible with human and animal systems; this allows the material to be implanted or manipulated in people and animals. The principal requirement for each biomedical alloy is that it be corrosion resistant when inserted into the body, and possess optimum mechanical properties. Bone has a modulus of elasticity (E) = 17 GPa (2.5 Msi - 2.5 mega pounds per square inch = 2.5 million psi). A corrosion rate of less than 0.01 mpy (mils per year) is required for a material to be considered a biomaterial. All types of corrosion have been observed in biomaterials in the body. Bone grows in response to applied stress. *Surface passivity* is the most important criterion for corrosion resistance. BioLAST is a new coating that passivates the surfaces of medical devices that come into contact with body fluids thus preventing the surface impurities from causing thrombosis. Further qualities of this new coating are its high degree of tissue compatibility and its possible resistance to bacterial infection. *Thrombosis* is a coagulation of blood in the heart or a blood vessel which forms a clot. The clot is caused by converting fibrinogen to fibrin. By coating the surface with *albumen fibrinogen,* adhesion will not occur and no thrombi can form at the device surface or farther downstream. Stainless steel and cobalt–chromium alloys depend on the presence of chromium and its ability to render the alloys passive. Over the long term, stainless steels are not sufficiently corrosion resistant. Additions of other elements enhance resistance to nonuniform types of corrosion such as pitting. (Refer to passive oxidation in Module 7.3 and prevention of corrosion in Module 4.4.4 for further information on the passivation mechanism.) Titanium and its alloys develop passivity without chromium. In addition, surface finish also may affect performance, as highly polished surfaces resist corrosion and wear more effectively than nonpolished surfaces.

Titanium is another example of a biomaterial that serves as implants for joint replacements and dental reconstruction (refer back to Figure 5-61). Figure 9-62 shows the many applications of titanium implants used in the human body. A coating process to improve bone growth around prosthetic devices substantially reduces the risk of implant failure. A thin layer of *hydroxyapatite (HA)* is first coated onto the implant. Then an ion implantation beam is focused on the surface, causing the HA to combine with the titanium which is strengthening the bond. The implant is then dipped into an HA solution filled with ceramic particles, and heated to form a hard coating around the initial implanted coating. The bone grows easily into the outer, porous HA layer.

New materials to repair broken and deformed bones that mimic the natural composition of bone have been formulated. This paste is injected into the joint for a speedy recovery without the normal heavy plaster casts. Synthetic polymers have been used as artificial skin to replace burned or otherwise damaged skin. The "plastic" skin, sometimes in conjunction with skin grafts, performs like real skin to protect the wound, allowing new skin to grow naturally. A new biomaterial called *Hedrocal* has been developed that can support injured bone areas. It is an open-celled lattice substance that appears similar to coral and cork, and is reinforced with tantalum (Ta). Having the properties of living bone, it can be implanted between two vertebra (any of the single bones or segments of the spinal column) where human bone will infiltrate the pores and form a true bone fusion across the joint. Human (living) bone consists of 75% hydroxyapatite (HA), which has the chemical formula $[Ca_5(PO_4)_3OH]$. The word *hydroxy* comes from the word *hydroxides,* which chemists use to describe compounds containing the polyatomic ion made up of oxygen and hydrogen, $-OH^{-1}$. For example, the compound NaOH, or lye, is called sodium hydroxide. *Apatite* is the name for a class of minerals that contain essentially calcium phosphate along with either fluorine, chlorine, or ions of hydroxy or carbonate. Human bone and teeth enamel are made principally from this chemical compound. Bacteria in the mouth produce acids from food trapped between teeth. These acids cause tooth enamel to dissolve and eventually produce cavities. Brushing the bacteria and food particles from the teeth and gums is one technique to reduce tooth decay. Recent studies indicate that the same bacteria may enter the bloodstream and contribute to the formation of plaque in the arterial arteries, ultimately causing heart attacks. In addition to practicing good dental hygiene, the use of fluoride-containing toothpaste results in the conversion of some hydroxyapatite into *fluorapatite* [$-Ca_5(PO_4)_3F$]. The crystalline structure of fluorapatite is more densely packed (less porous) than hydroxyapatite which makes it more resistant to penetration by acids in the mouth. This chemical reaction involves the replacement of the OH ion with the fluorine ion as seen by the above formula. This same replacement phenomenon occurs in the fossilization of bone where ancient bones are nearly all fluorapatite. A man-made ceramic derived from natural coral called **Pro Osteon**™ is being used to surgically repair bone defects. Provided in small blocks that can be cut and shaped, the surgeon can replace portions of bone with this ceramic replacement. It is sufficiently porous to allow tissue and bone to grow into its pores.

In Module 1, you were introduced to **Vitallium,** a biomaterial used for human prosthetics. The development of this alloy (cobalt, chromium, molybdenum, and nickel), first as a dental material (Figure 9-63) then as a material for orthopedic use (Figure 9-64), is typical of the synergy of engineering materials technology. Once Vitallium was found to be a suitable biomaterial during the 1920s and 1930s, then new methods of molding the alloy were sought. Because of the high melting temperature of the cobalt-based alloy, ethyl silicates were developed to serve as binders for the ceramic refractory to counter the problems of mold reaction and expansion. The process that evolved is called investment casting (see Module 5). During the 1930s the experience gained with casting Vitallium dentures and orthopedic prosthetics was refined and used to produce Vitallium and titanium components for jet engines for aircraft. That evolution has resulted in directional and single grained castings.

This synergy typifies the planned progress of materials science, engineering, and manufacturing. Wise and knowledgeable technologists, engineers, and technicians always keep informed about new developments around them. Events of serendipity may occur by chance, but the alert person can recognize their values.

9.14 ON TO THE FUTURE

We set the scene for this book with a high-tech mountain bicycle. When describing the many elements of engineering materials technology, we included customers' desires, designers' efforts to employ new materials and processes to meet the market demands, and manufacturers' methods of manufacturing to produce small lots of products with "agile manufacturing." Virtual reality, artificial intelligence, expert systems, multimedia computers, CAD-CAM, CIM, and robotics are among the evolving technologies and tools that will push engineering materials technology at an ever-increasing rate.

The morning paper, weekly magazines, and evening news bring you a constant barrage of materials evolutions and developments. You might read about the use of supercritical fluids to process plastics and other polymers in a more environmentally friendly method, compared to traditional polymer processing that gives off large volumes of pollutants. Or you may read how the economics of textiles is changed when the world's cotton supply is short and thus pushes up cotton textiles prices. To counter the rising price of fabrics, engineers look to recycled PET soda bottles as an alternative source for polyester and cotton blends. But the new polyester is not the coarse fabric of the 1970s leisure suit; rather, it is composed of "microfibers" that are finer, cooler, and softer than before. Watch for further developments on synthetic textiles as the free-electron laser (FEL, discussed in Section 9.4.2) moves out of the research lab into fabric manufacturing lines.

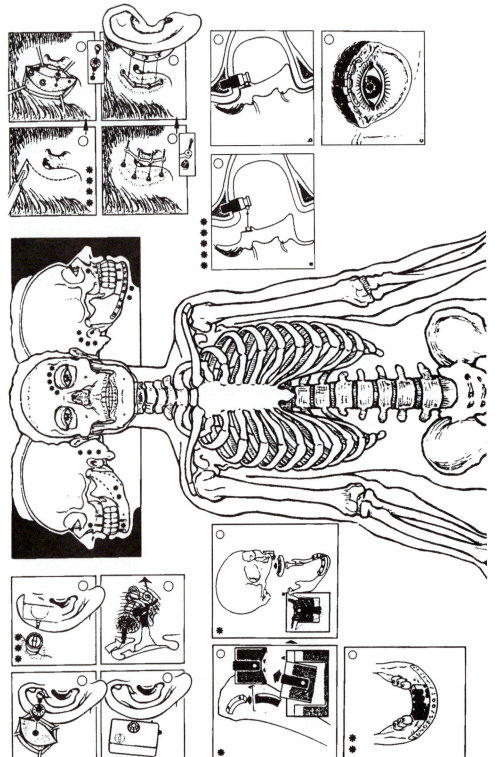

Figure 9-62 Because of its biocompatibility and ability to osseointegrate, titanium is used in many types of prostheses in humans. (Howmedica, Inc.) (*continued*)

Fredrik 81

© P-I Brånemark

TIBIA

Figure 9-62 (continued)

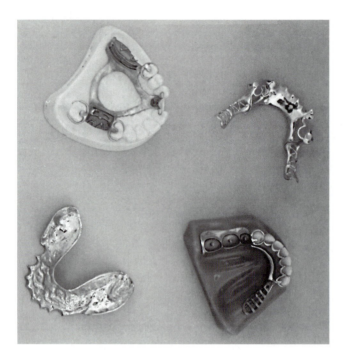

Figure 9-63 Vitallium dental castings. (Howmet)

Figure 9-64 Vitallium orthopedic castings. (Howmet)

9.14.1 Molecular Nanotechnology, Molecular Systems Engineering, or Molecular Manufacturing

Much of what has been covered in this module and many of the innovations throughout this book will rely on manipulating the very small nanostructures of materials to achieve the goals

Figure 9-65 High-resolution transmission electron micrograph of a typical area that reveals grain boundaries in a nanophase palladium. (NIST)

of advanced materials, such as smart materials and biomaterials. Electronic components such as transistors, capacitors, and diodes, with dimensions less than 0.25 micrometers, are emerging from the nation's laboratories. Units such as the micrometers are less useful as units of measure. The next smallest unit of measure is the ***nanometer (nm),*** defined as one-billionth of a meter. To emphasize its size, a human hair has a diameter that is 10 times the size of a nanometer. The diameter of most atoms is 0.1 to 0.4 nm. This new technology is being investigated for use in manufacturing and is described below in more detail.

Module 1 states that one constant of materials technology is constant change. It reminds one of the use of designer materials versus off-the-shelf materials, with their limited structures and properties. Module 2 recalls that materials will be produced in new ways as materials engineers work with manufacturing engineers and materials scientists to invent new materials processing. The Pause & Ponder for Module 3 asks the reader to recall the synthesis and processing capabilities of contemporary materials technology at the various scales of measurement, i.e., macro, micro, nano, and atomic. Module 3 also contains a reminder that both developed and newly developing countries are involved in improving or maintaining their competitiveness in the race for materials such as those listed in Table 3-1 under the nano and atomic categories.

The heading to this section describes, in present-day terms, the emerging field of engineering that uses nanoscale mechanical systems to guide the placement of reactive ***molecules,*** while building complex structures with atom-by-atom control. K. Eric Drexler has written extensively about the design, performance, and path of development of this system since 1981, when he published this untried concept.* He defines ***molecular manufacturing*** as the construction of objects to complex, atomic specifications using sequences of chemical reactions directed by nonbiological molecular machinery. ***Molecular nanotechnology*** describes the field, as a whole, comprising molecular manufacturing together with its techniques, its products, and their design and analysis. This new field is related to, yet distinct from, mechanical engineering, microtechnology (to be defined), chemistry, and molecular biology. As physics is related to engineering, so chemistry is related to molecular engineering. Molecular manufacturing applies the principles of mechanical engineering to chemistry. Figure 9-65 provides an example of the progress in molecular manufacturing.

Recent research into ***carbon nanotubes,*** rolled sheets of carbon just 1/50,000 the thickness of human hair, has turned up evidence that they are excellent conductors of electricity.

---

*Readers seeking further information on molecular nanotechnology may wish to contact the Foresight Institute regarding publications, conferences, and relevant research, such as that sponsored by the Institute for Molecular Manufacturing: Foresight Institute, Nanosystems information, PO Box 61058, Palo Alto, CA 94306. Tel: (415)-324-2490. Fax: (415)-324-2497. E-mail: foresight@cup.portal.com.

Nanotubes are formed by vaporized carbon that condenses into a series of hexagons. These hexagons naturally curl into hollow tubes. The twist of the tube determines its particular electronic properties. If the carbon sheet forming a nanotube is rolled up evenly, then the carbon molecules conduct electricity as if the sheet were a metallic substance. If the nanotube is rolled up in a twisted manner, then the nanotube acts as a semiconductor. Further, if the nanotubes are joined together end to end, then, their junction function acts as an electronic diode. A *diode* forms the basis for devices that convert alternating current into direct current. Further research is proceeding to construct *nanotransmitters* that could be built into electrical circuits.

Over the next 20 years, the development of this new manufacturing system for the bulk processing of materials and the fabrication of custom products will rely on the use of numerous small manufacturing systems working in parallel with locally available materials. The products will exhibit order-of-magnitude improvements in mechanical properties, and will be of high quality and low cost. Bulk processing will be attained when these machines have sufficient general capabilities to, first of all, manufacture copies of themselves and, second, be reprogrammed to manufacture finished products directly from raw materials, thus bypassing traditional fabricating techniques.

Advances in molecular nanotechnology since the introduction of the concept in 1981 include scanning probe microscopes (Module 3), design of proteins by molecular biologists, design of molecules to trap other molecules and ions, many scientific workstations and molecular modeling software that permit faster testing and construction of new designs, biomimetics (Module 3), and near-net-shape processes (Modules 3 and 5) that have eliminated the traditional secondary processes accompanying manufacturing. The next step is the development of new conceptual and mathematical tools to design and analyze a limited class of molecular manufacturing systems. The total costs of manufacture, excluding cost of development and distribution, will be almost wholly determined by the cost of materials. The relevant materials cost about 0.1\$/kg to 0.5\$/kg. Most manufactured products fall in the range between about 10\$/kg (i.e., automobiles, appliances) and 10^4/kg (i.e., aerospace vehicles).

9.14.2 Microtechnology/Microelectromechanical Systems (MEMS)

Much active research in engineering involves building electronic and mechanical systems on ever-smaller scales using microtechnologies. The micro scale (Module 3 and Tables 3-1 and 10-6) is, in terms of volume, 10^9 times larger than the nanometer scale, and microtechnologies provide no mechanism for gaining precise, molecular control of the surface and interior of a complex, three-dimensional structure. The materials involved can include diamond films, organic films, semiconductors, metal films, dielectrics, ferroelectrics, and piezoelectric films. This technology relies on processing techniques such as photolithographic pattern definition, plasma and wet etchings, physical and chemical vapor deposition, diffusion, and ion implantation. Some examples of the products of such processes are flat panel displays, solid-state sensors, and those related to microelectronics and optoelectronics industries. Refer to Table 9-4 to see how the free electron laser supports these emerging manufacturing processes. These techniques are essentially unrelated to those of molecular manufacturing, so a wide gap exists between *microtechnology* using the top-down approach (starting with large complex and irregular structures) and the chemistry of molecular manufacturing (nanotechnology) with its bottom-up path (starting with small, simple, and exact structures). In essence, microtechnology's goal is to make imprecise structures smaller, while the goal of molecular manufacturing is to make precise structures larger.

Micromachines are made from polysilicon, the same materials used in the manufacture of integrated circuits. As a result, the micromachines and integrated circuits can be constructed on one chip. Current production methods for quartz crystal timing devices and integrated circuits involve separate manufacturing steps for both the devices and the circuit, followed by assembly operations. The micromachines are embedded in a trench on a Si wafer. The wafers are used as a starting material for the conventional *complementary metal-oxide semiconductor (CMOS)* manufacturing process of integrated circuits. These circuits are built on the surface of the wafer, while the MEMS are sealed in the trench. The MEMS prototype timing fork ocillator functions as a clock source, generating frequencies of about 1 MHz with very low noise. Because of the low noise, the signals are constant, not disruptive, and ensure accuracy. This frequency provides the constant timing signals necessary for the digital electronics device to operate.

A minuscule combination lock called the Recordable Locking Device has been developed by Sandia National Laboratory to prevent unauthorized use of data on a computer's hard drive. Embedded onto a computer chip, the entire device is about the size of a shirt button. Devised by using **MEMS technology,** the device consists of a series of timing gears that move to an unlocked position when the operator enters the correct code. A series of six code wheels each less than 300 μm in diameter (about the size of the period at the end of this sentence) are driven by electrostatic drives that turn electrical impulses into mechanical motion. The lock "owner" sets a combination to any value from 1 to 1 million. The device "locks up" upon a wrong entry. The development phase is expected to conclude by the year 2001.

Sandia National Laboratories has developed a ***micro transmission*** for its similar-size micro engine. It has a 3 million to 1 gear ratio which theoretically can generate a force capable of moving a 1-pound object. The size of these mechanical devices is comparable to a grain of sand or about one-third the width of a human hair. Possible applications are aircraft sensors, optical switches for telephones, microsurgical instruments, and more effective airbag sensors for automobiles.

Battery technology has failed to miniaturize batteries sufficiently for use in small electronic and mechanical components. Scientists at MIT's Lincoln Laboratory have turned to MEMS devices to develop a turbine engine the size of a shirt button. Linked to a tiny electric generator, the turbine engine would be capable of producing 10 to 20 times as much power as the best chemical batteries. If successful, inexpensive miniature turbine generators could eventually be stamped out in large quantities similar to computer chips. Micromachinery fabricated from silicon would be limited in operation due to the high-temperature restrictions of silicon. Further research points toward the use of silicon carbide—a tougher, more temperature-resistant material.

Silicon micromachines, also called MEMS, are projected to find routine applications in industries such as automotive and medical diagnostics. A class of medical cutting tools etched from wafers of silicon has blades said to be 10 times sharper than metal tools. These tools, using ultrasonic vibrations, vibrate up to 200,000 times per second. Such technology could lead to greater precision for highly sensitive procedures such as cataract surgery and neurosurgery, and could be the basis for truly painless needles.

Three micromotors driving a microgear smaller in diameter than a human hair were reported to be fabricated from silicon by researchers at Sandia National Laboratories. This report typifies products being developed using microtechnology (Figures 9-66a–d). These devices were fabricated using both etching processes and silicon materials common to the microelectronics industry to facilitate their mass production. Each micromotor develops 0.5 μW of power rotating at a speed of 30 revolutions per second: sufficient to drive the larger gear (30 times larger than a microgear) at an angular velocity of one revolution per second. Open spaces in the large gear in some positions allow light to pass through, creating an optical shutter.

Figure 9-67 is a photograph of an Africanized, or killer, bee with a chip attached to its thorax. This solar-powered electronic package, with dimensions of 1 mm × 3 mm × 5 mm and weighing less than 50 mg, contains a micro-miniature infrared LED transmitter and tracking system. This product of microtechnology is being developed to allow entomologists to continuously track and study bees in the field to study their mating and foraging habits. When the chip was mounted on the bee, it was unable to fly. This illustrates how research often results in apparent solutions that do not succeed or that require further modifications. These results are normally discovered during field testing.

 EMT WEBLINK

Micro World with tutorials—*http://www.lbl.gov/MicroWorlds/*

(a)

Figure 9-66   (a) Three micromotors (rectangular structures, top and right), when energized by on–off voltages, power a micro gear (at center) to drive the larger gear. The motor consists of two silicon combs with a shuttle between them. The edges of the shuttle also have teeth that interdigitate with those of the stationary combs. When energized, the stationary combs alternately pull the shuttle via electrostatic interaction. (Ernest J. Garcia, Sandia National Laboratories)

(b)

Figure 9-66   (b) A closeup of the microgear (drive gear) meshing with the much larger optical-shutter gear. The attached shaft turns a drive gear in a quarter of a circle during the shaft's power stroke. Another micromotor, at right angles to the first, is timed to turn the microgear on the second quarter of its rotation. The two drives, alternating their force, convert reciprocating motion into rotary motion to drive the gear completely around. (Ernest J. Garcia, Sandia National Laboratories) (*continued*)

9.14.3 Challenges in Engineering Materials Technology

Throughout this book we have sought to transport you from the basic concepts of engineering materials technology, dealing with traditional materials, to advanced materials and future materials and processes technology. You have been exposed to new instruments, unique processes,

(c)

Figure 9-66 (c) No this is not Godzilla, but rather a spider mite (a miniscule, white fleck to the human eye) hanging out on a microlock mechanism. Note the scale key in the lower right corner shows 100 micrometers. (Sandia National Laboratories Intelligent Machining Initiative)

(d)

Figure 9-66 (d) Spider mite on mirror assembly. Imagine a machine so small that it is imperceptible to the human eye. Imagine working machines with gears no bigger than a grain of pollen. Imagine these machines being batch fabricated tens of thousands at a time, at a cost of only a few pennies each. Imagine a realm where the world of design is turned upside down, and the seemingly impossible suddenly becomes easy—a place where gravity and inertia are no longer important, but the effects of atomic forces and surface science dominate. The microdomain, a world now occupied by an explosive new technology known as MEMS (MicroElectroMechanical Systems) or, more simply micromachines. Sandia National Laboratories, motivated by a guiding vision for MEMS, has become a recognized leader in this emerging field. To learn more visit the MEMS site.

Figure 9-67 A micro-miniature infrared LED transmitter and tracking system is under development that will emit infrared pulses, detectable over a range of 1–2 km. The chip shown is proper size but it is not the actual transmitter. This chip was used to test the bee's ability to fly with the microchip on its back. (Ernest J. Garcia, Sandia National Laboratories)

and innovations in materials. At the same time, we have tried to help you develop problem-solving skills, learn to look around your environment, and observe what and how materials are currently being used and abused. We encouraged you to become familiar with the databases outside this book. Now we implore you to join a technical society such as ASM International, Society of Manufacturing Engineers, American Society for Testing and Materials, or any of the other groups who will assist you in keeping up with developments in our ever-expanding field! Student membership rates for these organizations are excellent, as are the benefits.

We encouraged you to consider the materials cycle and the consequences of each stage of the cycle. All who venture into technology must carry a concern for the environment. Selection of certain technologies and the associated materials issues to solve environmental problems is sometimes guided by politics and the perceptions of the general public, but the technologies selected may not result from the best use of the facts. An example is the mandating of electric-powered vehicles to relieve air pollution caused by internal combustion and gasoline-powered vehicles.

In Module 5 we discussed the emerging electric vehicles and the new batteries. As we suggested, battery technology has much maturing to do before becoming a strong option for most vehicle needs. Take the issue of weight comparison between gasoline-powered, internal combustion engines versus battery-powered motors. One gallon of gasoline weighs about 6 pounds. To produce the same energy as the 6 pounds of gasoline requires nearly 400 pounds of battery.

With the issue of air pollution, battery-powered vehicles also account for harmful emissions, albeit removed from the location of the electric vehicles. To produce electricity for recharging batteries, generating plants using fossil fuel waste as much as 65% of the energy to transform the fuel into electricity, and then there is another loss of 5–10% to wire the electricity down to the battery-charging receptacle. So generating electricity for battery charging not only creates harmful emissions, it is an inefficient use of fossil fuel.

Figure 9-68 "Smart coatings" may buy more time for the auto industry to improve power systems for vehicles that will be easier on our air quality. Smog-eating radiators use PremAir™ coatings to convert ozone (O_3)—the main component of smog—into breathable oxygen (O_2). (Ford Motor Company)

With the issue of batteries, materials science has yet to develop reliable, long-life batteries. Many other issues challenge the viability of battery-powered vehicles, including cost and fire hazards associated with current batteries. Experts feel that we should pursue more viable and cost-effective solutions for dealing with the many environmental issues of the internal combustion engine for vehicles, and wait for battery technology to mature before committing to electric vehicles.

An example of research aimed at cleaner internal combustion engines involves using a new coating on radiators (Figure 9-68) that actually offers the potential to clean the air as the engine runs rather than contributing to pollution. Such a development may buy more time for economical and practical improvements for vehicles powered by electricity, hydrogen, or other alternative fuels. The 2000 Volvo S80 sedans are equipped with the PremAir™ coated radiators.

The technically literate must enter into debates about the impact of technologies on our world and its inhabitants. Those who fear technology need to hear from those who understand it. To parallel the saying, "Those who don't know the lessons of history are doomed to repeat them," we can say, "Those who don't understand the fundamentals of technology may suffer from it and not enjoy its potential benefits."

The knowledge you have gained through this study should serve you well in everyday life and in your career. We would love to hear your comments on this book.

Good luck into the future!

SELF-ASSESSMENT

9-1. Would a material that has an electrical resistivity of $10^{12}\ \Omega \cdot cm$ make a good conductor or a good insulator?

9-2. Express a dielectric strength of 2000 V per mil in units of (a) V/cm; (b) V/m.

9-3. The parallel plates of an air-filled condenser are 1 mm apart. What is the plate's area (in square meters) if the capacitance is 1 nF?

9-4. Express the permitivity constant of free space in terms of picofarads per millimeter.

9-5. In selecting an optimum polymeric material from a table of dielectric strengths, discuss what effects a lack of knowledge of the thickness of the material selected would have if it were used in a capacitor.

9-6. In designing an electrical circuit, if the resistance is 16 Ω, the current is 7.5 A, and the resistivity is $5.6 \times 10^{-6}\ \Omega \cdot cm$, what is the conductivity?

9-7. A perfect dielectric has an angle δ of how many degrees?

9-8. What is the product of κ × tan δ called?

9-9. Name the three types of polarization discussed in this chapter.

9-10. A material that allows the passage of an electric field but does not conduct electrical current is known by what other name?

9-11. $BaTiO_3$ is chosen as a dielectric for a condenser. The area of the plates is 0.2 cm^2 and the separation between the plates is 0.05 cm. Find the capacitance.

9-12. Using Figure 9-15, the wavelength of red light is about 700 nm. What is the frequency of red light?

9-13. Explain why the electrical conductivity of semiconductor materials increases as temperature increases.

9-14. Iron and steel parts in electrical machinery, by necessity, must possess high permeability and preferably should be made of "soft" iron. What is the significance of this statement?

9-15. Give an example of the use of a metal with high residual magnetism.

9-16. Photoelectric cells are used in a variety of ways. Are solar batteries used in space such an application? If so, how efficient are these devices, in your opinion, in converting solar energy into electrical energy?

9-17. Describe two ways of producing X rays.

9-18. Distinguish among translucence, transparency, and transmittance.

9-19. Photoelasticity is sometimes defined as a visual, full-field technique for measuring stresses in plastic models of parts, usually before the parts are made. In relation to the material presented in the book, does this definition really define the term? Knowing that this stress analysis of plastic models uses a polariscope, where does photoelasticity play a part in the technique?

9-20. Thermal resistivity is as important as thermal conductivity. Illustrate both properties by citing an example of each from your own experience or knowledge.

9-21. Using magnetic chucks in a machining operation as an example, how does a knowledge of magnetic properties of metals pay big dividends?

9-22. Using Table 9-1 with room temperature of 68°F, determine the resistance of a copper wire with a length of 100 m and radius of 1 mm.

9-23. Monochromatic green light of frequency 6.2×10^{14} Hz produced by a laser consists of 5×10^{15} photons, which pass a point in 1 s. How much energy (joules) does each photon possess? (*Hint:* $E = \hbar f$.)

9-24. The minimum energy needed to extract electrons from inside a metal is known as the work function, *W*. *W* depends on the metal and its surface and is equal to the product of the threshold frequency, f_0, and Planck's constant, $\hbar$. If $W = 1.9$ eV, what is the minimum frequency of the light illuminating a metal surface that will release photoelectricity?

9-25. Yellow light with a wavelength of 5×10^{-7} m traveling in a vacuum enters a glass plate ($n = 1.5$). What is the velocity of yellow light in the glass plate in meters per second (m/s^{-1})?

9-26. Integrated circuits use many types of solid-state devices. Name three such devices.

9-27. What is the purpose of doping a semiconductor material?

9-28. Compare the energy required to raise a donor electron to the conduction band with the energy required to raise a valence electron.

9-29. Name one method by which an ideal semiconductor having no impurity atoms can be made to conduct electrons.

9-30. Once a donor electron has been raised into the conduction band by the receipt of energy from some external source, conduction occurs as a result of the combined actions of what charged bodies?

9-31. A Group VI element, selenium, is a semiconductor element. From what group in the periodic table would you choose a dopant to convert selenium to (a) a *p*-type material; (b) an *n*-type material?

9-32. Identify what are termed the majority carriers in an *n*-type semiconductor.

9-33. Explain what is meant by the term *biasing*.

9-34. The depletion zone in a *pn* junction differs in size when a forward bias is applied from a condition of zero bias. Specify what this difference is in terms of area.

9-35. What device operates in opposite fashion to a solar cell and produces light from electrical energy?

9-36. What creates the voltage in a typical solar cell that drives electrical current through the external circuit?

9-37. Reflection of sunlight from the surface layers of a photovoltaic cell is maximized by treating the surface with some form of antireflective coating. True or false? Explain.

9-38. Down through the ages, humans have dreamed of converting sunlight directly into electricity. How efficient is this process today? Define *efficiency* as used in this process.

9-39. Name four possible things that can happen to sunlight entering a solar cell.

9-40. Explain the terms *solar cell, module,* and *array.*

9-41. What is the major cause of poor efficiency of polycrystalline silicon cells in relation to single-crystal silicon cells?

9-42. Amorphous silicon used in handheld calculators is more efficient under fluorescent lighting than either single- or polycrystalline silicon. Name one characteristic that amorphous silicon has that explains this advantage.

9-43. Define the word *epitaxy* and explain why it is used in the semiconductor industry.

9-44. Both LED and LCD devices produce light. Which one requires more energy and consequently needs a battery source of power to function?

9-45. Optoelectronic devices all operate in accordance with a basic underlying scientific principle. Explain this principle.

9-46. The three types of liquid crystals are based on what characteristic of liquid crystals?

9-47. Name three influences that can change the structure of liquid crystals.

9-48. A homogeneous orientation of LC molecules can be accomplished in several ways. Name two such methods other than the use of an electric field.

9-49. In a reflective-type LC cell, the light source and the scattered light are on the same side of the cell. One glass plate is given a reflective coating. A transmitive-type cell's light source is on one side of the cell and the scattered light appears on the other. For this to happen, how are the glass plates modified, if at all? Draw a sketch of a transmitive-type cell similar to Figure 9-28.

9-50. Referring to Figure 9-50 to display the numeral "7," which segments need to be activated?

9-51. Referring to Figure 9-41, which axis (x or y) is known as the electric axis?

9-52. Name several applications of piezoelectric crystals.

9-53. The highest frequency of sounds that can be detected by the human ear is about how many cycles per second? Express in units of hertz.

9-54. A certain pickup for a record player converts the movements of a stylus into corresponding electrical signals. Other devices, such as a loudspeaker, use a similar element, a crystal. What material property peculiar to certain materials do these devices depend on to perform their functions?

9-55. Superconductors have great technological potential for many applications in this energy-dependent world. What is one difficulty that limits their development?

9-56. If a battery providing 1.5 volts of potential difference to light a small light bulb generates 300 mA of current flow, determine the resistance of the light bulb.

9-57. A copper wire 20 m in length is used to provide electricity to a device that offers 0.10 ohms of resistance. Find the diameter of the wire of sufficient size to carry this voltage and current. The resistivity of copper is $1.7 \times 10^{-8} \Omega$-m.

9-58. The maximum force on an electric wire is 4.14 N. The circuit carries a current of 25 A when placed across the poles of a circular magnet whose diameter is 22 cm. What is the strength of the magnetic field?

OBJECTIVE QUESTIONS

9-1. The dielectric constant for water.
  a. 7×10^6 Hz   **b.** 12×10^6 Hz   **c.** 78×10^6 Hz

9-2. A conductivity in ohm^{-1} cm^{-1} for a common metal would be _____ .
  a. 4,000   **b.** 100,000   **c.** 700,000

9-3. A dielectric is a nonconductive material whose polarization is caused by _____ .
  a. a magnetic field   **b.** an electric field   **c.** IR radiation

9-4. Dielectric constants are in demand by the _____ .
  a. machine tool industry   **b.** electronics industry   **c.** oil and gas industry

9-5. If the Mn atom loses two electrons, its paramagnetic properties _____ .

    a. stay the same   **b.** increase   **c.** decrease

9-6. The Na atom has an atomic number of 11. It is _____ .

    a. diamagnetic   **b.** paramagnetic   **c.** neither

9-7. The atomic number of Cl is 17, which tells us that Cl is _____ .

    a. diamagnetic   **b.** paramagnetic   **c.** ferromagnetic

9-8. A pair of electrons with opposing spins has _____ .

    a. a large magnetic field   **b.** a small magnetic field   **c.** no magnetic field

9-9. An example of a transition element.

    a. B   **b.** C   **c.** Sc

9-10. The wavelength of microwaves is about _____ .

    a. 10^{-5} nm   **b.** 30 m   **c.** 0.3 cm

9-11. The hysteresis loss (ergs/cm^3/cycle) for 4340 steel at 25°C is _____ .

    a. 2.7×10^4 **b.** 1.5×10^4 **c.** 11.4×10^4

9-12. High-energy protons are produced by _____ .

    a. alpha rays   **b.** X rays   **c.** beta rays

9-13. Electronic polarization deals with the polarity of _____ .

    a. atoms   **b.** ions   **c.** protons

9-14. The bending of light is known as _____ .

    a. reflection   **b.** refraction   **c.** incidence

9-15. The index of refraction for diamond is _____ .

    a. 0.67   **b.** 1.36   **c.** 2.42

9-16. Optical anisotropy refers to a material with _____ optical properties.

    a. good   **b.** poor   **c.** different

9-17. The size of the forbidden gap in conductors is _____ .

    a. large   **b.** zero   **c.** small

9-18. An electric current of one coulomb of charge passing a point in one second is known as a(n) _____ .

    a. micron   **b.** ampere   **c.** angstrom

9-19. The SI prefix μ (mu) stands for _____ .

    a. 10^{-6} in.   **b.** 10^{-6} **c.** 10^{-6} m

9-20. The *p*-side of a diode is called the _____ .

    a. cathode   **b.** anode   **c.** pentode

9-21. Solar cells are expensive primarily because of the cost of _____ .

    a. material   **b.** fabrication   **c.** tooling

9-22. A YCBO is a _____ .

    a. small yard tug   **b.** superconductor   **c.** computer memory code

9-23. High-T_c superconductors have an advantage over normal conductors because they offer no _____ to the flow of current.

    a. strain   **b.** resistance   **c.** dielectric   **d.** permeability

9-24. Which smart material might work a robot finger control or an antenna for a cellular phone because of its ability to remember its original shape after being deformed?

    a. MMC   **b.** piezoelectric   **c.** titanium   **d.** shape memory alloy

9-25. Biomimetics can help intelligent material systems technology by _____ .

    a. providing feedback and control   **b.** ensuring osseointegration

    c. aiding improved hybrids   **d.** prompting biotransfer

9-26. What is a main concern when selecting a material for bioengineering?

    a. Ability to resist heat

    b. Ability to provide feedback and control

    c. Compatibility with human and animal systems

    d. Good formability

REFERENCES & RELATED READINGS

ASM INTERNATIONAL. "Smart Materials Alter Shape of Adaptive Aircraft Wings," *Advanced Materials & Processes,* September 1995, p. 9.

CONSTANTZ, BRENT R. "Mineral Paste Surgically Injected to Repair Bones," Norian Corp., Cupertino, CA 95014, *Advanced Materials and Processes,* June 1995, p. 17.

DE NEUFVILLE, RICHARD, et al. "The Electric Car Unplugged," *Technology Review,* January 1996, pp. 30–36.

"Don't Let the Sun Shine In." Webasco Sunroofs of Maumee, OH, *Popular Science,* June 1994, p. 12.

DREXLER, K. ERIC. *Nanosystems—Molecular Machinery, Manufacturing, and Computation.* New York: John Wiley and Sons, Inc., 1992.

FISCHER, ARTHUR. "Superconductivity," *Popular Science,* April 1988, pp. 54–58.

FONASH, STEPHEN J. "Micro and Nanofabrication," Surface Engineering and Manufacturing Technology Center (SEMTC), U.S. Navy Manufacturing Science & Technology Program Report, February 1995, p. 3.

"Get Smart," *Compressed Air Magazine,* April/May 1995, pp. 36–41.

HAZEN, ROBERT M. "The New Alchemy," *Technology Review,* MIT, Bldg. W59, Cambridge, MA 02139, November/December 1994, pp. 24–26.

HUMMEL, ROLF E. *Electronic Properties of Materials,* 2nd ed. New York: Springer-Verlag, 1992.

KLIENFIELD, SONNY. *A Machine Called Indomitable.* New York: Time Books, 1985, pp. 92–97.

MCWHORTER, PAUL, JEFF SNIEGOWSKI, and ERNEST GARCIA. "Mighty Micrometers Drive Miniature Gears," *Advanced Materials and Processes,* December 1995, p. 6.

MALLARDI, JOSEPH L. *From Teeth to Jet Engines,* Howmet Corporation, 1992.

"Manufacturing Parts Drop by Drop," *Compressed Air Magazine,* March 1995, pp. 38–44.

Moving America: New Directions, New Opportunities, report to Congress, June 1990.

"Nanotechnology," *Compressed Air Magazine,* October/November 1994, pp. 16–23.

National MAGLEV Initiative: Annual Report—November 1992. Washington, DC: Federal Railroad Administration, 1992.

"Nimbus Information System." *CD-ROM Replication Guide.* Nimbus Manufacturing, Inc., Charlottesville, VA, 1995.

OLIVER, JOYCE. "Superconductors: The Heat Is On," *The Electron,* November/December 1988, pp. 12–17.

SARIKAYA, MEHMET, and AKSAY, ILHAN A. AKSAY (eds.). *Biomimetics: Design and Processing of Materials,* AIP Press, 1995.

SONG, KYO D. *Development of Microwave-driven Smart Materials Actuators.* NASA project overview, March 1, 2000.

"Stealth Drapes." Marvin Corp., Warroad, MN, *Popular Science,* July 1995, p. 10.

TUOMINEN, STEVE, and CRAIG WOJCIK. "Unique Alloys for Aerospace and Beyond," *Outlook Magazine,* Teledyne Wah Chang, Vol. 16, No. 2, 1995, pp. 1–6.

UENOHARA, MICHIYKI. "Electro-Ceramics and Applications in C&C Era," *Journal of Materials Education,* Vol. 10, No. 3, 1988, pp. 259–280.

WEBER, WILLIAM J. "Highly Conducting Electroceramics," Pacific Northwest Laboratory, October 1992.

10
Appendix Tables

TABLE 10-1A PERIODIC TABLE OF THE ELEMENTS

GROUP

Each element cell lists: Atomic Number, Atomic Weight, Oxidation States; Boiling Point (K), Melting Point (K), Density (g/cm³); Symbol; Electron Configuration; Name.

IA

#	Weight	Ox.	BP (K)	MP (K)	Density	Config	Name
1	1.00794	1	20.28	13.81	0.0899 †	$1s^1$	Hydrogen

IIA

#	Weight	Ox.	BP (K)	MP (K)	Density	Config	Name
3	6.941	1	1615	453.7	0.53	$1s^2 2s^1$	Lithium
4	9.01218	2	3243	1560	1.85	$1s^2 2s^2$	Beryllium
11	22.98977	1	1156	371.0	0.97	$[Ne]3s^1$	Sodium
12	24.305	2	1380	922	1.74	$[Ne]3s^2$	Magnesium

IIIA IVA VA VIA VIIA VIIIA

#	Weight	Ox.	BP (K)	MP (K)	Density	Config	Name
19	39.0983	1	1033	336.8	0.86	$[Ar]4s^1$	Potassium
20	40.078	2	1757	1112	1.55	$[Ar]4s^2$	Calcium
21	44.9559	3	3109	1814	2.99	$[Ar]3d^1 4s^2$	Scandium
22	47.87	4,3	3560	1935	4.54	$[Ar]3d^2 4s^2$	Titanium
23	50.9415	5,4,3,2	3650	2163	6.11	$[Ar]3d^3 4s^2$	Vanadium
24	51.996	6,3,2	2945	2130	7.19	$[Ar]3d^5 4s^1$	Chromium
25	54.9380	7,6,4,2,3	2235	1518	7.44	$[Ar]3d^5 4s^2$	Manganese
26	55.845	2,3	3023	1808	7.874	$[Ar]3d^6 4s^2$	Iron
27	58.9332	2,3	3143	1768	8.90	$[Ar]3d^7 4s^2$	Cobalt
37	85.4678	1	961	312.63	1.532	$[Kr]5s^1$	Rubidium
38	87.62	2	1655	1042	2.54	$[Kr]5s^2$	Strontium
39	88.9059	3	3611	1795	4.47	$[Kr]4d^1 5s^2$	Yttrium
40	91.224	4	4682	2128	6.51	$[Kr]4d^2 5s^2$	Zirconium
41	92.9064	5,3	5015	2742	8.57	$[Kr]4d^4 5s^1$	Niobium
42	95.94	6,5,4,3,2	4912	2896	10.22	$[Kr]4d^5 5s^1$	Molybdenum
43	(98)	7	4538	2477	11.5	$[Kr]4d^5 5s^2$	Technetium
44	101.07	2,3,4,6,8	4425	2610	12.37	$[Kr]4d^7 5s^1$	Ruthenium
45	102.9055	2,3,4	3970	2236	12.41	$[Kr]4d^8 5s^1$	Rhodium
55	132.9054	1	944	301.54	1.87	$[Xe]6s^1$	Cesium
56	137.33	2	2078	1002	3.59	$[Xe]6s^2$	Barium
57	138.9055	3 ★	3737	1191	6.15	$[Xe]5d^1 6s^2$	Lanthanum
72	178.49	4	4875	2504	13.31	$[Xe]4f^{14} 5d^2 6s^2$	Hafnium
73	180.9479	5	5730	3293	16.65	$[Xe]4f^{14} 5d^3 6s^2$	Tantalum
74	183.84	6,5,4,3,2	5825	3695	19.3	$[Xe]4f^{14} 5d^4 6s^2$	Tungsten
75	186.207	7,6,4,2	5870	3455	21.0	$[Xe]4f^{14} 5d^5 6s^2$	Rhenium
76	190.23	2,3,4,6,8	5300	3300	22.6	$[Xe]4f^{14} 5d^6 6s^2$	Osmium
77	192.22	2,3,4,6	4700	2720	22.6	$[Xe]4f^{14} 5d^7 6s^2$	Iridium
87	(223)	1	950	300		$[Rn]7s^1$	Francium
88	(226)	2	1413	973	5.0	$[Rn]7s^2$	Radium
89	(227)	3 ★★	3470*	1324	10.07	$[Rn]6d^1 7s^2$	Actinium
104	(261)					$[Rn]5f^{14} 6d^2 7s^2$	(Rutherfordium)
105	(262)					$[Rn]5f^{14} 6d^3 7s^2$	(Dubnium)
106	(263)					$[Rn]5f^{14} 6d^4 7s^2$	(Seaborgium)
107	(262)					$[Rn]5f^{14} 6d^5 7s^2$	(Bohrium)
108	(265)					$[Rn]5f^{14} 6d^6 7s^2$	(Hassium)
109	(266)					$[Rn]5f^{14} 6d^7 7s^2$	(Meitnerium)

* Estimated Values

KEY

- ATOMIC NUMBER
- ATOMIC WEIGHT (2)
- BOILING POINT, K
- MELTING POINT, K
- OXIDATION STATES (Bold most stable)
- SYMBOL
- DENSITY at 300 K (3) (g/cm³)
- ELECTRON CONFIGURATION
- NAME

#	Weight	Ox.	BP (K)	MP (K)	Density	Config	Name
30	65.39	2	1180	692.73	7.13	$[Ar]3d^{10} 4s^2$	Zinc

★ Lanthanides:

#	Weight	Ox.	BP (K)	MP (K)	Density	Config	Name
58	140.12	3,4	3715	1071	6.77	$[Xe]4f^1 5d^1 6s^2$	Cerium
59	140.9077	3,4	3785	1204	6.77	$[Xe]4f^3 6s^2$	Praseodymium
60	144.24	3	3347	1294	7.01	$[Xe]4f^4 6s^2$	Neodymium
61	(145)	3	3273	1315	7.22	$[Xe]4f^5 6s^2$	Promethium
62	150.36	3,2	2067	1347	7.52	$[Xe]4f^6 6s^2$	Samarium
63	151.964	3,2	1800	1095	5.24	$[Xe]4f^7 6s^2$	Europium

★★ Actinides:

#	Weight	Ox.	BP (K)	MP (K)	Density	Config	Name
90	232.0381	4	5060	2028	11.72	$[Rn]6d^2 7s^2$	Thorium
91	231.0359	5,4	4300*	1845	15.4	$[Rn]5f^2 6d^1 7s^2$	Protactinium
92	238.029	6,5,4,3	4407	1408	18.95	$[Rn]5f^3 6d^1 7s^2$	Uranium
93	(237)	6,5,4,3	4175*	912	20.2	$[Rn]5f^4 6d^1 7s^2$	Neptunium
94	(244)	6,5,4,3	3505	913	19.84	$[Rn]5f^6 7s^2$	Plutonium
95	(243)	6,5,4,3	2880	1449	13.7	$[Rn]5f^7 7s^2$	Americium

NOTES:
(1) Black — solid.
Red — gas.
Blue — liquid.
Outline — synthetically prepared.
(2) Based upon carbon-12. ( ) indicates most stable or best known isotope.
(3) Entries marked with daggers refer to the gaseous state at 273 K and 1 atm and are given in units of g/l.

VIII

2	4.00260
4.216
0.95
at 26 atm
0.1785†	**He**
1s²
Helium

IIIB	IVB	VB	VIB	VIIB

5	10.811 3	6	12.011 ±4,2	7	14.0067 ±3,5,4,2	8	15.9994 -2	9	18.99840 -1	10	20.1797
4275		5100*		77.344		90.188		85.0		27.10	
2365	**B**	3825*	**C**	63.15	**N**	54.8	**O**	53.55	**F**	24.55	**Ne**
2.34		2.26		1.251†		1.429†		1.696†		0.900†	
1s²2s²p¹		1s²2s²p²		1s²2s²p³		1s²2s²p⁴		1s²2s²p⁵		1s²2s²p⁶	
Boron		Carbon		Nitrogen		Oxygen		Fluorine		Neon	

13	26.98154 3	14	28.0855 4,2	15	30.97376 ±3,5,4	16	32.066 ±2,4,6	17	35.4527 ±1,3,5,7	18	39.948
2740		2630		553		717.82		239.18		87.45	
933.5	**Al**	1683	**Si**	317.3	**P**	392.2	**S**	172.17	**Cl**	83.95	**Ar**
2.70		2.33		1.82		2.07		3.214†		1.784†	
[Ne]3s²p¹		[Ne]3s²p²		[Ne]3s²p³		[Ne]3s²p⁴		[Ne]3s²p⁵		[Ne]3s²p⁶	
Aluminum		Silicon		Phosphorus		Sulfur		Chlorine		Argon	

IB	IIB

28	58.6934 2,3	29	63.546 2,1	30	65.39 2	31	69.723 3	32	72.61 4	33	74.9216 ±3,5	34	78.96 -2,4,6	35	79.904 ±1,5,7	36	83.80 0,2
3005		2840		1180		2478		3107		876 (subl.)		958		331.85		120.85	
1726	**Ni**	1356.6	**Cu**	692.73	**Zn**	302.92	**Ga**	1211.5	**Ge**	1090 (28 atm)	**As**	494	**Se**	265.95	**Br**	116	**Kr**
8.90		8.96		7.13		5.91		5.32		5.78		4.79		3.12		3.75†	
[Ar]3d⁸4s²		[Ar]3d¹⁰4s¹		[Ar]3d¹⁰4s²		[Ar]3d¹⁰4s²p¹		[Ar]3d¹⁰4s²p²		[Ar]3d¹⁰4s²p³		[Ar]3d¹⁰4s²p⁴		[Ar]3d¹⁰4s²p⁵		[Ar]3d¹⁰4s²p⁶	
Nickel		Copper		Zinc		Gallium		Germanium		Arsenic		Selenium		Bromine		Krypton	

46	106.42 2,4	47	107.868 1	48	112.41 2	49	114.82 3	50	118.710 4,2	51	121.760 ±3,5	52	127.60 -2,4,6	53	126.9045 ±1,5,7	54	131.29 0,2,4,6
3240		2436		1040		2350		2876		1860		1261		457.5		165.1	
1825	**Pd**	1235.08	**Ag**	594.26	**Cd**	429.78	**In**	505.12	**Sn**	903.91	**Sb**	722.72	**Te**	386.7	**I**	161.39	**Xe**
12.0		10.50		8.65		7.31		7.31		6.69		6.24		4.93		5.90†	
[Kr]4d¹⁰		[Kr]4d¹⁰5s¹		[Kr]4d¹⁰5s²		[Kr]4d¹⁰5s²p¹		[Kr]4d¹⁰5s²p²		[Kr]4d¹⁰5s²p³		[Kr]4d¹⁰5s²p⁴		[Kr]4d¹⁰5s²p⁵		[Kr]4d¹⁰5s²p⁶	
Palladium		Silver		Cadmium		Indium		Tin		Antimony		Tellurium		Iodine		Xenon	

78	195.08 2,4	79	196.9665 3,1	80	200.59 2,1	81	204.383 3,1	82	207.2 4,2	83	208.9804 3,5	84	(209) 4,2,6	85	(210) ±1,3,5,7	86	(222) 0,2
4100		3130		629.88		1746		2023		1837		—		610*		211.4	
2042.1	**Pt**	1337.58	**Au**	234.31	**Hg**	577	**Tl**	600.65	**Pb**	544.59	**Bi**	527	**Po**	575	**At**	202	**Rn**
21.45		19.3		11.85		11.35		11.35		9.75		9.3				9.73	
[Xe]4f¹⁴5d⁹6s¹		[Xe]4f¹⁴5d¹⁰6s¹		[Xe]4f¹⁴5d¹⁰6s²		[Xe]4f¹⁴5d¹⁰6s²p¹		[Xe]4f¹⁴5d¹⁰6s²p²		[Xe]4f¹⁴5d¹⁰6s²p³		[Xe]4f¹⁴5d¹⁰6s²p⁴		[Xe]4f¹⁴5d¹⁰6s²p⁵		[Xe]4f¹⁴5d¹⁰6s²p⁶	
Platinum		Gold		Mercury		Thallium		Lead		Bismuth		Polonium		Astatine		Radon	

110	(269)	111	(272)
=	**Uun**	=	**Uuu**
[Rn]5f¹⁴6d⁸7s¹*		[Rn]5f¹⁴6d⁹7s¹*	
(Ununnilium)		(Unununium)	

The A & B subgroup designations, applicable to elements in rows 4, 5, 6 and 7, are those recommended by the International Union of Pure and Applied Chemistry.

64	157.25	65	158.9253 3,4	66	162.50	67	164.9303 3	68	167.26 3	69	168.9342 3,2	70	173.04 3,2	71	174.967 3
3545		3500		2840		2968		3140		2223		1469		3668	
1585	**Gd**	1629	**Tb**	1685	**Dy**	1747	**Ho**	1802	**Er**	1818	**Tm**	1092	**Yb**	1936	**Lu**
7.90		8.23		8.55		8.80		9.07		9.32		6.97		9.84	
[Xe]4f⁷5d¹6s²		[Xe]4f⁹6s²		[Xe]4f¹⁰6s²		[Xe]4f¹¹6s²		[Xe]4f¹²6s²		[Xe]4f¹³6s²		[Xe]4f¹⁴6s²		[Xe]4f¹⁴5d¹6s²	
Gadolinium		Terbium		Dysprosium		Holmium		Erbium		Thulium		Ytterbium		Lutetium	

96	(247) 3	97	(247) 4,3	98	(251) 3	99	(252) 3	100	(257) 3	101	(258) 3	102	(259) 2,3	103	(262) 3
1620		—		1170*		1130*		1800*		1100*		1100*		1900*	
13.5	**Cm**	14*	**Bk**		**Cf**		**Es**		**Fm**		**Md**		**No**		**Lr**
[Rn]5f⁷6d¹7s²		[Rn]5f⁹7s²		[Rn]5f¹⁰7s²		[Rn]5f¹¹7s²		[Rn]5f¹²7s²		[Rn]5f¹³7s²		[Rn]5f¹⁴7s²		[Rn]5f¹⁴7s²7p¹	
Curium		Berkelium		Californium		Einsteinium		Fermium		Mendelevium		Nobelium		Lawrencium	

VWR Scientific Products

Side 1

Science Education

Catalog Number WLS-18806

P.O. Box 5229, Buffalo Grove, IL 60089-5229
1-800-727-4368 FAX 1-800-676-2540

TABLE 10-1B PERIODIC PROPERTIES OF THE ELEMENTS

1.9	2.0	2.1	2.2	2.3	2.4	2.5	2.6	2.7	2.8	2.9	3.0	3.1	3.2
59	63	67	70	74	76	79	82	84	86	88	89	91	92

The A & B subgroup designations, applicable to elements in rows 4, 5, 6 and 7, are those recommended by the International Union of Pure and Applied Chemistry.

NOTES: (1) For representative oxides (higher valence) of group. Oxide is acidic if color is red, basic if color is blue and amphoteric if both colors are shown. Intensity of color indicates relative strength.

(2) Cubic, face centered; cubic, body centered; cubic; hexagonal; rhombohedral; tetragonal; orthorhombic; monoclinic.

(3) At 300 K (27°C) (6) Generally at 293 K (20°C) (8) From density at 300 K (27°C) for liquid
(4) At boiling point   for polycrystalline material   and solid elements; values for gaseous elements
(5) At melting point (7) Quantum mechanical value   refer to liquid state at boiling point
                              for free atom

VWR Scientific Products
Science Education
P.O. Box 5229, Buffalo Grove, IL 60089-5229
1-800-727-4368 FAX 1-800-676-2540

Catalog Number WLS-18806

SIDE 2

TABLE 10-2 SYMBOLS OF THE ELEMENTS AND THEIR ATOMIC NUMBERS AND WEIGHTS[a]

Name	Symbol	Atomic Number	Atomic Weight[b]	Name	Symbol	Atomic Number	Atomic Weight[b]
Actinium	Ac	89	(227)	Lawrencium	Lr	103	(257)
Aluminum	Al	13	27.0	Lead	Pd	82	207.2
Americium	Am	95	(243)	Lithium	Li	3	6.94
Antimony	Sb	51	121.8	Lutetium	Lu	71	175.0
Argon	Ar	18	39.9	Magnesium	Mg	12	24.3
Arsenic	As	33	74.9	Manganese	Mn	25	54.9
Astatine	At	85	(210)	Mendelevium	Md	101	(256)
Barium	Ba	56	137.3	Mercury	Hg	80	200.6
Berkelium	Bk	97	(247)	Molybdenum	Mo	42	95.9
Beryllium	Be	4	9.01	Neodymium	Nd	60	144.2
Bismuth	Bi	83	209.0	Neon	Ne	10	20.2
Boron	B	5	10.8	Neptunium	Np	93	(237)
Bromine	Br	35	79.9	Nickel	Ni	28	58.7
Cadmium	Cd	48	112.4	Niobium	Nb	41	92.9
Calcium	Ca	20	40.1	Nitrogen	N	7	14.01
Californium	Cf	98	(251)	Nobelium	No	102	(254)
Carbon	C	6	12.01	Osmium	Os	76	190.2
Cerium	Ce	58	140.1	Oxygen	O	8	16.00
Cesium	Cs	55	132.9	Palladium	Pd	46	106.4
Chlorine	Cl	17	35.5	Phosphorus	P	15	31.0
Chromium	Cr	24	52.0	Platinum	Pt	78	195.1
Cobalt	Co	27	58.9	Plutonium	Pu	94	(242)
Copper	Cu	29	63.5	Polonium	Po	84	(210)
Curium	Cm	96	(247)	Potassium	K	19	39.1
Dysprosium	Dy	66	162.5	Praseodymium	Pr	59	140.9
Einsteinium	Es	99	(254)	Promethium	Pm	61	(147)
Erbium	Er	68	167.3	Protactinium	Pa	91	(231)
Europium	Eu	63	152.0	Radium	Ra	88	(226)
Fermium	Fm	100	(253)	Radon	Rn	86	(222)
Fluorine	F	9	19.0	Rhenium	Re	75	186.2
Francium	Fr	87	(223)	Rhodium	Rh	45	102.9
Gadolinium	Gd	64	157.3	Rubidium	Rb	37	85.5
Gallium	Ga	31	69.7	Ruthenium	Ru	44	101.1
Germanium	Ge	32	72.6	Samarium	Sm	62	150.4
Gold	Au	79	197.0	Scandium	Sc	21	45.0
Hafnium	Hf	72	178.5	Selenium	Se	34	79.0
Helium	He	2	4.00	Silicon	Si	14	28.1
Holmium	Ho	67	164.9	Silver	Ag	47	107.9
Hydrogen	H	1	1.008	Sodium	Na	11	23.0
Indium	In	49	114.8	Strontium	Sr	38	87.6
Iodine	I	53	126.9	Sulfur	S	16	32.1
Iridium	Ir	77	192.2	Tantalum	Ta	73	180.9
Iron	Fe	26	55.8	Technetium	Te	43	(99)
Krypton	Kr	36	83.8	Tellurium	Te	52	127.6
Lanthanum	La	57	138.9	Terbium	Tb	65	158.9

[a] The newest elements (rutherfordium, Rf-104, and hahnium, Ha-105) are not included.

[b] The values given in parentheses are mass numbers of the principal isotopes of unstable elements.

TABLE 10-2   (continued)

Name	Symbol	Atomic Number	Atomic Weight[b]	Name	Symbol	Atomic Number	Atomic weight[b]
Thallium	Tl	81	204.4	Vanadium	V	23	50.9
Thorium	Th	90	232.0	Xenon	Xe	54	131.3
Thulium	Tm	69	168.9	Ytterbium	Yb	70	173.0
Tin	Sn	50	118.7	Yttrium	Y	39	88.9
Titanium	Ti	22	47.9	Zinc	Zn	30	65.4
Tungsten	W	74	183.9	Zirconium	Zr	40	91.2
Uranium	U	92	238.0				

TABLE 10-3   GREEK SYMBOLS AND THEIR PRONUNCIATIONS

Γ	gamma (cap.)	η	eta (lc)
Δ	delta (cap.)	θ	theta (lc)
Θ	theta (cap.)	ϑ	theta (lc)
Λ	lambda (cap.)	κ	kappa (lc)
Ξ	xi (cap.)	λ	lambda (lc)
Π	pi (cap.)	μ	mu (lc)
Σ	sigma (cap.)	ν	nu (lc)
Υ	upsilon (cap.)	ξ	xi (lc)
Φ	phi (cap.)	π	pi (lc)
Ψ	psi (cap.)	ρ	rho (lc)
Ω	omega (cap.)	σ	sigma (lc)
α	alpha (lc)	τ	tau (lc)
β	beta (lc)	ϕ	phi (lc)
γ	gamma (lc)	φ	phi (lc)
δ	delta (lc)	χ	chi (lc)
ϵ	epsilon (lc)	ψ	psi (lc)
ζ	zeta (lc)	ω	omega (lc)

10.1 INTERNATIONAL SYSTEM OF UNITS AND UNIT CONVERSIONS

SI was created in 1960 by international agreement and represents a worldwide measurement system far superior to earlier measurement systems (gravitational) in expressing scientific and technical data. All gravitational systems using force as a fundamental dimension, including the American engineering system with its pound of mass a fundamental unit, are considered obsolete, and the changeover, though not mandatory, is proceeding at a steady pace, with some industries completing the changeover in record time. For most technology users, this means that they will continue to be confronted with problems that arise when two systems of measurement exist. Consequently, they must be very familiar with both systems and demonstrate ability to convert from one to another upon completion of their calculations.

Units.   SI units are grouped into three general classes: base or fundamental units, derived units, and supplementary units. Table 10-4 lists the base units as well as some

TABLE 10-4 NAMES AND SYMBOLS OF SI UNITS

Quantity	Name of Unit	Symbol	Expressed in Base Units Where Applicable
Base units			
Length	meter	m	
Mass	kilogram	kg	
Time	second	s	
Electric current	ampere	A	
Thermodynamic temperature	kelvin	K	
Luminous intensity	candela	cd	
Amount of substance	mole	mol	
Derived units			
Area	square meter	m^2	
Volume	cubic meter	m^3	
Frequency	hertz, cycle per second	Hz	s^{-1}
Density (mass)	kilogram per cubic meter	kg/m^3	
Velocity (linear)	meter per second	m/s	
Velocity (angular)	radian per second	rad/s	
Acceleration (linear)	meter per second squared	m/s^2	
Acceleration (angular)	radian per second squared	rad/s^2	
Force	newton, kilogram-meter per second squared	N	$kg \cdot m \cdot s^{-2}$
Permeability	henry per meter	H/m	$m \cdot kg \cdot s^{-2} \cdot A^{-2}$
Permittivity	farad per meter	F/m	$m^{-3} \cdot kg^{-1} \cdot s^4 \cdot A^2$
Pressure (mechanical stress)	pascal, newton per square meter	Pa	$N \cdot m^{-2}$
Kinematic viscosity	square meter per second	m^2/s	
Dynamic viscosity	newton-second per square meter	$N \cdot s/m^2$	$m^{-1} \cdot kg \cdot s^{-1}$
Work energy, quantity of heat	joule, newton-meter	$J, N \cdot m$	
Power	watt, joule per second	W, J/s	
Quantity of electricity, electric charge	coulomb	$C, A \cdot s$	
Potential difference, electromotive force	volt	V, W/A	
Electric field strength	volt per meter	V/m	$m \cdot kg \cdot s^{-3} \cdot A^{-1}$
Electric resistance	ohm	Ω, V/A	
Capacitance	farad	$F, A \cdot s/V$	
Magnetic flux	weber	$Wb, V \cdot s$	
Inductance	henry	$H, V \cdot s/A$	
Magnetic flux density	tesla	$T, Wb/m^2$	
Magnetic field strength	ampere per meter	A/m	
Magnetomotive force	ampere	A	
Luminous flux	lumen	lm	
Luminance	candela per square meter	cd/m^2	
Illuminance	lux	lx	
Wave number	1 per meter	m^{-1}	
Entropy	joule per kelvin	J/K	
Specific heat capacity	joule per kilogram kelvin	$J/(kg \cdot K)$	$m^2 \cdot s^{-2} \cdot K^{-1}$
Thermal conductivity	watt per meter kelvin	$W/(m \cdot K)$	$m \cdot kg \cdot s^{-3} \cdot K^{-1}$
Conductance	siemens	S, A/V	
Torque	newton-meter	N/m	$m^2 \cdot kg \cdot s^{-3} \cdot K^{-1}$
Supplementary units			
Plane angle	radian	rad	

TABLE 10-5   SPECIAL NAMES FOR UNITS

Quantity	Unit	Symbol	Formula
Frequency (of a periodic phenomenon)	hertz	Hz	$1/s$
Force	newton	N	$kg \cdot m/s^2$
Pressure, stress	pascal	Pa	N/m^2
Energy, work, quantity of heat	joule	J	$N \cdot m$
Power, radiant flux	watt	W	J/s
Quantity of electricity, electric charge	coulomb	C	$A \cdot s$
Electric potential, potential difference, electromotive force	volt	V	W/A
Capacitance	farad	F	C/V
Electric resistance	ohm	Ω	V/A
Conductance	siemens	S	A/V
Magnetic flux	weber	Wb	$V \cdot s$
Magnetic flux density	tesla	T	Wb/m^2
Inductance	henry	H	Wb/A
Luminous flux	lumen	lm	$cd \cdot sr$
Illuminance	lux	lx	lm/m^2

of the more common derived units used in this book. Special names are given to some of the derived units (see Table 10-5). For example, hertz (Hz) is the special name given to the SI unit for frequency. Units of force, stress, power, and energy also have special names and therefore need not be expressed in their base units. The amount of force required to accelerate 1 kilogram of mass 1 meter per second squared is given the special name of newton (N). Thus $1 \text{ N} = 1 \text{ kg} \cdot m/s^2$.

Prefixes.   Prefixes corresponding to powers of 10 are attached to the units discussed above in order to form larger or smaller units. In technical work the powers of 10 divisible by 3 are preferred. Table 10-6 contains the authorized prefixes.

TABLE 10-6   PREFIXES

Multiplication factor	Prefix	Symbol
1 000 000 000 000 000 000 = 10^{18}	exa	E
1 000 000 000 000 000 = 10^{15}	peta	P
1 000 000 000 000 = 10^{12}	tera	T
1 000 000 000 = 10^9	giga	G
1 000 000 = 10^6	mega	M
1 000 = 10^3	kilo	k
100 = 10^2	hecto	h
10 = 10^1	deka	da
0.1 = 10^{-1}	deci	d
0.01 = 10^{-2}	centi	c
0.001 = 10^{-3}	milli	m
0.000 001 = 10^{-6}	micro	μ
0.000 000 001 = 10^{-9}	nano	n
0.000 000 000 001 = 10^{-12}	pico	p
0.000 000 000 000 001 = 10^{-15}	femto	f
0.000 000 000 000 000 001 = 10^{-18}	atto	a

Rules for usage.   For standardized usage, the following rules should be observed:

1. Uppercase (capitals) and lowercase letters are never interchanged: kg, not KG.

2. The same symbol is used for plurals: N, not Ns; 14 meters or 14 m.

3. No space is left between the prefix and its unit symbol: GHz, not G Hz.

4. To form products, a raised dot is preferred (or a dot on a line): kN · m, or kN.m. The dot may be dispensed with if confusion is not created by its absence.

5. To form quotients, one solidus (an oblique line), a fraction line (horizontal), or a negative power is used to express derived units: m/s, $\frac{m}{s}$, or m · s^{-1}.

    Note: The solidus must not be repeated on the same line: m/s^2, not m/s/s. Also, kg/(m · s), $\frac{kg}{m \cdot s}$, or kg · m^{-1} · s^{-1}, but not kg/m/s. Note also the use of the parentheses to avoid ambiguity.

6. An exponent affixed to a symbol containing a prefix indicates that the multiple or submultiple of the unit is raised to the power expressed by the exponent.

$$1 \text{ mm}^3 = 10^{-9} \text{ m}^3, \text{ not } 10^{-3} \text{ m}^3$$
$$1 \text{ cm}^3 = 10^{-6} \text{ m}^3, \text{ not } 10^{-2} \text{ m}^3$$
$$1 \text{ cm}^{-1} = 10^2 \text{ m}^{-1}, \text{ not } 10^{-2} \text{ m}^{-1}$$

7. A period is used as a decimal marker. It is not used to separate groups of digits. A space is left for this purpose: 5 279 585 J, and 0.000 34 s.

8. Numbers are preferably expressed between the limits 0.1 and 1000, using the appropriate prefix to change the size of the unit: 5.23 GN.

9. For decimal numbers less than 1, the leading zero is never omitted: 0.625, not .625.

10. When units are written in words, they always start with lowercase letters except at the beginning of a sentence. If the unit is derived from the name of a person, the symbol is capitalized. Plurals of special names are written in the usual manner.

125 watts or 125 W
0.25 newton or 0.25 N
58.6 hertz or 58.6 Hz

11. A space or hyphen may be used to form the product expressed in words: newton - meters or newton meters.

12. For quotients, the word *per* may be used: newton per meter squared, kilogram per cubic meter.

13. The kelvin (symbol K) is the standard unit of temperature. In writing this absolute temperature, the word *degree* or its symbol (°) is not used: 472 K. In addition, K may be used to express an interval or a difference in temperature. Celsius temperature is expressed in degrees Celsius with symbol °C. The unit degrees Celsius is equal to the unit kelvin and may also be used to represent an interval or a difference of Celsius temperature: 25°C. Temperature in K = temperature in °C + 273.15.

Computations.   The SI system of units makes computations relatively simple because (1) a single unit is used to represent a particular physical quantity, (2) the system is coherent in that the factor of 1.0 replaces many conversion factors, and (3) SI is based on the decimal system.

Prior to computation, all prefixes should be replaced by their respective powers of 10. The final step is to select a suitable prefix to express the answer once the resulting answer is rounded to the appropriate number of significant digits.

Two examples illustrate the solution of typical problems using SI units and the preceding information.

Problem 1

Given: A metal rod under tensile load of 356 kN is allowed to withstand a unit stress of 110 NM/m².

Required: Find the diameter of the rod in millimeters.

Solution

1. Convert SI prefixes to powers of 10:

$$356 \text{ kN} = 356 \times 10^3 \text{ N}$$

$$110 \text{ MN/m}^2 = 110 \times 10^6 \text{ N/m}^2$$

2. Use the direct stress formula $(s = P/A)$ and solve for the area (A):

$$A = \frac{P}{s} = \frac{356 \times 10^3 \text{ N}}{110 \times 10^6 \text{ N/m}^2} = 3.24 \times 10^{-3} \text{ m}^2$$

3. Use the formula for circular area $\left(A = \frac{\pi}{4} D^2 \right)$ and solve for diameter (D):

$$D = \sqrt{\frac{4}{\pi} (3.24 \times 10^{-3} \text{ m}^2)}$$

or

$$D = \left[\frac{4}{\pi} (3.24 \times 10^{-3} \text{ m}^2) \right]^{1/2}$$

$$= \left[\frac{4}{\pi} (32.4 \times 10^{-4} \text{ m}^2) \right]^{1/2}$$

$$= 6.429 \times 10^{-2} \text{ m}$$

$$= 64.29 \times 10^{-3} \text{ m} = 64.3 \text{ mm (rounding up and using three significant digits)}$$

Problem 2

Given: Modulus of elasticity in tension (E) for steel is 29,120,000 psi.

Required: Express E in GPa.

Solution

1. Express E in terms of powers of 10:

$$E = 29.12 \times 10^6 \text{ psi}$$

TABLE 10-7 CONVERSIONS

Quantity	U.S. customary to SI
Acceleration, linear	$1 \text{ ft/s}^2 = 3.048 \times 10^{-1} \text{ m/s}^2$
Area	$1 \text{ in}^2 = 6.452 \times 10^2 \text{ mm}^2$
	$1 \text{ ft}^2 = 9.290 \times 10^{-2} \text{ m}^2$
Density (mass), ρ	$1 \text{ lb/in}^3 = 2.768 \times 10^4 \text{ kg/m}^3$
	$1 \text{ lb/ft}^3 = 1.602 \times 10^1 \text{ kg/m}^3$
Electric current, I	$1 \text{ ampere (A)} = 1 \text{ C/s}$
Energy, work	$1 \text{ Btu} = 1.055 \text{ kJ}$
	$1 \text{ in.} \cdot \text{lb} = 1.129 \times 10^{-1} \text{ J}$
	$1 \text{ ft} \cdot \text{lb} = 1.356 \text{ J}$
	$1 \text{ Btu} = 0.293 \text{ W} \cdot \text{h}$
Force	$1 \text{ lbf} = 4.448 \text{ N}$
	$1 \text{ kgf} = 9.807 \text{ N}$
Impulse	$1 \text{ lb} \cdot \text{s} = 4.448 \text{ N} \cdot \text{s}$
Length	$1 \text{ Å} = 1 \times 10^{-1} \text{ nm}$
	$1 \text{ microinch} = 2.540 \times 10^{-2} \text{ μm}$
	$1 \text{ mil} = 2.540 \times 10^1 \text{ μm}$
	$1 \text{ in.} = 2.540 \times 10^1 \text{ mm}$
	$1 \text{ ft} = 3.048 \times 10^{-1} \text{ m}$
Magnetic field strength, H	$1 \text{ oersted (Oe)} = 79.58 \text{ ampere turns per meter}$
Magnetic flux	$1 \text{ maxwell} \doteq 10^{-8} \text{ Wb}$
Magnetic flux density, B	$1 \text{ Wb/M}^2 = 1 \text{ T (tesla)}$
Modulus of elasticity, E	$1 \text{ lb/in}^2 = 6.895 \times 10^{-6} \text{ GPa}$
Moment of force, torque	$1 \text{ lb} \cdot \text{in.} = 1.130 \times 10^{-1} \text{ N} \cdot \text{m}$
	$1 \text{ lb} \cdot \text{ft} = 1.356 \text{ N} \cdot \text{m}$
Moment of inertia, I (of area)	$1 \text{ in}^4 = 4.162 \times 10^5 \text{ mm}^4$
Momentum, linear	$1 \text{ lb} \cdot \text{ft/s} = 1.383 \times 10^{-1} \text{ kg} \cdot \text{m/s}$
Power	$1 \text{ Btu/min} = 1.758 \times 10^{-2} \text{ kW}$
	$1 \text{ ft} \cdot \text{lb/min} = 2.259 \times 10^{-2} \text{ W}$
	$1 \text{ hp} = 7.457 \times 10^{-1} \text{ kW}$
Stress (pressure)	$1 \text{ lb/in}^2 = 6.895 \times 10^{-3} \text{ MPa}$
	$1 \text{ ksi} = 6.895 \text{ MPa}$
Temperature	$1°\text{F (difference)} = 0.555°\text{C}$
	$1.8 °\text{F} = 1°\text{C (difference)}$
Thermal expansion, linear coefficient, α	$\text{in./in./°F} = 1.8 \text{ K}^{-1}, \text{ K} = °\text{C} + 273.15$
Thermal conductivity, k	$1 \text{ Btu/ft} \cdot \text{hr} \cdot °\text{F} = 1.729 \text{ W/m} \cdot \text{K}$
Velocity, linear, v	$1 \text{ in./s} = 2.540 \times 10^1 \text{ mm/s}$
	$1 \text{ ft/s} = 3.048 \times 10^{-1} \text{ m/s}$
	$1 \text{ in./min} = 4.233 \times 10^{-1} \text{ mm/s}$
	$1 \text{ ft/min} = 5.080 \times 10^{-3} \text{ m/s}$
Velocity, angular, ω	$1 \text{ rev/min} = 1.047 \times 10^{-1} \text{ rad/s}$
Viscosity	$1 \text{ poise (P)} = 0.1 \text{ Pa} \cdot \text{s}$
Volume	$1 \text{ in}^3 = 1.639 \times 10^4 \text{ mm}^3$
	$1 \text{ ft}^3 = 2.832 \times 10^{-2} \text{ m}^3$
	$1 \text{ yd}^3 = 7.646 \times 10^{-1} \text{ m}^3$

$1 \text{ joule (J)} = 10^7 \text{ ergs} = 0.625 \times 10^{19} \text{ eV}$

$1 \text{ gauss (G)} = 10^{-4} \text{ Wb/m}^2 = 1 \text{ T}$

$1 \text{ weber (Wb)} = 1 \text{ T/m}^2$

$1 \text{ Å} = 10^{-10} \text{ m (obsolete)}$

$1 \text{ micrometer (micron)} = 10^{-6} \text{ m (obsolete)}$

TABLE 10-7A   ADDITIONAL USEFUL CONVERSIONS

1 W = 3.413 Btu/hr
1 W = 1 N·m·s⁻¹
1 W = 0.7376 ft·lb·s⁻¹
1 kW = 1.341 Hp
1 J = 0.2390 cal
1 cal = 4.186 J
1 Btu = 252 cal
1 J = 0.737562 lb·ft
1 J = 1 N·m = 1 W·s
1 Pa = 1 N/m²
1 N/m² = 1.450 lb/in² (psi)
1 m = 3.28 ft = 39.37 in.

$$°C = \frac{5}{9}(F - 32)$$

$$°F = \frac{9}{5}(F + 32)$$

K = °C + 273   (K = kelvin)
°R = °F + 460   (R = Rankine)
1 L = 1.06 qt
1 Pa·s = 10 poise

---

2. Locate conversion ratio (see Table 10-7):

$$1 \text{ psi} = 6.895 \times 10^{-3} \text{ MPa}$$

3. Express prefixes in terms of powers of 10 (see Table 10-6):

$$1 \text{ psi} = 6.895 \times 10^{3} \text{ Pa}$$

4. Multiply E by conversion ratio:

$$29.12 \times 10^{6} \text{ psi} \left(\frac{6.895 \times 10^{3} \text{ Pa}}{1 \text{ psi}} \right) = 200.8 \times 10^{9} \text{ Pa}$$

5. Express answer using required SI prefix (see Table 10-6):

200.8 GPa (using four significant digits)

REFERENCES

ASTM E 380: *Standard for Metric Practice*.
ANSI Z 210.1: *American National Standard for Metric Practice*.
ISO 1000: *SI Units and Recommendations for the Use of Their Multiples and of Certain Other Units*.
NBS Special Publication 330: *The International System of Units (SI)*.

TABLE 10-8 CONSTANTS

Quantity	Symbol	Value
Acceleration of gravity	g	$9.80 \text{ m} \cdot \text{s}^{-2}$
Atomic mass unit	amu	$1.66 \times 10^{-27} \text{ kg}$
Avogadro's number	N_A, N_0	$6.022 \times 10^{23} \text{ molecules/mole}$
Bohr radius	a_0	$5.292 \times 10^{-11} \text{ m}$
Electron charge	$q, -e$	$1.60 \times 10^{-19} \text{ C}$
Electron mass	m	$9.11 \times 10^{-31} \text{ kg}$
Electron volt	eV	$0.160 \times 10^{-18} \text{ J}$
Magnetic permeability of free space	μ_0	$4\pi \times 10^{-7} \text{ H/m} = 1.257 \times 10^{-6} \text{ H/m}$
Permittivity of free space	ϵ_0	$\dfrac{10^7}{4\pi c^2} = 8.854 \times 10^{-12} \text{ C/V} \cdot \text{m}$
Planck's constant	$\hbar$	$6.63 \times 10^{-34} \text{ J} \cdot \text{s}$
Speed of light (vacuum)	c	$3.00 \times 10^8 \text{ m} \cdot \text{s}^{-1}$

TABLE 10-9 REPRESENTATIVE PLASTICS

Common Name (Chemical Name)	ASTM Abbreviations	Trade Names	Common Structure	Grouping[a]	Typical Uses
ABS (acrylonitrile–butadiene–styrene)	ABS	Absinol, Abson, Cycolac, Royalite	Amorphous terpolymer	TP, EP	Pipe, toys, luggage, boat hulls, football helmets, gears, chrome-plated plumbing, and auto parts
Acetal (polyoxymethylene) (polymerized formaldehyde)	POM	Delrin, Celcon, Formaldafil	Highly crystalline homopolymer and copolymers	TP, EP	Gears, bearings, fan blades, shower heads, auto parts, and aerosol bottles
Acrylics [poly(methyl methacrylate)]	PMMA	Plexiglas, Lucite, Acrylite	Amorphous	TP, GP	Lenses, windows, signs, sculpture, light pipes, and skylights
(polyacrylonitrile)	PAN	Sayelle			

[a]TP, thermoplastic; TS, thermosetting; EP, engineering plastic; GP, general-purpose or special. ASTM, American Society for Testing and Materials.

[b]Engineering plastics.

TABLE 10-9 (continued)

Common Name (Chemical Name)	ASTM Abbreviations	Trade Names	Common Structure	Grouping[a]	Typical Uses
Alkyd plaskon (modified polyester resins)		Premix Dyal Glaskyd	Cross-link network	TS, GP	Coatings: enamel, lacquer and paint, molded electrical parts
Allylics (diallyl phthalate) (diallyl isophthalate) (diethylene glycol bisallyl/carbonate)	DAP DAIP CR39-allyl diglycol carbonate	Diall Poly-Dap	Cross-link network $\left[CH_2 = CH - CH_2 \right]_n$	TS, GP TS	Electronic parts, pump impellers, glass-fiber impregnate, dinnerware, watch crystals
Aminos (urea–formaldehyde) (melamine–formaldehyde)	UF MF	Plaskon Cymel	Cross-link network	TS, GP	Electrical parts, particle-board binders, coatings, dinnerware, paper impregnate
Cellulosics (cellulose acetate) (cellulose butyrate) (cellulose nitrate) (cellulose propionate) (ethyl cellulose)	CA CAB CN CAP EC	Tenite Uvex Nixonite Forticel Ethocel Methocel	Highly crystalline	TP, GP	Packaging film, pipe, optical frames, flashbulb shields, helmets, rollers

[a]TP, thermoplastic; TS, thermosetting; EP, engineering plastic; GP, general-purpose or special. ASTM, American Society for Testing and Materials.
[b]Engineering plastics.

TABLE 10-9 (continued)

Common Name (Chemical Name)	ASTM Abbreviations	Trade Names	Common Structure	Grouping[a]	Typical Uses
Epoxy	EP	Epi Rez Hysol	Cross-linked network or amorphous	TS, TP, GP	Coatings, fiber reinforcement, potting and encapsulating
Fluorocarbons (polytetrafluoroethylene)[b] (polychlorotrifluoroethylene)[b] [poly(vinylidene fluoride)]	FEP PTFE or TFE CTFE PVDF	Teflon Kelf Kynar	Highly crystalline	TP,[b] EP, GP	Seals, cookware, corrosion liquids hardware, chemical-processing equipment
Nylon (polyamide) (nylon 6) (nylon 6/6) (nylon 6/10) (nylon 6/12) (aramid)	PA	Nylon Celanese Zytel Plaskon Zytrel Nomex Kevlar	Crystalline	TP, EP	Bearings, tubing, gears, housings Parts exposed to moisture: nylon 6/10, nylon 6/12 Reinforcing fiber: aramid
Phonelic (phenol–formaldehyde)	PF	Bakelite Durite Resinox	Nylon 6 Cross-linked network	TS, GP	Electrical parts, housings, and binders

[a]TP, thermoplastic; TS, thermosetting; EP, engineering plastic; GP, general-purpose or special. ASTM, American Society for Testing and Materials.
[b]Engineering plastics.

TABLE 10-9 (continued)

Common Name (Chemical Name)	ASTM Abbreviations	Trade Names	Common Structure	Grouping[a]	Typical Uses
Phenoxy [poly(hydroxy ethers)]			Amorphous or cross-linked network	TP, TS, EP	Gas pipe, sports equipment, electrical housing, adhesives and coatings
Phenylene oxide [poly(phenylene oxide)]	PPO	Noryl	Crystalline	TP, EP	Auto trim, panels electrical housing, TV cabinets, pump parts
Polycarbonate	PC	Lexan Xenoy (blend)	Amorphous	TP, EP	Optical lenses, bullet-resistant windows, housings, cookers
Polyester [poly(ethylene terephthalate)] [poly(butylene terephthalate)] (aromatic polyesters)	PET (TS) PBT (TP)	Mylar Dacron Kodel Fortrel Laminac Selectron Gafite tp Ekonol	Amorphous or cross-linked network	TS, TP, GP, EP	Glass-fiber reinforcer, films, fibers

[a]TP, thermoplastic; TS, thermosetting; EP, engineering plastic; GP, general-purpose or special. ASTM, American Society for Testing and Materials.
[b]Engineering plastics.

TABLE 10-9 (continued)

Common Name (Chemical Name)	ASTM Abbreviations	Trade Names	Common Structure	Grouping[a]	Typical Uses
Polyimide	LaRC-ITP1 LaRC-CP1 UPILEX R$_1$	Kapton	Cross-linked network or amorphous	TS, TP, EP	Jet-engine vane bushings, seals, ball-bearing separators, high-temperature film, fiber matrix
Poly(amide/imide)		Torlon Tygon	Amorphous	TP, EP	Engineering-plastic gears, structural components, bearings, seals, and valves
Polyolefins			Low to high crystallinity or cross-linked network		
(polyethylene, low density)	LDPE	Alathon	polyethylene	TP, TS, GP, EP	PE: packaging, squeeze bottles, electrical insulation and tubing
(polyethylene, high density)[b]	HDPE	Dylan			
(polyethylene, ultra-high molecular weight)[b]	UHMWPE	Marlex			
(polypropylene)[b]	PP	Escon			PP: packaging, auto-battery cases, housings, electrical components and fan blades
(polyallomer)		Tenite			
(ethylene–vinyl acetate)	EVA				EVA: shoe soles and hypodermic syringes
[poly(ethylene terephthalate)]	PET	Amite Dacron Rynite	polypropylene		
(polybutylene)					
[poly(methyl pentene)]	TPX	Surlyn			
(ionomer)					

[a] TP, thermoplastic; TS, thermosetting; EP, engineering plastic; GP, general-purpose or special. ASTM, American Society for Testing and Materials.
[b] Engineering plastics.

TABLE 10-9 (continued)

Common Name (Chemical Name)	ASTM Abbreviations	Trade Names	Common Structure	Grouping[a]	Typical Uses
Poly(ether ether ketone)	PEEK	Victrex		EP	Crystalline polymer for high-temperatures and composite matrices
Poly(phenylene sulfide)	PPS	Ryton	Amorphous	TP, EP	Electrical terminal block and connectors, seals, gears
Polystyrene (styrene–acrylonitrile)	PS SAN	Styron Styrofoam Lustrex Dylite Tyril	Amorphous	TP, GP	Packaging, control knobs, TV cabinets, wood substitute, foam insulation
Polysulfone [poly(ether sulfone)] (polyphenylsulfone)		Udel Radel	Amorphous	TP, EP	Electrical insulators, auto-distributor caps, tubing, and aircraft cabin interiors
Polyurethane (isocyanate polyester or polyether)	PUR	Estane Flexane Texin Calspan	Amorphous or cross-linked network	TP, TS, EP	Reaction injection molded (RIM) foamed auto parts, solid tires, auto bumpers, synthetic leather

[a]TP, thermoplastic; TS, thermosetting; EP, engineering plastic; GP, general-purpose or special. ASTM, American Society for Testing and Materials.
[b]Engineering plastics.

TABLE 10-9 (continued)

Common Name (Chemical Name)	ASTM Abbreviations	Trade Names	Common Structure	Grouping[a]	Typical Uses
Silicones	SI	Silastic RTV Silicone RTV	Amorphous $\left[\begin{array}{c}CH_3 \\ -Si-O- \\ CH_3\end{array}\right]_n$	TS, GP	Room-temperature-vulcanizing (RTV) molds, fiber-matrix, electronic potting and encapsulating, heat seals
Vinyls [poly(vinyl chloride)] [poly(vinyl acetate)] [poly(vinyl alcohol)] [poly(vinyl butyral)] [poly(vinyl fluoride)] [poly(vinylidene chloride)]	PVC PVAc PVA PVB PVF PVDC	Vinylite Naugahyde Luxite Elvanol Butrar Tedlar Saran	Cross-linked network $\left[\begin{array}{c}H \ \ H \\ -C-C- \\ H \ \ Cl\end{array}\right]_n$ PVC $\left[\begin{array}{c}H \ \ H \\ -C-C- \\ H \ \ C{=}O \\ H-C-H \\ H\end{array}\right]_n$ PVAC	TS, TP, GP	Plastisol coating, upholstery, pipe, building trim, coatins, film

[a]TP, thermoplastic; TS, thermosetting; EP, engineering plastic; GP, general-purpose or special. ASTM, American Society for Testing and Materials.
[b]Engineering plastics.

TABLE 10-10 REPRESENTATIVE ELASTOMERS[a]

	NR	IR	SBR	IIR	BR	NBR
Abbreviation (ASTM):	NR	IR	SBR	IIR	BR	NBR
Common name:	Natural rubber	Isoprene	GRS or buna S	Butyl	Polybutadiene	Nitrile or buna N
Chemical name:	Natural polyisoprene	Polyisoprene	Styrene Butadiene	Isobutene Isoprene	Butadiene	Nitrile butadiene
(1) Tensile strength (kPa)	31	27.5	24	20.6	20.6	24
(2) Hardness Shore A	30–100	40–80	40–90	40–75	45–80	40–95
(3) Specific gravity	0.93	0.93	0.94	0.92	0.94	1.00
(4) Abrasion resistance	A	A	A	B	A	A
(5) Tear resistance	A	B	C	B	B	B
(6) Flexibility at low temperature	B	B	C	B	B	C
(7) Impact resistance	A	A	A	B	B	C
(8) Resiliency	A	A	B	C	A	B
(9) Creep	A	B	B	C	A	B
(10) ASTM/SAE type class	AA	AA	AA–BA	AA–BA	AA	BF, BG, PK, CH
(11) Maximum service temp. (°C)	70	70	100	100	70	100–125
(12) Heat-aging resistance	B–C	B–C	B	B–A	C	B
(13) Flame resistance	D	D	D	D	D	D
(14) Oil and gasoline resistance	X	X	X	X	C	A
(15) Oxidation resistance	C	B–C	C	C	C	C
(16) Ozone resistance	C	C–D	C–D	A	C–D	C–D
(17) Ultraviolet resistance	C–D	C–D	B–C	B	B–C	B–C
(18) Acid and base resistance	B–C	B–C	B–C	B	B–C	B
(19) Water absorption resistance	A	A	B	A	A	C
(20) Permeability to gases	C	C	C	A	C	B
(21) Electrical resistivity	A	A	A	A	A	D
(22) Adhesion to metals	A	A	A	B	A	A
(23) Trade names	—	Natsyn, Isoprene, Ameripol SN	K-Resin	Enjay Butyl, Petro-Tex Butyl	Diene, Ameripol CB	Paracril
(24) Typical uses	Tires, seals, bearings, couplings, shoe soles and heels	Same as natural rubber	Shock absorbers, belts, heels, sponges, gaskets, belts	Truck and auto tires, shock absorbers, inner tubes	Pneumatic tires, gaskets, seals, abrasion resistance belts	Gasoline, chemical and oil seals, gaskets and O-rings, belting

[a] A, excellent; B, good; C, fair; D, poor; X, not recommended.

TABLE 10-10 (continued)

	ACM	CR	CSM	FPM	EPDM	SI	UE	PTR
	Acrylate	Neoprene	Hypalom Chlorosulfanyl	Fluorocarbon Fluorinated	EPDM Ethylene	Silicone rubber	Urethane rubber Polyether	Thikol
	Polyacrylate	Chloropene	Polyethylene	hydrocarbon	propylene	Polysilicone	urethane	Polysulfide
(1)	17	27.5	27.5	17	20.6	4–10	34–55	4–10
(2)	40–90	40–90	50–95	60–90	30–90	25–80	35–100	20–80
(3)	1.10	1.23	1.12–1.28	1.45	0.86	1.14–2.05	1.06	1.34
(4)	B	B	B	A	B	D	B	D
(5)	C–B	C–B	B	B	C	B	B	D
(6)	D	C	B	D	B	A	C	C
(7)	D	B	C–B	B	B	D–C	B–A	D
(8)	C–B	A	C	C	B	D–A	C–A	C
(9)	C	B	C	B	C–B	C–A	C–A	D
(10)	DF, DH	BC, BE	CE	HK	AA, DA, CA	FC, FE, FK	AD, EC	AK
(11)	125	100	125	250	125	200–225	100	70
(12)	A	A	C	A	A	A	B–A	C–B
(13)	D	B	C	A	D	A	D	D
(14)	A	C	C	A	X	D–C	B	A
(15)	A	A	A	A	A	A	A	B
(16)	B	B	A	A	A	A	A	A
(17)	B	B	B	A	A	A	B	B
(18)	C	B	B	A	A	D	D	X
(19)	A	B	A	A	A	A	A	B
(20)	B	B	B	A	C	D	B	D
(21)	B	C	B	B	B	A	B	C
(22)	B	A	A	C–B	C	A	C	C–B
(23)	Hycar, Acrylon	Neoprene, Perbunanc	Hypalon	Viton, Proflo, Fluorel	Nordel, Epcar, Royalene	Adiprene, RTV, Silastic	Kalrez, Estane, Roylar	Thikol
(24)	Oil hose, colored and white parts, pressure and oil O-rings	Belts, hose, extruded goods, molded sheet, adhesives, chemical-tank liners	Laminated roofing, tarpaulins, reservoir and pond liners, diaphragms, shoe soles and heels, whitewall tires	Brake seals, ducting connectors, carburetors, needle tips, roll coverings	Garden and industrial hoses, belts, bike tires, electrical-wire insulation, paintable auto bumpers	Industrial tires and rolls, mining belts, die pads, gaskets and seals	Chemical O-ring seals, valve seats, gaskets, nuclear, oil, gas, hydraulic, and acid seals	Gasoline hose, printing rolls, caulking, adhesives and binders

TABLE 10-11 REPRESENTATIVE METALS

Material	Nominal Composition (Essential Elements) (%)		Form and Condition	Typical Mechanical Properties			
				Yield Strength (0.2% offset) (1000 psi)	Tensile Strength (1000 psi)	Elongation in 2 in. (%)	Hardness, Brinell or Rockwell
Copper CA 110 Sheet—ASTM B152 Rod—ASTM B124, B133 Wire—ASTM B1, B2, B3	Cu	99.90 min	Strip annealed Spring temper	10[a] 50[a]	32 55	45 4	40 HRF 60 HRB
Commercial bronze CA 220 Plate, sheet, strip, bar— ASTM B36 Wire—ASTM B 134	Cu Zn	90 10	Strip annealed Spring temper	10[a] 62[a]	37 72	45 3	53 HRF 78 HRB
Red brass CA 230 Strip, sheet, plate— ASTM B36 Wire—ASTM B134 Tube—ASTM B135	Cu Zn	85 15	Strip annealed Hard temper	15[a] 60[a]	40 75	50 7	50 135
Copper–nickel CA 715 Sheet—ASTM B122 Plate—ASTM B171 Tube—ASTM B111	Cu Ni Fe Mn	bal 30 0.55 0.5	Tube annealed	25[a]	60	45	45 HRB
Aluminum alloy Alclad 2024 Sheet and plate—ASTM B209	Core: 2024 Al Cu Mn Mg Cladding: Al	bal 4.5 0.6 1.5 99.3 min	Sheet annealed Heat annealed	11 42	26 64	20 19	— —
Aluminum alloy 3003 Sheet and plate—ASTM B209	Al Mn	bal 1.2	Sheet annealed Cold rolled	6 27	16 29	30 4	28 55
Aluminum alloy 5052 Sheet and plate—ASTM B209	Al Mg Cr	bal 2.5 0.25	Sheet annealed Cold rolled	13 37	28 42	25 7	47 77
Aluminum alloy 6061 Sheet and plate—ASTM B209	Al Si Cu Mg Cr	bal 0.6 0.25 1.0 0.25	Sheet annealed Heat treated	8 40	18 45	25 12	30 95
Aluminum alloy 707S Bar, rod, wire, and shapes—ASTM B221	Al Zn Cu Mg Cr	bal 5.6 1.6 2.5 0.3	Bar annealed Heat treated	15 73	33 83	16 11	60 150
Cast alluminum alloy 13 Castings—ASTM B85 Grade S12A	Al Si	bal 12.0	Die casting As cast	21	43	2.5	—

				Typical physical properties				
Density (lb/cu in)	Specific Gravity	Melting Point (°F)	Specific Heat (32 to 212°F) (Btu/lb/°F)	Thermal-expansion Coefficient (32 to 212°F) (in 10⁻⁶ in./in./°F)	Thermal Conductivity (32 to 212°F) (Btu/sq ft/ hr/°F/in.)	Electrical Resistivity (68°F) (ohms/cir mil ft)	Tensile Modulus of Elasticity (10⁶ psi)	Torsional Modulus of Elasticity (× 10⁶ psi)
0.322	8.91	1980	0.092	9.4	2512	10.3	17	6.4
—	—	—	—	9.8	—	—	—	—
0.318	8.80	1910	0.09	10.2	1308	23.6	17	6.4
—	—	—	—	—	—	—	—	—
0.316	8.75	1880	0.09	10.4	1104	28	17	6.4
—	—	—	—	—	—	—	—	—
0.323	8.94	2260	0.09	9.0	204	225	22	8.3
0.100	2.77	1180	0.23	12.6	1340	21	10.6	3.75
—	—	—	—	—	840	35	10.6	4.0
0.099	2.73	1210	0.23	12.9	1340	21	10.0	3.75
—	—	—	—	—	1070	26	10.0	3.75
0.097	2.68	1200	0.23	13.2	960	30	10.2	3.75
—	—	—	—	—	960	30	10.2	3.75
0.098	2.70	1205	0.23	13.0	1190	23	10.0	3.75
—	—	—	—	—	1070	26	10.0	3.75
0.101	2.80	1175	0.23	12.9	—	—	10.4	3.9
—	—	—	—	—	840	35	10.4	3.9
0.096	2.65	1080	0.23	11.5	870	34	10.3	3.85

TABLE 10-11   REPRESENTATIVE METALS

Material	Nominal Composition (Essential Elements) (%)		Form and Condition	Typical Mechanical Properties			
				Yield Strength (0.2% offset) (1000 psi)	Tensile Strength (1000 psi)	Elongation in 2 in. (%)	Hardness, Brinell or Rockwell
Magnesium alloy AZ 31B Plate and sheet—ASTM B90	Mg	bal	Sheet annealed	22	37	21	56
	Al	3.0					
	Zn	1.0	Hard sheet	32	42	15	73
	Mn	0.2 min					
Magnesium alloy AZ 80A Forgings—ASTM B91	Mg	bal	As forged	33	48	11	69
	Al	8.5					
	Zn	0.5	Forged and aged	36	50	6	72
	Mn	0.15 min					
Magnesium alloys AZ 91A and AZ 91B Castings—ASTM B94	Mg	bal					
	Al	9.0	Die cast	22	33	3	63
	Zn	0.7					
	Mn	0.2 min					
Titanium Ti-35A Forgings—ASTM B381 Sheet, strip, and plate—ASTM B265 Pipe—ASTM B337 Tubes—ASTM B338 Bars—ASTM B348	Ti	bal					
	C	0.08 max					
	Fe	0.12 max	Sheet annealed	30	40	30	135
	N₂	0.05 max					
	H₂	0.015 max					
Ti-6 Al-4 V alloy Sheet, strip, and plate—ASTM B265 Bar—ASTM B348 Forgings—ASTM B381	Ti	bal					
	Al	6.5	Sheet annealed	130	140	13	39 HRC
	V	4					
	C	0.08 max	Heat treated	165	175	12	—
	Fe	0.25 max					
	N₂	0.05 max					
	H₂	0.015 max					
Nickel 211 ASTM F290	Ni	95.0					
	Mn	4.75	Annealed	35	75	40	140
	C	0.10					
Nickel (cast)	Ni	95.6					
	Cu	0.5					
	Fe	0.5	As cast	25[a]	57	22	110
	Mn	0.8					
	Si	1.5					
	C	0.8					
Duranickel alloy 301	Ni	bal					
	Al	4.5					
	Si	0.55	Hot rolled and aged	132	185	28	330
	Ti	0.5					
	Mn	0.25					
	Fe	0.15					
	C	0.15					

				Typical Physical Properties				
Density (lb/cu in)	Specific Gravity	Melting Point (°F)	Specific Heat (32 to 212°F) (Btu/lb/°F)	Thermal-expansion Coefficient (32 to 212°F) (in 10^{-6} in./ in./°F)	Thermal Conductivity (32 to 212°F) (Btu/sq ft/ hr/°F/in.)	Electrical Resistivity (68°F) (ohms/cir mil ft)	Tensile Modulus of Elasticity (10^6 psi)	Torsional Modulus of Elasticity ($\times$ 10^6 psi)
0.064	1.77	1170	0.245	14.5	657	55	6.5	2.4
—	—	—	—	—	—	—	—	—
0.065	1.80	1130	0.25	14.5	522	87	6.5	2.4
—	—	—	—	—	—	—	—	—
0.065	1.80	1105	0.25	14.5	493	102	6.5	2.4
0.163	4.50	3063	0.124	4.8	108	336	14.9	6.5
0.160	4.42	3000	0.135	4.9	50	1026	16.5	6.1
—	—	—	—	—	—	—	—	—
0.315	8.73	2600	0.11	7.4	306	102	30	11
0.301	8.34	2550	0.13	8.85	410	125	21.5	—
0.298	8.75	2620	0.104	7.2	165	255	30	11

TABLE 10-11 REPRESENTATIVE METALS

Material	Nominal Composition (Essential Elements) (%)		Form and Condition	Typical Mechanical Properties			
				Yield Strength (0.2% offset) (1000 psi)	Tensile Strength (1000 psi)	Elongation in 2 in. (%)	Hardness, Brinell or Rockwell
Monel alloy 400 Rod and Bar—ASTM B164 Plate, sheet, and strip— ASTM B127 Tube—ASTM B165	Ni Cu Fe Mn Si C	bal 31.5 1.35 0.90 0.15 0.12	Rod hot-rolled annealed	30	79	48	125
Inconel Alloy 600 plate Sheet and strip—ASTM B168 Rod and bar—ASTM B166 Pipe and tube—ASTM B163 and B167	Ni Cr Fe Mn Si C Cu	bal 15.8 7.2 0.2 0.2 0.04 0.10	Rod hot-rolled annealed	36	90	47	150
Hastelloy Alloy W Wire—AMS 5786 Bar and forgings—AMS 5755A	Cr Mo Fe Ni	5 24.5 5.5 bal	Sheet annealed	53	123	55	—
Ingot iron	Fe	99.9 plus	Hot rolled	29	45	26	90
			Annealed	19	38	45	67
Wrought iron Forgings—ASTM A73	Fe Slag	bal 2.5	Hot rolled	30	48	30	100
Carbon steel—SAE 1020 ASTM A285	Fe C Mn Si	bal 0.20 0.45 0.25	Annealed	38	65	30	130
			Quenched and tempered at 1000° F	62	90	25	179
300 M alloy steel Bar and forgings—AMS 6416	Fe Mn Si Ni Cr Mo V C	bal 0.80 1.6 1.85 0.85 0.38 0.08 0.43	Hardened	240	290	10	535
Cast gray iron ASTM A48 class 30	C Si Mn Fe	3.4 1.8 0.8 bal	As cast	—	32	—	190
Malleable iron Castings—ASTM A47	C Si Mn Fe	2.5 1 0.55 max bal	Annealed	33	52	12	130

				Typical physical properties				
Density (lb/cu in)	Specific Gravity	Melting Point (°F)	Specific Heat (32 to 212°F) (Btu/lb/°F)	Thermal-expansion Coefficient (32 to 212°F) (in 10^{-6} in./in./°F)	Thermal Conductivity (32 to 212°F) (Btu/sq ft/hr/°F/in.)	Electrical Resistivity (68°F) (ohms/cir mil ft)	Tensile Modulus of Elasticity (10^6 psi)	Torsional Modulus of Elasticity ($\times\ 10^6$ psi)
0.319	8.84	2460	0.102	7.7	151	307	26	9.5
0.304	8.43	2600	0.106	7.4	103	620	31	11
0.325	9.03	2400	—	6.3	—	—	—	—
0.284	7.86	2795	0.108	6.8	490	57	30.1	11.8
—	—	—	—	—	—	—	—	—
0.278	7.70	2750	0.11	6.35	418	70	29	—
0.284	7.86	2760	0.107	6.7	360	60	30	—
—	—	—	—	—	—	—	—	—
0.283	7.84	2740	0.107	6.7	360	60	30	11.6
0.260	7.20	2150	—	6.7	310	400	14	—
0.264	7.32	2250	0.122	6.6	—	180	25	—

TABLE 10-11 REPRESENTATIVE METALS

Material	Nominal Composition (Essential Elements) (%)		Form and Condition	Typical Mechanical Properties			
				Yield Strength (0.2% offset) (1000 psi)	Tensile Strength (1000 psi)	Elongation in 2 in. (%)	Hardness, Brinell or Rockwell
Ductile iron (Nickle containing) Grade 60-40-18 castings—ASTM A536	Fe C Si Mn Ni	bal 3.6 2.3 0.5 0.75	Annealed	47	65	24	160
Ductile iron (Nickel containing) Grade 120-90-02 castings—ASTM A536	Fe C Si Mn Ni	bal 3.6 2.3 0.5 0.75	Oil quenched and tempered	120	140	4	325
Type 201 stainless steel (UNS 20100) Plate, sheet, and strip— ASTM A412 Bar—ASTM A429	Fe Cr Ni Mn N₂ C	bal 17 4.5 6.5 0.25 max 0.15 max	Strip annealed	55	115	60	90 HRB
Type 302 stainless steel (UNS 30200) Plate, sheet, and strip— ASTM A167 and A 240 Bar—ASTM A276 and A314 Wire—ASTM A313 Forgings—ASTM A473	Fe Cr Ni C	bal 18 9 0.15 max	Sheet annealed Cold rolled	40 up to 165	90 up to 190	50 5	85 HRB up to 40 HRC
Type 303 and 303 Se stainless steel (UNS 30323) Bar—ASTM A276 and A314 Forgings—ASTM A473	Fe Cr Ni S Se C	bal 18 9 0.15 min or 0.15 min 0.15 max	Bar annealed	35	90	50	160
Type 314 stainless steel (UNS 31400) Bar—ASTM A276 and A314	Fe Cr Ni Si C	bal 25 20 2.50 0.25 max	Bar annealed	50	100	45	180
Type 405 stainless steel (UNS 40500) Plate, sheet, and strip— ASTM A176 and A240 Tube—ASTM A268 Bar—ASTM A276 and A314	Fe Cr C Al	bal 12.5 0.08 max 0.20	Sheet annealed	40	65	25	75 HRB

Source: Courtesy of Inco.

				Typical Physical Properties				
Density (lb/cu in)	Specific Gravity	Melting Point (°F)	Specific Heat (32 to 212°F) (Btu/lb/°F)	Thermal-expansion Coefficient (32 to 212°F) (in 10^{-6} in./ in./°F)	Thermal Conductivity (32 to 212°F) (Btu/sq ft/ hr/°F/in.)	Electrical Resistivity (68°F) (ohms/cir mil ft)	Tensile Modulus of Elasticity (10^6 psi)	Torsional Modulus of Elasticity ($\times\ 10^6$ psi)
0.250	7.1	2150	—	6.2	276	399	24.5	9.3
0.252	7.2	2150	—	5.9	218	408	24.5	9.3
0.283	7.86	—	—	8.7	—	423	28.6	—
0.29 —	7.9 —	2590 —	0.12 —	9.6	113	432 —	28 —	12.5 —
0.29	—	2590	0.12	9.6	113	432	28	—
0.279	—	—	0.12	8.4	121	462	29	—
0.28	7.7	2790	0.11	6.0	—	360	29	—

TABLE 10-12 REPRESENTATIVE ULTRAHIGH-STRENGTH STEELS[b] (Courtesy Advanced Materials & Processes)

Steel and Composition[a] (%)	Tempering Temperature [°C (°F)]	Tensile Strength [MPa (10³ psi)]	Yield Strength [MPa (10³ psi)]	Elongation in 50 mm (2 in.)(%)	Reduction in Area (%)	Hardness[c]	Impact Energy[d] [J (ft·lbf)]
Medium-carbon low-alloy steels							
AISI/SAE 4130 (UNS G41300):[e] 0.28–0.33 C, 0.4–0.6 Mn, 0.2–0.35 Si, 0.8–1.1 Cr, 0.15–0.25 Mo	205 (400)	1765 (256)	1520 (220)	10	33	475 HB	18 (13) I
	425 (800)	1380 (200)	1170 (170)	16.5	49	375 HB	34 (25) I
	650 (1200)	965 (140)	830 (120)	22	63	270 HB	135 (100) I
AISI/SAE 4140 (UNS G41400):[f] 0.38–0.43 C, 0.75–1 Mn, 0.2–0.35 Si, 0.8–1.1 Cr, 0.15–0.25 Mo	205 (400)	1965 (285)	1740 (252)	11	42	578 HB	15 (11) I
	425 (800)	1450 (210)	1340 (195)	15	50	429 HB	28 (21) I
	650 (1200)	900 (130)	790 (114)	21	61	277 HB	112 (83) I
AISI/SAE 4340 (UNS G43400):[h] 0.38–0.43 C, 0.6–0.8 Mn, 0.2–0.35 Si, 0.7–0.9 Cr, 1.65–2 Ni, 0.2–0.3 Mo	205 (400)	1980 (287)	1860 (270)	11	39	53 HRC	20 (15) I
	425 (800)	1500 (217)	1365 (198)	14	48	46 HRC	16 (12) I
	650 (1200)	1020 (148)	860 (125)	20	60	31 HRC	100 (74) I
Alloy 300M (UNS K44220):[i] 0.4–0.46 C, 0.65–0.9 Mn, 1.45–1.8 Si, 0.7–0.95 Cr, 1.65–2 Ni, 0.3–0.45 Mo, 0.05 V min	205 (400)	2140 (310)	1650 (240)	7	27	54.5 HRC	21.7 (16) CVN
	425 (800)	1790 (260)	1480 (215)	8.5	23	45.5 HRC	13.6 (10) CVN

Material							
AISI/SAE 6150 (UNS G61500):[k] 0.48–0.53 C, 0.7–0.9 Mn, 0.2–0.35 Si, 0.8–1.1 Cr, 0.15–0.25 V	205 (400)	2050 (298)	1810 (263)	1	5	610 HB	—
	425 (800)	1585 (230)	1490 (216)	11	42	470 HB	14 (10) I
	595 (1100)	1150 (167)	1080 (157)	16	47	350 HB	28 (21) I
AISI/SAE 8640 (UNS G86400):[l] 0.38–0.43 C, 0.75–1 Mn, 0.2–0.35 Si, 0.4–0.6 Cr, 0.4–0.7 Ni, 0.15–0.25 Mo	205 (400)	1810 (263)	1670 (242)	8	25.8	55 HRC	11.5 (8.5) I
	425 (800)	1380 (200)	1230 (179)	10.5	46.3	44 HRC	27.8 (20.5) I
	650 (1200)	870 (126)	760 (110)	20.5	61	28 HRC	96.9 (71.5) I
Medium-alloy air-hardening steels							
H11 Mod (UNS T20811 Mod):[m] 0.37–0.43 C, 0.2–0.4 Mn, 0.8–1 Si, 4.75–5.25 Cr, 1.2–1.4 Mo, 0.4–0.6 V	540 (1000)	2010 (291)	1675 (243)	9.6	30.6	56 HRC	21 (15.5) CVN
	650 (1200)	1060 (154)	850 (124)	14.1	41.2	33 HRC	40 (29.5) CVN
AISI/SAE H13 (UNS T20813):[n] 0.32–0.45 C, 0.2–0.5 Mn, 0.8–1.2 Si, 4.75–5.5 Cr, 1.1–1.75 Mo, 0.8–1.2 V	525 (980)	1960 (284)	1570 (228)	13°	46.2	52 HRC	16 (12) CVN
	605 (1120)	1495 (217)	1290 (187)	15.4°	54	44 HRC	30 (22) CVN

TABLE 10-12 REPRESENTATIVE ULTRAHIGH-STRENGTH STEELS

Steel and Composition[a] (%)	Tempering Temperature [°C (°F)]	Tensile Strength [MPa (10³ psi)]	Yield Strength [MPa (10³ psi)]	Elongation in 50 mm (2 in.)(%)	Reduction in Area (%)	Fracture toughness, [K_I, MPa·m$^{1/2}$ (ksi·in.$^{1/2}$)] Hardness[c]	Impact Energy[d] [J (ft·lbf)]
High-fracture-toughness steels							
HP 9-4-30 (UNS K91283):[p] 0.29–0.34 C, 0.10–0.35 Mn, 0.2 Si max, 0.9–1.1 Cr, 7–8 Ni, 0.06–0.12 V, 4.25–4.75 Co	205 (400)[q]	1650–1790 (240–260)	1380–1450 (200–210)	8–12°	25–35	66–99 (60–90)	20–27 (15–20) CVN
	550 (1025)[r]	1520–1650 (220–240)	1310–1380 (190–200)	12–16°	35–50	99–115 (90–105)	24–34 (18–25) CVN
	Quenchant						
AF1410 (UNS K92571):[s] 0.13–0.17 C, 0.1 Mn max, 0.1 Si max, 1.8–2.2 Cr, 9.5–10.5 Ni, 0.9–1.1 Mo, 13.5–14.5 Co	Air	1680 (244)	1475 (214)	16	69	174 (158)	69 (51) CVN
	Oil	1750 (254)	1545 (224)	16	69	154 (140)	65 (48) CVN
	Water	1710 (248)	1570 (228)	16	70	160 (146)	65 (48) CVN

Source: Bruce A. Becherer and Thomas J. Witheford. "Heat Treating of Ultrahigh-Strength Steels," *ASM Handbook*, 10th ed., Vol. 4, *Heat Treating*, ASM International, Materials Park, Ohio, 1991, pp. 207–218.

[a]The ultrahigh-strength-steel category (1380 MPa, 200 × 10³ psi, minimum yield strength) also includes the 18% Ni maraging steels and a variety of stainless-type steels.

[b]Phosphorus and sulfur contents may vary with steelmaking practice, but usually do not exceed 0.035% P and 0.04% S.

[c]HB, Brinell; HRC, Rockwell C.

[d]I, Izod; CVN, Charpy V-notch.

[e]Round bars, 25-mm (1 in.) diameter; water quenched from 845 to 870°C (1550 to 1600°F).

[f]Round bars, 13-mm (0.5 in.) diameter; oil quenched from 845°C (1550°F).

[g]EAF-VAR plate, 25-mm (1 in.) thick; oil quenched from 870°C (1600°F), tempered 2 + 2 hr.

[h]Oil quenched from 845°C (1550°F).

[i]Round bars, 25-mm (1 in.) diameter; oil quenched from 860°C (1575°F).

[j]Bars normalized at 900°C (1650°F), oil quenched from 845°C (1550°F), D-6A was developed by Ladish Co. Inc., Cudahy, Wis.

[k]Round bars, 14-mm (0.55 in.) diameter; normalized at 870°C (1600°F), oil quenched from 860°C (1575°F).

[l]Round bars, 13.5-mm (0.53 in.) diameter; oil quenched from 830°C (1525°F).

[m]Longitudinal properties. Heat treatment: air cooled from 1010°C (1850°F), double tempered 2 + 2 hr.

[n]Longitudinal properties of round bars. Heat treatment: oil quenched from 1010°C (1850°F) double tempered 2 + 2 hr.

[o]Elongation in 4D.

[p]Oil quenched from 845°C (1550°F), refrigerated to −73°C (−100°F), double tempered, HP 9-4-30 was developed by Republic Steel Corp. (now LTV Steel Co.), Cleveland, Ohio.

[q]Hardness, 49–53 HRC.

[r]Hardness, 44–48 HRC.

[s]VIM/VAR plate, 50-mm (2 in.) thick. Heat treatment: 675° C (1250°F) for 8 hr, air cooled; 900°C (1650°F) for 1 hr, quenched; 830°C (1525°F) for 1 hr, quenched; refrigerated at −73°C (−100°F) for 1 hr; 510°C (950°F) for 1 hr, air cooled.

TABLE 10-13   PROPERTIES OF SELECTED CERAMICS

	Density (lb/in^3) (kg/m^3)	Hardness (M, Mohs) (K, Knoop)	Tensile Strength (psi) (MPa)	Thermal Conductivity (Btu · in./hr · ft^2 · °F) (W/m · K)		Coefficient of Thermal Expansion (10^{-6} F^{-1}) (10^{-6} K^{-1})
Alumina	0.14	M, 9	25,000	25°C	192–255	77–1830°F 4.3
(Al_2O_3)	3.8	K, 2500	172		27.7–36.7	298–1272 K 8.1
Beryllia	0.11	M, 9	23,000	25°C	1741	68–2550°F 5.28
(BeO)	2.92	K, 2000	159		250	293–1672 K 9.5
Boron carbide	0.087	M, 9	22,500	70°F	104–197	0–2550°F 1.73
(B_4C)	2.41	K, 2800	155		—	255–1672 K 3.1
Boron nitride	0.076	—	3,500	70°F	100–200	70–1800°F 4.17
(BN)	2.10		24.1			294–1255 K 7.5
Cordierite	0.065	M, 6.5	3,500	25°C	12–22	68–212°F 2.08
($2MgO \cdot 2Al_2O_3 \cdot 5SiO_2$)	1.8		24.1		1.7–3.2	293–373 K 3.7
Silicon carbide	0.11	M, 9	24,000	70°F	101	0–2552°F 2.4
(SiC)	3.17	K, 2500	165		—	225–1672 K 4.3
Steatite	0.09	M, 7.5	8,500	25°C	20–41	68–212°F 3.99
($MgO \cdot SiO_2$)	2.7	K, 1500	60		2.9–5.9	293–373 K 7.2
Zircon	0.13	M, 8	12,000	25°C	4.9–6.2	68–212°F 1.84
($ZrO_2 \cdot SiO_2$)	3.7		82.7			293–212 K 3.3

Source: Data adapted from *Machine Design,* '84 Materials Reference Issue 3M, and Ceramics Bulletin No. 757, *Materials Engineering,* '84 Materials Selection Issue.

Dielectric Constant at 10^6 Hz (Except as Noted)	Volume Resistivity	Compressive Strength (psi) (MPa)	Flexural Strength (psi) (MPa)		Impact Strength (in. · lb) (N · m)	Modulus of Elasticity (psi × 10^6) (GPa)	Safe Service Temperature (°F) (°C)
			70°F	2,250°F			
8.0–10.0	>10^{20}	340,000	48,000	31,000	6.5	50	3540
		2,344	311	214	0.73	379	1965
6.4–7.0	>10^{14}	260,000	33,000	—	—	47	4350
		1,793	228	—	—	324	2414
	—	420,000	44,000	—	—	65	1100
		2,896	303	—	—	448	611
4.1–4.8	—	45,000	—	—	—	7	3000
		310	—	—	—	48	1665
4.02–6.23	>10^{20}	40,000	8,000	—	2.5	7	2282
		276	55	—	0.28	48	1250
—	—	100,000	110,000	80,000	—	62	3200
		690	758	552	—	427	1776
5.9–6.3							
6 × 10^1 cps	>10^{20}	90,000	19,000	—	5.0	15	1832
(6 × 10^1 Hz)		620	131	—	0.56	103	1016
8.0–10.0	>10^{20}	100,000	22,000	—	5.5	23	2012
		690	152	—	0.62	159	1117

TABLE 10-14 REPRESENTATIVE GLASSES

U.S. Customary

Type	Color	Principal Use	Class	Corrosion Resistance			Thermal Expansion Multiply by 10^{-7} in./in./°F		Upper Working Temperatures (Mechanical Considerations Only)				Thermal Shock Resistance Plates 6 × 6 in.		
				Weathering	Water	Acid	32 to 572°F	77°F to Setting Point	Annealed		Tempered		Annealed		
									Normal Service (°F)	Extreme Service (°F)	Normal Service (°F)	Extreme Service (°F)	⅛ in. Thick (°F)	¼ in. Thick (°F)	½ in. Thick (°F)
(1) Soda lime	Clear	Lamp bulbs	I	3	2	2	52	58.3	230	860	428	482	149	122	95
(2) Potash soda lead	Clear	Lamp tubing	I	2	2	2	49.8	53.9	230	716	—	—	—	—	—
(3) Aluminosilicate ..	Clear	Electron tube	I	1	1	3	25.6	30	392	1202	752	842	257	212	158
(4) Borosilicate	Clear	Sealed-beam lamps	I	³1	³2	³2	20.4	21.2	446	860	500	500	320	266	294
(5) Borosilicate	Clear	General	I	2	2	2	18.9	20.6	446	842	482	482	320	266	194
(6) 96% Silica	Clear	High temp.	I	1	1	1	4.2	3.1*	1652	2192	—	—	—	—	—
(7) Fused silica	Clear	Optical	I	1	1	1	3.1	1.9*	1652	2012	—	—	572	338	266
(8) Glass-ceramic	White	Missle nose cones	II	—	1	4	31.7	—	1292	—	—	—			

Source: Courtesy of Corning Glass Works.

TABLE 10-14 (continued)

| | Thermal Stress Resistance °F | Viscosity Data | | | | Knoop Hardness, KHN$_{100}$ | Density (lb/ft³) | Young's Modulus (Multiply by 10⁶ psi) | Poisson's Ratio | Log$_{10}$ of volume resistivity (Ω·cm) | | | Dielectric properties at 1 MHz, 68°F | | | Refractive Index |
		Strain Point (°F)	Annealing Point (°F)	Softening Point (°F)	Working Point (°F)					77°F	482°F	662°F	Power Factor (%)	Dielectric Constant	Loss Factor (%)	
(1)	29	883	957	1285	1841	465	154	10.2	0.22	12.4	6.4	5.1	0.9	7.2	6.5	1.512
(2)	36	743	815	1166	1805	382	190.3	8.6	0.22	17+	10.1	8.0	0.12	6.7	0.8	1.560
(3)	47	1229	1310	1666	2134	514	164.7	12.5	0.24	17+	13.5	11.3	0.16	6.3	1.0	1.547
(4)	86	932	1011	1436	2133	—	140.4	9.3	0.19	18	8.1	6.6	0.45	4.85	2.18	1.476
(5)	94	892	973	1436	2188	442	139.8	9.0	0.20	17	9.4	7.7	0.18	4.5	0.79	1.473
(6)	396	1634	1868	2786	—	487	136	9.8	0.19	17+	9.7	8.1	0.04	3.8	0.15	1.458
(7)	515	1753	1983	2876	—	489	137.2	10.5	0.16	17+	11.8	10.2	0.001	3.8	0.0038	1.459
(8)	29	—	—	—	—	657	162.2	17.2	0.24	16.7	10.0	8.7	0.30	5.6	1.7	—

(Corning Glass Works)

TABLE 10-14 (continued)

SI

Type	Color	Principal Use	Class	Corrosion Resistance Weathering	Water	Acid	Thermal Expansion (Multiply by 10^{-7}) cm/cm/°C 0–300°C	25°C to Setting Point	Upper Working Temperatures (Mechanical Considerations Only) Annealed Normal Service (°C)	Extreme Service (°C)	Tempered Normal Service (°C)	Extreme Service (°C)	Thermal Shock Resistance Annealed Plates 15 × 15 cm 3.2 mm Thick (°C)	6.4 mm Thick (°C)	12.7 mm Thick (°C)
(1) Soda lime	Clear	Lamp bulbs	I	3	2	2	93.5	105	110	460	220	250	65	50	35
(2) Potash soda lead	Clear	Lamp tubing	I	2	2	2	89.5	97	110	380	—	—	65	50	35
(3) Aluminosilicate	Clear	Electron tube	I	1	1	3	46	54	200	650	400	450	125	100	70
(4) High lead	Clear	Solder sealing	II	1	1	4	84	92	100	300	—	—	—	—	—
(5) Borosilicate	Clear	General	I	2	2	2	34	37	230	450	250	250	160	130	90
(6) 96% Silica	Clear	High temp.	I	1	1	1	7.5	5.5*	900	1200	—	—	—	—	—
(7) Fused silica	Clear	Optical	I	1	1	1	5.5	3.5*	900	1100	—	—	200	170	130
(8) Glass-ceramic	White	Missile nose cones	II	—	1	4	57	—	700	—	—	—	—	—	130

TABLE 10-14   (continued)

	Thermal Stress Resistance °C	Knoop Hardness, KHN$_{100}$	Density (g/cm^3)	Young's Modulus (Multiply by 10^3 kg/mm^2)	Poisson's Ratio	Log$_{10}$ of Volume Resistivity (Ω · cm)			Dielectric Properties at 1 MHz, 20°C			Refractive Index
						25°C	250°C	350°F	Power Factor (%)	Dielectric Constant	Loss Factor (%)	
(1)	16	465	2.47	7.1	0.22	12.4	6.4	5.1	0.9	7.2	6.5	1.512
(2)	20	382	3.05	6.0	0.22	17+	10.1	8.0	0.12	6.7	0.8	1.560
(3)	26	514	2.64	8.8	0.24	17+	13.5	11.3	0.16	6.3	1.0	1.547
(4)	21	—	5.42	5.6	0.28	17+	10.6	8.7	0.22	15	3.3	1.86
(5)	52	442	2.24	6.3	0.20	17	9.4	7.7	0.18	4.5	0.79	1.473
(6)	220	487	2.18	6.9	0.19	17+	9.7	8.1	0.04	3.8	0.15	1.458
(7)	286	489	2.20	7.4	0.16	17+	11.8	10.2	0.001	3.8	0.0038	1.459
(8)	16	657	2.6	12	0.24	16.7	10.0	8.7	0.30	5.6	1.7	—

HARDNESS/TENSILE-STRENGTH CONVERSION

Hardness conversion chart for hardenable carbon and alloy steels: approximate relationship between hardnesses and tensile-strength. Conversions from one scale to another are made at the intercepts with the curve crossing the chart. For example, follow the horizontal line representing 400 Diamond Pyramid Hardness to its intersection with the conversion curve. From this point follow vertically downward for equivalent Rockwell C hardness (41), horizontally to the right for Brinell hardness (379) and tensile strength (187,000 psi), and vertically upward for the equivalent Shore hardness (55).

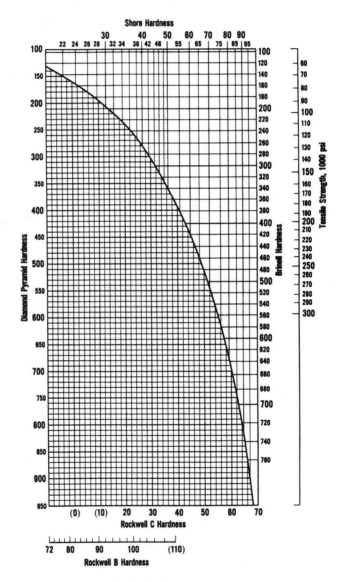

Source: Data from 1966 SAE handbook.

Index

Page numbers on which definitions appear are printed in **boldface.**

Index of Principles/Equations and Illustrative Problems

TABLE 10-7 CONVERSIONS

Quantity	U.S. customary to SI
Acceleration, linear	$1 \text{ ft/s}^2 = 3.048 \times 10^{-1} \text{ m/s}^2$
Area	$1 \text{ in}^2 = 6.452 \times 10^2 \text{ mm}^2$
	$1 \text{ ft}^2 = 9.290 \times 10^{-2} \text{ m}^2$
Density (mass), ρ	$1 \text{ lb/in}^3 = 2.768 \times 10^4 \text{ kg/m}^3$
	$1 \text{ lb/ft}^3 = 1.602 \times 10^1 \text{ kg/m}^3$
Electric current, I	$1 \text{ ampere (A)} = 1 \text{ C/s}$
Energy, work	$1 \text{ Btu} = 1.055 \text{ kJ}$
	$1 \text{ in.} \cdot \text{lb} = 1.129 \times 10^{-1} \text{ J}$
	$1 \text{ ft} \cdot \text{lb} = 1.356 \text{ J}$
	$1 \text{ Btu} = 0.293 \text{ W} \cdot \text{h}$
Force	$1 \text{ lbf} = 4.448 \text{ N}$
	$1 \text{ kgf} = 9.807 \text{ N}$
Impulse	$1 \text{ lb} \cdot \text{s} = 4.448 \text{ N} \cdot \text{s}$
Length	$1 \text{ Å} = 1 \times 10^{-1} \text{ nm}$
	$1 \text{ microinch} = 2.540 \times 10^{-2} \text{ } \mu\text{m}$
	$1 \text{ mil} = 2.540 \times 10^1 \text{ } \mu\text{m}$
	$1 \text{ in.} = 2.540 \times 10^1 \text{ mm}$
	$1 \text{ ft} = 3.048 \times 10^{-1} \text{ m}$
Magnetic field strength, H	$1 \text{ oersted (Oe)} = 79.58 \text{ ampere turns per meter}$
Magnetic flux	$1 \text{ maxwell} = 10^{-8} \text{ Wb}$
Magnetic flux density, B	$1 \text{ Wb/M}^2 = 1 \text{ T (tesla)}$
Modulus of elasticity, E	$1 \text{ lb/in}^2 = 6.895 \times 10^{-6} \text{ GPa}$
Moment of force, torque	$1 \text{ lb} \cdot \text{in.} = 1.130 \times 10^{-1} \text{ N} \cdot \text{m}$
	$1 \text{ lb} \cdot \text{ft} = 1.356 \text{ N} \cdot \text{m}$
Moment of inertia, I (of area)	$1 \text{ in}^4 = 4.162 \times 10^5 \text{ mm}^4$
Momentum, linear	$1 \text{ lb} \cdot \text{ft/s} = 1.383 \times 10^{-1} \text{ kg} \cdot \text{m/s}$
Power	$1 \text{ Btu/min} = 1.758 \times 10^{-2} \text{ kW}$
	$1 \text{ ft} \cdot \text{lb/min} = 2.259 \times 10^{-2} \text{ W}$
	$1 \text{ hp} = 7.457 \times 10^{-1} \text{ kW}$
Stress (pressure)	$1 \text{ lb/in}^2 = 6.895 \times 10^{-3} \text{ MPa}$
	$1 \text{ ksi} = 6.895 \text{ MPa}$
Temperature	$1°\text{F (difference)} = 0.555°\text{C}$
	$1.8 \text{ °F} = 1°\text{C (difference)}$
Thermal expansion, linear coefficient, α	$\text{in./in./°F} = 1.8 \text{ K}^{-1}, \text{ K} = °\text{C} + 273.15$
Thermal conductivity, k	$1 \text{ Btu/ft} \cdot \text{hr} \cdot °\text{F} = 1.729 \text{ W/m} \cdot \text{K}$
Velocity, linear, v	$1 \text{ in./s} = 2.540 \times 10^1 \text{ mm/s}$
	$1 \text{ ft/s} = 3.048 \times 10^{-1} \text{ m/s}$
	$1 \text{ in./min} = 4.233 \times 10^{-1} \text{ mm/s}$
	$1 \text{ ft/min} = 5.080 \times 10^{-3} \text{ m/s}$
Velocity, angular, ω	$1 \text{ rev/min} = 1.047 \times 10^{-1} \text{ rad/s}$
Viscosity	$1 \text{ poise (P)} = 0.1 \text{ Pa} \cdot \text{s}$
Volume	$1 \text{ in}^3 = 1.639 \times 10^4 \text{ mm}^3$
	$1 \text{ ft}^3 = 2.832 \times 10^{-2} \text{ m}^3$
	$1 \text{ yd}^3 = 7.646 \times 10^{-1} \text{ m}^3$

$1 \text{ joule (J)} = 10^7 \text{ ergs} = 0.625 \times 10^{19} \text{ eV}$

$1 \text{ gauss (G)} = 10^{-4} \text{ Wb/m}^2 = 1 \text{ T}$

$1 \text{ weber (Wb)} = 1 \text{ T/m}^2$

$1 \text{ Å} = 10^{-10} \text{ m (obsolete)}$

$1 \text{ micrometer (micron)} = 10^{-6} \text{ m (obsolete)}$

TABLE 10-7A   ADDITIONAL USEFUL CONVERSIONS

$1\text{ W} = 3.413\text{ Btu/hr}$
$1\text{ W} = 1\text{ N}\cdot\text{m}\cdot\text{s}^{-1}$
$1\text{ W} = 0.7376\text{ ft}\cdot\text{lb}\cdot\text{s}^{-1}$
$1\text{ kW} = 1.341\text{ Hp}$
$1\text{ J} = 0.2390\text{ cal}$
$1\text{ cal} = 4.186\text{ J}$
$1\text{ Btu} = 252\text{ cal}$
$1\text{ J} = 0.737562\text{ lb}\cdot\text{ft}$
$1\text{ J} = 1\text{ N}\cdot\text{m} = 1\text{ W}\cdot\text{s}$
$1\text{ Pa} = 1\text{ N/m}^2$
$1\text{ N/m}^2 = 1.450\text{ lb/in}^2\text{ (psi)}$
$1\text{ m} = 3.28\text{ ft} = 39.37\text{ in.}$

$°\text{C} = \dfrac{5}{9}(\text{F} - 32)$

$°\text{F} = \dfrac{9}{5}(\text{F} + 32)$

$\text{K} = °\text{C} + 273 \quad (\text{K} = \text{kelvin})$
$°\text{R} = °\text{F} + 460 \quad (\text{R} = \text{Rankine})$
$1\text{ L} = 1.06\text{ qt}$
$1\text{ Pa}\cdot\text{s} = 10\text{ poise}$

TABLE 10-8   CONSTANTS

Quantity	Symbol	Value
Acceleration of gravity	g	$9.80\text{ m}\cdot\text{s}^{-2}$
Atomic mass unit	amu	$1.66 \times 10^{-27}\text{ kg}$
Avogadro's number	N_A, N_0	$6.022 \times 10^{23}\text{ molecules/mole}$
Bohr radius	a_0	$5.292 \times 10^{-11}\text{ m}$
Electron charge	$q, -e$	$1.60 \times 10^{-19}\text{ C}$
Electron mass	m	$9.11 \times 10^{-31}\text{ kg}$
Electron volt	eV	$0.160 \times 10^{-18}\text{ J}$
Magnetic permeability of free space	μ_0	$4\pi \times 10^{-7}\text{ H/m} = 1.257 \times 10^{-6}\text{ H/m}$
Permittivity of free space	ϵ_0	$\dfrac{10^7}{4\pi c^2} = 8.854 \times 10^{-12}\text{ C/V}\cdot\text{m}$
Planck's constant	$\hbar$	$6.63 \times 10^{-34}\text{ J}\cdot\text{s}$
Speed of light (vacuum)	c	$3.00 \times 10^8\text{ m}\cdot\text{s}^{-1}$

TABLE 10-3   GREEK SYMBOLS AND THEIR PRONUNCIATIONS

Γ	gamma (cap.)	η	eta (lc)
Δ	delta (cap.)	θ	theta (lc)
Θ	theta (cap.)	ϑ	theta (lc)
Λ	lambda (cap.)	κ	kappa (lc)
Ξ	xi (cap.)	λ	lambda (lc)
Π	pi (cap.)	μ	mu (lc)
Σ	sigma (cap.)	ν	nu (lc)
Υ	upsilon (cap.)	ξ	xi (lc)
Φ	phi (cap.)	π	pi (lc)
Ψ	psi (cap.)	ρ	rho (lc)
Ω	omega (cap.)	σ	sigma (lc)
α	alpha (lc)	τ	tau (lc)
β	beta (lc)	ϕ	phi (lc)
γ	gamma (lc)	φ	phi (lc)
δ	delta (lc)	χ	chi (lc)
ϵ	epsilon (lc)	ψ	psi (lc)
ζ	zeta (lc)	ω	omega (lc)

VALUABLE SOFTWARE OFFER
35% DISCOUNT

MaterialTool
A Selection Guide of Materials
And Processes for Designers Version 2
Andrzej Jan Wroblewski
Suryanarayana Vanka

CD-ROM (0-13-934316-4)

Imagine a library in the palm of your hand

As an industrial design, architectural, or technology student, you are often faced with hundreds of materials and manufacturing processes - each with unique capabilities and uses. Imagine a single, comprehensive reference on your computer, which puts all the information you need about these materials and processes at your fingertips. One that can give you immediate access to material properties, typical applications, and commercial stock forms. One that can competently guide you through hundreds of manufacturing processes using hyper-linked text, active images, and interactive animations.

Stop imagining, and start believing

Wroblewski and Vanka's MaterialTool puts this and more within your reach with user-friendly, interactive software that catalogues over 1,000 materials and manufacturing processes - all fully supported with clear text descriptions, images, animations, property tables, hyperlinks, and references.

Coverage that's as comprehensive as it gets

Sections on Materials:
_ cover everything from acrylonitrile butadiene styrene to zirconium
_ provide tables of properties
_ present images showing visual characteristics
_ feature examples of typical applications
_ contain standard stock form information

Sections on Processes:
_ survey everything from abrading to zig-zag stitching
_ include self-paced animations
_ use images to show visual results of the process
_ offer tips for effective design
_ link related processes

Interaction that's rich with visuals...

You'll find superb graphics, visual explanations, and vibrant self-paced animations supporting the textual content. Plus self-paced tutorials; numerous interactive Material Property tables; standard stock form information for all material groups; literally thousands of interactive hyperlinks and images of typical applications; and references to videos, books, and websites for additional information.

Check out the demo packaged with this text.
Take advantage of this special offer and order today at a
35% discount.

To purchase your copy, mail this 35% off coupon to:
Prentice Hall Order Processing Department
P.O. Box 11073
Des Moines, IA 50336-1073

Please send me the CD-ROM listed below.

QUANTITY ISBN	TITLE	AUTHOR	PRICE EACH
0-13-934316-4	MaterialTool, version 2	Wroblewski/Vanka	Please call 1-800-282-0693 *

NAME _____

ADDRESS _____

CITY _____ STATE _____ ZIP _____

CHARGE MY ___VISA ___MASTER CARD ___AMERICAN EXPRESS ___DISCOVER CARD

CARD NUMBER _____ EXPIRATION DATE _____

Or ___PAYMENT ENCLOSED Customer's Signature

X_____

• Plus applicable sales tax, shipping and handling. (PRICES SUBJECT TO CHANGE WITHOUT NOTICE.)

• key M7896543, mpg 004